RECOMBINANT DNA
METHODOLOGY

SELECTED METHODS IN ENZYMOLOGY

Kivie Moldave (editor). RNA and Protein Synthesis, 1981

Daniel L. Purich (editor). Contemporary Enzyme Kinetics and Mechanism, 1983

Arthur Weissbach and Herbert Weissbach (editors). Methods for Plant Molecular Biology, 1988

P. Michael Conn (editor). Neuroendocrine Peptide Methodology, 1989

Ray Wu, Lawrence Grossman, and Kivie Moldave (editors). Recombinant DNA Methodology, 1989

Recombinant DNA Methodology

EDITED BY

Ray Wu
SECTION OF BIOCHEMISTRY
MOLECULAR AND CELL BIOLOGY
CORNELL UNIVERSITY
ITHACA, NEW YORK

Lawrence Grossman
DEPARTMENT OF BIOCHEMISTRY
THE JOHNS HOPKINS UNIVERSITY
SCHOOL OF HYGIENE AND PUBLIC HEALTH
BALTIMORE, MARYLAND

Kivie Moldave
DEPARTMENT OF BIOLOGY
UNIVERSITY OF CALIFORNIA, SANTA CRUZ
SANTA CRUZ, CALIFORNIA

ACADEMIC PRESS, INC.
Harcourt Brace Jovanovich, Publishers
San Diego   New York   Berkeley   Boston
London   Sydney   Tokyo   Toronto

Academic Press, Inc.
San Diego, California 92101

United Kingdom Edition published by
Academic Press Limited
24–28 Oval Road, London NW1 7DX

Library of Congress Cataloging-in-Publication Data

Recombinant DNA methodology / edited by Ray Wu, Lawrence Grossman,
    Kivie Moldave.
        p.   cm. -- (Selected methods in enzymology)
    Reprint of articles from volumes 68, 100, 101, 153, 154, and 155
of Methods in enzymology.
    Includes index.
    ISBN   0-12-765560-3   (alk. paper)
    1. Genetic engineering--Technique. 2. Recombinant DNA--Research-
-Methodology.   I. Wu, Ray. II. Grossman, Lawrence, Date.
III. Moldave, Kivie, Date. IV. Methods in enzymology.
V. Series
    [DNLM: 1. DNA, Recombinant--collected works.   QU 58 R3123]
QH442.R3828   1989
574.87'3282--dc19
DNLM/DLC
for Library of Congress   88-36438
                                                                      CIP

Printed in the United States of America
        90 91 92   9 8 7 6 5 4 3 2

Table of Contents

Section I. Enzymes in Recombinant DNA Research

Section II. Methods for Isolation, Purification, or Amplification of DNA

Section III. Vectors or Methods for Gene Cloning

Section IV. Vectors or Methods for Expression of Cloned Genes

Section V. Methods for Oligonucleotide-Directed Mutagenesis

Section VI. Miscellaneous Methods

List of Contributors

Article numbers are in parentheses following the names of contributors.
Affiliations listed are current.

CARL W. ANDERSON (38), *Biology Department, Brookhaven National Laboratory, Upton, New York 11973*

K. ARAI (12), *Department of Molecular Biology, DNAX Research Institute of Molecular and Cellular Biology, Palo Alto, California 94304*

AMY ARROW (40), *Biotix Inc., Commerce Park, Danbury, Connecticut 06810*

LESLIE BARNETT (15), *Medical Research Council (MRC) Molecular Genetics Unit, Cambridge CB2 2QH, England*

H. C. BIRNBOIM (8), *General Division, Ottawa Regional Cancer Centre, University of Ottawa, Ottawa, Ontario, Canada K1H 8L6*

GRANT A. BITTER (29), *AMGen, Thousand Oaks, California 91320*

ROBERT BLAKESLEY (2), *Bethesda Research Laboratories, Inc., Gaithersburg, Maryland 20877*

SYDNEY BRENNER (15), *Medical Research Council (MRC) Molecular Genetics Unit, Cambridge CB2 2QH, England*

JAMES R. BROACH (21), *Biology Department, Lewis Thomas Laboratory, Princeton University, Princeton, New Jersey 08544*

M. BROWNSTEIN (12), *Laboratory of Cell Biology, National Institute of Mental Health, National Institutes of Health, Bethesda, Maryland 20892*

CHARLES R. CANTOR (6), *Department of Genetics and Development, College of Physicians & Surgeons of Columbia University, New York, New York 10032*

JOHN CARBON (20), *Department of Biological Sciences, University of California, Santa Barbara, Santa Barbara, California 93106*

GEORGES F. CARLE (7), *Centre de Biochimie, Université de Nice, Nice, Cedex, France*

MALCOLM J. CASADABAN (23), *Department of Molecular Genetics and Cell Biology, Cummings Life Science Center, The University of Chicago, Chicago, Illinois 60637*

JOANY CHOU (23), *Department of Molecular Genetics and Cell Biology, Cummings Life Science Center, The University of Chicago, Chicago, Illinois 60637*

MAIR E. A. CHURCHILL (41), *Medical Research Council (MRC), Laboratory of Molecular Biology, Cambridge CB2 2QH, England*

LOUISE CLARKE (20), *Department of Biological Sciences, University of California, Santa Barbara, Santa Barbara, California 93106*

JOHN COLLINS (10), *Department of Genetics, Gesellschaft für Biotechnologische Forschung m.b.H., D-3300 Braunschweig, Federal Republic of Germany*

NICHOLAS R. COZZARELLI (1), *Division of Biochemistry and Molecular Biology, Department of Molecular and Cell Biology, University of California, Berkeley, Berkeley, California 94720*

RODERIC M. K. DALE (40), *Biotix Inc., Commerce Park, Danbury, Connecticut 06810*

RONALD W. DAVIS (17), *Department of Biochemistry, Stanford University School of Medicine, Stanford, California 94305*

BETH A. DOMBROSKI (41), *Department of Pediatrics, The Johns Hopkins University School of Medicine, Baltimore, Maryland 21205*

BERNARD S. DUDOCK (38), *Department of Biochemistry, State University of New York (SUNY) at Stony Brook, Stony Brook, New York 11794*

GUY D. DUFFAUD (26), *Graduate Department of Biochemistry, Brandeis University, Waltham, Massachusetts 02254*

KEVIN M. EGAN (29), *AMGen, Thousand Oaks, California 91320*

STEPHEN ELLEDGE (17), *Department of Biochemistry, Stanford University School of Medicine, Stanford, California 94305*

STEVEN G. ELLIOTT (29), *AMGen, Thousand Oaks, California 91320*

FRED A. FALOONA (9), *CETUS Corporation, Emeryville, California 94608*

WALTER FIERS (25), *Laboratory of Molecular Biology, State University of Ghent, B-9000 Ghent, Belgium*

ANDREW FIRE (36), *Department of Embryology, Carnegie Institute of Washington, Baltimore, Maryland 21210*

R. T. FRALEY (22), *Plant Molecular Biology Group, Biological Sciences Department, Corporate Research and Development Staff, Monsanto Company, St. Louis, Missouri 63198*

A. M. FRISCHAUF (16), *Molecular Analysis of Mammalian Mutations, The Imperial Cancer Research Fund, London WC2A 3PX, England*

HANS-JOACHIM FRITZ (31), *Institut für Molekulare Genetik, Universität Göttingen, D-3400 Göttingen, Federal Republic of Germany*

ROY FUCHS (2), *Biological Sciences Department, Monsanto Company, St. Louis, Missouri 63198*

JAMES C. GIFFIN (29), *AMGen, Thousand Oaks, California 91320*

A. GRAESSMANN (35), *Institut für Molekularbiologie und Biochemie der Freien Universität Berlin, D-1000 Berlin 33, Federal Republic of Germany*

LEONARD GUARENTE (27), *Department of Biology, Massachusetts Institute of Technology, Cambridge, Massachusetts 02139*

LI-HE GUO (4), *Shanghai Institute of Cell Biology, Academia Sinica, Shanghai, People's Republic of China*

J. B. GURDON (28), *CRC Molecular Embryology Group, Department of Zoology, University of Cambridge, Cambridge CB2 3EJ, England*

DOUGLAS HANAHAN (14), *Department of Biochemistry, Hormone Research Institute, University of California, San Francisco, San Francisco, California 94143*

N. PATRICK HIGGINS (1), *Department of Biochemistry, The University of Alabama at Birmingham, Birmingham, Alabama 35294*

DAVID E. HILL (34), *Applied bioTechnology Incorporated, Cambridge, Massachusetts 02142*

YEN-SEN HO (24), *Molecular Genetics Department, Smith Kline & French Laboratories, King of Prussia, Pennsylvania 19406*

R. B. HORSCH (22), *Plant Molecular Biology Group, Biological Sciences Department, Corporate Research and Development Staff, Monsanto Company, St. Louis, Missouri 63198*

CHU-LAI HSIAO (20), *Division of Microbiology, Wyeth Laboratories, Philadelphia, Pennsylvania 19101*

P. C. HUANG (33), *Department of Biochemistry, The Johns Hopkins University, School of Hygiene and Public Health, Baltimore, Maryland 21205*

MASAYORI INOUYE (26), *Department of Biochemistry, University of Medicine and Dentistry of New Jersey at Rutgers, Robert Wood Johnson Medical School, Piscataway, New Jersey 08854*

MATTHEW O. JONES (29), *AMGen, Thousand Oaks, California 91320*

LAURANCE KAM (41), *University of Pennsylvania, School of Medicine, Philadelphia, Pennsylvania 19104*

JONATHAN KARN (15), *Medical Research Council (MRC), Laboratory of Molecular Biology, Cambridge CB2 2QH, England*

M. KAWAICHI (12), *Department of Medical Chemistry, Kyoto University, Kyoto 606, Japan*

H. J. KLEE (22), *Plant Molecular Biology Group, Biological Sciences Department, Corporate Research and Development Staff, Monsanto Company, St. Louis, Missouri 63198*

RAYMOND A. KOSKI (29), *AMGen, Thousand Oaks, California 91320*

WILFRIED KRAMER (31), *Institut für Molekulare Genetik, Universität Göttingen, D-3400 Göttingen, Federal Republic of Germany*

THOMAS A. KUNKEL (32), *Laboratory of Molecular Genetics, National Institute of Environmental Health Sciences, National Institutes of Health, Research Triangle Park, North Carolina 27709*

F. LEE (12), *Department of Molecular Biology, DNAX Research Institute of Molecular and Cellular Biology, Palo Alto, California 94304*

H. LEHRACH (16), *Genome Analysis, The Imperial Cancer Research Fund, London WC2A 3PX, England*

A. LOYTER (35), *Department of Biological Chemistry, Institute of Life Sciences, The Hebrew University of Jerusalem, Jerusalem 91904, Israel*

MICHAEL McCLELLAND (3), *Department of Biochemistry and Molecular Biology, The University of Chicago, Chicago, Illinois 60637*

JAMES L. MANLEY (36), *Department of Biological Sciences, Columbia University, New York, New York 10027*

PAUL E. MARCH (26), *Department of Biochemistry, University of Medicine and Dentistry of New Jersey at Rutgers, Robert Wood Johnson Medical School, Piscataway, New Jersey 08854*

ANNE MARMENOUT (25), *Innogenetics, Zwijnaarde, Belgium*

ALFONSO MARTINEZ-ARIAS (23), *Department of Zoology, University of Cambridge, Cambridge CB2 3EJ, England*

WILLIAM C. MERRICK (37), *Department of Biochemistry, Case Western Reserve University, School of Medicine, Cleveland, Ohio 44106*

MATTHEW MESELSON (14), *Department of Biochemistry and Molecular Biology, Harvard University, Cambridge, Massachusetts 02138*

JOACHIM MESSING (11), *Waksman Institute of Microbiology, Rutgers, The State University of New Jersey, Piscataway, New Jersey 08855*

D. A. MORRISON (13), *ABC Research Department, National Defense Research Institute, S-901 82 Umea, Sweden*

KARY B. MULLIS (9), *6767 Neptune Place, Apartment 4, La Jolla, California 92307*

ANDREW W. MURRAY (19), *Department of Biochemistry and Biophysics, University of California, San Francisco, San Francisco, California 94143*

N. MURRAY (16), *Department of Molecular Biology, University of Edinburgh, Edinburgh EH9 3JR, Scotland*

H. OKAYAMA (12), *Research Institute for Microbial Diseases, Osaka University, Suita, Osaka 565, Japan*

ARNOLD R OLIPHANT (34), *Department of Medical Informatics, Division of Genetic Epidemiology, University of Utah, Salt Lake City, Utah 84108*

MAYNARD V. OLSON (7), *Department of Genetics, Washington University School of Medicine, St. Louis, Missouri 63110*

RICHARD PINE (33), *Laboratory of Molecular Cell Biology, The Rockefeller University, New York, New York 10021*

INGO POTRYKUS (39), *Institute for Plant Sciences, CH-1892 Zurich, Switzerland*

A. RAZIN (35), *Department of Cellular Biochemistry, The Hebrew University, Hadassah Medical School, Jerusalem 91010, Israel*

ERIK REMAUT (25), *Laboratory of Molecular Biology, State University of Ghent, B-9000 Ghent, Belgium*

JOHN D. ROBERTS (32), *Laboratory of Molecular Genetics, National Institute of Environmental Health Sciences, National Institutes of Health, Research Triangle Park, North Carolina 27709*

S. G. ROGERS (22), *Plant Molecular Biology Group, Biological Sciences Department, Corporate Research and Development Staff, Monsanto Company, St. Louis, Missouri 63198*

MARTIN ROSENBERG (24), *Biopharmaceutical Research & Development, Smith Kline & French Laboratories, King of Prussia, Pennsylvania 19406*

RODNEY J. ROTHSTEIN (18), *Department of Genetics and Development, College of Physicians & Surgeons of Columbia University, New York, New York 10032*

STEPHANIE W. RUBY (19), *Division of Biology, California Institute of Technology, Pasadena, California 91125*

MARK SAMUELS (36), *Biology Department, Princeton University, Princeton, New Jersey 08544*

STUART K. SHAPIRA (23), *The Childrens Hospital, Boston, Massachusetts 02115*

PHILLIP A. SHARP (36), *Center for Cancer Research, Massachusetts Institute of Technology, Cambridge, Massachusetts 02139*

ALLAN SHATZMAN (24), *Biological Process Sciences, Smith Kline & French Laboratories, King of Prussia, Pennsylvania 19406*

RAYMOND D. SHILLITO (39), *Biotechnology Research, CIBA-GEIGY Corporation, Research Triangle Park, North Carolina 27709*

GUUS SIMONS (25), *N.I.Z.O., 6710 Ede, The Netherlands*

CASSANDRA L. SMITH (6), *Departments of Microbiology and Psychiatry, College of Physicians & Surgeons of Columbia University, New York, New York 10032*

MICHAEL SMITH (30), *Biotechnology Laboratory, The University of British Columbia, Vancouver, British Columbia, Canada V6T 1W5*

MICHAEL SNYDER (17), *Department of Biochemistry, Stanford University School of Medicine, Stanford, California 94305*

EDWIN SOUTHERN (5), *Medical Research Council (MRC) Mammalian Genome Unit, Edinburgh EH9 3JT, Scotland*

J. WILLIAM STRAUS (38), *Biology Department, Vassar College, Poughkeepsie, New York 12601*

KEVIN STRUHL (34), *Department of Biological Chemistry, Harvard Medical School, Boston, Massachusetts 02115*

DOUGLAS SWEETSER (17), *Whitehead Institute for Biomedical Research, Cambridge, Massachusetts 02142*

JACK W. SZOSTAK (19), *Department of Molecular Biology, Massachusetts General Hospital, Boston, Massachusetts 02114*

THOMAS D. TULLIUS (41), *Department of Chemistry, The Johns Hopkins University, Baltimore, Maryland 21218*

A. VAINSTEIN (35), *Department of Horticulture, Faculty of Agriculture, The Hebrew University of Jerusalem, Rehovot 76-100, Israel*

JEFFREY VIEIRA (11), *Department of Microbiology and Immunology, Stanford University School of Medicine, Stanford, California 94305*

M. P. WICKENS (28), *Department of Biochemistry, University of Wisconsin-Madison, Madison, Wisconsin 53706*

RAY WU (4), *Section of Biochemistry, Molecular and Cell Biology, Cornell University, Ithaca, New York 14853*

T. YOKOTA (12), *Department of Molecular Biology, DNAX Research Institute of Molecular and Cellular Biology, Palo Alto, California 94304*

RICHARD A. YOUNG (17), *Whitehead Institute for Biomedical Research, Cambridge, Massachusetts 02142, and Department of Biology, Massachusetts Institute of Technology, Cambridge, Massachusetts 02139*

RICHARD A. ZAKOUR (32), *Biotech Research Laboratories, Inc., Rockville, Maryland 20850*

MARK J. ZOLLER (30), *Department of Cardiovascular Research, Genentech, Inc., South San Francisco, California 94080*

Preface

Recombinant DNA methods are revolutionary techniques that allow the cloning of a single gene—isolation of a gene in large amounts—from a pool of millions of genes. These techniques also allow specific modification of the isolated genes or their regulator regions for analysis or for reintroduction into cells for the production of large amounts of specific RNA or protein molecules. These powerful new methods lead to unprecedented possibilities for solving complex biological problems or for producing new and better products in the fields of medicine, agriculture, and industry.

This volume of *Selected Methods in Enzymology* includes important contributions from Volumes 68, 100, 101, 153, 154, and 155 of *Methods in Enzymology*. The selection of articles was based mainly on the following criteria: inclusion in the article of extensive description or theoretical discussion of important methods and specific information still up-to-date and useful.

Where a cross reference is given to a volume and paper in this series, it refers to the *Methods in Enzymology* series. Where only volumes and paper numbers are referred to, the volumes too are those in the *Methods in Enzymology* series.

We thank the authors and the staff of Academic Press who made this volume possible.

RAY WU
LAWRENCE GROSSMAN
KIVIE MOLDAVE

Contents of Methods in Enzymology
Volumes 68, 100, 101, 153, 154, and 155

Section IV. Vehicles and Hosts for the Cloning of Recombinant DNA

Section V. Screening and Selection of Cloned DNA

Section VI. Detection and Analysis of Expression of Cloned Genes

Volume 100

Section I. Use of Enzymes in Recombinant DNA Research

Section II. Enzymes Affecting the Gross Morphology of DNA

A. Topoisomerases Type I

B. Topoisomerases Type II

Section V. Analytical Methods for Gene Products

Section VI. Mutagenesis: *In Vitro* and *in Vivo*

Volume 101

Section I. New Vectors for Cloning Genes

Section II. Cloning of Genes into Yeast Cells

Section III. Systems for Monitoring Cloned Gene Expression

A. Intact Cell Systems

B. Introduction of Genes into Mammalian Cells

C. Cell-Free Systems; Transcription

D. Cell-Free Systems; Translation

Volume 153

Section I. Vectors for Cloning DNA

Section II. Vectors for Expression of Cloned Genes

Volume 154

Section I. Methods for Cloning cDNA

Section II. Identification of Cloned Genes and Mapping of Genes

Section III. Chemical Synthesis and Analysis of Oligodeoxynucleotides

Volume 155

Section I. Restriction Enzymes

Section I

Enzymes in Recombinant DNA Research

[1] DNA-Joining Enzymes: A Review

By N. PATRICK HIGGINS and NICHOLAS R. COZZARELLI

The first DNA-joining enzymes identified were the DNA ligases. They join together DNA chains by transmuting the high-energy pyrophosphate linkage of a nucleotide cofactor into a phosphoester bond between 5'-phosphoryl and 3'-hydroxyl termini. The impetus for the nearly simultaneous discovery of DNA ligase in bacteria[1-3] and in bacteriophage-infected cells[4,5] was the recognition that DNA fragments were joined together during genetic recombination.[6,7] Shortly thereafter, Okazaki and co-workers[8] showed that ligase played the key role in DNA replication of joining the nascent small pieces of DNA at the replication fork. Ligase has since been found to participate in the synthesis and repair of DNA in a variety of organisms.[9-11]

The termini of DNA strands to be joined by ligase must in general be abutted by base pairing to a complementary strand. This ensures proper alignment and preservation of the nucleotide sequence of DNA; thus, the reaction can be considered *conservative*. Template-dependent synthesis and faithful conservation of DNA coding information is also the hallmark of most DNA polymerases. There is, however, increasing interest in *radical* or template-independent reactions. Since they are not disciplined by a complementary strand, they generally change the nucleotide sequence at the site of reaction. Polynucleotide phosphorylase and terminal deoxynucleotidyltransferase are examples of radical polymerases that generate

[1] M. Gellert, *Proc. Natl. Acad. Sci. U.S.A.* **57,** 148 (1967).

[2] B. M. Olivera, and I. R. Lehman, *Proc. Natl. Acad. Sci. U.S.A.* **57,** 1426 (1967).

[3] M. L. Gefter, A. Becker, and J. Hurwitz, *Proc. Natl. Acad. Sci. U.S.A.* **58,** 240 (1967).

[4] B. Weiss, and C. C. Richardson, *Proc. Natl. Acad. Sci. U.S.A.* **57,** 1021 (1967).

[5] N. R. Cozzarelli, N. E. Melechen, T. M. Jovin, and A. Kornberg, *Biochem. Biophys. Res. Commun.* **28,** 578 (1967).

[6] M. Meselson, and J. J. Weigle, *Proc. Natl. Acad. Sci. U.S.A.* **47,** 869 (1961).

[7] G. M. Kellenberger, M. L. Zichichi, and J. J. Weigle, *Proc. Natl. Acad. Sci. U.S.A.* **47,** 869 (1964).

[8] K. Sugimoto, T. Okazaki, and R. Okazaki, *Proc. Natl. Acad. Sci. U.S.A.* **60,** 1356 (1968).

[9] I. R. Lehman, *Science* **186,** 790 (1974).

[10] K. A. Nasmyth, *Cell* **12,** 1109 (1977).

[11] L. H. Johnston, and K. A. Nasmyth, *Nature (London)* **274,** 891 (1978).

new sequences at the 3'-hydroxyl terminus of DNA molecules.[12-14] T4-induced RNA ligase is a radical ligase that catalyzes the uninstructed joining of both RNA and DNA chains. Although a role for T4 RNA ligase in the joining of nucleic acids *in vivo* has not been demonstrated, some presumably radical enzymes must circularize the single-stranded RNA viroids in plant tissues[15,16] and splice precursor RNA chains to form functional messages in eukaryotes.[17] T4 DNA ligase at high concentrations also carries out a template-independent joining of molecules with base-paired ends, so that it can operate in either a conservative or radical mode.

The term "ligase" will be reserved for proteins that join the preexisting termini of polynucleotide chains. This is the basis for the common assay for RNA and DNA ligases—conversion of a terminal 5'-^{32}P-labeled phosphoryl to a form resistant to digestion with bacterial alkaline phosphatase. However, several enzymes have been discovered that join polynucleotides in a fundamentally different manner. The substrate for these enzymes is not the ends but continuous stretches of nucleic acids which are broken and rejoined in a concerted fashion. These breakage–reunion (B-R) enzymes[98a] require no cofactor to supply energy for reforming the DNA backbone bond. Presumably, the energy released by breakage is stored in a covalent enzyme–DNA intermediate and utilized for joining. A major class of B-R enzymes catalyze the interconversion of topological isomers of DNA and are designated topoisomerases.[18] The first enzyme isolated in this class of proteins was *Escherichia coli* ω protein, and other examples are the bacterial DNA gyrases and the DNA untwisting enzymes from eukaryotic cells. The integration of bacteriophage λ DNA involves breaking at least four strands of duplex DNA and resealing the ends to different partners without alteration in the winding number. The reaction does not require any energy cofactor. This enzyme and the topoisomerases isolated to date are conservative B-R enzymes. However, the φX174 *cisA*-coded protein is a site-specific B-R enzyme which apparently is not template-instructed, and there may be additional radical B-R enzymes. Several models for movement of insertion elements have proposed radical joining reactions at the terminus of the transposing

[12] S. Gillam, K. Waterman, and M. Smith, *Nucleic Acids Res.* **2**, 613 (1975).
[13] K. Kato, J. M. Gonçalves, G. E. Houts, and F. J. Bollum, *J. Biol. Chem.* **242**, 2780 (1967).
[14] R. Roychoudhury, E. Jay, and R. Wu, *Nucleic Acids Res.* **3**, 863 (1976).
[15] T. O. Diener, *Annu. Rev. Microbiol.* **28**, 23 (1974).
[16] H. L. Sänger, G. Klotz, D. Riesner, H. J. Gross, and A. K. Kleinschmidt, *Proc. Natl. Acad. Sci. U.S.A.* **73**, 3852 (1976).
[17] F. Crick, *Science* **204**, 264 (1979).
[18] J. C. Wang, and L. F. Liu, in "Molecular Genetics" (J. H. Taylor, ed.), Part III, pp. 65–88. Academic Press, New York, 1979.

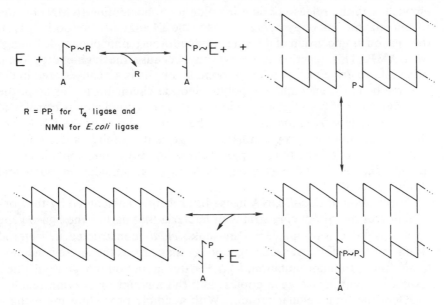

R = PP$_i$ for T$_4$ ligase and
NMN for *E. coli* ligase

FIG. 1. Reaction mechanism of DNA ligase (E).

region[19,20]; these reactions might be carried out by ligases, B-R enzymes, or a combination of both. This article is a brief review of the four classes of enzymes that ligate polynucleotide chains *in vitro*, the radical and conservative ligases and B-R enzymes. Since the conservative enzymes are more often summarized, emphasis will be placed on the enzymes that carry out template-independent reactions.

DNA Ligase

The *E. coli* and bacteriophage T4-induced enzymes are the most thoroughly investigated DNA ligases, and the work up to 1974 has been summarized in the review by Lehman.[9] DNA ligase from *E. coli* is a single polypeptide chain of molecular weight 74,000, and bacteriophage T4 induces a DNA ligase having a single chain with a molecular weight of 68,000. Phosphodiester bond synthesis is coupled to the cleavage of a pyrophosphate bond in nicotinamide adenine dinucleotide (NAD) for the bacterial enzyme and in ATP for the phage enzyme (Fig. 1). Two different covalent intermediates have been identified as partial reaction products for each enzyme. The first step of the reaction is the adenylylation of free

[19] N. D. F. Grindley and D. J. Sherratt, *Cold Spring Harbor Symp. Quant. Biol.* **43**, 1257 (1978).
[20] J. A. Shapiro, *Proc. Natl. Acad. Sci. U.S.A.* **76**, 1933 (1979).

enzyme with the release of nicotinamide mononucleotide (NMN) for the *E. coli* enzyme and of pyrophosphate for the T4 enzyme. For both ligases, the epsilon amino group of a lysine residue forms a phosphoamide linkage with AMP. This reaction is noteworthy because the high-energy bond necessary to drive phosphodiester bond synthesis is actually stored in the enzyme before it encounters a polynucleotide chain. From this step, the reaction paths of both ligases are identical. The second step is transfer of the AMP moiety from the enzyme to the 5'-phosphoryl of the DNA substrate, recreating a pyrophosphate linkage and thereby preserving the high-energy bond. The third step of the reaction is the nucleophilic attack of the adjacent 3'-hydroxyl group to form a phosphodiester bond and eliminate AMP.

Purification of *E. coli* DNA ligase has recently been aided by the construction of overproducing strains of bacteria that harbor the ligase gene on a plasmid or episome.[21,22] Panasenko *et al.* constructed a bacterial strain lysogenic for a transducing λ phage that carries the ligase gene next to an "up" promoter mutation, *lopll*.[21] After induction, phage replication results in many ligase gene copies, and the amount of enzyme reaches 2–3% of the total cellular protein. With a simple procedure involving a single chromatographic step, 30 mg of homogeneous enzyme was prepared from 120 g of these cells.

Reliable schemes for the preparation of homogeneous T4 DNA ligase are available.[23,24] Recently a simpler protocol has been reported to yield nearly pure enzyme after only two chromatographic steps.[25] A significant advantage of preparing the T4 ligase is that at least three other valuable T4-induced enzymes can be concurrently purified from the same batch of infected cells. Panet *et al.*[24] devised a purification procedure for the simultaneous isolation of polynucleotide kinase, DNA ligase, and DNA polymerase. RNA ligase can also be interdigitated; the enzyme elutes after DNA ligase on DEAE-cellulose and does not bind to phosphocellulose.

A number of different assays have been used to monitor DNA ligases. These include joining the cohesive ends of bacteriophage λ,[1,3] sealing nicks in natural DNA molecules,[4] intermolecular joining of short homopolymers annealed to a template strand,[2] linking polynucleotides to the ends of oligonucleotides immobilized on cellulose under the instruction of

[21] S. M. Panasenko, R. J. Alazard, and I. R. Lehman, *J. Biol. Chem.* **253,** 4590 (1978).

[22] K. Borck, J. D. Beggs, W. J. Brammar, A. S. Hopkins, and N. E. Murray, *Mol. Gen. Genet.* **146,** 199 (1976).

[23] B. Weiss, A. Jacquemin-Sablon, T. R. Live, G. C. Fareed, and C. C. Richardson, *J. Biol. Chem.* **243,** 4543 (1968).

[24] A. Panet, J. H. van de Sande, P. C. Loewen, H. G. Khorana, A. J. Raae, J. R. Lillehaug, and K. Kleppe, *Biochemistry* **12,** 5045 (1973).

[25] K.-W. Knopf, *Eur. J. Biochem.* **73,** 33 (1977).

a complementary strand,[5] circularization of the self-complementary polymer poly[d(A-T)],[26] pyrophosphate exchange,[27,28] and formation of an acid-precipitable enzyme–adenylate complex.[23,25,29]

DNA ligases catalyze a variety of inter- and intramolecular joining reactions, and some examples are given in Table I. The reactions catalyzed by the *E. coli* enzyme are asterisked. *Escherichia coli* ligase joins oligo(dT) that is base-paired to poly(dA); it also joins oligo(dA) base-paired to poly(dT), but this is a much less favorable reaction. The self-complementary poly[d(A-T)] can fold back upon itself so as to oppose the ends and ligase seals these into closed loop molecules (circles). Joining of the 5'-phosphoryl of a DNA chain to the 3'-hydroxyl of an RNA molecule has been shown in two ways.[30] First, oligo(A) was joined to the 5'-phosphoryl of oligo(dA) on a poly(dT) template strand. Second, ligase can circularize poly[d(A-T)]pU in which UMP occupies the 3'-hydroxyl terminal position.

T4 DNA ligase is a much more permissive enzyme than *E. coli* ligase. In addition to the reactions mentioned above, it catalyzes a number of ligations that cannot as yet be performed with the *E. coli* enzyme. Kleppe *et al.*[31] and Fareed *et al.*[32] found that ligation of oligo(dT) annealed to poly(A) took place at a few percent of the rate occurring with oligo(dT)·poly(dA). The phage enzyme will also act as an RNA ligase in the strict sense and join RNA to RNA, but this is a very unfavorable reaction. The joining of oligo(I), oligo(C), or oligo(U) in the presence of their complementary polyribonucleotide partners occurs at rates at least two orders of magnitude lower than with DNA substrates.[28] Joining of the 5'-phosphoryl end of an RNA molecule to the 3'-hydroxyl of a DNA chain was first noted by Westegard *et al.*[33] who found that T4 ligase would seal in an RNA fragment used to prime DNA synthesis on the ϕX174 viral strand. Similarly, Nath and Hurwitz[30] showed that T4 ligase joined chains of $(pA)_3(pdA)_n$ when hybridized to poly(dT), albeit at a low rate. The closed circular form of ColE1 DNA purified from chloramphenicol-treated *E. coli* can contain ribonucleotide segments.[34] If this RNA is a remnant of the primer for DNA synthesis, then some *E. coli* ligase must

[26] P. Modrich and I. R. Lehman, *J. Biol. Chem.* **245**, 3626 (1970).

[27] V. Sgaramella and S. D. Ehrlich, *Eur. J. Biochem.* **86**, 531 (1978).

[28] H. Sano and G. Feix, *Biochemistry* **13**, 5110 (1974).

[29] S. B. Zimmerman and C. K. Oshinsky, *J. Biol. Chem.* **244**, 4689 (1969).

[30] K. Nath and J. Hurwitz, *J. Biol. Chem.* **249**, 3680 (1974).

[31] K. Kleppe, J. H. van de Sande, and H. G. Khorana, *Proc. Natl. Acad. Sci. U.S.A.* **67**, 68 (1970).

[32] G. C. Fareed, E. M. Wilt, and C. C. Richardson, *J. Biol. Chem.* **246**, 925 (1971).

[33] O. Westegaard, D. Brutlag, and A. Kornberg, *J. Biol. Chem.* **248**, 1361 (1973).

[34] D. G. Blair, D. B. Clewell, D. J. Sherratt, and D. R. Helinski, *Proc. Natl. Acad. Sci. U.S.A.* **69**, 2518 (1972).

TABLE I

EXAMPLES OF DNA LIGASE REACTIONS[a]

Type of reaction	Substrate	Template	Linkages formed	References
DNA–DNA	Oligo(dT)*	Poly(dA)	dTpdT	2, 30–32, 35
DNA–DNA	Oligo(dA)*	Poly(dT)	dApdA	32
DNA–DNA	Oligo[d(A-T)]*	Self-complementary	dTpdA and dApdT	27, 30, 32
DNA–DNA	Oligo(dT)	Poly(A)	dTpdT	28, 32, 35
DNA–DNA	(pdT)$_{11}$pdC	Poly(dA)	dCpdT	36
DNA–DNA	Termini of exonuclease III-digested DNA	?	Complementary strand, terminal cross-links	37
DNA–DNA	Blunt-ended duplex DNA	None	Phosphodiester bonds between duplex ends	40
RNA–DNA	Oligo(dA) plus oligo(A)*	Poly(dT)	ApdA	30
RNA–DNA	Oligo[d(A-T)]pU*	Self-complementary	UpdA	30
RNA–DNA	Oligo(dT) plus oligo(U)	Poly(dA)	UpdT	30
RNA–RNA	Oligo(A)	Poly(dT)	ApA	28, 31, 32
RNA–RNA	Oligo(I)	Poly(C)	IpI	28
RNA–RNA	Oligo(C)	Poly(I)	CpC	28
RNA–RNA	Oligo(U)	Poly(A)	UpU	28
DNA–RNA	(pA)$_3$(pdA)$_n$	Poly(dT)	dApA	30

[a] All reactions listed are catalyzed by T4 ligase, but *E. coli* ligase carries out only the asterisked reactions. The convention for "Type of reaction" and "Linkages formed" is that the molecule contributing the 3'-hydroxyl group is on the left and the moiety supplying the 5'-phosphoryl is on the right.

be able to join RNA to DNA even though the known *E. coli* ligase has not yet demonstrated this capacity *in vitro*.

Extensive base pairing is not required for T4 DNA ligase to join nucleotide chains. Harvey and Wright[35] showed that T4 ligase joined oligonucleotides in the presence of their complement at temperatures above their T_m. For example, the rate of joining of $(pdT)_9$ of the presence of poly(dA) was optimal at 25°, and ligation was found at temperatures as high as 37°. Yet, in the absence of ligase at least, at 25° no complex between nonathymidylate and poly(dA) was observed by gel chromatography. It is quite possible that in these studies the enzyme stabilized a transient duplex since the complementary strand is required for the ligation. However, a clear example of a radical ligation was the early observation of Tsiapolis and Narang[36] that T4 ligase joined molecules with a terminal mismatch. The polymer $(dpT)_{11}pdC$ hybridized to poly(dA) was joined to form products 400 nucleotides long. This type of reaction could provide an important method for creating site-specific mutations, and further investigations should be made. Another reaction of T4 ligase proceeding with an imperfectly paired substrate was reported by Weiss.[37] A mixture of T4 DNA ligase and *E. coli* exonuclease III cross-links the termini of duplex DNA producing molecules that rapidly renatured after denaturation. A mixture of T7-induced enzymes including ligase also carries out terminal cross-linking.[38] The mechanism suggested for these reactions was exonuclease digestion of the 3'-end to expose a sequence of two to three nucleotides complementary to the 5'-end of the sister strand, base pairing which created a terminal loop and apposed the 5'- and 3'-termini, and finally ligation.

The most emphatically radical of all the T4 ligase reactions was discovered by Sgaramella *et al.*[39,40] who found that it would carry out the intermolecular joining of DNA substrates at completely base-paired ends. This reaction, called blunt-end joining, is a powerful method for structuring DNA molecules *in vitro* (see below) but proceeds much less readily than nick sealing. Blunt-end joining is not linearly dependent on enzyme concentration but increases greatly at higher enzyme levels. Although T4 RNA ligase does not catalyze blunt-end joining, it markedly stimulates

[35] C. L. Harvey and R. Wright, *Biochemistry* **11**, 2667 (1972).

[36] C. M. Tsiapolis and S. A. Narang, *Biochem. Biophys. Res. Commun.* **39**, 631 (1970).

[37] B. Weiss, *J. Mol. Biol.* **103**, 669 (1976).

[38] P. Sadowski, A. McGreer, and A. Becker, *Can. J. Biochem.* **52**, 525 (1974).

[39] V. Sgaramella, J. H. van de Sande, and H. G. Khorana, *Proc. Natl. Acad. Sci. U.S.A.* **67**, 1468 (1970).

[40] V. Sgaramella and H. G. Khorana, *J. Mol. Biol.* **72**, 493 (1972).

the reaction, particularly at low DNA ligase concentrations.[41] As much as a 20-fold increase in the rate was observed at low DNA ligase concentrations, and the stimulation was specific for blunt-end joining because RNA ligase had only a marginal influence on the joining of cohesive ends. In the presence of RNA ligase, T4 DNA ligase had about the same turnover rate (1 mol/min per mol of ligase) for blunt-end and cohesive-end joining. The apparent K_m for blunt ends in the presence or absence of RNA ligase was 50 μM (in terms of 5′-termini) which was two orders of magnitude higher than for a nicked substrate. Joining of the base-paired ends of HaeIII restriction enzyme fragments of ColE1 DNA was examined in detail.[41] The reaction was highly efficient with all sizes of molecules participating in the reaction; linear molecules longer than ColE1 DNA were generated, and 10% of the molecules were circular. The circles were on average smaller than the linear products. However, in a more dilute reaction, 40% of the product molecules were circular.[42] Therefore, reactions in which an intermolecular product is desired should be carried out at high substrate concentrations, whereas circularization is favored by low substrate concentrations and short polymers. The precision of the blunt-end joining reaction was demonstrated by the ability of the joined products to be re-digested with HaeIII restriction enzyme.[41,42]

Poor blunt-end joining has been reported with preparations of T4 DNA ligase active in sealing nicks, even though it has been proved that the identical enzyme is operative in both reactions.[41] There are several possible reasons for this observation. First, the degree of contamination with T4 RNA ligase sharply influences the blunt-end joining reaction. Second, the reaction exhibits bimolecular reaction kinetics at low substrate concentrations[43] and a high K_m for the substrate. Therefore, dilute reactions may result in disappointing yields. Third, in the absence of RNA ligase, the blunt-end joining reaction is not linear with respect to the DNA ligase concentration and requires much larger amounts of enzyme (10- to 30-fold) than sealing nicks.[26,41]

The use of DNA ligase in de novo construction of DNA of defined sequence has been pioneered by Khorana and his co-workers.[44] An example of the strategy employed was synthesis of the duplex DNA corresponding to yeast alanine tRNA. Overlapping polydeoxyribonucleotide segments

[41] A. Sugino, H. M. Goodman, H. L. Heyneker, J. Shine, H. W. Boyer, and N. R. Cozzarelli, J. Biol. Chem. 252, 3987 (1977).

[42] M. Mottes, C. Morandi, S. Cremaschi, and V. Sgaramella, Nucleic Acids Res. 4, 2467 (1977).

[43] K. V. Deugau and J. H. van de Sande, Biochemistry 17, 723 (1978).

[44] H. G. Khorana, K. L. Agarwal, H. Büchi, M. G. Caruthers, N. K. Gupta, K. Kleppe, A. Kumar, E. Ohtsuka, U. L. RajBhandary, J. H. van de Sande, V. Sgaramella, T. Terao, H. Weber, and T. Yamada, J. Mol. Biol. 72, 209 (1972).

representing the entire two strands of the DNA were synthesized by chemical methods. Their lengths ranged from 8 to 12 nucleotides, and the overlapping regions were at least 4 nucleotides long. The last step employed DNA ligase to stitch the base-paired fragments together. T4 DNA ligase has the advantage of requiring a smaller overlapping sequence than the *E. coli* enzyme.

DNA ligase is routinely used in the cloning of DNA to join large DNA fragments containing complementary and antiparallel single-strand extensions commonly called cohesive or sticky ends; the techniques have recently been reviewed by Vosberg.[45] There are three general approaches to creating cohesive ends. The simplest method is to use a type II restriction endonuclease that introduces staggered scissions at unique sequences. A second method uses terminal deoxynucleotidyltransferase to generate single-stranded tails of either poly(dA) and poly(dT) or poly(dG) and poly(dC) at the 3'-hydroxyl terminus of different populations of DNA molecules. In the third method, short duplexes containing a restriction nuclease cleavage site called "linkers" are joined to DNA by T4 ligase-catalyzed blunt-end joining. Linkers have been constructed for the *Eco*RI, *Hin*dIII, *Bam*I, *Hpa*II, *Mbo*I, and *Pst*I restriction enzymes.[46–49] The decanucleotide duplex

$$\begin{array}{c} \text{pCCGGATCCGG} \\ \text{GGCCTAGGCCp} \end{array}$$

has sites for the *Mbo*I ($\downarrow$GATC), *Hpa*II (C$\downarrow$CGG), and *Bam*I (G$\downarrow$GATCC) nucleases (arrows indicate the site of cleavage).[48] After digesting the product with one of the restriction enzymes that cleaves the linkers, the product can be bound by cohesive-end joining to any cloning vehicle that has one of these restriction sites. Examples of genetic sequences cloned using linkers are the *lac* operator region,[46–48,50] rat insulin,[51] human chorionic somatomammotropin,[52] and rat growth hor-

[45] H.-P. Vosberg, *Hum. Genet.* **40,** 1 (1977).

[46] K. J. Marians, R. Wu, J. Stawinski, T. Hozumi, and S. A. Narang, *Nature (London)* **263,** 744 (1976).

[47] H. L. Heynecker, J. Shine, H. M. Goodman, H. W. Boyer, J. Rosenberg, R. E. Dickerson, S. A. Narang, K. Itakura, S. Liu, and A. D. Riggs, *Nature (London)* **263,** 748 (1976).

[48] C. P. Bahl, K. J. Marians, R. Wu, J. Stawinski, and S. A. Narang, *Gene* **1,** 81 (1976).

[49] R. H. Scheller, R. E. Dickerson, H. W. Boyer, A. D. Riggs, and K. Itakura, *Science* **196,** 177 (1977).

[50] C. P. Bahl, R. Wu, and S. A. Narang, *Gene* **3,** 123 (1978).

[51] A. Ullrich, J. Shine, J. Chirgwin, R. Pictet, E. Tischer, W. J. Rutter, and H. M. Goodman, *Science* **196,** 1313 (1977).

[52] J. Shine, P. H. Seeburg, J. A. Martial, J. D. Baxter, and H. M. Goodman, *Nature (London)* **270,** 494 (1977).

mone.[53] The cI gene of bacteriophage λ was cloned by direct insertion of the sequence into the plasmid pCR11 with blunt-end joining.[54]

The E. coli and T4-induced DNA ligases have been paradigms for similar enzymes in other cells. Bacillus subtilis contains a DNA ligase requiring NAD,[55] like the E. coli enzyme, and bacteriophage T7 induces an ATP-dependent ligase.[56] DNA ligases have also been found in a large number of animal and plant cell types, and these have recently been reviewed by Soderhall and Lindahl.[57] Mammalian cells contain two different enzymes designated I and II.[58-63] DNA ligase I accounts for most of the ligase activity in dividing but not resting cells, and it fluctuates according to growth phase. The enzyme has a native molecular weight of approximately 200,000 based on sedimentation and gel filtration data.[58,64,65] DNA ligase I has an apparent K_m for ATP of 0.2–1.5 μM.[65-67] Like the E. coli and T4 enzymes, it forms an adenylylated enzyme intermediate[58] and joins the 5'-phosphoryl terminus of DNA to the 3'-hydroxyl end of a DNA (or less efficiently RNA) chain in a duplex.[68] DNA ligase II is not inhibited by antibodies directed against DNA ligase I.[58] DNA ligase II is also smaller, having a molecular weight of about 100,000,[58,60,62] and its apparent K_m for ATP of 40–100 μM is two orders of magnitude higher than that of DNA ligase I. Mitochondria contain a DNA ligase which is very similar to DNA ligase II in molecular size and in K_m for ATP.[69]

[53] P. H. Seeburg, J. Shine, J. A. Martial, J. D. Baxter, and H. M. Goodman, Nature (London) 270, 486 (1977).
[54] K. Backman, M. Ptashne, and W. Gilbert, Proc. Natl. Acad. Sci. U.S.A. 73, 4174 (1976).
[55] P. J. Laipis, B. M. Olivera, and A. T. Ganesan, Proc. Natl. Acad. Sci. U.S.A. 62, 289 (1969).
[56] A. Becker, G. Lyn, M. Gefter, and J. Hurwitz, Proc. Natl. Acad. Sci. U.S.A. 58, 1996 (1967).
[57] S. Soderhall and T. Lindahl, FEBS Lett. 67, 8 (1976).
[58] S. Soderhall and T. Lindahl, J. Biol. Chem. 250, 8438 (1975).
[59] S. Soderhall and T. Lindahl, Biochem. Biophys. Res. Commun. 53, 910 (1973).
[60] H. Teraoka, M. Shimoyachi, and K. Tsukada, FEBS Lett. 54, 217 (1975).
[61] D. G. Evans, S. H. Ton, and H. M. Kier, Biochem. Soc. Trans. 3, 1131 (1975).
[62] S. B. Zimmerman and C. J. Levin, J. Biol. Chem. 250, 149 (1975).
[63] S. Soderhall, Nature (London) 260, 640 (1976).
[64] P. Beard, Biochim. Biophys. Acta 269, 385 (1972).
[65] G. C. F. Pedrali Noy, S. Spadari, G. Ciarrocchi, A. M. Pedrini, and A. Falaschi, Eur. J. Biochem. 39, 343 (1973).
[66] U. Bertazzoni, M. Mathelet, and F. Campagnari, Biochim. Biophys. Acta 287, 404 (1972).
[67] H. Young, S. H. Ton, L. A. F. Morrice, R. S. Feldberg, and H. M. Keir, Biochem. Soc. Trans. 1, 520 (1973).
[68] E. Bedows, J. T. Wachsman, and R. I. Gumport, Biochemistry 16, 2231 (1977).
[69] C. J. Levin and S. B. Zimmerman, Biochem. Biophys. Res. Commun. 69, 514 (1976).

RNA Ligase

RNA ligase consists of a single 41,000-dalton polypeptide chain[70,71] encoded by bacteriophage T4 gene *63*.[72] It was discovered in T4-infected cells by Silber *et al.*[73] as an enzyme that catalyzed the circularization of single-stranded poly(A). RNA ligase has since been shown to carry out inter- as well as intramolecular joining of RNA and DNA chains. In the terminology widely employed for RNA ligase, the 5'-phosphoryl-terminated moiety is called the donor and the 3'-hydroxyl-terminated portion is termed the acceptor. The mechanism of RNA ligase-catalyzed reactions is analogous to that of DNA ligase (Fig. 1), except for the important distinction that a complementary strand is not required[74]; the abutment of ends is carried out at acceptor and donor binding sites on the enzyme. Like T4 DNA ligase, the energy cofactor is ATP. The first step in the reaction is formation of an adenylylated enzyme. This is a stable intermediate that can be easily identified by polyacrylamide gel electrophoresis in the presence of sodium dodecyl sulfate (SDS).[75] The effect of adenylylation is striking, because the decrease in electrophoretic mobility of the protein is that expected of an increase in molecular weight of 4000, although the actual increase is only 350. Adenylylation is reversible, so the enzyme catalyzes pyrophosphate exchange with ATP.[74] Transfer of AMP from the enzyme to the 5'-phosphoryl of the donor terminus requires the presence of an acceptor end.[76] The 3'-hydroxyl terminus of the acceptor may affect a conformational change necessary for adenyl transfer—the enzyme does not cock the trigger unless a target is in sight! The activated donor molecule does not necessarily join in the acceptor which stimulated its formation. The term *acceptor exchange* was coined for cases where the activated intermediate was formed in the presence of one acceptor but ligated to a different acceptor.[76] The adenylylated donor intermediate is difficult to detect with RNA substrates but accumulates in reactions with a DNA acceptor.[76] Evidently a DNA acceptor is better at evoking donor adenylylation than ligation. Proof that the adenylylated donor is a true reaction intermediate was obtained by purifying the compound and showing that it could be joined to an acceptor in the absence of

[70] T. J. Snopek, A. Sugino, K. Agarwal, and N. R. Cozzarelli, *Biochem. Biophys. Res. Commun.* **68,** 417 (1976).

[71] J. A. Last and W. F. Anderson, *Arch. Biochem. Biophys.* **174,** 167 (1976).

[72] T. J. Snopek, W. B. Wood, M. P. Conley, P. Chen, and N. R. Cozzarelli, *Proc. Natl. Acad. Sci. U.S.A.* **74,** 3355 (1977).

[73] R. Silber, V. G. Malathi, and J. Hurwitz, *Proc. Natl. Acad. Sci. U.S.A.* **69,** 3009 (1972).

[74] J. W. Cranston, R. Silber, V. G. Malathi, and J. Hurwitz, *J. Biol. Chem.* **249,** 7447 (1974).

[75] N. P. Higgins, A. P. Geballe, T. J. Snopek, A. Sugino, and N. R. Cozzarelli, *Nucleic Acids Res.* **4,** 3175 (1977).

[76] A. Sugino, T. J. Snopek, and N. R. Cozzarelli, *J. Biol. Chem.* **252,** 1732 (1977).

ATP with the stoichiometric release of AMP.[76,77] In addition, many ade-nylylated synthetic compounds are donor substrates in the absence of ATP (see below).[78]

RNA ligase can be easily purified in large amounts. The most specific enzyme assay is the one originally used by Silber *et al.*[73]—cyclization of [5'-^{32}P]poly(A) which renders the ^{32}P resistant to digestion with alkaline phosphatase. At later stages in the purification, the enzyme can also be detected by exchange of labeled pyrophosphate with ATP[74] and by forma-tion of an acid-precipitable enzyme–adenylate complex with [^{3}H]ATP.[25] The kinetics of the appearance of RNA ligase after T4 infection is highly unusual in that the enzyme is synthesized throughout the latent period. For maximum enzyme yield, harvest of infected cells should be delayed as long as possible. Moreover, a seven-fold increase in the amount of RNA ligase results from infection with T4 double-mutant strains carrying one mutation in the *regA* gene and another mutation in a gene required for DNA synthesis.[75] Using this rich source of enzyme, Higgins *et al.*[75] pre-pared physically homogeneous, nuclease-free enzyme in high yield. There are several other purification procedures yielding RNase-free enzyme of good purity,[70,71,73] but these preparations contain a contaminating DNA exonuclease that removes 5'-mononucleotides from the 3'-terminus of single-stranded substrates. This nuclease level, although small compared to the ligase activity, represents a major obstacle for DNA-joining reac-tions that often require high levels of enzyme. Chromatography on DNA–agarose removes this exonuclease.[75]

RNA ligase is a versatile reagent for the radical construction of nucleic acids. Intermolecular reactions leading to high-yield synthesis of oligori-bonucleotides with a defined sequence have been described by several groups.[77,79-83] Self-polymerization of acceptors can be prevented by using polynucleotides with hydroxyls at both the 5'- and 3'-ends, and several strategies have been employed to limit self-reaction of the donor. One method is to use a large excess of acceptor.[77,79] However, this may re-quire unacceptably high acceptor concentrations, since intramolecular

[77] E. Ohtsuka, S. Nishikawa, M. Sugiura, and M. Ikehara, *Nucleic Acids Res.* **3**, 1613 (1976).

[78] T. E. England, R. I. Gumport, and O. C. Uhlenbeck, *Proc. Natl. Acad. Sci. U.S.A.* **74**, 4839 (1977).

[79] G. C. Walker, O. C. Uhlenbeck, E. Bedows, and R. I. Gumport, *Proc. Natl. Acad. Sci. U.S.A.* **72**, 122 (1975).

[80] J. J. Sninsky, J. A. Last, and P. T. Gilham, *Nucleic Acids Res.* **3**, 3157 (1976).

[81] O. C. Uhlenbeck and V. Cameron, *Nucleic Acids Res.* **4**, 85 (1977).

[82] E. Ohtsuka, S. Nishikawa, R. Fukumoto, S. Tanaka, A. F. Markham, M. Ikehara, and M. Sugiura, *Eur. J. Biochem.* **81**, 285 (1977).

[83] Y. Kikuchi, F. Hishinuma, and K. Sakaguchi, *Proc. Natl. Acad. Sci. U.S.A.* **75**, 1270 (1978).

joining is generally a favored reaction for donors long enough to circularize (six to eight nucleotides). Donors with a phosphoryl at both the 3'- and 5'-termini are used efficiently with equimolar acceptor concentrations,[81] and Kikuchi et al. recently synthesized on a milligram scale an octanucleotide segment of bacteriophage Qβ coat protein gene using this method.[84] Periodate oxidation[85] and addition of methoxyethyl[80] and ethoxymethylidene[86] groups have been used to block the 3'-hydroxyl of donors. The latter methods are easily reversible, and Ohtsuka et al.[86] used this approach to synthesize a heptadecanucleotide corresponding to bases 61–77 of E. coli tRNA^{fMet}. The smallest monoaddition donors are the mononucleoside 3',5'-bisphosphates pCp, pAp, pUp, pGp, pdCp, pdAp, pdTp, pdUp, and pdGp. They add to the 3'-terminus of a number of different acceptors including tRNA, mRNA, and double- and single-stranded viral RNA.[87,88] These studies are examples of the important application of RNA ligase technology to the precise engineering of naturally derived RNA molecules. In addition, "mutant" yeast tRNA^{Phe} was synthesized by Kaufmann and Littauer[89] by joining two half-molecules in which the Y base at position 37 had been removed. More recently, Meyhack et al.[90] constructed six model substrates of the B. subtilis 5 S rRNA precursor to test the substrate recognition requirements of the M5 processing nuclease.

DNA is about as good a donor as RNA for RNA ligase-catalyzed reactions but is a poorer acceptor. Snopek et al.[70] showed that a number of short synthetic oligodeoxyribonucleotides participated in intra- and intermolecular reactions. Cyclization of $(pdT)_n$ from 6 to 30 nucleotides long was observed, and at the optimum chain length of 20 the rate was $\frac{1}{10}$ that of cyclization of poly(A).[76] Sugino et al.[76] showed that 2",3'-deoxynucleotides such as dideoxythymidylate (d_2T) were effective 3'-blocking groups to prevent self-joining of DNA donors. Intermolecular joining of the donor $(pdC)_8pd_2T$ with $dA(pdA)_5$ was observed. Facilitation of donor activation by acceptor exchange improved the reaction-addition of ApApA resulted in a 10-fold stimulation of DNA-to-DNA joining, and the final yield represented an acceptable 13%. As a donor for intermolecular joining with the ribonucleotide acceptor $A(pA)_{\overline{20}}$, $(dpT)_{12-18}pd_2T$

[84] Y. Kikuchi and K. Sakaguchi, Nucleic Acids Res. 5, 591 (1978).
[85] G. Kaufmann and N. R. Kallenbach, Nature (London) 254, 452 (1975).
[86] E. Ohtsuka, S. Nishikawa, A. F. Markham, S. Tanaka, T. Miyake, T. Wakabayashi, M. Ikehara, and M. Sugiura, Biochemistry 17, 4894 (1978).
[87] G. A. Bruce and O. C. Uhlenbeck, Nucleic Acids Res. 5, 3665 (1978).
[88] T. E. England and O. C. Uhlenbeck, Nature (London) 275, 560 (1978).
[89] G. Kaufmann and U. Z. Littauer, Proc. Natl. Acad. Sci. U.S.A. 71, 3741 (1974).
[90] B. Meyhack, B. Pace, O. C. Uhlenbeck, and N. R. Pace, Proc. Natl. Acad. Sci. U.S.A. 75, 3045 (1978).

was even better than periodate-oxidized $(pA)_{\overline{20}}$. Completing the set of possible combinations, $dA(pdA)_5$ served as an acceptor for oxidized $(pA)_{\overline{20}}$, although it was roughly $\frac{1}{10}$ as effective as ApApA. Finally, Hinton et al.[91] have reported that oligodeoxyribonucleotide acceptors can be joined to 2'-deoxyribonucleoside 3',5'-bisphosphates.

RNA ligase also joins 3',5'-hydroxyl ribohomopolymers to large natural DNA substrates, generating single-strand extensions at the 5'-termini.[92] This reaction, called 5'-tailing, is surprisingly insensitive to the size and structure of the DNA donor. Double- and single-stranded DNA fragments from 10 to 6000 nucleotides long can be joined quantitatively to a number of oligoribonucleotide acceptors. The DNAs tested thus far include ColE1 DNA digested with EcoR1 or HaeIII restriction endonucleases which produce cohesive and base-paired ends, respectively, heat-denatured HaeIII-generated ColE1 DNA fragments, full-length linear ϕX174 viral DNA, and a self-complementary synthetic decanucleotide. The acceptor requirements are much more fastidious. The most efficient acceptor is $A(pA)_5$, but $I(pI)_5$ and $C(pC)_5$ are also utilized well, whereas $U(pU)_5$ is inert. The shortest acceptor joined is a trinucleoside diphosphate, and the optimum length is six nucleotides, but polymers between 10 and 20 nucleotides long are ligated at half the maximal rate. The 5'-tailing reaction provides an efficient radical method for generating extensions of predetermined length and base composition at the 5'-ends; it complements the terminal deoxynucleotidyltransferase-mediated addition of either ribo- or deoxyribonucleotides to the 3'-hydroxyl terminus.[12,14] On duplex DNA substrates the 5'-tailing product provides a natural template for reverse transcriptase, and therefore it would be possible to build duplex structures at the 5'-end of long molecules.

RNA ligase is capable of performing an amazing variety of reactions, but it is not totally permissive. As donors, it accepts molecules as diverse as a 3',5'-mononucleoside bisphosphate and the 5'-base-paired end of duplex DNA thousands of base pairs long. There is even more latitude for the already activated donors of the general formula A-5'PP-X, since X does not even need to be a purine or pyrimidine. England et al.[78] have found that ADP ribose, diphospho-CoA, FAD, and even ADP cyanoethanol, along with AppA, AppC, AppG, AppU, and AppT, can be joined to oligoribonucleotides. The acceptor requirements are much stricter. The base, sugar, and chain length have all been shown to influence both the rate and the final yield of joining in a number of systems. The smallest acceptor is a trinucleoside diphosphate. A ribose moiety at the 3'-terminus is preferred by usually a factor of 10 or more to a deoxyribose. Oligo(A) at

[91] D. M. Hinton, J. A. Baez, and R. I. Gumport, Biochemistry 17, 5091 (1978).
[92] N. P. Higgins, A. P. Geballe, and N. R. Cozzarelli, Nucleic Acids Res. 6, 1013 (1979).

the acceptor terminus provides the best substrates, followed by oligo(I) and oligo(C); oligo(U) is generally very poor. However, this may be an oversimplification. Kikuchi et al.[83] found that $U(pU)_3$ would join efficiently to pCp but not at all to pAp, and there are other examples where it is not simply the acceptor that is important but the particular combination of acceptor and donor.[86]

The enzyme has thus far proved to have an unmeasurably high K_m for both the donor and acceptor termini. Therefore, the reaction increases proportionately with substrate concentration. This is sometimes an important factor in deciding whether or not a reaction with a poor substrate like a DNA acceptor can profitably be attempted. With a high concentration of molecules, there is a good chance for success; but if the molecule is extremely large, then the prospects are diminished. Fortunately large amounts of pure enzyme are easily obtainable and can be used to drive some sluggish reactions.

It is ironic that with the wealth of RNA ligase reactions demonstrated in vitro there is no evidence for an involvement in nucleic acid metabolism in vivo other than the weak indication provided by the initiation of its synthesis early after infection. However, unlike the enzymes known to be required for nucleic acid metabolism, it continues to be produced throughout the latent period. Mutants that have no detectable RNA ligase activity seem to have no abnormality in RNA or DNA metabolism but are blocked in the essential step of attachment of tail fibers to phage heads, notably the last step in virus development.[72] This morphogenetic step is efficiently catalyzed by RNA ligase in vitro.[72] Here is an intriguing formal analogy to the RNA ligase facilitation of blunt-end joining by DNA ligase where its proposed role is also the apposition of participating macromolecular entities. The unmeasurably high K_m for all nucleic acids tested suggests that its natural nucleic acid substrate, if it exists at all, may yet remain to be identified.

In eukaryotes, joining of RNA is clearly of widespread importance. The direct transcript is generally longer than the functional RNA.[93,94] An activity has been identified that cleaves a precursor at specific sites and joins the ends to splice the molecules into a functional unit. O'Farrell et al.[95] and Knapp et al.[96] have found that a yeast extract carries out site-

[93] D. F. Klessig, Cell 12, 9 (1977).
[94] S. M. Berget, C. Moore, and P. A. Sharp, Proc. Natl. Acad. Sci. U.S.A. 74, 3171 (1977).
[95] P. Z. O'Farrell, B. Cordel, P. Valenzuela, W. J. Rutter, and H. M. Goodman, Nature (London) 274, 438 (1978).
[96] G. Knapp, J. S. Beckmann, P. F. Johnson, S. A. Fuhrman, and J. Abelson, Cell 14, 221 (1978).

specific cleavage and rejoining of precursors of several tRNAs. Blanchard *et al.*[97] have also reported evidence for *in vitro* splicing of adenovirus-2 mRNA in HeLa cell extracts. The mechanism of this reaction is as yet undetermined, but the requirement for ATP in the yeast system may signal that a ligase analogous to RNA ligase is needed for the joining process. The enzymology of circular RNA production in plants is unknown.

B-R Enzymes

In 1971 Wang[98] purified an enzyme from *E. coli,* the ω protein, that removed superhelical turns from covalently closed DNA molecules. The reaction must involve breakage of backbone bonds, rotation about the helix axis opposite the sense of the supertwists, and reunion of the backbone. Nonetheless, ω had no demonstrable nuclease or ligase activity and required no energy cofactor. Moreover, since relaxation could occur gradually, ω must prevent free rotation of the transiently broken ends of DNA. To account for these results Wang[98] proposed that the phosphodiester bond energy released by DNA strand rupture was stored in a high-energy complex of enzyme and DNA, and that the dissociation of this complex was concomitant with reformation of the DNA backbone. This model accounted for the lack of ligase activity in ω, because the substrate for the joining component is not nicked DNA but instead a transient nick is formed in continuous regions of DNA by the enzymes. The ends of the nick do not rotate freely because one end is covalently bound to the enzyme and the adjacent other end could easily be impeded by the large enzyme. The requirement for an exogneous energy source is circumvented by the high-energy intermediate. This should be compared to the DNA ligase reaction where the energy extracted from the cofactor in the first step is stored in subsequent reaction intermediates.

The breakage-reunion[98a] (B-R) model is probably correct and has heralded new classes of proteins that play important roles in the replication, recombination, repair, and transcription of DNA. One class is the topoisomerases that carry out the interconversion of topological isomers of

---

[97] J. M. Blanchard, J. Weber, W. Jelinek, and J. E. Darnell, *Proc. Natl. Acad. Sci. U.S.A.* **75,** 5344 (1978).

[98] J. Wang, *J. Mol. Biol.* **55,** 523 (1971).

[98a] We prefer the term *breakage-reunion* to *nicking-closing* because the interruption in the DNA need not be a nick but could be across both strands.

DNA. Examples include ω proteins, bacterial DNA gyrases, and eukaryotic untwisting enzymes.[18] B-R proteins whose primary activity is not the interconversion of topological isomers of DNA include the ϕX174 *cisA* protein and the λ *int* protein. The relaxation complexes of small plasmids studied by Helinski and co-workers[99] may be intermediate in B-R reactions, since denaturation of relaxation complexes and topoisomerases under defined conditions leads to DNA strand rupture with nucleic acid covalently linked to protein.[100–104]

ω proteins from *E. coli* and *Micrococcus luteus* have been purified to homogeneity and consist of single 110,000- and 120,000-dalton polypeptides, respectively.[101,105,106] Both enzymes require Mg^{2+} and act efficiently on highly negatively twisted (underwound) DNA substrates. Covalently closed DNAs containing few negative or positive superhelical turns are relaxed very poorly. With single-stranded circular DNA substrates these proteins catalyze the formation of topologically knotted rings.[105,107] In this reaction a region of the circle must be passed between the ends of the interrupted DNA backbone prior to ligation. Another reaction demonstrating the same principle is the reassociation of single-stranded complementary circular DNA molecules into a covalently closed duplex structure.[108] The high-energy reaction intermediate poised for resealing contains ω covalently linked to the 5′-phosphoryl of the transiently broken DNA chain.[101]

The rat liver DNA untwisting enzyme has also been extensively purified and consists of a single polypeptide of about 65,000 daltons.[109,110] This protein and other similar eukaryotic activities are set apart from the prokaryotic relaxing enzymes by the lack of a divalent metal ion requirement, the efficient removal of positive as well as negative superhelical coils, and, in the key reaction intermediate, the covalent attachment of pro-

[99] D. G. Blair and D. R. Helinski, *J. Biol. Chem.* **250**, 8785 (1975).

[100] D. G. Guiney and D. R. Helinski, *J. Biol. Chem.* **250**, 8796 (1975).

[101] R. E. Depew, L. F. Liu, and J. C. Wang, *J. Biol. Chem.* **253**, 511 (1978).

[102] J. J. Champoux, *Proc. Natl. Acad. Sci. U.S.A.* **74**, 3800 (1977).

[103] A. Sugino, C. L. Peebles, K. N. Kreuzer, and N. R. Cozzarelli, *Proc. Natl. Acad. Sci. U.S.A.* **74**, 4767 (1977).

[104] M. Gellert, K. Mizuuchi, M. H. O'Dea, T. Itoh, and J. Tomizawa, *Proc. Natl. Acad. Sci. U.S.A.* **74**, 4772 (1977).

[105] V. T. Kung and J. C. Wang, *J. Biol. Chem.* **252**, 5398 (1978).

[106] R. Hecht and H. W. Thielman, *Nucleic Acids Res.* **4**, 4235 (1978).

[107] L. F. Liu, R. E. Depew, and J. C. Wang, *J. Mol. Biol.* **106**, 439 (1976).

[108] K. Kirkegaard and J. C. Wang, *Nucleic Acids Res.* **5**, 3811 (1978).

[109] W. Keller, *Proc. Natl. Acad. Sci. U.S.A.* **72**, 4876 (1975).

[110] J. J. Champoux and B. L. McConaughy, *Biochemistry* **15**, 4638 (1976).

tein to a 3'-phosphoryl DNA terminus.[102] Like ω, the rat liver enzyme can intertwine single-stranded covalently closed circular molecules to form a closed duplex molecule.[111] This unusual activity may signal an involvement of untwisting enzymes and ω in the homologous recombination of topologically constrained DNA molecules.[108,111] The positive coil-relaxing activity of untwisting enzymes, histones, and an assembly factor are required for the efficient *in vitro* synthesis of chromatin-like material,[112] and this has been suggested as the mechanism for generation of negative superhelical turns in eukaryotic DNA molecules.[113]

DNA gyrase, which was discovered by Gellert *et al.*[114] as a host activity required for the *in vitro* integrative recombination of relaxed bacteriophage λ DNA molecules, introduces negative superhelical coils into relaxed DNA. The enzyme requires ATP and Mg^{2+} and is stimulated by spermidine. Temperature-sensitive mutants have been isolated that have temperature-sensitive gyrase activity.[115,116] Shift of these mutants to the nonpermissive temperature arrests DNA synthesis, as does addition of the gyrase inhibitor nalidixate or novobiocin. The enzyme is also important for transcription, recombination, and repair of cellular DNA.[114–118] For *E. coli* gyrase, the subunits designated A and B contain the 105,000-dalton product of the *nalA* gene and the 95,000-dalton product of the *cou* gene, respectively.[103,104,119,120] The *M. luteus* enzyme is composed of two polypeptides, α and β, with molecular weights of 115,000 and 97,000, respectively.[121] Gyrase can be purified as a complex of A and B proteins or in the form of subunits which can be mixed to reconstitute activity.[119] In the absence of ATP the enzyme relaxes negative superhelical turns in DNA. Thus, like other topoisomerases, ATP is not required to break and rejoin

[111] J. J. Champoux, *Proc. Natl. Acad. Sci. U.S.A.* **74**, 5328 (1977).
[112] R. A. Laskey, B. M. Honda, A. D. Mills, and J. T. Finch, *Nature (London)* **275**, 416 (1978).
[113] J. E. Germond, B. Hirt, P. Oudet, M. Gross-Bellard, and P. Chambon, *Proc. Natl. Acad. Sci. U.S.A.* **72**, 1843 (1975).
[114] M. Gellert, K. Mizuuchi, M. H. O'Dea, and H. A. Nash, *Proc. Natl. Acad. Sci. U.S.A.* **73**, 3872 (1976).
[115] K. N. Kreuzer, K. McEntee, A. P. Geballe, and N. R. Cozzarelli, *Mol. Gen. Genet.* **167**, 129 (1978).
[116] C. L. Peebles, N. P. Higgins, K. N. Kreuzer, A. Morrison, P. O. Brown, A. Sugino, and N. R. Cozzarelli, *Cold Spring Harbor Symp. Quant. Biol.* **43**, 41 (1978).
[117] C. L. Smith, K. Kubo, and F. Imamoto, *Nature (London)* **275**, 424 (1978).
[118] J. B. Hays and S. Boehmer, *Proc. Natl. Acad. Sci. U.S.A.* **75**, 4125 (1978).
[119] N. P. Higgins, C. L. Peebles, A. Sugino, and N. R. Cozzarelli, *Proc. Natl. Acad. Sci. U.S.A.* **75**, 1773 (1978).
[120] M. Gellert, M. H. O'Dea, T. Itoh, and J. Tomizawa, *Proc. Natl. Acad. Sci. U.S.A.* **73**, 4474 (1976).
[121] L. F. Liu, and J. C. Wang, *Proc. Natl. Acad. Sci. U.S.A.* **75**, 2098 (1978).

the DNA backbone bonds. Hydrolysis of ATP is not even required for introduction of negative supercoils, since the nonhydrolyzable ATP analog, adenyl-5′-y1-imidodiphosphate, promotes limited supercoiling.[122] The covalent intermediate of DNA gyrase differs from those of ω and rat liver untwisting enzymes in two important and interesting ways. First, denaturation of gyrase leads to site-specific breaks[103,104]; breakage by the relaxing enzyme is not random but occurs at many more sites. Second, gyrase breakage is across both strands of DNA, and protein becomes covalently attached to the two resulting 5′-phosphoryl groups which are separated by a four-base stagger.[123] The enzyme thus generates a four-base cohesive end which can be repaired *in vitro* with *E. coli* DNA polymerase I.[123]

The *cisA* protein coded by φX174 is a polypeptide of 60,000 daltons, which is required for virus replication.[124,125] The proposed role for this enzyme in the replication of closed duplex (RF) DNA is as follows. Upon binding to the supercoiled RF molecule it produces a site-specific break at approximately residue 4300 of the positive strand by forming a covalent bond with the 5′-phosphoryl.[126] This is like the intermediate proposed by Wang for ω, but in this case the demonstrable covalent bond and rupture of the strands is a normal event and not tied to enzyme denaturation. The *cisA* protein interacts with *rep* unwinding protein, helix-destabilizing protein, and DNA polymerase in the unwinding of the duplex concomitant with DNA replication.[127] Chain growth proceeds with the *cisA*-bound terminus locked at the replication fork.[128] Upon complete traverse of the circle, *cisA* protein cleaves the new origin and rejoins the two original 5′- and 3′-ends, releasing a single-strand circle. Thus, like topoisomerases, *cisA* protein is a B-R enzyme but one with a protracted coupling period. It may be a radical enzyme, since a template may not be required; but if so, unlike the situation in the radical ligases we have considered, the nucleotide sequence of the substrate is conserved.

The site-specific integration of bacteriophage λ DNA into host chromosomal sequences requires the *int* gene product. Recently this protein

[122] A. Sugino, N. P. Higgins, P. O. Brown, C. L. Peebles, and N. R. Cozzarelli, *Proc. Natl. Acad. Sci. U.S.A.* **75**, 4838 (1978).

[123] A. Morrison and N. R. Cozzarelli, *Cell* **17**, 175 (1979).

[124] S. Eisenberg, J. F. Scott, and A. Kornberg, *Proc. Natl. Acad. Sci. U.S.A.* **73**, 1594 (1976).

[125] J.-E. Ikeda, A. Yudelevich, and J. Hurwitz, *Proc. Natl. Acad. Sci. U.S.A.* **73**, 2269 (1976).

[126] S. Eisenberg and A. Kornberg, *J. Biol. Chem.* **254**, 5328 (1979).

[127] J. R. Scott, S. Eisenberg, L. L. Bertsch, and A. Kornberg, *Proc. Natl. Acad. Sci. U.S.A.* **74**, 193 (1977).

[128] S. Eisenberg, J. Griffith, and A. Kornberg, *Proc. Natl. Acad. Sci. U.S.A.* **74**, 3198 (1977).

was purified and shown to have a molecular weight of 40,000.[129] The purified *int* gene product is a site-specific DNA-binding protein and must be supplemented with host proteins to carry out recombination. The reaction proceeds in the absence of Mg^{2+} or an energy cofactor, but spermidine is required, as in the presence of negative supercoils in the DNA substrate under some reaction conditions. Interestingly, the superhelical turns present in the substrate are not lost during the reaction, which means that the four ends that are broken and joined to separate partners are physically restrained during the complete reaction sequence.[130] Although there is no evidence as yet for a covalent protein–DNA intermediate in the integration reaction, the lack of an energy cofactor and the conservation of the supercoiling of the DNA substrate presupposes a coupled B-R mechanism.

Perspectives

Enzymes that carry out joining of DNA chains can be grouped into two distinct categories—ligases and B-R enzymes (Table II). The former join the ends of polynucleotides and require an energy source. Both the *E. coli* and the phage T4-coded DNA ligases are primarily conservative, i.e., template-directed enzymes, and play essential roles in DNA metabolism. However, T4 DNA ligase can also act radically and alter the nucleotide sequence of the substrate, as in blunt-end joining and ligation of mismatched termini. T4 RNA ligase is strictly a radical enzyme. Unfortunately, it is unknown whether the radical properties of ligases are biologically important.

B-R enzymes, on the other hand, carry out a coupled breakage and reunion of nucleic acids at contiguous regions via covalent protein–DNA intermediates and do not require energy input. However, a cofactor may be used to drive an endergonic reaction, as in the generation of negative supercoils by DNA gyrase. The role of B-R enzymes in the replication and recombination of DNA is becoming increasingly clear. DNA gyrase maintains the negative superhelical tension in DNA for replication and transcription, and for integration of bacteriophage λ. The *cisA* protein initiates the replication of ϕX174 closed duplex DNA. Although not yet proved to be a B-R reaction, λ integration has the properties anticipated—no energy requirement for joining, and conservation of topological structure of the DNA throughout the reaction. The integration of phage mu, a transposon, may share common host factors with the λ-

[129] Y. Kikuchi and H. A. Nash, *J. Biol. Chem.* **253,** 7149 (1978).
[130] K. Mizuuchi, M. Gellert and H. A. Nash, *J. Mol. Biol.* **121,** 375 (1978).

TABLE II

JOINING ENZYMES[a]

Enzyme	Classification	Function	Genetic loci	Protomer molecular weight	Type of reaction intermediate	Energy cofactor
E. coli DNA ligase	Ligase, conservative	Replication and repair of DNA	lig	74,000	Protein–5′-pA	NAD
T4 DNA ligase	Ligase, conservative and radical	Replication and repair of viral DNA	T4 gene 30	68,000	Protein–5′-pA	ATP
T4 RNA ligase	Ligase, radical	Tail fiber attachment	T4 gene 63	41,000	Protein–5′-pA	ATP
E. coli ω	B-R, conservative	?	?	110,000	Protein–5′-pDNA	None
DNA untwisting enzyme	B-R, conservative	Chromatin assembly	?	65,000	Protein–3′-pDNA	None
E. coli DNA gyrase	B-R, conservative	Replication, transcription, repair, and recombination of DNA	nalA and cou	105,000 and 95,000	Protein–5′-pDNA	ATP
φX174 cisA protein	B-R, radical?	Initiation of closed duplex φX174 DNA replication	cisA	60,000	Protein–5′-pDNA	None
λ int protein	B-R, conservative	Integration and excision of λ DNA	λ int and host factors	40,000 and unknown	?	None

[a] Conventions are defined in the footnote to Table I.

integration system (H. Nash, personal communication). However, a short sequence of the host DNA is duplicated in the same orientation immediately adjacent to the two ends of some inserted transposable elements.[19] Therefore, if a B-R enzyme is responsible for insertion, it is likely to be radical.

The enzymes with the widest use for joining nucleic acids are the two T4 ligases. These have complementary substrate specificity. With T4 DNA and RNA ligases available, virtually all possible combinations of inter- and intramolecular ligation of single- and double-stranded RNA and DNA substrates can be performed. Their utility in constructing recombinant DNA is manifest.

[2] Guide to the Use of Type II Restriction Endonucleases

By ROY FUCHS and ROBERT BLAKESLEY

Type II restriction endonucleases are DNases that recognize specific oligonucleotide sequences, make double-strand cleavages, and generate unique, equal molar fragments of a DNA molecule. By the nature of their controllable, predictable, infrequent, and site-specific cleavage of DNA, restriction endonucleases proved to be extremely useful as tools in dissecting, analyzing, and reconfiguring genetic information at the molecular level. Over 350 different restriction endonucleases have been isolated from a wide variety of prokaryotic sources, representing at least 85 different recognition sequences.[1,2] A number of excellent reviews detail the variety of restriction enzymes and their sources,[2,3] their purification and determination of their sequence specificity,[4,5] and their physical properties, kinetics, and reaction mechanism.[6] Here we provide a summary, based on the literature and our experience in this laboratory, emphasizing the practical aspects for using restriction endonucleases as tools. This review focuses on the reaction, its components and the conditions that affect enzymic activity and sequence fidelity, methods for terminating the reaction, some reaction variations, and a troubleshooting guide to help identify and solve restriction endonuclease-related problems.

The Reaction

Despite the diversity of the source and specificity for the over 350 type II restriction endonucleases identified to date,[1,2] their reaction conditions are remarkably similar. Compared to other classes of enzymes these conditions are also very simple. The restriction endonuclease reaction (Table I) is typically composed of the substrate DNA incubated at 37° in a solution buffered near pH 7.5, containing Mg^{2+}, frequently Na^+, and the selected restriction enzyme. Specific reaction details as found in the liter-

[1] R. Blakesley, in "Gene Amplification and Analysis," Vol. 1: "Restriction Endonucleases" (J. G. Chirikjian, ed.), p. 1. Elsevier/North-Holland, Amsterdam, 1981.

[2] R. J. Roberts, Nucleic Acids Res. 10, r117 (1982).

[3] J. G. Chirikjian, "Gene Amplification and Analysis," Vol. 1: "Restriction Endonucleases." Elsevier/North-Holland, Amsterdam, 1981.

[4] R. J. Roberts, CRC Crit. Rev. Biochem. 4, 123 (1976).

[5] This series, Vol. 65, several articles.

[6] R. D. Wells, R. D. Klein, and C. K. Singleton, in "The Enzymes" (P. D. Boyer, ed.), 3rd ed., Vol. 14, Part A, p. 157. Academic Press, New York, 1981.

RECOMBINANT DNA
METHODOLOGY

TABLE I
GENERALIZED REACTION CONDITIONS FOR
RESTRICTION ENDONUCLEASES

	Reaction type	
Conditions	Analytical	Preparative
Volume	20–100 μl	0.5–5 ml
DNA	0.1–10 μg	10–500 μg
Enzyme	1–5 units/μg DNA	1–5 units/μg DNA
Tris-HCl (pH 7.5)	20–50 mM	50 mM
MgCl$_2$	5–10 mM	10 mM
2-Mercaptoethanol	5–10 mM	5–10 mM
Bovine serum albumin	50–500 μg/ml	200–500 μg/ml
Glycerol	<5% (v/v)	<5% (v/v)
NaCl	As required	As required
Time	1 hr	1–5 hr
Temperature	37°	37°

ature for the more frequently used enzymes are listed in Table II. Note that in most cases these data do not represent optimal reaction conditions.

By convention, a unit of restriction endonuclease activity is usually defined as that amount of enzyme required to digest completely 1 μg of DNA (usually of bacteriophage lambda) in 1 hr.[4] This definition was chosen for convenience, since the useful, readily measurable end result of a restriction endonuclease reaction is completely cleaved DNA. However, a unit defined in this manner measures enzyme activity by an end point rather than by the classical initial rate term. Thus, traditional kinetic arguments based upon substrate saturating (initial rate) conditions cannot be applied to restriction endonucleases defined in this (enzyme saturating) manner.

One reason why there are few proper kinetic data on restriction endonucleases lies in the difficulty in measuring restriction enzyme activities during the linear portion of the reaction when using the standard enzyme assay.[7] The strong emphasis placed on their use as research tools in molecular biology rather than on investigation of their biochemical properties also contributed to the deficiency. Hence we lack good experimental data on conditions for optimal activity. For most newly isolated restriction endonucleases, assay buffers were selected for convenience during enzyme isolation rather than for optimal reactivity. These conditions have persisted as dogma. Thus, the implied precision and unique-

[7] P. A. Sharp, B. Sugden, and J. Sambrook, *Biochemistry* **12**, 3055 (1973).

ness of these values, e.g., pH 7.2 vs pH 7.4, is frequently without experimental basis. In fact, where investigated, restriction endonucleases usually show relatively broad activity profiles for the various reaction parameters.[8-10]

The fact that restriction endonucleases are active under a variety of conditions indicates that, similar to other nucleases, they are rather hardy enzymes. From an enzymologist's viewpoint, these enzymes can be mishandled and still demonstrate activity. But to achieve reproducible, efficient, and specific DNA cleavages, certain factors concerning restriction enzyme reactions should be considered. From our experience the most important factors for proper restriction endonuclease use are (a) the purity and physical characteristics of the substrate DNA; (b) the reagents used in the reaction; (c) the assay volume and associated errors; and (d) the time and temperature of incubation.

In the following sections each of these reaction parameters is discussed in detail. General conclusions are drawn in order to provide the researcher a framework in which properly to use restriction endonucleases. However, one must always be cognizant of the fact that each restriction endonuclease represents a unique enzymic protein. Any kinetic or biochemical generalization applied to the over 350 restriction enzymes will find exceptions.

DNA

The single most critical component of a restriction endonuclease reaction is the DNA substrate. DNA products generated in the reaction are directly affected by the degree of purity of the DNA substrate. Improperly prepared DNA samples will be cleaved poorly, if at all, producing partially digested DNA. In addition to DNA purity, other DNA-associated parameters that affect the products of the restriction endonuclease reaction include: DNA concentration, the specific sequence at and adjacent to the recognition site (including nucleotide modifications), and the secondary/tertiary DNA structure. Physical data pertaining to the DNA to be cleaved, if known, can guide one in choosing appropriate reaction conditions or prereaction treatments. Conversely, the response of a DNA of unknown physical properties to a standard restriction endonuclease digest can suggest certain characteristics of the DNA, e.g., the extent of methylation (see below).

---

[8] R. W. Blakesley, J. B. Dodgson, I. F. Nes, and R. D. Wells, J. Biol. Chem. **252,** 7300 (1977).

[9] P. J. Greene, M. S. Poonian, A. L. Nussbaum, L. Tobias, D. E. Garfin, H. W. Boyer, and H. M. Goodman, J. Mol. Biol. **99,** 237 (1975).

[10] B. Hinsch and M.-R. Kula, Nucleic Acids Res. **8,** 623 (1980).

TABLE II

REACTION CONDITIONS FOR CERTAIN RESTRICTION ENDONUCLEASES SELECTED FROM THE LITERATURE

Enzyme	Temperature (°C)	pH	Tris-HCl (mM)	$MgCl_2$ (mM)	NaCl (mM)	2-Mercapto-ethanol (mM)	Notes	Reference[*]
AluI	37	7.9	6	6	—	6	—	1
AsuI	37	7.5	20	10	100	—	—	2
AvaI	37	7.5	20	10	100	20	—	3
AvaII	37	7.5	20	10	100	20	—	3
BalI	37	7.9	6	6	—	6	—	4
BamHI	37	7.3	10	13[a]	50–100[a]	—	—	5
BamHI·1[b]	37	8.5	20	10	—	2	c	6
BclI	50[a]	7.4	12	12	—	0.5 mM DTT[b]	—	7
BglI	30[a]	9.5[a]	20 mM GOH[b]	20[a]	150[a]	7	d, e	8
BglII	30[a]	9.5[a]	20 mM GOH[b]	10[a]	—	—	e, f	8
BspI	37	8[a]	25	20[a]	50[a]	—	—	9
BstI	37–50[a]	7–9.5[a]	100	0.5–2[a]	—	—	g	10
BstI*[b]	37	9	100	>10	—	—	h	10
Bst1503I	65[a]	7.8[a]	10[a]	0.2[a]	—	6.6	—	11
BsuI	37	7.4[a]	10	10[a]	150[a]	1 mM DTT[b]	i	12
BsuI*[b]	37	8.5	25	10	—	5	—	12
ClaI	37	7.4	6	6	50	6	—	13
DdeI	37	7.5	100	5	100	—	j	14
DpnI	37	7.5	50	5	50	—	k, l	15
DpnII	37	7.5	50	5	50	—	k, l	15
EcaI	37	8	10	10	—	—	l	16
EcoRI	37	7.1–7.5[a]	100	5[a]	50[a]	—	—	17
EcoRI*[b]	37	8.5	25	2	—	—	e	18
EcoRII	37	7.4	25	5	—	—	—	19
FnuDII	37	7.9	6	6	50–150[a]	6	—	20
HaeII	37	7.9	6	6	—	6	—	21

Enzyme	Temperature (°C)	pH				0.5 mM DTT		Reference
Hae III	70[a]	7.5[a]	50[a]	5[a]	—	—	—	22
Hga I	37	7.6	10	5[a]	—	7	m	23
Hgi AI	30–45[a]	7.5–8.5[a]	10	10	100–150[a]	10	—	24
Hha I	37	7.9	6	6	—	6	—	25
Hinc II	37	7.9	10	6.6	60	6	e, n	26
Hind III	37	8.5[a]	10	10[a]	60[a]	7	j	27
Hinf I	37	7.5	6.6	10	50	6.6	k, o	28
Hpa I	45[a]	7.7–8.1[a]	10	5[a]	—	10	k, p	29
Hpa II	37	7.5	10	10	—	10	—	29
Hph I	37	7.4	10	10	6 mM KCl	10	—	30
Kpn I	37	7.9	6	6	—	6	—	31
Mbo I	37	7.9	6	6	—	6	—	32
Mbo II	37	7.9	6	6	—	6	—	32
Mla I	40	7.4	6.7	6.7	60 mM KCl	6.7	q	33
Msp I	37	8.0	20	10	5	10	k	34
Nci I	37	7.5	6	6	6	6	k	35
Ngo II	55[a]	8.5[a]	100	1[a]	20	—	—	36
Pst I	37	7.4	6.6	6.6	50	6	—	37
Pvu I	37	7.9	6	6	—	6	o	38
Pvu II	37	7.9	6	6	—	6	l	38
Rsa I	34[a]	7.9[a]	10	6	—	0.5 mM DTT[b]	—	39
Rsh I	30–37[a]	7.9	10	6	—	0.5 mM DTT[b]	—	40
Sal I	37	7.9	6	6	—	6	—	41
Sau3A I	30	7.5	6	15	60	6	l	42
Sau96 I	30	7.4	6	15	60	6	—	43
Sma I	37	7.5	10	10	50	—	r	44
Sph I	37	7.3–7.8[a]	6	6[a]	50[a]	6	e, s	45
Sst I	37	7.5	10	10	100	10	t	46
Stu I	37	7.9	10	10	100	—	—	47
Taq I	37	7.4	10	10	—	10	—	48
Tha I	60[a]	7.4	10	10	—	—	—	49
Tth I	60[a]	7.5–8.5[a]	20	5	50	10	—	50
Tth111, I	65[a]	7.4	8	8[a]	50[a]	8	—	51

(continued)

TABLE II (continued)

Enzyme	Temperature (°C)	pH	Tris-HCl (mM)	MgCl$_2$ (mM)	NaCl (mM)	2-Mercapto-ethanol (mM)	Notes	Reference*
Tth111, II	65^a	7.4	6	6	120–150^a	6	r	52
XbaI	37	7.9	6	6	—	6	—	53
XhoI	37	7.9	6	6	—	6	—	54
XmaI	37	7.9	6	6	—	6	—	44
XorII	37	7.4	6	12–24^a	—	6	e	55

^{a} Optimal condition.

^{b} Abbreviations: DTT, dithiothreitol; GOH, glycine-NaOH; BamHI.1, BstI*, BsuI*, EcoRI*, the secondary, "star" activities of BamHI, BstI, BsuI, and EcoRI, respectively.

^{c} In addition, 36% (v/v) glycerol and >20 × excess of enzyme are needed.

^{d} Activity is stimulated twofold with 200 mM NaCl.

^{e} Mn^{2+} can substitute for Mg^{2+}.

^{f} Zn^{2+} inhibits activity.

^{g} Activity is inhibited 50% by 50 mM NaCl.

^{h} In addition, >5% glycerol is needed.

^{i} In addition, 25% glycerol and 20–40 × excess of enzyme are needed.

^{j} In addition, 500 µg of bovine serum albumin are needed per milliliter.

^{k} In addition, 100 µg of bovine serum albumin are needed per milliliter.

^{l} Activity is inhibited by ≧100 mM NaCl.

^{m} Active to 500 mM KCl.

^{n} Activity is inhibited by >250 mM NaCl or pH <7.

^{o} Active to 200 mM NaCl.

^{p} Activity is inhibited by >60 mM NaCl.

^{q} In addition, 50 µg of bovine serum albumin are needed per milliliter.

^{r} Active at 70°.

^{s} Active to 300 mM NaCl.

^{t} Activity is inhibited by >200 mM NaCl.

* Key to references:

1. R. J. Roberts, P. A. Myers, A. Morrison, and K. Murray, J. Mol. Biol. **102**, 157 (1976).

2. S. G. Hughes, T. Bruce, and K. Murray, *Biochem. J.* **185**, 59 (1980).

3. S. G. Hughes, and K. Murray, *Biochem. J.* **185**, 65 (1980).

4. R. E. Gelinas, P. A. Myers, G. A. Weiss, R. J. Roberts, and K. Murray, *J. Mol. Biol.* **114**, 433 (1977).

5. B. Hinsch, and M. Kula, *Nucleic Acids Res.* **8**, 623 (1980).

6. George, R. W. Blakesley, and J. G. Chirikjian, *J. Biol. Chem.* **255**, 6521 (1980).

7. A. H. A. Bingham, T. Atkinson, D. Sciaky, and R. J. Roberts, *Nucleic Acids Res.* **5**, 3457 (1978).

8. T. A. Bickle, V. Pirrotta, and R. Imber, this series, Vol. 65, p. 132.

9. P. Venetianer, this series, Vol. 65, p. 109.

10. C. M. Clarke, and B. S. Hartley, *Biochem. J.* **177**, 49 (1979).

11. J. F. Catterall, and N. E. Welker, this series, Vol. 65, p. 167.

12. S. Bron, and W. Horz, this series, Vol. 65, p. 112.

13. H. Mayer, R. Grosschedl, H. Schutte, and G. Hobom, *Nucleic Acids Res.* **9**, 4833 (1981).

14. R. A. Makula, and R. B. Meagher, *Nucleic Acids Res.* **8**, 3125 (1980).

15. S. A. Lacks, this series, Vol. 65, p. 138.

16. G. Hobom, E. Schwarz, M. Melzer, and H. Mayer, *Nucleic Acids Res.* **9**, 4823 (1981).

17. R. A. Rubin, and P. Modrich, this series, Vol. 65, p. 96.

18. B. Polisky, P. Greene, D. E. Garfin, B. J. McCarthy, H. M. Goodman, and H. W. Boyer, *Proc. Natl. Acad. Sci. U.S.A.* **72**, 3310 (1975).

19. S. G. Hattman, and S. Hattman, *J. Mol. Biol.* **98**, 645 (1975).

20. A. C. P. Lui, B. C. McBride, G. F. Vovis, and M. Smith, *Nucleic Acids Res.* **6**, 1 (1979).

21. R. J. Roberts, J. B. Breitmeyer, N. F. Tabachnik, and P. A. Myers, *J. Mol. Biol.* **91**, 121 (1975).

22. R. W. Blakesley, J. B. Dodgson, I. F. Nes, and R. D. Wells, *J. Biol. Chem.* **252**, 7300 (1977).

23. M. Takanami, *Methods Mol. Biol.* **7**, 113 (1974).

24. N. L. Brown, M. McClelland, and P. R. Whitehead, *Gene* **9**, 49 (1980).

25. R. J. Roberts, P. A. Myers, A. Morrison, and K. Murray, *J. Mol. Biol.* **103**, 199 (1976).

26. A. Landy, E. Ruedisueli, L. Robinson, C. Foeller, and W. Ross, *Biochemistry* **13**, 2134 (1974).

27. H. O. Smith and G. M. Marley, this series, Vol. 65, p. 104.

28. K. N. Subramanian, S. M. Weissman, B. S. Zain, and R. J. Roberts, *J. Mol. Biol.* **110**, 297 (1977).

29. J. L. Hines, T. R. Chauncey, and K. L. Agarwal, this series, Vol. 65, p. 153.

30. D. G. Kleid, this series, Vol. 65, p. 163.

31. J. Tomassini, R. Roychoudhury, R. Wu, and R. J. Roberts, *Nucleic Acids Res.* **5**, 4055 (1978).

32. R. E. Gelinas, P. A. Myers, and R. J. Roberts, *J. Mol. Biol.* **114**, 169 (1977).

33. M. Duyvesteyn, and A. de Waard, *FEBS Lett.* **111**, 423 (1980).

34. O. J. Yoo, and K. L. Agarwal, *J. Biol. Chem.* **255**, 10559 (1980).

TABLE II (*continued*)

35. R. Watson, M. Zuker, S. M. Martin, and L. P. Visentin, *FEBS Lett.* **118**, 47 (1980).

36. D. J. Clanton, W. S. Riggsby, and R. V. Miller, *J. Bacteriol.* **137**, 1299 (1979).

37. D. I. Smith, F. R. Blattner, and J. Davies, *Nucleic Acids Res.* **3**, 343 (1976).

38. T. R. Gingeras, L. Greenough, I. Schildkraut, and R. J. Roberts, *Nucleic Acids Res.* **9**, 4525 (1981).

39. S. P. Lynn, L. K. Cohen, S. Kaplan, and J. F. Gardner, *J. Bacteriol.* **142**, 380 (1980).

40. S. P. Lynn, L. K. Cohen, J. F. Gardner, and S. Kaplan, *J. Bacteriol.* **138**, 505 (1979).

41. J. R. Arrand, P. A. Myers, and R. J. Roberts, *J. Mol. Biol.* **118**, 127 (1978).

42. J. S. Sussenbach, C. H. Monfoort, R. Schiphof, and E. E. Stobberingh, *Nucleic Acids Res.* **3**, 3193 (1976).

43. J. S. Sussenbach, P. H. Steenbergh, J. A. Rost, W. J. van Leeuwen, and J. D. A. van Embden, *Nucleic Acids Res.* **5**, 1153 (1978).

44. S. A. Endow and R. J. Roberts, *J. Mol. Biol.* **112**, 521 (1977).

45. L. Y. Fuchs, L. Covarrubias, L. Escalante, S. Sanchez, and F. Bolivar, *Gene* **10**, 39 (1980).

46. A. Rambach, this series, Vol. 65, p. 170.

47. H. Shimotsu, H. Takahashi, and H. Saito, *Gene* **11**, 219 (1980).

48. S. Sato, C. A. Hutchison, III, and J. I. Harris, *Proc. Natl. Acad. Sci. U.S.A.* **74**, 542 (1977).

49. D. J. McConnell, D. Searcy, and G. Sutcliffe, *Nucleic Acids Res.* **5**, 1979 (1978).

50. A. Venegas, R. Vicuna, A. Alonso, F. Valdes, and A. Yudelevich, *FEBS Lett.* **109**, 156 (1980).

51. T. Shinomiya and S. Sato, *Nucleic Acids Res.* **8**, 43 (1980).

52. Shinomiya, M. Kobayashi, and S. Sato, *Nucleic Acids Res.* **8**, 3275 (1980).

53. B. S. Zain, and R. J. Roberts, *J. Mol. Biol.* **115**, 249 (1977).

54. T. R. Gingeras, P. A. Myers, J. A. Olsen, F.A. Hanberg, and R. J. Roberts, *J. Mol. Biol.* **118**, 113 (1978).

55. R. Y. H. Wang, J. G. Shedlarski, M. B. Farber, D. Kuebbing, and M. Ehrlich, *Biochim. Biophys. Acta* **606**, 371 (1980).

Depending upon the subsequent use of the cleaved DNA, the demands on the purity of the DNA may vary. Generally, RNA and/or DNA contamination does not significantly interfere with the apparent restriction reaction rate as measured by digest completion. This is in spite of the fact that nonspecific binding to nucleic acids reduces the effective concentration of a restriction endonuclease. Contaminating nucleic acids more often interfere by obscuring the detection or selection of reaction products. For example, positive clones screened by rapid lysis methods[11] may be difficult to identify if the insert DNA excised by restriction endonuclease cleavage migrates in the same region as the intense broad tRNA band upon agarose gel electrophoresis. In such cases, treatment with DNase-free RNase or purification with a quick minicolumn using RPC-5 ANA-LOG[12] is recommended. On the other hand, sequencing protocols, e.g., the M13mp7 dideoxy method,[13] require highly purified DNA as restriction cleavage products. Protein contaminations are tolerated in a restriction reaction as long as the products eventually are protein-free. It should be noted, however, that the presence of other nucleases will reduce the integrity of the product, whereas proteins tightly bound to the DNA may lessen or block the cleavage reaction. DNAs are customarily deproteinized by phenol extraction prior to restriction endonuclease treatment.

Compounds involved in DNA isolation should be rigorously removed by dialysis or by ethanol precipitation and drying prior to addition of the DNA sample to the restriction endonuclease reaction. For example, Hg^{2+}, phenol, chloroform, ethanol, ethylene(diaminetetraacetic) acid (EDTA), sodium dodecyl sulfate (SDS), and NaCl at high levels interfere with restriction reactions, and some can alter the recognition specificity of restriction endonucleases. Drugs frequently used in DNA studies, e.g., actinomycin and distamycin A,[14] also influence restriction endonuclease activity.

In a typical reaction, the restriction endonuclease is in considerable molar excess of the substrate DNA. Therefore, consideration of DNA concentration usually is not required. In fact, it was necessary to dilute HaeIII[8] or BamHI[15] approximately 1000-fold from typical unit assay conditions in order to observe a substrate cleavage rate proportional to the

[11] R. W. Davis, M. Thomas, J. Cameron, T. P. St. John, S. Scherer, and R. A. Padgett, this series, Vol. 65, p. 404.

[12] J. A. Thompson, R. W. Blakesley, K. Doran, C. J. Hough, and R. D. Wells, this series, Vol. 100, p. 368.

[13] J. Messing, R. Crea, and P. H. Seeburg, *Nucleic Acids Res.* **9**, 309 (1981).

[14] V. V. Nosikov, E. A. Braga, A. V. Karlishev, A. L. Zhuze, and O. L. Polyanovsky, *Nucleic Acids Res.* **3**, 2293 (1976).

[15] J. George, unpublished results, 1981.

amount of enzyme added to the reaction. Further, caution must be exercised when attempting to extrapolate the amount of enzyme required for a complete digest based upon the number of recognition sites in a particular DNA. Preliminary observations using the enzyme-saturated, end point-dependent unit assay indicates that apparently no general correlation exists between recognition site density and restriction enzyme units required.[16]

By exception, the concentration of the substrate DNA did influence the apparent reaction rate for HindIII under enzyme-saturating conditions. A typical reaction for unit determination contains 1μg of lambda DNA in a 50-μl reaction volume (20 μg/ml). One unit, but not 0.5 unit, of HindIII completely cleaves 1 μg of lambda DNA. One unit of HindIII also completely cleaves 4 μg (80 μg/ml) of lambda DNA under these conditions.[16] This peculiar response in HindIII activity cannot be attributed to enzyme : DNA concentration ratios, but is assumed to reflect the absolute DNA concentration dependence of HindIII. In contrast to the increased HindIII activity in the presence of increased DNA, 10 units of HpaI, KpnI, or Sau3AI proved to be insufficient to cleave completely 4 μg (80 μg/ml) of lambda DNA in a 15-hr reaction.[16] This phenomenon may be attributed to the viscosity produced by high concentrations of high molecular weight DNA (e.g., lambda DNA), which can inhibit enzyme diffusion and, therefore, inhibit some enzyme activities. These apparently anomalous results point out that one cannot directly compare units determined by titrating enzyme with those obtained by titrating (changing the concentration of) DNA. Further, DNA concentrations near or below the K_m of a restriction enzyme (1–10 nM^6) could also inhibit apparent enzyme cleavage. However, for lambda DNA the K_m is approximately 1000-fold less than the concentration used in the standard reaction for unit determination. From these observations it is recommended that the DNA concentration be at or near that used in the unit assay reaction for the particular restriction endonuclease.

Restriction endonucleases probably show their greatest sensitivity to the DNA sequence. Obviously, the sequence of the recognition site is essentially invariant, as this distinguishes type II restriction endonucleases from other nucleases. The stringent sequence requirement frequently can be relaxed by alterations of the reaction environment, generating the "star" activity (see below) observed for a number of enzymes, EcoRI being the most notable. Sequences adjacent to the recognition site also influence the rate of cleavage. A nearly 10-fold difference in reaction rate was observed between two of the EcoRI sites in lambda DNA.[17] A

[16] This laboratory, unpublished results, 1981.
[17] M. Thomas and R. W. Davis, J. Mol. Biol. 91, 315 (1975).

TABLE III
EFFECT OF BASE ANALOG SUBSTITUTIONS IN DNA ON RESTRICTION
ENDONUCLEASE ACTIVITY

Enzyme	Recognition sequence	Relative activity of base analogs[a-c]				
		HMC	GHMC	U	HMU	BrdU
BamHI[d]	GGATCC		−	++	+	+
EcoRI[d-f]	GAATTC	++	−	++	+	+
HaeII[d]	PuGCGCPy		−		+	+
HhaI[d]	GCGC		−		++	
HindII[d,e]	GTPyPuAC	−	−	+	+	
HindIII[d-f]	AAGCTT	−	−	+	+	+
HpaI[d,g]	GTTAAC		−	+	+	+
HpaII[d]	CCGG		−		++	+
MboI[g]	GATC					Enhanced 5-fold

[a] Activity symbols: ++, full activity; +, diminished activity; −, no activity; blank, not tested.

[b] In these studies, HMC or GHMC were in place of cytosine, while U, HMU, or BrdU replaced thymidine in the tested DNAs.

[c] Abbreviations used: HMC, 5-hydroxymethylcytosine; GHMC, glucosylated 5-hydroxymethylcytosine; U, uridine; HMU, 5-hydroxymethyluridine; BrdU, 5-bromodeoxyuridine; Py, pyrimidine; Pu, purine.

[d] K. L. Berkner and W. R. Folk, J. Biol. Chem. 254, 2551 (1979).

[e] D. A. Kaplan and D. P. Nierlich, J. Biol. Chem. 250, 2395 (1975).

[f] M. A. Marchionni and D. J. Roufa, J. Biol. Chem. 253, 9075 (1978).

[g] J. Petruska and D. Horn, Biochem. Biophys. Res. Commun. 96, 1317 (1980).

similar effect was reported for PstI.[18] In addition, thymine substituted by 5-bromodeoxyuridine prevented cleavage of some SmaI sites in the DNA tested, even though the 5-bromodeoxyuridine was not part of the canonical recognition sequence (CCCGGG).[19]

Nucleotide changes within the recognition sequence more directly affect the restriction endonuclease reaction (Tables III and IV). For EcoRI, cleavage was unaffected by 5-hydroxymethylcytosine substitution for cytosine[20] or by the absence or the presence of the 2-amino group of guanine.[21] Glycosylation of 5-hydroxymethylcytosine, however, made the DNA resistant to cleavage by EcoRI as well as by HpaI, HindII, HindIII, BamHI, HaeII, HpaII and HhaI.[22] Substitution of thymine with

[18] K. Armstrong and W. R. Bauer, Nucleic Acids Res. 10, 993 (1982).
[19] M. A. Marchionni and D. J. Roufa, J. Biol. Chem. 253, 9075 (1978).
[20] P. Modrich and R. A. Rubin, J. Biol. Chem. 252, 7273 (1977).
[21] D. A. Kaplan and D. P. Nierlich, J. Biol. Chem. 250, 2395 (1975).
[22] K. L. Berkner and W. R. Folk, J. Biol. Chem. 254, 2551 (1979).

5-hydroxymethyluridine diminished activities of enzymes with AT-containing sites, whereas a differential effect was observed for uridine and 5-bromodeoxyuridine substitutions.[22] Methylation of nucleotides within restriction endonuclease recognition sequences, occurring almost exclusively as 5-methylcytosine or N^6-methyladenine, prevented most

TABLE IV

METHYLATED DNAs AS SUBSTRATES FOR RESTRICTION ENDONUCLEASES[a]

Enzyme	Sequences containing 5-methylcytosine or N^6-methyladenine[b]		References
	Cleaved	Not cleaved	
Aos II	—	GPu^{m}CGPyC	*d, e*
Ava I	—	CPy^{m}CGPuG	*f*
Ava II	—	GG(A)C^{m}C (T)	*g*
Bst NI	C^{m}C(A)GG (T)	—	*h*
Eco RII	—	C^{m}C(A)GG (T)	*h, i*
Hae II	—	PuG^{m}CGCPy	*e, f*
Hae III	GGC^{m}C	GG^{m}CC	*j, k*
Hap II	—	C^{m}CGG	*e, l*
Hha I	—	G^{m}CGC	*f, j*
Hpa II	^{m}CCGG	C^{m}CGG	*j, l*
Msp I	C^{m}CGG	^{m}CCGG	*l, m*
Pst I	—	^{m}CTGCAG	*n*
Pvu II	—	^{m}CAGCTG	*n*
Sal I	—	GT^{m}CGAC	*d, e*
Sma I	—	CC^{m}CGGG	*e, u*
Taq I	T^{m}CGA	—	*o*
Xho I	—	CT^{m}CGAG	*d, e*
Xma I	CC^{m}CGGG	—	*u*
Bam HI	GG^{m}ATCC	—	*g, p*
Bgl II	AG^{m}ATCT	—	*p*
Dpn I	G^{m}ATC^c	—	*o, q*
Dpn II	—	G^{m}ATC	*q*
Eco RI	—	GA^{m}ATTC	*r*
Fnu EI	G^{m}ATC	—	*s*
Hind II	—	GTPyPu^{m}AC	*k*
Hind III	—	^{m}AAGCTT	*k*
Hpa I	—	GTTA^{m}AC	*t*
Mbo I	—	G^{m}ATC	*p, s*
Mbo II	—	GAAG^{m}A	*g*
Sau 3AI	G^{m}ATC	—	*o, p*
Taq I	—	TCG^{m}A	*d, o*

enzymes from cleaving. In Table IV are listed the responses of a variety of restriction enzymes to DNA methylation. Several enzymes were found to vary in their response to hemimethylated DNAs, where only one of the two strands is methylated (Table IV).[23]

Modification of all or the vast majority of certain base types within the DNA of certain bacteriophages has, as expected, more drastic effects on the ability and rate of restriction endonuclease cleavage than modifications that occur solely within the recognition sequences described above.

[23] Y. Gruenbaum, H. Cedar, and A. Razin, *Nucleic Acids Res.* **9**, 2509 (1981).

---

[a] The enzymes *Bst*NI, *Hinc*II, *Hinf*I, *Hpa*I, and *Taq*I have been reported to cleave hemimethylated DNA (i.e., only one DNA strand contains ^mC). In addition *Msp*I, *Sau*3A, and *Hae*III nick the unmethylated strand of the hemimethylated DNA [R. E. Streeck, *Gene* **12**, 267 (1980); and Y. Gruenbaum, H. Cedar, and A. Razin, *Nucleic Acids Res.* **9**, 2509 (1981)].

[b] Abbreviations used: —, not determined; Pu, purine; Py, pyrimidine; ^mC, 5-methylcytosine; ^mA, N^6-methyladenine.

[c] Methylation is required for cleavage.

[d] L. H. T. van der Ploeg and R. A. Flavell, *Cell* **19**, 947 (1980).

[e] M. Ehrlich and R. Y. H. Wang, *Science* **212**, 1350 (1981).

[f] A. P. Bird and E. M. Southern, *J. Mol. Biol.* **118**, 27 (1978).

[g] K. Bachman, *Gene* **11**, 169 (1980).

[h] S. Hattman, C. Gribbin, and C. A. Hutchison, III, *J. Virol.* **32**, 845 (1979).

[i] M. S. May and S. Hattman, *J. Bacteriol.* **122**, 129 (1975).

[j] M. B. Mann and H. O. Smith, *Nucleic Acids Res.* **4**, 4211 (1977).

[k] P. H. Roy and H. O. Smith, *J. Mol. Biol.* **81**, 427 (1973).

[l] C. Waalwijk and R. A. Flavell, *Nucleic Acids Res.* **5**, 3231 (1978).

[m] T. W. Sneider, *Nucleic Acids Res.* **8**, 3829 (1980).

[n] A. P. Dobritsa and S. V. Dobritsa, *Gene* **10**, 105 (1980).

[o] R. E. Streeck, *Gene* **12**, 267 (1980).

[p] B. Dreiseikelman, R. Eichenlaub, and W. Wackernagel, *Biochim. Biophys. Acta* **562**, 418 (1979).

[q] S. Lacks and B. Greenberg, *J. Biol. Chem.* **250**, 4060 (1975).

[r] A. Dugaiczyk, J. Hedgepeth, H. W. Boyer, and H. M. Goodman, *Biochemistry* **13**, 503 (1974).

[s] A. C. P. Lui, B. C. McBride, G. F. Vovis, and M. Smith, *Nucleic Acids Res.* **6**, 1 (1979).

[t] L.-H. Huang, C. M. Farnet, K. C. Ehrlich, and M. Ehlich, *Nucleic Acids Res.* **10**, 1579 (1982).

[u] H. Youssoufian and C. Mulder, *J. Mol. Biol.* **150**, 133 (1981).

When 30 type II restriction endonucleases were separately incubated with *Xanthomonas oryzae* phage XP12 DNA, all cytosine residues of which are modified to 5-methylcytosine, only *Taq*I cleaved efficiently. When bacteriophage T4 DNA, which contains only 5-hydroxymethylcytosine, but not cytosine, was tested, again only *Taq*I cleaved, although inefficiently. The complete substitution of thymine residues with 5-hydroxymethyluracil in the genome of *Bacillus subtilis* phages SP01 and PBS1 either had no effect or for, some of the restriction enzymes, only reduced cleavage efficiency. The substitution of thymine by phosphogluconated or glucosylated 5-(4′,5′-dihydroxy)pentyluracil in *B. subtilis* phage SP15 DNA precluded cleaving by most of the restriction endonucleases tested.[24] *Dde*I, *Taq*I, *Tha*I, and *Bst*NI did cleave this DNA very poorly. Complete nucleotide substitutions cause drastic alterations not only in the recognition sequences for these restriction enzymes, but also in the secondary and tertiary DNA structures.

The proximity of the recognition site to the terminus of a DNA can also influence cleavage. *Hpa*II and *Mno*I required at least one base preceding the 5′ end of the recognition sequence for cleavage.[25] The minimal duplex hexanucleotide recognition sequences for *Eco*RI (GAATTC), *Bam*HI (GGATCC), and *Hin*dIII (AAGCTT) were resistant to cleavage. However, *Eco*RI will cleave if the sequence is extended by one base to GAATTCA.[26] On the other hand, when *Hha*I (GCGC) cleaved poly(dG-dC), about 85% of the product was the limit tetranucleotide.[27]

Secondary and tertiary structure of the recognition/cleavage site also affects the restriction endonuclease reaction rate. Restriction enzymes typically require the substrate cleavage site to be in a duplex form for cleavage as shown for *Hae*III,[8] *Eco*RI,[9] and *Msp*I.[28] *Hin*dIII apparently requires at least two uninterrupted turns of the double helix for cleavage.[26] Certain restriction endonucleases (*Bsp*RI, *Hae*III, *Hha*I, *Hinf*I, *Mbo*I, *Mbo*II, *Msp*I, and *Sfa*I) will cleave "single-stranded" viral DNAs of bacteriophages φX174, M13, or f1 whose cleavage sites are in the duplex form. Even though *Hpa*II was reported to cleave a single strand,[29] there is no conclusive evidence that a bona fide single-stranded restriction site is cleaved. The fact that certain enzymes do not cleave the "single-stranded" viral DNAs indicates that properties in addition to the DNA

[24] L.-H. Huang, C. M. Farnet, K. C. Ehrlich, and M. Ehrlich, *Nucleic Acids Res.* **10,** 1579 (1982).
[25] B. R. Baumstark, R. J. Roberts, and U. L. RajBhandary, *J. Biol. Chem.* **254,** 8943 (1979).
[26] Y. A. Berlin, N. M. Zvonok, and S. A. Chuvpilo, *Bioorg. Khim.* **6,** 1522 (1980).
[27] R. J. Roberts, P. A. Myers, A. Morrison, and K. Murray, *J. Mol. Biol.* **103,** 199 (1976).
[28] O. J. Yoo, and K. L. Agarwal, *J. Biol. Chem.* **255,** 10559 (1980).
[29] K. Horiuchi, and N. D. Zinder, *Proc. Natl. Acad. Sci. U.S.A.* **72,** 2555 (1975).

recognition sequence are required for restriction endonucleolytic cleavage (for review, see Wells and Neuendorf[30]).

Cleavage of RNA · DNA hybrid molecules were described for several restriction endonucleases.[31] The fate of the RNA was not followed, but presumably RNA was degraded to small oligonucleotides. This would not be surprising since restriction endonucleases are frequently not assayed for, or purified from, ribonucleases. It is difficult unequivocally to conclude that true RNA · DNA hybrids were cleaved, since the remaining DNA strand could potentially self-hybridize, as in the "single-stranded" viral DNAs, to provide the appropriate duplex substrate. This must await further experimentation.

Another DNA structural variant frequently encountered in restriction endonuclease reactions is superhelicity. Generally, larger amounts of restriction enzyme are required to cleave supercoiled plasmid or viral DNAs completely than for linear DNA. A comparison of the relative cleavage efficiencies for several supercoiled and linear DNAs are presented in Table V. If a supercoiled DNA (e.g., pBR322 plasmid DNA) is first linearized with a restriction endonuclease or relaxed with topoisomerase,[32] frequently less enzyme is needed for complete cleavage (Table VI).

Reagents

The components of a restriction endonuclease buffer system should be of the highest quality available. Contaminants, e.g., heavy metals in buffer components, should be looked for and avoided. Reagents should be free of enzyme activities, especially nucleases. Filter or heat-sterilize all reagent stocks, then store frozen and replace frequently in order to maintain quality and integrity. For convenience several of the reagents can be mixed together as a 10-fold concentrated stock solution. When added to the final reaction mixture, an appropriate single dilution into sterile water is made. These precautions will help to ensure the desired quality in the DNA product of the reaction.

A number of buffers are available to maintain the assay pH between 7 and 8. Tris(hydroxymethyl)aminomethane (Tris), the most widely used and least noxious, has a large temperature coefficient that should be considered when preparing and using this buffer. The pH of Tris buffers also

[30] R. D. Wells, and S. K. Neuendorf, in "Gene Amplification and Analysis," Vol. I: "Restriction Endonucleases" (J. G. Chirikjian, ed.), p. 101. Elsevier/North-Holland, Amsterdam, 1981.

[31] P. L. Molloy, and R. H. Symons, Nucleic Acids Res. 8, 2939 (1980).

[32] J. LeBon, C. Kado, L. Rosenthal, and J. G. Chirikjian, Proc. Natl. Acad. Sci. U.S.A. 74, 542 (1977).

TABLE V
RELATIVE ACTIVITIES OF CERTAIN RESTRICTION
ENDONUCLEASES ON SEVERAL DNA SUBSTRATES[a]

Enzyme[d]	Enzyme units required for complete cleavage of specified DNA[b,c]				
	Lambda	Ad-2	pBR322	ϕX174RF	SV40
BamHI	1	2	3	—	4
EcoRI	1	1	2.5	—	3
HhaI	1	10	4	1	2
HindIII	1	3	2.5	—	10
HinfI	1	1	1	1	1
HpaII	1	1	2	1	10
PstI	1	2	1.5	1	1
PvuII	1	4	4	—	4
Sau3AI	1	2	2.5	—	1.5
TaqI	1.5	1	10	1	0.5
XorII	1	1	>10	—	—

[a] H. Belle Isle, unpublished results, 1981.

[b] Activity was measured by incubation of 1 μg of the specified DNA with various amounts of the respective restriction endonucleases under appropriate standard reaction conditions. These values represent the minimum number of units of enzyme required for complete digestion of the specified DNA as monitored by agarose gel electrophoresis [P. A. Sharp, B. Sugden, and J. Sambrook, *Biochemistry* **12**, 3055 (1973)]. Enzyme activity units are defined as the minimum amount of enzyme required to digest completely 1 μg of lambda (or ϕX174 RF for TaqI, or Ad-2 for XorII) DNA under standard reaction conditions.

[c] Abbreviations used: lambda, bacteriophage lambda CI857 Sam7; Ad-2, Adenovirus type 2; pBR322, supercoiled plasmid pBR322; ϕX174 RF, supercoiled bacteriophage ϕX174 replicative form; SV40, supercoiled simian virus 40; and —, recognition sequence for this enzyme not present in this DNA.

[d] All enzymes and DNAs were from Bethesda Research Laboratories, Inc.

varies with concentration and should therefore be reset upon dilution. Glycine is useful as a restriction endonuclease buffer for reactions at pH >9. Phosphate is an excellent buffer for assays between pH 6.0 and 7.5 and has a minimal temperature coefficient. But phosphate buffers should be used only if no subsequent enzyme reactions are to be performed that

TABLE VI
EFFECT OF DNA SUPERHELICITY ON RESTRICTION
ENZYME ACTIVITY[a]

Enzyme[c]	Enzyme units required for complete cleavage[b]	
	Supercoiled pBR322 DNA	Linear pBR322 DNA[d]
BamHI	2	1
EcoRI	2.5	1
HindIII	2.5	2.5
SalI	7.5	3

[a] H. Belle Isle, unpublished results, 1981.

[b] Activity was measured by incubation of 1 μg of pBR322 DNA with various amounts of the respective restriction endonucleases under appropriate standard reaction conditions. These values represent the minimum number of units of enzyme required for complete digestion of the DNA as monitored by agarose gel electrophoresis [P. A. Sharp, B. Sugden, and J. Sambrook, *Biochemistry* 12, 3055 (1973)]. Enzyme activity units are defined as the minimum amount of enzyme required to digest completely 1 μg of lambda DNA under standard reaction conditions.

[c] All enzymes and DNAs were from Bethesda Research Laboratories, Inc.

[d] Linear form III pBR322 DNA was prepared by incubation of supercoiled form I DNA with PstI, followed by phenol extraction and ethanol precipitation.

are inhibited by the phosphate ion, e.g., DNA end-labeling[33] or ligation.[34] Typical methods of phenol extraction or ethanol precipitation will not significantly reduce the phosphate ion content in a DNA sample. Dialysis or multiple ethanol precipitations with 2.5 M ammonium acetate are, on the other hand, effective. Citrate and other biological buffers that chelate Mg^{2+} cannot be used.

The selected buffer concentration must be sufficient to maintain the proper pH of the final reaction mixture. Buffer concentrations greater than 10 mM are recommended to provide the appropriate buffering capacity under conditions where the pH of most distilled water supplies are low. In addition, the reaction pH should not be altered when a relatively large volume of an assay component, e.g., the DNA substrate, is added. In general, the reaction rate is not significantly affected by the concentra-

[33] G. Chaconas and J. H. van de Sande, this series, Vol. 65, p. 75.
[34] A. W. Hu, manuscript in preparation (1982).

tion of Tris buffer above 10 mM e.g., HaeIII demonstrated $<20\%$ variance in reactivity between 15 and 120 mM.[8]

Many restriction enzymes have significant activity over a rather broad pH range. HaeIII has an activity optimum at pH 7.5, but retains at least 50% of its activity when assayed at 1.5 units above or below pH 7.5.[8] Some other enzymes studied, BstI,[35] HaeII,[8] HgiAI,[36] HhaI,[8] NgoII,[37] SphI,[38] and TthI[39] showed similar profiles. Selected enzymes such as EcoRI are sensitive to altered pH. Not only does EcoRI activity significantly decrease,[40] but an altered activity (see Secondary Activity below) appears when the pH is increased from 7.2 to 8.5.[41] Thus, the pH should be maintained at the recommended value by a buffer with adequate capacity.

Type II restriction endonucleases require Mg^{2+} as the only cofactor. Complete chelation of Mg^{2+} by EDTA can thus effectively stop the reaction. Restriction enzyme activities are relatively insensitive to the Mg^{2+} concentration; similar rates are observed from 5 to 30 mM.[8,42] Similar to other nucleic acid enzymes, some restriction endonucleases accept Mn^{2+} as a substitute for Mg^{2+}, although with varying results. EcoRI and HindIII change their recognition specificity with such replacement.[43,44] HaeIII is approximately 50% as active with $MnCl_2$ as with $MgCl_2$,[8] while XorII[42] and TthI[39] are equally active with Mg^{2+} or Mn^{2+}.

Whereas EcoRI functions, although inefficiently, with other divalent cations (Mn^{2+}, Co^{2+}, Zn^{2+}), Mg^{2+} cannot be replaced by other divalent cations (Cu^{2+}, Ba^{2+}, Cr^{2+}, Co^{2+}, Zn^{2+}, and Ni^{2+}) in the HaeIII reaction.[8] BamHI showed secondary "star" activity when Zn^{2+} or Co^{2+} replaced Mg^{2+} at pH 6, but no activity at pH 8.5.[15] BspI is quite active with Mn^{2+}, but completely inhibited with Zn^{2+}.[45] It is unclear at this point whether metal ions such as Zn^{2+} contribute to restriction endonuclease structural

[35] C. M. Clarke and B. S. Hartley, *Biochem. J.* **177,** 49 (1979).
[36] N. L. Brown, M. McClelland, and P. R. Whitehead, *Gene* **9,** 49 (1980).
[37] D. J. Clanton, W. S. Riggsby, and R. V. Miller, *J. Bacteriol.* **137,** 1299 (1979).
[38] L. Y. Fuchs, L. Covarrubias, L. Escalante, S. Sanchez, and F. Bolivar, *Gene* **10,** 39 (1980).
[39] A. Venegas, R. Vicuna, A. Alonso, F. Valdes, and A. Yuldelevich, *FEBS Lett.* **109,** 156 (1980).
[40] R. A. Rubin and P. Modrich, this series, Vol. 65, p. 96.
[41] B. Polisky, P. Greene, D. E. Garfin, B. J. McCarthy, H. M. Goodman, and H. W. Boyer, *Proc. Natl. Acad. Sci. U.S.A.* **72,** 3310 (1975).
[42] R. Y.-H. Wang, J. G. Shedlarski, M. B. Farber, D. Kuebbing, and M. Ehrlich, *Biochim. Biophys. Acta* **606,** 371 (1980).
[43] T. I. Tikchonenko, E. V. Karamov, B. A. Zavizion, and B. S. Naroditsky, *Gene* **4,** 195 (1978).
[44] M. Hsu and P. Berg, *Biochemistry* **17,** 131 (1978).
[45] P. Venetianer, this series, Vol. 65, p. 109.

stability as demonstrated with other nucleic acid enzymes, such as *Escherichia coli* DNA polymerase I.[46] Unless metal chelators such as EDTA, EGTA, or *o*-phenanthroline are present in the reaction, one need not be concerned about adding to the reaction metal ions other than Mg^{2+} for the activity or fidelity of restriction endonucleases.

Restriction endonucleases show a wide diversity in their responses to ionic strength (Table II). Most enzymes do not absolutely require specific monovalent cations, but rather are stimulated by the corresponding ionic strength. *Sma*I, however, does have an absolute requirement for K^+.[16] Many enzymes are stimulated by 50–100 mM NaCl or KCl (e.g., *Sph*I,[38] *Mlu*I[47]), whereas others are drastically inhibited at concentrations >20 mM (e.g., *Fok*I,[47] *Hin*dII,[48] and *Fnu*DI[49]). Other cations, e.g., NH_4^+, can in some cases provide the stimulating ionic strength.[50] Loss of restriction enzyme activity (Table VII) and recognition specificity[41,51] can result from inappropriate monovalent cation concentrations. Recommended concentrations (see Table II or VII) should therefore be closely followed. Special caution also should be used in selecting the appropriate buffers for multiple enzyme digestions (see Other Reaction Considerations).

Sulfhydryl reagents such as 2-mercaptoethanol and dithiothreitol are routinely used in restriction enzyme reactions. Historically, 2-mercaptoethanol was added to restriction enzyme preparations and reactions as a general precaution based on the labilities of other nucleic acid enzymes. Nath demonstrated that not all restriction endonucleases require such reagents.[52] *Bgl*II, *Eco*RI, *Hin*dIII, *Hpa*I, *Sal*I, and *Sst*II activities are insensitive, and *Ava*I, *Bam*HI, *Pvu*I, and *Sma*I activities are inhibited by the sulfhydryl reactive compounds *p*-mercuribenzoate and *N*-ethylmaleimide. In other studies, *Hpa*I and *Hpa*II,[53] and *Eco*RI[43] were unaffected, whereas the "star" activity of *Eco*RI (*Eco*RI*)[43] was sulfhydryl sensitive. Where not required, the sulfhydryl reagents should be omitted from the reaction to prevent stabilization of possible contaminating activities. When used, only freshly prepared stocks of 2-mercaptoethanol and dithiothreitol at final reaction concentrations of no greater than 10 and 1.0 mM, respectively, should be employed.

Bovine serum albumin (BSA) or gelatin is frequently used in restric-

[46] A. Kornberg, "DNA Replication." Freeman, San Francisco, California, 1980.
[47] H. Sugisaki and S. Kanazawa, *Gene* **16**, 73 (1981).
[48] H. O. Smith and G. M. Marley, this series, Vol. 65, p. 104.
[49] A. C. P. Lui, B. C. McBride, G. F. Vovis, and M. Smith, *Nucleic Acids Res.* **6**, 1 (1979).
[50] D. I. Smith, F. R. Blattner, and J. Davies, *Nucleic Acids Res.* **3**, 343 (1976).
[51] R. A. Makula and R. B. Meagher, *Nucleic Acids Res.* **8**, 3125 (1980).
[52] K. Nath, *Arch. Biochem. Biophys.* **212**, 611 (1981).
[53] J. L. Hines, T. R. Chauncey, and K. L. Agarwal, this series, Vol. 65, p. 153.

TABLE VII

RESTRICTION ENDONUCLEASE ACTIVITY IN CORE BUFFER[a]

Enzyme[d]	Relative enzyme activity (percent) in core buffer with[b,c]		
	0 mM NaCl	50 mM NaCl	100 mM NaCl
AccI	200	100	50
AluI	140	100	40
AvaI	75	125	40
AvaII	150	125	50
BalI	27	14	5
BamHI	67	117	33
BclI	120	120	80
BglI	33	67	89
BglII	50	88	100
BstEII	40	160	120
CfoI	50	20	5
ClaI	86	43	3
DdeI	50	150	200
DpnI	133	133	117
EcoRI[e]	10	10	10
EcoRII	50	75	100
HaeII	200	100	25
HaeIII	114	100	50
HhaI	33	56	67
HincII	25	75	100
HindIII	160	200	120
HinfI	100	100	100
HpaI	25	12	5
HpaII	71	43	7
KpnI	67	33	6
MboI	75	100	50
MboII	50	30	10
MnlI	85	95	75
MspI	100	33	17
NciI	67	22	6
PstI	100	125	100
PvuII	33	67	50
SalI	10	25	150
Sau3AI	40	20	10
Sau96	53	80	53
SmaI[f]	0	0	0
SphI	10	20	40
SstI	57	71	29

TABLE VII (*continued*)

Enzyme[d]	Relative enzyme activity (percent) in core buffer with[b,c]		
	0 mM NaCl	50 mM NaCl	100 mM NaCl
*Taq*I	12	25	50
*Tha*I	83	67	33
*Xba*I	100	70	50
*Xho*I	117	150	150
*Xma*III	50	67	33
*Xor*II	50	25	5

[a] A. MarSchel, unpublished results, 1981.

[b] Core buffer is 50 mM Tris-HCl (pH 8.0), 10 mM $MgCl_2$, 1 mM dithiothreitol, 100 μg of bovine serum albumin per milliliter, and an appropriate amount of NaCl.

[c] Enzyme activity was measured by incubation of an appropriate DNA with various amounts of the respective endonuclease in the standard reaction buffer, in core buffer, or in core buffer supplemented with NaCl. The standard buffer was either that listed in Table II or, in some cases, the listed buffer as modified by this laboratory to give greater activity. One enzyme unit is defined as the minimum amount of enzyme required to digest completely 1 μg of lambda (or adenovirus type 2 for *Bcl*I, *Eco*RII, *Sal*I, *Sau*96I, *Sma*I, *Sst*I, *Xba*I, *Xho*I, *Xma*III, and *Xor*II; ϕX174 RF for *Taq*I; SV40 form I for *Mbo*I, and *Mbo*II; or pBR322 for *Dpn*I and *Mnl*I) DNA as monitored by agarose gel electrophoresis [P. A. Sharp, B. Sugden, and J. Sambrook, *Biochemistry* **12**, 3055 (1973)]. The unit concentration of each enzyme determined in the core buffer, or in core buffer with NaCl is listed as a percentage of the unit concentration determined in the standard buffer (designated 100% activity).

[d] All enzymes and DNAs were from Bethesda Research Laboratories, Inc.

[e] *Eco*RI has a narrow pH optimum range for enzyme activity. When the pH was lowered to 7.2, the following relative enzymic activities were obtained: 44%, 89%, and 67% in core buffer supplemented with 0, 50, and 100 mM NaCl, respectively.

[f] *Sma*I has an absolute requirement for K^+, which is absent from the core buffer. When the core buffer contains 15 mM KCl, the following relative enzymic activities were obtained: 50%, 25%, and 7% in core buffer supplemented with 0, 50, and 100 mM NaCl, respectively.

tion endonuclease preparations to stabilize enzyme activity in long-term incubation or storage. *Hpa*I and*Hpa*II are quite unstable when the protein concentration is <20 μg/ml.[53] Addition of exogenous proteins protects the restriction endonucleases from proteases, nonspecific adsorption, and harmful environmental factors such as heat, surface tension, and chemicals, that cause denaturation. Only sterile solutions of nuclease-free BSA or heavy metal-free gelatin should be added to restriction enzyme reactions. In general, little harm results from addition of these proteins to the reaction. Occasionally, excess BSA binding to DNA causes band smearing during gel electrophoresis. This is eliminated by the addition of SDS to the sample followed by heating to 65° for 5 min prior to sample loading.

The importance of water quality should not be overlooked. Glass-distilled water free of ions and organic compounds should be used for all buffers and reaction components. Deionized water is satisfactory provided the content of organic material is not significant.

Glycerol added to restriction endonuclease stocks stabilizes the enzymes and prevents freezing at low temperature (−20°) during long-term storage. A number of restriction enzymes show reduced recognition specificity in the presence of glycerol (see below). In general, restriction enzyme reactions should contain <5% (v/v) glycerol (final concentration).

Core Buffer System

Many laboratories stock a large panel of individual buffer systems appropriate for the many restriction endonucleases in use (see Table II). Identification of one or a few primary buffer systems that would take advantage of the similarities of the restriction enzymes, while reflecting as closely as possible the optima for each enzyme, would provide a valuable convenience for restriction endonuclease use. For example, reaction conditions for the enzymes reported by Roberts and co-workers (e.g., *Alu*I, *Bal*I, and *Xho*I; see reviews by Roberts[2,4]) were based on a single buffer system, the "6/6/6" [6 mM Tris-HCl (pH 7.9), 6 mM MgCl$_2$, and 6 mM 2-mercaptoethanol]. Although this system suffices for those enzymes, it can be improved upon by consideration of more recent data on restriction endonuclease reactions.

In application of several facts described in the preceding section, we devised a basic assay system, the "core buffer" [50 mM Tris-HCl (pH 8.0), 10 mM MgCl$_2$, 1 mM dithiothreitol, and 100 μg of BSA per milliliter] to which is added 0, 50, or 100 mM NaCl depending upon an individual enzyme's greatest activity. In Table VII are compared the relative activities for a number of commonly used restriction endonucleases assayed both in this core buffer system and in the standard buffer. More than 60% of the enzymes tested were at least 80% as active in the core buffer as in the standard buffer. In fact, 34% of the enzymes exhibited higher activity in the core buffer, demonstrating that many of the standard buffers are suboptimal. As expected for any class of enzymes this large, some enzymes are not amenable to the core buffer reaction conditions. *Bal*I, *Hpa*I, *Sau*3A, and *Sph*I lost more than 50% of their activities under the core buffer conditions. These enzymes should continue to be used as described in Table II. Although the present core buffer system fails to identify the optima that are obtained from initial rate studies, it permits a rational consolidation and a practical solution to the variety of buffer systems currently in use.

Volume

Although restriction endonucleases exhibit activity over wide concentration ranges, the reaction volume should be carefully selected. Analytical reactions (<50 μl) are especially susceptible to significant concentration errors. Pipetting of small volumes of reaction components can introduce significant error, especially when using repeating pipettes outside their tolerance limits. Viscous solutions, e.g., the restriction endonuclease stocks, are especially difficult to dispense accurately in volumes of less than 5 μl. Significant variation in the extent of reaction can be observed with inadvertent delivery of insufficient enzyme. Positive displacement or calibrated glass micropipettes are recommended for measuring critical volumes. Alternatively, samples should be diluted so that ≥ 5 μl can be pipetted.

Component concentrations in small volume reactions (<50 μl) can also be altered significantly during incubation. This is especially apparent in long-term (>1 hr) or high-temperature ($>37°$) incubations, which evaporate a considerable percentage of the water. Reactions in capped microfuge tubes can trap the water, but the collected moisture should be centrifuged into the reaction volume occassionally. Overlayering the reaction volume with mineral oil for high-temperature incubations offers another solution; however, one must be careful during retrieval of reaction products.

Large-volume reactions (>0.5 ml) can on occasion fail to give complete DNA digestion. Scaled-up reactions should take into account final DNA and enzyme concentrations. Viscous DNA solutions inhibit enzyme diffusion and can significantly reduce apparent enzyme activity. For troublesome digestions, sometimes 20 0.5-ml reactions are more successful than a single 10-ml reaction.

Incubation Time and Temperature

The restriction endonuclease unit presents a practical, though unusual, enzyme activity definition based on complete digestion of the substrate. Frequently used is the equation

$$a \ (\mu\text{g of DNA cleaved}) = b \ (\text{units of enzyme}) \times c \ (\text{hours of incubation})$$

This equation is sometimes useful as a guide, but extrapolation of incubation time or amount of enzyme from this definition can lead to erroneous results. For example, one unit of restriction enzyme may or may not represent sufficient enzyme molecules to cleave 2 μg of DNA completely in 2 hr of incubation. The extrapolation assumes that the enzyme activity is stable and linear over the entire incubation period. From our experience not all restriction enzymes remain completely active at their reaction temperature for 1 hr or longer. An exception is *Bal*I, which remains active for at least 16 hr.[54] Although long (overnight) incubations can occasionally be successful in saving on the amount of restriction enzyme used, it is not recommended. From experience, contaminating nonspecific exonucleases and endonucleases usually survive better than the specific restriction endonucleases in long incubations. Thus, even slight nonspecific nuclease contamination can, given enough time, destroy the precision and uniqueness of fragments generated by restriction enzyme cleavage. The most reliable results are obtained by maintaining reaction conditions as defined for unit activity.

Most restriction endonuclease activities are determined at 37°. Restriction enzymes isolated from thermophilic bacteria are more stable and more active at temperatures higher than 37°.[35,39,55] Selected restriction endonucleases were studied to ascertain the effect of assay temperature on their activity. *Eco*RI is inactive above 42°,[9] while *Bam*HI loses significant activity at 55°.[10] Curiously, *Hae*III (from a nonthermophilic bacterium) is fully active at 70°, whereas its companion enzyme *Hae*II is inactivated above 42°.[8] High temperature reactivity of restriction endonucleases can be used advantageously, e.g., as probes of DNA secondary structure[8] or for suppression of contaminating enzymic activities.[55] Below 37° most restriction endonucleases remain active, although at reduced rates. Thus, DNA cleavage will occur once all necessary reaction components are present, even though the reaction vessel remains on the bench top or in an ice bath. For example, *Eco*RI was demonstrated to cleave a

[54] R. E. Gelinas, P. A. Myers, G. A. Weiss, R. J. Roberts and K. Murray, *J. Mol. Biol.* **114**, 433 (1977).
[55] S. Sato, C. A. Hutchison III, and J. I. Harris, *Proc. Natl, Acad. Sci. U.S.A.* **74**, 542 (1977).

duplex octanucleotide in the temperature range from 5° to 30°.[9] Hence, the order of addition of components to the reaction mixture should place the enzyme last, at which point the reaction is deemed to have started.

Stopping Reactions

Restriction endonuclease reactions can be stopped by one of several different methods. The method chosen depends upon the subsequent use of the DNA products. For reactions performed solely for the purpose of analyzing the DNA fragments by gel electrophoresis, chelation of Mg^{2+} by EDTA is an effective method to terminate cleavage. If desired, the reaction can be reestablished readily by addition of excess Mg^{2+}. Dissociation and/or denaturation of the restriction endonuclease by adding 0.1% SDS also stops the reaction. For ease and efficiency we add to the reaction one-tenth volume of a solution containing 50% (v/v) glycerol, 100 mM Na$_2$ EDTA (pH 8), 1% (w/v) SDS, and 0.1% (w/v) bromophenol blue. Incubation of this mixture at 65° for 5 min just prior to gel application ensures distinct, reproducible DNA fragment patterns by dissociating bound proteins (e.g., BSA) and reducing DNA·DNA associations, such as the "sticky ends" of lambda DNA.

When the products of the reaction are to be used subsequently for kinasing, ligation, or sequencing, the reaction can be terminated, in some cases, by heat inactivation of the enzyme, or more reliably by phenol extraction of the DNA fragments. Some enzymes such as EcoRI[9] or HaeII[8] are irreversibly inactivated by exposure to 65° for 5 min, whereas TthI[39] and HindIII[48] remain active after this treatment. Therefore, we suggest extraction of the DNA from the reaction mixture with an equal volume of phenol freshly saturated with 0.1 M Tris-HCl (pH 8). An ether extraction to remove the residual phenol is followed by two consecutive precipitations of the DNA with one-half volume of 7.5 M ammonium acetate and two volumes of ethanol for 30 min at −70°. Suspension of the DNA in appropriate buffer provides restriction fragments free of restriction reaction components, phenol, and the enzyme.

Detection of Reaction Products

Total DNA Mass

Upon completion of a restriction endonuclease reaction, the DNA fragments are typically separated by agarose or polyacrylamide gel electrophoresis.[7,56] Usually, the resolved fragments are detected by direct

[56] E. Southern, this volume [5].

staining. Fluorescence of ethidium bromide bound to DNA is the most frequently used method to observe the DNA fragments. In agarose gels a sensitivity of about 20 ng per band is expected. Native, single-stranded DNA and RNA will also fluoresce, but with relatively less intensity. Methylene blue, acridine orange, and Stains-All[57] also can be used. Since Stains-All employs 50% formamide as a solvent, this stain is very useful for detecting DNA fragments in gels run under denaturing conditions.[58] Ethidium bromide, on the other hand, stains very poorly, if at all, under these conditions. Uniform radioactive labeling of DNA also provides a means to detect the total mass of each DNA band by autoradiography.[56]

The fragments of a DNA generated by a restriction enzyme reaction are equimolar with respect to one another. Thus, detection of DNA by mass provides a direct correlation between stain intensity and fragment length. Conversely, the relative molarities of restriction fragments of known lengths can be determined from their relative intensities. If used quantitatively a standard curve must be employed, as the linear relationship between intensity and mass is valid only over a narrow range.[59] Note also that especially small DNA fragments (<75 base pairs) may be difficult to detect by this method.

Total DNA Ends

The intensity of radioactively end-labeled DNA restriction fragments following gel electrophoretic separation and autoradiography[56] is molarity dependent. In contrast to the mass-dependent measurement, this method visualizes each DNA fragment equally, regardless of size. Short oligonucleotides are easily detectable by this technique. End-labeling methods are also several orders of magnitude more sensitive than direct staining. The 5'-phosphate end generated by almost all restriction endonucleases[4] (*Nci*I was found to generate 5'-hydroxyl and 3'-phosphate ends[34]) is radioactively labeled (^{32}P) by treatment of the fragments with alkaline phosphatase followed by incubation with polynucleotide kinase and [γ-^{32}P]ATP.[33] Alternatively, the 3' end is labeled by one of several enzymic procedures.[5]

Detection of Specific DNA Sequences

A specific DNA sequence can be detected among a complex mixture of DNA sequences by using a radiolabeled DNA or RNA probe comple-

---

[57] A. E. Dahlberg, C. W. Dingman, and A. C. Peacock, *J. Mol. Biol.* **41**, 139 (1969).
[58] T. Maniatis and A. Efstratiadis, this series, Vol. 65, p. 299.
[59] A. Prunell, this series, Vol. 65, p. 353.

mentary to the desired DNA sequence. Southern or blot hybridization[56] is highly sensitive and specific, capable of detecting a single specific DNA sequence in the midst of a tremendous excess of nonspecific DNA sequences. Restriction endonuclease fragments radiolabeled by nick translation or radiolabeled synthetic polynucleotides can serve as effective hybridization probes.

Other Reaction Considerations

In addition to the conditions described above, other reaction parameters pertinent to the use of restriction endonucleases and the generated products include (a) the extent of methylation of the DNA substrate and the selection of the appropriate restriction endonucleases to cleave methylated DNA; (b) those conditions that elicit the expression of secondary (star) activities of specific restriction endonucleases; (c) the parameters required to generate partial digestion of DNAs; (d) the ability to perform multiple digestions; and (e) the level of contaminating endonuclease and exonuclease activities.

Methylation

In bacterial systems, methylation usually occurs at either an adenine residue (N-6 position) or a cytosine residue (5 position) within the recognition sequence(s) for the specific endogenous restriction endonuclease(s).[60] Methylation of eukaryotic DNA is almost exclusively restricted to the 5 position of cytosine and primarily (>90%) to the cytosine residues present in the dinucleotide CpG.[61] Findings in eukaryotic systems have suggested the involvement of methylation in numerous functions which include: transcriptional regulation, differentiation, influence of chromosomal structure, DNA repair and recombination, and designation of sites for mutation (reviewed by Ehrlich and Wang[60]). The sensitivity of restriction endonuclease cleavage to methylation (Table IV) can be used advantageously to deduce the patterns and the extent of methylation in DNA. For example, the differential reactivity of the isoschizomers MspI and HpaII to ^{m}CG was used to identify gross tissue specific differences in methylation patterns and, more important, to identify the methylation status of cleavage sites within a specific gene or genetic region.[62]

---

[60] M. Ehrlich and R. Y.-H. Wang, Science 212, 1350 (1981).
[61] A. Razin and A. D. Riggs, Science 210, 604 (1980).
[62] C. Waalwijk and R. A. Flavell, Nucleic Acids Res. 5, 4631 (1978).

Secondary (Star) Activity of Restriction Endonucleases

Secondary (star) activity of a restriction endonuclease refers to the relaxation of the strict canonical recognition sequence specificity resulting in the production of additional cleavages within a DNA. One prominent example is EcoRI, where the usual hexanucleotide sequence (GAATTC) is reduced to a tetranucleotide sequence (AATT) for EcoRI* (EcoRI "star").[41] Several parameters responsible either individually or in combination for generating star activities include (a) glycerol concentration; (b) ionic strength; (c) pH; (d) the presence of organic solvents; (e) divalent cations; and (f) high enzyme-to-DNA ratios. Restriction enzymes that have been shown to express secondary activities under these conditions are listed in Table VIII (and in Table II). Cleavage in the presence of a high glycerol concentration in the reaction mixture represents the most commonly recognized factor associated with secondary activities.

At restriction enzyme-to-DNA ratios of 50 units/μg, glycerol concentrations as low as 7.5% (v/v) can cause the generation of star activities.[63,64] At lower enzyme-to-DNA ratios (10 units/μg), glycerol concentrations of 20% (v/v) or greater are required before restriction enzyme star activities are observed.[63,64] Relatively low levels of organic solvents such as dimethyl sulfoxide (DMSO), ethanol, ethylene glycol, and dioxane, can also produce similar losses in cleavage specificity.[43,63,64] DMSO at concentrations of 1–2% (v/v) in the final reaction mixture can cause star activities.[43,63]

For restriction enzymes that require high salt concentrations (100 mM or greater) in the reaction mixture, a reduction in the salt concentration can result in the generation of secondary activities.[16] BamHI, for example, at enzyme-to-DNA ratios of 100 units/μg cleaves pBR322 at one site in reactions containing 100 mM NaCl, at two sites in reactions containing 50 mM NaCl, and at eight sites in reactions in the absence of NaCl.[15,63] Additional factors such as the substitution of Mn^{2+} for Mg^{2+} as the divalent cation has also been reported to stimulate star activities of both EcoRI[43] and HindIII.[44] Increasing the assay pH from pH 7.5 to 8.5 also increases EcoRI* activity.[41]

Although the generation of secondary activities can provide restriction enzymes with new sequence specificities (no isoschizomers are known for EcoRI*) that may prove to be useful in some instances, these activities rarely result in complete or equal cleavage of all possible secondary recognition sites. Therefore, to eliminate or minimize the expression of restriction enzyme secondary activities, all restriction enzyme assays

[63] J. George, R. W. Blakesley, and J. G. Chirikjian, J. Biol. Chem. 255, 6521 (1980).
[64] E. Malyguine, P. Vannier, and P. Yot, Gene 8, 163 (1980).

TABLE VIII
Reaction Conditions That Induce Secondary "Star" Activity in Certain Restriction Endonucleases

Enzyme	Alterations of standard reaction conditions[a]	References
*Ava*I	A, B, D	*b, c*
*Bam*HI	A, B, C, D, E, H	*b, c, d, e*
*Bst*I	B, D	*b, f*
*Bsu*I	B, D, F	*g*
*Eco*RI	A, B, D, E, F	*b, h, i, j*
*Hae*III	B, D	*b*
*Hha*I	B, D, G	*b, e*
*Hin*dIII	E	*k*
*Hpa*I	A, B, D	*b, c*
*Pst*I	A, B, D, G	*c, e*
*Pvu*II	B, D	*l*
*Sal*I	A, B, D, G	*b, c, e*
*Sst*I	B, D, G	*b, e*
*Sst*II	B, D	*b*
*Xba*I	B, D, G	*b, e*

[a] Abbreviations used: A, ethylene glycol (45%); B, glycerol (12–20%); C, ethanol (12%); D, high enzyme : DNA ratio (>25 units/μg); E, Mn^{2+} substituted for Mg^{2+}; F, pH 8.5; G, dimethyl sulfoxide (8%); and H, absence of NaCl.

[b] J. George and J. G. Chirikjian, *Proc. Natl. Acad. Sci. U.S.A.* **79,** 2432 (1982).

[c] K. Nath and B. A. Azzolina, *in* "Gene Amplification and Analysis," Vol. 1: "Restriction Endonucleases" (J. G. Chirikjian, ed.), p. 113. Elsevier/North-Holland, Amsterdam, 1981.

[d] J. George, R. W. Blakesley, and J. G. Chirikjian, *J. Biol. Chem.* **255,** 6521 (1980).

[e] E. Malyguine, P. Vannier, and P. Yot, *Gene* **8,** 163 (1980).

[f] C. M. Clarke and B. S. Hartley, *Biochem. J.* **177,** 49 (1979).

[g] K. Heininger, W. Horz, and H. G. Zachau, *Gene* **1,** 291 (1977).

[h] B. Polisky, P. Greene, D. E. Garfin, B. J. McCarthy, H. M. Goodman, and H. W. Boyer, *Proc. Natl. Acad. Sci. U.S.A.* **72,** 3310 (1975).

[i] C. J. Woodbury, Jr., O. Hagenbuchle, and P. H. von Hippel, *J. Biol. Chem.* **255,** 11534 (1980).

[j] T. I. Tikchonenko, E. V. Karamov, B. A. Zavizion, and B. S. Naroditsky, *Gene* **4,** 195 (1978).

[k] M. Hsu, and P. Berg, *Biochemistry* **17,** 131 (1978).

[l] H. Belle Isle, unpublished results, 1981.

should be performed under the recommended standard assay conditions especially in regard to pH, ionic strength, and divalent cation concentration. The amount of glycerol introduced into the assay should be kept below 5% (v/v), and prolonged incubation with high enzyme-to-DNA ratios should be avoided. In addition, the introduction of additional components via the DNA substrate, especially DNA previously exposed to organic solvents, can be minimized by dialyzing the DNA prior to restriction enzyme cleavage.

Partial Digestion of DNA

Partial digestion refers to incomplete cleavage of the DNA, observed as fragments of higher molecular weight than the final cleavage products. These usually disappear by increasing incubation time or the amount of enzyme added. When DNA fragments generated by restriction endonuclease cleavage (e.g., Sau3A) are used in "shotgun" cloning experiments, partial digestion of the DNA substrate is frequently desirable. Under this condition internal recognition sequences for the selected restriction endonuclease remain intact at a frequency nearly dependent upon the amount of enzyme added and the incubation condition used. Partial digestions also could be obtained by substitution of other divalent cations (e.g., Mn^{2+} or Zn^{2+} [65]) for Mg^{2+} (see Table II) to slow the reaction or by addition of DNA binding ligands, such as actinomycin[14,66] and 6,4'-diamidino-2-phenylindole.[67] Each of these methods, however, is nonrandom, showing a hierarchy of cleavage rates for the various sites within the DNA. A more effective technique to generate random partial digests is partially to methylate the DNA prior to restriction endonuclease cleavage.[68]

Multiple Digestions

Mapping analysis or isolation of particular DNA fragments frequently requires the digestion of DNA by more than one restriction endonuclease. When sufficient quantities of DNA are available, the safest procedure for multiple digestion involves independent restriction enzyme digestions separated by phenol extraction and ethanol precipitation. However, when DNA substrate quantities are limited and where the selected restriction endonucleases have similar assay requirements (e.g., pH, [Mg^{2+}], [NaCl], buffer), two consecutive or simultaneous digestions can proceed with no buffer alterations. This consideration was important in establishing the

[65] T. A. Bickle, V. Pirrotta, and R. Imber, this series, Vol. 65, p. 132.
[66] M. Goppelt, J. Langowski, A. Pingoud, W. Haupt, C. Urbanke, H. Mayer, and G. Maass, *Nucleic Acids Res.* **9,** 6115 (1981).
[67] J. Kania and T. G. Fanning, *Eur. J. Biochem.* **67,** 367 (1976).
[68] E. Ferrari, D. J. Henner, and J. A. Hoch, *J. Bacteriol.* **146,** 430 (1981).

core buffer system (see the section The Reaction). But even where identical reaction conditions are recommended for two enzymes, digestions should be performed consecutively, rather than simultaneously, to ensure that each enzyme cleaves completely. When double digestions require restriction enzymes with different recommended assay conditions, each reaction should be performed under its optimal conditions. For example, to perform a *Kpn*I, *Hin*fI double digestion where both enzymes have identical assay requirements except for NaCl concentration (Tables II, and VII), one should first cleave to completion with *Kpn*I in *Kpn*I assay buffer, then increase the NaCl concentration and cleave with *Hin*fI. For enzymes with significantly different pH, buffer, salt, or Mg^{2+} requirements, the assay buffer can be changed effectively and the DNA quantitatively recovered by a 2- to 3-hr dialysis in a microdialyzer prior to digestion with the second restriction endonuclease. Use of the recommended reaction conditions for each restriction enzyme ensures production of the appropriate restriction enzyme fragments.

Contaminating Activities

Because restriction endonucleases are used essentially as reagents in DNA cleaving reactions, they need to be free of inhibitors and contaminating activities that could interfere with either the cleavage analysis or the subsequent use of the cleaved DNA products for cloning, sequencing, etc. Two general classes of contaminating activities prevail: first, other endonuclease activities that could alter the number, the size, and the termini of fragments produced; second, exonuclease activities that could remove nucleotides from either the 3' or 5' ends of the resultant fragments and inhibit subsequent ligation and labeling experiments. Commercially available restriction enzymes are routinely characterized for and purified away from both types of nuclease contamination.

In addition to exonucleases that specifically degrade the 3' and/or 5' ends of double-stranded DNA, we have identified in several restriction endonuclease preparations a 3' exonuclease activity specific for single-stranded DNA. Thus, DNA fragments with 3' extended single-strand ends (e.g., *Hae*II and *Kpn*I) are readily degraded by this contaminating activity. Potential problems arising from contaminating exo- and endonucleases can be reduced by using the highest quality of restriction endonuclease available, the minimum quanity of enzyme required for complete digestion of the DNA, and the recommended assay conditions.

Troubleshooting Guide

In Table IX are listed a number of the common problems encountered when using restriction endonucleases, a probable cause, and a suggested

TABLE IX
TROUBLESHOOTING GUIDE

Problem	Probable causes	Suggested solutions
No cleavage	Inactive restriction enzyme	Check enzyme activity on unit substrate DNA.
	Presence of inhibitor, e.g., SDS, phenol, EDTA	Precipitate DNA twice with 1/2 volume of 7.5 M ammonium acetate plus 2 volumes of ethanol, or dialyze DNA sample.
	Nonoptimal reaction composition or temperature (thermophiles)	Prepare fresh buffer, check assay temperature.
	Inadequate gel separation of DNA substrate and fragments (e.g., SsrI cleavage of lambda DNA)	Lower the percentage of gel, or do double restriction enzyme digestion (e.g., EcoRI digested lambda DNA as substrate for SsrI).
	DNA methylation (e.g., EcoRII not cleaving pBR322 DNA)	Mix test DNA and unit substrate DNA, then cleave with selected enzyme; use isoschizomer insensitive to DNA methylation; replicate plasmid in mec⁻ dam⁻ *E. coli* host.
	DNA unmethylated (DpnI requires methylated DNA)	Replicate plasmid in mec⁺ dam⁺ *E. coli* host. Use isoschizomer that cleaves unmethylated DNA (e.g., use Sau3AI rather than DpnI)
	Other DNA modification	To identify, mix unit substrate DNA and test DNA, then cleave both with the selected restriction enzyme.
	Impure DNA	To detect, compare ability to cleave test DNA and unit substrate DNA. Remove impurities with an RPC-5 ANALOG column or by precipitating twice with 2 volumes of ethanol in the presence of 1/2 volume of 7.5 M ammonium acetate at −70° for 30 min.
	DNA has no recognition sequences for selected restriction enzyme.	Confirm restriction enzyme activity and lack of inhibitors as above. Do 10-fold excess units of enzyme. Cleave test DNA with several other restriction enzymes to ensure that impurities in the DNA or DNA methylation are not responsible for lack of cleavage.
Partial cleavage	Loss of restriction enzyme activity	Use 5- to 10-fold excess restriction enzyme. Check for conditions that cause enzyme activity loss, see below.

	Cause	Remedy
	Incorrectly diluted enzyme	Do unit titration assay or redilute from fresh enzyme stock.
	Presence of inhibitor(s), e.g., SDS, phenol, EDTA, or plasticizer from microfuge tubes	See above.
	Improper reaction conditions	Prepare fresh assay buffer, check assay temperature. Try overnight digestion; add 0.01% Triton X-100 to increase stability during incubation. Determine activity on unit substrate DNA.
	Test DNA requires more restriction enzyme for complete cleavage than unit substrate DNA (e.g., see Table V).	Do 5- to 10-fold excess restriction digest.
	DNA impure	See above.
	Methylation of only a subset of the recognition sequences (e.g., XbaI cleavage of lambda DNA)	See above.
	Other DNA modifications	See above.
	Portion of substrate DNA left unreacted on side of microfuge tube	Centrifuge 1-2 sec in Eppendorf centrifuge prior to incubation.
	Pipetting error (especially of viscous solutions)	Use positive displacement pipettors; dilute so >5 µl are pipetted.
	Annealed DNA ends (e.g., lambda DNA)	Heat DNA at 65° for 5 min prior to gel electrophoresis.
	Denaturation of restriction enzyme by assay reagents, temperature, and vortexing	Use recommended reaction conditions and temperature; avoid vigorous vortexing.
	Differences in nucleotide sequences adjacent to recognition site	Detect by cleaving a mixture of test DNA and unit substrate DNA; use 5- to 10-fold excess enzyme.
	Loss of restriction enzyme activity upon dilution	Dilute only into recommended storage buffer. Use immediately; diluted enzyme usually does not store well. Do 5- to 10-fold excess digest to obtain complete cleavage.
Persistent partial	Partial methylation (e.g., XbaI cleavage of lambda DNA)	See above
	Differences in nucleotide sequence adjacent to recognition site	See above
Difficulty cleaving supercoiled DNA	DNA structure	Relax DNA with topoisomerase I, then cleave; linearize plasmid first with another restriction enzyme; or use excess restriction enzyme.

(continued)

TABLE IX (*continued*)

Problem	Probable causes	Suggested solutions
Failure to obtain expected enzyme activity	Assayed on DNA different from that used to determine unit activity	Use DNA for unit assay for quantitation of enzyme activity.
	Pipetting errors (especially of viscous solutions)	Dilute so >5 μl can be added to reaction.
	Concentration of bovine serum albumin (BSA)	Use recommended BSA concentration in assay and storage buffers
	Loss of enzyme activity	See below.
More than expected number of DNA fragments	Restriction enzyme "star" activity	Detect by appearance of extra DNA fragments produced by cleavage of DNA used for unit determination; check assay conditions, especially glycerol concentration, or Mn^{2+} for Mg^{2+}; precipitate DNA twice with half volume of 7.5 M ammonium acetate and 2 volumes of ethanol at −70° for 30 min; minimize quantity of enzyme used.
	Presence of second restriction enzyme	Detect second activity by comparing restriction digest pattern to that expected for the DNA used for unit determination.
	Test DNA contaminated with another DNA	Detect by DNA minus enzyme on gel; assay other restriction enzymes on the same DNA substrate; purify test DNA from contaminant by either gel electrophoresis or RPC-5 ANALOG column chromatography.
No DNA observed	DNA quantitation in error (e.g., RNA contamination)	Treat DNA preparation with 100 μg/ml DNase-free RNase, phenol extract, then either dialyze or precipitate twice with ethanol.
	Nonspecific precipitation in reaction	Dialyze DNA or precipitate DNA twice with ethanol prior to assay.
Rapid loss of restriction enzyme activity upon storage	Improper storage temperature	Store enzymes in recommended storage buffer plus 50% glycerol at −20° in a non-frost-free freezer.

Problem	Cause	Recommendation
	Enzymes stored diluted.	Store restriction enzymes only in concentrated form.
	Incorrect storage buffer (e.g., EcoRI in Tris-HCl)	Use recommended storage buffer; check for pH changes with temperature.
	Low protein concentration	Store enzymes with 500 μg/ml nuclease-free BSA.
Diffuse DNA bands after gel electrophoresis	Protein binding to DNA	Heat cleaved DNA at 65° for 5 min in the presence 0.1% SDS prior to loading the gel.
	Exonuclease contamination	Detect by monitoring acid-soluble material after incubation of enzyme with radioactively labeled DNA; minimize the quantity of restriction enzyme used and/or incubation time.
Large precipitates after ethanol precipitation of DNA	MgPO$_4$ precipitation	Dialyze DNA after precipitation; use buffer with little or no phosphate; add excess EDTA to the DNA prior to precipitation to chelate the magnesium.
Poor ligation efficiency	High phosphate or salt carry-over from restriction digest	Dialyze DNA fragments after restriction digest; remove phosphate or salt with small molecular sieve column or multiple ethanol precipitations.
	Incomplete removal of restriction enzyme	Extract DNA after restriction digest with equal volumes of buffered phenol, chloroform, and ether, then precipitate with ethanol.
	Incomplete removal or inactivated bacterial alkaline phosphatase	Extract DNA after phosphatase treatment with buffered phenol, chloroform, and ether. Precipitate with ethanol.
	ATPase contamination	Extract the DNA with an equal volume of phenol, chloroform, ether, then precipitate with ethanol.
	Ligation of blunt ends	Use excess T4 DNA ligase (1–2 units per picomole of free ends).
	Exonuclease contamination	Detect as above; minimize amount of enzyme or incubation time.
	Unstable buffer components	Prepare fresh ligation buffer.
Poor kinase efficiency	Phosphate carry-over	Remove phosphate by either dialysis, molecular sieve, or RPC-5 ANALOG column.
	Self-annealing of GC-rich ends.	Heat DNA at 65° for 5 min prior to kinase reaction.

solution. This list is not necessarily complete, nor are the solutions unique, but the guide is intended to be a quick reference for effectively utilizing restriction endonucleases as tools for molecular biology.

Acknowledgments

We acknowledge Drs. H. Belle Isle, D. Appleby, A. Hu, and A. MarSchel of this laboratory for their contributions of unpublished data. We express our appreciation to P. Hammond for typing the manuscript and to Drs. D. Rabussay, J. George, J. A. Thompson, J. Kane, and D. Hendrick for their valuable comments and suggestions.

[3] Site-Specific Cleavage of DNA at 8-, 9-, and 10-bp Sequences

By MICHAEL MCCLELLAND

Introduction

Site-specific cleavage of DNA at 8-, 9-, and 10-bp sequences has been reported.[1,2] This technique relies on a restriction enzyme, *Dpn*I, which only cuts the sequence GATC when both strands are methylated at adenine[3,4];

$$5' \cdots GmA \ T \ C \cdots 3'$$
$$3' \cdots C \ TmA \ G \cdots 5'$$

*Dpn*I does not cut the DNA of most species because they lack this methylated sequence.

Using the sequence-specific modification methylases M.*Taq*I (TCG^mA),[5] M.*Mbo*II (GAAGmA), and M.*Cla*I (ATCG^mAT) enables selective cleavage of DNA by *Dpn*I at TCGATCGA, GAAGATCTTC, and ATCGATCGAT, respectively, in DNA which is otherwise uncuttable by *Dpn*I (Fig. 1).[1,2] When M.*Mbo*II is used in conjunction with M.*Cla*I, cleavage by *Dpn*I occurs at the four 10-bp sequences GAAGATCTTC, GAAGATCGAT, ATCGATCTTC, and ATCGATCGAT. Thus M.*Mbo*II/ M.*Cla*I sites occur as frequently as a 9-base recognition sequence.[2] Other methylation-dependent *Dpn*I cleavage systems can be envisioned. These are listed in Table I.[6–10]

Cleavage systems of 8, 9, and 10 bp produce average fragment sizes of 65,000, 250,000 and 1,000,000 bp, respectively. Fragments of this size can

[1] M. McClelland, L. Kessler, and M. Bittner, *Proc. Natl. Acad. Sci. U.S.A.* **81,** 983 (1984).

[2] M. McClelland, M. Nelson, and C. Cantor, *Nucleic Acids Res.* **13,** 7171 (1985).

[3] S. Lacks and B. Greenberg, *J. Mol. Biol.* **114,** 153 (1980).

[4] G. E. Geier and P. Modrich, *J. Biol. Chem.* **254,** 1408 (1979).

[5] Unless otherwise indicated, the sequence of only one strand is shown, oriented 5' to 3'. ^mA represents 6-methyladenine.

[6] H. van Ormondt, J. A. Lautenberger, S. Linn, and A. deWaard, *FEBS Lett.* **33,** 177 (1973).

[7] M. McClelland and M. Nelson, *Nucleic Acids Res.* **15,** r219 (1987).

[8] M. Nelson, C. Christ, and I. Schildkraut, *Nucleic Acids Res.* **12,** 5165 (1984).

[9] K. Kita, N. Hiraoka, A. Oshima, S. Kadonishi, and A. Obayashi, *Nucleic Acids Res.* **13,** 8685 (1985).

[10] R. J. Roberts, *Nucleic Acids Res.* **13,** r165 (1985).

RECOMBINANT DNA
METHODOLOGY

A   NO METHYLATION

```
5' N N N G A T C N N N 3'
3' N N N C T A G N N N 5'     no 6mA


              |
              |            m
              | Dpn I (GA\TC)
              |

           NO CUTTING
```

B   HEMIMETHYLATION

```
5' N T C G A T C N N N 3'
3' N A G C T A G N N N 5'


              |                m
              | M.Taq I (TCGA)
              |

         m
5' N T C G A T C N N N 3'
3' N A G C T A G N N N 5'     only one 6mA
     m


              |
              |            m
              | Dpn I (GA\TC)
              |

           NO CUTTING
```

C   DOUBLE METHYLATION

```
5' N T C G A T C G A N 3'
3' N A G C T A G C T N 5'


              |                m
              | M.Taq I (TCGA)
              |

         m         m
5' N T C G A T C G A N 3'
3' N A G C T A G C T N 5'
     m         m


              |
              |            m
              | Dpn I (GA\TC)
              |

         m                m
5' N T C G A    pT C G A N 3'
3' N A G C Tp      A G C T N 5'
     m                m
```

CUTTING only AT THE EIGHT
BASE PAIR SEQUENCE TCGATCGA

FIG. 1. M.TaqI TCGmA-dependent DpnI GmA\TC cleavage at the 8-bp sequence
TCGATCGA.

now be separated electrophoretically.[11,12] Potential applications for rare cleavage systems include demonstrating physical linkage between distant DNA markers, generating subchromosomal libraries, and megabase chromosome walking.[1,2]

DNA Preparation

For most applications of methods involving cleavage of DNA at rare sites it is necessary that the DNA remain unsheared throughout the procedure since the desired product molecules may be up to millions of base pairs long. Thus DNA is prepared by a method which protects it from shear during purification, methylation, and cleavage. The insert method, developed by Schwartz, in which cells are suspended in a block of agarose, is described elsewhere.[11] A similar method using agarose microbeads first devised by Cook et al. for the study of chromatin[13,14] can be adapted to the preparation of high-molecular-weight DNA. Beads have the advantage of a very large surface area and short surface-to-center distance when compared to inserts, and they can also be pipetted without DNA shearing. However, inserts are sometimes easier to manipulate and if the DNA to be prepared contains small fragments or plasmids these may diffuse out of beads more rapidly than out of inserts.

Microbead Preparation

1. Washed cells are placed in the appropriate buffer: phosphate-buffered saline for mammalian cells or 10 mM Tris (pH 8.0), 100 mM EDTA, and 10 mM EGTA (to inhibit Ca^{2+}-dependent nucleases) for yeast and *Escherichia coli*. Living cells are suspended in 1% low-gelling-temperature (LGT) agarose at 38–40° by gentle swirling. (Addition of EDTA to agarose before melting results in agarose aggregates.) A total DNA concentration of up to 100 μg/100 μl is acceptable. For a mammalian cell this is 5×10^7 cells/ml; for yeast, 10^9 cells/ml; and for *E. coli*, 10^{10} cells/ml.

Note: The major technical problem with the use of *Dpn*I is that the batch of agarose can influence the subsequent enzymatic reactions, thus, it is advisable to use a high-quality agarose and to do a pilot experiment with agarose and GmATC-methylated DNA to ensure the batch will allow cleavage with *Dpn*I and any other nucleases to be employed. Substrates for *Dpn*I cleavage include plasmids from *dam*+ strains of *E. coli* which are methylated at GmATC *in vivo*. The quality of the agarose can be im-

[11] D. Schwartz and C. R. Cantor, *Cell* 37, 67 (1984).
[12] G. F. Carle and M. V. Olsen, *Nucleic Acids Res.* 12, 5647 (1984).
[13] P. R. Cook, *EMBO J.* 3, 1837 (1984).
[14] D. A. Jackson and P. R. Cook, *EMBO J.* 4, 913 (1985).

TABLE I

AVAILABLE AND POTENTIAL CLEAVAGE STRATEGIES[a]

Methylase(s)	Recognition sequence	Effective size (base pairs)
*M.*Mbo*II/M.*Taq*I	GAA<u>GATC</u>GA	7.7
M.*Acc*III/M.*Taq*I	TCCG<u>GATC</u>GA	7.9
M.*Eco*B/M.*Taq*I	AGCAN₅T<u>GATC</u>GA	7.9
M.*Nru*I/M.*Taq*I	TCGC<u>GATC</u>GA	7.9
M.*Taq*I/M.*Taq*II	TC<u>GATC</u>GGTC	7.9
M.*Taq*I/M.*Xba*I	TC<u>GATC</u>TAGA	7.9
*M.*Taq*I/M.*Taq*I	TC<u>GATC</u>GA	8.0
*M.*Cla*I/M.*Mbo*II	ATC<u>GATC</u>TTC	9.0
M.*Acc*III/M.*Cla*I	TCCG<u>GATC</u>GAT	9.7
M.*Acc*III/M.*Mbo*II	TCCG<u>GATC</u>TTC	9.7
M.*Cla*I/M.*Nru*I	ATC<u>GATC</u>GCGA	9.7
M.*Cla*I/M.*Taq*II	ATC<u>GATC</u>GGTC	9.7
M.*Cla*I/M.*Xba*I	ATC<u>GATC</u>TAGA	9.7
M.*Mbo*II/M.*Nru*I	GAA<u>GATC</u>GCGA	9.7
M.*Mbo*II/M.*Taq*II	GAA<u>GATC</u>GGTC	9.7
M.*Mbo*II/M.*Xba*I	GAA<u>GATC</u>TAGA	9.7
M.*Nru*I/M.*Taq*II	TCGC<u>GATC</u>GGTC	9.7
M.*Eco*B/M.*Cla*I	AGCAN₅T<u>GATC</u>GAT	9.9
M.*Eco*B/M.*Mbo*II	AGCAN₅T<u>GATC</u>TTC	9.9
*M.*Cla*I/M.*Cla*I	ATC<u>GATC</u>GAT	10.0
*M.*Mbo*II/M.*Mbo*II	GAA<u>GATC</u>TTC	10.0
M.*Acc*III/M.*Nru*I	TCCG<u>GATC</u>GCGA	11.0
M.*Acc*III/M.*Taq*II	TCCG<u>GATC</u>GGTC	11.0
M.*Acc*III/M.*Xba*I	TCCG<u>GATC</u>TAGA	11.0
M.*Nru*I/M.*Xba*I	TCGC<u>GATC</u>TAGA	11.0
M.*Taq*II/M.*Xba*I	GACC<u>GATC</u>TAGA	11.0
M.*Acc*III/M.*Eco*B	TCCG<u>GATC</u>AN₅TGCT	11.7
M.*Eco*B/M.*Nru*I	AGCAN₅T<u>GATC</u>GCGA	11.7
M.*Eco*B/M.*Taq*II	AGCAN₅T<u>GATC</u>GGTC	11.7
M.*Eco*B/M.*Xba*I	AGCAN₅T<u>GATC</u>TAGA	11.7
M.*Acc*III/M.*Acc*III	TCCG<u>GATC</u>CGGA	12.0
M.*Nru*I/M.*Nru*I	TCGC<u>GATC</u>GCGA	12.0
M.*Taq*II/M.*Taq*II	GACC<u>GATC</u>GGTC	12.0
M.*Xba*I/M.*Xba*I	TCTA<u>GATC</u>TAGA	12.0
M.*Eco*B/M.*Eco*B	AGCAN₅T<u>GATC</u>AN₅TGCT	14.0

[a] Available and potential cleavage systems using *Dpn*I and either one or two site-specific methylases are presented. Each system uses a methylase which overlaps GATC by two or more bases. More complicated schemes employing more than two methylases can be envisioned but are not included here. Where there is more than one recognition sequence for *Dpn*I, only one of the combinations is shown. For instance, the system based on M.*Cla*I and M.*Mbo*II has four recognition sequences but only the sequence ATCGATCTTC is shown.

proved by purification with DEAE.[13] Also note that if a significant proportion of chromosomes are in the process of replication they may contribute a background of fragmented molecules, although, we do not find the use of replication inhibitors increases the average size of DNA prepared.

2. The cells, suspended in melted agarose, are placed in twice the volume of mineral oil, also at 38–40°, in a clean Corex tube and shaken in one continuous burst at medium power (800 cycles/min) on a vortex mixer at room temperature for 30 sec. An emulsion is formed.[13,14]

3. After cooling at 4°, the 100-μm beds of agarose are mixed with a twofold-volume excess of buffer and spun at 2000 rpm in a bench-top centrifuge. The aqueous phase containing the beads is decanted. Microbead aggregates may have formed. These are dispersed in a large-volume excess of buffer then the beads are reprecipitated by spinning at 2000 rpm in a bench-top centrifuge. The beads should be of relatively uniform diameter ranging from 25 to 100 μm.[13,14] It is advisable to practice on plain agarose until proficient.

4. In general, the beads are incubated in a 50-fold-volume excess with a cell wall-disrupting enzyme. For cells without walls such as mammalian

---

The effective recognition sequence size for each strategy is obtained by adding together the cleavage frequencies of each recognition sequence then taking the log (base 4) of the reciprocal. For example, for M.ClaI/M.MboII this is calculated from the frequency of GAAGATCTTC (1/4^{10}) + GAAGATCGAT (1/4^{10}) + ATCGATCTTC (1/4^{10}) + ATCGATCGAT (1/4^{10}) = (1/4^9). Effective size = 9 bp. A more complicated example is the M.ClaI/M.NruI system which includes the 10-base M.ClaI/M.ClaI sequence ATCGATCGAT, the 11-base M.ClaI/M.NruI sequences ATCGATCGCGA and TCGCGATCGAT, and the 12-base M.NruI/M.NruI sequence TCGCGATCGCGA. This gives an overall cutting frequency equivalent to 9.7 bp.

* The cutting systems designated with an asterisk have been demonstrated.[1,2] The others are theoretical, based on known restriction recognition sequences. M.HphI (GGTGA) has been excluded because it is now known to be a cytosine methylase (R. Feehery and M. Nelson, unpublished results). Schemes using M.EcoB have not been tried but are likely to work as M.EcoB has the correct specificity (TGCTN$_8$TG^{m}A).[6] AccIII, NruI, and XbaI do not cleave DNA methylated at adenine in their recognition sequences,[7-9] a prerequisite if the corresponding methylase is to have adenine specificity. Isoschizomers of MboII include NsuI. Isoschizomers of NruI include AmaI and Sbo13.[10] Isoschizomers of DpnI include CfuI, NmuEI, NmuDI, and NsuDI (R. Camp, P. Hurlin, and I. Schildkraut, unpublished results).

It is likely that there are other methylation-dependent restriction systems with different sequence specificities. These should yield a large number of new, rare cutting specificities.

cells, skip to step 5. The cell wall of the yeast *Saccharomyces cerevisiae* can be removed by the following solution at 37° for 3 hr[11]:

  25 μg/ml Zymolyase 5000 (Kirin Breweries)
  500 mM EDTA
  10 mM EGTA
  7% 2-mercaptoethanol

  For *E. coli* use the following solution at 37° for 1 hr:

  lysozyme (100 μg/ml)
  100 mM EDTA
  10 mM EGTA
  10 mM Tris · HCl (pH 8.0)
  0.1% Sarkosyl

The beads are pelleted by centrifugation.

5. The beads from step 3 or 4 are then placed in a solution containing lysis buffer:

  2 mg/ml proteinase K
  1% SDS
  500 mM EDTA
  10 mM EGTA
  10 mM Tris · HCl (pH 8.5)

Preincubate at 65° for 30 min to remove nuclease contaminants.

Note: In some species with cell walls it may not always be necessary to include step 4. DNA will be released by electrophoresis from some cells with walls when treated with only proteinase K and SDS.

6. The beads are incubated for 8–15 hr at 50° in a rotary water bath. Proteinase K works up to 65° but 50° is the highest temperature at which most low-melting-temperature agarose can be incubated without melting.

7. The beads are repeatedly washed in sterile 10 mM EDTA and 10 mM Tris (pH 8.0). The first few washes should contain 1 mM PMSF, which is an inhibitor of serine protease such as proteinase K. The PMSF is prepared as a fresh 100 mM stock in isopropyl alcohol. If necessary, the quality of the beads or inserts can be improved by electrophoresis in TE at 10 V per cm for 1 hr. This removes small DNA fragments, whereas those over one million base pairs remain in the agarose. Beads can be stored in 10× TE or in proteinase K buffer for many months.

Note: In this and all subsequent steps care should be taken to maintain nuclease-free conditions and to use double-distilled water.

DNA Methylation

The methylases employed in this method do not require divalent cations, thus they can be used in the presence of EDTA and EGTA to minimize nuclease cleavage during the methylation step. All the methylases will work in 0 to 100 mM NaCl buffers and employ 80 μM S-adenosylmethionine (SAM) as the methyl donor. The purification of DNA methylases is described in *Methods in Enzymology*, Vol. 155 (see Nelson and McClelland [5]). M.*Taq*I (TCG^{m}A) for *Dpn*I cleavage at TCGATCGA and M.*Cla*I (ATCG^{m}AT) for *Dpn*I cleavage at ATCGATCGAT are both available from New England Biolabs.

1. Preincubate the agarose beads at 50° for 1 min in 10 volumes of filter-sterilized methylase buffer:

> 100 mM Tris (pH 8.0)
> 10 mM EDTA
> 1 mM EGTA
> 10 μg/ml BSA (nuclease free)
> 5 mM dithiothreitol
> (0–100 mM NaCl)

2. Replace buffer with two volumes of fresh buffer. Methylation is performed in a volume twice that of the beads in buffer containing 80 μM SAM and at least 5 U of methylase per microgram of DNA. Incubation is performed for 2 hr, and then the buffer is changed and the procedure repeated for 2 hr (or overnight if the methylase is highly purified).

Note: The methylases are stored in 50% glycerol. A glycerol concentration in excess of 5% may inhibit some methylases. Also note that the efficiency of incorporation of methyl groups can be ascertained either by using tritiated SAM [in this case only 2 μM SAM (84 Ci/mmol) is used], or by monitoring cleavage with the corresponding restriction endonuclease. This latter method requires a larger amount of DNA.

DNA Cleavage

1. The methylated DNA beads or inserts are treated for 4–24 hr at 50° with a solution containing lysis buffer:

> 2 mg/ml proteinase K
> 1% SDS
> 500 mM EDTA
> 10 mM EGTA
> 10 mM Tris · HCl (pH 8.5)

Preincubated at 65° for 30 min to remove nuclease contaminants.
The lysis buffer is removed by repeated buffer changes with shaking in

100 mM Tris (pH 8.0)
1 mM EDTA
1 mM EGTA

at 50° and then replaced twice with restriction buffer (vendor's recommended conditions).

DpnI buffer is as follows:

150 mM NaCl
100 mM Tris · HCl (pH 8.0)
10 mM MgCl$_2$
100 μg/ml BSA

2. Digestion is performed in twice the bead volume using 5 U of DpnI/
μg DNA for 3 hr at 37° with shaking. It is usually advisable then to change
the restriction buffer and redigest for a further 3 hr. DpnI is available from
New England Biolabs and Boerhinger Mannheim Biochemicals.

3. At this point certain procedures may require radioactive labeling of
the DNA. This is most easily achieved with T4 DNA polymerase. The
restriction buffer is replaced with an equal volume of T4 buffer:

33 mM Tris · acetate (pH 7.9)
66 mM potassium acetate
10 mM MgCl$_2$
5 mM DTT
100 μg/ml BSA

Change after 10 min. Then 100 mM dTTP and dCTP and 2 μCi of ^{32}P-
labeled dATP and dGTP (800 Ci/mmol) are added along with 5 U of T4
polymerase per microgram of DNA. The reaction is continued for 10 min
at 37°. The reaction is stopped by the addition of an equal volume of

100 mM EDTA
10 mM Tris (pH 8.0)
1% SDS

Excess label can be removed by washing and centrifugation of the
beads.

DNA Electrophoresis

Beads can be sealed into the wells of a 0.5 to 1.5% agarose gel by
layering 100 μl of normal-gelling-temperature agarose at 50° on top of the
beads after they are placed in the well.

The type of gel depends on the electrophoresis method. For pulsed gel electrophoresis[11,12] either Tris · borate EDTA or Tris · acetate electrophoresis buffer can be used. TEA may increase the DNA carrying capacity of the gel, allowing more DNA to be electrophoresed without smearing.

Double-Decker Gels

Double-decker gels can be useful if two identical DNA patterns are required from PGE gels. After electrophoresis, one layer of a double-decker gel can be stained for UV visualization, dried for autoradiography, or blotted to nitrocellulose while the other identical layer remains as a source of intact DNA. This method is of particular use for PFG because of the variability in PFG DNA separation patterns even for gels run under ostensibly identical conditions.

Double-decker gels are made as follows:

1. Melted agarose at the appropriate concentration is prepared and kept sealed in a 60° water bath.

2. Half the volume of agarose necessary for constructing the gel is poured into a horizontal gel former with well-forming combs in place. This layer is allowed to gel.

3. The second layer is poured on top of the first from the stoppered bottle kept at 60°.

4. When both layers are gelled they can be treated like a normal gel and loaded with the samples of interest. However, in order for the DNA sample to be distributed in the well such that it will migrate both into the top and bottom layers it is advisable to use (a) microbeads mixed with a volume of 1% low-melting-temperature agarose at 50° equal to 80% of the well volume and load immediately; or (b) inserts placed with their longest edge up and then sealed with 1% LMT agarose at 50°; or (c) a liquid solution of DNA mixed with melted 1% LMT agarose at 50° with a total load volume 80% of the well volume and load immediately.

No special precautions are necessary when handling these gels. In fact, the two layers are held firmly together and only a very close inspection will reveal the interface.

After electrophoresis the two layers can be separated with a broad flat spatula. A corner is cut from the gel, at a distance from the region containing the separated DNA, revealing a visible interface between the layers. The spatula is inserted carefully between the layers. Once the spatula has slid a few centimeters into the gel gentle lateral pressure will cause it to slide along the interface between the layers. It is possible to slice through the wells and gel plugs without damage if the spatula is inserted deep between the layers before approaching the wells. It is necessary to slide over the whole surface of contact between the layers before attempting to separate the layers, otherwise the gel will rip. Generally, two layers with an identical DNA pattern can be produced at the first attempt.

Applications include identifying a DNA of interest in one layer then aligning this information with the second layer and obtaining the DNA intact from this layer.

1. The position of a band of interest can be determined from an ethidium-stained layer on a UV box. The DNA band is then cut from the other layer. This band will be ethidium free and has escaped exposure to UV light.

2. The untreated layer can be placed on top of an autoradiogram from an end-labeled digest or from a Southern blot of the other layer and the band of interest cut out. Southern blots are performed by standard methods.[15]

Large DNAs obtained in this manner from pulsed gels may be electroeluted by pulsed electrophoresis and cleaved, labeled, or ligated. Alternatively, they can be cleaved, labeled, or even ligated *in situ* or after electrotransfer to BA85 paper.[15-17] If these steps are to be performed *in situ* the gel should be cast from a high-quality agarose such as Seakem Gel Technology Grade. The agarose slice containing the desired band must be repeatedly soaked in a large volume of the appropriate buffer before enzyme reactions are performed. In some cases where the DNA does not have to remain intact the agarose can be melted and diluted many fold to prevent resolidification and thereby allow manipulation to be done in liquid rather than gel slices.

The double-decker gels can also be used for analysis of DNA or proteins after conventional electrophoresis or PFG in two dimensions. For example, in principle, DNA can be cut with *Not*I, end labeled, electrophoresed by PFG, then the lane removed and the DNA cut on BA85 paper[16,17] or *in situ* with *Eco*RI.[18] The lane is placed at 90° to a homogeneous field and re-electrophoresed. One layer can be autoradiographed to show the *Not*I/*Eco*RI fragments and the other layer will supply fragments for cloning.

We now have the ability to produce large DNA fragment sizes, separate them, and obtain the DNA fragment of interest intact and partially

[15] M. McClelland, R. Jones, Y. Patel, and M. Nelson, *Nucleic Acids Res.* **15**, in press.
[16] M. McClelland, unpublished results.
[17] S. K. Poddar and J. Maniloff, *Gene* **49**, 93 (1986).
[18] S. G. Fischer and L. S. Lerman, this series, Vol. 68, p. 183.

purified for further manipulation. This should eventually allow the development of useful techniques such as the production of subchromosomal libraries, megabase chromosome walking, and cloning in extremely large vectors.[1,2,15,19]

Acknowledgments

I thank Mike Nelson for invaluable discussions and Robert Benezra and Randy Morse for critical reading of the manuscript. I thank Charles Cantor in whose laboratory much of this work was performed. This work was supported by a Lucille P. Markey Scholar Award.

[19] M. Nelson and M. McClelland, *in* "Gene Amplification and Analysis" (J. G. Chirikjian, ed.), Vol. 5. Elsevier/North-Holland, New York, 1987.

[4] Exonuclease III: Use for DNA Sequence Analysis and in Specific Deletions of Nucleotides

By LI-HE GUO and RAY WU

Exonuclease III of *Escherichia coli* catalyzes the sequential hydrolysis of mononucleotides from the 3' termini of duplex DNA molecules.[1] Using a high ratio of exonuclease III to DNA ends and moderate concentration of salt (90 mM KCl), digestion of DNA is relatively synchronous, removing approximately 10 nucleotides per minute from each 3' terminus[2,3] at room temperature.

We have developed an improved enzymic method for DNA sequence analysis based on the partial digestion of duplex DNA with exonuclease III to produce DNA molecules with 3' ends shortened to varying lengths. After exonuclease III treatment, 3' ends are extended and labeled by repair synthesis[3] using the dideoxynucleotide chain termination method of Sanger *et al.*[4] The exonuclease III method (Procedures 8 and 9, and after Procedure 6 under DNA Sequencing) is the only sequencing method that can sequence both strands of a cloned DNA fragment over 1000 base pairs in length without prior gel fractionation of the fragments. We also describe a method for making deletions of nucleotides in DNA employing exonuclease III. Comparison is made between exonuclease III and *Bal*31 nuclease for making specific deletions.

In order to increase the ease and flexibility of cloning and sequencing any gene based on the exonuclease III method, we have constructed a family of pWR plasmids. They were derived from plasmid pUR222 of Rüther *et al.*[5] by deleting about 530 base pairs between the *amp* region and the *lac* region to produce pWR1. This deletion resulted in a fourfold increase in the copy number of the plasmid. A *Hin*dIII site was next inserted into the polylinker region of pWR1. The resultant pWR2 plasmid has eight unique restriction sites in the *lacZ'* gene, which can be used for cloning genes for sequencing by the exonuclease III method described here. Genes cloned in pWR2 can also be sequenced by other DNA sequencing methods[4,6] (procedures 7 and 10).

[1] C. C. Richardson, I. R. Lehman, and A. Kornberg, *J. Biol. Chem.* **239**, 251 (1964).
[2] R. Wu, G. Ruben, B. Siegel, E. Jay, P. Spielman, C. D. Tu, *Biochemistry* **15**, 734 (1976).
[3] L. Guo and R. Wu, *Nucleic Acids Res.* **10**, 2065 (1982).
[4] F. Sanger, S. Nicklen, and A. R. Coulson, *Proc. Natl. Acad. Sci. U.S.A.* **74**, 5463 (1977).
[5] U. Rüther, M. Koenen, K. Otto, and B. Müller-Hill, *Nucleic Acids Res.* **9**, 4087 (1981).
[6] A. M. Maxam and W. Gilbert, this series, Vol. 65, p. 499.

RECOMBINANT DNA
METHODOLOGY

Properties of pUR222

The *lacZ'* gene encodes 59 amino acid residues of the α-peptide of β-galactosidase, a number that is sufficient for α-peptide activity.[8-11] Therefore bacteria harboring a plasmid that includes this region with its operator and promoter should make blue colonies on indicator plates containing isopropylthiogalactoside (IPTG) and 5-bromo-4-chloroindolyl-β-D-galactoside (X-gal). If an exogenous DNA fragment is inserted between the *lac* promoter and the *lacZ'* gene, or into the *lacZ'* gene, bacteria harboring this plasmid should give rise to white colonies. This makes the selection easy for colonies carrying inserts.[5,8-11]

Rüther *et al.*[5] constructed a multipurpose plasmid, pUR222, which contains six unique cloning sites (*Pst*I, *Sal*I, *Acc*I, *Hind*II, *Bam*HI, and *Eco*RI) in a small region of its *lacZ'* gene. Bacteria harboring recombinant plasmids generally give rise to white colonies, whereas those containing only pUR222 form blue colonies on indicator plates. DNA cloned into this plasmid can be labeled and sequenced directly, after cutting with the proper restriction enzymes, using the procedure of Maxam and Gilbert[6]; it is not necessary to isolate the labeled fragment that is to be sequenced.

Construction and Properties of pWR Plasmids

Figure 1 shows the construction of pWR2, one of five related pWR plasmids. There are two *Pvu*II sites in pUR222, one located in the *lacZ'* gene and the other preceding the *lac* promoter. To remove the latter, plasmid pUR222 was first partially digested with restriction enzyme *Pvu*II. After partial digestion to cut only one of the two *Pvu*II sites, the linear pUR222 DNA was digested with *Bal*31 nuclease (Procedure 14) to delete about 530 base pairs. After ligation and transformation (Procedures 3 and 4), a number of blue colonies were obtained, and plasmid DNA was prepared from several of them. The DNA was analyzed by digestion with *Pvu*II and *Eco*RI and a colony with only one *Pvu*II site located in the *lacZ'* gene was selected.

For inserting a *Hind*III site[7] into the polylinker region, this plasmid was digested with *Eco*RI and *Bam*HI, and the resulting cohesive ends were filled in by using the large fragment of DNA polymerase I in the presence of four dNTPs (Procedure 10b). After electrophoresis (Proce-

[7] C. P. Bahl, K. J. Marians, R. Wu, J. Stawinsky, and S. A. Narang, *Gene* **1,** 81 (1976).
[8] B. Gronenborn and J. Messing, *Nature (London)* **272,** 375 (1978).
[9] J. Messing, B. Gronenborn, B. Müller-Hill, P. H. Hofschneider, *Proc. Natl. Acad. Sci. U.S.A.* **74,** 3642 (1977).
[10] U. Rüther, *Mol. Gen. Genet.* **178,** 475 (1980).
[11] A. Ullmann, F. Jacob, and J. Monod, *J. Mol. Biol.* **24,** 339 (1967).

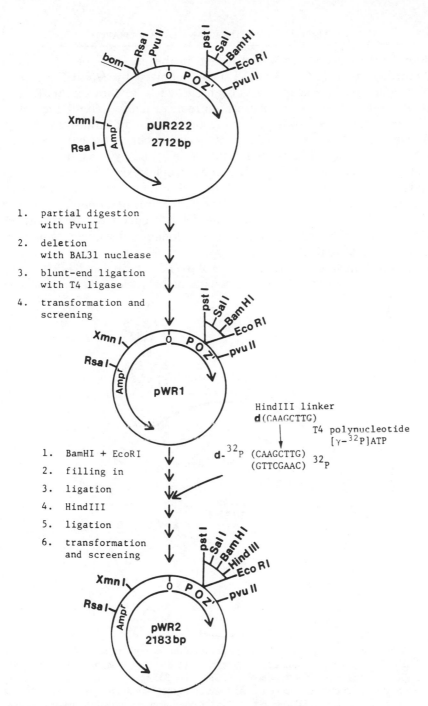

FIG. 1. Schematic representation of plasmid pWR2 construction.

dure 2), the plasmid DNA was ligated in the presence of a 5'-labeled *Hind*III adaptor, d(C-A-A-G-C-T-T-G), followed by *Hind*III digestion (Procedure 3). This adaptor has been chosen to restore the *Eco*RI and *Bam*HI sites. After electrophoresis to remove unused *Hind*III adaptor, the DNA was ligated to form a circle and was used to transform *E. coli* (Procedures 3 and 4). Transformants were screened for blue colonies; DNA was isolated from them and futher analyzed by separate digestion with *Eco*RI, *Hind*III, or *Bam*HI and electrophoresis. Five related plasmids were selected as being useful, and all of them gave very high copy number. One of them, pWR2, has the new features of a unique *Pvu*II site in the *lacZ'* gene, and a unique *Hind*III site added to the polylinker region. These sites are available for DNA cloning and sequencing in addition to sites in pUR222.[5]

Figure 2 shows the restriction map of pWR2. This plasmid is 2183 base pairs (bp) in length and consists of two functional regions. One region (355

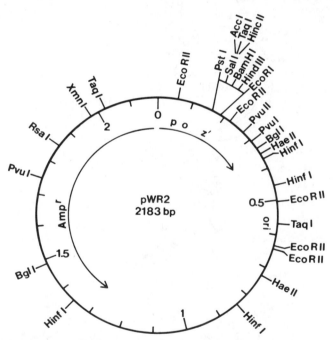

FIG. 2. Restriction map of pWR2. The positions of the *lac* region and the *Amp*^r region are indicated. Their junction is numbered as 0 (zero). The first base pair (bp) in the *lac* region (see Fig. 3) is designated nucleotide No. 1. There are 355 bp in the *lac* region, which is interrupted by the polylinker sequence, and 883 bp in the *Amp*^r region. Bacteria harboring recombinant plasmids with DNA inserts within the polylinker generally form white colonies, whereas those containing only pWR2 give rise to blue colonies on indicator plates.

bp) contains the *lac* promoter and operator and part of the *lacZ'* gene. The other region (1828 bp) contains the ampicillin-resistant gene (*amp* or *bla*) and the origin of replication (ori) of the plasmid. There are 10 unique restriction enzyme recognition sites in pWR2. Eight of them are located with the *lacZ'* gene, and two within the β-lactamase (*amp* or *bla*) gene. The locations of the restriction enzyme sites are tabulated in Table I. The junction of the *lac* region and the *Amp*[r] region is taken as the zero position on the physical map (see Figs. 2 and 3). The first base pair in the *lac* region is designated nucleotide No. 1.

In plasmid pWR2 seven unique restriction sites (*Pst*I, *Sal*I, *Acc*I, *Hinc*II, *Bam*HI, *Hin*dIII, and *Eco*RI) are located between codon No. 4 (thr) and No. 6 (ser) of the *lacZ'* gene (Fig. 4), and an eighth, *Pvu*II is in codon No. 35 (ser). DNA cloned into the *Bam*HI site of pWR2 can be directly sequenced by the chemical method. This is not possible with pUR222.[5] The other pWR plasmids (Fig. 4) differ from pWR2 in the number of unique restriction sites in the polylinker region.

We have observed that bacteria carrying pWR2 without inserts in the polylinker region are capable of growing in either M9 minimal medium or L broth medium up to 1000 μg of ampicillin per milliliter. However, bacteria harboring recombinant plasmids with inserts in the polylinker region grow more slowly and can tolerate only up to 25 μg of ampicillin per milliliter. The reason for this decrease in resistance is not clear but could be related to the fact that 530 bp have been deleted between the *bla* promoter and the *lacZ'* gene in pWR2. In the *Amp*[r] region of pWR2, there are only 27 base residues immediately preceding the start codon (ATG) of the β-lactamase gene (*bla*) (see Fig. 3, lower strand). This sequence is too short for the efficient functioning of the *bla* promoter. Efficient expression probably depends on another promoter within or downstream from the *lacZ'* gene, which is interrupted by insertion of a DNA fragment into the polylinker region.

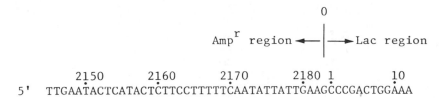

FIG. 3. DNA sequence of the intercistronic region between the *bla* gene (*Amp*[r] region) and the *lac* region in pWR plasmids. ATG start codon for translation of β-lactamase is underlined. Sequences with homology to 16 S RNA are indicated by dashed underlines.

TABLE I
pWR2: LOCATIONS OF RESTRICTION RECOGNITION SEQUENCES

Enzymes[a]	Number of cleavage sites	Locations							
AccI	1	173							
AcyI	1	1907							
AluI	11	48	143	186	282	419	645	781	1038
		1559	1659	1722					
AsuI	5	301	1412	1491	1508	1730			
AvaII	2	1508	1730						
BamHI	1	178							
BbvI	12	165	260	333	382	400	819	884	887
		1093	1421	1610	1787				
BglI	2	342	1489						
DdeI	4	752	1161	1327	1867				
EcoRI	1	190							
EcoRII	5	65	232	503	624	637			
FnuDII	4	524	1105	1435	1928				
Fnu4HI	18	166	261	334	383	401	404	522	677
		820	885	888	1094	1422	1611	1761	1788
		1883	2112						
GdiII	1	1758							
HaeI	3	492	503	955					
HaeII	2	355	725						
HaeIII	9	200	302	492	503	521	955	1413	1493
		1760							
HgaI	3	579	1157	1907					
HgiAI	3	795	1956	2041					
HgiCI	2	60	1318						
HgiEII	1	1058							
HhaI	13	26		333	354	387	657	724	824
		998	1107	1500	1593	1930			
HincII	1	174							
HindIII	1	184							
HinfI	4	377	452	848	1364				
HpaII	10	88	684	831	857	1047	1451	1485	1552
		1622	1904						
HphI	5	1221	1448	1844	2070	2085			
MboII	7	300	355	1127	1218	1973	2051	2160	
MnlI	9	295	366	592	649	916	1316	1397	1527
		1733							
MstI	2	332	1590						
PstI	1	170							
PvuI	2	313	1739						
PvuII	1	282							
RruI	1	1849							
RsaI	1	1849							

TABLE I (continued)

Enzymes[a]	Number of cleavage sites	Locations							
SalI	1	172							
Sau3AI	15	178	310	1043	1118	1129	1137	1215	1227
		1332	1673	1691	1737	1995	2012	2048	
TaqI	3	173	577	2020					
XmnI	1	1970							

[a] The following enzymes do not digest pWR2. These sites are useful for sequencing the cloned gene if present.

AvaI, BalI, BclI, BglII, BstEII, ClaI, EcoRV, HpaI, KpnI, NcoI, NruI, SacI, SacII, SmaI, SphI, StuI, Tth111I, XbaI, XhoI, XmaIII

Soberón et al.[12] showed that the deletion of the DNA region between the origin of replication and the *tet* gene of pBR322 did not affect the cloning capacity of the resulting plasmids. However, since a DNA region like the ColE1 relaxation site (*bom*) is deleted in these plasmids, they are superior EK2 vectors. Use of these vectors should permit lowering the degree of physical containment by at least one level.[13] The relaxation site (*bom*) in the pWR plasmids has also been deleted (Fig. 1), and therefore they are also superior vectors from the safety point of view.

Twigg and Sherratt[14] reported the construction of ColE derivatives in which deletion of a nonessential DNA region carrying the ColE1 relaxation site resulted in an increased copy number. In the construction of pWR plasmids, we also deleted this region from pUR222, and found that the copy number of pWR2 is at least four times higher than that of pUR222 or pBR322. The high yield of pWR2 and its derivatives allows the application of a rapid isolation procedure, yielding plasmid of sufficient purity and amount for DNA sequencing. The purity of the plasmid is high owing to a high ratio of plasmid DNA to chromosomal DNA.

[12] X. Soberón, L. Covarrubias, and F. Bolivar, *Gene* **9**, 287 (1980).
[13] L. Covarrubias, L. Cervantes, A. Covarrubias, X. Soberón, I. Vichido, A. Blanco, Y. M. Kupersztoch-Portnoy, and F. Bolivar, *Gene* **13**, 25 (1981).
[14] A. J. Twigg and D. Sherratt, *Nature (London)* **283**, 216 (1980).

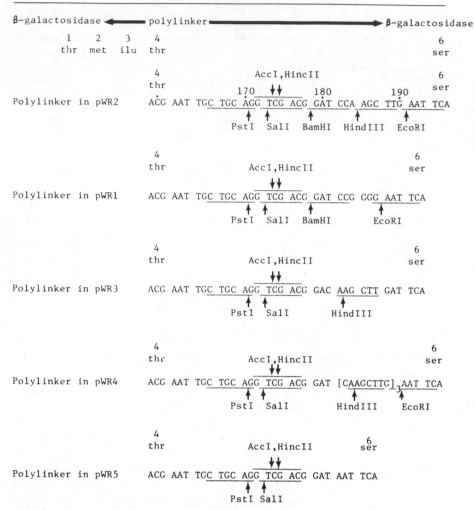

FIG. 4. The nucleotide sequences of polylinker inserted in the *lacZ'* gene region in pWR2 and its derivatives. The polylinkers are located between amino acid codon 4 (*thr*) and 6 (*ser*) of β-galactosidase gene.

Solutions, Buffers, Media, Gel Formulas, and Commercial Sources

Solutions

1. Chloroform-isoamyl alcohol (24 : 1, v/v), for extraction
2. Phenol saturated with 1 *M* Tris-HCl, pH 8, for extraction
3. 0.1 *M* EDTA–1.5 *M* NaOAc, for terminating enzyme reactions

4. 40% acrylamide stock solution for gel electrophoresis, 38% acrylamide–2% bisacrylamide in water, stable for at least half a year when stored at 4°
5. 5% ammonium persulfate for gel electrophoresis, stable for at least a year when stored at 4°
6. 0.3% each of bromophenol blue and xylene cyanole FF in 60% glycerol, for gel electrophoresis
7. 0.3% each of bromophenol blue and xylene cyanole FF in 10 M urea, for gel electrophoresis
8. 1 N NaOH–100 mM EDTA, for gel electrophoresis
9. 0.2 N NaOH–10 mM EDTA for gel electrophoresis
10. 1% agarose in TA or TB buffer plus 0.5 μg of ethidium bromide per milliliter

Buffers

11. 10 × T4 ligase buffer: 500 mM Tris-HCl, pH 7.6, 50 mM MgCl$_2$, 50 mM dithiothreitol, 5 mM ATP
12. 10 × TA buffer for gel electrophoresis: 400 mM Tris-HCl, pH 7.9, 30 mM NaOAc, 10 mM EDTA
13. 10 × TB buffer for gel electrophoresis: 500 mM Tris-borate, pH 8.3, 10 mM EDTA
14. TE buffer: 10 mM Tris-HCl, pH 8, 1 mM EDTA
15. 10 × RE buffer for restriction enzyme digestion and DNA polymerase repair synthesis: 500 mM Tris-HCl, pH 7.6, 500 mM KCl, 100 mM MgCl$_2$, 100 mM dithiothreitol
16. 10 × exonuclease III buffer: 660 mM Tris-HCl, pH 8, 770 mM NaCl, 50 mM MgCl$_2$, 100 mM dithiothreitol
17. 10 × λ exonuclease buffer: 500 mM Tris-HCl, pH 9.5, 20 mM MgCl$_2$, 30 mM dithiothreitol
18. 10 × terminal transferase buffer for tailing DNA: 1 M sodium cacodylate, pH 7, 10 mM CoCl$_2$, 2 mM dithiothreitol
19. 10 × S1 buffer for removal of single-stranded DNA: 0.5 M NaOAc, pH 4, 0.5 M NaCl, 60 mM ZnSO$_4$
20. 5 × *Bal*31 buffer for deletion of a segment of DNA with *Bal*31 nuclease: 100 mM Tris-HCl, pH 8, 60 mM MgCl$_2$, 60 mM CaCl$_2$, 3 M NaCl, 5 mM EDTA
21. Lysozyme solution for preparation of plasmid DNA: 2 mg/ml crystalline lysozyme, 50 mM glucose, 25 mM Tris-HCl, pH 8, 10 mM EDTA. Prepare daily from crystalline lysozyme, and stock solutions of other components; store at 4°
22. Lysozyme solution for mini preparation of plasmid DNA: 5 mg of crystalline lysozyme per milliliter in 25 mM Tris-HCl, pH 8
23. RNase A solution for preparation of plasmid DNA: 10 mg of

RNase A per milliliter in 50 mM NaOAc, pH 4.8, heated at 90° for 5 min prior to use
24. Triton solution for preparation of plasmid DNA: 0.3% Triton X-100, 150 mM Tris-HCl, pH 8, 200 mM EDTA
25. 30% polyethylene glycol (PEG) 6000–1.8 M NaCl for preparation of plasmid DNA

Media

26. YT broth contains (per liter): 8 g of (Bacto)tryptone, 5 g of (Bacto) yeast extract, 5 g of NaCl (if used in plates, add 15 g of agar), and deionized water. After autoclaving, if necessary, add ampicillin (10–20 μg/ml final concentration) after the solution has cooled to below 50°. 2 × YT broth (per liter): 16 g tryptone, 10 g yeast extract, and 5 g NaCl.
27. 20 × M9 salt: 140 g of Na_2HPO_4, 60 g of KH_2PO_4, 10 g of NaCl, 20 g of NH_4Cl, and H_2O to 1 liter
28. M9 medium containing IPTG and X-gal is made as follows: Agar, 15 g in 900 ml of water, is autoclaved. To the mixture add 50 ml of 20 × M9 salt, 20 ml of 20% casamino acids, 10 ml of 20% glucose, 5 ml of thiamin (1 mg/ml), 1 ml of 1 M $MgSO_4$, 0.1 ml of 1 M $CaCl_2$ (the components have been separately autocalved). After the mixture has cooled to below 50°, add ampicillin (20 μg/ml final concentration), IPTG (40 mg in 1.5 ml H_2O) and X-gal (40 mg in 1.5 ml of N,N-dimethylformamide). If using M9 medium for liquid culture, agar, IPTG, and X-gal should be omitted.

Polyacrylamide Gel Formulas

29. Volumes required for preparing polyacrylamide gels of various size and thickness:

Length (cm)	Width (cm)	Thickness (mm)	Volume required (ml)
40	35	3	500
40	20	3	260
40	35	2	400
40	20	2	220
40	35	1.5	300
40	35	0.6	100
40	35	0.4	70
80	35	0.4	150

30. Volumes of essential components needed to prepare polyacrylamide gels of different concentration:

	Percent concentration of polyacrylamide gel								
	3	4	5	6	8	10	12	15	20
Components	Volume (in ml) needed[a]								
40% acrylamide/ Bis[b] (19:1) mixture	15	20	25	30	40	50	60	75	100
10 × Tris-borate buffer	20	20	20	20	20	20	20	20	20
Distilled water	162.8	157.8	152.8	147.8	137.8	127.8	117.8	102.8	77.8
TEMED	0.2	0.2	0.2	0.2	0.2	0.2	0.2	0.2	0.2
5% $(NH_4)_2S_2O_4$	2	2	2	2	2	2	2	2	2

[a] Final volume = 200 ml.
[b] Bis = bisacrylamide.

31. Quantities of ingredients necessary for preparing denaturing (containing 8 M urea) acrylamide/bis mixture (19:1) of different polyacrylamide concentration:

	Percent concentration			
	5	8	15	20
Ingredients	Volume and weight required			
40% acrylamide/Bis (19:1) mixture (ml)	62.5	100	187.5	150
Urea, ultrapure (g)	240.24	240.24	240.24	126
10 × Tris-borate buffer (ml)	50	50	50	30
Distilled water (ml)	200	162.5	75	21.3
Total final volume (ml)	500	500	500	300
Final molarity of urea	8 M	8 M	8 M	7 M

32. Polynucleotide chain length corresponding to the migration rate of dye markers (bromophenol blue, and xylene cyanole FF) in different polyacrylamide concentrations of plain and 8 M urea-containing gel:

Percent concentration of plain and 8 M urea-containing polyacrylamide gel		Polynucleotide chain length, corresponding to the migration rate of dye	
		Bromophenol blue	Xylene cyanole FF
Plain			
Acrylamide	Acrylamide/Bis		
8%	29:1	20	125
5%	19:1	55	170
4%	19:1	83	400
4%	29:1	100	430
8 M Urea			
Acrylamide	Acrylamide/Bis		
20%	19:1	10	30
20%	29:1	12	35
15%	19:1	12	35
15%	29:1	14	48
12%	19:1	16	45
10%	19:1	18	55
8%	19:1	20	73
8%	29:1	23	90
5%	19:1	31	130–140
4%	19:1	50	250–300
3%	19:1	80	410

We thank Robert Yang for information on gels used for items 29–32.

Commerical Sources of Reagents, Isotopes, and Enzymes

$[\gamma\text{-}^{32}P]$ATP (>2000 Ci/mmol): The Radiochemical Centre, Amersham
Agarose: Bethesda Research Laboratories (BRL)
Acrylamide: Bio-Rad Laboratories
Adenosine triphosphate: P-L Biochemicals
Bisacrylamide: Bio-Rad Laboratories
Bromophenol blue: BDH Chemicals, Ltd.
5-Bromo-4-chloroindodyl-β-D-galactoside (X-gal): Sigma Chemical Co.
Bacto-tryptone: Difco Laboratories
Bacto-yeast extract: Difco Laboratories
Bovine pancreatic RNase A: Sigma Chemical Co.
Casamino acids: Difco Laboratories
*Bal*31 nuclease: BRL
$[\alpha\text{-}^{32}P]$dNTPs (410 Ci/mmol): The Radiochemical Centre, Amersham
$[^{35}S]$Deoxyadenosine 5′-[α-thio]triphosphate (600 Ci/mmol): New England Nuclear

dNTPs: P-L Biochemicals
ddNTPs: P-L Biochemicals
Exonuclease III: BRL
λ exonuclease: New England BioLabs
DNA polymerase (Klenow fragment): New England BioLabs
HindIII linker (d-CAAGCTTG): Collaborative Research, Inc.
Isopropylthiogalactoside (IPTG): Sigma Chemical Company
MacConkey agar: Difco Laboratories
Polyethylene glycol 6000: J. T. Baker Chemical Corp.
Restriction enzymes: New England BioLabs and BRL
Reverse transcriptase: Life Science Incorporated
S1 nuclease: Sigma Chemical Co.
Sodium cacodylate: Fisher Scientific
T4 DNA ligase: New England BioLabs
T4 polynucleotide kinase: New England BioLabs
Terminal transferase: P-L Biochemicals
Thiamin hydrochloride (B_1): Calbiochem
Triton X-100: Sigma Chemical Co.
Xylene cyanole: McIB Manufacturing Chemists

Cloning into pWR Plasmids

The major steps underlying genetic engineering technology include isolation and specific cleavage of DNA, ligation of DNA fragments to a cloning vector, transformation and selection of the desired clone, confirming the cloned gene by physical mapping and DNA sequencing, and expression of the cloned gene. Usually several vectors are needed to serve these functions. However, it is most convenient if a single vector can serve all these functions. A plasmid has been constructed to serve all these functions. This plasmid, pWR2, has eight unique restriction enzyme sites in *lacZ'* gene (see Fig. 4) for convenient cloning of different DNA fragments. The desired clone can be readily selected by color change of clones from blue to colorless, and the insert DNA can be directly sequenced using different methods. The cloned gene can be expressed by using the *lac* promoter in this plasmid.

Plasmid pWR2 provides a wide selection of unique restriction sites for cloning exogenous DNA fragments. Table II lists these restriction sites and shows how these sites can be chosen for cloning of a large variety of restriction fragments and for sequencing the cloned DNA using either the exonuclease III method[3] or the chemical method.[6]

Sometimes there are problems with cloning of blunt-ended DNA into the *Hinc*II site in pWR2. We found that, when plasmid DNA digested

TABLE II
CLONING AND SEQUENCING IN PLASMID pWR2

Restriction fragments to be cloned	Cloning sites	Sequencing			
		Exo III method restriction enzyme		Chemical method restriction enzyme	
		1st cut	2nd cut	1st cut and labeling	2nd cut
EcoRI EcoRI*	EcoRI	PvuII	PstI	PvuII	PvuI or BglI
		HindIII	EcoRI	HindIII	BamHI
HindIII	HindIII	EcoRI	PstI	PvuII	PvuI or BglI
		SalI	EcoRI	BamHI	SalI
BamHI BglII	BamHI	EcoRI	PstI	HindIII	EcoRI
BclI Sau3A XhoII		SalI	EcoRI	SalI	PstI
SalI XhoI (AvaI)	SalI	EcoRI	PstI	HindIII	EcoRI
		PstI	EcoRI		
AccI AsuII ClaI	AccI	EcoRI	PstI	HindIII	EcoRI
HpaII TaqI		PstI	EcoRI		
PstI	PstI	EcoRI	PstI	HindIII	EcoRI
		XmnI	EcoRI		
Blunt-ended fragments	HincII	EcoRI	PstI	HindIII	EcoRI
		PstI	EcoRI		
	PvuII	PvuI	EcoRI	EcoRI	HindIII
		EcoRi	PvuI or BglI		

with HincII (Lot 11214, BRL) was blunt-end ligated to the HincII-digested DNA, there were a number of white clones that did not carry DNA inserts. The lacZ' gene had probably been inactivated by exonuclease contaminating the HincII. However, the white clones with DNA inserts can be distinguished from those without insert by the greater ampicillin resistance of the latter. Gardner et al.[15] also met with the same problems with blunt-ended cloning in M13mp7, but they were able to overcome this problem by using a particular batch of HincII (BRL, Lot 2651).

Procedures for Isolating, Digesting, Cloning, and Sequencing of DNA

Procedure 1. Restriction Enzyme Digestion of Plasmid DNA or Other DNA. To 0.5–3 μg of DNA, add 1–10 units of a restriction enzyme and

[15] R. C. Gardner, A. J. Howarth, P. Hahn, M. Brown-Lendi, R. J. Shepherd, and J. Messing, *Nucleic Acids Res.* **9,** 2871 (1981).

1 μl of 10 × RE buffer. Adjust the volume to 10 μl with distilled water and incubate at 37° for 30–60 min. For shotgun cloning, 0.5 μg of vector and 3 μg of DNA containing the desired insert fragment are digested together and then used directly for Procedure 3.

Procedure 2. Isolation of DNA Fragments by Gel Electrophoresis. To isolate a particular DNA fragment to be cloned, use Procedure 1 on 3 μg of the DNA carrying the desired fragment. Add 2 μl of dyes (0.3% bromophenol blue and xylene cyanole in 60% glycerol) to the reaction mixture. The sample is electrophoresed on a horizontal gel apparatus (10 × 8.2 cm) containing 1% low melting point agarose (BRL) plus 0.5 μg of ethidium bromide per milliliter in TB buffer at 60 V and 2 mA. (Warning: current should not be over 3 mA or 7.5 V/cm.) Cut out the band to be cloned with a razor blade and put it into an Eppendorf tube (1.5 ml). After melting the gel by heating at 70° for 2 min, measure the volume and add 0.1 volume of 5 M NaCl. Vortex and continue heating at 70° for 3 min. Extract the solution twice with an equal volume of phenol saturated with 1 M Tris-HCl (pH 8) and once with an equal volume of chloroform–isoamyl alcohol (24 : 1, v/v). Extract the aqueous phase several times with a large volume of *n*-butanol to remove ethidium bromide and to reduce the volume of the solution. Add 2.5 volumes of ethanol to the above solution, chill at −70° for 5 min, and centrifuge at 12,000 rpm (Eppendorf centrifuge) for 5 min. If there is a white salt precipitate at the bottom of the tube, add a little distilled water to dissolve it and repeat the ethanol precipitation step. Remove the supernatant and add 200–500 μl of ethanol to rinse the DNA pellet. Centrifuge the tube for 3 min and remove the supernatant. Dry the DNA pellet under vacuum for 5 min:

Procedure 3. Ligation of Insert into Vector. To vector plus linearized insert DNA (in 10 μl, from Procedure 1) or an isolated insert fragment plus linearized vector in 10 μl (from Procedure 1 plus 2), add 10 μl of 0.1 M EDTA–1.5 M NaOAc and 30 μl of TE buffer. Extract the mixture once with phenol saturated with 1 M Tris-HCl (pH 8) and once with chloroform–isoamyl alcohol (24 : 1, v/v). To the aqueous phase add 125 μl of ethanol. Chill the mixture at −70° for 5 min and centrifuge for 5 min. To the DNA pellet add 200 μl of ethanol and centrifuge for 3 min. Dry the plasmid DNA pellet under vacuum for 5 min. Resuspend the sample in 8 μl of H_2O and 1 μl of 10 × ligation buffer. After vortexing, add 1 μl of T_4 DNA ligase (3 units, New England BioLabs) to the mixture and incubate at 4° for 4–18 hr.

Procedure 4. Transformation

a. To the ligation mixture from Procedure 3, add 200 μl of *E. coli* $F^-Z^-\Delta M15recA$[5] or JM101,[8] made competent and stored frozen according to Morrison,[16] thawed on ice for 15 min. Incubate at 0° for 30 min. Heat at 42° for 2 min. Add 1 ml of 2× YT broth (no ampicillin) to the mixture and incubate at 37° for 60 min. Plate 100 μl of serially diluted samples in 10 mM NaCl on petri dishes containing (per milliliter) M9 agar medium plus 20 μg of ampicillin, 5 μg of thiamin, and 40 μg each of IPTG and X-gal or YT agar medium plus 20 μg/ml ampicillin, 40 μg/ml each of IPTG and X-gal. Incubate the plates at 37° for 15–24 hr, and select those bacteria that give white colonies.

b. To the ligation mixture from Procedure 3, add 40 μl of 0.1 M CaCl₂ and 200 μl of competent cells as in Procedure 4a. Incubate on ice for 60 min. Heat at 42° for 2 min, and transfer the mixture to 5 ml of YT broth. Shake at 37° for 4–5 hr. Transfer the culture into a 50-ml sterile centrifuge tube, and centrifuge at 8000 rpm for 5 min. Pour off the supernatant and resuspend the pellet in 1 ml of 10 mM NaCl. Plate 100 μl of serially diluted samples on petri dishes as described above. This slightly longer procedure yields two or three times more transformants.

Procedure 5. Large-Scale Preparation of Plasmid.[17] Inoculate a single colony in 20 ml of YT broth and shake overnight at 37°. Transfer the culture to 1 liter of M9 medium or YT medium plus 25 μg of ampicillin per milliliter for bacteria harboring plasmid pWR2, or 10 μg of ampicillin per milliliter for those harboring recombinant plasmids with inserted DNA in the polylinker region. Shake at 37° until A_{600} reaches 1.0. Add 150 mg of chloramphenicol and continue shaking at 37° overnight. Pour the culture into four 500-ml centrifuge bottles and incubate in ice for 15 min. Centrifuge at 8000 rpm for 10 min. Resuspend the pellets in 50 ml of lysozyme solution. Incubate on ice for 30 min. Add 1 ml of RNase A solution and 24 ml of Triton solution. Incubate on ice for 30 min, and centrifuge in a Beckman 50.2 Ti rotor at 3000 rpm for 1 hr at 4°. Transfer the supernatant into four 50-ml phenol-resistant plastic tubes with tight-fitting caps

---

[16] D. A. Morrison, this volume [13].

[17] Plasmid DNA preparations may be contaminated with DNases or inhibitors of restriction enzymes. It is recommended that the following tests be carried out for each plasmid preparation. Three samples (1 μg of plasmid in each) are used for testing: sample (a), no incubation; sample (b), incubation with only restriction enzyme buffer at 37° for 3 hr; sample (c), incubation with a restriction enzyme and buffer at 37° for 3 hr (HindIII or PstI are more sensitive to inhibitors than many other enzymes). After incubation, the samples are loaded on a mini-agarose gel (10 × 8.2 cm, see Procedures 2 and 12) for 1.5–2 hr. Sample (a) gives the percentage of form I plasmid DNA. Sharp bands in samples (b) and (c) indicate the lack of contaminating DNases, and smearing of DNA bands indicates the presence of DNases. Sample (c) serves to test whether an inhibitor of the restriction enzyme is present. Contaminated DNase may be removed by adding 3 M NaOAc to the plasmid DNA followed by ethanol precipitation. It may be necessary to carry out another phenol extraction and ethanol precipitation.

(Sarstedt, No. 60·547) and extract twice with phenol saturated with 1 M Tris-HCl (pH 8) and once with chloroform–isoamyl alcohol (24 : 1, v/v). Transfer the aqueous phase into a 500-ml centrifuge bottle. Add 8 ml of 3 M NaOAc and 200 ml of ethanol. The mixture is chilled at −70° for 20 min and centrifuged at 10,000 rpm for 30 min. Remove the supernatant and dry the DNA pellet under vacuum for 15 min. Resuspend the DNA pellet in 50 ml of TE buffer, and then add 20 ml of 30% PEG-6000–1.8 M NaCl. Incubate at 4° overnight. Centrifuge at 10,000 rpm for 20 min and remove the supernatant. Resuspend the plasmid DNA pellet in 5 ml of 0.3 M NaOAc and transfer into a 50-ml centrifuge tube. To the DNA solution add 13 ml of ethanol, chill at −70° for 15 min, and centrifuge at 10,000 rpm for 15 min. The DNA pellet is washed once with 20 ml of ethanol, dried, and resuspended in 2 ml of TE buffer. The final concentration of plasmid DNA is usually around 1 μg/μl. The total yield of plasmid pWR2 DNA using this chloramphenicol amplification procedure is very high, about 2 mg per liter of culture using pWR2 or its derivatives.

Procedure 6. Mini Preparation of Plasmid.[17] Transfer a single colony into 5 ml of YT broth or M9 medium with ampicillin as in Procedure 5 or streak on one-eighth of a petri dish containing YT broth agar medium plus 10 μg of ampicillin per milliliter. Incubate at 37° overnight. Pellet the bacteria at 10,000 rpm for 5 min or scrape up bacteria from the plate and transfer them into a 1.5-ml Eppendorf tube. Resuspend the cell pellet in 200 μl of cold TE buffer and 5 μl of 0.5 M EDTA (pH 8). Add 50 μl of lysozyme solution. Incubate on ice for 15 min; add 5 μl of RNase solution and 120 μl of Triton solution. Incubate on ice for 15 min, and centrifuge for 15 min. Extract the supernatant once with phenol and once with chloroform–isoamyl alcohol. To the aqueous phase add 30 μl of 3 M NaOAc and 700 μl of ethanol, chill at −70° for 10 min, and centrifuge for 5 min. Wash once with 1 ml of ethanol. The DNA pellet is dried and resuspended in 100 μl of TE buffer followed by addition of 40 μl of 30% polyethylene glycol 6000–1.8 M NaCl. Put on ice for 4 hr or at 4° overnight. Centrifuge for 15 min, and resuspend the plasmid DNA pellet in 40 μl of 0.3 M NaOAc followed by addition of 100 μl of ethanol. Centrifuge and wash the precipitate once with ethanol. Dry the plasmid DNA pellet and resuspend in 20 μl of TE buffer. The DNA concentration is usually around 1 μg/μl. Usually the DNA sample is pure enough for DNA sequencing using any one of three methods (procedures 7, 8, and 10).[3,4,6]

DNA Sequencing

There are three methods for sequencing a DNA fragment inserted in pWR2 without prior isolation and purification of the DNA fragment. They include the exonuclease III method of Guo and Wu,[3] the primer extension method of Sanger et al.,[4] and the chemical method of Maxam and Gilbert.[6]

One limitation with Sanger's method[4] is the requirement of single-stranded DNA as a template for hybridization with the added primer. One can overcome this limitation by cloning the DNA into a M13 phage[8] and isolate single-stranded DNA. One can use an alternative method[18] for sequencing double-stranded DNA by first linearizing a plasmid DNA with a restriction enzyme at a site far away from the DNA insert to be sequenced. The DNA is then heated at 100° for 3 min in the presence of a synthetic primer, followed by quenching the mixture to 0°. Subsequent steps are the same as the chain terminator procedure for DNA sequencing.[4]

Two universal primers (16-mers) have been designed and are being synthesized for sequencing any DNA fragments cloned into the polylinker region of pWR2. Primer I has the sequence of 5′ d(A-C-C-A-T-G-A-T-T-A-C-G-A-A-T-T), which can bind to the region between nucleotides 149 and 164 (see Fig. 8) and be extended rightward to sequence one strand of the cloned gene (see Fig. 6). Primer II has the sequence of 5′ d(C-A-C-G-A-C-G-T-T-G-T-A-A-A-A-C), which is complementary to nucleotides 221–206 and can bind to pWR and be extended leftward to sequence the other strand of the cloned gene. Primer II can be replaced by the 19-mer synthetic primer, 5′ d(T-T-G-T-A-A-A-A-C-G-A-C-G-G-C-C-A-G-T), for sequencing DNA cloned into M13mp2, mWJ22, or mWJ43.[19]

One potential problem with Maxam and Gilbert's method[6] is that a double-stranded DNA sometimes cannot be sequenced to give clean gel patterns because of breaks or gaps in the DNA. In these cases, it is best to separate the two strands of DNA prior to carrying out base-specific chemical cleavage reaction. Since strand separation procedure is time-consuming, we have made an improvement on the method of Rüther et al.[5] to overcome this difficulty. The 5′ ends of a plasmid linearized in the polylinker are first labeled with polynucleotide kinase and [γ-^{32}P]ATP followed by cutting at another restriction site in the polylinker region to produce a very short and a very long fragment, each labeled at the polylinker end. The single-end labeled fragments are digested with E. coli exonuclease III to destroy the very short piece and convert part of the long piece into single-stranded DNA before degradation by base-specific

[18] R. B. Wallace, M. J. Johnson, S. V. Suggs, Ken-ichi Miyoshi, R. Bhatt, and K. Itakura, Gene 16, 21 (1981).

[19] R. Wu, L. Lau, H. Hsiung, W. Sung, R. Brousseau, and S. A. Narang, Miami Winter Symp. 17, 419 (1980).

chemical reaction. This will give more reproducible sequencing results than the method employing double-stranded fragments.

Each of the above two methods can be used to sequence both strands of a short DNA fragment cloned in the polylinker region of pWR2, provided that the DNA is shorter than 400 bp. For sequence analysis of a longer DNA fragment the exonuclease III method[3] can be used, since it can determine the DNA sequences not only from both ends near the sites of cloning, but also from one or several restriction sites located within the cloned DNA (see Table II and Fig. 6). It is very likely that there are one or more restriction site(s) available in the middle of the DNA to be sequenced, and these sites can be used for the second digestion (e.g., X or Y in Fig. 6). The sequence can then be read from the second site within the cloned DNA. It should be pointed out that the second site within the cloned gene need not be a unique restriction site in the recombinant plasmid as long as other sites of the same enzyme are sufficiently distant so that no other labeled fragment falls into the size range of the labeled fragment to be sequenced.[3] If these sites are close to an identical site within the cloned gene to be utilized for the second digestion, Procedure 9 can be used to block one end of the DNA from exonuclease III digestion (see Fig. 7). If a site within the cloned gene is unique on the recombinant plasmid, the site can serve either as the second site or the first site.

All three sequencing methods can be used for sequence analysis of a gene cloned into pWR2. The primer method is the most simple and rapid for short DNA fragments (below 400 bp), but the exonuclease III method is the most useful, since it can determine sequences of both short and long DNA fragments (up to 2000 long) cloned into the plasmid. The major advantage is that a long DNA need not be fragmented as much for cloning and sequence analysis. For example, a 5000 bp DNA can be digested to give 3 or 4 fragments and each one cloned into pWR2 for complete sequence analysis by the exonuclease III method. In contrast, in the M13 method, a 5000 bp DNA needs to be digested to give about 15 fragments, and each of these needs to be cloned into M13mp8 or M13mp9 in both orientations to make a total of 30 clones. After clone selection and sequencing, additional effort is needed to line up the 15 fragments to give the linear sequence.

One requirement for the success of the exonuclease III method is the quality of restriction enzymes for the second digestion, which must be free of contaminating enzymes, so that the 5' ends will remain intact. The restriction enzyme can be tested for contaminating enzymes according to the method described in Procedure 8. After PvuII and exonuclease III digestion of pWR2 DNA followed by labeling with DNA polymerase I and second digestion with EcoRI, the reaction mixture is heated at 70° for

10 min to terminate the reaction. One portion of the sample will be saved as the control, the other portion will be incubated with the restriction enzyme to be tested at 37° for 30 min. After electrophoresis, if the gel pattern of the 92 bp long *Eco*RI-*Pvu*II fragment in the test sample is as clear as the control, the restriction enzyme is considered to be pure and useful for sequence analysis.

Procedure 7. Primer Method[4,18] *for DNA Sequencing.* A recombinant plasmid DNA can be first linearized using one of several sites (e.g., *Pvu*II) for sequencing the inserted DNA from primer I or primer II (see Fig. 6). The restriction enzyme digestion of plasmid DNA is the same as in Procedure 1, except that 0.4 pmol of DNA instead of 3 μg is used and incubated at 37° for 15–30 min. After incubation, add 1 μl of synthetic (16-mer) primer I or II (8–10 pmol). Transfer the mixture into a capillary, seal, heat at 100° for 3 min, and then quench at 0° for several minutes. Transfer the mixture into an Eppendorf tube containing [α-^{32}P]dATP (8 μCi, 410 Ci/mmol) that has been dried down. Add 3 μl of 1 × RE buffer into the tube. After vortexing and centrifuging, pipette aliquots of 3 μl into 4 Eppendorf tubes marked A, G, C, and T, respectively. To each tube add 1 μl of the appropriate ddNTP–dNTP mix (see Table III) and 1 μl of DNA polymerase (Klenow fragment 0.2–0.6 unit). Incubate at 23° for 10 min. To each reaction mixture, add 1 μl of 0.5 mM dATP (for chase) and incubate at 23° for an additional 10 min. Stop the reactions by 1 μl of 1 N NaOH–0.1 M EDTAand 4 μl of 10 M urea containing 0.3% bromophenol blue and xylene cyanole dyes. After standing at room temperature (23°) for 5 min, load the samples on a sequencing gel.[3,4,6] Alternatively, stop the reactions by ethanol precipitation as described in procedure 8(c).

Procedure 8. Exonuclease III Method for DNA Sequencing. There are two options for exonuclease III sequencing methods.[3] Here we describe only Method II using chain terminator. The principle of this method for

TABLE III
COMPOSITION OF ddNTP–dNTP MIXTURE[a]

Mix	ddATP (mM)	ddGTP (mM)	ddCTP (mM)	ddTTP (mM)	dATP (μM)	dGTP (μM)	dCTP (μM)	dTTP (μM)
ddATP–dNTP	0.6				2.5	500	500	500
ddGTP–dNTP		2			1	40	500	500
ddCTP–dNTP			2		1	500	40	500
ddTTP–dNTP				4	1	500	500	40

[a] Dideoxy- and deoxyribonucleoside triphosphate are dissolved in 1× RE buffer (50 mM Tris-HCl, pH 7.6, 50 mM KCl, 10 mM MgCl$_2$, 10 mM dithiothreitol). The mixtures can be stored at −20° and used for at least 3 months.

DNA sequencing is illustrated in Fig. 5. The restriction sites available in pWR2 for the exonuclease III sequencing method are tabulated in Table II.

To sequence a long DNA (e.g., 1000 bp) cloned in the polylinker region of pWR2, Fig. 6 gives a feasible strategy. Both X and Y sites are

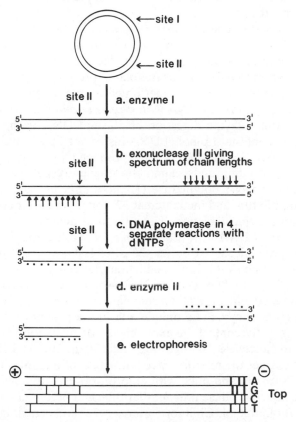

FIG. 5. The principle of the exonuclease III method for sequencing DNA. A fragment of DNA to be sequenced is cloned into a plasmid (represented by a double circle) between restriction sites I and II. In this example, the plasmid is first digested with restriction enzyme I to give a linear DNA. The digestion by exonuclease III (step b) gave a family of molecules with 3′ ends shortened. These molecules are an ideal template-primer system in which the shortened strands with 3′ ends serve as the primers. The digested DNA in step b is distributed in four tubes, and the 3′ ends of the DNA are extended (step C) using the chain termination method.[4] After digestion with a second restriction enzmye (step d), the DNA fragments are fractionated on a denaturing polyacrylamide gel (step e). The shorter fragments are well separated on the lower part of the gel (left-hand side of gel), and the longer fragments are retained near the top of the gel. Four lanes represent DNA fragments terminated with each of the four different ddNTPs.

available for sequencing both strands of DNA. If not all the requirements listed in the table for Fig. 6 can be met, one end of the restricted DNA can be blocked and the other end digested with exonuclease III, according to the strategy shown in Fig. 7 and described in Procedure 9. If X or Y site is unique in the recombinant plasmid, it may also serve for the first digestion to linearize the plasmid DNA and to obtain sequence information on both sides of these sites.

It is easy to determine the restriction sites in the cloned fragment by digesting it (either after gel separation or in the intact plasmid) with several restriction enzymes and running gel electrophoresis. The number of sites and the approximate size of fragments produced by the restriction enzymes are important, but an exact physical map is not necessary.

a. Digestion of plasmid DNA with a restriction enzyme. The mixture for the restriction enzyme digestion of plasmid DNA is the same as in Procedure 1, except that 1 pmol of DNA instead of 3 μg is used. If there is a need to remove 3' protruding ends produced by restriction-enzyme digestion (3' protruding ends cannot be efficiently cut by exonuclease III), 1–2 units of DNA polymerase (Klenow enzyme) are added to the same reaction mixture and incubated at 37° for 15 min after completion of the restriction enzyme digestion. The mixture is heated at 70° for 10 min.

b. Digestion of DNA with exonuclease III. To the above mixture add 4 μl of 10 × exonuclease III buffer, 26 μl of H_2O, and 1–4 μl of exonuclease III (see Table IV). It is advisable that the exonuclease III digestion be carried out beyond the second restriction site (i.e., the incubation time should be about 10 min longer than calculated). If the second site to be cut with a restriction enzyme is a unique site in the plasmid DNA, it is not necessary to strictly control the exonuclease III digestion. After completion of the exonuclease III digestion, add 10 μl 0.1 M EDTA–1.5 M NaOAc and extract the mixture once with 50 μl of phenol saturated with 1 M Tris-HCl (pH 8) and once with chloroform–isoamyl alcohol (24 : 1). Transfer the aqueous phase into another tube and precipitate the DNA by adding 130 μl of ethanol. Chill for 5 min at −70° and centrifuge at 0° for 5 min. Remove the supernatant and add 200 μl of ethanol to rinse the DNA pellet. Centrifuge for 3 min and remove the supernatant. Dry the DNA pellet under vacuum for 5 min.

FIG. 6. A strategy for sequencing a long DNA fragment cloned in the polylinker region of pWR2 by using the exonuclease III method. The upper part of this figure shows a DNA fragment 1000 base pairs in length cloned in *Bam*HI site of the polylinker region. There are two restriction sites (X and Y) within the insert. If the restriction sites X and Y meet all the requirements listed in the table for this figure, both strands of the inserted DNA can be sequenced using the exonuclease III method without necessity of isolation and strand separation of DNA. Dashed arrows show directions of exonuclease III digestion of an inserted DNA; solid arrows show directions of DNA sequencing.

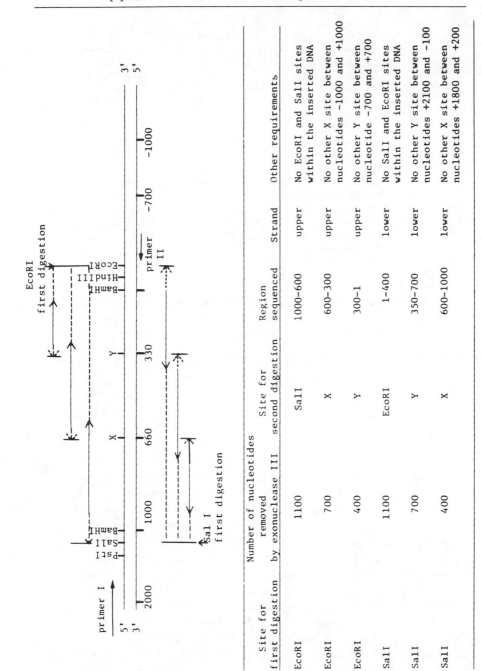

Site for first digestion	Number of nucleotides removed by exonuclease III	Site for second digestion	Region sequenced	Strand	Other requirements
EcoRI	1100	SalI	1000–600	upper	No EcoRI and SalI sites within the inserted DNA
EcoRI	700	X	600–300	upper	No other X site between nucleotides −1000 and +1000
EcoRI	400	Y	300–1	upper	No other Y site between nucleotide −700 and +700
SalI	1100	EcoRI	1–400	lower	No SalI and EcoRI sites within the inserted DNA
SalI	700	Y	350–700	lower	No other Y site between nucleotides +2100 and −100
SalI	400	X	600–1000	lower	No other X site between nucleotides +1800 and +200

TABLE IV

DIGESTION OF DNA WITH EXONUCLEASE III[a]

Number of nucleotides removed from each end of DNA	Incubation (min)	Exo III (U/pmol DNA)
100–250	10–25	15–20
250–500	25–50	20–25
500–750	50–75	25–30
750–1000	75–100	30–35
1000–1500	100–150	35–45

[a] To remove approximately 10 nucleotides per minute from each 3' end of DNA, the concentration of *Escherichia coli* exonuclease III (ExoIII) needed is given in the table. A unit of exonuclease III is the amount of enzyme that liberates 1 nmol of mononucleotides from a sonicated DNA substrate in 30 min at 37°. BRL ExoIII, Lot 2429, was used.

c. Labeling of partially digested DNA and second restriction enzyme digestion. Resuspend the DNA pellet in 18 μl of H_2O and 2 μl of 10 × RE buffer, and heat at 70° for 5 min followed by a brief centrifugation (sometimes the heating may be omitted). Pipette 14 μl into a tube containing [α-^{32}P]dATP (8 μCi, 410 Ci/mmol) that has been dried down. Pipette aliquots of 3 μl into four tubes marked A, G, C, and T, respectively. Add to each tube 1 μl of the appropriate ddNTP–dNTP mix (see Table III) and 1 μl of DNA polymerase (Klenow enzyme) (0.2–0.6 unit). Incubate at 37° or 23° for 10 min and then chase by adding 1 μl of 0.5 mM dATP and incubate for 5 min. To each tube add 1 μl of a restriction enzyme (1–2 units) and incubate at 37° for 10 min. Stop the reactions by addition of 1 μl of 1 N NaOH–0.1 M EDTA and 4 μl of 10 M urea containing 0.3% bromophenol blue and xylene cyanole dyes. Alternatively, stop the reactions by the addition of 1 μl of tRNA (10 μg/μl), 3 μl of 0.1 M EDTA–1.5 M NaOAc, and 25 μl of ethanol. After chilling and centrifugation, resuspend the labeled DNA pellets in 3 μl of 0.2 N NaOH–10 mM EDTA and 3 μl of 10 M urea containing 0.3% dyes. Let the preparation stand at room temperature (23°) for 5 min, then load the samples on a sequencing gel.[3,4,6]

Procedure 9. Methods to Digest and Sequence Only a Selected End of a Double-Stranded DNA Using Exonuclease III. *Escherichia coli* exonuclease III can digest a linear duplex DNA from the 3' ends of both strands (see Fig. 5). DNA molecules with 3' recessed or blunt ends are good substrates for exonuclease III digestion, but DNA with 3' protruding ends are not efficiently digested. If the 3' protruding end is long (e.g., >20

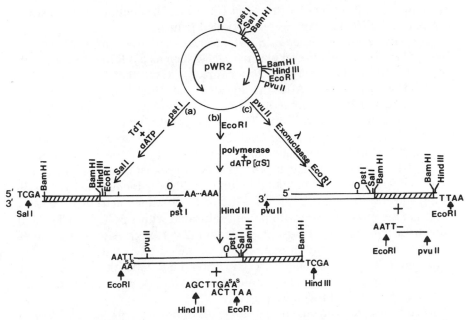

FIG. 7. Three methods for blocking one end of plasmid DNA from *Escherichia coli* exonuclease III digestion. For sequencing a DNA fragment, shown with hatched lines, which is cloned into the *Bam*HI site in the polylinker region of pWR2, one end of the linearized DNA can be blocked in one of three ways: (*a*) tailing one end of DNA with calf thymus terminal transferase (TdT); (*b*) introducing an α-thio nucleotide into one end of DNA with DNA polymerase I; (*c*) producing a long protruding 3' end of DNA with λ exonuclease.

nucleotides), it would be resistant to exonuclease III digestion. Therefore a DNA with a recessed or blunt end at one end and a long 3' single strand at the other end can be digested only at one end (e.g., an inserted DNA fragment shown with hatched lines in Fig. 7 can be digested with exonuclease III only from one 3' end). Another strategy for blocking the 3' end at one end of a DNA fragment is to introduce a 2'-deoxynucleoside 5'-O-(1-thiophosphate)[20] into one of the 3' ends of DNA fragment using *E. coli* DNA polymerase I (Fig. 7). This modified 3' end is resistant to exonuclease III digestion.[21] The application of these procedures to DNA sequencing is shown in Fig. 7. The procedures are also useful for making

[20] T. A. Kunkel, F. Eckstein, A. S. Mildvan, R. M. Koplitz, and L. A. Loeb, *Proc. Natl. Acad. Sci. U.S.A.* **78**, 6734 (1981).
[21] S. D. Putney, S. J. Benkovic, and P. R. Schimmel, *Proc. Natl. Acad. Sci. U.S.A.* **78**, 7350 (1981).

asymmetrical deletion of a stretch of DNA (see the section on application of plasmid pWR for expression of cloned genes).

   Procedure 9a. Addition of a poly(dA) tail to 3' ends of a DNA fragment (Fig. 7). Digestion of a plasmid DNA such as pWR2 with restriction enzyme *Pst*I is the same as in Procedure 1, except that 1 pmol of DNA instead of 3 μg is used. After the digestion, add 3 μl of 0.1 M EDTA–1.5 M NaOAc and 35 μl of ethanol to the reaction mixture, chill at $-70°$ for 5 min, and centrifuge for 5 min. The DNA pellet is rinsed with 50 μl of ethanol, centrifuged for 3 min, and dried down under vacuum for 5 min. The DNA pellet is resuspended in 2 μl of 10× terminal transferase buffer, 1 μl of 1 mM dATP, terminal deoxynucleotidyltransferase (10 units), and distilled water is added to a final volume of 20 μl.[22] Incubate at 30° for 30 min, and add 10 μl of 0.1 M EDTA–1.5 M NaOAc and 30 μl of TE buffer. Extract the mixture once with phenol saturated with 1 M Tris-HCl (pH 8) and once with chloroform–isoamyl alcohol (24 : 1, v/v). After ethanol precipitation, washing, and drying, digest the tailed DNA with a restriction enzyme followed by exonuclease III. Extend and label the 3' ends using the chain terminator procedure, and digest with a second restriction enzyme according to Procedure 8.

   Procedure 9b. Introduction of an α-phosphorothioate nucleotide into one end of DNA to block that end from exonuclease III digestion (Fig. 7). Digestion of a plasmid DNA such as pWR2 with *Eco*RI is the same as in Procedure 1, except that 1 pmol instead of 3 μg of DNA is used. After digestion, add 12–24 pmol of dATP [α-S], 1 μl of 1 mM dTTP and 1 μl of DNA polymerase I (Klenow enzyme, 0.5–1 unit) for repair of *Eco*RI-digested ends. Reverse transcriptase may be used to replace Klenow enzyme for repair of restriction enzyme-digested ends. The final volume of the reaction mixture is 12–14 μl, and the incubation is carried out at 37° for 15 min. Add 5 μl of 0.1 M EDTA–1.5 M NaOAc. After ethanol precipitation, washing, and drying, digest the linear plasmid DNA with its modified 3' ends with *Hin*dIII to produce two fragments, each with a modified 3' end. This modified end is resistant to exonuclease III digestion, but the other 3' end can be digested and then labeled using the chain-terminator procedure.

   Procedure 9c. Producing long 3' protruding ends with λ exonuclease (Fig. 7). Digestion of a plasmid DNA such as pWR2 with *Pvu*II enzyme is the same as in Procedure 1, except that 1 pmol instead of 3 μg of DNA is used. After *Pvu*II digestion, add 5 μl of 10× λ buffer, 35 μl of H_2O, and 2 units of λ exonuclease. Incubate at 6° or 23° for 30 min. Under these conditions, λ exonuclease removes 50–60 nucleotides (but no more) from

[22] G. Deng and R. Wu, *Nucleic Acids Res.* **9**, 4173 (1981).

each 5' end of a DNA fragment. After digestion, add 10 μl of 0.1 M EDTA–1.5 M NaOAc to the reaction mixture and extract once with an equal volume of phenol and chloroform–isoamyl alcohol, respectively. After ethanol precipitation, washing, and drying, digest the DNA with long 3' protruding tails with EcoRI or other restriction enzyme to produce two fragments, each with only one 5' protruding or even end. This end alone can be digested by exonuclease III and labeled using the chain termination procedure.

Some Problems and Solutions for Use of the Exonuclease III Method

The rate of exonuclease III digestion can be controlled conveniently and averages about 10 nucleotides per minute at each end of DNA at 23° if the salt concentration in the digestion solution is 90 mM and the appropriate ratio of exonuclease III (units) to DNA (pmol) is used. The rate of exonuclease III digestion of DNA falls off with time; therefore, for removal of a large number of nucleotides the ratio of the enzyme to DNA must be increased for compensation (Table IV). In order to control the salt concentration during exonuclease III digestion, the salt used for the restriction enzyme cleavage of plasmid DNA must be taken into consideration. The rate of exonuclease III digestion was 10, 7.5, or 5 nucleotides per minute at 90 mM, 105 mM, or 125 mM salt, respectively.

We found that labeled oligonucleotides shorter than 15 residues usually give rise to very weak bands, even though exonuclease III digestion goes beyond the site to be cut by the second restriction enzyme. This may be due to the possibility that, after the digestion with the second enzyme, the short oligonucleotides dissociates from the template and is rapidly degraded.[23]

The purity of the restriction enzyme used for the second digestion is especially important because the common 5' end of DNA fragments to be separated on the gel is produced by the enzyme. Contamination by exonucleases or other endonucleases would give extraneous bands in the gel pattern. Restriction enzymes that recognize six nucleotide sequences seem better than those that recognize four nucleotide sequences. Occasionally, there are some extraneous bands in one lane of gel pattern, especially in the A reaction lane when labeled dATP was used, but not in the other three lanes. The selection of another restriction enzyme for the second digestion may solve the problem.

It is recommended that the exonuclease III digestion be carried out beyond the site to be cut with the second restriction enzyme, by increas-

[23] A. J. H. Smith, this series, Vol. 65, p. 560.

ing the incubation time by about 10 min. Otherwise, some extraneous bands in the gel may appear, especially in the region for short oligonucleotides.

Other possible problems, causes, and solutions are tabulated in Table V. Any ambiguity in sequence analysis of one strand of DNA often can be resolved by determining the complementary sequence in the same region or by using different DNA sequencing methods described in this paper. As a rule, to be assured of absolute reliability, the sequence of both strands of a DNA molecule must be determined.

Procedure 10. Chemical Method for DNA Sequencing

a. Labeling plasmid DNA at the 5' ends. The restriction sites available in pWR2 for chemical sequencing of DNA are tabulated in Table II. Plasmid DNA linearized according to Procedure 1 is labeled at the 5' end by polynucleotide kinase and [γ-^{32}P]ATP as in the method of Maxam and Gilbert,[6] except that calf intestinal alkaline phosphatase (1 unit per 1 pmol of DNA) is used to remove the terminal phosphates and incubation is carried out at 60° for 30 min. After labeling the 5' ends, the digestion by a second restriction enzyme produces two fragments each labeled only at one end (a very small and a very large fragment). It may be of advantage to digest the DNA by exonuclease III to produce single strands (Procedure 8a and b) before carrying out the base-specific chemical cleavage reactions.

b. Labeling plasmid DNA at 3' recessive ends.[24] Plasmid DNA is linearized with a restriction enzyme according to Procedure 1. After digestion, transfer the reaction mixture into another Eppendorf tube containing [α-^{32}P]dNTP (5–10 μCi, 410 Ci/mmol) that has been dried down. After resuspension, add 1 μl of DNA polymerase (Klenow enzyme, 0.5–1 unit), and incubate at 23° for 10 min. For labeling *Eco*RI-digested ends of DNA, add 1 μl of 1 m*M* dATP and continue the incubation for 5 min. Heat the reaction mixture at 70° for 10 min to inactivate the polymerase, add 6–12 units of another restriction enzyme, and incubate at 37° for 30–60 min. Add 10 μl of 0.1 *M* EDTA–1.5 *M* NaOAc and 30 μl of TE buffer to stop the reaction. Extract DNA once with phenol saturated with 1 *M* Tris-HCl (pH 8) and once with chloroform–isoamyl alcohol (24:1, v/v). To the aqueous phase add 125 μl of ethanol, chill at −70° for 5 min and centrifuge for 5 min. Wash the DNA pellet with 200 μl of ethanol and dry under vacuum.

Base-specific chemical cleavage reactions are carried out according to the method of Maxam and Gilbert.[6]

---

[24] R. Wu, *J. Mol. Biol.* **51,** 501 (1970).

*Labeling a DNA Fragment Cloned in pWR2 as Probe for
  Molecular Hybridization*

A DNA fragment cloned in pWR2 or its derivatives can serve as a probe for molecular hybridization. There are two methods for the synthesis of radioactive labeled probes with specific activities of at least 10^8 cpm per microgram of DNA. Method I uses primer I or primer II and extends it to get single-stranded labeled DNA (see Procedure 7). Single-stranded probes are free from interference by the complementary strand, and thus are more efficient in hybridizing to the target DNA. Method II makes double-stranded labeled DNA in which only one strand is labeled. The principle is the same as the exonuclease III method for DNA sequencing (see Fig. 5 and Procedure 8). These methods can produce radioactive probes with higher specific activity than those obtained by the method of nick translation.

Procedure 11. Method I for Labeling DNA. Linearize the recombinant plasmid DNA, add synthetic primer, heat, and quench the restriction enzyme digest as in Procedure 7. Transfer the DNA mixture into an Eppendorf tube containing [α-^{32}P]dNTP (50–200 μCi, 410 Ci/mmol) that has been dried. After resuspension, add 1 μl of 1 mM each of three other cold dNTPs and 1 μl of DNA polymerase I (Klenow enzyme, 0.5–1 unit). Incubate at 23° for 15 min and add 1 μl of 1 mM cold dNTP corresponding to the labeled dNTP; continue the incubation at 23° for 5 min. Stop the reaction by addition of 5 μl of 0.1 M EDTA–1.5 M NaOAc. After ethanol precipitation, washing, and drying, the labeled DNA can be used as a hybridization probe.

Procedure 12. Method II for Labeling DNA. Digest the recombinant plasmid DNA (1 μg) in RE buffer with 2–4 units of a restriction enzyme as in Procedure 1. Incubate at 37° for 30–60 min. To the mixture (10 μl), add 1 μl of 0.1 M KCl and the appropriate amount of exonuclease III, and incubate at 23° for appropriate time (see Table IV). Stop the reaction by heating the sample at 80° for 10 min. Transfer the sample into an Eppendorf tube containing [α-^{32}P]dNTP (50–200 μCi), which has been dried down. After resuspension, add 1 μl of 1 mM each of three other cold dNTPs and 1 μl of DNA polymerase I (Klenow enzyme, 0.5–1 unit). Incubate at 37° for 10 min, and add 1 μl of 1 mM cold dNTP corresponding to the radioactive nucleotide and 1 μl of a second restriction enzyme (2–4 units). Incubate at 37° for 15 min, and stop reaction by addition of 5 μl of 0.1 M EDTA–1.5 M NaOAc. After ethanol precipitation, washing and drying, the labeled DNA fragment (one strand labeled) is electrophoresized in 1% low melting point agarose as in Procedure 2 or regular agarose. If using regular agarose for isolation of labeled DNA frag-

TABLE V

DIAGNOSIS AND CORRECTION OF PROBLEMS

Problem	Probable cause	Solution
A high background of gel bands	(a) Contamination of chromosomal DNA in plasmid DNA sample	(a), (b), and (c) Further purify plasmid DNA by phenol and chloroform extraction, by adding NH_4OAc to 3 M and reprecipitate[17] with 2 volumes of EtOH, by agarose gel electrophoresis, or by CsCl banding.
	(b) A lot of nicks in plasmid DNA (a high concentration of form II DNA)	
	(c) Plasmid DNA preparation is contaminated with DNase.	
	(d) The restriction enzyme for the first digestion of DNA is contaminated with DNase, especially if a large excess of enzyme is used.	(d) Digest plasmid DNA with a purer enzyme or a different restriction enzyme.
	(e) Labeled $[\alpha\text{-}^{32}P]dNTP$ is dirty.	(e) Get a clean $[\alpha\text{-}^{32}P]dNTP$.
Band intensity in upper part of the gel is stronger than that in the lower part of the gel	(a) The incubation time of exonuclease digestion is too short.	(a) Carry out a longer incubation.
	(b) The salt concentration of exnuclease III digestion solution is too high.	(b) Dilute the salt concentration (K^+ and Na^+ ions) to 90 mM.
	(c) The specific activity of exonuclease III is lower than expected.	(c) Increase the amount of enzyme.
	(d) The ratio of ddNTP to dNTP is too low.	(d) Adjust the ratio of ddNTP to dNTP.
A band pattern in which every band appears as a doublet or triplet	(a) The second restriction enzyme has a contaminating exonuclease activity.	(a) The second restriction enzyme digestion should be carried out with a purer enzyme or incubation time should be shortened.
	(b) DNA polymerase had not been inactivated completely.	(b) After repair labeling, heat at 70° for a longer time prior to the second restriction enzyme digestion.

Extraneous gel bands or bands across all the lanes of the gel in a particular region of the sequence	(a) If the primers generated by exonuclease III digestion are capable of forming some internal secondary structure as a self-priming template, it may be labeled during the priming reaction and appears as extra bands after the second restriction enzyme cut. (b) The first or second restriction enzyme(s) may be contaminated with a double strand specific endonuclease. (c) The band(s) may be due to a pileup of DNA polymerase at a particular sequence.	(a) Heat the sample at 70° for 5 min prior to addition of cold and hot dNTPs mix and DNA polymerase; or the incubation of exonuclease III digestion should be carried out for a longer time. Using a different lot of DNA polymerase may help. (b) Use a new batch of restriction enzyme(s). (c) Use the chemical method of sequencing[6] for this portion of the sequence.
Band patterns obscure and superimposed upon one another	(a) The plasmid DNA to be sequenced may be contaminated with other DNA. (b) The base composition of the template DNA has a very asymmetric G : C ratio. (c) The rate of exonuclease III digestion of DNA is too rapid, so that the digestion is not synchronous. The primers generated in this way are not good for priming.	(a) Further purify the plasmid DNA. (b) Adjust the ratio of dideoxy- to deoxynucleoside triphosphates. (c) Lower the rate of ExoIII digestion by increasing the salt (K^+ or Na^+) concentration to 90 mM or decreasing the amount of ExoIII.
A sudden compression or increase in the spacing of bands, then returns to the normal band spacing	Due to the formation of secondary structure that moves abnormally during electrophoresis.	Run the gel at higher voltages or use dITP instead of dGTP for incorporation.
A large bubble across all the lanes in the upper part of the gel	The salt concentration of samples is too high.	Precipitate the DNA with ethanol prior to loading the gel.

TABLE VI
EXPRESSION OF A GENE CLONED IN pWR2

Number of base pairs preceding the start codon within a gene to be cloned in pWR2	Sites in which a gene may be cloned for expression
$3n$	Filled-in *Acc*I
	S1 nuclease-treated *Eco*RI
	S1 nuclease-treated *Hin*dIII
	S1 nuclease-treated *Bam*HI
	S1 nuclease-treated *Sal*I
	S1 nuclease-treated *Pst*I
$3n + 1$	*Hin*cII
	*Pvu*II
$3n + 2$	S1 nuclease-treated *Acc*I
	Filled-in *Eco*RI
	Filled-in *Hin*dIII
	Filled-in *Bam*HI
	Filled-in *Sal*I

ment, TA buffer and 60 V (10 mA) are used. The labeled DNA fragment is eluted from the agarose gel by electrophoresis.[25]

Application of Plasmid pWR for Expression of Cloned Genes

If a DNA fragment carrying a protein-coding sequence is inserted into one of the unique restriction sites in the *lacZ'* gene of pWR and the resultant reading frame is in phase with the *lacZ'* gene, a protein fused with β-galactosidase can be produced in bacteria. Owing to the high copy number of the pWR plasmid and the few proteins encoded by the plasmid, the fusion protein can be readily identified by gel electrophoresis. If the inserted gene is not more than about 300 nucleotides long and the reading frame is in phase with the *lacZ'* gene, the colonies in question are light blue, showing that some fused α-peptide of β-galactosidase has been produced. It is easy to recognize and select these colonies in which the cloned gene is expressed as a protein fused to the α-peptide. The sites in β-galactosidase gene of pWR2 for insertion of blunt-ended DNA fragments to produce fused proteins are shown in Table VI.

For a gene cloned in the polylinker region, if the product of a single protein rather than a fused protein is desired, the stretch of nucleotides between the Shine–Dalgaro (SD) sequence of α-peptide and the polylinker (see Fig. 8), or between the SD sequence and the start codon (ATG)

[25] R. C.-A. Yang, J. Lis, and R. Wu, this series, Vol. 68, p. 176.

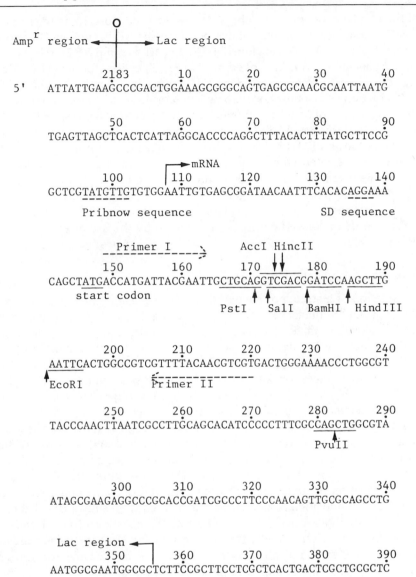

FIG. 8. Nucleotide sequence of *lac* region in pWR2. Two dashed arrows represent two primers and the direction of chain extension. The first primer, d(A-C-C-A-T-G-A-T-T-A-C-G-A-A-T-T), can bind to the region between nucleotides 149 and 164, and the second primer, d(C-A-C-G-A-C-G-T-T-G-T-A-A-A-A-C), is complementary to nucleotides 221–206. Each primer can be used to determine the sequence of a specific strand of a DNA fragment cloned in the polylinker region.

of the cloned gene, may be deleted by use of Procedure 13 or 14. For the deletion of a stretch of nucleotides between the SD region and the poly-linker (for example, a gene cloned in the *Eco*RI site in pWR2), the recombinant plasmid DNA is linearized by *Bam*HI digestion, followed by blocking these ends from exonuclease III digestion using Procedure 9b (introduction of α-phosphorothioate nucleotides into these ends). Then *Pst*I digestion followed by exonuclease III treatment is carried out by using Procedures 8a and 8b to remove the stretch of nucleotides between positions 140 and 170 (Fig. 8). The resultant plasmid DNA is digested with *Hind*III and treated with S_1 nuclease using Procedure 13. The plasmid DNA is next self-ligated using Procedure 3. For the deletion of a stretch of nucleotides between the SD region and the start codon of the cloned gene, the recombinant plasmid DNA is first linearized with a restriction enzyme (*Pst*I or *Sal*I or *Bam*HI or *Hind*III) followed by the use of Procedure 14 (*Bal*31).

Deletion of a Specific Region of DNA

Deletion of a stretch of nucleotides from a plasmid DNA with exonuclease III and S_1 nuclease using Procedures 8a, 8b, and 13 usually is more easily controlled than with *Bal*31 nuclease (Procedure 14). *Escherichia coli* exonuclease III digests DNA from the 3′ ends with almost no base specificity. Under appropriate conditions of salt and enzyme concentration, the rate of digestion is about 10 nucleotides per minute at each end of DNA at 23°.[2,3] Exonuclease III digestion followed by S_1 nuclease[2] is useful for making deletions if the objective is to get a family of deletions covering all sizes. On the other hand, the rate of *Bal*31 nuclease digestion depends to a great extent upon the DNA sequence and the type of the terminus. For example, Fig. 9a shows that the patterns of *Bal*31 digestions of *Sma*I- and *Ava*I-treated pYT2 DNA[3] are different. The rate of *Bal*31 digestion of *Ava*I ends is faster than that of *Sma*I ends, although their base composition is almost the same. Figure 9b shows *Bal*31 digestion of *Hha*I-treated pYT2 DNA. The rate of digestion of *Hha*I ends is faster than that of *Ava*I ends. Degradation by *Bal*31 of the G-rich region in the 5′ → 3′ strand of pYT2 DNA is slow, so that G bands on the gel pattern in Fig. 9b are much more intense. The advantage of *Bal*31 is that the rate of digestion is relatively fast, so that it is convenient for deleting a very large stretch of DNA (such as 2000 base pairs). The disadvantage is that not all sizes of DNA are represented in the final digestion mixture. The rate of digestion of an A : T-rich region can be as much as 100-fold greater than that of a G-rich region. The rate of digestion is also sequence specific. We found that among the G-containing dinucleotide sequences, the rate follows the pattern of TG > GT = CG > AG > GC = GA > GG,

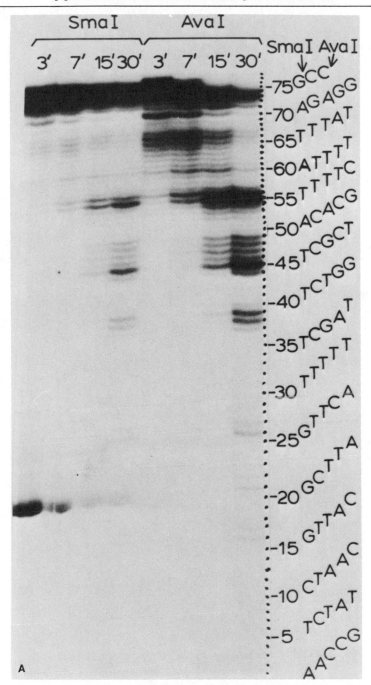

FIG. 9. (A) autoradiogram of *Bal*31 nuclease digestions of *Sma*I- and *Ava*I-treated pYT2 DNA. (B) Autoradiogram of *Bal*31 nuclease digestion of *Hha*I-treated pYT2 DNA.

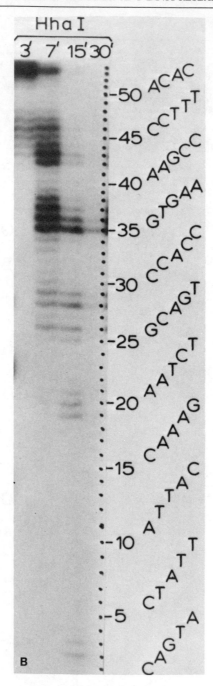

FIG. 9B. See legend on p. 107.

and the majority of the DNA termini generated by *Bal*31 are G-C base pairs.

Procedure 13. Deleting a Stretch of Nucleotides with Exonuclease III and S_1 Nuclease. Plasmid DNA is linearized with a restriction enzyme using Procedure 1 and then digested with exonuclease III using Procedures 8a and 8b, but not extracted with phenol and chloroform–isoamyl alcohol. After completion of exonuclease III digestion, immediately add 4.5 μl of 10 × S_1 nuclease buffer (2) and S_1 nuclease (20 units per microgram of DNA). Incubate at 23° for 15 min. Add 10 μl of 0.1 *M* EDTA–1.5 *M* NaOAc and 25 μl of TE buffer, extract the mixture once with 50 μl of phenol and once with 50 μl of chloroform–isoamyl alcohol (24 : 1, v/v). Precipitate the DNA in the aqueous phase with ethanol, rinse DNA pellet once with ethanol and dry under vacuum. The DNA is then self-ligated using Procedure 3.

Procedure 14. Deleting a Stretch of Nucleotides with Bal31 Nuclease. Plasmid DNA is linearized with a restriction enzyme using Procedure 1. To this, add 8 μl of 5 × *Bal*31 buffer, 20 μl of water, *Bal*31 enzyme (1 unit of enzyme per microgram of DNA), and incubate at 23°. Between 5 and 10 bp are removed per minute from each end of *Sma*I-cut DNA or over 200 bp of *Hind*III-cut or *Eco*RI-cut DNA. After completion of the *Bal*31 digestion, add 2 μl of 0.5 *M* EDTA to stop the reaction. Extract the mixture with 50 μl of phenol and 50 μl of chloroform–isoamyl alcohol. Precipitate the DNA from the aqueous phase with ethanol, rinse the DNA pellet once with ethanol, and dry under vacuum.

Acknowledgments

We thank Ullrich Rüther for the pUR222 plasmid and the *E. coli F⁻Z⁻ΔM15recA*. We are grateful to J. Yun Tso and Robert Yang for valuable help and discussion. This work was supported by Research Grant GM29179 from the National Institutes of Health.

Section II

Methods for Isolation, Purification, or Amplification of DNA

[5] Gel Electrophoresis of Restriction Fragments

By EDWIN SOUTHERN

Analytical Procedures

There are many designs of apparatus that can be used for the electrophoresis of DNA in agarose or polyacrylamide gels.[1-3] Some of them are very easy to make from glass or Perspex, and a simple apparatus can give excellent results (Fig. 1). For both agarose and polyacrylamide gels, glass has advantages over Perspex: Perspex and other plastics have a higher coefficient of thermal expansion than glass and distort more when molten agarose is poured into them; Perspex and many other plastics inhibit the polymerization of acrylamide. Agarose gels thinner than about 3 mm are fragile and difficult to handle, but polyacrylamide gels can be used at thicknesses of less than 1 mm.

Horizontal slab gels are normally used for agarose gels below 0.5%. Some workers prefer to use horizontal gels for all applications, because they are easier to use, but in our hands vertical gels give sharper bands.

Agarose gels are simply prepared by adding the powder to cold electrophoresis buffer and bringing the solution to a boil while stirring or shaking. The solution is cooled to a temperature a few degrees above the setting point and poured into the apparatus or mold. Several buffers are employed; Tris-acetate or -phosphate at about 50 mM and pH 7.5–8 with 1 mM EDTA can be used, as can sodium phosphate, provided the buffer is circulated between the anode and cathode compartment during the run. Discontinuous buffer systems and stacking gels are not normally used. Denatured DNA can be separated in polyacrylamide gels containing formamide,[4] or in alkaline agarose gels.[3]

Before loading restriction digests of DNA the reaction is usually stopped by adding EDTA and/or heating the mixture to 65°. A dense solvent and a marker dye are usually also added. Glycerol or sucrose is often used to increase, the density, but these low-molecular-weight solutes cause streaming of the sample up the sides of the loading slot, which leads to U-shaped bands. Ficoll added to 1–2% avoids this effect. Bromphenol blue or orange G can be used as a visible marker. The latter travels faster and causes less quenching of the ethidium bromide (EtBr) fluorescence.

[1] F. W. Studier, *J. Mol. Biol.* **79**, 237 (1973).

[2] T. M. Shinnick, E. Lund, O. Smithies, and F. R. Blattner, *Nucleic Acids Res.* **2**, 1911 (1975).

[3] M. W. McDonell, M. N. Simon, and F. W. Studier, *J. Mol. Biol.* **110**, 119 (1977).

[4] T. Maniatis, A. Jeffrey, and H. van de Sande, *Biochemistry* **14**, 3787 (1975).

RECOMBINANT DNA
METHODOLOGY

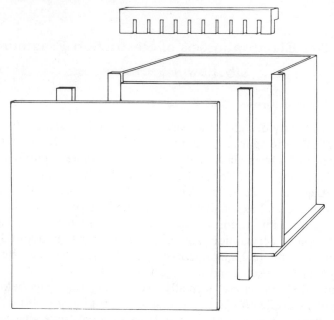

FIG. 1 Vertical gel electrophoresis apparatus. This design is typical of many based on the original apparatus of Studier.[1] The box is made of Perspex, apart from the front, which is best made of glass. The "comb" used to cast loading slots is also made of plastic, as are the spacers; many plastics inhibit polymerization of acrylamide, but polyvinyl chloride can be used with these gels. The front glass plate is held about 3 mm above the base and clamped in position. Molten agarose is poured into the gaps to a height of 1–2 cm, while the apparatus is tilted back. When this has set, the rest of the molten agarose, or acrylamide monomers, is poured in, and the comb inserted into the top to mold the slots. The box is filled with electrophoresis buffer and placed in a vessel containing the same buffer. Electrodes made of platinum wire are placed one inside the box and the other outside along the front.

The maximum load of DNA that can be applied to the gel depends on the average size of the fragments and whether the mixture is simple or complex. As a rough guide, 50 μg/cm^2 of an *Eco*RI or *Hin*dIII digest of a complex DNA is close to the maximum that can be separated with high resolution on agarose gels. Higher loading is possible with lower-molecular-weight DNA, with single-stranded DNA, and with polyacrylamide gels. Lower loading might be necessary with simple DNAs such as digests of viral DNA, or with mixtures of higher average molecular weight.

Gel Dimensions

For rapid analysis gels about 80 mm long may give adequate separation and can be run in approximately an hour at a voltage gradient of ca.

10 V/cm; for many purposes 3-mm-wide slots give bands that are wide enough. However, for accurate size measurement and high resolution 400-mm-long gels are often used. The width of the slot also has a considerable effect on resolving power; wider slots provide the eye with more information to discern fine detail in the band pattern. Thus, 20-mm slots give noticeably better resolution than 10-mm slots. Both separation and resolution are affected by the voltage gradient. Low-molecular-weight fragments diffuse and are thus best separated at fairly high-voltage gradients. Large fragments, however, diffuse very slowly, and best resolution is achieved by a low-voltage gradient.[3] It may be necessary to use a voltage gradient less than 0.5 V/cm and carry out the separation for several days to resolve large DNA fragments.[5]

Detection of DNA Fragments

EtBr may be included in the gel (added while the gel is molten) and running buffers; it has only a small effect on the mobility of DNA. However, it is not advisable to include it with high loads of DNA, because the faster-moving DNA can mop up all the EtBr, leaving the slow-moving molecules unstained. For accurate quantitative work it is advisable to stain the DNA after electrophoresis is complete. About $\frac{1}{2}$ hr using a concentration of 0.5 μg/ml is standard practice.[6] However, these conditions do not saturate the DNA in a 3-mm thick gel, and longer staining may give higher sensitivity. Background fluorescence due to unbound EtBr in the gel can be reduced by destaining, but extensive washing removes the ethidium bound to the DNA.

Photography

Fluorescence of the DNA–EtBr complex is stimulated by illumination with uv light of 254, 300, or 366 nm.

The gel may be illuminated from the side, with an angle of incidence of about 5°–10°. Placing the gel on black Perspex gives a cleaner background than placing it on glass or clear Perspex which may itself fluoresce in the uv light. Alternatively, the gel may be illuminated from below (transillumination). With 254-nm lamps the gel is placed directly on the uv pass filter that covers the lamp, but with 300- and 366-nm lamps the gel can be placed on a sheet of uv transparent Perspex (Plexiglas 218, Rohm and Haas). This material is transparent to wavelengths above 300 nm but is opaque to 254 nm. Transillumination gives higher intensity than incident

[5] W. L. Fangman, *Nucleic Acids Res.* **5,** 653 (1978).
[6] P. A. Sharp, B. Sugden, and J. Sambrook, *Biochemistry* **12,** 3055 (1973).

illumination, but more light from the lamps has to be eliminated by filters on the camera.

The short wavelength gives high sensitivity and is most commonly used, but it causes damage to the DNA during the period of photographic exposure. Less damage is caused by 300-nm lamps which also give high sensitivity,[7] but visible light emitted by these lamps is more difficult to filter out than that from 254-nm lamps and requires an interference filter on the camera to remove red light. If the interference filter cuts off both above and below the emission maximum of 590 nm, it can be used without a red absorption filter. Depending on the power of the lamp, the concentration of DNA, and the sensitivity of the film, exposure times may vary from a few seconds up to half an hour. Exposures longer than a few minutes lead to reciprocity failure in film response. This has the effect of suppressing the background, giving a cleaner appearance to the picture, but faint bands are also underrepresented and may be missed under these conditions. Reciprocity failure can be overcome by prefogging the film with a brief exposure from an attenuated electronic flash, sufficient to give a fog level of about 0.05 OD_{600} on the developed film.

Lamps with an output at 366 nm give about $\frac{1}{10}$ the excitation of shorter wavelengths[7] but, as they cause little damage to the DNA, they are used to illuminate gels from which DNA is to be recovered. They can also be used to illuminate gels still contained in the electrophoresis apparatus.

Large-format film gives better results than miniature film, and most laboratories use either Polaroid, Ilford FP4, or Kodak Tri X. Because the wavelength is close to the red end of the visible spectrum, the film's sensitivity to EtBr fluorescence is not directly related to its speed rating, and we have found that fast emulsions are no more sensitive than those listed here. Sensitivity can be increased by prefogging as described above and by doubling the recommended development time. For quantitative work prefogging improves the linearity of film response to faint images, but standard development should be used.

A standard giving a variety of band sizes and intensities, such as a restriction digest of a simple DNA, should be included in gels used for quantitative analysis, and the intensity of the bands in the standard should span the full range of those to be measured. The film density in the bands is measured in a microdensitometer such as the Joyce–Loebl. A calibration curve is plotted from peak areas of the bands in the standard, against their size. Ideally this should be a straight line. If it is not, corrections must be applied to measurements of unknowns. Of course, measurements taken from gels are rarely used to give absolute amounts but are used to measure relative amounts of fragments in a mixture.

[7] C. F. Brunk and L. Simpson, *Anal. Biochem.* **82,** 455 (1977).

Film Detection of Radioactive DNA

Radioautography of ^3P *and* ^{125}I. The wet gel may be placed against the x-ray film, with a thin plastic film such as Saran Wrap in between. For long exposures the DNA should be fixed by immersing the gel in 5% acetic acid. Drying the gel gives sharper resolution. Dilute agarose gels are easily dried to a thin film supported by a paper backing. The gel is laid on a piece of Saran Wrap and covered with a piece of hard filter paper. Absorbent paper towels weighted down with a heavy plate are laid on top of this. Most of the liquid is squeezed out in about an hour, leaving the DNA trapped in a thin film of gel which adheres to the paper. This can be radioautographed without further drying. Both agarose and polyacrylamide gels can be completely desiccated using an apparatus (Fig. 2) that applies pressure, heat, and suction to the gel. Heat should not be applied to agarose gels before most of the liquid has been removed by suction. Suction is then continued while the apparatus is heated over a boiling water bath or hot plate. Polyacrylamide gels become brittle when dry, unless glycerol permeates the gel before drying.

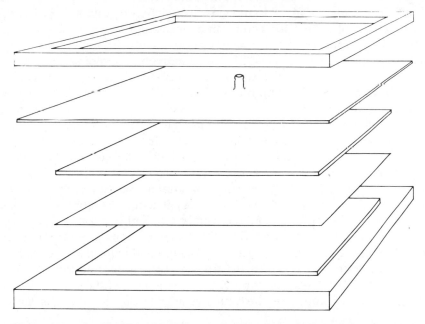

FIG. 2. Apparatus for drying gels. The gel is laid on a film of polythene on the baseplate and overlaid with a sheet of wet filter paper and a sheet of porous polythene. A sheet of rubber with a nozzle is clamped over the sandwich, and suction applied. When most of the liquid has been removed, the assembly is placed on a steam bath or hot plate until the gel has dried down to a thin film attached to the filter paper.

Indirect radioautography enhances the sensitivity to ^{32}P about 10-fold and to ^{125}I about 15 to 50-fold.[8] The x-ray film is fogged to an OD_{600} of 0.1–0.2 and laid over the dry gel with the fogged side uppermost. A calcium tungstate image intensification screen is laid over the film, and exposure is carried out at $-70°$. Film development is normal. After exposure, film should be unwrapped while still cold; if it is allowed to warm up, pressures that cause physical fogging may be generated. Tritium can be detected by allowing 2,5-diphenyloxazole PPO to permeate the gel after replacing the water in the gel by an organic solvent. The gel is soaked in methanol for 1 hr and then in a solution of PPO (10% w/w). The gel is then dried, without heating, and exposed at $-70°$ to prefogged x-ray film.[9]

Detection of Specific Sequences

It is often useful to be able to identify a particular sequence in DNA fragments separated by gel electrophoresis. There are currently a number of methods available to achieve this purpose. Restriction fragments that bind to a protein may be separated as a complex by filtration through a membrane filter, and the fragment identified by electrophoresis after elution from the membrane.[10] Radioactive restriction fragments that hybridize to a particular RNA molecule can be measured by gel electrophoresis after nonhybridized parts of the fragments have been digested with the single-strand-specific nuclease S1.[11] Alternatively, nonradioactive restriction fragments, after they have been separated by agarose gel electrophoresis, can be probed with radioactive RNA or DNA to mark the position of specific sequences. Two methods have been developed for hybridizing a radioactive probe to restriction fragments after they have been separated by electrophoresis in agarose gels. Both methods start by denaturing the DNA in the gel by soaking it in alkali. (This step can be omitted if the DNA fragments were denatured before electrophoresis.) Alkali is then neutralized by soaking the gel in a neutral or slightly acidic buffer. At this point, the gel may be dried down to a paper-thin membrane, trapping most of the DNA within it, and the membrane used as a matrix for subsequent hybridization.[2] Alternatively, DNA may be transferred from the gel to a sheet of cellulose nitrate, retaining the original pattern.[12] The sheet of cellulose nitrate is laid against the gel, and solvent blotted through it by stacking absorbent paper on top. DNA is carried out of the gel by the flow of solvent and trapped in the cellulose nitrate paper, which is subse-

[8] R. A. Laskey and A. D. Mills, *Eur. J. Biochem.* **56,** 335 (1975).
[9] R. A. Laskey and A. D. Mills, *FEBS Lett.* **82,** 314 (1977).
[10] V. Pirrotta, *Nucleic Acids Res.* **3,** 1747 (1976).
[11] A. J. Berk and P. A. Sharp, *Cell* **12,** 721 (1977).
[12] E. M. Southern, *J. Mol. Biol.* **98,** 503 (1975).

quently used for hybridization using well-established methods. After hybridization, bands of radioactivity are best detected by radioautography or autofluorography.

Standard Procedure for Transfer of DNA from Agarose Gels

Step 1. Prepare the transfer apparatus (Fig. 3). This consists of a tray filled to the brim with 20× SSC, a glass plate supported on two sides of the tray, and a thick pad of filter paper soaked in 20× SSC draped over the glass plate with two ends dipping into the solution in the tray.

Step 2. Place the gel on a glass or plastic sheet and immerse it a solution of 1.5 *M* NaCl and 0.5 *N* NaOH for 15–30 min, occasionally rocking the tray.

Step 3. Carefully decant the alkaline solution, or draw it off at a water pump, and rinse the gel with water.

Step 4. Soak the gel in a neutralizing solution (e.g., 3 *M* sodium acetate, pH 5.5, or 2 *M* NaCl in 1 *M* Tris-HCl adjusted to pH 5.5) for 30 min.

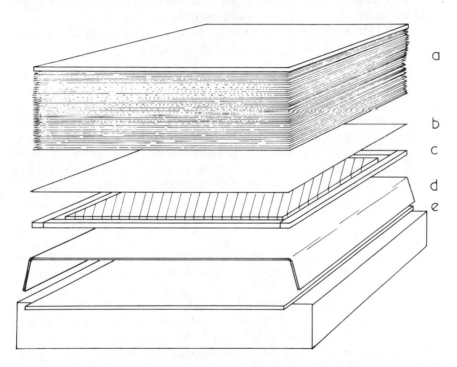

FIG. 3. Transfer of DNA from flat gels to cellulose nitrate paper. (a) Stack of paper towels weighted with a glass plate. (b) Cellulose nitrate paper. (c) Gel surrounded by plastic strips which support the edges of the cellulose paper and the towels. (d) Wad of thick filter paper that dips into the tray of 20× SSC. (e) Tray of 20× SSC with a glass plate to support the wad of wet filter paper.

Step 5. Slide the gel carefully from the plate on to the pad of filter paper, taking care to avoid trapping air beneath it.

Step 6. Cover the paper around the gel with a layer of waterproof film (e.g., thin polythene or Saran Wrap). This helps to prevent the absorbent paper, which may sag down, from becoming saturated.

Step 7. Set pieces of plastic sheet around the gel with a gap of 2 mm away from the gel. The pieces should stand about the same height as the gel.

Step 8. Squeegee excess liquid from the surface of the gel.

Step 9. Take a sheet of cellulose nitrate paper (e.g., Millipore HAWP Schleicher and Schuell BA85 or Sartorius) and wet it by floating it on water. The sheet should be at least 1 cm longer and wider than the gel. Handle the sheet with forceps or wear gloves.

Step 10. Lay the cellulose nitrate sheet on the gel, taking care not to trap air beneath it. The edges of the sheet should be supported by the pieces of plastic around the gel.

Step 11. Soak a piece of filter paper in 20× SSC and lay it on top of the cellulose nitrate, taking care to avoid trapping air beneath it.

Step 12. Stack absorbent paper (e.g., paper towels) on top of the filter paper and weigh it down lightly with a glass plate.

The rate of transfer out of the gel depends on DNA size, and gel concentration and thickness. The usual practice is to leave the transfer overnight. Large fragments (above 10 kb) may not be completely transferred from gels in this period; they can be transferred more effectively if they are broken down by irradiation with shortwave uv before denaturation. The time taken to photograph a gel in 254-nm light is enough to introduce an appreciable number of breaks.[7] Small fragments do not absorb to cellulose nitrate in low salt. However, sonicated DNA (M_r ca. 10^5) absorbs completely in 20× SSC.[12]

Hybridization

After the transfer is complete, the position and orientation of the gel are marked on the cellulose nitrate sheet with a ballpoint pen, and the sheet is thoroughly rinsed with 2× SSC. It is then baked at 80° in a vacuum oven for 2 hr. The sheet can be stored for many months after baking. Before hybridization the sheet is treated with Denhardt's solution.[13] (Some workers use 2× or even 10× usual strength, and it is an advantage to include also some denatured nonhomologous DNA.) Denhardt's solution is prepared by dissolving in 3× SSC, Ficoll (MW 400,000), polyvinylpyrollidone (MW 360,000), and bovine serum albumin, each to a concentration of 0.02% (w/v). The cellulose nitrate sheet is steeped in this solu-

[13] D. T. Denhardt, *Biochem. Biophys. Res. Commun.* **23,** 641 (1966).

tion at 65° for a few hours. The hybridization probe dissolved in the same solvent including 0.2% sodium dodecyl sulfate (SDS) is then introduced. Various methods have been devised for carrying out the hybridization of large sheets in the smallest possible volumes:

1. The sheet is wrapped around a rod which is a close fit inside a cylinder. Enough liquid is introduced to fill the space around the filter.

2. The sheet is wrapped around the inside of a cylinder or test tube which is sealed and placed on a roller or a platform that can be rotated in a horizontal plane. Enough liquid is introduced to form a continuous film over the filter as the cylinder is rotated.

3. The filter is sealed in a polythene bag, or plastic box, with enough liquid to cover it liberally. The bag or box is submerged in a water-filled shaking bath and shaken during hybridization.

4. The filter is wetted with hybridization mixture and then submerged in oil or fixed to a glass plate which is then sealed with Saran Wrap.

The aim of all these methods is to give uniform covering of the sheet with a small volume. To avoid high backgrounds it is important that the sheet be covered with a generous amount of liquid and that no areas dry out. Background is less of a problem if formamide is included in the hybridization mixture, but this solvent has the disadvantage that it slows down the rate of hybridization and also causes losses of DNA from the cellulose nitrate. Air coming out of solution can cause points of high background, which are avoided by deaerating the hybridization mixture. Sweat can cause absorption of the probes, so the sheet should not be handled with bare fingers.

The time required for hybridization depends on the concentration and sequence complexity of the probe, as well as on the temperature, solvent, and salt concentration. Usually it is not necessary to saturate all the hybridizing sequence on the filter with radioactive probe to give a readily detectable signal. Most commonly, hybridization is carried out at about 65°–70° in 2–6× SSC for a period of 1–48 hr.

Many methods of radioactive labeling can be used to make radioactive hybridization probes. The following have been used to hybridize to transfers: ^{3}H- or ^{32}P-labeled cDNA, nick-translated DNA, RNA terminally labeled with ^{32}P, ^{125}I-labeled RNA, tRNA charged with radioactive amino acids, and RNAs labeled *in vivo* with ^{3}H or ^{32}P.

After hybridization, the sheet is washed in a large volume of dilute salt solution at a high temperature for an hour or more; 0.1–2× SSC at 60°–70° are the conditions most commonly used. The high-stringency wash reduces the background of probe sticking nonspecifically to the filter and to nonhomologous sequences in the DNA. However, it should not be used in the detection of mismatched duplexes or small fragments.

RNase treatment (20 μg/ml in 2× SSC at room temperature for 20 min) can be used to reduce background when the probe is RNA; usually this treatment is not necessary.

After washing the filter it is dried and exposed to x-ray film as described for dried gels. For the detection of ^{3}H the sheet is first impregnated with PPO by dipping it into a 20% solution of PPO in toluene and allowing the solvent to evaporate while the sheet is held horizontal.

There are several problems associated with using the technique quantitatively. The method is rarely used to give an absolute measure of the amount of hybridization but to compare the extent of hybridization with different bands; even so there are problems. First, large fragments may not transfer completely from the gel to the cellulose nitrate. Second, tandemly repeated sequences such as satellite DNAs may reassociate in the gel to form networks that do not transfer. Third, the efficiency of hybridization of fragments falls off sharply below an M_r of 5×10^5. Fourth, the probe may "overhang" the ends of the fragment being detected and give an overrepresentation of its proportion (when the probe is RNA, such ends can be trimmed off with RNase, but enzymes specific for single-stranded DNA remove all DNA from the filter). Fifth, film detection methods other than direct radioautography do not give linear film response with very low amounts of radioactivity even with presensitized film. Many of these problems can be overcome by the use of appropriate internal standards. But in their absence, it should be apparent that the method can give only a rough measure of the amount of a fragment.

Estimation of Size from Mobility

Graphical Methods. There are two graphical methods of relating DNA size to mobility in gels. The log of the size may be plotted against mobility, or the size may be plotted against the reciprocal of the mobility. Both methods give a straight-line relationship in the small size range but a curve in the large size range. The curvature, in both cases, is more marked when the separation is carried out with high-voltage gradients. For accurate measurements in the high-molecular-weight range, separation should be carried out with low-voltage gradients.

If a plot of size L against the reciprocal of mobility $1/m$ gives a curve, this may be corrected to a straight line by plotting L against $1/(m - m_0)$, where m_0 is a factor calculated as described in the next section.

Calculation of Size from Mobility

Graphical methods of relating data introduce unnecessary errors. When the relationship can be represented by a simple equation, calculation is more accurate and often more convenient. The relationship

$$L = (k_1/m) + k_2 \qquad (1)$$

is accurate over a fairly wide range provided the separation was carried out with a low-voltage gradient. When a plot of L versus $1/m$ is curved in the range of the measurements, it can be made to fit to a straight line of the form

$$L = k_1/(m - m_0) + k_2 \qquad (2)$$

where m_0 is a correction factor determined by imposing the condition that the three lines joining three points all have the same slope. Three points are chosen, corresponding to size standards L_1, L_2, and L_3 with mobilities m_1, m_2, and m_3. The three values of L should span the range in which measurements are to be made. The value of m_0 that determines that these three points are joined by a straight line is given by

$$m_0 = \frac{m_3(L_2 - L_3)(m_2 - m_1) - m_1(L_1 - L_2)(m_3 - m_2)}{(L_2 - L_3)(m_2 - m_1) - (L_1 - L_2)(m_3 - m_2)} \qquad (3)$$

Rearrangement gives a form that is easier to compute with a simple calculator:

$$m_0 = \frac{m_3 - m_1((L_1 - L_2)/(L_2 - L_3) \times (m_3 - m_2)/(m_2 - m_1))}{1 - ((L_1 - L_2)/(L_2 - L_3) \times (m_3 - m_2)/(m_2 - m_1))} \qquad (4)$$

The terms in parentheses are the same and need only be calculated once.

To calculate the equation of the line [Eq. (2)], k_1 and k_2 are calculated from

$$k_1 = \frac{L_1 - L_2}{1/(m_1 - m_0) - 1/(m_2 - m_0)}$$

$$k_2 = L_1 - k_1/(m_1 - m_0)$$

Equation (2) can now be used to calculate the sizes of DNA fragments in the range between L_1 and L_3.

Size Standards

The sequences of some small DNAs have been completely determined, and the positions of a large number of restriction endonuclease sites are therefore precisely known. Fragments produced from these DNAs give the most accurate size standards in the range 0–5000 nucleotide pairs. Probably the most convenient of these DNAs to prepare in bulk is that of the plasmid pBR322[14] (Table II). The sizes of a number of restric-

---

[14] J. G. Sutcliffe, *Nucleic Acids Res.* **5,** 2721 (1978).

TABLE I
SIZES OF RESTRICTION FRAGMENTS OF PHAGE λ DNA[a]

Fragment	EcoRI[b]	HindIII[b]	EcoRI plus HindIII[b]	BglII[c]	AvaI[c]
A	21.8	23.7	21.8	22.8	15.9
B	7.52	9.46	5.24	13.6	8.8
C	5.93	6.75 (6.61)	5.05	9.8	6.1
D	5.54	4.26	4.21	2.3	4.6 (two fragments)
E	4.80	2.26	3.41	0.46	4.1
F	3.41	1.98	1.98		1.8
G		0.58	1.90		1.61
H			1.71 (1.57)		1.55
I			1.32		
J			0.93		
K			0.84		
L			0.58		

[a] All values are in kilobases and the full size of λ DNA is taken as 49 kb.
[b] Data from Phillipsen and Davies.[15] Values are for wild-type λ with values for λ cI857 in parentheses.
[c] Data from unpublished measurements of B. Smith on λ cI857.

tion fragments of phage λ DNA (Table I) have been measured in the electron microscope to an accuracy of about 1%. These together with the full-sized phage DNA cover the range 0.5–49 kb.[15] (Note that the strain of λ often used for bulk production, λ cI857, gives some restriction fragments differing in size from those of the wild type.) A continuous range of size standards is produced from a polymer series. Such a series can be produced by partial digestion with a restriction endonuclease of a tandemly repeated sequence such as a satellite DNA,[16] or alternatively by polymerization with ligase of a molecule that has sticky ends. Polymers of a small plasmid such as λ dV (3.1 kb) give markers in the moderate to high size range. Polymers of phage λ DNA would extend the range of accurate markers beyond 100 kb.

Measurement of Mobility

It is difficult to measure the position of bands stained with EtBr directly on the gel, and mobilities are usually taken from photographs. For the greatest accuracy, the photograph should be traced in a microdensitometer at sufficient scale expansion that the distance of the peaks from the origin can be measured accurately. Alternatively, measurements can be taken from a photographic enlargement. The accuracy of these measurements should be better than 1% if the full accuracy of the method is required.

[15] P. Phillipsen Kramer and R. Davies, *J. Mol. Biol.* **123,** 371 (1978).
[16] E. M. Southern, *J. Mol. Biol.* **94,** 51 (1975).

TABLE II
THE SIZES OF THE RESTRICTION FRAGMENTS OF pBR322[a,b]

HaeIII	HpaII	AluI	Hinfl	TaqI	ThaI	HhaI	HaeII	MboI
587	622	910	1631	1444	581	393	1876	1374
540	527	659	517	1307	493	347	622	665
504	404	655	506	475	452	337	439	358
458	309	521	396	368	372	332	430	341
434	242	403	344	315	355	270	370	317
267	238	281	298	312	341	259	227	272
234	217	257	221	141	332	206	181	258
213	201	226	220		330	190	83	207
192	190	136	154		145	174	60	105
184	180	100	75		129	153	53	91
124	160	63			129	152	21	78
123	160	57			122	151		75
104	147	49			115	141		46
89	147	19			104	132		36
80	122	15			97	131		31
64	110	11			68	109		27
57	90				66	104		18
51	76				61	100		17
21	67				27	93		15
18	34				26	83		12
11	34				10	75		11
7	26				5	67		8
	26				2	62		
	15					60		
	9					53		
	9					40		
						36		
						33		
						30		
						28		
						21		

[a] Data from Sutcliffe.[14]
[b] These sizes (in base pairs) do not include any extension which may be left by the particular enzyme.

Mobilities can, of course, be measured directly from radioautographs of dried gels or hybridized transfers. But nonradioactive standards are often used, and their mobilities must be taken from the photograph of the ethidium stain. In this case great care must be taken, in determining the photographic reduction of the ethidium-stained gel, to convert the mobilities of the standards to the same scale as the radioautograph.

Preparative Gel Electrophoresis

Recovery of DNA from Gel Slices

Method 1. Gel slices are shaken overnight in a buffer containing SDS. The liquid is filtered through a membrane filter or extracted with phenol, and the DNA is recovered by ethanol precipitation. This method is most effective with polyacrylamide gels and small DNA fragments. Agarose gels are usually crushed before treatment.

Method 2. Agarose is dissolved in about 10 volumes of 5 *M* sodium perchlorate[16] at 60° or in saturated potassium iodide solution at room temperature. Enough hydroxyapatite (HAP), as a solid or a thick slurry, is added to absorb the DNA and, after brief shaking, the HAP is centrifuged down. The HAP is resuspended in perchlorate or iodide solution to wash away traces of agarose and then washed in 0.12 *M* potassium phosphate buffer. A small (1- to 2-ml) column of Sephadex G50, swollen in a buffer that is appropriate for storing the DNA, is prepared in a disposable syringe or pipette. The HAP is layered carefully on top of the Sephadex. DNA is eluted from the HAP with 1 *M* potassium phosphate, from which it is separated as it passes through the Sephadex.[17] Radioactive DNA is readily detected in the fractions. Nonradioactive DNA can be detected by mixing a small drop with EtBr solution and examining under uv light.

Method 3. Agarose is dissolved in saturated potassium iodide. The solution should not contain more than 0.1–0.2% agarose and should be adjusted to a density of about 1.5 g/ml ($n_{20}^D = 1.421$). The solution is centrifuged in a SW rotor at about 200,000 *g* for 20–40 hr at 20°. DNA bands at about 1.465 g/ml, and agarose above 1.55 g/ml.[18] If EtBr (20 μg/ml) is included in the gradient, the band of DNA can be seen in longwave uv light. The band is drawn off (extracted three times with *n*-butanol if EtBr is present), and potassium iodide removed by dialysis.

Method 4. Slices of gel are placed in a dialysis bag containing a dilute buffer such as 100 m*M* sodium acetate and 1.0 m*M* EDTA, which is laid in a shallow layer of the buffer between two electrodes. Direct current is passed between the electrodes for about $\frac{1}{2}$ hr, and the polarity is reversed for about 10 sec to detach the DNA from the surface of the membrane. The buffer is then drawn from the dialysis bag.[3]

Alternatively, elution may be carried out in a glass or plastic tube which has one end sealed with a knotted dialysis tube. The tube is held in a standard electrophoresis apparatus. After overnight elution, and a

---

[17] H. F. Tabak and R. A. Flavell, *Nucleic Acids Res.* **7**, 2321 (1978).
[18] N. Blin, A. V. Gabain, and H. Bujard, *FEBS Lett.* **53**, 84 (1975).

10-sec polarity reversal, the buffer containing DNA is dripped from the end of the dialysis tube.

Method 5. DNA can be eluted from horizontal slab gels without cutting the gel into pieces. A trough is cut in front of the band of DNA and filled with a suspension of HAP. The current is switched on so that the DNA runs into the HAP, where it is absorbed.[17] The HAP is then eluted as described in method 2.

Method 6. Pieces of gel are placed between two layers of Parafilm and frozen. The frozen sandwich is squeezed between finger and thumb while the gel thaws. The drop of liquid exuded by this treatment contains a high proportion of the DNA.[19]

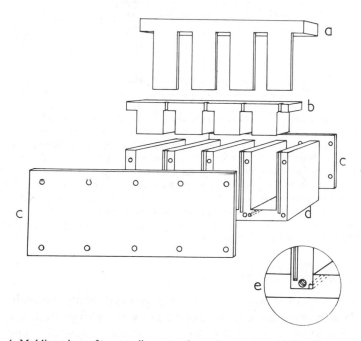

FIG. 4. Molding pieces for a small preparative gel apparatus. (a) Slot formers, which are placed behind the cover. (b) Cover, which can be placed at varying heights for gels of different sizes; this piece is left in position after the gel has set and during the run. (c) Front and back which are removed after the gel has set. (d) Main body; the grooves down the front of the uprights form seatings for the tubing to the pump used to collect fractions. (e) Detail showing the grooves down the front and the channel through the base. The channel maintains the height of the buffer in the sample collection slot at the same level as that in the rear buffer compartment. These channels are sealed off during fraction collection by a gate at the back of the apparatus (see Fig. 6).

[19] R. W. J. Thuring, J. P. M. Sanders, and P. Borst, *Anal. Biochem.* **66,** 213 (1975).

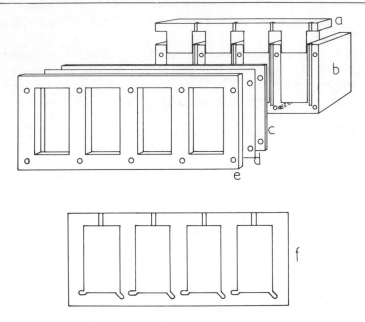

FIG. 5. Front assembly of a small preparative gel apparatus. (a) Cover, shown here in the highest position. (b) Main body. (c) Spacer, which holds the dialysis membrane 1.5–2 mm away from the gel and forms the chambers from which fractions are collected; it is grooved (see detail) with three channels in each sector—two at the bottom to connect the buffer in the chamber to the collecting tube and to the channel running through to the back of the apparatus and one at the top to let air in when the sample is pumped out. (d) Dialysis membrane. (e) Clamping frame which holds the front assembly to the apparatus; nylon screws are used to avoid corrosion. (f) Detail showing grooves in the face of the spacer; the "windows" in the spacer should be about 1 mm narrower than the width of the gel to prevent the gel from sliding forward against the membrane.

Special Preparative Gel Apparatus

Elution of DNA from pieces cut out of gels is difficult and tedious and, when a large number have to be collected or when a large amount of DNA is to be fractionated, it is better to use a special preparative apparatus. Two designs are shown here. One is for fractionation of up to 50 mg of DNA, and the other for fractionation of smaller amounts. The second apparatus can be used to fractionate several samples at the same time. Both operate on the same principle, the differences being only in the size and geometry.

Small Apparatus. The smaller apparatus (Figs. 4–6) has several channels—four in the illustration. Each channel holds a block of gel, cast *in situ,* with a loading slot molded into it. A dialysis membrane is held 1.5 mm away from the end of the gel, leaving a space from which the

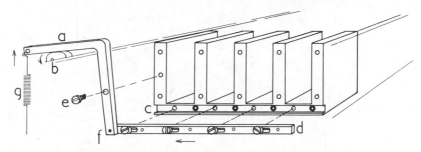

FIG. 6. Rear assembly for small preparative gel apparatus. (a) Lever which moves the gate across at each fraction collection. (c) Channel to seat the gate (d); the light circles represent threaded holes to take the nylon screws that clamp the gate to the rear of the bases. The dark circles represent rubber O rings placed in the ends of the channels that run through the base; these O rings protrude slightly to seal against the front face of the gate. (d) Gate, which is held, with light pressure, against the O rings by the nylon screws; the screws pass through slots in the gate to allow movement across the O rings; at each fraction collection, the motor-drive cam (b) raises the lever (a), moving the gate to the left; the channels are then sealed off while the fraction is pumped out of the collection chambers; when these chambers are completely emptied, the cam returns to its rest position; the spring (g) returns the gate to the right, placing the holes in register with the channels and allowing buffer from the cathode compartment to flow into the collection chambers. (e) Fulcrum of the lever. (f) Hinge holding the lever to the gate. (g) Return spring.

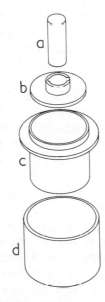

FIG. 7. Molding pieces for annular gel. The pieces are constructed of Perspex or other plastic. (a) Mold for central hole. (b) Cover which is left in position after gel has set. (c) Mold for loading slot. (d) Outer molding piece.

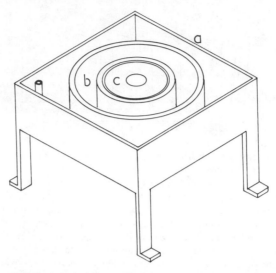

FIG. 8. Gel in position after casting. (a) Main vessel. (b) Outer electrode. (c) Gel, shown here without cover, which should be left in position.

samples are collected. This space is connected by a channel running underneath the gel to the buffer in the cathode compartment; thus the buffer in the space is maintained at a constant level. When the sample is pumped out of the space, the connection is closed off by a gate which is pushed over the ends of the grooves at the cathode side (Fig. 6). The gate is operated by a motor-driven cam and lever. The space between the gel and the membrane is also connected to a pump which removes the samples to a fraction collector.

Large Apparatus. The large-scale gel (Figs. 7–11) is cast as an annulus, with the loading slot running around the outside and a hole in the center to take the anode. The cathode is a helix of platinum wire wound on the inside of a wide tube that surrounds the gel; the anode is a helix wound around a "bobbin" that holds a tube of dialysis membrane (Fig. 9). The bobbin electrode is clamped in the base of the apparatus at its center and with a gap of about 2 mm between the membrane and the gel (Fig. 10). This space is connected to the cathode buffer by a tube that can be closed by a magnetic valve; it is also connected to the pump used to collect samples. Another pump is used to circulate the buffer continuously during the run. Buffer is pumped from a reservoir through the base of the bobbin electrode, up through the space between the membrane and the anode out through the top of the bobbin electrode, and then through a tube to circulate around the space between the outside of the gel and the cathode (Fig. 11). An overflow tube to the reservoir keeps the level of buffer constant at the same height as the gel.

FIG. 9. Central electrode assembly. (a) Collar which crimps dialysis membrane to upper flange. (b) Nozzles for hoses carrying buffer flow. (c) Terminal. (d) Central core, made of a tough plastic such as nylon, with platinum winding. (e) Lower end, threaded to take knurled nut. (f) Lower flanges, grooved to take O rings that hold bottom end of membrane.

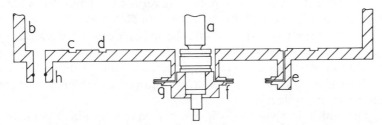

FIG. 10. Detail of base of main vessel. (a) Central electrode. (b) Side wall. (c) Groove to seat outer electrode. (d) Groove to seat outer molding piece. (e and f) Nipples to take tubing that connects outer buffer to collecting slot; the tubing is closed off by a magnetic valve during sample collection. (g) Nipple for tube connection to fraction collector. (h) Seating for buffer overflow tube.

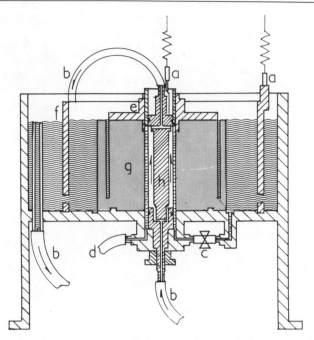

Fig. 11. Cross section through complete assembly. (a) Termini. (b) Hoses carrying circulating buffer. (c) Magnetic valve. (d) Outlet to pump to fraction collector. (e) Cover. (f) Outer electrode. (g) Gel. (h) Central electrode.

The operating cycle is controlled by the fraction collector and slave timer. The interval between fractions is set on the fraction collector. At each changeover, a pulse from the fraction collector (taken from, say, the event marker) activates the slave timer which makes the following switches over a period of 2 min:

1. The polarity of the dc supply is reversed for $\frac{1}{2}$ min to release the DNA from the membrane into the collection slot.

2. The dc supply is switched off, the magnetic pinch cock closes the tube between the outside and the inside compartments, or the motor-driven cam moves the gate across the channel, and the pump to the fraction collector is started. One minute is allowed to completely clear the sample from the collection compartment(s).

3. The pump is switched off, and the magnetic pinch cock or gate opened ($\frac{1}{2}$ min) so that the collection compartment(s) refills with buffer.

4. The cycle is restarted by switching on the dc supply with normal polarity.

The slave timer may be electronic or electromechanical. Suitable timers are commercially available, and the design described by Brownstone[20] can also be used with this apparatus.

Casting the Gel

Small Apparatus. (1) A plate is screwed onto each end of the gel container, and the cover lowered to give the desired height to the gel (Fig. 4). (2) Agarose dissolved in electrophoresis buffer (100 mM Tris-acetate, pH 7.8, 10 mM EDTA) and cooled to just above the setting point is poured into as many channels as are needed for the separation, and any bubbles trapped beneath the cover are removed by tilting the apparatus. (3) The slot former is lowered into the gel behind the cover, and the gel left to set. (4) The slot former and the end plates are removed; any loose bits of agarose are cleared away.

Large Apparatus. (1) The molding pieces that form the outside of the gel and the central hole are placed in position (Fig. 7), and the joins sealed with a little molten agarose. (2) When this has set, gel is poured to the required height. The slot former and the cover are then lowered into position. (3) When the gel has set (it is advisable to leave it overnight), the slot former, the central rod, and the outer molding piece are removed, and loose bits of agarose cleared away. The cover is left in place.

Setting Up the Apparatus

Small Apparatus. (1) A moist dialysis membrane is sandwiched between the spacer and the clamping frames (Fig. 5). Holes are pierced in the membrane at the positions of the holes in the frame. (2) This assembly is screwed against the front of the gel container, the membrane being stretched as the screws are tightened. (3) Excess membrane is trimmed off, the gate is attached to the back, and the whole assembly placed in position in the main vessel. (4) Tubes to the pump are pushed into their grooves. (5) The vessel is filled with buffer to the height of the gel and the sample is loaded.

When the sample has run into the gel, the buffer level can be raised above the top of the gel to prevent it from drying out. Saran Wrap or a polythene sheet is floated on the surface of the buffer to stop evaporation.

Large Apparatus. (1) The collar of the central electrode and an O ring are slipped over the top. (2) A length of moist dialysis tubing is fed over the lower end of the bobbin, under the collar and O ring, and over the top end of the bobbin. (3) The O ring is forced into its seating in the lower

---

[20] A. Brownstone, *Anal. Biochem.* **70,** 572 (1976).

flange to clamp the lower end of the dialysis tube in place. (4) The collar is then moved up, while the dialysis tube is pulled tight, to clamp the tube against the O ring seated in the upper flange. The distance between the lower end of the collar is adjusted so that the end of the collar protrudes just below the cover on the gel. (5) Excess dialysis tubing is trimmed off, and the electrode is placed in position in the center of the gel and clamped to the base with the knurled nut. (6) Tubing to carry the circulating buffer is attached to the top and bottom of the electrode assembly. (7) With the connection between the cathode buffer and the collecting slot closed, buffer circulation is started. (8) Electric resistance between the two electrodes should be high. If it is not, it is likely that the membrane is touching the gel because it is distorted or too distended. Reducing the flow rate or refitting the membrane may break the contact. Alternatively, conduction between the electrodes may be caused by leakage through the dialysis membrane filling the gap between it and the gel. This can be checked by switching on the collection pump.

If all is well, the collection slot is filled with buffer by unclamping the connecting tube. The sample is loaded into the peripheral slot and run into the gel. The buffer level is then raised to prevent the slot from drying out.

The preparative apparatuses described here have high resolving power and, to exploit this fully, it is necessary to collect a large number of fractions. For example, if a restriction enzyme digest of a complex DNA is fractionated into about 50 fractions, there is little overlap between fractions. Resolution depends on a number of factors which include the gel concentration, path length, time between fractions, and accuracy with which the apparatus is constructed. In the annular apparatus the path length is not readily varied, and this is one disadvantage of this design. However, there is an approximately linear relationship between the molecular weight of the DNA fragment and the time required for its elution from the apparatus. This linear relationship holds over a wide range of fragment sizes and gel concentrations between 0.4 and 2.4%. But if the molecular weight is plotted against the elution time for different gel concentrations, the slope of the line decreases with increasing gel concentration. An important consequence of this is that increasing the gel concentration has the same effect as increasing the path length. Thus, to increase separation in either gel apparatus it is not necessary to increase the size of the gel; it is enough to increase the gel concentration. There is, however, an upper limit to the gel concentration that can be used, because as the gel concentration is increased the total time for a separation increases and this could become unacceptably long. A consequence of increasing resolution by increasing either gel concentration or path length is that the

DNA concentration in the fractions is reduced, and this makes recovery and analysis of the fraction more difficult.

Some guidance as to the appropriate choice of gel concentration for a given separation is provided by the data in Fig. 12.

Analysis of Fractions

The average concentration of DNA in the fraction collected from the annular apparatus is about 100 μg/ml, and from the smaller apparatus about 30 μg/ml, concentrations that are high enough to read in a spectro-

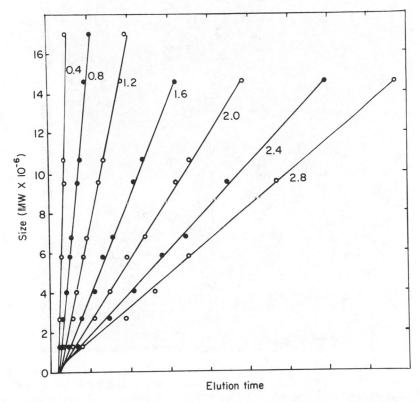

FIG. 12. Dependence of elution time on fragment size and gel concentration. The time taken for a fragment to be eluted from a preparative gel is approximately proportional to its size. Elution time for any fragment size increases with gel concentration as shown. Relative elution times only are shown here, because other factors such as gel path length and voltage gradient also affect elution time. Data for the 1.2% gel were taken from the preparative gel separation shown in Fig. 13. Data for other gel concentrations were calculated from relative mobilities in analytical gel (unpublished experiments of E. M. Southern and R. West).

Fraction Number

8 18 28 38 48 55 60 64 67   70 73 76 79 82 84   86 88 90 92

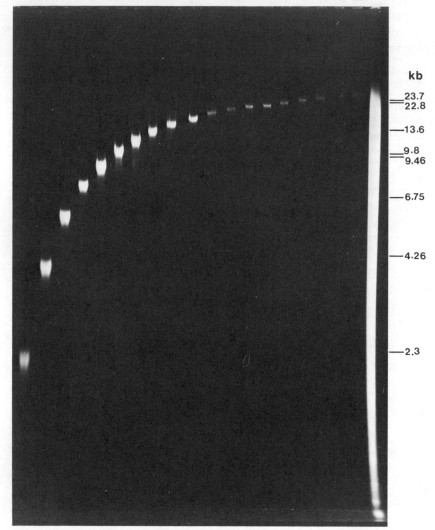

kb

—23.7
—22.8

—13.6

—9.8
—9.46

—6.75

—4.26

—2.3

FIG. 13. Fractionation of a *Hin*dIII digest of rabbit DNA on the annular preparative gel apparatus. A *Hin*dIII digest of rabbit liver DNA (6 mg) dissolved in 100 mM sodium acetate, pH 7.0, containing 1% Ficoll and a little orange G was applied to a 1.2% agarose gel 4 cm high with a 12.5 cm diameter. The central hole had a diameter of 3 cm, giving a radial path length of 4.75 cm for the separation. Twenty-four volts were applied across the electrodes. Fractions (ca. 9 ml) were collected according to the following schedule: 54 fractions at 20 min, 10 at 40 min, 20 at 60 min, and at 90 min thereafter. The photograph shows an analytical agarose gel (0.8%). Fractions of 5 μl each separated by about 3 hr were applied to this gel which was stained in EtBr and photographed as normal. The track on the right shows the original sample that was applied to the preparative gel. Size standards were a *Hin*dIII and a *Bgl*II digest of λ cI857 DNA. DNA was recovered from each fraction by adding 2 volumes of ethanol and centrifuging after keeping at −20° for 1 hr.

photometer. However, we have found that the uv absorption of buffers containing Tris rises throughout the run, probably because the base is oxidized. Analysis of the fractions by gel electrophoresis is more informative, and a few microliters of each fraction is enough for this analysis (Fig. 13). Where many fractions are to be analyzed it is convenient to combine them. For example, 100 fractions can be analyzed in 10 slots of a slab gel by combining portions of every tenth fraction.

Recovery and Further Purification of Fragments

Fractions may be concentrated approximately 10-fold by adding 2 volumes of s-butanol, removing the organic phase, and extracting the s-butanol from the aqueous phase with chloroform. Alternatively, the DNA may be precipitated with ethanol, provided the running buffer was sodium or Tris-acetate. Occasionally it is found that the DNA recovered from gels is resistant to the actions of enzymes such as restriction endonucleases or ligase. It is assumed that the inhibition is due to impurities leached from the agarose. A variety of methods have been used to remove these inhibitors, such as phenol extraction, HAP fractionation, and banding in cesium chloride.

[6] Purification, Specific Fragmentation, and Separation of Large DNA Molecules

By CASSANDRA L. SMITH and CHARLES R. CANTOR

Three new techniques have been recently developed that allow the fractionation and analysis of DNA molecules on a size scale much larger than previously possible. These methods permit routine handling of DNAs up to 1.5 million base pairs (bp). The first technique involves the preparation of unbroken genomic DNAs inside agarose gels. The second involves digestion of such DNAs in agarose with restriction nucleases that produce discrete, large fragments. The third technique, pulsed-field gel (PFG) electrophoresis, allows the size separation of DNAs ranging from 10,000 bp (10 kb) to more than 1.5 million bp (1.5 Mb). Here, these new techniques will be described and examples of their applicability to the analysis of bacterial genomes and unicellular eukaryotic genomes will be demonstrated. The reader is referred elsewhere for a more detailed discussion of how the three techniques are applied when handling higher eukaryotic genomes.[1]

Preparation of Intact DNA

Ordinary DNA preparative procedures are carried out in solution. In a typical DNA preparation, if the cells have walls these are first removed by appropriate enzymatic treatment. The resulting spheroblasts or protoplasts are then broken open by destruction of their cell membranes with detergents and a metal chelator (EDTA). This produces a complex mixture of DNA, RNA, and proteins. Treatment of the mixture with proteases and RNases may remove some of the unwanted components. Additional proteins are removed by chemical extraction of the DNA solution with phenol, and the DNA is concentrated by alcohol precipitation and centrifugation. To obtain clean DNA preparations that do not have contaminants which interfere with subsequent analytical or preparatory procedures, it is frequently necessary to repeat some of these steps. It is also essential that, during preparation, DNAs are not exposed unnecessarily to DNases.

Here, our goal is the preparation of genomic DNAs in as near to an

---

[1] C. L. Smith, S. K. Lawrance, G. A. Gillespie, C. R. Cantor, S. M. Weissman, and F. S. Collins, this series, Vol. 151 [35].

TABLE I

SIZES OF CHROMOSOMAL DNAs

DNA	Size (Mb)
Human 1	250
Human 21	50
Schizosaccharomyces pombe	3–12
Saccharomyces cerevisiae	0.22–1.3
Trypanosomes	0.05–0.1
	0.3–0.7
	0.2–3
Plasmodium	0.5–4
Leishmania	0.3–2
Blue-green bacteria	12
Pseudomonas	10
Bacillus subtilis	5
Escherichia coli	4.6
Salmonella typhimurium	4
Neisseria gonorrhoeae	1.8
Mycoplasma	0.75–1.5
G phage	0.63
T4 phage	0.18
λ phage	0.05

intact state as possible. The problem in such a preparation is the enormous size of these molecules. Table I summarizes the sizes estimated for typical chromosomal DNAs from a variety of organisms. While a few chromosomes are less than 1 Mb, such as the smaller yeast and unicellular parasitic protozoa chromosomes, most are larger. For example, Drosophila chromosomes are composed of single DNA molecules up to 100 Mb long.[2] Human chromosomes are also thought to be composed of single DNA molecules. These will range in size from 50 to 250 Mb long. It is impossible to use standard procedures to obtain large DNA molecules, such as intact chromosomal DNAs. Large DNA molecules are exquisitely sensitive to DNase damage. Special precautions must be taken to suppress cellular DNases and to avoid contaminating the samples with DNases in the laboratory. More importantly, large DNA molecules are subject to shear damage that is proportional to the square of their molecular weights. The handling described above for a standard solution preparation of DNA always breaks DNA molecules into pieces less than 10^6 bp long. Thus, laboratories using typical solution methods for preparing DNA are routinely breaking these molecules into random pieces, usually

[2] R. Kavenoff, L. C. Klotz, and B. H. Zimm, *Cold Spring Harbor Symp. Quant. Biol.* **38,** 1 (1973).

less than few hundred kilobases. This is acceptable because subsequent standard molecular biological techniques usually involve working with pieces of DNA that are much smaller still and range only up to 20 kb long. However, it is obvious that in order to work with DNA molecules the size of intact chromosomes, one must be able to prevent the type of shear damage that occurs readily when such molecules are put into solution.

Recently, a very simple technique for purifying, handling, and storing large DNA pieces without breaking them was described.[3] This technique is even adaptable to field studies since it requires only a few chemicals and simple equipment and permits the collection of a large number of samples in a short period of time. The samples are stable at 50° for days, at room temperature for months, and at 4° indefinitely. This means that samples can be shipped without refrigeration. The samples are easily manipulated by transferring them to new solutions or loading them onto electrophoresis gels.

The original procedure was described for the isolation of chromosomes from yeast cells.[3] The procedure involves purifying chromosomes from freshly grown yeast cells in solid agarose blocks, called inserts. The agarose protects the DNA from shear breakage. Freshly grown cells have little DNA destruction by endogenous DNases and therefore more intact chromosomes. A typical yeast insert contains about 100 million cells. This is equal to about 10 μg of DNA. Since only 1–2 μg is needed per electrophoresis analysis, a single insert can be used for several electrophoresis runs.

Yeast DNA in Agarose Inserts

Cells are grown overnight in YPD (10 g yeast extract, 20 g dextrose, and 20 g Bacto peptone) and washed in 50 mM EDTA (pH 7.5) and resuspended in the same buffer at twice the cell concentration wanted for each insert. Then 2 ml of the yeast suspension is mixed with 0.04 ml of 2 mg/ml Zymolase 5000 [in 10 mM NaPO$_4$ (pH 7.4) + 50% glycerol] and 2 ml of 1% liquified low gelling-temperature agarose (Seaplaque FMC, Inc.) and is brought to 45–50°. The mixture is put into molds and allowed to cool and solidify. The mold is designed to make small, 0.05- to 0.1-ml, rectangular inserts. Typical molds are 2 × 5 × 10 mm and can make 100 inserts at a time. After the insert has solidified it is pushed out of the mold into a solution of 0.5 M EDTA (pH 7.5) + 7.5% 2-mercaptoethanol, 5 ml per 20 inserts. (Yes, this is really 0.5 M EDTA and 7.5% 2-mercaptoethanol.) This treatment removes the cell wall material and makes spheroplasts.[4]

[3] D. Schwartz and C. R. Cantor, *Cell* **37**, 67 (1984).
[4] G. D. Lauer, T. M. Robert, and L. Klotz, *J. Mol. Biol.* **114**, 507 (1977).

The inserts are incubated overnight at 37°, and then incubated for 2 days in a solution (ESP) containing detergent (1% sodium lauroyl sarcosine), a very high concentration of EDTA (0.5 M, pH 9–9.5), and 1 mg/ml proteinase K. A total volume of 5 ml is used per 20 inserts. Inserts can then be stored in ESP indefinitely at 4°. It is also safe to ship the samples in ESP at ambient temperature.

The detergent is needed to destroy the cell membrane. The EDTA not only helps destroy the cell membrane, but also inactivates DNases and prevents aggregation of the DNA molecules. The proteinase K destroys and digests proteins such as DNases that are present *in vivo* and which bind to DNA and interfere with the electrophoretic separation of chromosomes described in the next section.

Bacterial DNA in Agarose Inserts

The above procedure has been modified successfully to include bacteria, which have a very different cell wall composition from that of yeast.[5] Freshly grown bacterial cells are washed and resuspended in 1 M NaCl and 10 mM Tris · Cl (pH 7.6) at typically 1.8×10^9 cell/ml. The suspension is mixed with an equal volume of 1% agarose and allowed to solidify in a mold as described above. The inserts are incubated overnight at 37° with gentle shaking, in a solution [EC-lysis, containing 6 mM Tris · Cl (pH 7.6), 1 M NaCl, 100 mM EDTA (pH 7.5), 0.5% Brij-58, 0.2% deoxycholate, 0.5% Sarkosyl, 1 mg/ml lysozyme and 20 μg/ml RNase], using approximately 5 ml EC-lysis per 20 inserts. The inserts are then treated with ESP as described above for yeast and stored in ESP until needed for further use. The basic procedures can be extended to a whole variety of organisms by incubating cells in inserts in solutions used to remove the cell wall material of the particular type of organism. Even whole flies (*Drosophila*) have successfully yielded high-molecular-weight DNA by these procedures. A simplified version of the yeast procedure has been used successfully to obtain intact chromosomes from cells lacking cells walls, for example, from humans, mice, and protozoa.[5-7] This simply involves incubation in ESP. For human samples 10^6 tissue culture cells are used for each agarose insert, providing about 10 μg of DNA. The samples obtained can and have been used for a variety of recombinant DNA techniques in addition to chromosomal electrophoresis analysis. A

[5] C. L. Smith, P. Warburton, A. Gaal, and C. R. Cantor, *Genet. Eng.* **8**, 45 (1986).
[6] L. H. T. Van der Ploeg, D. C. Schwartz, C. R. Cantor, and P. Borst, *Cell* **37**, 77 (1984).
[7] D. J. Kemp, L. M. Corcoran, R. L. Coppel, H. D. Stahl, A. E. Bianco, G. V. Brown, and R. F. Anders, *Nature (London)* **315**, 347 (1985).

slightly different agarose technique for obtaining high-molecular-weight DNA from mammalian cells was recently described.[8] This yields DNA entrapped in suspensions of small agarose beads.

Pulsed-Field Gel Electrophoretic Separation of Intact Chromosomal
  DNA Molecules

Electrophoresis techniques developed in the last 50 years have allowed very effective separation of proteins and nucleic acids on the basis of molecular weight. However, conventional DNA electrophoresis is only able to separate pieces of DNA smaller than 20 kb long. A piece of DNA 20 kb long represents only about 0.02% of an average human chromosome, 1.4×10^8 bp long. The development of pulsed-field gel electrophoresis now allows one to work with pieces of DNA that are 200 times the previous size limit, i.e., 4-Mb pieces of DNA that are 3% the size of an average human chromosome[9] and 8% the size of the smallest human chromosome.[10] The intact chromosomal DNAs of yeast and unicellular parasitic protozoa are within the current size limits of PFG electrophoresis and have been successfully separated by this technique. Figures 1 and 2 show separated yeast chromosomal DNAs which range in size from 0.2 to 1.5 Mb. Recently, separations of chromosomal DNAs of the malarial parasite, *Plasmodium falciparum*, have been reported; some of these are larger than the largest yeast chromosomal DNA and the largest is estimated to be about 4 million bp long.[7] However, the absence of established length standards above 1.5 Mb makes firm estimates of the current operating limits of PFG electrophoresis quite tenuous. Intact human chromosomes (50–250 Mb) have not yet been separated by PFG electrophoresis. However, there is no compelling reason why the technique might not eventually be extendable to separation of DNA molecules in this size range.

In solution, in the presence of an electrical field, all DNA molecules move with the same average velocity, because both the charge on the DNA and frictional resistance of the DNA molecules are directly proportional to size. During gel electrophoresis a DNA molecule twice the size of another will feel twice the pull but also twice the drag. Fractionation of DNA molecules during conventional gel electrophoresis occurs because

[8] P. R. Cook, *EMBO J.* **3,** 1837 (1984).
[9] D. C. Schwartz, W. Saffran, J. Welsh, R. Haas, M. Goldenberg, and C. R. Cantor, *Cold Spring Harbor Symp. Quant. Biol.* **47,** 189 (1983).
[10] J. de Grouchy and C. Turleau, "Clinical Atlas of Human Chromosomes," 2nd Ed., p. 373. Wiley, New York, 1984.

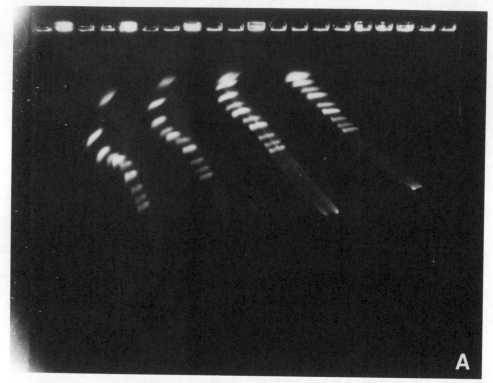

FIG. 1. Separation of yeast chromosomes by PFG electrophoresis using a single inhomogeneous (si) field in a 20-cm apparatus. DNA samples were prepared in 0.5% agarose inserts and loaded onto 1% agarose gels. PFG electrophoresis was carried out for 24 hr using the single inhomogeneous configuration indicated in Fig. 5. Applied voltages were 325 V in the vertical direction and 115 V in the horizontal direction. Pulse times were 80 sec (A) and 30 sec (B). Samples were loaded in the four wells which appear brightest in A and in 8 wells (at top left) in B.

molecules moving in an electrical field are sieved through an agarose or acrylamide matrix.[11] Agarose is used for DNA in the size range of 500 bp to 20 kb because it has large pores. Acrylamide is used for the size range of 1 to 700 bp because it has smaller pore sizes. Both of these matrices have a wide range of pore sizes. A small molecule will fit through all the pores and thus can travel through the gel in a relatively straight line. A

[11] C. R. Cantor and P. Schimmel, "Biophysical Chemistry," p. 576. Freeman, San Francisco, California, 1983.

FIG. 1B.

large molecule will be unable to enter most pores and will have to travel via a circuitous and much longer route. Thus its net translational velocity through the gel will be much smaller.

DNA molecules above 20 kb long usually cannot be separated by gel electrophoresis because they are larger than the pore size of the matrix. However, DNA size is a bit misleading because DNA molecules are not rigid and unperturbable. They are, on the average, spherical Gaussian coils and can travel through a gel matrix by deforming their shape in order to pass through the pores (Fig. 3). Thus, in an electrical field in a gel, large DNA molecules appear to behave like highly extended coils, oriented with the long axis of the coil parallel to the electrical field. The motion of such a deformed coil through a gel under the influence of the electrical field is thought to be like extrusion of plastic through a die. Large molecules adapt their size and shape to match the gel pores. The gel pores

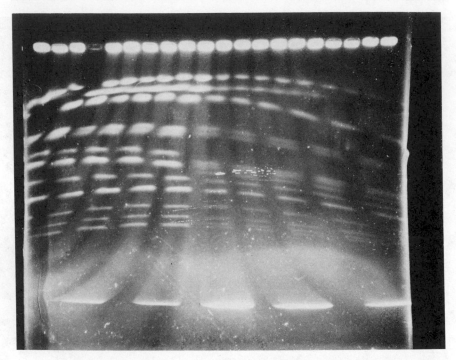

FIG. 2. Separation of yeast chromosomes by PFG electrophoresis using a double inhomogeneous (di) field with a 20-cm 1% LE agarose gel plate in a 36-cm apparatus. DNA samples were prepared and loaded as described in Fig. 1. PFG electrophoresis was carried out for 42 hr using the double inhomogeneous configuration shown in Fig. 4. Applied voltage was 450 V. Pulse time was 90 sec. Two different yeast strains were used, D273-10B/A1 for the left five lanes and DBY782 for the right five lanes. The gel was photographed along the diagonal to give the illusion that the bands had run straight.

determine the surface area of the molecules, hence they determine the frictional pull exerted per unit length of DNA migrating through the gel. All large molecules travel at the same velocity, independent of size, because there is no sieving effect of the gel matrix. The velocity is the same for all molecules because charge differences will be countered by the frictional pull of the different size molecules.

In standard gel electrophoresis the electrical field is applied constantly in one direction (Fig. 3, top). In PFG electrophoresis the electrical field is applied alternately in two directions (Fig. 3, bottom, and Fig. 4). This change in field direction forces DNA molecules to try, alternately, to orient themselves in two directions. The timing of the actual field change is effectively instantaneous relative to DNA motion. Thus a DNA coil

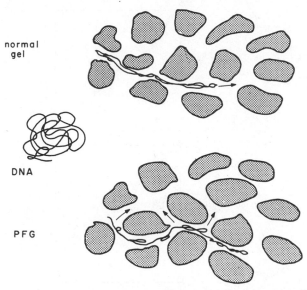

normal
gel

DNA

PFG

Fig. 3. Separation of DNA by ordinary and PFG electrophoresis. DNA traveling in a matrix during standard electrophoresis is oriented parallel to the field and is sieved through the matrix (top). DNA traveling in the alternating fields used during PFG electrophoresis is forced to spend most of its time reorienting itself perpendicular to its long axis (bottom).

with its long axis parallel to the previous field now has to deform itself along a roughly perpendicular axis in order to travel through the matrix (Fig. 3, bottom). The field is then changed back to the first orientation and the molecule is again forced to change orientation. The time spent in each direction is called the pulse time. If the pulse time is too long, the DNA molecule will effectively reorient itself rapidly to the new field and move by ordinary electrophoresis (Fig. 4, bottom). No size fractionation will occur even though there are alternating electrical fields because molecules of all sizes will have the same ordinary electrophoretic velocity in both directions. If the pulse time is too short the molecule will not have time to reorient at all (Fig. 4, top). Instead, it will see the average electrical field between the two directions and will move along this field by ordinary electrophoresis. Again there will be no size fractionation. At intermediate pulse times molecules spend most of their time reorienting and as a result effective size fractionation occurs (Fig. 4, middle).

The optimal pulse time for fractionation will be different for different length molecules. The time it takes for a molecule to reorient itself in a new electrical field will depend on its length. In general, larger molecules

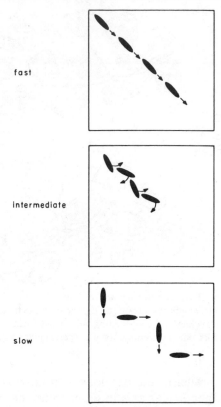

FIG. 4. The separation of DNA molecules by PFG electrophoresis depends on the pulse time. When the pulse time is too long the DNA molecules completely reorient themselves along their long axes and no separation occurs (top). When the pulse time is too short, the molecules only see the average electrical field (bottom). At intermediate pulse times the molecules spend most of their time trying to reorient (middle).

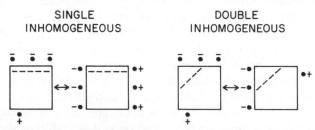

FIG. 5. Examples of two PFG electrophoresis electrode configurations. A typical single inhomogeneous electrode configuration is shown along with the current standard double inhomogeneous electrode configuration. Gel slots used for sample loading are indicated by dashes.

will be optimally fractionated at longer pulse times. At long pulse times, small molecules can completely reorient and will not be fractionated. In contrast, at short pulse times large molecules will only see the average field. This is illustrated in Fig. 1 with yeast chromosomes. PFG electrophoresis with short pulse times (30 sec) only resolves the first three lower molecular-weight chromosomes ranging in size from 250 to 600 kb (Fig. 1, bottom), while longer pulse times (80 sec) are required to separate the larger chromosomes (Fig. 1A). Separation of yeast chromosomes using a 90-sec pulse time in a different PFG electrode configuration is shown in Fig. 2. (The different electrode configurations are shown in Fig. 5 and are discussed below.)

Pulse times used with PFG electrophoresis can be tuned such that maximal separation of particular sized molecules is obtained. However, typical running conditions result in fairly high resolution of DNA fragments over a wide molecular-weight range. In Fig. 6, some lanes contain a bacteriophage λ DNA ladder, where each successively slower moving band in the lane consists of molecules increased in size by a constant increment, 42.5 kb. The resolution between bands with a constant increment in molecular weight is roughly constant or even increases with a molecular weight. This ability to achieve high resolution over a wide range may be due to the presence of the field inhomogeneity discussed below.

A variety of different types of apparatus have been used successfully for PFG electrophoresis. These differ basically in the size of the apparatus, the shape of the electrical fields, the way in which the electrodes are constructed, and the way in which the apparatus is maintained at constant temperature. The earliest apparatus described was quite small with typical running gel areas of only 10 × 10 cm. Two basic electrical field shapes were used: a homogeneous field like that generated by pairs of parallel, ordinary, linear electrodes in conventional electrophoresis, and an inhomogeneous electrical field using, for example, a linear anode and a single-point cathode. It was found empirically[7] that the highest resolution was achieved when both alternate fields were double inhomogeneous (di) configuration, but quite excellent results could also be achieved by combining a single inhomogeneous (si) field with a homogeneous field (si configuration).[3] Examples of the actual configurations used are shown in Fig. 5.

Experience has shown that the di configuration gives aesthetically very pleasing results when the samples are loaded at a 45° angle, as shown in Fig. 5.[12] Better performance is definitely obtained with both configura-

[12] G. F. Carle and M. V. Olson, *Nucleic Acids Res.* **12**, 5647 (1984).

FIG. 6. PFG separation of *Not*I fragments of different isolates of *E. coli*. Samples were prepared as described in text. PFG electrophoresis was carried out using the di configuration with a 20-cm gel in a 36-cm apparatus at 450 V for 42 hr using 25-sec pulse times. Samples were loaded in the 10 inner wells. The two extreme outside lanes contain intact yeast chromosomal DNA. Next to these are 42.5-kb monomers and various multimers of λ with DNA used as length standards. The inner six lanes contain different strains of *E. coli* DNA digested with the restriction nuclease *Not*I.

tions as the size of the apparatus increases. A 20-cm gel is the largest gel conveniently handled. However, placing such a gel at a 45° angle in a 33-cm apparatus allows almost all of the gel to be used as an effective working area for separation. Further improvements in resolution are seen when the size of the apparatus is increased further to 55 cm. In general, a constant field strength of 10 V/cm is recommended and so larger apparatus means overall higher voltages. Running times for effective separations range from 4 to 6 hr in a 10-cm apparatus, to 24 to 40 hr in a 28-cm apparatus, to 60 to 72 hr in a 55-cm apparatus.

The effect of electrical field shape on PFG electrophoresis performance is only beginning to be understood. For high-resolution separation the critical variable appears to be to the angle between the two alternate electrical fields. Thus, with the di configuration, while the two electrode arrays are perpendicular, the actual electrical fields generated by the arrays are biased toward the point cathodes. Thus the actual angles between the alternate fields tend to range from more than 90° to almost 180° in the running gel. Calculations combined with experiments show clearly that angles less than 90° give extremely poor PFG electrophoresis performance, while angles above 120° all appear quite effective. Thus, one major role of the inhomogeneous fields is to produce large angles. A second role of the fields is to produce an angle gradient. With typical successful di electrode configurations, all molecules experience a continuous increase in angle between alternate fields as they move along the diagonal of the apparatus. A larger angle inevitably means less net translational motion along the diagonal. Thus any band of identical molecules becomes sharpened as it moves, because the molecules at the leading edge are always moving slower than the molecules at the trailing edge. This band sharpening helps provide the extraordinary resolution seen in optimal PFG experiments.

The inhomogeneous fields appear to play a third role in expediting PFG separation across a wide size range simultaneously. Smaller molecules move faster through the gel and reach regions of larger angle and lower net electrical field. In contrast, larger molecules remain in regions of the gel where stronger fields and smaller angles occur. The pulse time required for optimal separation depends not only on DNA size but also on field strength and angle. It turns out that by using inhomogeneous fields one creates the same effect as a program of different pulse times so that both large and small DNAs find themselves, at least during some portion of the PFG electrophoresis run, under conditions where the effective local pulse time affords high resolution separation in their size range.

Most workers today favor the di configuration. In addition to its aesthetic appeal (bands appear to be running straight), it has been observed that there is less breakage of large chromosomes when the di configuration is used. Current thinking is that one possible size limitation of PFG electrophoresis is due to shear breakage of DNAs larger than 2 Mb. This may be avoided by lowering the field strength, by using support materials with larger and more regular pore sizes, and by using more homogeneous fields or smaller field gradients. Alternately, some way of partially condensing DNA during PFG electrophoresis might assist extension to higher molecular-weight DNA. However, the optimal experimental PFG configuration is still not known for any DNA size range.

Current standard PFG electrophoretic analyses use the electrode configurations shown in Fig. 5 and produce separation patterns typified by those in Figs. 1 and 2. From 10 to 20 samples are loaded per gel, with 0.1–1.0 μg of bacterial DNA or 1–10 μg of mamnalian DNA loaded per sample.

PFG Electrophoresis

A typical pulsed-field gel electrophoretic separation would proceed as described below, using the LKB Pulsaphore apparatus with a di configuration (LKB Producktor, Box 305, S-161, 26 Bromma, Sweden). A 1% agarose 20-cm running gel is cast in a frame in the usual way. A typical gel thickness is 5 mm. The gel comb is placed parallel to one edge of the square gel. The exact position of the comb can influence the results. Generally satisfactory performance is obtained when the comb is located 3 cm in from the top of the gel. DNA samples in inserts are loaded into the 10 central wells and pressed up against the edges. Where necessary, inserts can be sliced into smaller pieces with a glass cover slip. Small insert slices can be glued in place with liquid low-gelling agarose. It is quite important that the size of the gel inserts is not larger than the size of the wells.

The running gel is placed into the electrophoresis apparatus at a 45° angle with sufficient TBE buffer [0.1 M Tris base, 0.1 M boric acid, and 0.4 mM EDTA (pH 8.2)] at 15° to cover the gel, and is equilibrated at 15° prior to electrophoresis. Electrophoresis is started by applying pulsed fields. Some trial and error is inevitable to optimize separations in a particular apparatus. However, a generally good set of starting conditions is as follows: (1) electrode configurations—a single anode, placed one-fourth of the way in from the corner of the box, in the general di geometry shown in Fig. 5, and at least three cathodes, roughly evenly spaced; (2) applied electrical field strength—10 V/cm of box; (3) careful temperature control during the entire electrophoretic run at 15°, by circulating the buffer continuously over the running gel and either passing the buffer through an external cooling bath[3] or passing it over internal cooling coils in the PFG apparatus itself, as in the LKB apparatus; (4) pulse times—chosen as illustrated above; unless the characteristics of the sample are known in advance, we recommend three pilot runs at, typically, 30-, 60-, and 120-sec pulse times in a 33-cm PFG apparatus; and (5) running time—40 hr or approximately 1.3 hr/cm of box size. After the electrophoresis has run for the desired time period, the power is shut off. The sample can be stored for many days with no band broadening since the molecules are far too large to show any significant diffusion. The gels can be stained with ethidium bromide and photographed in the normal way. Details of recommended procedures for blotting PFG samples are given elsewhere.[1]

Specific Fragmentation of High-Molecular-Weight DNA

Although the intact chromosomes of bacteria are only 0.75–20 Mb, they do not migrate in any electrophoresis system because of their circular topology. It is believed that circular molecules get impaled on the agarose rods so that they can no longer move. In order to work with these chromosomes it is necessary to convert them to linear DNA molecules similar to those of higher organisms. The procedures used to cut them at specific places are the same types of procedures that would be required to cut the extremely large chromosomes of mammals into pieces that can be separated by PFG electrophoresis. Although feasibility studies of cutting DNA into discrete, large fragments have been undertaken with both bacterial and mammalian chromosomes, the bacterial samples are easier and faster to work with and permit direct visualization of DNA fragments by ethidium bromide staining. Much of the technology has been worked out with bacteria and then, subsequently, extended to both mouse and human samples.

Type II restriction endonucleases are enzymes that cut DNA at specific sequences. Most restriction enzymes have 6- or 4-bp recognition sequences. Simple statistics shows that, for DNAs with 50% AT, a complete digest with these enzymes will yield DNA fragments of average sizes of 4096 or 256 bp, respectively. These size pieces are amenable to current recombinant DNA technology manipulations. The ability to generate, separate, and modify such defined small fragments of DNA was an essential step in developing the field of recombinant DNA. PFG electrophoresis now enables the separation of much larger DNA molecules. Hence, recombinant DNA technologies must now be extended to generate and manipulate large fragments.

Three restriction enzymes, *Sfi*I, *Not*I, and *Rsr*II, tend to cut DNA less often than other restriction enzymes because their recognition sequence is larger than 6 bp long.[13,14] Other enzymes cut mammalian DNA less frequently than their 6-bp recognition sequence would predict because their recognition sequence contains one or more CG sequences. In higher eukaryotes, this sequence occurs at only one-fourth the frequency expected from the base composition.[15] Cutting of mammalian DNA by particular restriction nucleases may also be inhibited when endogenous methylases modify the recognition sequence.[16] Two restriction enzymes

[13] Q. Bo-Qin and I. Shildkraut, *Nucleic Acids Res.* **12,** 4507 (1984).
[14] C. D. O'Connor, E. Metcalf, C. J. Wrighton, T. J. R. Harris, and J. R. Saunders, *Nucleic Acids Res.* **12,** 6701 (1984).
[15] G. G. Lennon and N. W. Fraser, *J. Mol. Evol.* **19,** 286 (1983).
[16] M. McClelland and M. Nelson, *Nucleic Acids Res.* **13,** r201 (1985).

(*Not*I and *Sfi*I) with 8-bp cutting sites are available commercially. In a random DNA sequence these enzymes would be expected to cut large molecular-weight DNA into pieces averaging 64-kb. These DNA pieces are above the conventional size limits of ordinary electrophoresis and the size limits of the DNA usually purified in solution by ordinary techniques. However, fragments of this size are amenable to analysis by PFG electrophoresis and can be obtained from the digests of DNA preparations described in the preceding section. In fact, the identification of restriction enzymes with rare cutting sites has been hindered by the fact that neither a substrate nor a means to separate large fragments has previously been available. The use of bacterial insert DNA preparations should now facilitate the search for other restriction enzymes with infrequently occurring recognition sequences (see below).

A typical result derived by cutting the *E. coli* chromosome with the restriction endonuclease *Not*I is shown Fig. 6. The extreme outside lanes contain yeast samples, while the next inner lane on each side contains a size standard of λ phage DNA molecules tandemly annealed. The six inner lanes are loaded with inserts made from different strains of *E. coli* that have been washed and digested with *Not*I, as described below. The results show that the *E. coli* chromosome is cut into about 20 fragments ranging in size from 40 to over 500 kb and which seem to have an average size around 200 kb rather than the theoretical value of 64 kb. *Sfi*I digests of *E. coli* result in slightly more numerous fragments but these are still much larger on average than the expected theoretical distribution. It is not clear why there are so few *Not*I and *Sfi*I restriction sites on the *E. coli* chromosome.

The procedure used for obtaining the results shown in Fig. 6 and for those discussed below with mammalian DNA is as follows.

Restriction Nuclease Digestion of DNA in Agarose

Each DNA insert block is prepared for digestion by first being treated twice with 1 ml of 1 mM phenylmethylsulfonyl fluoride (PMSF) in TE [10 mM Tris · Cl and 0.1 mM EDTA (pH. 7.5)] buffer by gentle rotation at room temperature for 2 hr. PMSF inactivates the proteinase present in the lysis solution. The inserts are further washed three times with an additional 1 ml of TE buffer without PMSF. Inserts can be safely stored at this stage for at least several weeks. For digestion, the inserts are moved into Eppendorf tubes containing 20 units of restriction enzyme per 1 μg of DNA in 250 μl of the recommended reaction buffer containing fresh sulfhydryl reagent and supplemented with 100 μg/ml bovine serum albumin. The inserts are incubated 2–16 hr at 37° with gentle shaking. The reaction buffer is aspirated off and each insert is incubated for 2 hr at 50° in 1 ml ES (ES = ESP without proteinase K). The ES is removed and 250 μl ESP is added and the samples are incubated an additional 2 hr at 50° before loading onto a gel. We have found that all proteins must be removed from DNA in order to obtain good PFG separation of fragments. Slices of insert containing typically 0.2 μg of DNA each are loaded onto 1% agarose gels and run in di configuration.

Similar procedures are effective with mouse cells[16a] and human cells.[1] Mammalian genomes are about 1000-fold larger than bacterial genomes (3 × 10⁹ bp). The size (or complexity) of the genome prevents complete resolution of most of the pieces. Therefore, unlike the *E. coli* digestions shown in Fig 6, ethidium bromide staining of large genomes digested with restriction enzymes usually shows mostly smears instead of bands. Specific restriction fragments of mammalian DNA must be identified by blotting and hybridization.[1]

For most potential applications of PFG electrophoresis, as long as one has the technology to separate the fragments generated, the larger the fragments available the better. Besides screening directly for restriction enzymes with rare cutting sites, a variety of ways are being developed to cut DNA into large, defined pieces. A new technique, originally shown to work on small plasmids,[17] also looks feasible for *E. coli* DNA.[17a] This procedure takes advantage of the fact that the *Dpn*I restriction enzyme, which has a 4-bp recognition sequence (GATC), will only cut GATC sequences when the As on both strands are methylated. This methylation can be carried out *in vitro* by using a DNA methylase, *MTaq*I, which recognizes the overlapping sequence TCGA. *Dpn*I sites are only generated when two M*Taq*I sites appear tandemly in a DNA sequence. Thus a 4-bp-specific enzyme is converted to one with 8-bp specificity, as shown below (methylated bases are indicated by an asterisk).

TCGA*TCGA*
*ATCT*ATCT

Besides restriction endonucleases there are other enzymes that cut DNA at specific sequences. Some of these, such as the λ phage *ter*, have very large 100- to 200-bp recognition sequences.[18,19] They have the poten-

[16a] C. L. Smith, J. Berman, G. Yancopoulous, C. R. Cantor, and F. Alt, unpublished observations.

[17] M. McClelland, L. G. Kessler, and M. Bittner, *Proc. Natl. Acad. Sci. U.S.A.* **81**, 983 (1984).

[17a] C. L. Smith *et al.*, unpublished results.

[18] A. Becker and M. Gold, *Proc. Natl. Acad. Sci. U.S.A.* **75**, 4199 (1978).

[19] M. Gold and A. Becker, *J. Biol. Chem.* **258**, 14619 (1983).

tial to generate very large fragments from mammalian chromosomes. Other potential cutting protocols involve the *in vitro* use of the sophisticated proteins the cell normally uses to promote chromosomal rearrangements. These include the endonuclease responsible for mating-type switching in yeast[20] and the enzymes involved in recombination.[21]

Applications of Large DNA Technology to the Study of Organisms with Simple Genomes

Organisms such as yeast and parasitic protozoa have chromosomal DNAs in the size range currently amenable to PFG electrophoretic separations. Thus the technology described above provides direct cytogenetic analysis simply by electrophoresis. The number and size of chromosomes is revealed directly from stained, gel-separated DNA. All cloned genes can be assigned directly to chromosomes by simple Southern blots. Single chromosome libraries can be constructed by electroelution of the DNA of individual separated chromosomes followed by fragmentation and cloning into an appropriate vector.

In addition to these static measurements of genome organization, technology to handle large DNA molecules allows simple monitoring and analysis of genome dynamics. Chromosomal rearrangement and transposition studies using PFG technology have been carried out with yeast[3,22] and a variety of protozoa.[6,7,23,24] Using PFG technology it was shown that expression of specific trypanosome variant surface glycoprotein (VSG) genes is regulated by genomic rearrangements. Genomic rearrangements regulating gene expression occur throughout nature and are part of the normal life cycles of a variety of organisms,[25] and PFG should find wide applicability in detecting these events. It has also become apparent that genomic rearrangements are responsible for a variety of tumors[26,27] and

[20] R. Kostriken, J. N. Strathern, A. J. S. Klar, J. B. Hicks, and F. Heffron, *Cell* **35**, 167 (1983).
[21] P. Howard-Flanders, S. W. West, and A. Stasiak, *Nature (London)* **309**, 215 (1984).
[22] P. Heiter, C. Mann, M. Snyder, and R. W. Davis, *Cell* **40**, 381 (1985).
[23] L. H. T. Van der Ploeg, A. W. C. A. Cornelissen, P. A. M. Michels, and P. Borst, *Cell* **39**, 213 (1984).
[24] L. H. T. Van der Ploeg, A. W. C. A. Cornelissen, J. D. Barry, and P. Borst, *EMBO J.* **3**, 3109 (1984).
[25] J. A. Shapiro, "Mobile Genetic Elements," p. 668. Academic Press, New York, New York, 1983.
[26] M. Schwab, G. Ramsay, K. Alitalo, H. E. Varmus, J. M. Bishop, T. Matinsson, G. Levan, and A. Levan, *Nature (London)* **315**, 345 (1985).
[27] Y. Tsujimoto, E. Jaffe, J. Cossman, J. Gorham, P. C. Nowell, and C. M. Croce, *Nature (London)* **315**, 340 (1985).

PFG technology should be equally useful to study the rearrangement of mammalian genomes.

Finer analysis of genome organization and dynamics is possible by cutting chromosomal DNAs with restriction nucleases that yield large fragments, as described above. It should be possible to order these fragments by several different methods. If a detailed genetic map is available, simple Southern blotting will place most large fragments in order. Known genetic rearrangements, such as insertions, deletions, or inversions, can aid in this process. If no map is available, different single and double restriction nuclease digests can be assembled into a map by overlap, just as in conventional restriction map construction. Size estimates of DNAs from PFG analysis appear to be sufficiently accurate to allow this approach.

More powerful methods are potentially available. One of the most attractive uses junction probes, which are clones containing rare cutting sites, such as *Not*I sites.[5] Hybridization of a Southern blot of a *Not*I-cut genomic DNA with such a probe would detect two large DNA fragments. These must be adjacent. Hence, repeating this process with a complete set of *Not*I junction probes will yield the complete ordered arrangement of the distances between *Not*I sites. Where more than one *Not*I fragment of the same size exists, ambiguities in identifying particular large fragments can be resolved in relatively straightforward ways. Then the established data will constitute a complete *Not*I restriction map.

Once a physical map is available, genes can be located to within the resolution of the map by simple blotting and hybridization. Genome rearrangements can be mapped in a similar way. The power of the method is the speed and lack of ambiguity of direct physical analysis relative to most genetic methods. In addition, it is clearly applicable to organisms for which little or no classical genetic analysis is currently feasible.

Acknowledgment

This work was supported by grants from the NIH, GM 14825 and CA 39782.

[7] Orthogonal-Field-Alternation Gel Electrophoresis

By Georges F. Carle *and* Maynard V. Olson

Introduction

Until recently, mixtures of large DNA molecules were difficult to analyze. At about the size of bacteriophage λ (48.5 kb), standard chromatographic, sedimentation, and electrophoretic methods of separating DNA molecules begin to fail. Although the effective size range of some of these methods can be extended by using extreme conditions—for example, by carrying out velocity sedimentation at low rotor speeds[1] or electrophoresis at low voltages in dilute gel matrices[2,3]—none of these approaches has proved sufficiently powerful to attract widespread use. This situation changed abruptly in 1983, when Schwartz *et al.* reported that the electrophoretic mobilities of DNA molecules up to at least several hundred kilobase pairs become strongly size dependent when they are electrophoresed in the presence of two alternately applied, approximately perpendicular electric fields.[4] Use of this technique has allowed the analysis of many previously uncharacterized DNA molecules, such as the intact chromosomal DNA molecules of yeast and several protozoans.[5-8]

In this article, we present detailed protocols for separating large DNA molecules by orthogonal-field-alternation gel electrophoresis (OFAGE) and discuss some of the practical and theoretical issues that affect this technique. Sufficient details are also provided to allow the assembly of an apparatus, which is functionally identical to one whose characteristics have been previously described.[7]

Sample Preparation

As the size of DNA molecules increases, it becomes progressively more difficult to keep them intact. Both mechanical shear and nucleolytic degradation must be controlled to produce acceptable OFAGE samples.

[1] T. D. Petes and W. L. Fangman, *Proc. Natl. Acad. Sci. U.S.A.* **69**, 1188 (1972).
[2] P. Serwer, *Biochemistry* **19**, 3001 (1980).
[3] W. L. Fangman, *Nucleic Acids Res.* **5**, 653 (1978).
[4] D. C. Schwartz, W. Saffran, J. Welsh, R. Haas, M. Goldenberg, and C. R. Cantor, *Cold Spring Harbor Symp. Quant. Biol.* **47**, 189 (1982).
[5] D. C. Schwartz and C. R. Cantor, *Cell* **37**, 67 (1984).
[6] L. H. T. Van der Ploeg, D. C. Schwartz, C. R. Cantor, and P. Borst, *Cell* **37**, 77 (1984).
[7] G. F. Carle and M. V. Olson, *Nucleic Acids Res.* **12**, 5647 (1984).
[8] G. F. Carle and M. V. Olson, *Proc. Natl. Acad. Sci. U.S.A.* **82**, 3756 (1985).

As a rule, some degree of special handling becomes necessary as soon as molecules are appreciably larger than 50 kb. It is not difficult, however, to prepare intact molecules of 200–300 kb from most biological sources using standard methods of DNA purification. Even such procedures as phenol extraction, alcohol precipitation, and micropipetting can be applied to DNA molecules in this size range as long as nucleolytic degradation is controlled by the liberal use of EDTA, detergent, and proteases, and mechanical agitation is kept to a minimum. By avoiding phenol extraction, alcohol precipitation, and micropipetting, molecules as large as 500 kb can be prepared in standard buffers.[7]

To handle intact DNA molecules larger than 500 kb, however, it has proved necessary to prepare DNA samples by *in situ* lysis of cells or spheroplasts in a semisolid matrix. The two most widely used methods employ an agarose matrix, either as a solid plug[5] or in the form of microbeads.[9] We have had good success preparing yeast samples with both techniques. The microbead preparations are more rapid and are particularly convenient when samples must be prepared simultaneously from many different strains. Microbeads have the disadvantage, however, that at least in the case of yeast, the results are only satisfactory when cells are embedded at relatively low cell densities. Consequently, when maximal DNA concentrations are required, we emloy solid plugs. Our adaptations of the microbead and solid-plug protocols, as applied to yeast, are given below:

Solid-Plug Protocol

   Solutions

     SCE buffer: 1.0 M sorbitol
                  0.1 M sodium citrate
                  0.06 M EDTA
                  (overall pH adjusted to 7 with HCl)

The following volumes are suitable for preparing 10 solid-plug samples, poured in 60 × 20-mm Petri plates.

     Solution I: 10 ml SCE buffer
                  0.5 ml 2-mercaptoethanol
                  10 mg zymolyase 60,000 (Miles)
     Solution II: 46 ml 0.5 M EDTA (pH 9.0)
                  0.5 ml 1 M Tris · HCl (pH 8.0)
                  4 ml 2-mercaptoethanol

[9] P. R. Cook, *EMBO J.* **3,** 1837 (1984).

Solution III: 45 ml 0.5 *M* EDTA (pH 9.0)
5 ml 10% sodium *N*-laurylsarcosinate
0.5 ml 1 *M* Tris · HCl (pH 8.0)
50 mg proteinase K (Boehringer Mannheim)

Sample Preparation

1. Grow cells overnight to late log phase in 100 ml of YPD broth (for 1 liter use 10 g yeast extract, 20 g peptone, and 20 g glucose; see note *a* below).

2. Harvest the cells by centrifugation at 4° in 50-ml bottles.

3. Resuspend the pellets in 20 ml 0.05 *M* EDTA (pH 7.5) at 0° and pool them in one 50-ml bottle.

4. Rinse the cells once by repeating steps 2 and 3.

5. Discard the supernatant and resuspend the cells in SCE to achieve a concentration of 0.1 g/ml (wet weight). Let the suspension equilibrate at room temperature for a few minutes.

6. Mix: 3 ml cells
1 ml Solution I
5 ml 1% low-gelling agarose (Sigma type VII) in 0.125 *M* EDTA (pH 7.5), cooled to 42°

7. Pour into a 60 × 20-mm Petri dish (see note *b* below).

8. Allow the mixture to solidify at room temperature and overlay with 5 ml of Solution II. Place the Petri dish in a sealed plastic bag and incubate overnight at 37° with gentle agitation.

9. Replace overlay with 5 ml of Solution III, reseal the plastic bag, and incubate the plate overnight at 50° with gentle agitation (see note *c* below).

10. Replace the overlay with 5 ml of 0.5 *M* EDTA (pH 9.0) and store at 4°.

Gels are loaded by cutting a slice of the chilled agarose with a razor blade and gently teasing the slice into a well formed by a comb (1.5 mm thick) that forms one continuous slot across the gel. If multiple samples are loaded, they are simply lined up one after another in the slot.

Notes on Solid-Plug Protocol

a. A 100-ml overnight culture will produce 1 to 2 g of wet cells (depending on the strain), which is enough to make two to three plates containing 0.3–0.4 g of wet cells each. The sharpest bands are obtained with as little as 0.2 g of wet cells/plate. When more than 0.8 g of cells/plate are used, the background becomes intense and the sharpness of the bands is adversely affected.

b. For step 6, the mixture of cells and agarose should be homogeneous. Clumping of the agarose, which may occur if the temperature of the cell suspension or the agarose is too low, must be avoided.

c. The agarose usually clears visibly in 2–3 hr due to lysis of the cells; once clearing is complete, the samples can be used without further incubation.

d. Samples can be stored at 4° for several months, but they often give the sharpest bands and lowest background (particularly for the largest chromosomes) when less than 2 weeks old.

Microbead Protocol

Solutions

TE8: 10 mM Tris · HCl (pH 8.0), 1 mM EDTA (pH 8.0)
Spheroplasting buffer (for 2 ml of beads):
   3 ml SCE
   2 ml 0.5 M EDTA (pH 9.0)
   1 mg zymolyase 60,000 (Miles)
   0.25 ml 2-mercaptoethanol
Lysis buffer (for 2 ml of beads):
   4.5 ml 0.5 M EDTA (pH 9.0)
   0.05 ml 1 M Tris · HCl (pH 8.0)
   0.5 ml 10% sodium N-laurylsarcosinate
   5 mg proteinase K (Boehringer Mannheim)

Sample Preparation

1. Grow cells overnight in 5 ml of YPD broth (see note a below).
2. Mix in a 50-ml Erlenmeyer flask:
   4 ml of culture
   10 ml paraffin oil kept at 42°
   1 ml 2.5% low-gelling agarose kept at 50°
3. Mix vigorously on a vortex at maximum speed for about 1 min until a fine emulsion is obtained.
4. Cool rapidly in an ice-water bath while swirling the flask (see note b below).
5. Transfer the cold emulsion to a 50-ml polystyrene tube and add 20 ml TE8.
6. Spin in a table-top centrifuge (2500 rpm for 5 min).
7. Discard paraffin oil and supernatant, resuspend the beads in 30 ml TE8 and centrifuge as in step 6.
8. Resuspend the beads in 10 ml SCE, transfer to a 15 ml polypropylene tube, and centrifuge as in step 6.
9. Discard supernatant, add 5 ml spheroplasting buffer, and incubate at 37° for 1 hr on a rotating drum. Centrifuge as in step 6.

10. Discard supernatant, add 5 ml lysis buffer, and incubate at 50° for 1 hr on a rotating drum. Centrifuge as in step 6.

11. Discard supernatant and replace with 1 ml 0.5 M EDTA (pH 9.0). Store at 4° (see note c below).

With the above protocol, 0.08 ml of beads is sufficient to visualize all the yeast chromosomes when loaded into a 12 × 1.5-mm well.

Notes on Microbead Protocol

a. The overnight culture can be concentrated for poorly growing strains but we have found this step unnecessary in most cases. This protocol will yield approximately 2 ml of microbeads.

b. Making a good emulsion and cooling it quickly are the critical steps in generating microbeads that are small enough to be easily pipetted. If the preparation of beads that are too large to pipet readily is a common problem, the use of an ice–salt bath for the initial cooling step (~20 sec) is recommended.

c. The stability of the bead preparations on storage is similar to that of plugs (see above).

Electrophoresis

OFAGE results are influenced by an unusual number of interdependent variables. Because small changes in the experimental conditions—particularly the switching cycle, the voltage, and the temperature—can have large and often unexpected effects, it is advisable to use thoroughly tested conditions during the initial characterization of samples. In our laboratory, the large majority of routine applications are carried out using just three sets of standard conditions, all of which employ the same apparatus. Because the electric-field geometry is the most critical of the experimental variables, it is particularly desirable to leave the electrodes in fixed positions. All the examples that we discuss presuppose that the geometry of our published apparatus has been duplicated exactly (Fig. 1). Even minor changes in the positioning of the electrodes or in the overall size and shape of the electrophoresis chamber can have large effects. Our three standard sets of conditions are summarized below and are illustrated in Figs. 2 and 3. Figure 3 also illustrates the use of OFAGE to assign a previously unmapped gene to a chromosome (CDC3, chromosome XII).

It should be noted that the relative migration of chromosomes IV and XII is very sensitive to the electrophoretic conditions. In Fig. 3, chromosome XII migrates ahead of IV, whereas under other conditions it mi-

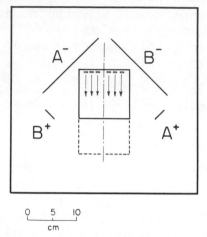

FIG. 1. A schematic diagram of the electrophoresis chamber. A more complete diagram is given in Ref. 7. The dotted line represent a gel plate longer than the regular 10 × 10-cm plate. The longer plate is sometimes useful in preserving smaller fragments; it should be noted that "forward" components of the electric field extend well past the ends of the electrodes in the region between A⁺ and B⁺. Since the critical measurements concern the position of the electrodes relative to the gel, as opposed to the sides of the box, the precise electrode positions are given below using an x–y coordinate system that intersects at the center of the box and has axes parallel to the sides of the box. The $(+,+)$ quadrant is the back right corner of the box, when the box is viewed from the side where the negative electrodes converge (i.e., it contains electrode B⁺). In this system, the coordinates of the eight electrode ends are as follows, in units of millimeters: A⁺ electrode $(-120, +24)$ and $(-101, +43)$; B⁺ electrode $(+120, +24)$ and $(+101, +43)$; B⁻ electrode $(-13, -130)$ and $(-130, -13)$; and A⁻ electrode $(+13, -130)$ and $(+130, -13)$. The gel should be positioned so that the center of the well lies at position $(0, -54)$; therefore, the gel plate is not precisely in the center of the box. When using a longer gel plate (10 × 18 cm), it is important to position the well at the same location as in a regular 10 × 10-cm plate. The position of the well relative to the electrode has a strong effect on the banding pattern.

grates behind IV. This behavior is probably characteristic of all DNA molecules as large as XII, which is almost certainly the largest yeast chromosome.[10] When characterizing DNA molecules larger than 1–2 Mb, a monotonically inverse relationship between size and mobility should not be assumed.

Broad Size-Range Conditions (50 to >1000 kb, Fig. 2A)

1.5% agarose

0.5× TBE (1× TBE = 90 mM Tris base, 90 mM boric acid, 2.5 mM Na$_2$H$_2$EDTA, overall pH not adjusted but expected to be approximately 8.2)

[10] R. K. Mortimer and D. Schild, *Microbiol. Rev.* **49**, 181 (1985).

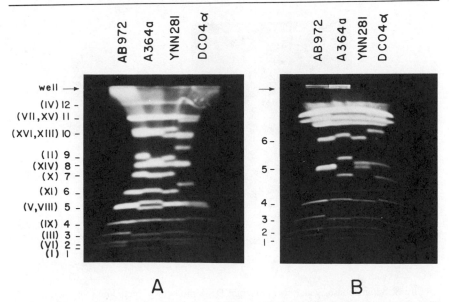

FIG. 2. Separation of yeast chromosomal DNA molecules under broad size-range conditions. (A) Data for the four strains AB972, A364a, YNN281, and DCO4α under the broad size-range conditions (40-sec switching). (B) The separation of the smaller bands has been increased by using the small-molecule conditions (30-sec switching). The band numbering and assignments refer to our standard strain AB972[8] in both panels. Not all the bands in the other strains have been analyzed, but we have determined assignments for the following resolved doublets that allow resolution of pairs of chromosomes that comigrate in AB972: strain A364a resolves band 5 into 5A (chromosome VIII) and 5B (chromosome V), strain YNN281 resolves band 10 into 10A (chromosome XIII) and 10B (chromosome XVI), and strain DCO4α resolves band 11 into 11A (chromosome XV) and 11B (chromosome VII). In all cases the A band is the one with the higher mobility.

40- or 50-sec switching interval
300 V
18 hr
13°

Small-Molecule Conditions (20 to 500 kb, Fig. 2B)

1% agarose
0.5× TBE
30-sec switching interval
300 V
18 hr
13°

Large-Molecule Conditions (>1000 kb, Fig. 3A)

0.4% agarose

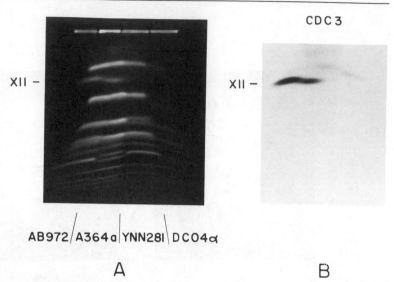

FIG. 3. Separation of yeast chromosomal DNA molecules, including chromosome XII, under large-molecule conditions. Data are for the same four yeast strains analyzed in Fig. 2. (A) The ethidium bromide-stained gel; (B) hybridization with a DNA probe specific for CDC3 shows that this previously unmapped yeast gene is on chromosome XII. Reproducible detection of chromosomes as large as XII requires low agarose concentration (0.4%) and lower voltage (250–200 V) than is normally employed. Band sharpness is somewhat diminished by the low agarose concentration.

0.5× TBE
50- to 60-sec switching interval
200–250 V
≥18 hr
13°

Apparatus Construction

The functionally critical feature of an OFAGE apparatus is the positioning of the electrodes in relation to the gel and the perimeter of the electrophoresis chamber. Detailed measurements on these points are presented in Ref. 7 and Fig. 1. Other features of the construction are straightforward. The main gel box is constructed from clear acrylic (e.g., Plexiglas, a product of Rohm & Haas) that is 1/4 inch thick. The internal dimensions of the electrophoresis chamber are 15 × 15 × 2 inches. Underneath the two ends of the electrophoresis chamber that are parallel to the wells of the gel are mixing manifolds for buffer recirculation. These manifolds have a square cross-section with internal dimensions of 3/4 × 1 inch (width × depth) and run the full 15-inch length of the chamber. Ports for buffer intake and outflow are centered along the 15-inch length of the manifolds. Each manifold is connected to the main electrophoresis chamber by a set of six holes drilled in the bottom of the chamber; the holes have diameters of 1/4 inch, are inset 5/8-inch from the edge of the chamber, and have center-to-center spacings of 2 1/4 inches. This arrangement gives relatively uniform flow of buffer, parallel to the direction of electrophoretic migration, across the whole width of the electrophoresis chamber. The electrophoresis chamber should also be equipped with some type of three-point leveling system. It is important for the apparatus to be level for two reasons: first, levelness favors uniform buffer flow and consequently uniform temperature, and second, it provides the uniform buffer depth that is essential for establishment of the correct electric field. Finally, the electrodes should be installed in the bottom of the box by drilling small holes at the electrode end points; these holes should be sealed with some removable glue, such as a silicone cement, since the electrodes may need periodic replacement (see Troubleshooting). For the electrodes, 26-gauge, 100% platinum wire (0.4 mm diameter) is recommended; the design specified in Fig. 1 requires about 20 inches of wire, including allowance for making connections to the leads.

The buffer loop involves a peristaltic pump capable of maintaining a flow of 250 ml/min (e.g., a Cole Parmer Masterflex T-7553-00 drive with a T-7019-21 head using tubing with an inside diameter of 5/16 inch) in series with a heat exchanger. A simple heat exchanger can be built from two concentric rolls of polyethylene tubing formed from 15-m sections of tubing with an outside diameter of 6.4 mm and an inside diameter of 4.3 mm; these rolls can be formed into 23 loops with diameters of approximately 20 cm, which are immersed in a passive water bath that simply serves to facilitate heat conduction between the electrophoresis buffer loop and the coolant loop. Low-density polyethylene is the preferred material for the heat-exchanger tubing because it has higher heat conductance than other common flexible polymers. Water is pumped through the coolant loop from a refrigerated water bath that is equipped for external circulation and has the capacity to serve as a steady-state heat sink for a load of at least 75 W (e.g., a Neslab Instruments Model RTE-9B).

The central component of the electrical circuitry is a double-pole, double-throw power relay (e.g., Magnecraft W199ABX-14). One pole of the relay switches the negative DC output from the power supply between the two negative electrodes, while the other pole plays the analogous role for the positive output. One set of electrodes (e.g., the A set in Fig. 1) is activated when line voltage is applied to the relay's coil, while the other set of electrodes (e.g., the B set) is activated when it is not. A programmable laboratory timer (e.g., Chrontrol Model CT-4) is used to control the periodic application of line voltage to the relay's coil.

Apparatus Safety

An apparatus of the general design described here poses some special safety problems that go beyond those associated with standard electrophoresis equipment. Although the voltages and current used in OFAGE experiments are lower than those associated with many DNA sequencing protocols, they are higher and substantially more dangerous than those normally employed when separating DNA fragments on agarose gels. There is the added hazard of recirculation of electrically hot buffer through a buffer loop that involves a number of plumbing connections. Failure of the tubing in the peristaltic pump head (see Troubleshooting) is only one of several ways that buffer spills can occur that pose a potential electrical shock hazard. It is recommended that the electrophoresis chamber, the pump, and all tubing connections be placed in an outer containment box with a safety interlock that prevents opening of the box while power is going either to the pump or to the gel. Although the containment box should be watertight it should not be airtight: there should be good venting at the top to prevent accumulations of the hydrogen gas that is generated during electrophoresis. Detailed plans for a containment box that is suitable for the apparatus described here are available from the authors.

Troubleshooting

Electrophoresis Conditions. OFAGE involves an element of "brinksmanship" in which the standard electrophoresis conditions border closely on disaster. It is probable that this situation is a direct reflection of the underlying fractionation principle, which we believe is largely based on the retardation of large DNA molecules by interactions with the gel matrix that are only barely reversible. Specifically, immediately after a switching event, when inappropriately aligned molecules are subjected to forces in the new field direction, it is likely that local segments of the molecules penetrate the gel matrix along many paths that are counterproductive with respect to overall molecular translocation. Under optimum conditions, this process gives rise to size-dependent retardation of large molecules and good fractionation. Under only slightly more extreme conditions, the same process causes irreversible entanglement of large molecules in the gel matrix.

The phenomenon of irreversible loss of the largest molecules is readily demonstrated by varying critical aspects of the electrophoresis conditions. For example, Fig. 4A shows an OFAGE separation of the yeast chromosomes carried out at $0.1\times$ TBE, rather than the conventional $0.5\times$ TBE; all other conditions are identical to the "broad size-range conditions" specified above. The photograph of the ethidium bromide-stained gel in Fig. 4A clearly illustrates the selective loss of the larger molecules (compare with Fig. 2A). The fates of these molecules were followed by carrying out Southern hybridization using probes for chromosomes V (band 5), XIII (band 10), and IV (band 12). The results, in Fig. 4B, confirm the impression given by the stained gel that the loss of bands associated with the larger chromosomes is due to a smearing of the corresponding sequences between the normal band positions and the well. At one extreme, chromosome V is only slightly affected, while at the other, chromosome IV barely enters the gel. Chromosome XIII, which is of intermediate size, shows intermediate behavior: a faint band is detectable in the normal position for band 10, but most of the hybridization signal is in a broad smear that extends all the way back to the well. We interpret these results as indicating that there is a size-dependent risk at each switching cycle of molecules becoming irreversibly entangled in the gel matrix.

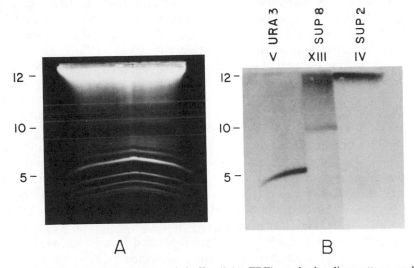

FIG. 4. Effect of low-ionic-strength buffer ($0.1\times$ TBE) on the banding pattern, under broad size-range conditions (50-sec switching). DNA from the gel in A was transferred to a single sheet of nitrocellulose. This sheet was then cut into three strips that were separately hybridized to three chromosome-specific probes (URA3 for chromosome V, SUP8 for chromosome XIII, and SUP2 for chromosome IV). The filter strips were then positioned in their original alignment before the autoradiogram in B was exposed. Similar results were obtained under high-voltage (400 V) or high-temperature (21°) conditions.

Under the particular low-salt electrophoresis conditions illustrated in Fig. 4, the risk of entanglement is small for chromosome V, intermediate for XIII, and so high for IV that none of the molecules migrates any significant distance.

Results indistinguishable from those shown in Fig. 4 can be obtained by increasing the voltage to 400 V or the temperature to 21°, while keeping all other broad size-range conditions unchanged. It is self-evident that a higher voltage will increase the likelihood of irreversible entanglement, while a full understanding of the effects of temperature and salt are less readily predictable from first principles. These variables affect the flexibility, the degree of condensation, and the extent of charge shielding of the DNA, all of which may be expected to influence the probability of irreversible entanglement. Predictions about the directions of the effects, however, must be made cautiously both because the matrix is also affected by changes in the temperature and ionic strength and because entanglement is a complex process that involves the relative ease of segmental penetration of and extrication from inappropriate paths.

From a troubleshooting standpoint, it is important to distinguish between the loss of large molecules due to electrophoretic effects such as those discussed above, and their loss during sample preparation. Samples in which the DNA has been mildly degraded, either due to shear or nucleolytic activity, also show the preferential loss of large molecules. The only reliable method of distinguishing between the two phenomena is by a Southern analysis, as was illustrated in Fig. 4A. Electrophoretic entanglement causes bands to smear behind their normal positions, while degradation causes them to smear ahead of them. In either case, the distinction can only be made reliably when the probed molecule is displaying intermediate behavior (i.e., some members of the population are forming the normal band, while other members are affected by degradation or entanglement). It should also be borne in mind that degraded or entangled molecules can be spread out over such large regions of the gel that they may only be detectable when a Southern experiment is carried out at high sensitivity; a low-sensitivity Southern experiment is likely to reveal only the molecules that are concentrated in a band, even if the intensity of the band has been seriously diminished by one or more of the above processes.

While we have frequently encountered the loss of large molecules due to irreversible entanglement in the gel matrix, we have seen no evidence for electrophoretic shearing, a phenomenon that would be expected to have effects similar to those of degradation. Loss of large molecules at high-voltage gradients has been observed by others and interpreted to be due to electrophoretic shearing.[11] Although we cannot exclude the possibility that electrophoretic shearing occurs under some experimental conditions, it has been our consistent experience that excessive voltage gradients, as well as high temperatures or low salt concentrations, cause bands to smear behind their normal positions as predicted for entanglement.

Apparatus. Long-term maintenance of an OFAGE apparatus requires attention to some details that are rarely problems with conventional electrophoresis equipment. Most of our difficulties have involved three components: the electrodes, the peristaltic-pump tubing, and the timer. With respect to the electrodes, we find that even 100% platinum dissolves slowly in the electrophoresis buffer under standard OFAGE conditions, and the electrodes may require replacement after approximately 1000 hr of use. Corrosion of the electrodes, as expected, is concentrated in regions where oxygen production is most vigorous at some stage in the switching cycle. Specifically, the ends of the "negative" electrodes that are nearest the wells are affected because they act as anodes during their passive cycle, and the ends of the "positive" electrodes that are farthest from the wells are affected because they are the sites of the most vigorous oxidative activity during their active cycle. Undoubtedly, one of the components of the running buffer—probably the Tris—facilitates platinum oxidation by forming a coordination complex with the liberated Pt(II). Presumably, the problem is observed in an OFAGE apparatus, but not in conventional electrophoresis chambers, because the electrolytic activity is concentrated in particular regions of the electrodes during OFAGE.

With respect to peristaltic-pump tubing, it is commonplace to encounter leaks in the tubing within the pump head when a pump is run continuously for long periods. We recommend Norprene Food-grade tubing (Barnant Company, Barrington, IL), which has already lasted 10 times longer, in our lab, than standard silicone tubing and is still performing well. An alternative solution to this problem is to use a nonperistaltic pump whose working parts are made from nonconducting materials (e.g., the L-1239 electrical isolation pump manufactured by Micropump Corp., Concord CA).

Finally, in our experience, typical microprocessor-controlled laboratory timers are somewhat unreliable. If irreproducible results are obtained, it is wise to monitor the switching cycle at regular intervals during an experiment. Particularly because OFAGE experiments are often shut down automatically at the end of a run, a gross malfunction of the timer earlier in the experiment can easily escape detection.

[11] L. H. T. Van der Ploeg, A. W. C. A. Cornelissen, P. A. M. Michels, and P. Borst, *Cell* **39**, 213 (1984).

Discussion

We conclude with a few comments about apparatus design and some likely future directions of large-DNA electrophoresis. The most perplexing issue in this area concerns the field geometry. As discussed above, the electrophoretic results are extremely sensitive to this variable. Changes of even a few millimeters in the positioning or lengths of the electrodes— particularly the short electrodes—have substantial effects. The resolution of the larger yeast chromosomes is much more sensitive to electric-field geometry than is the behavior of the smaller ones. It should be emphasized that the overall field geometry is a composite of several effects: the lengths and positioning of the active electrodes are obviously important, but so is the presence of the "passive electrodes," as well as the overall size and shape of the electrophoresis chamber. An important point about apparatus design is that the final field geometry in all current large-DNA electrophoresis apparatuses has been arrived at empirically, and the real question about any given apparatus is not how logical its design is, but how well it works. For example, the "distorting" effects of the passive electrodes are not inherently bad. Indeed, in our apparatus, they contribute to the overall field geometry, which is what has been empirically optimized; if the passive-electrode effects were eliminated, other compensating changes in the design would have to be made in order to achieve comparable electrophoretic results.

Nonetheless, we expect to see movement toward simpler field geometries, if only because the use of more uniform fields that intersect at constant angles would make the results less sensitive to the exact design of a given apparatus. We are convinced from our experience with a variety of experimental apparatuses that excellent results can be obtained using entirely uniform electric fields. What is less obvious, at present, is which of several possible methods of producing uniform fields will prove to be the best overall instrumental design.

A design for an OFAGE-like apparatus that produces uniform fields that intersect at 120° has recently been published.[12] This design can be implemented as an upgrade to the apparatus described here simply by replacing the electrodes and modifying the external electrical circuitry. Preliminary experience in our laboratory with an OFAGE apparatus that has been modified in this way has been highly favorable.

There is also the likelihood that many applications of large-DNA electrophoresis will be taken over by more radically different methods, such as the newly discovered technique of field-inversion gel electrophoresis

[12] G. Chu, D. Vollrath, and R. W. Davis, *Science* **234**, 1582 (1986).

(FIGE), which can be carried out using an ordinary electrophoresis chamber with only two electrodes.[13] Given the fluid state of current technology, researchers with specific applications of large-DNA electrophoresis will have to decide how to strike a balance between using the most advanced technology and simply getting the job done. OFAGE offers a well-tested experimental system whose idiosyncracies are now well known. For this reason alone, it offers an attractive solution to many experimental problems. In our own laboratory, for example, despite the availability of several different experimental apparatuses—each of which can outperform OFAGE in one or another respect—we still use a standard OFAGE apparatus for many purposes. Particularly when a well-calibrated overview of a broad size range is required, as for example when mapping new yeast genes or examining the electrophoretic karyotypes of new yeast strains, we have found it difficult to improve upon.

Acknowledgments

We would like to acknowledge A. Link for supplying one of the OFAGE photographs, and J. Pringle for providing a probe for CDC3. This work was supported in part by Grant GM28232 from the National Institutes of Health, and by a graduate research assistantship to G.C. from the Washington University Division of Biology and Biomedical Sciences.

[13] G. F. Carle, M. Frank, and M. V. Olson, *Science* **232,** 65 (1986).

[8] A Rapid Alkaline Extraction Method for the Isolation of Plasmid DNA

By H. C. BIRNBOIM

Plasmids are double-stranded circular DNA molecules that have the property of self-replication, independent of chromosomal DNA. In bacteria, they carry genes that may specify a variety of host properties. In recent years naturally occurring plasmids have been modified to produce new plasmids, which are used as cloning vehicles in recombinant DNA research. Although the presence of a plasmid in a bacterial cell may be detected genetically as a change in phenotype (e.g., resistance to a particular antibiotic), often it is necessary to isolate plasmid DNA for molecular studies, such as size determination, restriction enzyme mapping, and nucleotide sequencing, or for the construction of new hybrid plasmids. The degree of purification required will depend upon the intended use. Highly purified material can be prepared by the "cleared lysate" method, which involves a long period of centrifugation in a dye–CsCl gradient.[1] Less purified plasmid DNA is often satisfactory for recombinant DNA studies, and a large number of shorter and simpler methods have been developed (see Birnboim and Doly[2] and references therein). This chapter describes one such method that uses an alkaline extraction step. It is rapid enough to be used as a screening method, permitting 50–100 or more samples to be extracted in a few hours. The DNA is sufficiently pure to be digestible by restriction enzymes, an important advantage for screening. A preparative version that allows isolation of larger quantitites of more highly purified material is also described.

Principle

Isolation of plasmid DNA requires that it be separated from host-cell chromosomal DNA as well as other macromolecular components. Alkaline extraction exploits the covalently closed circular (CCC) nature of plasmid DNA and the very high molecular weight of chromosomal DNA. When a cell extract is exposed to conditions of alkaline pH in the range 12.0–12.6, linear (chromosomal) DNA will denature but CCC DNA will not.[3] pH adjustment is simplified by using glucose as a buffer. On neutral-

[1] D. B. Clewell and D. R. Helinski, *Proc. Natl. Acad. Sci. U.S.A.* **62**, 1159 (1969).

[2] H. C. Birnboim and J. Doly, *Nucleic Acids Res.* **7**, 1513 (1979).

[3] P. H. Pouwels, C. M. Knijnenburg, J. van Rotterdam, and J. A. Cohen, *J. Mol. Biol.* **32**, 169 (1968).

RECOMBINANT DNA
METHODOLOGY

izing the extract in the presence of a high concentration of salt, precipitation of chromosomal DNA occurs. We presume this is because interstrand reassociations occur at multiple sites owing to the very high molecular weight of the DNA, which then leads to the formation of an insoluble DNA network.[4] CCC DNA remains in the soluble fraction. The bulk of cellular RNA and protein will also precipitate under these conditions if protein is first complexed with an anionic detergent, sodium dodecyl sulfate (SDS). By combining reagents appropriately, precipitation of most of the chromosomal DNA, RNA, and protein can be accomplished in a single step.

Materials and Reagents

Bacterial Strains and Plasmids. For the experiments to be described here as an illustration of the method, *Escherichia coli* K12 Strain RR1 containing a dimeric form of plasmid pBR322 was used.[5] Plasmids containing *Eco*RI fragments of coliphage T4 in *E. coli* K12 strain K802 were from E. Young, University of Washington.

Equipment for the Screening Method

Bench-top centrifuge capable of generating r.c.f. of 8000–13,000 *g,* such as the Eppendorf Model 5412 microcentrifuge. This accommodates 12 polypropylene tubes (1.5 ml capacity).

Racks to support 60 centrifuge tubes; these can be constructed by drilling holes 11 mm in diameter in sheets of 2 mm-thick aluminum.

Pasteur pipette, drawn out to a fine tip and flame-polished is useful in aspirating supernatants after centrifugation.

Repetitive pipettor capable of delivering 0.1–0.2 ml volumes is also helpful when large numbers of samples are to be processed.

Equipment for the Preparative Method

Refrigerated preparative centrifuge such as a Sorval RC-5

Rotors to accommodate 50-ml and 250-ml bottles

Solutions

Lysozyme solution: glucose (50 m*M*), CDTA (10 m*M*), and Tris-HCl (25 m*M*) (pH 8.0); the solution can be kept for many weeks at room temperature. Lysozyme is added at a concentration of 1 mg/ml shortly before use and dissolved by gentle mixing. The chelating agent, cyclohexane diaminetetraacetic acid (CDTA), is available from Sigma or Aldrich Chemical Company and can be rendered

[4] R. J. Britten, D. E. Graham, and B. R. Neufeld, this series, Vol. 29, p. 363.
[5] F. Bolivar, R. L. Rodriguez, P. J. Greene, M. C. Betlach, H. L. Heyneker, and H. W. Boyer, *Gene* **2,** 95 (1977).

colorless if necessary by treatment with charcoal. EDTA can be substituted, but CDTA has the following advantages over EDTA: at a concentration of 1 mM or higher, it appears to inhibit growth of microorganisms, allowing solutions to be stored at room temperature; it chelates metal ions much more effectively; it is more soluble in alcoholic and acidic solutions.

Alkaline SDS. This solution contains 0.2 N NaOH, 1% SDS. It is used to lyse cells and denature chromosomal DNA. Because of the buffering capacity of glucose, which is added in the first solution, proper pH for denaturation can be obtained without use of a pH meter. The shelf life at room temperature is about a week or longer, depending upon the source of SDS; reagent grade sodium dodecyl sulfate from BDH Chemicals has been satisfactory.

High-salt solution: 3 M potassium acetate, 1.8 M formic acid. For 100 ml of solution, use 29.4 g of potassium acetate and 5 ml of 90% formic acid; store at room temperature. If turbid, the solution should be clarified by filtration. It is used to neutralize the alkali used in the previous step and provide conditions under which chromosomal DNA, RNA, and protein will precipitate. Formic acid was substituted for acetic acid, used in the earlier procedure,[2] to make this solution easier to prepare and easier to adjust to pH 8 in later steps. Potassium acetate was substituted for sodium acetate because it is slightly more effective in precipitating denatured chromosomal DNA and SDS–protein complexes. Potassium acetate has also been used by other workers.[6]

Acetate–MOPS: 0.1 M sodium acetate, 0.05 M MOPS, adjusted to pH 8.0 with NaOH. It is stable at room temperature if stored over a drop of chloroform. MOPS (morpholinopropanesulfonic acid) is from Sigma.

CDTA–Tris: 1 mM CDTA, 10 mM Tris-HCl (pH 7.5); stable indefinitely at room temperature

Additional Solutions Required for the Preparative Method

LiCl solution: 5 M LiCl, 0.05 M MOPS, adjusted to pH 8.0 with NaOH. Filter through 0.45-μm membrane filter if turbid.

Ribonucleases: Ribonuclease A stock solution, 1 mg/ml in 5 mM Tris-HCl (pH 8.0); ribonuclease T1 stock solution, 500 units/ml in 5 mM Tris-HCl (pH 8.0). Both are available from Sigma, and solutions are heated at 80° for 10 min after preparation to inactivate contaminating deoxyribonuclease, if present. Store at −20°.

[6] D. Ish-Horowicz and J. F. Burke, *Nucleic Acids Res.* **9,** 2989 (1981).

CG-50 Ion-exchange resin: Amberlite CG-50 (200–400 mesh), a carboxylic acid-type cation-exchange resin. It is prepared by washing with 1 N HCl, water, 1 N NaOH, water and is finally equilibrated with 10 mM MOPS, 1 mM CDTA at pH 8. Fines are removed during washing. It is stored as a 50% (v/v) slurry.

Screening Method

Clones for plasmid extraction are grown to saturation in small volumes (2–3 ml) of medium such as L broth in the presence of the appropriate antibotic. Alternatively, single colonies (about 4 mm in diameter) from an agar plate can be used. The number that can be screened is usually limited by the number of slots available in an agarose gel electrophoresis apparatus; extracting 60 samples at a time is convenient. All manipulations and centrifugations (in a microcentrifuge) are at room temperature unless otherwise indicated.

1. One-half milliliter of each culture is transferred to a 1.5-ml Eppendorf tube for extraction; the remainder is stored at $-20°$ for future use after addition of an equal volume of 80% (v/v) glycerol. If single colonies are to be used, they can be scraped with a sterile toothpick and suspended in 0.5 ml of water. Tubes are centrifuged for 15 sec. Longer times may make the pellet difficult to suspend. The supernatant is carefully removed using a fine-tip aspirator, and the cell pellet is loosened from the wall by vortexing two tubes together on a mixer while allowing their tips to clatter.

2. Each pellet is suspended thoroughly in 0.1 ml of lysozyme solution by vortexing immediately after each addition. The suspension is held at $0°$ for 5 min.

3. Alkaline SDS (0.2 ml) (at room temperature) is added. The sample is mixed gently by inversion several times and held at $0°$ for 5 min. The lysate should become almost clear initially, but will become cloudy on standing as SDS precipitates.

4. High-salt solution (0.15 ml) (at room temperature) is added. It is again mixed gently and held at $0°$ for 15 min. A curdlike precipitate will form. Centrifuge for 2 min.

5. Part of the supernatant (0.35 ml) is transferred into another tube, care being taken to avoid disturbing the bulky pellet. Cold ethanol (0.9 ml) is added to the sample, which is held at $-20°$ for 15 min, then centrifuged for 1 min. Each tube should be oriented in the centrifuge so that the position of the pellet (which may not be obvious) will be known in order that it not be disturbed as the supernatant is removed. The pellet is dis-

solved in 0.1 ml of acetate–MOPS and reprecipitated with 0.2 ml of ethanol. This washing step should be repeated once more if the DNA is to be digested with enzymes or used for transformation. The final pellet is suspended in 0.04 ml of water or CDTA–Tris. It is suitable for analysis by gel electrophoresis directly or after digestion by restriction enzymes, or can be used for transformation of other cells. One-fifth of each sample is usually sufficient to give a discernible band on a gel.

Comments on the Screening Procedure. The method has been used successfully with other *E. coli* strains, with other bacteria, and with very large plasmids. Some cells lyse adequately in alkaline SDS without prior lysozyme treatment. The times indicated at each step are approximate minimum times and appear not to be very critical. Gentle handling at steps 3 and 4 helps to eliminate chromosomal DNA (presumably by preserving its high molecular weight), but some loss of plasmid may occur. RNA present in samples prepared for screening can obscure short fragments of DNA, which may be released by restriction nuclease treatment. If this is anticipated, samples dissolved in CDTA–Tris can be incubated with ribonuclease A (50 μg/ml) for 10 min prior to addition of concentrated buffer for restriction-enzyme digestion. Note that ribonuclease can activate trace amounts of endonuclease I (if present) by eliminating tRNA, a powerful inhibitor of the enzyme.[7]

Preparative Method

The first steps of the preparative method are similar to those of the screening method, and the reagents are identical. Either a nonamplified or an amplified culture can be used. Advantages of the latter are that smaller volumes of reagents at the early steps are needed and the yield (per liter of culture) appears to be higher. pBR322 in *Escherichia coli* can be amplified as follows. Twenty milliliters of an overnight culture is used to inoculate 1 liter of L broth containing 100 μg of ampicillin per milliliter. When the optical density at 600 nm reaches 0.8–1.0, chloramphenicol (170 μg/ml) is added and incubation continued for 18–20 hr. Alternatively, a nonamplified, overnight culture can be used. The volumes of extraction solutions given below are for 1 liter of culture of amplified cells.

1. Cells are harvested by centrifugation in 250- or 500-ml bottles at 6000 g for 10 min at 0°. Higher speeds may make the pellets difficult to suspend. The pellets are brought up in about 50 ml of water and recentrifuged. The cell pellet is first suspended as well as possible in 1 ml of

[7] I. R. Lehman, G. G. Roussos, and E. A. Pratt, *J. Biol. Chem.* **237**, 819 (1962).

glucose–CDTA–Tris at 0°, then 9 ml of cold glucose–CDTA–Tris (containing 10 mg of lysozyme) is added. For nonamplified cells, four times the amount of glucose–CDTA–Tris and lysozyme are used. The cell suspension is kept at 0° for 30 min.

2. Twenty milliliters of alkaline SDS (at room temperature) is added. The mixture is stirred fairly gently with a glass rod until nearly homogeneous and clear. It is kept at 0° for 10 min, then 15 ml of high-salt solution is added. The mixture is stirred a little more vigorously than before for several minutes until a coarse white precipitate forms. After standing at 0° for 30 min, the precipitate is removed by centrifugation at 12,000 g for 10 min at 0°. For nonamplified cells, the volumes used are four times those indicated.

3. The supernatant is transferred into another tube, and 2 volumes of cold ethanol is added. The precipitate of nucleic acids that forms on standing at −20° for 20 min is collected by centrifugation and dissolved in 5 ml of acetate–MOPS (12 ml for nonamplified cells). Nucleic acids are again precipitated with 2 volumes of ethanol and dissolved in 2 ml of water for both amplified and nonamplified cells. Up to this stage, the procedure is very similar to that used for screening.

4. The volume of solution is measured, and an equal volume of LiCl solution is added. The sample is held at 0° for 15 min, and the heavy precipitate that forms is removed by centrifugation at 12,000 g for 10 min at 0°. The clear supernatant is heated at 60° for 10 min, and a small amount of additional precipitate that may form is removed by centrifugation as before. Plasmid DNA is precipitated by the addition of 2 volumes of cold ethanol to the supernatant solution. After holding for 15 min at −20°, the precipitate is collected by centrifugation and redissolved in 2.5 ml of acetate–MOPS. After another ethanol precipitation, plasmid DNA (with some contaminating low molecular weight RNA at this stage) is dissolved in 2 ml of CDTA–Tris.

5a. Contaminating RNA is removed by treatment with ribonuclease and precipitation with isopropanol as follows. Ribonucleases A and T1 are added to 10 μg and 5 units per milliliter, respectively. After incubation at 37° for 15 min, 0.04 ml of 10% SDS and 2 ml of acetate–MOPS are added. Four milliliters of isopropanol is added dropwise with mixing, and the precipitate of plasmid DNA that forms after 15 min at room temperature is collected by centrifugation at 20°. Acetate–MOPS (2 ml) is added and plasmid DNA (which may be difficult to dissolve after isopropanol precipitation) is reprecipitated with 2 volumes of ethanol. Plasmid DNA is dissolved in 2 ml of CDTA–Tris and can be stored either frozen or at 4° over a drop of chloroform.

5b. An alternative choice of procedures at this point, which avoids steps 5a and 6, involves binding and elution of plasmid DNA from glass powder.[8]

6. In *E. coli* strains containing endonuclease I (*endA*$^+$), traces of this enzyme may survive previous treatments. The enzyme is readily eliminated by binding to an ion-exchange resin as follows.[7] Plasmid DNA (2 ml) from the previous step is combined with 2 ml of CG-50 slurry and 0.08 ml of LiCl solution. The mixture is gently shaken at room temperature for 30 min, then centrifuged to recover the supernatant. The pellet is washed with 1 ml of acetate–MOPS. The supernatants are combined, and purified plasmid DNA is precipitated with ethanol and dissolved in 2 ml of CDTA–Tris.

Additional Purification Where Required. Some preparations of plasmid DNA contain what appears to be cell wall carbohydrate material (that is, it gives a positive reaction in a phenol–sulfuric acid test[9]). If present, it can be detected after step 5 by its faintly colloidal appearance or by the present of insoluble material. Much of it can be removed by vortexing the sample for 2 min with a few drops of chloroform and then centrifuging at 12,000 g for 15 min at 20°. Plasmid DNA remains in the aqueous phase. Traces of chromosomal DNA can be removed by repeating the alkali denaturing step,[8] by extraction with phenol under acidic conditions,[10] by centrifugation to equilibrium in ethidium bromide/CsCl gradients,[11] or by chromatography on acridine yellow.[12]

Characterization of Plasmid DNA by Agarose Gel Electrophoresis

Hybrid Plasmids Extracted by the Screening Method. Twenty-eight individual clones of hybrid plasmids containing fragments of coliphage T4 DNA inserted into pBR322 were prepared by alkaline extraction as an illustration of the method (Fig. 1). The results are typical of those that may be expected with the screening procedure. Bands of plasmid DNA are detectable in nearly every slot, although the intensity of the individual band varies. Thirty samples can be handled readily on a 20 cm-wide gel. Overall, the mobility of the CCC form of the plasmid should provide a good indication of the size of the inserted fragment.

[8] M. A. Marko, R. Chipperfield, and H. C. Birnboim, *Anal. Biochem.* **121**, 382 (1982).
[9] G. Ashwell, this series, Vol. 8, p. 85.
[10] M. Zasloff, G. D. Ginder, and G. Felsenfeld, *Nucleic Acids Res.* **5**, 1139 (1978).
[11] R. Radloff, W. Bauer, and J. Vinograd, *Proc. Natl. Acad. Sci. U.S.A.* **57**, 1514 (1967).
[12] W. S. Vincent, III and E. S. Goldstein, *Anal. Biochem.* **110**, 123 (1981).

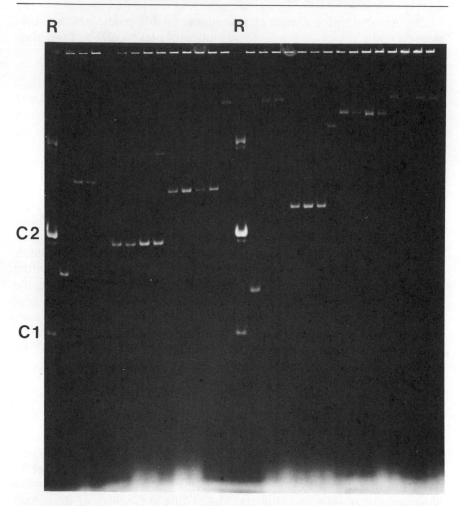

FIG. 1. Screening of coliphage T4–pBR322 recombinant plasmids by gel electrophoresis of alkali-extracted plasmid DNA. A total of 28 clones, representing 7 different hybrid plasmids, was extracted as described in the text, and 10 μl of each extract was applied to a vertical 0.8% agarose gel slab (0.3 × 20 × 20 cm); the electrophoresis buffer contained 40 mM Tris base, 20 mM sodium acetate, 2 mM EDTA, adjusted to pH 7.8 with acetic acid. Electrophoresis was carried out at room temperature for 15 hr with an applied voltage of 55 V. Gels were stained for 30 min at room temperature with ethidium bromide (1 μg/ml in water) and photographed through a Kodak No. 24 filter with 300 nm UV illumination [C. F. Brunk and L. Simpson, *Anal. Biochem.* **82,** 455 (1977)] using Polaroid type 665 pos/neg film. UV lamps were obtained from Fotodyne, Inc., New Berlin, Wisconsin. R is a reference mixture of the monomer and dimer form of pBR322 DNA. Other designations are described in the legend to Fig. 2.

Preparative Method. This version of the method is illustrated by preparing dimeric pBR322 plasmid DNA from a 1-liter culture of amplified cells. Samples were taken after steps 3, 4, and 5 of the purification procedure and are shown in slots A, B, and C, respectively, of Fig. 2. The principal differences between samples that can be seen on the gel are the elimination of intensely staining material (RNA) near the bottom of the gel and a small increase in the amount of open-circular form after ribonuclease treatment in slot C. After step 5, the preparation is virtually free of RNA fragments, as determined by Sephadex G-50 chromatography.[8] Very little contaminating chromosomal DNA can be seen on this agarose gel, but a little more can be expected if nonamplified cells were used instead as a source of plasmid DNA. Contaminating chromosomal DNA is seen more readily in some slots of Fig. 1 as material that remains at the origin or is distributed diffusely near the top of the gel.

Identification of Bands. Plasmid DNA in extracts can exist in a number of different forms that can make bands on a gel sometimes difficult to identify unambiguously. It is necessary to be aware that a plasmid may exist as a dimer or even higher multimer and that DNA can be present in the CCC form (with varying number of superhelical turns), the open-circular (nicked) form, or the linear form. Fortunately, the complexity is not usually a problem, since all these molecular species should give rise to identical fragments after restriction nuclease digestion. The principal band seen in Fig. 2 is the CCC dimer of pBR322. This can be deduced because a small amount of the CCC monomer can be seen, and because both a linear monomer form generated by complete *Hin*dIII digestion and a linear dimer form resulting from partial digestion appear (see Fig. 4). Other slowly migrating bands are presumed to be higher multimers. In addition to the molecular species discussed, a small amount of another form can be generated as a result of the alkaline extraction step. This is the "irreversibly denatured" form, which can be seen as a faint band running ahead of the CCC band in Fig. 2, slot A. It is removed at the next step of purification or on digestion with small amounts of nuclease S1, a single-strand specific nuclease (Fig. 3). The nicked form and a small amount of the linear form are also produced by S1, so this enzyme can be used to help in the identification of minor bands.

Digestion with Restriction Endonuclease. An important next step in the characterization of plasmids is often an examination of the fragments produced by digestion with different restriction enzymes. If a large number of samples is to be digested, such as in screening, susceptibility of plasmid DNA in the crude extract to restriction enzymes is an essential feature of a rapid extraction method. An example of the digestibility of pBR322 at different stages of purification is shown in Fig. 4. Slots A–C

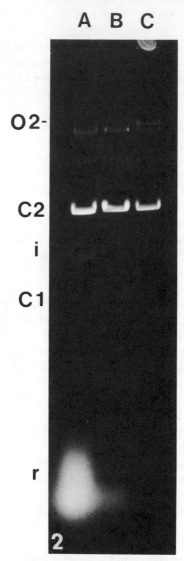

FIG. 2. Purification of dimeric pBR322 DNA by the preparative method. Samples (corresponding to 0.15 ml of culture) were taken after steps 3, 4, and 5 and applied to an agarose gel (slots A–C, respectively). Electrophoresis was carried out as in Fig. 1. The designation of bands in this and other figures is as follows: r, ribosomal and transfer RNA; i, irreversibly denatured form of CCC DNA; C, covalently closed circular form of plasmid DNA; O, open circular form; L, linear form; 1 and 2 refer to the monomeric and dimeric forms.

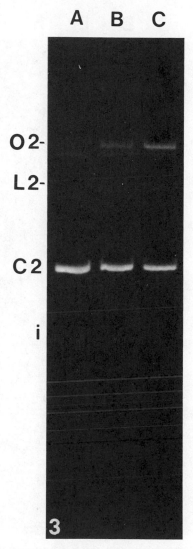

FIG. 3. Treatment of plasmid DNA with nuclease S1. A sample of pBR322 DNA taken after step 3 (see Fig. 2) was treated lightly with nuclease S1 to assist in identifying minor bands. Undigested DNA is in slot A; the effects of treatment for 10 min at 37° with 30 units/ ml (slot B) or 60 units/ml (slot C) of nuclease S1 are shown. The enzyme was from Miles Laboratories, and the buffer contained sodium acetate (50 mM), ZnCl$_2$ (1 mM), pH 4.5; the DNA concentration was approximately 20 μg/ml. On treatment with nuclease S1, the faint band of irreversibly denatured CCC DNA disappeared and the open-circular and linear forms were generated.

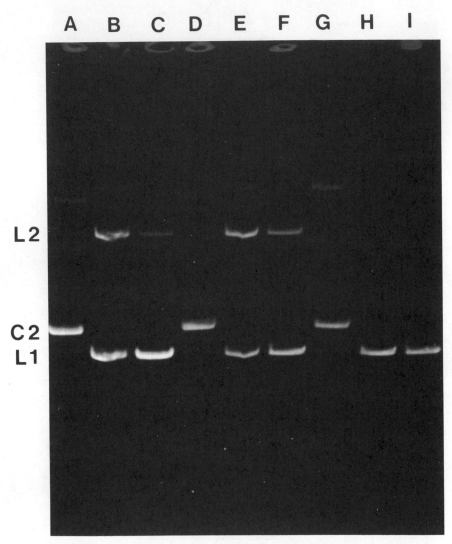

FIG. 4. Treatment of plasmid DNA with restriction nuclease HindIII. Undigested samples in slots A, D, and G correspond to samples in slots A, B and C, respectively, of Fig. 2. Each sample, taken at successively later stages in the purification procedure, was treated with either 0.35 unit (B, E, H), or 0.70 unit (C, F, I) of HindIII (from Boehringer) at 37° for 30 min, then analyzed by gel electrophoresis as in Fig. 1. Designation of bands is as in Fig. 2.

contain plasmid DNA that corresponds to the purity obtained in the screening method. The CCC-dimer form is quite readily cleaved to the linear monomer form. Not unexpectedly, highly purified plasmid DNA in slot G is more readily digested than the less pure samples in slots A and D.

Conclusions

Alkaline extraction has proved to be a useful method for the isolation of covalently closed circular DNA from bacterial cells. It is sufficiently simple and reliable to be useful for screening clones containing hybrid plasmids and also for the preparation of more highly purified plasmid DNA.

Acknowledgments

I thank Dr. J. D. Childs for helpful suggestions and discussion and for permitting the use of the experiment shown in Fig. 1, and Dr. E. Young for supplying plasmids of coliphage T4 DNA in pBR322.

[9] Specific Synthesis of DNA *in Vitro* via a Polymerase-Catalyzed Chain Reaction

By KARY B. MULLIS and FRED A. FALOONA

We have devised a method whereby a nucleic acid sequence can be exponentially amplified *in vitro*. The same method can be used to alter the amplified sequence or to append new sequence information to it. It is necessary that the ends of the sequence be known in sufficient detail that oligonucleotides can be synthesized which will hybridize to them, and that a small amount of the sequence be available to initiate the reaction. It is not necessary that the sequence to be synthesized enzymatically be present initially in a pure form; it can be a minor fraction of a complex mixture, such as a segment of a single-copy gene in whole human DNA. The sequence to be synthesized can be present initially as a discrete molecule or it can be part of a larger molecule. In either case, the product of the reaction will be a discrete dsDNA molecule with termini corresponding to the 5' ends of the oligomers employed.

Synthesis of a 110-bp fragment from a larger molecule via this procedure, which we have termed polymerase chain reaction, is depicted in Fig. 1. A source of DNA including the desired sequence is denatured in the presence of a large molar excess of two oligonucleotides and the four deoxyribonucleoside triphosphates. The oligonucleotides are complementary to different strands of the desired sequence and at relative positions along the sequence such that the DNA polymerase extension product of the one, when denatured, can serve as a template for the other, and vice versa. DNA polymerase is added and a reaction allowed to occur. The reaction products are denatured and the process is repeated until the desired amount of the 110-bp sequence bounded by the two oligonucleotides is obtained.

During the first and each subsequent reaction cycle extension of each oligonucleotide on the original template will produce one new ssDNA molecule of indefinite length. These "long products" will accumulate in a linear fashion, i.e., the amount present after any number of cycles will be linearly proportional to the number of cycles. The long products thus produced will act as templates for one or the other of the oligonucleotides during subsequent cycles and extension of these oligonucleotides by polymerase will produce molecules of a specific length, in this case, 110 bases long. These will also function as templates for one or the other of the oligonucleotides producing more 110-base molecules. Thus a chain reac-

```
|----------------------------- 110-bp -----------------------------------|

                                                          3'-PC04
                                               EXTENDS<---ccacttgcacctacttcaac
                                                          ||||||||||||||||||||
---------------------------ACACAACTGTGTTCACTAGC---------    ---    GGTGAACGTGGATGAAGTTG---------------------
Human betaglobin region
---------------------------TGTGTTGACACAAGTGATCG---------    ---    CCACTTGCACCTACTTCAAC-----------------
                           ||||||||||||||||||||
                           acacaactgtgttcactagc---->EXTENDS
                                     3'-PC03
                                                      | Polymerase +
                           CYCLE 1                    V dNTPs

---------------------------TGTGTTGACACAAGTGATCG---------    ---    ccacttgcacctacttcaac
                           ::::::::::::::::::::::::::::::||||||||||||||||||||||||||||||||||||
---------------------------ACACAACTGTGTTCACTAGC---------    ---    GGTGAACGTGGATGAAGTTG---------

---------------------------TGTGTTGACACAAGTGATCG---------    ---    CCACTTGCACCTACTTCAAC-----------------
                           ||||||||||||||||||||||||||||||    ||||||||||||||||||||
                           acacaactgtgttcactagc---------    ---    GGTGAACGTGGATGAAGTTG-----------------

                                                      | Denature,
                                                      V Anneal

---------------------------TGTGTTGACACAAGTGATCGT--------    ---    ccacttgcacctacttcaac
                           ||||||||||||||||||||
                           acacaactgtgttcactagc---->EXTENDS
                                     3'-PC03                          3'-PC04
                                                             EXTENDS<---ccacttgcacctacttcaac
                                                                        ||||||||||||||||||||
---------------------------ACACAACTGTGTTCACTAGC---------    ---    GGTGAACGTGGATGAAGTTG-----------

---------------------------TGTGTTGACACAAGTGATCG---------    ---    CCACTTGCACCTACTTCAAC-----------------
                           ||||||||||||||||||||
                           acacaactgtgttcactagc---->EXTENDS
                                     3'-PC03                          3'-PC04
                                                             EXTENDS<---ccacttgcacctacttcaac
                                                                        ||||||||||||||||||||
                           acacaactgtgttcactagc---------    ---    GGTGAACGTGGATGAAGTTG-----------------

                                                      | Polymerase +
                           CYCLE 2                    V dNTPs

---------------------------TGTGTTGACACAAGTGATCG---------    ---    ccacttgcacctacttcaac
                           ||||||||||||||||||||||||||||||    ||||||||||||||||||||||||||||||||||||||||||||
                           acacaactgtgttcactagc---------    ---    GGTGAACGTGGATGAAGTTG

---------------------------TGTGTTGACACAAGTGATCG---------    ---    ccacttgcacctacttcaac
-----------||||||||||||||||||||||||||||||||||||||    ---    |||||||||||||||||||||||||||||||||||||
---------------------------ACACAACTGTGTTCACTAGC---------    ---    GGTGAACGTGGATGAAGTTG--------------- ---

---------------------------TGTGTTGACACAAGTGATCG---------    ---    CCACTTGCACCTACTTCAAC------------- ---
                           ||||||||||||||||||||||||||||||    ||||||||||||||||||||||||||||||||||||||||||||
                           acacaactgtgttcactagc---------    ---    GGTGAACGTGGATGAAGTTG-----------------

                           TGTGTTGACACAAGTGATCG---------    ---    ccacttgcacctacttcaac
                           ||||||||||||||||||||||||||||||    ||||||||||||||||||||||||||||||||||||||||||||
                           acacaactgtgttcactagc---------    ---    GGTGAACGTGGATGAAGTTG--------------

                                                      | Denature,
                                                      V Anneal

---------------------------TGTGTTGACACAAGTGATCG---------    ---    ccacttgcacctacttcaac
                           ||||||||||||||||||||
                           acacaactgtgttcactagc---->EXTENDS                  3'-PC04
                                     3'-PC03                          EXTENDS<---ccacttgcacctacttcaac
                                                                                ||||||||||||||||||||
                           acacaactgtgttcactagc---------    ---    GGTGAACGTGGATGAAGTTG

---------------------------TGTGTTGACACAAGTGATCG---------    ---    ccacttgcacctacttcaac
                           ||||||||||||||||||||
                           acacaactgtgttcactagc---->EXTENDS                  3'-PC04
                                     3'-PC03                          EXTENDS<---ccacttgcacctacttcaac
                                                                                ||||||||||||||||||||
---------------------------ACACAACTGTGTTCACTAGC---------    ---    GGTGAACGTGGATGAAGTTG---------- ---

---------------------------TGTGTTGACACAAGTGATCG---------    ---    CCACTTGCACCTACTTCAAC-----------------
                           ||||||||||||||||||||
                           acacaactgtgttcactagc---->EXTENDS                  3'-PC04
                                     3'-PC03                          EXTENDS<---ccacttgcacctacttcaac
                                                                                ||||||||||||||||||||
                           acacaactgtgttcactagc---------    ---    GGTGAACGTGGATGAAGTTG-----------------

                           TGTGTTGACACAAGTGATCG---------    ---    ccacttgcacctacttcaac
                           ||||||||||||||||||||
                           acacaactgtgttcactagc---->EXTENDS                  3'-PC04
                                     3'-PC03                          EXTENDS<---ccacttgcacctacttcaac
                                                                                ||||||||||||||||||||
                           acacaactgtgttcactagc---------    ---    GGTGAACGTGGATGAAGTTG-----------------

                                                      | Polymerase +
                           CYCLE 3                    V dNTPs

---------------------------TGTGTTGACACAAGTGATCG---------    ---    ccacttgcacctacttcaac
                           ||||||||||||||||||||||||||||||    ||||||||||||||||||||||||||||||||||||||||||||
                           acacaactgtgttcactagc---------    ---    GGTGAACGTGGATGAAGTTG

                           TGTGTTGACACAAGTGATCG---------    ---    ccacttgcacctacttcaac
                           ||||||||||||||||||||||||||||||    ||||||||||||||||||||||||||||||||||||||||||||
                           ACACAACTGTGTTCACTAGC---------    ---    GGTGAACGTGGATGAAGTTG

---------------------------TGTGTTGACACAAGTGATCG---------    ---    ccacttgcacctacttcaac
                           ||||||||||||||||||||||||||||||    ||||||||||||||||||||||||||||||||||||||||||||
                           acacaactgtgttcactagc---------    ---    GGTGAACGTGGATGAAGTTG

---------------------------TGTGTTGACACAAGTGATCG---------    ---    ccacttgcacctacttcaac
-----------||||||||||||||||||||||||||||||||||||||    ---    |||||||||||||||||||||||||||||||||||||
---------------------------ACACAACTGTGTTCACTAGC---------    ---    GGTGAACGTGGATGAAGTTG--------------- ---

---------------------------TGTGTTGACACAAGTGATCG---------    ---    CCACTTGCACCTACTTCAAC----------------- ---
                           ||||||||||||||||||||||||||||||    ||||||||||||||||||||||||||||||||||||||||||||
                           acacaactgtgttcactagc---------    ---    GGTGAACGTGGATGAAGTTG-----------------

                           TGTGTTGACACAAGTGATCG---------    ---    ccacttgcacctacttcaac
                           ||||||||||||||||||||||||||||||    ||||||||||||||||||||||||||||||||||||||||||||
                           acacaactgtgttcactagc---------    ---    GGTGAACGTGGATGAAGTTG-----------------

                           TGTGTTGACACAAGTGATCG---------    ---    ccacttgcacctacttcaac
                           ||||||||||||||||||||||||||||||    ||||||||||||||||||||||||||||||||||||||||||||
                           acacaactgtgttcactagc---------    ---    GGTGAACGTGGATGAAGTTG-----------------

                           TGTGTTGACACAAGTGATCG---------    ---    ccacttgcacctacttcaac
                           ||||||||||||||||||||||||||||||    ||||||||||||||||||||||||||||||||||||||||||||
                           acacaactgtgttcactagc---------    ---    GGTGAACGTGGATGAAGTTG-----------------
```

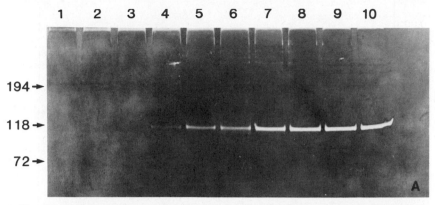

FIG. 2. (A) Reactions were performed as in Method I. DNA target was pBR328::βA, oligonucleotides were PC03 and PC04 at 10 μM, and dNTPs were labeled with α-^{32}P at 500 Ci/mol. After each synthesis cycle 10-μl aliquots were removed and these (lanes 1–10) were analyzed on a 14% polyacrylamide gel in 90 mM Tris–borate and 2.5 mM EDTA at pH 8.3 and 24 V/cm for 2.5 hr. The completed gel was soaked 20 min in the same buffer with the addition of 0.5 μg/ml ethidium bromide, washed with the original buffer, and photographed in UV light using a red filter. The numbers on the left margin indicate the sizes of DNA in base pairs. (B) The 110-bp fragment produced was excised from the gel under UV light and the incorporated ^{32}P counted by Cerenkov radiation. An attempt to fit the data to an equation of the form pmol/10 μl = $0.01[(1 + y)^N - yN - 1]$, where N represents the number of cycles and y the fractional yield per cycle, was optimal with $y = 0.619$. (C) The 8-μl aliquots from the tenth cycle of a reaction similar to the above were subjected to restriction analysis by addition of 1 μl BSA (25 mg/ml) and 1 μl of the appropriate enzyme (undiluted, as supplied by the manufacturer); reacted at 37° for 15 hr; PAGE was performed as above. (1) 1 μg ϕX174/HaeIII digest, (2) no enzyme, (3) 8 units HinfI, (4) 0.5 units MnlI, (5) 2 units MstII, (6) 3.5 units NcoI. The numbers on the left margin indicate the sizes (in base pairs) of DNA.

tion can be sustained which will result in the accumulation of a specific 110-bp dsDNA at an exponential rate relative to the number of cycles.

Figure 2 demonstrates the exponential growth of the 110-bp fragment beginning with 0.1 pmol of a plasmid template. After 10 cycles of polymerase chain reaction, the target sequence was amplified 100 times. The data have been fit to a simple exponential curve (Fig. 2B), which assumes that the fraction of template molecules successfully copied in each cycle remains constant over the 10 cycles. This is probably not true; however, the precision of the available data and our present level of sophistication in fully understanding the several factors involved do not seem to justify a more elaborate mathematical model. This analysis results in a calculated yield per cycle of about 62%. Amplification of this same 110-bp fragment

FIG. 1. The polymerase chain reaction amplification of a 110-bp fragment from the first exon of the human β-globin gene.

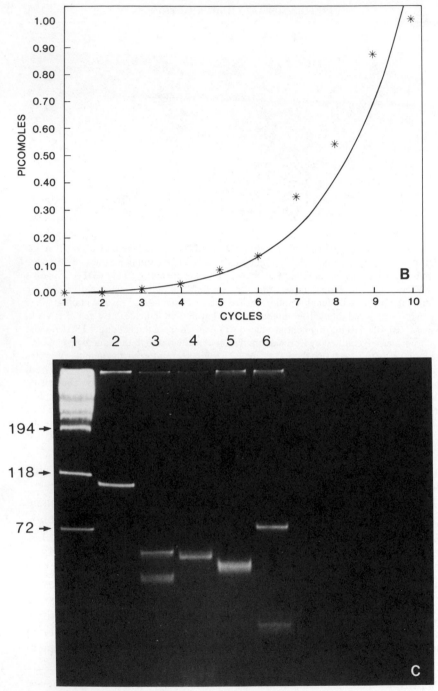

FIG. 2B and C. See legend on p. 191.

starting with 1 μg total human DNA (contains approximately 5 × 10⁻¹⁹ mol of the target sequence from a single-copy gene) produced a 200,000-fold increase of this fragment after 20 cycles. This corresponds to a calculated yield of 85% per cycle.[1] This yield is higher than that in the first example in which the target sequence is present at a higher concentration. It is likely that when the target DNA is present in high concentrations, rehybridization of the amplified fragments occurs more readily than their hybridization to primer molecules.

Materials and Methods

Oligonucleotides were synthesized using an automated DNA synthesis machine (Biosearch, Inc., San Rafael, California) using phosphoramidite chemistry. Synthesis and purification were performed according to the directions provided by the manufacturer.

Oligodeoxyribonucleotides	Designed to produce	From template
FF02 CGCATTAAAGCTTATCGATG	75 bp with FF03	pBR322
FF03 TAGGCGTATCACGAGGCCCT		
FF05 CTTCCCCATCGGTGATGTCG	500 bp with FF03	pBR322
FF05 CCAGCAAGACGTAGCCCAGC	1000 bp with FF03	pBR322
KM29 GGTTGGCCAATCTACTCCCAGG		
KM30 TAACCTTGATACCAACCTGCCC	240 bp with KM29	Globin DNA
KM38 TGGTCTCCTTAAACCTGTCTT	268 bp with KM29	Globin DNA
KM47 AATTAATACGACTCACTATAGGGAGA- TAGGCGTATCACGAGGCCCT	As FF03 plus 26 bp	pBR322
PC03 ACACAACTGTGTTCACTAGC		
PC04 CAACTTCATCCACGTTCACC	110 bp with PC03	Globin DNA
PC05 TTTGCTTCTGACACAACTGTGTTCACTAGC		
PC06 GCCTCACCACCAACTTCATCCACGTTCACC	130 bp with PC05	Globin DNA
PC07 CAGACACCATGGTGCACCTGACTCCTG		
PC08 CCCCACAGGGCAGTAACGGCAGACTTCTCC	58 bp with PC07	Globin DNA

Plasmid pBR328::BA, containing a 1.9-kb insert from the first exon of the human β-globin A allele, and pBR328 :: βS, representing the β-globin S allele, were kindly provided by R. Saiki.

Restriction enzymes were purchased from New England Biolabs, Beverly, Massachusetts. Klenow fragment of *Escherichia coli* DNA polymerase was purchased from United States Biochemical Corp., Cleveland,

[1] R. Saiki, S. Scharf, F. Faloona, K. Mullis, G. Horn, H. Erlich, and N. Arnheim, *Science* **230**, 1350 (1985).

Ohio, and was the product of a Klenow fragment clone rather than an enzymatic cleavage of DNA polymerase I.

Acrylamide was from Bio-Rad Laboratories, Richmond, California; deoxyribonucleoside triphosphates were from Sigma Chemical Co., St. Louis, Missouri.

NuSieve agarose was purchased from FMC Corporation. Gels were prepared by boiling the appropriate amount of agarose in 90 mM Tris–borate at pH 8.3, 2.5 mM in EDTA, and containing 0.5 μg/ml ethidium bromide. Poured into horizontal trays, the gels were ~0.5 cm thick, 10 cm long, and were run for 60–90 min at 10 V/cm submerged in the buffer described above. From 4 to 6% NuSieve agarose gels provide separations comparable to 10–15% polyacrylamide; they are considerably easier to cast and load and can be monitored while running with a hand-held UV light. Prior to photography, gels are soaked in water for 20 min to remove unbound ethidium bromide.

The following method is representative of a number of PCR protocols which have been successfully utilized. Specific variations on this procedure are noted in the figure legends and several are summarized below.

Polymerase Chain Reaction: Method I

Dissolve 0.1 pmol pBR322 (1 nM) and 300 pmol each of oligonucleotides FF02 and FF03 (3 μM) (see Diagram 1), and 150 nmol of each deoxynucleoside triphosphate (1.5 mM) in 100 μl 30 mM Tris–acetate (pH 7.9), 60 mM sodium acetate, 10 mM dithiothreitol, and 10 mM magnesium acetate. The solution is brought to 100° for 1 min, and is cooled to 25° for 30 sec in a waterbath. Add 1.0 μl containing 5 units of Klenow fragment of

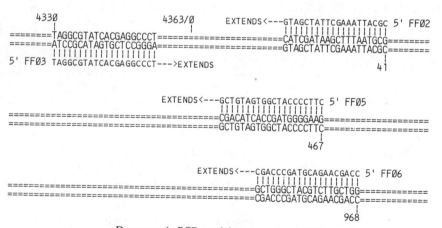

DIAGRAM 1. PCR model systems in pBR322.

E. coli DNA polymerase I and allow the reaction to proceed for 2 min at 25°, after which the cycle of heating, cooling, adding enzyme, and reacting is repeated nine times.

Method I (Summary of Above)

Target DNA: 0.1 pmol
Oligonucleotides: 3 μM, 20-mers
Buffer: 100 μl 30 mM Tris–acetate (pH 7.9) 60 mM sodium acetate, 10 mM Magnesium acetate, and 10 mM DTT
dNTPs: 1.5 mM
Enzyme: 5 units Klenow fragment
Cycles: Number: 10
        Denaturation: 100°, 1 min
        Primer hybridization: 25°, 30 sec
        Reaction: 25°, 2 min

Method II (Nested Primer Sets)

Target DNA: 10 μg human DNA (0.5×10^{-5} pmol)
Oligonucleotides: 2 μM, outer set: 20-mers; inner set: 27-mer and 30-mer
Buffer: 100 μl 30 mM Tris–acetate (pH 7.9), 60 mM sodium acetate, 10 mM magnesium acetate, and 10 mM DTT
dNTPs: 1.0 mM
Enzyme: 2 units Klenow fragment
Cycles: Following 20 cycles of amplification with the outer-set primers, a 10-μl aliquot of this reaction was diluted into a further 100-μl reaction mixture containing the inner-set primers and 10 more cycles were performed.
        Denaturation: 100°, 1 min
        Primer hybridization: 25°, 1 min
        Reaction: 25°, 2 min

Method III[1]

Target DNA: 1 μg to 20 ng human DNA (0.5×10^{-6} to 1×10^{-8} pmol)
Oligonucleotides: 1 μM, 20-mers
Buffer: 100 μl 10 mM Tris–chloride (pH 7.5), 50 mM sodium acetate, and 10 mM magnesium chloride
dNTPs: 1.5 mM
Enzyme: 1 unit Klenow fragment
Cycles: Number: 20–25
        Denaturation: 95°, 5 min, first cycle

95°, 2 min, subsequent cycles
Primer hybridization: 30°, 2 min
Reaction: 30°, 2 min

Method IV[2]

Target DNA: 1 µg human DNA (0.5×10^{-6} pmol)
Oligonucleotides: 1 µM, 20–28-mers
Buffer: 100 µl 30 mM Tris–acetate (pH 7.9), 60 mM sodium acetate, 10 mM Magnesium acetate
dNTPs: 1.5 mM
Enzyme: 1 unit Klenow fragment
Cycles: Number: 20
       Denaturation: 95°, 2 min
       Primer hybridization: 37°, 2 min
       Reaction: 37°, 2 min

Method V[2]

As Method IV except
Buffer: 10% DMSO added to Method IV buffer
Cycles: Number: 27

Method VI[3]

Target DNA: 5 ng human DNA containing target + 250 ng human DNA deleted for target, or 1 µg human DNA containing an unknown amount of HTLV-III viral DNA sequence
Oligonucleotides: 1 µM, 15–18-mers
Buffer: 100 µl 10 mM Tris–chloride (pH 7.5), 50 mM sodium chloride, and 10 mM magnesium chloride
dNTPs: 1.5 mM
Enzyme: 1 unit Klenow fragment
Cycles: Number: 20–25
       Denaturation: 95°, 2 min
       Primer hybridization: 25°, 2 min
       Reaction: 25°, 2 min

Specificity of the Amplification Reaction

This process has been employed to amplify DNA segments from 24 to 1000 bp in length using template DNA ranging in purity from a highly

[2] S. Scharf, G. Horn, and H. Erlich, submitted for publication.
[3] S. Kwok, D. Mack, K. Mullis, B. Poiesz, G. Ehrlich, D. Blair, A. Friedman-Kien, and J. J. Sninsky, submitted for publication.

purified synthetic single-stranded DNA to a totally unpurified single-copy gene in whole human DNA. Despite the low stringency of the hybridizations the specificity of the overall reaction is intrinsically high, probably due to the requirement that two separate and coordinated priming events occur at each cycle. Beginning with purified plasmid DNA as initial template and pairs of primers intended to produce fragments in the range of 200 bp or less, homogeneous products have usually been observed. Using similar templates, but primers chosen to amplify larger fragments, longer reaction times are required and considerable production of DNA fragments other than that intended is observed (Fig. 3). These by-products are usually smaller than the intended product and can be accounted for by "mispriming" events wherein the 3' end of one of the primers interacts with a region of partial homology within the sequence of the primary product (see Diagram 2). The probability for synthesis of a by-product representing a subfragment of the primary product is higher than the probability for synthesis of a by-product representing some different sequence in the original reaction for two reasons. First, the concentration of the primary product becomes relatively high during the reaction; and second, any single "mispriming" on a molecule of primary product will result in the production of a new molecule, which like the primary product will contain two primer sites. (A primer "site" in this context would be either a region complementary to one of the primers or a region containing one of the primers, which would in successive cycles produce a sequence complementary to it.) The synthesis of multiple DNA fragments is thus more likely if the intended fragment is large and the final desired concentration of the product is high. The ~225-bp by-product of the amplification of a 500-bp fragment from pBR322 depicted in Fig. 3B can be ac-

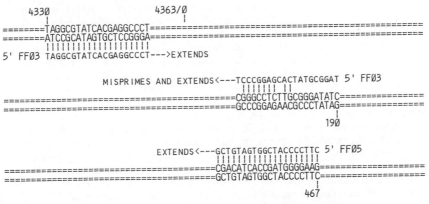

DIAGRAM 2. Probable second priming site on pBR322 for FF03.

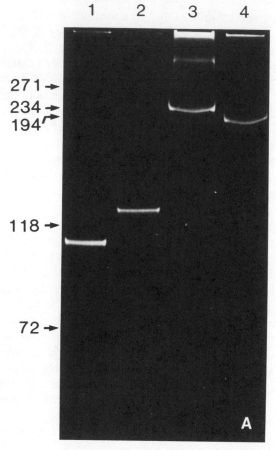

FIG. 3. (A) Reactions were performed as in Method I. DNA target was pBR328 :: βA, oligonucleotides were (1) PC03 and PC04, (2) PC05 and PC06, (3) KM29 and KM38 (reaction time was 20 min), (4) KM29 and KM30; DNA target was pBR328 :: βS digested with *Mst*II prior to the reaction. This plasmid is cut several times by *Mst*II but not within the sequence to be amplified by KM29 and KM30. A similar reaction with pBR328 :: βA which is cut within the target sequence yields no amplified product. The numbers on the left margin indicate the sizes (in base pairs) of DNA. (B) Reactions were performed as in Method I, except reaction times were 20 min per cycle at 37°. Oligonucleotides were FF03 and FF05. Final product was rehybridized for 15 hr at 57°. Electrophoresis was on a 4% NuSieve agarose gel. The numbers on the left margin indicate the sizes (in base pairs) of DNA. (C) (1) Reactions were performed as in Method I. Oligonucleotides were FF02 and FF03. The tenth reaction cycle was terminated by freezing and an 8-μl aliquot was applied to a 4% NuSieve agarose gel visualized with ethidium bromide. (2) Reactions were the same as in (1) except that the oligonucleotides used were FF02 and KM47, which were designed to produce a 101-bp fragment, 26 nucleotides of which are not present in pBR322. The numbers on the left margin indicate the sizes (in base pairs) of DNA. (D) (1) Reactions were performed as in (B). Oligonucleotides were FF03 and FF06. (2) Same as (1) except that KM47 was substituted for FF03. The numbers on the left margin indicate the sizes (in base pairs) of DNA.

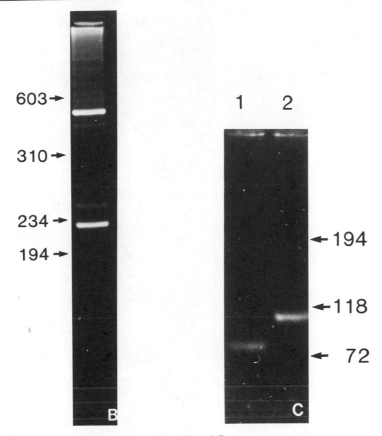

FIG. 3B and C.

counted for by a second priming site for FF03 in which 9 out of 11 of the 3′ nucleotides of FF03 find a match within the amplified product.

In Vitro Mutations

"Mispriming" can be usefully employed to make intentional *in vitro* mutations or to add sequence information to one or both ends of a given sequence. A primer which is not a perfect match to the template sequence but which is nonetheless able to hybridize sufficiently to be enzymatically extended will produce a product which contains the sequence of the primer rather than the corresponding sequence of the original template. When this product in a subsequent cycle is template for the second primer the extension product produced will be a perfect match to the first primer

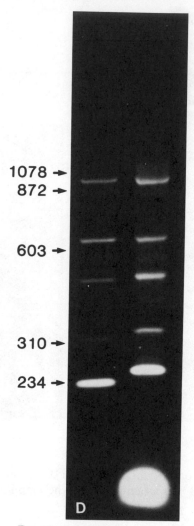

1078 →
872 →

603 →

310 →

234 →

D

FIG. 3D. See legend on p. 198.

and an *in vitro* mutation will have been introduced. In further cycles this mutation will be amplified with an undiminished efficiency since no further mispaired primings are required.

A primer which carries a noncomplementary extension on its 5′ end can be used to insert a new sequence in the product adjacent to the template sequence being copied. In Fig. 3C, lane 2, a 26-bp T7 phage promoter has been appended to a 75-bp sequence from pBR322 by using an oligonucleotide with 20 complementary bases and a 26-base 5' extension. The procedure required less than 2 hr and produced 2 pmol of the relatively pure 101-bp fragment from 100 fmol of pBR322. Similarly in Fig. 3D, the T7 promoter has been inserted adjacent to a 1000-bp fragment from pBR322.

Scharf *et al.*,[2] in order to facilitate the cloning of human genomic fragments, inserted restriction sites onto the ends of amplified sequences by the use of primers appropriately mismatched on their 5' ends.

Detection of Minute Quantities of DNA

A microgram of human DNA contains 5×10^{-19} moles of each single-copy sequence. This is ~300,000 molecules. Detection of single-copy sequences in whole human DNA or other similarly complex mixtures of nucleic acids presents a problem which has only been successfully approached using labeled hybridization probes.

Saiki *et al.*,[1] by combining a PCR amplification with a labeled hybridization probe technique, have significantly reduced the time and uncertainty involved in determining the sequence of a single base pair change in the human genome from only a microgram of DNA. They performed a 20-cycle amplification, which required less than 2 hr, and achieved a 200,000-fold increase in the level of a 110-bp sequence in the first exon of the β-globin gene. Once amplified the sequence was relatively simple to analyze.

We attempted to amplify the same 110-bp fragment to a slightly higher level so as to enable visual detection via ethidium bromide staining of a gel. For fragments in this size range, 100 fmol gives rise to a clearly visible band, thus, 0.1 aliquot of a 200,000-fold amplification of 10 μg of human DNA should be sufficient. And so it is; however, control experiments with DNA from a cell line harboring a β-globin deletion indicated that the 110-bp fragment produced was not exclusively representative of the β globin locus. That is, fragments of ~110 bp were being amplified even though no β-globin sequences were present. On the chance that whatever was causing this "background" might not share extensive homology with β-globin in the central 60 nucleotides of this 110-bp region, we attempted to increase the specificity of the process by introducing a second stage of amplification using a second set of primers nested within the first (see Diagram 3). By requiring four separate priming events to take place, we were thus able to amplify approximately 2,000,000-fold and readily detect a β-globin-specific product (Fig. 4).

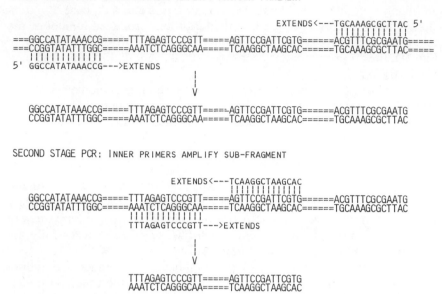

FIRST STAGE PCR: OUTER PRIMERS AMPLIFY PRIMARY FRAGMENT

DIAGRAM 3. Nested primer sites, which enable a second layer of specificity. (The sequences here are only examples and have no particular significance.)

The wild-type β-globin allele can be distinguished from the sickle-type allele by the presence of a site for the restriction enzyme *Dde*I. Thus, *Dde*I treatment of the DNA prior to amplification, or of the amplified product subsequent to amplification, will serve to distinguish between these two allelles.

Scharf *et al.*,[2] beginning with 1 μg of human DNA and oligonucleotides 26 and 28 nucleotides in length that were designed to amplify a 240-bp region of the *HLA-DQ-α* gene after 27 cycles of PCR, were able to visualize the predicted fragment via ethidium staining of an agarose gel. In contrast to our results with β-globin, controls with *HLA*-deleted cell lines revealed that this single-stage amplification was specific for the intended target.

Similar amplifications of other human loci have resulted in varying degrees of specificity and efficiency. No simple explanations for this variability, based on, for example, oligomer size, target size, sequence, and temperature, have been forthcoming; however, the number of examples of attempted amplifications of different human sequences is still small.

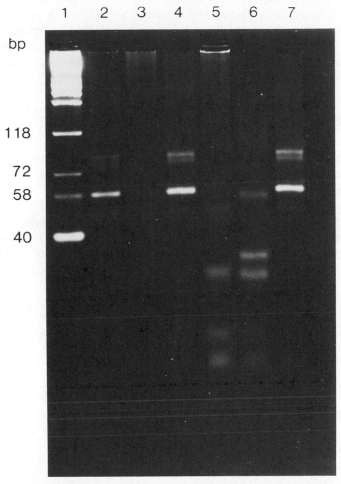

FIG. 4. Reactions were performed as in Method II, and 8-µl aliquots (representing 80 ng of unamplified DNA) were subjected to electrophoresis on a 4% NuSieve agarose gel stained with ethidium bromide. Oligonucleotides were PC03 and PC04, followed by PC07 and PC08 (the nested set). DNA target was as follows: lane (2), human DNA homozygous for the wild-type β-globin allele; lane (3), as in (2) but treated prior to amplification with *Dde*I, which cleaves the intended target and prevents amplification; lane (4), human DNA homozygous for the sickle β-globin allele treated prior to amplification with *Dde*I, which for this allele does not cleave the intended target; lane (5), salmon sperm DNA. Following the final amplification an aliquot of the reaction in (2) was subjected to cleavage with *Dde*I, which should convert the 58-bp wild-type product into 27- and 31-bp fragments [lane (6)]; an aliquot of the reaction in (4) was similarly treated with *Dde*I after amplification [lane (7)]. The 58-bp product from the sickle allele, as expected, contains no *Dde*I site. The numbers on the left margin indicate the sizes (in base pairs) of DNA.

Kwok *et al.*[3] have demonstrated that DNA sequences present at less than one copy per human genome can be successfully amplified and detected. Using an isotopic detection system they were able to identify β-globin sequences in as little as 5 ng of human DNa and have demonstrated sequences of HTLV-III in cell lines derived from patients affected with AIDS.

The polymerase chain reaction has thus found immediate use in developmental DNA diagnostic procedures[1,3] and in molecular cloning from genomic DNA[2]; it should be useful wherever increased amounts and relative purification of a particular nucleic acid sequence would be advantageous, or when alterations or additions to the ends of a sequence are required.

We are exploring the possibility of utilizing a heat-stable DNA polymerase so as to avoid the need for addition of new enzyme after each cycle of thermal denaturation; in addition, it is anticipated that increasing the temperature at which the priming and polymerization reactions take place will have a beneficial effect on the specificity of the amplification.

Acknowledgments

We wish to acknowledge the interest and support of Thomas White, and we would like to thank Corey Levenson, Lauri Goda, and Dragan Spasic for preparation of oligonucleotides; Randy Saiki, Stephen Scharf, Glenn Horn, Henry Erlich, Norman Arnheim, and Ed Sheldon for useful discussions regarding the amplification of human sequences; and Denise Ramirez for assistance with the manuscript.

Section III

Vectors or Methods for Gene Cloning

[10] *Escherichia coli* Plasmids Packageable *in Vitro* in λ Bacteriophage Particles

By JOHN COLLINS

It has recently been shown that *in vitro* λ bacteriophage packaging systems can be used for the efficient production of hybrid bacteriophage from DNA prepared by fusion of a bacteriophage vector and foreign DNA fragments *in vitro*.[1,2] The *in vitro* packaging system appears to be insensitive to the DNA built on adjacent to the small λ region required for packaging. This is illustrated by the fact that most of the λ DNA, including all regions required for λ replication and lysogeny, can be replaced by plasmid DNA and efficient packaging will still occur. The packaging is then dependent on the entire molecule being of a certain minimum size and on the presence of concatemeric forms induced by *in vitro* restriction endonuclease cleavage and ligation at high DNA concentrations[3,4] (Fig. 1).

The conditions required for efficient packaging of plasmids carrying the λ *cos* site are described below, with particular emphasis on conditions leading to the selection of hybrids carrying large pieces of foreign DNA subsequent to *in vitro* recombination. The *in vitro* packaging system is based on that developed by Hohn[1,3–5] with some modifications, but it is likely that the plasmid vectors described for use with this system (cosmids)[3,4] can also function well with other λ *in vitro* packaging systems, as long as the criteria for preparation of a good DNA substrate are observed. A selection of 12 plasmid vectors that can be used for cloning with over nine restriction endonucleases that generate cohesive ends is described. Vectors of different molecular weight are also included, since they may be used for selective cloning of foreign DNA fragments in a particular size range.

Plasmids usable with this λ packaging technique are referred to below as cosmids, since the λ region necessary for recognition by the λ packaging system is called the λ *cos* site. As demonstrated below, the particular advantage of the system over the λ vectors is the small size of the vector, which consequently allows packaging of large foreign DNA frag-

[1] H. Hohn and K. Murray, *Proc. Natl. Acad. Sci. U.S.A.* **74,** 3259 (1977).

[2] N. Sternberg, D. Tiemeier, and L. Enquist, *Gene* **1,** 255 (1977).

[3] J. Collins and B. Hohn, *Proc. Natl. Acad. Sci. U.S.A.* **75,** 4242 (1978).

[4] J. Collins and H. J. Brüning, *Gene* **4,** 85 (1978).

[5] B. Hohn, this series, Vol. 68, p. 299.

RECOMBINANT DNA
METHODOLOGY

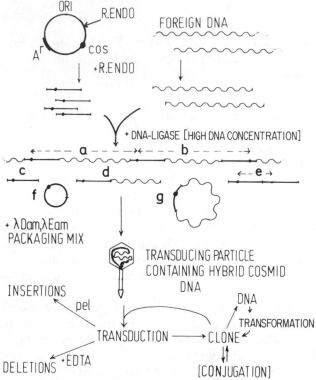

FIG. 1. A scheme illustrating a method of gene cloning with plasmids containing a selectable marker (e.g., antibiotic resistance, A^r), the cohered λ cohesive ends (cos), a replication origin (ORI), and a site (↓) for a restriction endonuclease (R. ENDO). A plasmid having such properties is called a cosmid. The cosmid is cleaved to linear form with a restriction endonuclease. The foreign DNA to be cloned is cleaved partially with an enzyme generating the same cohesive ends (fragments in the 16–30 Md size range). The DNAs are mixed and ligated. The use of high DNA concentrations during this ligation increases the formation of molecules of types a, b, and e. The DNA forms most likely to be packaged and subsequently lead to formation of a transductant are a and e, where the distance between the *cos* sites is about 30 Md. Packaging of the hybrid DNA takes place by cleavage of two consecutive, similarly oriented *cos* sites and the subsequent encapsulation of the intervening DNA in the λ particle. For small vectors the likelihood of producing transductants of the vector alone (from molecules e and f or higher polymers) is low. The packaging takes place in an *in vitro* packaging mix,[5] and the transducing particles formed are used to inject λ-sensitive *su*$^-$ *E. coli*. The hybrid plasmid is injected, circularizes, and replicates as a normal plasmid without expression of any λ function. The isolated clones can now be further manipulated genetically (1) by repackaging the hybrid cosmids by superinfection with helper phages (efficiency low); (2) by conjugation, when the plasmid is mobilizable, by introduction of a sex factor; or (3) by transformation of purified supercoiled DNA. The use of repackaging could be used, presumably, in conjunction with standard methods used in genetics to select for insertions or deletions on the basis of the effect of DNA size (1) on particle stability, enhanced by EDTA to select for deletions, or (2) on penetration during transduction enhanced on infection of *E. coli pel*$^-$ hosts to select for insertions. Reproduced by permission of Elsevier North Holland Biomedical Press (Collins and Brüning[4]).

ments. The main advantages of the cosmid system over plasmid cloning vectors are (1) the high efficiency of hybrid clone formation; (2) the fact that essentially only hybrid plasmids are formed, thereby obviating the need for screening or selecting for hybrid clones or pretreating vector DNA to prevent vector ring closure; and (3) the fact that the system selects particularly for large fragments in contrast to plasmid transformation systems which give a bias toward the cloning of smaller fragments.

The development of new vectors can be rapid, since the regions required for the principal properties of the cosmid (replication origin, selective marker, single restriction site, and *cos* site) are small, and vectors may be easily developed to meet specific cloning requirements.

Principle of the Method

For DNA to be efficiently transported into the *Escherichia coli* recipient cell through *in vitro* packaging, two main criteria can be distinguished:

1. Packaging of the DNA into the λ particle, which is dependent on ATP and on the presence of suitable recognition sites distributed along the DNA to be packaged (see below).

2. Injection of the packaged DNA into the recipient cell, which is dependent on the size of the packaged DNA, being efficient in the approximate range 23–32 Md.[3,4,6]

DNA that can be used for packaging may have one of the following forms[7]: (1) linear molecules having λ cohesive ends (m and m'), or (2) concatemeric molecules with the fused (cohered) cohesive end site (*cos* site consisting of m plus m') dispersed along the concatemer.

Molecules apparently not used either *in vivo*[8-13] or *in vitro*[14-16] as a efficient substrate for packaging are circular molecules with a single *cos* site. Suitable concatemeric substrates may be produced from them by recombination *in vitro*[17] or *in vivo*[11,18] but at a low frequency. The exact extent of the region in the vicinity of the *cos* site required for packaging is

[6] M. Feiss, R. A. Fisher, M. A. Crayton, and C. Egner, *Virology* **77**, 281 (1977).
[7] T. Hohn and I. Katsura, *Curr. Top. Microbiol. Immunol.* **78**, 69 (1977).
[8] J. Szpirer and P. Brachet, *Mol. Gen. Genet.* **108**, 78 (1970).
[9] F. W. Stahl, K. D. McMillan, M. M. Stahl, R. E. Malone, Y. Nozu, and V. E. A. Russo, *J. Mol. Biol.* **68**, 57 (1972).
[10] L. W. Enquist and A. Skalka, *J. Mol. Biol.* **75**, 185 (1973).
[11] M. Feiss and T. Margulies, *Mol. Gen. Genet.* **127**, 285 (1973).
[12] D. Freifelder, L. Chud, and E. E. Levine, *J. Mol. Biol.* **83**, 503 (1974).
[13] P. Dawson, A. Skalka, and L. D. Simon, *J. Mol. Biol.* **93**, 167 (1975).
[14] B. Hohn, B. Klein, M. Wurtz, A. Lustig, and T. Hohn, *J. Supramol. Struct.* **2**, 302 (1974).
[15] B. Hohn, *J. Mol. Biol.* **98**, 1975).
[16] M. Syvanen, *J. Mol. Biol.* **91**, 165 (1975).
[17] M. Syvanen, *Proc. Natl. Acad. Sci. U.S.A.* **71**, 2496 (1974).
[18] K. Umene, K. Shimada, and Y. Takagi, *Mol. Gen. Genet.* **159**, 39 (1978).

not well defined, but about 178 base pairs spanning this site have been se-quenced[19] and show unusual palindromic regions occurring up to 60 base pairs away from the axis of the cohesive ends.

A suitable substrate for packaging is derived by the ligation of restriction endonuclease-cleaved foreign and vector DNA at high DNA concentrations (Fig. 1). The number of possible assemblies of vector and/or foreign DNA is large (molecules a–g in Fig. 1). From this random ligation molecules of forms a and e will be cleaved out at the *cos* sites and packaged *in vitro* into λ heads. This having occurred, the assembly of the bacteriophage tail and foot takes place to give a complete particle able to inject the encapsulated cosmid or cosmid hybrid DNA into an appropriate recipient *Escherichia coli*. This injection of the packaged DNA only occurs with reasonable efficiency if the DNA is in the size range 23–31 Md, i.e., a large hybrid or a multimer of the vector. This leads to automatic production of clones containing mostly large hybrid plasmids after subsequent selection for the antibiotic resistance carried on the plasmid vector. A further reduction in the percentage of nonhybrid vector (multimeric form) can be obtained by using large vectors (16–20 Md), small vectors (4–7 Md), or temperature-sensitive replication mutants with a low copy number (Table I).

On injection, the cosmid hybrid DNA, which is in a linear form, having λ cohesive ends (m and m'), circularizes by reforming the *cos* site (m plus m'). DNA replication now continues under control of the plasmid replicon. Since no bacteriophage functions are expressed, no bacteriophage particles can be formed without further infection with a λ helper phage.

The *in vitro* packaging system described here[14] consists of a mixture of induced cultures of λ *Dam Sam* and λ *Eam Sam* lysogens, with suppressor minus backgrounds, which are only able to complement each other in λ prehead formation *in vitro* on lysis of the cells. Lysis is induced by warming up the cells. During the packaging of exogenous (cosmid hybrid) DNA the DNA of the bacteriophage used to produce the packaging mix is also packaged to some extent. For this reason the adsorption and injection of the packaged DNA must be carried out in a suppressor minus *E. coli* host. If the λ DNA present in the packaging mix is inactivated by uv irradiation,[5] a suppressor plus recipient strain can be used. In addition, to prevent degradation of foreign DNA cloned into the hybrids, a recipient should be used that has a defective host restriction system.

[19] B. P. Nichols and J. E. Donelson, *J. Virol.* **26**, 429 (1978).

TABLE I
Properties of Cosmid Cloning Vectors

Vector cosmid	720	703ΔBglII	703ΔHindIII	703ΔPstI	720ΔBglII-3	720ΔBglII-2	720ΔBglII-2 ΔPstI	74	75-65	75-58	76	77	78	79
MW (Megadalton)	16	7.2	11.2	7.6	13.3	10	7.2	10.5	8.2	7.6	4.0	6.3	6.8	4
Selection	rif	E1	E1	E1	rif	rif	E1	Ap	Ap	Ap	Ap	Ap Tc	pro^+	Ap Tc
EcoRI G^AATTC				+				+	+	+	+	+	+	+
BamHI G^GATCC								+	+	+			+	−
BglII A^GATCT		+		+	+	+	+	+	+			+		
SalI G^TCGAC		+		+		+	+				+	−	+	+
PstI C TGCA^G				+		+						+	−	+
ClaI A T^CGAT												+		
XmaI C^CCGGG	+	+	+		+	+								
HindIII A^AGCTT	+			+							+	−	+	−
KpnI GGTAC^C						+								
Copy N°	15	15	15	15	15	15	15	15	5	5	5	30	5	30
Temp. sensit. (ts)									ts	ts	ts		ts	
mob	+	+	+	+	+	+	+	+	+	−	−	−	−	−

[a] Cosmid vectors of the designated molecular weights can be selected for with rifampicin (Rif), ampicillin (Ap), tetracycline (Tc), or colicin E1 (E1). Where a single cutting site for a restriction endonuclease exists this is indicated by a plus sign under the selection still available and a minus sign under the resistance inactivated by insertion at that site. The KpnI site in pJC720 Δ BglII-2 destroys rifampicin resistance, leaving only E1 immunity for selection. The plasmid copy number in LB broth at 30° is indicated below the list of restriction enzymes and their cut sites on the DNA. Temperature-sensitive plasmid replication leads to loss of the plasmid at 42° and a lower copy number at 30°; both properties help reduce the background of nonhybrid clones in cloning experiments. Mobilizability (Mob^+ or mob^-) is indicated in the bottom row and refers to loss of plasmid functions required for plasmid mobilization in sex factor-promoted conjugation (but not to loss of the plasmid conjugational transfer origin). Restriction endonuclease maps of these plasmids are given in Figs. 3 and 4. pJC77, a hybrid of SauI fragments from pJC75-58 and pBR322[29] is still being tested for its properties as a cloning vector. pJC78 is a vector designed to have exceptionally good containment qualities. In addition to the containment qualities detailed for pJC75-58 it has no homology with the E. coli chromosome or known R factors. The proB and C alleles it carries are derived from Methylomonas[30] pHC79 is made from a pBR322 SauI-induced rearrangement, which has a single BglII site (pJC80, 2.95 Md) located between the end of the tetracycline resistance gene and the origin of replication. A 1.1 Md BglII fragment carrying the cos site from λ Charon 3A was inserted at this BglII site.[31] The small cos fragment is easily reisolatable from pHC79, and this should facilitate the construction of further cosmid derivatives. Reproduced by permission of Elsevier North Holland Biomedical Press (Collins and Brüning[4]).

Materials and Reagents

Spermidine hydrochloride (Serva); putrescine (Merck); adenosine 5'-triphosphate (Boehringer); BamHI (Boehringer); HindIII (Boehringer); DNA ligase (Boehringer); BglII (gift from W. Rüger); SauI (gift from Streeck); SalI, EcoRI, PstI, and ClaI (gifts from H. Mayer and H. Schütte).

Preparation of DNA

Plasmid vector DNA was prepared from 1 liter L Broth (1% Bacto-tryptone, 0.5% yeast extract, 0.5% NaCl) cultures grown at 30° either overnight in stationary phase or after a 12-hr treatment with chloramphenicol (150 μg/ml added at $OD_{550} = 1.0$). This latter method of plasmid amplification in the absence of protein synthesis is possible since all vectors described here are derived from ColE1. Cleared lysates were prepared after the method of Katz et al.[20] using a solution of lysozyme, EDTA, and Triton X-100. The plasmid DNA was then further purified by CsCl–ethidium bromide centrifugation (Ti60 Beckman rotor) [20-ml gradients at 36,000 rpm for 36 hr at 15° followed by recentrifugation of the pooled lower (plasmid supercoil) band in a Ti50 rotor at 36,000 rpm for 36 hr at 15°]. After extraction of the ethidium bromide with three washes of iso-propanol at 0° and dialysis against three changes of 10 mM Tris, pH 8.0, and 0.1 mM EDTA buffer the plasmid DNA was concentrated by precipitation with 2 volumes ethanol in the presence of 100 mM NaCl at $-20°$ for 10 hr. All DNA preparations were stored at concentrations of 200–1000 μg/ml in sterile 10 mM Tris, pH 8.0, and 0.1 mM EDTA at $-20°$. Repeated freezing and thawing were avoided by keeping small aliquots.

High-molecular-weight chromosomal DNAs are prepared from cells or tissues by published procedures shown to give molecular weights in excess of 30 Md. During purification RNase and CsCl ultracentrifugation are included. DNA preparations are kept unfrozen in 10 mM Tris, pH 8.0, and 0.1 mM EDTA.

Restriction Endonucleases and DNA Cleavage

The endonucleases listed in Table II,[21] which generate cohesive ends, may be used for cloning with the vectors described.[22]

[20] L. Katz, D. T. Kingsbury, and D. R. Helinski, J. Bacteriol. **114,** 577 (1973).
[21] Recognition sites: R. Roberts, personal communication; Crit. Rev. Biochem. **4,** 123 (1976).
[22] H. Mayer and G. Hobom, personal communication.

TABLE II
ENDONUCLEASES USED FOR CLONING[a]

Vector cleaved with:	Foreign DNA cleaved with:
HindIII (A ↓ AGCTT)	HindIII
EcoRI (G ↓ AATTC)	EcoRI (EcoRI*)
SalI (G ↓ TCGAC)	SalI, XhoI (C ↓ TCGAG), AvaI (C ↓ PyCGPuG)
BamHI (G ↓ GATCC) ⎱	⎰ BamHI, SauI (↓ GATC)
BglII (A ↓ GATCT) ⎰	⎱ BclI (T ↓ GATCA), BglII (A ↓ GATCT)
PstI (CTGCA ↓ G)	PstI
XmaI (C ↓ CCGGG)[b]	XmaI
ClaI[22] (AT ↓ CGAT)[c]	ClaI, TaqI (T ↓ CGA), HpaII (C ↓ CGG)
KpnI (GGTAC ↓ C)	KpnI

[a] Restriction endonuclease digestions were carried out in the following buffer: 10 mM MgCl$_2$, 10 mM NaCl, 10 mM Tris-HCl, pH 7.8, with the exception of KpnI for which 6 mM MgCl$_2$, 6 mM NaCl, 6 mM Tris-HCl, pH 7.5, 10 mM dithiothreitol, 100 μg/ml BSA was used.

[b] XmaI is included only on a theoretical basis, since it is currently considered highly unstable, is not commercially available, and is appreciably contaminated with nonspecific endonucleases. Any plasmid carrying a single site for any restriction endonuclease not cutting within the λ cos site or within the genes used for selection or for replication can be used as a packageable cloning vector (cosmid, Figs. 1 and 2, Table I). It is not known if this cloning technique works with ligation of flush-molecules as are derived from HincII or SmaI cleavage for example. In view of the high degree of ligation required at high DNA concentrations it is not to be expected that this would be particularly efficient.

[c] The enzyme ClaI recently purified from a Caryophon species (H. Mayer and H. Schütte, personal communication) has been shown to have the recognition sequence ATCGAT and to produce the same two base pair sticky ends as TaqI (T ↓ CGA) and HpaII (C ↓ CGG) (H. Mayer, personal communication). The particular advantage of using "multicut" enzymes, with only 4-bp recognition sites, for the production of gene banks is discussed later.

Methods

The details of the λ packaging method used are described by Hohn,[5] and the application of the packaging system to a specific cloning problem using λ vectors is described elsewhere.[23]

The description of the packaging system as applied to cosmids is therefore confined to the vectors so far available and to variables that have an effect on the efficiency of hybrid formation.

[23] R. Lenhard-Schuller, B. Hohn, C. Brack, M. Hirama, and S. P. Tonegawa, Proc. Natl. Acad. Sci. U.S.A. 75, 4709 (1978).

Bacteria and Plasmids

Preparation of the plasmids pJC720 and pJC703 has been described,[24,25] as has the derivation of their deletions.[4] Preparation of the ampicillin-resistant pJC74 by cloning *Pst*I fragments from R1*drd*19 into pJC703 *Pst*I (Figs. 2 and 3) has been described.[4] Temperature-sensitive derivatives of pJC74 are shown in Fig. 3, where their formation is also outlined.[4] The formation of pJC76 is described in the legend for Fig. 4.[26] pJC77, mentioned in Table I, has not been fully mapped but is one of a number of derivatives from a pBR322[27,28] in which *Sau*I fragments from pJC75-58 have been inserted into a *Sau*I site of pBR322.

CC309 ColE1+ (*thr leu thy*) was used for colicin E1 production and carries a chromosomal mutation leading to temperature inducibility of colicin. JC411 R1*drd*16, ColE2+ (*his pro arg met*) was obtained from D. Helinski. HB101 (*pro leu r⁻ m⁻*) from B. Hohn was used as a recipient for cosmid transduction after packaging.

Properties of Cosmid Cloning Vectors

A summary of the main properties of a number of cosmids is shown in Table I.[4,29-31]

The molecular weight of the plasmid determines the maximum amount of foreign DNA that can be cloned (approximately a 31-Md cosmid), the average size of the DNA cloned (approximately a 26-Md cosmid), and the smallest size of the DNA that can be cloned (approximately a 19-Md cosmid). Hybrids as small as 10 Md have been obtained but with a low frequency, and it is not clear whether or not they were segregated by deletion from larger plasmids.

The Selection for Cosmid Hybrids. Rifampicin selection must be carried out with some care in view of the low rifampicin resistance (30 μg/ml) of strains carrying these vectors. First, a relatively long expression time ($1\frac{1}{2}$–2 hr) is needed after the transduction-adsorption step, in LB broth (1% Bacto-tryptone, 0.5% Difco yeast extract, 0.5% NaCl), before plating. The plates must not be wet and should be incubated *in the dark* for 48 hr before counting. Plates were made using 30 μg/ml rifam-

---

[24] J. Collins, N. P. Fiil, P. Jørgensen, and J. D. Friesen, *Control Ribosome Synth., Proc. Alfred Benzon Symp., 9th Munksgaard, Copenhagen 1976* p. 356 (1977).

[25] J. Collins, *Curr. Top. Microbiol. Immunol.* **78,** 121 (1978).

[26] J. Collins, in preparation.

[27] F. Bolivar and K. Backman, this series, Vol. 68, p. 245.

[28] F. Bolivar, R. L. Rodriguez, M. C. Betlach, and H. W. Boyer, *Proc. Natl. Acad. Sci. U.S.A.* **74,** 5265 (1977).

[29] F. Bolivar, R. L. Rodriguez, P. J. Greene, M. C. Betlach, H. L. Heyneker, H. W. Boyer, J. H. Crosa, and S. Falkow, *Gene* **2,** 95 (1977).

[30] J. Collins and H. Berger, in preparation.

[31] B. Hohn and J. Collins, submitted to *Proc. Natl. Acad. Sci.*

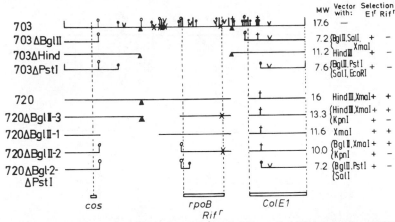

FIG. 2. Restriction maps of deletions of the ColE1 rpoB (rif^R) plasmids pJC703 and pJC720 usable as cosmids.[4] The vectors of the indicated molecular weight were derived by direct deletion from pJC703 or pJC720 with the indicated enzymes. The selection available (E1 immunity or rifampicin resistance) in conjunction with particular restriction enzymes is shown at the right. Reproduced by permission of Elsevier North Holland Biomedical Press (Collins and Brüning[4]).

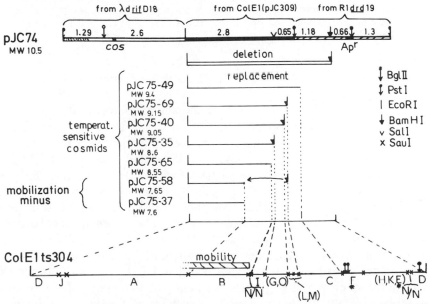

FIG. 3. The derivation of temperature-sensitive derivatives of pJC74, in which the EcoRI-BamHI fragment of pJC74 was replaced by EcoRI-SauI fragments from the temperature-sensitive ColE1 ts304.[4] The location of conjugational mobilization functions and the SauI fragments of ColE1 ts304 are shown below. SauI fragment P (20 bp) has not been placed. Fragments E and F are also indistinguishable. The dotted region to the left of the cos site is a region of E. coli chromosomal DNA from λ dRifD18. Reproduced by permission of Elsevier North Holland Biomedical Press (Collins and Brüning[4]).

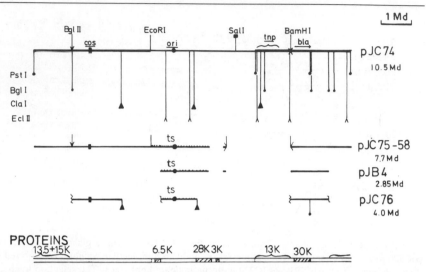

FIG. 4. Restriction map of three temperature-sensitive derivatives of pJC74. pJB4 was obtained as a spontaneous deletion from pJC75-58 during experiments to cause deletions with *Bgl*I. pJC76 was obtained by introducing a *Bgl*II-*Cla*I fragment from pJC74 into pJB4 cut with *Bam*HI and *Cla*I. The signs   and   indicate the ends joined to each other. The lower part of the diagram shown the positions of seven proteins coded for by pJC74. pJB4 makes only the 30,000-molecular-weight β-lactamase (Ap^R).[26] pJB4 and pJC76 have lost ColE1 immunity.

picin, 1.5% agar, 0.5% NaCl, 0.5% Difco yeast extract, and 1% Bacto-tryptone.

Selection for E1-immune colonies is made as follows. First, 0.1 ml of an overnight LB broth culture of CC309, a temperature-sensitive *E. coli* ColE1⁺ in which colicin synthesis is induced at a high temperature (lethal at 42°), is spread on LB broth plates and incubated overnight at 37°. The surface growth is scraped off with a glass rod, and the plates inverted over 3 drops of chloroform (in a watch glass) for ½ hr on the bench, so as to sterilize the plates. These plates are then used for the colicin selection. Not more than 10⁶ cells are added per plate. The plates are incubated at 42° so as to reduce the background of E1^R mutants occurring frequently as a result of lipopolysaccharide overproduction. Colonies are picked off after 36 hr. Retesting for sensitivity to colicin E2 at 42° is used later to establish the identity of the colonies as *E. coli* and to distinguish E1 immunity (plasmid-coded) from colicin E resistance (E1 plus E2 resistance, usually a chromosomal mutation or contaminant), if required. Sensitivity to colicin E2 is tested on simple cross-streak plates on which an *E. coli* ColE2⁺ strain has been grown, the culture scraped off, and the plate chloroformed.

Selection for ampicillin resistance (50 μg/ml) and tetracycline resistance (20 μg/ml) was carried out on LB broth plates after a 30-min (Ap^R) or a 60-min expression time (Tc^R).

Copy Number. For pJC720, and the deletions of this plasmid and pJC703, copy numbers have been estimated roughly to be similar to that of ColE1 (15 copies per cell). For the temperature-sensitive plasmids the copy number has been measured more accurately by subsequent segregation kinetics at 42°.[4] Copy numbers are 5.3 at 30°, 4.5 at 34°, and 2.1 at 37° after 12 generations at these temperatures. Tetramers and trimers of pJC75-58 have been isolated and the copy numbers estimated as 1.4 and 1.8, respectively, after 12 generations at 30°; however, the populations contain 55% (trimer) and 75% (tetramer) ampicillin-sensitive cells which have lost the plasmid. This leads to formation by the polymeric plasmids of small colonies on plates after 48 hr. This instability of vector multimers leads to a reduction in the nonhybrid background in cloning (see below).

Mobilizability. All the cosmids described here are derived from ColE1 which can be mobilized during conjugation promoted by a sex factor. pJC75-58, pJC76, and pJC77 have deletions in the region for a protein required for mobilization.[4,28] pJC75-58 and pJC76 probably still have the site at which mobilization starts and could perhaps still be complemented for transfer by a double infection with ColD and a sex factor.[32] The sexual properties of the plasmids are important in the context of the discussion of their use as EKII (German B2) safety vectors.

At present no recombinant DNA experiments may be carried out with sexually competent strains. Authorization to use cosmids of the pJC75-58, pJC76, or pJC77 type as safety vectors will largely depend on the production of a safe E. coli λ-adsorbing "incapacitated" host, such as those currently being developed in Curtiss's laboratory.[33]

DNA Preparation. Plasmid DNA was prepared by CsCl–ethidium bromide centrifugation of cleared lysates of strain *E. coli* C1 (*ser$^-$*) carrying the required plasmid vector. Starting with 5-liter cultures, two rounds of CsCl–ethidium bromide gradient ultracentrifugation were used to concentrate the DNA to about 500–1000 μg/ml. The DNA was finally dialyzed in 10 mM Tris-HCl and 0.1 mM EDTA, pH 7.0. The *E. coli* chromosomal DNA to be cloned was prepared by lysozyme–sodium diodecyl sulfate lysis of the cells, protease K and RNase treatment, and phenol–chloroform extraction. The detailed method varies according to the source of the DNA. It should be noted that the following criteria are important: The DNA must be of very high molecular weight, i.e., 30–50 Md or larger, and free of nucleases, and the DNA concentration

[32] G. Dougan, M. Saul, G. Warren, and D. Sherratt, *Mol. Gen. Genet.* **158**, 325 (1978).
[33] R. Curtiss, III, *Annu. Rev. Microbiol.* **30**, 507 (1976).

should be accurately known. If the DNA is relatively free of RNA and protein, the uv absorption at 260 nm (A_{260}) can be used (1 mg/ml DNA = A_{260} 20). DNA from λ b515 b519 cI857 Sam7 was used as a control DNA for the efficiency of the packaging system using a E. coli su recipient for absorption and plaque titration.[5] Efficiencies of 10^7-10^8 plaque-forming units (PFU) per microgram of λ DNA should be obtained before using the packaging mix for cosmid packaging.

DNA Restriction and Ligation

Partial digests of the chromosomal DNA were made by serial dilution of the restriction enzyme (HindIII or BglII) and incubation for 30 min at 37°. The reactions were stopped by an incubation for 10 min at 70°, and the vector DNA was digested to completion. To check the molecular weight of DNA accurately in this high-molecular-weight range 0.4% agarose gels were used in an electrophoresis apparatus (H. Hözel Technik) in which no hydrostatic pressure was exerted on the gel (electrophoresis buffer—0.09 M Tris, 2.5 mM EDTA, 0.09 M boric acid, pH 8.4). Slab gels 3 mm thick, 13 cm wide, and 10 cm deep were run at 180 V for 3 hr. The linear vector and λ DNA were used for molecular weight markers.

The DNAs were mixed to give the indicated (Table III) ratios of vector to chromosomal DNA, over a ninefold range, at a final concentration of 300 μg/ml. If required, the DNAs were concentrated by precipitation in 70% ethanol, 100 mM NaCl, 0.1 mM EDTA, and 10 mM Tris, pH 7.8, for 4 hr at −20° and resuspension in 10 mM Tris-HCl, pH 7.8, and 0.1 mM EDTA after desiccation of the pellet. Resuspension of the DNA at these high DNA concentrations requires at least a ½-hour shaking, being easier if carried out with digested DNA.

The final mix for ligation (20–50 μl) was adjusted to 10 mM $MgCl_2$ (no more than 10 mM NaCl) and 10 mM Tris-HCl, pH 7.5, heated to 70° for 5 min, cooled slowly at room temperature, and placed on ice for 30 min. Dithiothreitol (10 mM), bovine serum albumin (BSA, 100 μg/ml, filter-sterilized), 100 μM ATP, and 0.2 unit of T4 DNA ligase (Boehringer) were added per 5 μg DNA. Ligation was continued for at least 8 hr at 8°. The DNA can be left in this state for several days and then frozen at −20° for several months without affecting the yields of clones subsequent to packaging.

Packaging. The in vitro packaging mix used was essentially that developed by Hohn,[5] in which uv irradiation of the packaging mix and addition of DNase-containing mix to the packaged mix after the initial packaging step were omitted. The packaging mix contained heat-induced E. coli N205 (λ imm434 cIts b2 red3 Eam4 Sam7) and N205 (λ imm434 cI75ts b2

TABLE III

Exp.	Av. M.W. E.coli DNA	Vector E.coli	DNA conc. in ligat.			Colonies		Prototrophs		
			Vector	E.coli	Total	per 2µl DNA	per µg E.coli DNA	pro+	leu+	
			µg/ml							
1	7·10^6	3	225	75	300	$6.0 \cdot 10^4$	$4.0 \cdot 10^5$	11/4150(0.26%)	13/4150(0.31%)	pJC75-58 × BglII
2		1	150	150	300	$1.1 \cdot 10^4$	$3.6 \cdot 10^4$	3/1841 (0.16%)	10/1841 (0.54%)	
3		0.33	75	225	300	$2.4 \cdot 10^2$	$5.2 \cdot 10^2$	–	–	
4	20·10^6	3	225	75	300	$8.6 \cdot 10^3$	$5.8 \cdot 10^4$	0/1260(<.08%)	8/1260(0.62%)	
5		1	150	150	300	$5.8 \cdot 10^3$	$1.9 \cdot 10^4$	0/ 961(<.1%)	2/ 961 (0.21%)	
6		0.33	75	225	300	$2.3 \cdot 10^2$	$5.1 \cdot 10^2$	–	–	
7	–	–	225	–	225	$4.8 \cdot 10^2$	–	0/1008(<.1%)	0/1008(<.1%)	
8	–	–	75	–	75	$2.4 \cdot 10^1$	–	–	–	
9	10·10^6	3	225	75	300	$8.7 \cdot 10^3$	$5.8 \cdot 10^4$	13/5000(0.26%)	0/5000(<.02%)	pJC720 × HindIII
10		1	150	150	300	$1.3 \cdot 10^3$	$4.3 \cdot 10^3$	–	–	
11		0.33	75	225	300	$3.0 \cdot 10^2$	$6.6 \cdot 10^2$	–	–	
12	30·10^6	3	225	75	300	$1.1 \cdot 10^4$	$7.1 \cdot 10^4$	7/5000(0.14%)	1/5000(0.02%)	
13		1	150	150	300	$1.5 \cdot 10^3$	$4.8 \cdot 10^3$	–	–	
14		0.33	75	225	300	$3.5 \cdot 10^2$	$7.7 \cdot 10^2$	–	–	
15	–	–	225	–	225	$4.0 \cdot 10^3$	–	0/1200(<.08%)	0/1200(<.08%)	
16	–	–	150	–	150	$1.8 \cdot 10^3$	–	–	–	
17	–	–	75	–	75	$8.1 \cdot 10^2$	–	–	–	

[a] Comparison of packaging efficiencies with cosmids pJC75-58 (7.65 Md) and pJC720 (16 Md) as a function of the average molecular weight of the foreign DNA added to the ligation mix and as a function of the ratio of vector to foreign DNA. pJC75-58 and pJC720 were cut to completion with BglII and HindIII, respectively. The foreign DNA was cut partially with the appropriate enzyme to the average molecular weight indicated in the second column, as estimated by 0.4% agarose gel electrophoresis. The ratio of vector to foreign (E. coli) DNA is given in the second column. As indicated in columns 4–6, the DNA concentration during ligation was kept constant at 300 µg/ml, except for the controls where vector DNA was ligated alone. Two-microliter aliquots were packaged and adsorbed to HB101. The number of cosmid-containing colonies, ampicillin-resistant (pJC75-58, 30°) or 30 µg/ml rifampicin-resistant (pJC720, 37°), is indicated in column 7, and the number of clones obtained per microgram of foreign DNA is shown in column 8. The last two columns indicate the frequency of pro+ and leu+ prototrophs among the cosmid "transductants," as tested by replica-plating the indicated number of antibiotic-resistant colonies on minimal medium. Analysis of plasmids carried by some of these hybrids indicated that the pro+ allele was carried on a 16-Md BglII fragment and the leu+ allele on a 18.5-Md BglII fragment, and that the leu+ allele had at least two HindIII sites in or close to the gene.[4] Analysis of the size of the plasmid in small, cleared lysates of isolated colonies (about 30 from each experiment) indicated that in experiment 1, 90%; experiment 2, 80%; experiment 4, 80%; experiment 5, 70%; experiment 9, 80%; and experiment 12, 70% of the plasmids were larger than the vector plasmids (average molecular weight about 26 Md). Supercoiled DNA from 5000 colonies from experiments 1 and 4 were further analyzed, as shown in Fig. 5.

red3 Dam15Sam7) in the following buffer: 40 mM Tris-HCl, pH 8.0, 10 mM spermidine hydrochloride, 10 mM putrescine hydrochloride, 0.1% mercaptoethanol, and 7% dimethyl sulfoxide. This mix had been distributed in 20-μl portions in 1.5-ml capped plastic centrifuge tubes (Eppendorf), frozen in liquid nitrogen, and stored up to 2 months at $-65°$. Just before use the mix was transported in liquid nitrogen to the bench and placed in ice for about 3 min; 1 μl 38 mM ATP was then added to the still frozen mix. A few seconds later the ligated DNA sample (1–15 μl, usually 5 μl) was added and mixed during thawing, which took place immediately. The amount of DNA added per 20 μl packaging mix was usually 1 μg, but increasing this to 4.5 μg still gave approximately the same hybrid yield per microgram of DNA. After incubation of the packaging mix at 37° (or 25°) for 30 min, DNase was added (10 μg/ml) and MgCl$_2$ (10 mM). When the thick pellet was again liquid (2–10 min at 37°), 0.5 ml of phage dilution buffer[5] (40 mM Na$_2$HPO$_4$, 20 mM KH$_2$PO$_4$, 80 mM NaCl, 20 mM NH$_4$Cl, 0.1 mM CaCl$_2$, 10 mM MgCl$_2$, 1 mM MgSO$_4$) and a drop of chloroform were added. After a 2-min centrifugation at 5000 g the supernatant was removed and used as a bacteriophage suspension for transduction of *E. coli* HB101, which had been grown to late exponential phase (A_{550} = 1.0) in L Broth containing 0.5% maltose.[5] Plating and selection were carried out as described above.

After selection for ampicillin resistance (pJC75-58) or rifampicin resistance (pJC720) colonies were replica-plated onto minimal medium lacking either proline or leucine, so as to test for complementation of these auxotrophies in HB101 by the cloned DNA.

Results

The Production of Escherichia coli Gene Banks with Cosmids pJC720 and pJC75-38[3,4]

Cosmid Transduction Frequency without Added Foreign DNA. It can be seen from the transduction frequencies obtained in experiments 7, 8, and 15–17 (Table III) that pJC720 gives a higher background of (nonhybrid) colonies than pJC75-58 when religated in the absence of added high-molecular-weight DNA. Other vectors, in the size range 10–12 Md and lacking the temperature-sensitive phenotype, give even higher backgrounds, sometimes reaching 10^5 colonies per microgram of vector DNA.

Effect of Added High-Molecular-Weight Foreign DNA. In experiments 1, 2, 4, 5, 9, and 12 (Table III) a clear increase in the number of transductant colonies was produced by the addition of foreign DNA of the indicated molecular weight. As discussed below, this is due to the formation of large hybrid plasmids which now make up the majority of the transductant population.

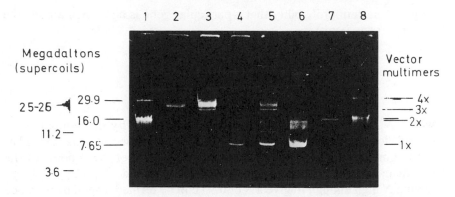

FIG. 5. Electrophoresis with 0.4% agarose gel of supercoiled DNA from gene banks produced in experiments 1 and 4 (Table III). Marker supercoil DNAs are at positions 1 and 8, pCOS10 (29.9 Md), upper band, position 6, pJC75-58 (7.65 Md), and position 7, pJC720 (16.0 Md). Positions 2 and 3 contain 100 and 400 ng, respectively, of supercoiled DNA isolated from 5000 antibiotic-resistant colonies from experiment 1 (Table III). Positions 4 and 5 contain 100 and 400 ng, respectively, of DNA from 5000 antibiotic-resistant colonies from Experiment 4 (Table III). The very faint bands at 3.6 and 11.2 Md are due to a small plasmid (and its hybrid with the vector DNA) which was present in the original foreign *E. coli* DNA cloned. Reproduced by permission of Elsevier North Holland Biomedical Press (Collins and Brüning[4]).

Effect of the Vector/Foreign DNA Ratio. The ratio of vector to foreign DNA is seen to have a marked effect in that the higher the ratio the higher the production of hybrids per input foreign DNA. This conforms with the hypothesis that the formation of sandwich-type molecules (a in Fig. 1) are required for packaging and the subsequent efficient production of hybrid DNA clones.

The Frequency of Hybrids among the Transduced Clones. The proportion of clones carrying large DNA fragments can be estimated either by physical measurement (Fig. 5) or by a genetic test in which one determines what percentage of the total population carries a particular gene.[34]

According to the physical measurements, i.e., gel electrophoresis of the supercoiled DNA from a mixture of 5000 colonies from experiments 1 and 4, the majority of the plasmids are found in the size range 25–26 Md, although in experiment 4 at least half of the DNA appears to be of the same size as the vector DNA, with small numbers of dimers and somewhat more in the trimeric form. In experiment 1, the average size of foreign DNA per hybrid is estimated to be about 26 − 7.6 = 18.4 Md.

According to Clarke and Carbon[34] the number of hybrid clones needed

[34] L. Clarke and J. Carbon, *Proc. Natl. Acad. Sci. U.S.A.* **72**, 4361 (1975).

to give a certain probability that a particular gene is contained among the hybrids (gene bank) is

$$N = \frac{\ln (1 - P)}{\ln \{1 - [(L - X)/M]\}}$$

where N is the number of clones, P is the probability of finding the required clone, L is the average size of the fragments cloned, X is the size of the fragment (gene) screened for, and M is the size of the genome of the organism from which the cloned DNA was obtained. A group of clones having N high enough to give a P of 0.95 can be termed a gene bank, since it should contain clones carrying DNA representative of the whole donor genome. Considering $X = 0$, $M = 2.5 \times 10^9$, and $L = 18 \times 10^6$, $N = 415$; i.e., in this calculation one would expect at least 0.24% of the hybrid clones to carry any particular gene. When a large number of clones are screened, the number of specific hybrids per total clone number should approach L/M^{15}, i.e., $18 \times 10^6/2.5 \times 10^9 = 0.72\%$.

This criterion for a gene bank appears from the results in experiment 1 (Table III) to have been reached in this example; i.e., the genetic data support the physical data that on average the clones contain 18×10^6 daltons of foreign DNA apiece.

Preferential Exclusion of Certain Fragments. It is also apparent from these genetic analyses that another effect is being demonstrated, namely, the preferential exclusion of some fragments, depending on the restriction enzyme use for cloning, the size of the vector, or the extent of digestion of the foreign DNA. For example, in the *Hin*dIII cloning experiments the *leu* gene is almost excluded, and using partially cut *Bgl*II DNA leads to a selective reduction in the frequency of *pro+* hybrids. Analysis of the distribution of restriction endonuclease sites in the region of the *pro* and *leu* genes leads to the interpretation that large restriction fragments (> 16 Md) will be eliminated from partial digests and that genes cut frequently will be absent from more complete digests and also reduced in frequency in partial digests if the neighboring fragments are of high molecular weight.[4]

Comments

The optimum efficiency of cloning is very high, e.g., in the experiment described 4×10^5 hybrid clones per microgram of foreign DNA, but efficiencies of 6×10^5 have been obtained. Thus in a single 20-μl packaging experiment approximately 10^6 hybrid clones are attainable, where the average size of the DNA insert is about 18 Md. This is considerably in excess of the number required for a gene bank from higher eukaryotes.

As demonstrated by the example above, the optimum yield can only be obtained by using the correct vector/foreign DNA ratio and with vector–foreign DNA fragments in the correct size range. The high DNA concentration for DNA ligation is also extremely important.

Complete cutting of the vector DNA is not important, as the contribution by uncut vector supercoils is insignificant.[4] One can therefore use the shortest incubation conditions possible so as to protect the cohesive ends formed from attack by exonucleases which may be contaminating the restriction enzyme used.

The use of alkaline phosphatase treatment[35] (to prevent ring closure of the vector DNA and cause the formation of hybrids with untreated DNA) is unnecessary in view of the size requirement for packaging and transduction, which essentially ensures hybrid formation. A useful application of the alkaline phosphatase treatment seems to be removal of the 5'-phosphates from the DNA fragments to be cloned. This ensures the formation of hybrids in which the cloned DNA is derived only from DNA fragments that were contiguous on the original genome. This should help remove much of the ambiguity from fine-structure mapping involving cloning of DNA adjacent to previously cloned fragments.[36]

Mention should be made of the results from earlier experiments[3] in which cosmid hybrids were constructed containing the same fragment repeated a number of times. Apart from interest in this finding as a method of gene amplification, the main observation was that the repeat units were stable and that *no* derivatives were found containing fragments in palindromic orientation (i.e., with a twofold axis of inverted symmetry). In addition, it was found that in these experiments, designed to produce repeats, nearly 10% of all hybrids carried small deletions. This points to the possibly highly unstable nature of palindromic structures (or inverted repeats?) in *E. coli,* an effect that could present serious difficulties in the cloning of large DNA fragments from higher eukaryotes.

The combinations of restriction enzymes available for cloning in conjunction with cosmids are shown in Table I, and a list of cosmids containing single sites for these enzymes is presented in Table I. As demonstrated in the experiment below, the use of restriction enzymes with six-base-pair recognition sites can lead to the exclusion of very large fragments or very small fragments, because of the random distribution of cutting sites and the size limitations imposed by the packaging system. The small fragments may be cloneable by using partial digests but, if they are

---

[35] A. Ullrich, J. Shine, J. Chirgwin, R. Pietet, E. Tischer, W. J. Rulter, and H. M. Goodman, *Science* **196,** 1313 (1977).

[36] A. Royal, A. Garapin, B. Cami, F. Perrin, J. L. Mandel, M. LeMeur, F. Brégégère, F. Gannon, J. P. LePennec, P. Chambon, and P. Kouilsky. *Nature* **279,** 125 (1979).

adjacent to very large fragments, they will still be eliminated. A simple way to overcome this problem in the production of complete gene banks is the use of partial digests with enzymes cutting more frequently (i.e., with four-base recognition sites or degenerate six-base sites), such as *Sau*I (for *Bam*HI or *Bgl*II vectors), *Ava*I (for *Sal*I vectors), and *Taq*I and *Hpa*II (for *Cla*I vectors). This procedure should give rise to a very random collection of fragments and to the cloning of regions representative of the whole genome, in a manner similar to the shearing and tailing procedures used in conjunction with other cloning techniques.[34] This method has been used to clone a gene of biotechnological interest from a *Klebsiella* strain, in pJC75-58, using *Sau*I partial digestion of the foreign DNA and *Bgl*II digestion of the vector. This method was successful after failures with *Hin*dIII and *Eco*RI partial and complete digests (J. Collins and H. Mayer, unpublished results).

Last, I should like to point out that the procedure for producing the packaging mix was developed entirely by Hohn.[5]

Acknowledgments

This report on attempts to optimize some of the conditions for cosmid packaging is based on original experiments conducted during a close collaboration between Barbara Hohn and myself.[3] I wish to thank her for sharing with me the belief that such a system could be developed and for her collaboration in bringing that belief to a practical and fruitful conclusion, in addition to her help in critically reading the manuscript.

[11] Production of Single-Stranded Plasmid DNA

By JEFFREY VIEIRA and JOACHIM MESSING

Introduction

In the study of gene structure and function, the techniques of DNA analysis that are efficiently carried out on single-strand (ss) DNA templates, such as DNA sequencing and site-specific *in vitro* mutagenesis, have been of great importance. Because of this, the vectors developed from the ssDNA bacteriophages M13, fd, or f1, which allow the easy isolation of strand-specific templates, have been widely used. While these vectors are very valuable for the production of ssDNA, they have certain negative aspects in comparison to plasmid vectors (e.g., increased instability of some inserts, the minimum size of phage vectors). Work from the laboratory of N. Zinder showed that a plasmid carrying the intergenic region (IG) of f1 could be packaged as ssDNA into a viral particle by a helper phage.[1] This led to the construction of vectors that could combine the advantages of both plasmid and phage vectors.[2] Since that time a number of plasmids carrying the intergenic region of M13 or f1 have been constructed with a variety of features.[3]

A problem that has been encountered in the use of these plasmid/phage chimeric vectors (plage) is the significant reduction in the amount of ssDNA that is produced as compared to phage vectors. Phage vectors can have titers of plaque-forming units (pfu) of 10^{12}/ml and give yields of a few micrograms per milliliter of ssDNA. It might then be expected that cells carrying both a plage and helper phage would give titers of 5×10^{11}/ml for each of the two. However, this is not the case due to interference by the plage with the replication of the phage.[4] This results in a reduction in the phage copy number and, therefore, reduces the phage gene products necessary for production of ssDNA. This interference results in a 10- to 100-fold reduction in the phage titer and a level of ss plasmid DNA particles of about 10^{10} colony forming units (cfu) per milliliter.[1] Phage mutants that show interference resistance have been isolated.[4,5] These mutants can increase the yield of ss plasmid by 10-fold and concurrently

---

[1] G. P. Dotto, V. Enea, and N. D. Zinder, *Virology* **114,** 463 (1981).

[2] N. D. Zinder and J. D. Boeke, *Gene* **19,** 1 (1982).

[3] D. Mead and B. Kemper, *in* "Vectors: A Survey of Molecular Cloning Vectors and Their Uses" (R. Rodriguez and D. T. Denhardt, eds.). Butterworth, Boston, Massachusetts, 1986.

[4] V. Enea and N. D. Zinder, *Virology* **122,** 222 (1982).

[5] A. Levinson, D. Silver, and B. Seed, *J. Mol. Appl. Genet.* **2,** 507 (1984).

RECOMBINANT DNA
METHODOLOGY

increase the level of phage by a similar amount. Whether wild-type (wt) phage or an interference-resistant mutant is used as helper the yield of plasmid ssDNA is usually about equal to that of the phage,[3] and as the plasmid size increases the ratio shifts to favor the phage.[5] In order to increase both the quantitative and qualitative yield of the plasmid ssDNA, a helper phage, M13KO7, has been constructed that preferentially packages plasmid DNA over phage DNA. In this chapter, M13KO7 will be described and its uses discussed.

M13 Biology

Certain aspects of M13 biology and M13 mutants play an important role in the functioning of M13KO7, so a short review of its biology is appropriate.[6,7] M13 is a phage that contains a circular ssDNA molecule of 6407 bases packaged in a filamentous virion which is extruded from the cell without lysis. It can infect only cells having an F pili, to which it binds for entering the cell. The phage genome consists of 9 genes encoding 10 proteins and contains an intergenic region of 508 bases. The proteins expressed by the phage are involved in the following processes: I and IV are involved in phage morphogenesis, III, VI, VII, VIII, and IX are virion proteins, V is an ssDNA binding protein, X is probably involved in replication, and II creates a site-specific (+) strand nick within the IG region of the double-stranded replicative form (RF) of the phage DNA molecule at which DNA synthesis is initiated.

Phage replication consists of three phases: (1) ss–ds, (2) ds–ds, and (3) ds–ss. The ss–ds phase is carried out entirely by host enzymes. For phases 2 and 3, gene II, which encodes both proteins II and X, is required for initiating DNA synthesis; all other functions necessary for synthesis are supplied by the host. The DNA synthesis initiated by the action of the gene II protein (gIIp) leads to both the replication of the ds molecule and the production of the ssDNA that is to be packaged in the mature virion. The phage is replicated by a rolling circle mechanism that is terminated by gIIp cleaving the displaced (+) strand at the same site and resealing it to create a circular ssDNA molecule. Early in the phage life cycle this ssDNA molecule is converted to the ds RF but later in the phage life cycle gVp binds to the (+) strand, preventing it from being converted to dsDNA and resulting in it being packaged into viral particles. The assembly of the virion occurs in the cell membrane where the gVp is replaced by the

[6] D. T. Denhardt, D. Dressler, and D. S. Ray (eds.), "The Single-Stranded DNA Phages." Cold Spring Harbor Lab., Cold Spring Harbor, New York, 1978.
[7] N. D. Zinder and K. Horiuchi, *Microbiol. Rev.* **49**, 101 (1985).

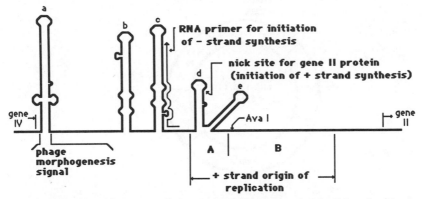

FIG. 1. The M13 intergenic region is schematically presented. It is 508 nucleotides long and is situated between genes II and IV. Potential secondary structure is represented by hairpin structures a–e.[8] Important functional regions are also shown.

gVIIIp and the other virion proteins as the phage particle is extruded from the cell.

The IG structure contains regions important for four phage processes[8–10]: (1) The sequences necessary for the recognition of an ssDNA by phage proteins for its efficient packaging into viral particles; (2) the site of synthesis of an RNA primer that is used to initiate (−) strand synthesis; (3) the initiation; and (4) the termination of (+) strand synthesis. In Fig. 1 the IG, which has the potential to form five hairpin structures, is represented schematically and important regions designated. Most important to the functioning of M13KO7 is the origin of replication of the (+) strand. The origin consists of 140 bp and can be divided into two domains. Domain A, about 40 bp, is essential for replication and contains the recognition sequence for gIIp to create the nick that initiates and terminates replication of the RF. Domain B is about 100 bp long and acts as an enhancer for gIIp to function at domain A. The effect of domain B can be demonstrated by the fact that a disruption or deletion of it will decrease phage yield by 100-fold.[9] Two types of mutants, a qualitative mutation from M13mp1[11] and two quantitative ones from R218 and R325,[12] that compensate for the loss of a functional domain B have been analyzed. The qualitative mutant from mp1, which has an 800-bp insertion within B,

[8] H. Schaller, *Cold Spring Harbor Symp. Quant. Biol.* **45,** 177 (1978).

[9] G. P. Dotto, K. Horiuchi, and N. D. Zinder, *J. Mol. Biol.* **172,** 507 (1984).

[10] G. P. Dotto and N. D. Zinder, *Virology* **130,** 252 (1983).

[11] J. Messing, B. Gronenborn, B. Muller-Hill, and P. H. Hofschneider, *Proc. Natl. Acad. Sci. U.S.A.* **74,** 3642 (1977).

[12] G. P. Dotto and N. D. Zinder, *Proc. Natl. Acad. Sci. U.S.A.* **81,** 1336 (1984).

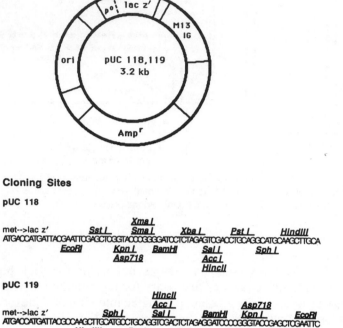

Fig. 2. Structure of pUC 118 and 119 and the DNA sequence of the unique restriction enzyme sites within the sequence encoding the *lacZ* peptide.

consists of a single G-to-T substitution that changes a methionine (codon 40) to an isoleucine within the gIIp.[13] This change allows the mp1gIIp to function efficiently enough on an origin consisting of only domain A to give wild-type levels of phage. In R218 and R325 the loss of a functional domain B is compensated for by mutations that cause the overproduction of a normal gIIp at 10-fold normal levels.[12,13] Even though a wild-type gIIp works very poorly on a domain B-deficient origin, the excess level of gIIp achieves enough initiation of replication to give normal levels of phage.

pUC 118 and 119

All ss plasmid DNA vectors carry a phage intergenic region. The entire complement of functions necessary for the packaging of ssDNA

---

[13] G. P. Dotto, K. Horiuchi, and N. D. Zinder, *Nature (London)* **311**, 279 (1984).

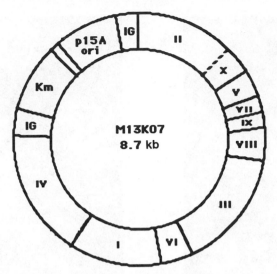

FIG. 3. Structure of M13KO7.

into viral particles will work *in trans* on an IG region. The vectors used in the experiments described here are pUC 118 and 119 (Fig. 2). They are pUC 18 and 19,[14] respectively, with the IG region of M13 from the *Hgi*AI site (5465) to the *Dra*I site (5941) inserted at the unique *Nde*I site (2499) of pUC. The orientation of the M13 IG region is such that the strand of the *lac* region that is packaged as ssDNA is the same as in the M13mp vectors.

M13KO7

M13KO7 (Fig. 3) is an M13 phage that has the gene II of M13mp1 and the insertion of the origin of replication from p15A[15] and the kanamycin-resistance gene from Tn 903[16] at the *Ava*I site (5825) of M13. With the p15A origin, the phage is able to replicate independent of gIIp. This allows the phage to overcome the effects of interference and maintain adequate genome levels for the expression of proteins needed for ssDNA production when it is growing in the presence of a plage. The effect of the addition of the plasmid origin is shown in Fig. 4B. The insertion of the p15A origin and the kanamycin-resistance gene separates the A and B

[14] J. Norrander, T. Kempe, and J. Messing, *Gene* **26,** 101 (1983).
[15] G. Selzer, T. Som, T. Itoh, and J. Tomizawa, *Cell* **32,** 119 (1983).
[16] N. D. F. Grindley and C. M. Joyce, *Proc. Natl. Acad. Sci. U.S.A.* **77,** 7176 (1980).

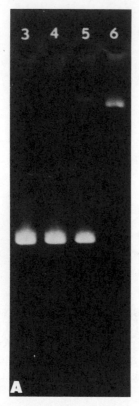

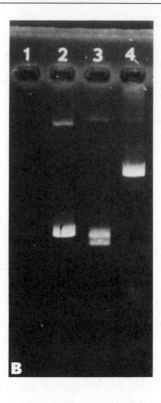

FIG. 4. In all gel lanes 40 μl of the supernatant fraction after centrifugation of the culture was mixed with 6 μl of SDS gel-loading buffer and loaded on the gel. (A) Lane 3: pUC 118 with M13KO7 as helper phage. Plasmid titer is 5×10^{11} cfu/ml, phage titer is 8×10^9 pfu/ml. Lane 4: pUC 119 with M13KO7 as helper phage. Plasmid titer is 6×10^{11} cfu/ml, phage titer is 8×10^9 pfu/ml. Lane 5: pUC 119 with M13KO19 (similar to KO7, but with a deletion of domain B of the phage origin of replication) as helper phage. Lane 6: M13KO7. (B) Lane 1: pUC 119 with an M13mp8 phage carrying the kanamycin gene, but no plasmid origin of replication, as helper phage. Lane 2: pUC 119 with M13KO19 as helper phage. Lane 3: pUC 19 with the M13 IG region in the same location as 119, but in the opposite orientation. Lane 4: pUC 118 with 2.5-kb insert.

domains of the phage origin of replication, creating an origin that is less efficient for the functioning of the mp1 gIIp than the wild-type origin carried by the plage. This, plus the high copy number of pUC, leads to the preferential packaging of plasmid DNA into viral particles. The mp1 gIIp functions well enough on the altered origin when M13KO7 is grown by itself to produce a high titer of phage for use as inoculum for the production of ss plasmid.

Materials and Reagents

Strains

MV1184: *ara,*Δ*(lac–pro)*, *strA*, thi, (ϕ80Δ*lacIZ*ΔM15),Δ*(srl–recA)* 306::Tn*10*(tet^r); F': *traD*36, *proAB*, *lacI^{q}Z*Δm15)

Media

2× YT (per liter): 16 g Difco Bacto tryptone, 10 g Difco Bacto yeast extract, 5 g NaCl, 10 mM KPO$_4$, pH 7.5

2× YT plates: 15 g Difco Bacto agar added to 1 liter of 2× YT

YT soft agar (per liter): 8 g Difco Bacto tryptone, 5 g yeast extract, 5 g NaCl, 7 g agar

M9 plates: For 1 liter of 10× M9 salts: combine 60 g Na$_2$HPO$_4$, 30 g KH$_2$PO$_4$, 0.5 g NaCl, 10 g NH$_4$Cl dissolved in H$_2$O to a final volume of 970 ml and autoclave. After autoclaving add 10 ml of a sterile 1 M MgSO$_4$ solution and 20 ml of a sterile 0.05 M CaCl$_2$ solution. For 1 liter of plates autoclave 15 g of agar in 890 ml. After autoclaving add 100 ml 10× M9 salts, 10 ml of a 20% glucose solution, and 1 ml of a 1% thiamin solution

Solutions

SDS gel loading buffer: 0.05% bromphenol blue, 0.2 M EDTA, pH 8.0, 50% glycerol, 1% SDS

TE buffer: 10 mM Tris–HCl, pH 8.0, 1 mM EDTA, pH 8.0

Growth of M13KO7

M13KO7 exhibits some instability of the insert during growth, but this does not create a problem if it is propagated correctly. The procedure for the production of M13KO7 is the following. M13KO7 supernatant is streaked on a YT agar plate and then 4 ml of soft agar, to which 0.5 ml of a culture of MV 1184 (OD$_{600}$ > 0.8) has been added, is poured across the plate from the dilute side of the streak toward the more concentrated side. After 6–12 hr of incubation at 37° single plaques are picked and grown individually in 2–3 ml of YT containing kanamycin (70 μg/ml) overnight. The cells are then pelleted by centrifugation, and the supernatant is used as inoculum of M13KO7. The phage in the supernatant will remain viable for months when stored at 4°.

Production of ss Plasmid DNA

For the production of ss plasmid DNA it is important that a low-density culture of plage-containing cells, infected with M13KO7, be grown for 14–18 hr with very good aeration. The medium that is used is

$2\times$ YT supplemented with 0.001% thiamin, 150 μg/ml ampicillin, and, when appropriate, 70 μg/ml kanamycin. Commonly used methods are the following:

1. A culture of MV1184 (pUC 118/119) in early log phase is infected with M13KO7 at a multiplicity of infection (moi) of 2–10 and incubated at 37° for 1 hr and 15 min. The infection should be carried out on a roller or a shaker at low rpm. After this time the cells are diluted, if necessary, to an $OD_{600} < 0.2$ and kanamycin is added to a final concentration of 70 μg/ml. The culture is then grown for 14–18 hr at 37°. Culture conditions are usually 2–3 ml in an 18-mm culture tube on a roller or 5–10 ml in a 125-ml culture flask on a shaker at 300 rpm. Pellet the cells by centrifugation (8000 g, 10 min) and remove the supernatant to a fresh tube. Add one-ninth of the supernatant volume of 40% PEG and of 5 M sodium acetate and mix well. Place on ice 30 min and pellet the viral particles by centrifugation (8000 g, 10 min) and pour off the supernatant. Remove the remaining supernatant with a sterile cotton swab. Resuspend the pellet in 200 μl TE buffer by vortexing. Add 150 μl of TE-saturated phenol (pH 7) and vortex for 30 sec. Add 50 μl of CHCl$_3$, vortex, and centrifuge for 5 min (Brinkman Eppendorf centrifuge). Remove the aqueous layer to a fresh tube and repeat phenol/CHCl$_3$ extraction. Remove the aqueous layer to a fresh tube and add an equal volume of CHCl$_3$, vortex, and centrifuge for 5 min. Remove the aqueous layer to another tube and add 3 vol of ether. Vortex well and centrifuge briefly. Remove the ether, add one-twentieth the volume of 3 M sodium acetate (pH 7), and precipitate the DNA with 2.5 vol of ethanol at −70° for 30 min and then pellet by centrifugation. Once the pellet is dry it can be resuspended in TE and used in the same manner as has been previously described for the use of M13 ssDNA templates.[17]

2. For the screening of plasmid for inserts a colony selected from a plate is added to 2–3 ml of medium containing M13KO7 (~10^7/ml) and grown at 37° for a few hours. Kanamycin is then added and the cultures are incubated for 14–18 hr at 37°. The cells are then pelleted and 40 μl of supernatant is mixed with 6 μl of loading buffer and electrophoresed on a 1% agarose gel, stained with ethidium bromide, and viewed with UV illumination.

Discussion

The use of M13KO7 for the production of ss plasmid DNA normally gives titers of cfu of 10^{11}–5×10^{11}/ml and phage titers 10- to 100-fold lower

[17] J. Messing, this series, Vol. 101, p. 20.

(Fig. 4A). Plasmids containing inserts as large as 9 kb have been packaged as ssDNA without a significant loss in yield (M. McMullen and P. Das, personal communication) and instability has not been a problem. It has been observed that some clones, irregardless of size, give reduced levels of ssDNA. This reduction in yield has been both dependent (M. McMullen, personal communication) and independent (J. Braam, personal communication) of the orientation of the insert. M13KO7 has given high yields of ssDNA from pUC-derived vectors, but when it was used as a helper phage with pZ150,[19] a vector constructed from pBR 322, the yield of ssDNA was not significantly different from the yield given by other helper phages. Whether this is due to the lower copy number of pBR as compared to pUC or to some effect of the vector structure is not known. It has been noted that the position and orientation of the IG region within the plasmid can affect its packaging as ssDNA. An example is shown in Fig. 4B (lane 3). This plasmid has the IG region inserted in the same position but the opposite orientation as compared to pUC 119/118, and always gives two bands. However, if the IG region, in the opposite orientation of 118/119, is inserted within the polycloning sites of a pUC vector, the resulting plasmid yields a single band after gel electrophoresis (data not shown). A large variation in the yield of ss plasmid DNA has been seen between different bacterial strains. MV 1184 (derived from JM 83) and MV 1190 (derived from JM 101) have given satisfactory yields. MV 1304 (derived from JM 105) gives much reduced yields and JM 109 undergoes significant lysis when it contains both plasmid and phage.

Acknowledgments

We would like to thank B. McClure, R. Zagursky, M. Berman, and D. Mead for valuable discussions. We also thank M. Volkert for the MV bacterial strains and Claudia Dembinski for help in preparing this manuscript. This work was supported by the Department of Energy, Grant #DE-FG05-85ER13367.

[18] M. J. Zoller and M. Smith, this volume [30].
[19] R. J. Zagursky and M. L. Berman, *Gene* **27,** 183 (1984).

[12] High-Efficiency Cloning of Full-Length cDNA; Construction and Screening of cDNA Expression Libraries for Mammalian Cells

By H. OKAYAMA, M. KAWAICHI, M. BROWNSTEIN, F. LEE, T. YOKOTA, and K. ARAI

cDNA cloning constitutes one of the essential steps to isolate and characterize complex eukaryotic genes, and to express them in a wide variety of host cells. Without cloned cDNA, it is extremely difficult to define the introns and exons, the coding and noncoding sequences, and the transcriptional promoter and terminator of genes. Cloning of cDNA, however, is generally far more difficult than any other recombinant DNA work, requiring multiple sequential enzymatic reactions. It involves *in vitro* synthesis of a DNA copy of mRNA, its subsequent conversion to a duplex cDNA, and insertion into an appropriate prokaryotic vector. Due to the intrinsic difficulty of these reactions as well as the inefficiency of the cloning protocols devised, the yield of clones is low and many of clones are truncated.[1]

The cloning method developed by Okayama and Berg[2] circumvents many of these problems, and permits a high yield of full-length cDNA clones regardless of their size.[3-6] The method utilizes two specially engineered plasmid DNA fragments, "vector primer" and "linker DNA." In addition, several specific enzymes are used for efficient synthesis of a duplex DNA copy of mRNA and for efficient insertion of this DNA into a plasmid. Excellent yields of full-length clones and the unidirectional insertion of cDNA into the vector are the result. These features not only facilitate cloning and analysis but are also ideally suited for the expression of functional cDNA.

To take full advantage of the features of this method, Okayama and

---

[1] A. Efstratiadis and L. Villa-Komaroff, *in* "Genetic Engineering" (J. K. Setlow and A. Hollaender, eds.), Vol. 1, p. 1. Plenum, New York, 1979.

[2] H. Okayama and P. Berg, *Mol. Cell. Biol.* **1,** 161 (1982).

[3] D. H. Maclennan, C. J. Brandl, B. Korczak, and N. M. Green, *Nature (London)* **316,** 696 (1985).

[4] L. C. Kun, A. McClelland, and F. H. Ruddle, *Cell* **37,** 95 (1984).

[5] K. Shigesada, G. R. Stark, J. A. Maley, L. A. Niswander, and J. N. Davidson, *Mol. Cell. Biol.* **5,** 1735 (1985).

[6] S. M. Hollenberg, C. Weinberger, E. S. Ong, G. Cerelli, A. Oro, R. Lebo, E. B. Thompson, M. G. Rosenfeld, and R. M. Evans, *Nature (London)* **318,** 635 (1985).

RECOMBINANT DNA
METHODOLOGY

Berg[7] have modified the original vector. The modified vector, pcD, has had SV40 transcriptional signals introduced into the vector primer and linker DNAs to promote efficient expression of inserted cDNAs in mammalian cells. Construction of cDNA libraries in the pcD expression vector thus permits screening or selection of particular clones on the basis of their expressed function in mammalian cells, in addition to regular screening with hybridization probes.

Expression cloning has proven extremely powerful if appropriate functional assays or genetic complementation selection systems are available.[8–14] In fact, Yokota et al.[11,12] and Lee et al.[13,14] have recently isolated full-length cDNA clones encoding mouse and human lymphokines without any prior knowledge of their chemical properties, relying entirely on transient expression assays using cultured mammalian cells. Similar modifications have been made to promote the expression of cDNA in yeast, thereby permitting yeast mutant cells to be used as possible complementation hosts.[15,16]

In this chapter, we describe detailed procedures for the construction of full-length cDNA expression libraries and the screening of the libraries for particular clones based on their transient expression in mammalian cells. Methods for library transduction and screening based on stable expression are described in Vol. 151 of *Methods in Enzymology*. If expression cloning is not envisioned, the original vector[2] or one described by others[17] can be used with slight modifications of the procedure described below.

---

[7] H. Okayama and P. Berg, *Mol. Cell. Biol.* **2,** 280 (1983).

[8] D. H. Joly, H. Okayama, P. Berg, A. C. Esty, D. Filpula, P. Bohlen, G. G. Johnson, J. E. Shivery, T. Hunkapiller, and T. Friedmann, *Proc. Natl. Sci. Acad. U.S.A.* **80,** 477 (1983).

[9] D. Ayusawa, K. Takeishi, S. Kaneda, K. Shimizu, H. Koyama, and T. Seno, *J. Biol. Chem.* **259,** 1436 (1984).

[10] H. Okayama and P. Berg, *Mol. Cell. Biol.* **5,** 1136 (1985).

[11] T. Yokota, F. Lee, D. Rennick, C. Hall, N. Arai, T. Mosmann, G. Nabel, H. Cantor, and K. Arai, *Proc. Natl. Acad. Sci. U.S.A.* **81,** 1070 (1985).

[12] T. Yokota, N. Arai, F. Lee, D. Rennick, T. Mosmann, and K. Arai, *Proc. Natl. Acad. Sci. U.S.A.* **82,** 68 (1985).

[13] F. Lee, T. Yokota, T. Otsuka, L. Gemmell, N. Larson, L. Luh, K. Arai, and D. Rennick, *Proc. Natl. Acad. Sci. U.S.A.* **82,** 4360 (1985).

[14] F. Lee, T. Yokota, T. Otsuka, P. Meyerson, D. Villaret, R. Coffman, T. Mosmann, D. Rennick, N. Roehm, C. Smith, C. Zlotnick, and K. Arai, *Proc. Natl. Acad. Sci. U.S.A.* **83,** 2061 (1986).

[15] G. L. McKnight and B. C. McConaughy, *Proc. Natl. Acad. Sci. U.S.A.* **80,** 4412 (1983).

[16] A. Miyajima, N. Nakayama, I. Miyajima, N. Arai, H. Okayama, and K. Arai, *Nucleic Acids Res.* **12,** 6639 (1984).

[17] D. C. Alexander, T. D. McKnight, and B. G. Williams, *Gene* **31,** 79 (1984).

Methods

Clean, intact mRNA is prepared from cultured cells or tissue by the guanidine thiocyanate method[18] followed by two cycles of oligo(dT)–cellulose column chromatography. The purified mRNA is then reverse transcribed by the avian myeloblastosis enzyme in a reaction primed with the pcD-based vector primer, a plasmid DNA fragment that contains a poly(dT) tail at one end and a HindIII restriction site near the other end (Figs. 1 and 2).[7] The vector also contains the SV40 poly(A) addition signal downstream of the tail site as well as the pBR322 replication origin and the β-lactamase gene. Reverse transcription results in the synthesis of a cDNA : mRNA hybrid covalently linked to the vector molecule (Fig. 3). This product is tailed with oligo(dC) at its 3' ends and digested with HindIII to release an oligo(dC) tail from the vector end and to create a HindIII cohesive end. The C-tailed cDNA : mRNA hybrid linked to the vector is cyclized by addition of DNA ligase and a pcD-based linker DNA—an oligo(dG)-tailed DNA fragment with a HindIII cohesive end (this linker contains the SV40 early promoter and the late splice junctions) (Figs. 1 and 2). Finally, the RNA strand is converted to DNA by nick-translation repair catalyzed by Escherichia coli DNA polymerase I, RNase H, and DNA ligase. The end product, a closed circular cDNA recombinant, is transfected into a highly competent E. coli host to establish a cDNA clone library.

In the steps that have just been enumerated, double-stranded, full-length DNA copies of the original mRNAs are efficiently synthesized and inserted into the vector to form a functional composite gene with the protein coding sequence derived from the cDNA and the transcriptional and RNA processing signals from the SV40 genome. To screen for or select a particular clone on the basis of the function it encodes, the library is acutely transfected or stably transduced into cultured cells. Procedures for stable transduction are described in Chap. [32] of Vol. 151 of Methods in Enzymology.

Preparation of mRNA

Successful construction of full-length cDNA libraries depends heavily on the quality of the mRNA preparation. The use of intact, uncontaminated mRNA is essential for generating full-length clones. Messenger RNA prepared by the guanidine thiocyanate method[18] satisfies the above

[18] J. M. Chilgwin, A. E. Przybyla, R. J. MacDonald, and W. J. Rutter, Biochemistry 18, 5294 (1978).

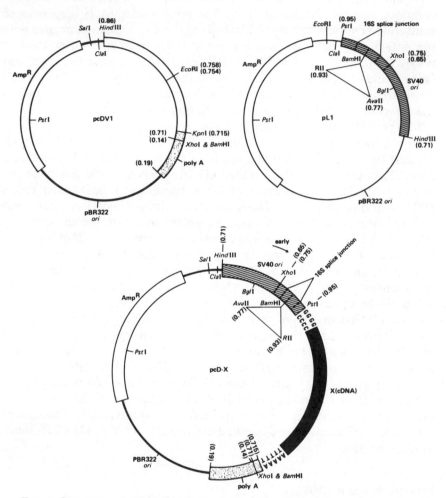

FIG. 1. Structure and component parts of the pcD vector and its precursor plasmids, pcDV1 and pL1. The principal elements of the pcD vector are a segment containing the SV40 replication origin and the early promoter joined to a segment containing the 19 S and 16 S SV40 late splice junctions (hatched area); the various cDNA inserts flanked by dG/dC and dA/dT stretches that connect them to the vector (solid black area); a segment containing the SV40 late polyadenylation signal [poly(A)] (stippled area); and the segment containing the pBR322 β-lactamase gene and the origin of replication (thin and open area). pcDV1 and pL1 provide the pcD-based vector primer and linker DNA, respectively. For the preparation of the vector primer and linker DNA, see Methods sections and Fig. 2.

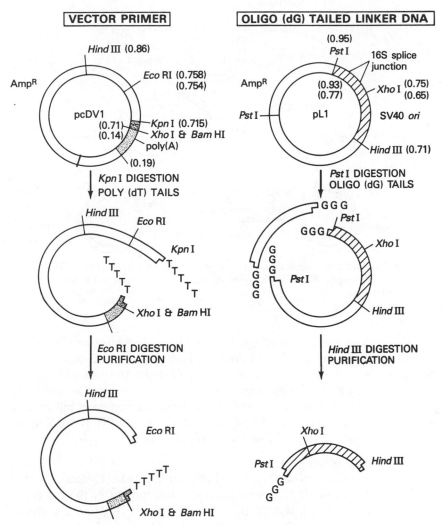

FIG. 2. Preparation of vector primer and linker DNAs.

criteria and is reverse transcribed efficiently. It has successfully been used for cloning a number of cDNAs.[3-14] The method described below is a slight modification of the original method that ensures complete inactivation of RNases through all the steps of RNA isolation.

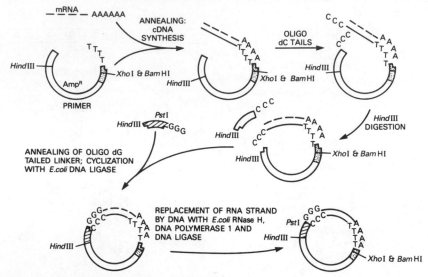

FIG. 3. Enzymatic steps in the construction of pcD–cDNA recombinants. The designations of the DNA segments are as described in Fig. 1. For experimental details and comments, see Methods.

Reagents

All solutions are prepared using autoclaved glassware or sterile disposable plasticware, autoclaved double-distilled water and chemicals of the finest grade. Solutions are sterilized by filtration through Nalgen 0.45 μm Millipore filters and subsequently by autoclaving (except as noted). In general, treatment of solutions with diethyl pyrocarbonate is not recommended since residual diethyl pyrocarbonate may modify the RNA, resulting in a marked reduction in its template activity.

5.5 M GTC solution:
   5.5 M guanidine thiocyanate (Fluka or Eastman-Kodak), 25 mM sodium citrate, 0.5% sodium lauryl sarcosine. After the pH is adjusted to 7.0 with NaOH, the solution is filter-sterilized and stored at 4°. Prior to use, 2-mercaptoethanol is added to a final concentration of 0.2 M.
4 M GTC solution: 5.5 M solution diluted to 4 M with sterile distilled water.
CsTFA solution:
   cesium trifluoroacetate (density 1.51 ± 0.01 g/ml), 0.1 M ethylenediaminetetraacetic acid (EDTA) (pH 7.0). Prepared with cesium trifluoroacetate (2 g/ml) (CsTFA, Pharmacia) and 0.25 M EDTA (pH 7.0).

TE: 10 mM Tris–HCl (pH 7.5), 1 mM EDTA.
1 M NaCl
2 M NaCl
1 M acetic acid, filter sterilized
Oligo(dT)–cellulose (Collaborative Research), Type 3.
> Resins provided by some other supplier may not be useful because mRNA purified on these often has significant template activity for reverse transcription without addition of primer. This is likely to be due to contamination of the mRNA with oligo(dT) leached from the resin.

TE/NaCl: a 1 : 1 mixture of TE and 1 M NaCl.
Ethidium bromide stock solution: 10 mg/ml water, stored at 4°.
Yeast tRNA stock solution: 1 mg/ml, dissolved in sterile water.

Procedure

Step 1. Extraction of Total RNA. Approximately 2–4 × 10^8 cells or 1–3 g of tissue are treated with 100 ml of the 5.5 M GTC solution. Cultured cells immediately lyse but tissue generally requires homogenization to facilitate lysis. The viscous lysate is transferred to a sterile beaker, and the DNA is sheared by passing the lysate through a 16- to 18-gauge needle attached to a syringe several times until the viscosity decreases. After removal of cell debris by a brief low speed centrifugation, the lysate is gently overlaid onto a 17-ml cushion of CsTFA solution in autoclaved SW28 centrifuge tubes and centrifuged at 25,000 rpm for 24 hr at 15°.

After centrifugation, the upper GTC layer and the DNA band at the interface are removed by aspiration. The tubes are quickly inverted, and their contents are poured into a beaker. Still inverted, they are placed on a paper towel to drain for 5 min, and then the bottom 2 cm of the tube is cut off with a razor blade or scalpel; the remainder is discarded. After the bottom of the tube is removed, the cup that is formed is turned over again and placed on a bed of ice. The RNA pellet is dissolved in a total of 0.4 ml of the 4 M GTC solution. After insoluble materials are removed by brief centrifugation in an Eppendorf microfuge, the RNA is precipitated as follows: 10 μl of 1 M acetic acid and 300 μl of ethanol are added to the solution and chilled at −20° for at least 3 hr. The RNA is pelleted by centrifugation at 4° for 10 min in a microfuge. The RNA pellet is dissolved in 1 ml of TE, and the insoluble material is removed by centrifugation. One hundred microliters of 2 M NaCl and 3 ml of ethanol are added to the solution. The RNA is precipitated by centrifugation after chilling at −20° for several hours. The RNA may be stored as a wet precipitate.

Step 2. Oligo(dT)–Cellulose Column Chromatography. Poly(A)$^+$ RNA is separated from the total RNA by oligo(dT)–cellulose column chromatography.[19] Oligo(dT)–cellulose is suspended in TE, and the fines are removed by decantation. A column 1.5 cm in height is made in an autoclaved Econocolumn (0.6 cm diameter) (Bio-Rad), washed with several column volumes of TE, and equilibrated with TE/NaCl. The RNA pellet is dissolved in 1 ml of TE. It is then incubated at 65° for 5 min, quickly chilled on ice, and 1 ml of 1 M NaCl is added. The RNA sample is applied to the column and the flowthrough is applied again. The column is washed with 5 bed volumes of TE/NaCl and eluted with 3 bed volumes of TE. One-half milliliter fractions of TE are collected. The RNA eluted is assayed by the spot test.

Small samples (0.5–3 μl) from each fraction are mixed with 20 μl of 1 μg/ml ethidium bromide (freshly prepared from the stock solution). The mixture is spotted onto a sheet of plastic wrap placed on a UV light box; ethidium bromide bound to RNA in the positive fractions emits a red–orange fluorescence. Fractions containing poly(A)$^+$ RNA are combined, incubated at 65° for 5 min, and chilled on ice. After adding an equal volume of 1 M NaCl, the sample is reapplied to the original column that has, in the meantime, been washed with TE and reequilibrated with TE/NaCl. The column is washed and eluted as above. Poly(A)$^+$ RNA eluted from the column is precipitated by adding 0.2 volume of 2 M NaCl and 3 volumes of ethanol, chilling on dry ice for 30 min, and centrifuging in a microfuge at 4°. The RNA pellet is dissolved in 20 μl of TE. The RNA concentration is determined by the spot test, as described above, using $E. coli$ tRNA solutions of known concentrations as standards (the assay can be used to measure between 100 and 400 ng of RNA). Ethanol is added to a final concentration of 50%, and the solution is stored at $-20°$. Generally 20–30 μg of poly(A)$^+$ RNA is obtained from 3 × 10^8 cells or 1 g of tissue. This RNA is more than enough for making a library.

Prior to use, the RNA should be tested. Reverse-transcribe 2 μg of RNA using 0.5 μg of oligo(dT) primer in place of the vector primer under the conditions described in the cDNA Cloning section "Pilot-scale reaction." Calculate the percent conversion to cDNA from the amounts of cDNA synthesized and RNA used [(μg of cDNA synthesized/μg of RNA used) × 100]. Generally 15–20% of the poly(A)$^+$ RNA prepared by this method can be converted to cDNA. If the number is considerably smaller than this, the RNA should not be used.

Comments. Fresh tissue or healthy cells should be used for the preparation of mRNA. Messenger RNA from mycoplasma-infected cells is often partially degraded. The use of such RNA leads to a failure to generate

[19] H. Aviv and P. Leder, *Proc. Natl. Acad. Sci. U.S.A.* **69,** 1408 (1972).

full-length clones. A great deal of caution should be taken to prevent the contamination of glassware and solutions by RNase. Solutions should be freshly prepared each time from stock buffers and solutions that have been guarded against contamination. As long as samples are carefully handled, wearing gloves is not necessary. The use of sodium dodecyl sulfate (SDS) in oligo(dT)–cellulose column chromatography is not recommended since residual SDS may inactivate reverse transcriptase.

Preparation of Vector Primer and Linker DNAs

The pcD-based vector primer and linker DNAs are prepared from pcDV1 and pL1, respectively, using the enzymatic treatments and purification procedures illustrated in Fig. 2. Briefly, pcDV1 is linearized by *Kpn*I digestion, and poly(dT) tails are added to the ends of the linear DNA with terminal transferase. One tail is removed by *Eco*RI digestion. The resulting large fragment is purified by agarose gel electrophoresis and subsequent oligo(dA)–cellulose column chromatography, and used as the vector primer. Untailed or uncut DNA, which produces significant background colonies, is effectively removed by the column purification step.

The best cloning results have been obtained with vector primer having 40 to 60 dT residues per tail. The reaction conditions that allow the addition of poly(dT) tails of this size vary with the lot of transferase and the preparation of DNA. Therefore, optimization of tailing conditions should be established for each preparation with a pilot-scale reaction.

Excessive digestion of DNA with restriction endonucleases should be avoided to minimize nicking of DNA by contaminating nucleases. The ends at nicks serve as effective primers for terminal transferase as well as for reverse transcriptase.[20] The resulting homopolymer tails and branching structures at nicks will reduce cloning efficiency.

Oligo(dG)-tailed linker DNA is prepared by *Pst*I digestion of pL1 DNA. Oligo(dG) tails of 10–15 dG residues are added to the ends. After *Hin*dIII digestion, the tailed fragment that contains the SV40 sequences is purified by agarose gel electrophoresis.

Reagents

*Kpn*I, *Eco*RI, *Pst*I, and *Hin*dIII (New England Biolabs)
Terminal deoxynucleotidyltransferase from calf thymus (Pharmacia)
Oligo(dA)–cellulose (Collaborative Research)
Loading buffer: 1 M NaCl, 10 mM Tris–HCl (pH 7.5), 1 mM EDTA.

---

[20] T. Nelson and D. Brutlag, this series, Vol. 68, p. 43.

10× *Kpn*I buffer: 60 mM NaCl, 60 mM Tris–HCl (pH 7.5), 60 mM MgCl$_2$, 60 mM 2-mercaptoethanol, 1 mg bovine serum albumin (BSA) (Miles, Pentex crystallized)/ml.

5× *Eco*RI buffer: 0.25 M NaCl, 0.5 M Tris–HCl (pH 7.5), 25 mM MgCl$_2$, 0.5 mg BSA/ml.

10× *Pst*I buffer: 1 M NaCl, 0.1 M Tris–HCl (pH 7.5), 0.1 M MgCl$_2$, 1 mg BSA/ml.

10× *Hind*III buffer: 0.5 M NaCl, 0.5 M Tris–HCl (pH 8.0), 0.1 M MgCl$_2$, 1 mg BSA/ml.

10× terminal transferase buffer:

1.4 M sodium cacodylate, 0.3 M Tris base, 10 mM CoCl$_2$. Adjusted to pH 7.6 at room temperature (the pH of the 10-fold diluted solution will be 6.8 at 37°); the buffer is filter-sterilized and stored at 4°.

1 mM dithiothreitol (DTT)

[*methyl*-1′,2′-^{3}H]dTTP and [8-^{3}H]dGTP (New England Nuclear)

NaI solution:

90.8 g of NaI and 1.5 g of Na$_2$SO$_3$ are dissolved in a total 100 ml of distilled water. Insoluble impurities are removed by filtering through Whatman No. 1 filter paper, and 0.5 g of Na$_2$SO$_3$ is added to the solution to protect the NaI from oxidation.

Glass bead suspension:

200 ml of silica-325 mesh (a powdered flint glass obtainable from ceramic stores) is suspended in 500 ml of water. The suspension is stirred with a magnetic stirrer at room temperature for 1 hr. Coarse particles are allowed to settle for 1 hr, and the supernatant is collected. Fines are spun down in a Sorval centrifuge and resuspended in 200 ml of water. An equal volume of nitric acid is added, and the suspension is heated almost to the boiling point in a chemical hood. The glass beads are sedimented by centrifugation and washed with water until the pH is 5–6. The beads are suspended in a volume of water equal to their own volume and stored at 4°.

Ethanol wash solution: 50% ethanol, 0.1 M NaCl, 10 mM Tris–HCl (pH 7.5), 1 mM EDTA.

TE: 10 mM Tris–HCl (pH 7.5), 1 mM EDTA.

SDS/EDTA stop solution: 5% sodium dodecyl sulfate, 125 mM sodium EDTA (pH 8.0).

BPB/XC: 1% bromophenol blue, 1% xylene cyanol, 50% glycerol.

25× TAE: 121 g of Tris base, 18.6 g of disodium EDTA, 28.7 ml of acetic acid in total 1 liter of water.

Phenol/chloroform: a 1 : 1 (by volume) mixture of water-saturated phenol and chloroform.

Procedure

For a diagram of this procedure, see Fig. 2. To determine optimal conditions, pilot reactions are carried out at each step. In all the procedures described below, submicroliter amounts of enzyme are pipeted with a 1-μl Hamilton syringe connected to an autoclaved Teflon tubing.

Preparation of Vector Primer

Step 1. KpnI Digestion of pcDV1. pcDV1 DNA is prepared by the standard Triton X-100–lysozyme extraction method followed by two cycles of CsCl equilibrium gradient centrifugation.[21]

A small-scale *Kpn*I digestion is carried out at 37° in a 20-μl reaction mixture containing 2 μl of 10× *Kpn*I buffer, 20 μg pcDV1 DNA, and 20 units of *Kpn*I enzyme. One-half microliter aliquots are removed every 20 min up to 2 hr and mixed with 9 μl of TE, 1 μl of SDS/EDTA stop solution, and 1 μl of BPB/XC. The aliquots are then analyzed by agarose gel (1%) electrophoresis in 1× TAE, and the minimum time required for at least 95% digestion of the plasmid is determined.

Large-scale digestion is performed with 500–600 μg of DNA in a proportionally scaled-up reaction mixture for the determined time. The reaction is terminated by adding 0.1 volume of SDS/EDTA stop solution. The mixture is extracted twice with an equal volume of phenol/chloroform. One-tenth volume of 2 M NaCl and 2 volumes of ethanol are added to the aqueous solution, and the solution is chilled on dry ice for 20 min and centrifuged at 4° for 15 min in an Eppendorf microfuge. The pellet is dissolved in TE, and the ethanol precipitation is repeated once more. The pellet is dissolved in water (not TE) to make a solution of approximately 4 μg/μl. The DNA concentration is determined by measuring its absorbance at 260 nm. A small sample is analyzed by gel electrophoresis to check the completion of digestion.

Step 2. Poly(dT) Tailing. A pilot tailing reaction is carried out in a 10 μl mixture containing 1 μl of 10× terminal transferase buffer, 1 μl of 1 mM DTT, 20 μg of *Kpn*I-digested DNA (19 pmol of DNA ends), 1 μl of 2.5 mM [³H]dTTP (250–500 dpm/pmol), and 20 units of terminal transferase. The mixture is warmed to 37° prior to addition of the enzyme. After 5, 10, 15, 20, and 30 min of incubation, 1-μl aliquots are taken with Drummond microcapillary pipettes and mixed with 50 μl of ice-cold TE containing 10 μg of plasmid DNA carrier. The DNA is precipitated by addition of 50 μl of 20% trichloroacetic acid (TCA) to each tube and

[21] L. Katz, D. T. Kingsbury, and D. R. Helinski, *J. Bacteriol.* **114**, 577 (1973).

collected on Whatman GF/C glass filter disks (2.4 cm diameter). The filters are washed 4 times with 10 ml of 10% TCA and rinsed with 10 ml of ethanol. After being dried in a oven, the radioactivity on the filter is measured in a toluene-based scintillation fluid. The average number of dT residues per DNA end is calculated based on the total counts incorporated, the counting efficiency of ^{3}H on a glass filter, the specific activity of the [^{3}H]dTTP used, and the number of DNA ends (19 pmol). Figure 4 shows the typical time course of the tailing reaction. The rate of dT incorporation decreases after 10 min of incubation and levels off at around 15 min. The incubation time that results in the formation of 40–50 dT long tails is determined.

Large-scale tailing is then carried out in a 200-μl reaction mixture containing 400 μg of DNA under the conditions determined by doing the pilot reaction. The reaction is terminated by adding 20 μl of the SDS/ EDTA stop solution. After two extractions with phenol/chloroform, the DNA is precipitated by adding 20 μl of 2 M NaCl and 400 μl of ethanol as described above. The ethanol precipitation is repeated once more, and the pellet is dissolved in 100 μl of TE.

Step 3. EcoRI Digestion. A miniscale *Eco*RI digestion is performed in a 5-μl reaction volume containing 2 μl of 5× EcoRI buffer, 2.5 μl (8–9 μg) of poly(dT)-tailed DNA, and 10 units of *Eco*RI. The incubation time required for at least 95% digestion of DNA is determined by analyzing the products on agarose gels as described above after incubation times of 30– 90 min. The remainder of the DNA (100 μl) is digested in a 200 μl reaction mixture under the conditions determined in the pilot study.

After the reaction is stopped with 20 μl of the SDS/EDTA solution and

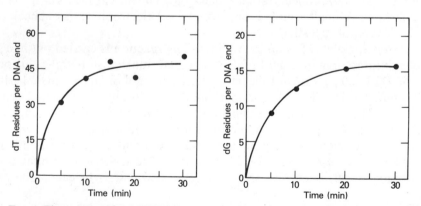

FIG. 4. Time course of poly(dT) (left) and oligo(dG) (right) tailings. Experimental details are described in Methods.

20 μl of BPB/XC, the product is purified by preparative agarose gel (1%, 20 × 25 × 0.6 cm) electrophoresis in 1× TAE buffer. The gel is stained with ethidium bromide (1 μg/ml), and the area of the gel containing the vector primer (the larger DNA fragment) is cut out with a razor blade and sliced into pieces. The gel is weighed and transferred to a plastic bench-top centrifuge tube, to which 2 ml of NaI solution is added for each gram of gel. The tube is placed in a 37° water bath and incubated with occasional vigorous shaking until the gel is completely solubilized. Approximately 0.3 ml of glass bead suspension (0.8 ml/mg DNA) is added, and the solution is cooled on ice and incubated at 4° for 1–2 hrs with gentle rocking. The glass beads, which bind DNA, are recovered by brief centrifugation and washed with 1.5 ml of ice-cold NaI solution once and then twice with 1.5 ml of ice-cold ethanol wash solution. The washed beads are suspended in 1 ml of TE and incubated at 37° for 30 min to dissociate the DNA. The TE is separated from the beads by brief centrifugation in a microfuge; then the beads are extracted once more with 1 ml of TE. Both extracts are pooled, and, after several brief centrifugations to remove residual fine glass particles, the DNA is recovered by ethanol precipitation. This step has 50–80% yield.

Step 4. Oligo(dA)–Cellulose Column Chromatography. Oligo(dA)–cellulose column chromatography is used to remove untailed DNA. All solutions should be RNase free. Oligo(dA)–cellulose is suspended in loading buffer and packed in a column (0.6 cm diameter × 2.5 cm height). The column is washed with several bed volumes of sterile distilled water and equilibrated with the loading buffer at 0–4°. The DNA pellet is dissolved in 1 ml of the loading buffer, cooled on ice and applied to the column. After the column is washed with several bed volumes of the buffer at 0–4°, the bound DNA is eluted with sterile distilled water at room temperature. One milliliter fractions are collected. Small samples are removed from each fraction and the radioactivity is counted in an aqueous scintillation fluid. Radioactive fractions are combined, and the DNA is recovered by ethanol precipitation. The pellet is dissolved in TE to give a solution of about 3 μg/μl. Based on the radioactivity and the amount of DNA recovered, the average length of the poly(dT) tails can be redetermined. The overall yield of vector primer is 30–40%.

The vector primer solution is adjusted to a concentration of 1.2–1.5 μg/μl by adding an equal volume of ethanol, and stored at −20°.

Preparation of Oligo(dG)-Tailed Linker DNA

Step 1. PstI Digestion of pL1. A pilot reaction is performed in a 10-μl solution containing 1 μl of 10× PstI buffer, 5 μg of pL1 DNA, and 5–10 units of PstI. The minimal time for 95% digestion is determined as described above. Two hundred micrograms of pL1 DNA is digested in a 400-μl reaction mixture under the conditions determined. After termination of the reaction with 30 μl of SDS/EDTA solution, the mixture is extracted twice with phenol/chloroform, and the digested DNA is recovered by ethanol precipitation. After being washed with 70% ethanol, the pellet is dissolved in 60 μl of sterile distilled water (not TE). The DNA concentration must accurately be determined by measuring the absorbance at 260 nm.

Step 2. Oligo(dG) Tailing. The optimal conditions for dG tailing are determined in a pilot reaction as above. The reaction mixture (10 μl) contains 1 μl each of 10× terminal transferase buffer, 1 mM DTT, and 1 mM [^{3}H]dGTP (500–1000 dpm/pmol), 12 μg of the PstI-cut pL1, and 12 units of terminal transferase. The mixture is preincubated for 5–10 min at 37° before adding enzyme. After incubation times ranging from 10 to 60 min, 1-μl samples are taken with Drummond microcapillary pipets, and TCA-insoluble counts are determined as described earlier. From the counts and the amount of DNA used, the size of the dG tails is determined. A typical time course is shown in Fig. 4. Under the conditions used, the dG tailing reaction is self-limiting and stops after 10–15 dG residues are added, perhaps due to the formation of double-strand structures within and/or between tails. As a result of the formation of concatenated DNA by base-pairing between tails, the solution becomes more viscous as the reaction proceeds.

The preparative-scale tailing reaction is carried out with 120 μg of DNA in a 100-μl reaction volume. The reaction is terminated with 10 μl of SDS/EDTA solution. After several extractions with phenol/chloroform, the tailed DNA is recovered by ethanol precipitation. The pellet is washed with 70% ethanol and dissolved in 62 μl of TE (approximately 2 μg DNA/μl).

Step 3. HindIII Digestion. The tailed DNA is digested with HindIII. After determining the conditions to be used by means of a pilot reaction (10 μl) containing 1 μl of 10× HindIII buffer, 2 μl (4 μg) of the tailed DNA, and 20 units of HindIII, the rest of the DNA is digested in a 300-μl reaction mixture. The reaction is stopped, and the product is separated by preparative agarose gel (1.8%) (20 × 25 × 0.6 cm) electrophoresis as described above. The tailed linker fragment (450 bp) is recovered by the glass bead method as described above, except that a somewhat larger amount of glass bead suspension and a longer incubation (12–14 hr) at 4° are required to ensure complete adsorption of the DNA to the beads. The DNA recovered is precipitated with ethanol and dissolved in TE to give a concentration of 0.5–1 μg/μl. The DNA concentration and the precise size of the tails are determined from the absorbance at 260 nm and the radioactivity of the solution. After adjusting the concentration to 0.6–1.0 pmol DNA/μl of TE, an equal volume of ethanol is added, and the linker solution is stored at $-20°$.

Comments. Contamination of the vector primer and linker DNA solutions by agarose or impurities in the agarose will strongly inhibit the reactions that these DNAs are involved in. DNA recovered by the glass bead method[22] is very clean, and we have not encountered any difficulty in using it. DNA recovered by other methods may not be clean enough. Gel buffers containing borate should not be used; borate inhibits solubilization of agarose gel by NaI. As a result of the formation of high molecular weight DNA by base-pairing between dG-tails, the yield of linker DNA from agarose gels is generally poor.

cDNA Cloning

Steps in the construction of cDNA–plasmid recombinants are illustrated in Fig. 3.

Step 1. cDNA Synthesis

mRNA is reverse-transcribed with the avian myeloblastosis enzyme in a reaction primed with poly(dT)-tailed vector molecules, resulting in the synthesis of cDNA : mRNA hybrids covalently linked to the vector molecules. The use of high concentrations of deoxynucleoside triphosphates (2 mM each)[23] is designed to inhibit the enzyme-associated RNase H activity. The latter enzyme activity cleaves off the 5' end of RNA on the hybrid in an exonucleolytic fashion, thereby inducing the formation of hairpin structures at the single-stranded 3' ends of the cDNA. Such structures induce cloning artifacts.

Reagents

Poly(A)$^+$ RNA, 10 μg (20% template activity, see section Preparation of mRNA).
Vector primer DNA, 3.2 μg
Avian myelobastosis reverse transcriptase (Seikagaku or Bio-Rad) RNase free.

[22] B. Vogelstein and D. Gillespie, *Proc. Natl. Acad. Sci. U.S.A.* **76**, 615 (1979).
[23] D. L. Kacian and J. C. Mayers, *Proc. Natl. Acad. Sci. U.S.A.* **73**, 2191 (1976).

After being received, the enzyme is aliquoted and stored in liquid nitrogen. Repeated freezing and thawing inactivate the enzyme.

10× reaction buffer:

500 mM Tris–HCl (pH 8.5 at 20°), 80 mM MgCl$_2$, 300 mM KCl, 3 mM dithiothreitol (DTT).

Prepared from autoclaved, filtered 1 M stock solutions of Tris–HCl and the salts, and filter-sterilized 1 M dithiothreitol.

20 mM dNTP: a mixture of 20 mM dGTP, dATP, and dTTP.

20 mM dCTP

Prepared in sterile distilled water and neutralized with 0.1 N NaOH (final pH 6–6.5). The pH is monitored by applying small aliquots to narrow-range pH paper with a capillary pipet. The solution is filter-sterilized with a small Millipore filter, and the concentration is determined by measuring absorbance at 260 nm.

20 mM [α-^{32}P]dCTP (400–700 cpm/pmol):

Approximately 100 μCi of [α-^{32}P]dCTP (200–5000 Ci/mmol) are dried down under reduced pressure and dissolved in 15 μl of 20 mM dCTP.

10% trichloroacetic acid (TCA)

0.25 M EDTA (pH 8.0), filter-sterilized and autoclaved after preparation.

10% SDS

Phenol/chloroform

4 M ammonium acetate

Procedure

Pilot-scale reaction. Prior to large-scale cDNA synthesis, a small-scale reaction should always be performed to test the mRNA and other reagents. The reaction is carried out as described below but in a 15-μl total reaction volume. Under these conditions, the number of effective mRNA template molecules is about twice the number of primer molecules used (generally only 15–20% of the RNA molecules in the final preparation are effective as templates). After a 30-min incubation, a 1-μl aliquot is removed with a Drummond glass capillary pipet and added to 20 μl of 10% TCA. TCA-insoluble species are collected on a 0.45 μm Millipore filter, type HA, and the radioactivity is measured. From the radioactive counts, the specific activity of the ^{32}P-dCTP, and the amount of vector primer (1 μg = 0.5 pmol as primer) used, the average size of the cDNA can be estimated assuming that 100% of the primer is utilized for the synthesis. Generally the average size of the cDNA falls between 1.0 and 1.2 kb (120–150 pmol dCTP incorporated/μg vector primer). The production of long cDNA (4–6 kb) can be confirmed by phenol extraction and ethanol precipitation of an aliquot of the reaction mixture followed by electrophoresis of the product on denaturing agarose gels (0.7–1%) and autoradiography.

If the estimated size of the cDNA is considerably shorter than expected or if the production of long cDNA is not detected, check the mRNA, the reverse transcriptase, and the vector primer by using globin mRNA (commercially available), reverse transcriptase from another lot or source, and/or oligo(dT) primer as controls.

Large-scale cDNA synthesis. Approximately 6–7 μg of mRNA (20% template activity, see Preparation of mRNA) is ethanol-precipitated in a small Eppendorf microfuge tube (never let the RNA dry completely), and dissolved in 16 μl of 5 m*M* Tris–HCl (pH 7.5). The solution is heated at 65° for 5 min and transferred to 37°. The reaction is immediately initiated by addition of 3 μl each of 20 m*M* dNTP and 20 m*M* ^{32}P-dCTP, 2 μl of vector primer (1.2 μg/μl), 3 μl of 10× reaction mixture, and 3 μl of reverse transcriptase (15 U/μl). After a 30-min incubation, the mixture is terminated with 3 μl of 0.25 *M* EDTA (pH 8.0) and 1.5 μl of 10% SDS. At the end of reaction, a 1-μl aliquot is taken with a Drummond glass capillary pipet and precipitated with 10% TCA, and the radioactivity incorporated is determined as above to monitor the cDNA synthesis. The reaction mixture is extracted with 30 μl of phenol/chloroform twice. The aqueous phase is transferred to another tube, and 35 μl of 4 *M* ammonium acetate and 140 μl of ethanol are added. The solution is chilled on dry ice for 15 min and then warmed to room temperature with occasional vortexing to dissolve free deoxynucleoside triphosphates that have precipitated. As it warms up, the solution clears. The cDNA product is then precipitated by centrifugation in an Eppendorf microfuge for 15 min at 4°. The pellet is dissolved in 35 μl of TE and reprecipitated with ethanol as above. This ethanol precipitation step may be repeated once more to ensure complete removal of free deoxynucleoside triphosphates. Finally the pellet is rinsed with ethanol prior to the next step. The yield of product after three ethanol precipitations is 70–80%.

Comments. The use of clean, intact mRNA, fresh RNase-free reverse transcriptase, and a well-prepared vector primer is essential for the efficient synthesis of long cDNA. Inclusion of RNase inhibitors in the reaction mixture is of little value as long as clean enzyme is used, and may have an adverse effect.

Step 2. C-Tailing

Tails 10–15 dC residues long are added to the 3' ends of the cDNA and vector by calf thymus terminal transferase. RNA is not a substrate for this enzyme. Addition of poly(A) to the reaction mixture prevents preferential tailing of unutilized vector molecules, thereby minimizing cloning of vector molecules with no inserts.

Reagents

Terminal deoxynucleotidyltransferase from calf thymus:
Pharmacia, minimal nuclease grade. Stored in liquid nitrogen; once thawed, the preparation should be stored at $-20°$ but for no longer than 1 month. The enzyme is unstable.

$10\times$ reaction buffer:
See Preparation of Vector Primer and Linker DNAs.

$2mM$ [α-^{32}P]dCTP (4000–7000 cpm/pmol):
Approximately 100 μCi of [α-^{32}P]dCTP (200–5000 Ci/mmol) is dried down under reduced pressure and dissolved in 15 μl of 2 mM dCTP.

1 mM dithiothreitol (DTT)

Poly(A) (Miles), 0.15 μg/μl:
Prepared with sterile water. The average size of the chain is 5–6 S.

Procedure. The pellet is dissolved in 12 μl of water, and 2 μl each of $10\times$ reaction buffer, poly(A) (0.15 μg/μl), and 1 mM DTT and 0.6 μl of 2 mM ^{32}P-dCTP are added. After vortexing, the solution is preincubated at $37°$ for 5 min. A 1-μl aliquot is taken with a Drummond microcapillary pipet and precipitated in 20 μl of 10% TCA as described earlier. The reaction is started by the addition of 2 μl of terminal transferase (10–15 units/μl) and lasts for 5 min. Just before stopping the reaction, another 1-μl aliquot is taken and precipitated with TCA as above. The reaction is terminated with 2 μl of 0.25 M EDTA (pH 8.0) and 1 μl of 10% SDS. The mixture is extracted with 20 μl of phenol/chloroform, the aqueous phase is collected, and the product is ethanol-precipitated twice in the presence of 2 M ammonium acetate as in Step 1. The pellet is rinsed with ethanol.

The average size of C-tails formed, where A is the total counts of cDNA formed in the RT reaction (in cpm), B the TCA-insoluble counts in the first aliquot (in cpm), C the TCA-insoluble counts in the second aliquot (in cpm), D the specific activity of the 2 mM ^{32}P-dCTP (in cpm/pmol), and $18 \times B/A$ the recovery of cDNA synthesized or vector primer, is calculated as follows:

$$\text{Average size of C-tails} = \frac{(C - B) \times A}{2 \times D \times B \times 1.2 \text{ pmol (vector primer)}}$$

Formation of C-tails with an average length of 8–15 dC residues should be aimed at. If the length of the tail is much shorter, the binding of the G-tailed linker (see below) to the C-tailed cDNA will not be very strong, and cyclization will be inefficient. If the C-tail is too long (>20–25), expression in mammalian cells will be adversely affected.

To set up tailing conditions for your own lot of enzyme, a pilot reaction should be carried out first at one-half scale with the product of Step 1. Changes in incubation time and/or enzyme amount may be necessary.

Step 3. HindIII Restriction Endonuclease Digestion

Cleavage with *Hin*dIII removes the unwanted C-tail from the vector end and creates a sticky end for the G-tailed linker DNA to bind. *Hin*dIII does not cleave DNA : RNA hybrids efficiently, and we have not had any problem in cloning cDNA containing multiple *Hin*dIII sites.[5]

Using the reaction conditions described below, the *Hin*dIII cleavage is often relatively poor (30–50% digestion). This may be due to inhibition by the large amounts of free mRNA or by salts carried over from the previous step. For the best results, the use of a clean, fresh, active enzyme is necessary. Even with an excess of enzyme it is difficult to obtain complete digestion; that is not necessary in any case and should not be attempted because it will surely lead to a loss of some cDNA clones containing *Hin*dIII sites. The extent of digestion can be roughly estimated by looking at the small fragment (about 500 bp) cleaved off from the vector end on 1.5% agarose gels.

Reagents

*Hin*dIII restriction endonuclease (New England Biolabs), 20 units/μl
10× reaction buffer:
500 mM NaCl, 500 mM Tris–HCl (pH 8.0), 100 mM MgCl$_2$, 1 mg bovine serum albumin (BSA) (Miles, Pentex, crystallized)/ml. Prepared from sterilized stock solutions.

Procedure. The pellet is dissolved in 26 μl of sterilized water, and 3 μl of 10× reaction mixture is added. After brief vortexing, 0.7 μl of *Hin*dIII (20 U/μl) is added to the tube, and it is placed in a 37° water bath for 1 hr. The reaction is stopped with 3 μl of 0.25 M EDTA (pH 8.0) and 1.5 μl of 10% SDS. The mixture is extracted twice with phenol/chloroform, and the product is precipitated twice with ethanol in the presence of ammonium acetate as described above. The pellet is rinsed with ethanol, dissolved in 10 μl of TE, and stored at −20° after addition of 10 μl ethanol (total 20 μl) to prevent freezing. The product is stable for several years under these conditions.

*Comments. Hin*dIII enzyme preparations that have been stored at −20° for more than 2–3 months may not cleave the C-tailed cDNA : mRNA–vector though they will restrict pure plasmids. Use fresh enzyme.

Step 4. Oligo(dG)-Tailed Linker DNA-Mediated Cyclization and Repair of RNA Strand

The HindIII-cut, C-tailed cDNA : mRNA–vector is cyclized by DNA ligase and the oligo(dG)-tailed linker DNA that bridges the C-tail and the HindIII end of the hybrid–vector. The RNA strand is then replaced by DNA in a nick-translation repair reaction: RNase H introduces nicks in the RNA, DNA polymerase I nick-translates utilizing the nicks as priming sites, and ligase seals all the nicks. The end products are closed circular cDNA–vector recombinants.

Reagents

HindIII-cut, C-tailed cDNA : mRNA–vector (Step 3)
dG-tailed linker DNA (0.3 pmol/μl)
5× hybridization buffer:
    50 mM Tris–HCl (pH 7.5), 5 mM EDTA, 500 mM NaCl. Stored at −20°.
5× ligase buffer:
    100 mM Tris–HCl (pH 7.5), 20 mM MgCl$_2$, 50 mM (NH$_4$)$_2$SO$_4$, 500 mM KCl, 250 μg BSA/ml. Stored at −20°.
2 mM dNTP: mixture of dATP, dGTP, and dTTP (2 mM each).
2 mM dCTP
10 mM βNAD
E. coli DNA ligase (Pharmacia), nuclease free
E. coli DNA polymerase (Boehringer Mannheim), nuclease free
E. coli RNase H (Pharmacia), nuclease free
    Contamination of the E. coli enzyme preparations by endonuclease specific for double- or single-stranded DNA can be detected by digestion of supercoiled pBR322 or single-stranded ϕX174 DNAs under the conditions specified below for each enzyme followed by analysis of their degradation by agarose gel electrophoresis.

Procedure. One microliter of the HindIII-digested, C-tailed mRNA : cDNA–vector solution (50% ethanol/TE) is added to a 1.5-ml Eppendorf tube along with 0.08 pmol of oligo(dG)-tailed linker DNA (about 1.5-fold excess over C-tailed ends), 2 μl of 5× hybridization buffer, and enough water to yield a final volume of 10 μl. The tube is placed in a 65° water bath for 5 min, then placed in a 43° bath for 30 min and transferred to a bed of ice. Eighteen microliters of 5× ligase buffer, 70.7 μl of water, and 1 μl of 10 mM NAD are added to the tube, which is then incubated on ice for 10 min. After the addition of 0.6 μg of DNA ligase, the tube is gently vortexed and incubated overnight in a 12° water bath. (The ligase is fairly labile and must be handled with care. It is reasonably stable when stored at −20°.)

After the cyclization step, the RNA strand must be replaced by DNA. The following are added to the tube: 2 μl each of 2 mM dNTP and 2 mM dCTP, 0.5 μl of 10 mM NAD, 0.3 μg of DNA ligase, 0.25 μg of DNA polymerase, and 0.1 unit of RNase H. The mixture is gently vortexed and incubated for 1 hr at 12°, then 1 hr at room temperature. The product is frozen at −20°.

Comments. It is imperative to use pure and active enzymes for this step. Nicking of the cDNA strand by nucleases before the RNA strand is replaced by DNA will completely destroy the recombinant. Before attempting a large-scale transfection, the cyclized product should be tested in a small-scale transfection. One microliter of the product should yield 1–3 × 10⁴ colonies (this product is only 5–10% as active in transfecting cells as a corresponding amount of intact pBR322).

Preparation of Competent DH1 Cells

To make large cDNA libraries with the expenditure of reasonable amounts of mRNA, highly competent cells are required. We routinely prepare competent DH1 cells with transfection efficiencies of 3 × 10⁸ to 10⁹ colonies per microgram pBR322 DNA. The method described below is a modification of Hanahan's procedure.[24] It is somewhat simpler than the original, quite reliable, and can be used to prepare the large quantity of DH1 cells needed to construct big libraries.

Reagents

Double-distilled water or Milli Q water (not regular deionized water) should be used to prepare SOB, SOC, and FTB reagents.

SOB medium:
   2% Bactotryptone (Difco), 0.5% yeast extract (Difco), 10 mM NaCl, 2.5 mM KCl, 10 mM MgCl$_2$, 10 mM MgSO$_4$.
All of the components except the magnesium salts are mixed and autoclaved. The solution is cooled, made 20 mM in Mg^{2+} with a 2 M Mg^{2+} stock solution (1 M MgCl$_2$ + 1 M MgSO$_4$), and filter-sterilized.

[24] D. H. Hanahan, *J. Mol. Biol.* **166,** 557 (1983).

Freeze–thaw buffer (FTB):

10 mM CH$_3$COOK, 45 mM MnCl$_2$, 10 mM CaCl$_2$, 3 mM hexamine cobalt chloride, 100 mM KCl, 10% glycerol.

Adjusted to pH 6.4 with 0.1 M HCl, filter-sterilized, and stored at 4°.

Dimethyl sulfoxide (DMSO):

MCB, spectrograde. A fresh bottle of DMSO should be used for best results. Alternatively, the content of a fresh bottle of DMSO can be aliquoted, stored at −20°, and thawed just before use.

Procedure. Frozen stock DH1 cells are thawed, streaked on an LB-broth agar plate, and cultured overnight at 37°. About 10–12 large colonies are transferred to 1 liter of SOB medium in a 4-liter flask, and grown to an OD$_{600}$ of 0.7 at 18–20°, with vigorous shaking of the flask (200–250 rpm). The flask is removed from the incubator and placed on ice for 10 min. The culture is transferred to two 500-ml Sorvall centrifuge bottles and spun at 3000 g for 10 min at 4°. The pellet is resuspended in 330 ml of ice-cold FTB, incubated in an ice bath for 10 min, and spun down as above. The cell pellet is gently resuspended in 80 ml of FTB, and DMSO is added with gentle swirling to a final concentration of 7%. After incubating in an ice bath for 10 min, between 0.5 and 1 ml of the cell suspension is aliquoted into Nunc tissue culture cell freezer tubes and immediately placed in liquid nitrogen.

The frozen competent cells can be stored in liquid nitrogen for 1–2 months without a significant loss of competency. Prior to use, each preparation of competent cells should be assayed using standard plasmids, such as pBR322.

Transfection of DH1 Cells

Described below is a typical protocol for establishing a cDNA library containing 1–2 × 10^6 clones in *E. coli*. Depending on the size of the library desired, one should scale the reaction up or down.

Reagents

SOC medium: SOB with 20 mM glucose.

Prepare a 2 M filter-sterilized glucose stock, add the glucose after making complete SOB medium, filter-sterilize, and store at room temperature.

Procedure. Four milliliters of competent cells are thawed at room temperature and placed in an ice bath as soon as thawing is complete. A maximum of 60 μl of the cyclized cDNA plasmid is added to the 4 ml of cells. The cells are then incubated in an ice bath for 30 min. Four hundred microliters of the transfected cell suspension is dispensed into each of 10 Falcon 2059 tubes in a bed of ice. They are then incubated in a 43° water bath for 90 sec and transferred to an ice bath. After 1.6 ml of SOC is added, the tubes are placed in a 37° incubator for 1 hr and shaken vigorously. The above transfection steps are repeated once more with another 4 ml of competent cells until a total 120 μl of the cyclized product is finally used. The entire suspension of transfected cells is then combined and transferred to 1 liter of L broth containing 50 μg/ml ampicillin. To determine the number of independent clones generated, a 0.1–0.2 ml aliquot is removed, mixed with 2.5 ml of L-broth soft agar at 43°, and plated on LB-broth agar containing ampicillin. Colonies are counted after an overnight incubation. The rest of the culture is grown to confluency at 37° and aliquoted in Nunc tubes. The tubes are stored at −70° or in liquid nitrogen after addition of DMSO (7%).

Comments. Pilot transfections with cyclized plasmid should always be undertaken before attempting to make a large library. A big water bath should be used to do large-scale transfections or else the water temperature will fall. We commonly observe a decrease of 50% in transfection efficiency from that predicted by pilot assays when we do large-scale transfections as above. The final cyclized product (Step 4, above) is only 5–10% as potent in transfecting cell as intact pBR322. (MC1061 *recA* cells also give a transformation efficiency comparable to DH1 cells.)

Transient Expression Screening of a cDNA Library

This screening method[10] relies on transient expression of cDNA clones in mammalian cells. The approach does not require any prior knowledge of the protein product itself. One only needs a specific biological or enzymatic assay for the presence of the protein in the cells or medium. For the techniques to work, the activity sought must be attributable to a single gene product.

In addition to the library of cDNA clones to be transfected, there are two components necessary for transient expression screening. First, one needs an appropriate recipient cell line. Because the pcD vector carries the SV40 early promoter and origin of replication, COS cells have been used as hosts. These cells contain an origin-defective SV40 genome and constitutively produce T antigen, a viral gene product needed to direct DNA replication initiating at the SV40 origin.[25] These cells are capable of greatly amplifying the number of DNA molecules taken up by the trans-

[25] Y. Gluzman, *Cell* **23**, 175 (1981).

fected cells, increasing the amount of gene product synthesized. The second requirement is an efficient method for introducing DNA into the COS cells. Based on previous studies, DEAE-dextran has been used to effectively introduce DNA into recipient cells. Immediately after introducing DNA into the cells they are treated for 3 hr with chloroquine. This has been found to stimulate levels of expression 5- to 10-fold.[26] Secreted products accumulate in the cell supernatants, and these are harvested 72 hr after the transfection and assayed. Using this procedure, Yokota *et al.*[12] and Lee *et al.*[13,14] have directly isolated full-length cDNA clones for mouse interleukin 2, human granulocyte–macrophage colony-stimulating factor, and mouse B-cell-stimulating factor-1 from concanavalin A-activated T-cell cDNA libraries.

Expression cloning of transiently expressed cDNAs can be used successfully provided that a sensitive assay is available and that the clones of interest are present in reasonable abundance (>0.01%). This procedure should be adaptable to screening protocols involving immunological detection of either intracellular or cell surface gene products. If one is screening for rare cDNA clones (<0.01%), prior enrichment of the library to be screened will probably prove necessary.

Reagents

COS cells[25]:
   The cells are grown on tissue culture plates in Dulbecco modified Eagle's medium (DME) supplemented with 10% fetal calf serum (FCS), L-glutamine (2 mM), and antibiotics (penicillin/streptomycin). They are passed every 3 days by splitting 1 : 10 using a solution of trypsin–EDTA (see below).
Growth medium: DME containing 10% FCS, L-glutamine, penicillin/streptomycin.
Tris-buffered serum-free medium (Tris–SFM): DME buffered with 50 mM Tris–HCl (pH 7.4).
Collecting medium: DME containing 4% FCS, L-glutamine, penicillin/streptomycin.
DEAE–dextran stock solution:
   20 mg/ml (Pharmacia) in sterile water. Do not autoclave or filter.
Chloroquine stock solution:
   10 mM in 50 mM Tris–HCl (pH 7.4), 140 mM NaCl. Filter-sterilize.

Procedure. The main library is split into sublibraries containing an appropriate number of cDNA clones (50–100), and plasmid DNA is pre-

[26] H. Luthman and G. Magnusson, *Nucleic Acids Res.* **11,** 1295 (1983).

pared from the sublibraries. (The size of the sublibraries depends on the sensitivity of the assay for the desired gene product.)

Recipient cells are prepared from confluent plates of COS cells. The medium is removed by suction, and the plates are washed once with phosphate-buffered saline (PBS). One milliliter of trypsin–EDTA is added to each plate, and the plates are incubated at 37° for about 5–10 min to allow the cells to detach. Nine milliliters of DME with 10% FCS is added to each plate, and the cells are released and resuspended by gentle pipeting. Cells are then replated at 10^6 cells/plate in 10 ml of growth medium and incubated at 37° overnight. The medium is removed, and the plates are washed twice with Tris–SFM or PBS. Four milliliters of Tris–SFM containing 80 μl of DEAE–dextran (20 mg/ml) and 10–50 μg of plasmid DNA is added to each plate followed by incubation at 37° for 4 hr. The medium is removed, and the plates are washed with Tris–SFM or PBS. Five milliliters of DME containing 2% FCS, L-glutamine, and 100 μM chloroquine are added to each plate. After incubation at 37° for 3 hr, the plates are washed twice with Tris–SFM or PBS, fed with 4 ml of collecting medium, and incubated at 37°. The supernatants are harvested after 72 hr for assay.

Trouble Shooting Chart

Problems in library construction	Possible causes

1. Low yield of colonies	A. cDNA synthesis
	   i. Heavily nicked vector primer—Prepare with clean enzymes
	B. C-tailing
	   i. Contamination of dNTP from previous step—Try three ethanol precipitations
	   ii. Unstable or inactive enzyme
	   iii. Endonuclease contamination in enzyme
	   iv. Incomplete phenol extraction or incomplete removal of phenol or ethanol after cDNA synthesis
	C. HindIII digestion
	   i. Inactive or unstable enzyme
	   ii. Endonuclease contamination in enzyme
	   iii. Incomplete phenol extraction or incomplete removal of phenol or ethanol after C-tailing
	D. Cyclization and repair
	   i. Inactive ligase or polymerase
	   ii. Nuclease contamination in one of the three enzyme
	   iii. Linker DNA contaminated with agarose
	E. Transfection
	   i. Incompetent DH1 cells

(continued)

Trouble Shooting Chart (*continued*)

Problems in library construction	Possible causes

2. Few plasmids contain inserts	A. cDNA synthesis
	    i. mRNA with low template activity
	        a. Contamination by ribosomal RNA—Use high flow rate for loading a sample and washing column in oligo(dT)–cellulose chromatography
	        b. mRNA contaminated by impurities that inhibit reverse transcriptase—Prepare new one
	    ii. mRNA contaminated by oligo(dT)—Use highest quality oligo(dT)–cellulose; extensively wash column before use
	    iii. Vector primer contaminated by oligo(dA)—Use highest quality oligo(dA)–cellulose; extensively wash column before use
	    iv. Vector primer has T-tails that are too long or too short
	    v. Inactive or unstable reverse transcriptase
	B. C-tailing
	    i. Degraded poly(A)

3. Inserts are short	A. cDNA synthesis
	    i. mRNA degraded or contaminated with impurities—Prepare new mRNA
	    ii. Inactive or RNase-contaminated reverse transcriptase
	B. C-tailing
	    i. Endonuclease-contaminated terminal transferase

[13] Transformation and Preservation of Competent Bacterial Cells by Freezing

By D. A. MORRISON

The preparation of microbial cultures competent for transformation by added DNA, whether using one of the naturally transformable species or an artificial treatment to render cells permeable to DNA, is typically a multistep procedure occupying at least the greater part of one working day. In many applications of transformation where the actual transformation step is done infrequently, and the properties of the transformed cell lines themselves are the principal objects of study, preparation of competent cells for each experiment presents little difficulty. However, in work where transformation is performed daily, as in studies of uptake or recombination mechanisms or where transformation is used as an analytical tool for the bioassay of DNA samples, a ready supply of competent cells of reproducible and known properties can be of great convenience, with the daily repetitive work reduced from hours to minutes. The preservation of competent or precompetent cultures by freezing has long been exploited in studies using *Bacillus subtilis*,[1] *Streptococcus pneumoniae*,[2] and *Haemophilus influenzae*.[3] This chapter presents and discusses a method that has given reasonable success, in this and other laboratories, for *Escherichia coli* cells rendered susceptible to DNA by $CaCl_2$ treatment, a procedure adapted slightly from that described by Lederberg and Cohen[4] with the addition of glycerol to allow freezing of the competent cells.

Preparation and Freezing of Competent Cells

Materials
1. 1.0 M $CaCl_2$
2. 1.0 M $MgCl_2$
3. Glycerol, reagent grade
4. Distilled water
5. L broth: 10 g Bacto-tryptone, 5 g Bacto yeast extract, 5 g NaCl, and 1 g glucose per liter, adjusted to pH 7.0 with 1 N NaOH[5]

(All stock solutions are sterilized in the autoclave.)

[1] D. Dubnau and R. Davidoff-Abelson, *J. Mol. Biol.* **56**, 209 (1971).
[2] M. S. Fox and M. K. Allen, *Proc. Natl. Acad. Sci. U.S.A.* **52**, 412 (1964).
[3] L. Nickel and S. H. Goodgal, *J. Bacteriol.* **88**, 1538 (1964).
[4] E. M. Lederberg and S. N. Cohen, *J. Bacteriol.* **119**, 1072 (1974).
[5] E. S. Lennox, *Virology* **1**, 190 (1955).

RECOMBINANT DNA
METHODOLOGY

Procedure. L broth is inoculated with RR1 cells, usually from a stock of titer 10^9/ml stored at $-82°$ in broth supplemented with 10% glycerol, at an initial density of about 10^7/ml. Two 500-ml portions are incubated in 2-liter Erlenmeyer flasks with aeration at 37°. When the culture reaches $OD_{550} = 0.5$ (about 5×10^8/ml), it is combined in a 4-liter flask and swirled vigorously in a salt–ice water bath for 3 min to bring the temperature to $0°–5°$. The culture is divided among five centrifuge bottles and sedimented in a Sorvall GSA rotor in the cold at 8000 rpm for 8 min. The pellets are gently resuspended, by repeated flushing with a 10-ml pipette, in a total of 250 ml of ice-cold 0.1 M $MgCl_2$ and assembled in one bottle. With gentle agitation this is accomplished in 5–10 min. The cells are sedimented as before. The pellet is resuspended gently in 250 ml cold 0.1 M $CaCl_2$, kept at $0°$ for 20 min, and then sedimented in the cold as before. Finally, the pellet of $CaCl_2$-treated cells is resuspended in 43 ml of 0.1 M $CaCl_2$ mixed with 7 ml of glycerol. Resuspension of pellets is facilitated by repeated flushings with a 10-ml pipette with a wide tip.

The suspension of competent cells is distributed into cold screw-cap tubes in volumes (0.1–10 ml) determined by the projected application. This is most conveniently carried out in a cold room with a chilled pipette or sterile repeating syringe dispenser. The entire set of capped tubes is drained of water and ice briefly and placed in a bath of liquid nitrogen, or of acetone or alcohol chilled with pieces of solid CO_2. After 5 min, they are removed to a freezer for storage at $-82°$.

Transformation

For use, a tube of frozen cells is allowed to thaw in an ice-water bath for 10 min. Cells are then distributed, in the cold, into tubes containing DNA samples in small volumes of buffer or 0.1 M $CaCl_2$, kept at $0°$ for 30 min more, placed in a 42° bath for 2 min, and diluted into 50 volumes of broth at 37° to reduce the calcium concentration and allow resumption of growth.

Comments

1. Tubes of a batch of competent cells prepared by this method have been used without loss of transformability after 15 months at $-82°$.

2. When thawed and maintained on ice, cells lose transformability slowly, retaining 50% of initial activity after 3 hr.

3. The concentration of glycerol does not appear to be critical; 10 or 20% solutions give similar results.

4. Although the freezing and thawing step itself does not reduce the viability of treated cells, it often renders them more sensitive to the $CaCl_2$

and following treatments of the transformation procedure, resulting in a lower final survival. This is partially compensated for by increasing the number of cells used.

5. The CaCl$_2$ treatment procedure described by S. Kushner[6] may also be adapted for cryogenic preservation. Incorporation of 15% glycerol in all buffers, as well as in the dilution broth, preservation of washed cells by freezing in the first washing buffer (0.01 M morpholinopropane sulfonic acid, 0.01 M RbCl) before calcium treatment, and increasing the number of cells used in the final reaction volume by a factor of 10 give reproducible results similar to those described here.

Assay

In the case of transformation involving a drug resistance marker, the products of transformation may be measured by a pour plate procedure taking advantage of diffusion-limited exposure of cells to the selective drug. The plates are designed with a drug-free zone separating the cell-containing layer from the drug agar layer, and the drug concentration is adjusted so that inhibitory levels are not reached in the cell-bearing zone until phenotypic expression is complete.

Materials
  1. L broth
  2. L broth containing 2% agar, maintained at 52°
  3. 13 × 100 mm slip-cap tubes maintained at 49° in a heating block and containing 1.5 ml each of L agar
  4. 15 × 60 mm petri dishes

Procedure. Cells are diluted appropriately in L broth, and a volume to be plated is added to L broth in a 13 × 100 mm tube to a final volume of 1.5 ml at room temperature. A 3-ml base layer of L agar is allowed to harden in each petri dish. The 1.5-ml cell sample is poured into a tube of molten agar, mixed, and poured onto the base layer. When this has hardened (about 1–2 min), a third 3-ml layer of L agar is pipetted over the cell-containing layer. Finally, for selection of drug-resistant transformants, an additional (fourth) 3-ml layer of L agar containing the drug is applied and allowed to harden. After overnight incubation at 37°, colonies appear as small spheres or disks, about ½–2 mm in diameter, depending on the time of incubation and the total colony density on the plate. They may be counted conveniently under a dissecting microscope fitted with a square-ruled eyepiece micrometer disk.

---

[6] S. R. Kushner, *in* "Genetic Engineering" (H. W. Boyer, ed.) p. 17. North-Holland Publ., Amsterdam, 1978.

Comments

1. The amount of drug added in the final agar layer must be chosen with care for each genetic marker and each recipient strain. The appropriate range may be determined by plating a set of identical samples of a transformed culture, within a few minutes after initial dilution into broth, on a series of plates, and incorporating a series of drug concentrations in the final agar layers. For the tetracycline resistance gene of pMB9 in RR1 recipient cells, for example, it is observed that drug levels from 15 to 60 μg/ml are satisfactory. Below 15 μg/ml, transformants are difficult or impossible to detect amid a background of small drug-sensitive colonies. Above 60 μg/ml even transformants fail to form colonies, because of drug effects on expression and/or growth.

2. The events occurring after transformation of a thawed frozen competent culture of strain RR1 and dilution into broth are illustrated in Fig. 1.[7] The viable cells resume growth and division very quickly. The number of plasmid-bearing cells also begins to increase promptly, in parallel with the total viable population. There is no segregation delay, often observed in other transformation systems. This result, obtained with the diffusion-limited, double-overlay plating method described here, is compared with determinations of the number of drug-resistant cells made by plating samples directly in drug agar. It is clear that plasmid-bearing cells do not immediately express the resistance phenotype; about a 1-hr incubation in broth is required before all transformants can survive drug exposure.

3. The net result of the process described in the preceding paragraph is that, by the time of full phenotypic expression of the transformed gene, each initial transformant is represented by several descendants. The immediate double-overlay plating method allows full phenotypic expression and also ensures that all the cells descended from a single transformant are confined to a single colony; each colony thus represents an independent transformation event if plated soon after the transformation heat pulse.

4. The plating method described combines a number of features of value for routine or high-volume work. With a standard $7\times-30\times$ dissecting microscope, the number of colonies that may be quantitated on a single small petri dish covers a very wide range, up to 50,000 with calibration of the eyepiece grid at several magnifications. Samples may be assayed immediately after transformation, as the drug plate design allows

[7] D. Morrison, *J. Bacteriol.* **132**, 349 (1977).

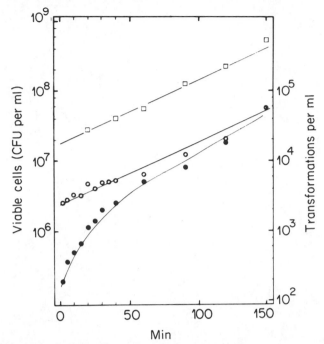

FIG. 1. Segregation and phenotypic expression of tetracycline resistance in an RR1 culture transformed with pMB9 DNA after preservation at −82°. Samples of the transformed culture taken at the indicated times after dilution into L broth were plated without drug (□), by the diffusion-limited double-overlay method (○), and by direct embedding in drug agar (●) (Morrison,[7] reproduced with permission).

phenotypic expression after plating. The amount of agar used per assay is small. For assay of many samples, all broth and agar volumes may be dispensed into tubes or onto plates with a sterile repeating syringe, such as the Cornwall Pipetter. Selective plates need not be prepared in advance. Finally, cells from individual colonies are easily retrieved with a toothpick or Pasteur pipette.

[14] Plasmid Screening at High Colony Density

By Douglas Hanahan and Matthew Meselson

Bacterial plasmid vectors are widely employed in the isolation, amplification, mutagenesis, and analysis of DNA sequences. A number of applications involve locating the products of rare events. Such identification can be considerably facilitated by the ability to screen for specific plasmids at high colony density. This chapter updates and expands upon methodology devised for that purpose.[1-3]

Principle

This procedure involves establishing bacterial colonies directly upon nitrocellulose filters laid on agar plates and replicating the distributions onto other nitrocellulose filters. Maintaining and replicating colonies on a nitrocellulose support allows very large numbers to be readily manipulated (at least 10^5 colonies per 82 mm in diameter filter). Colonies can be established under one set of conditions and then the distribution (or a replica of it) transferred easily to another.

Initially, a primary colony distribution is created. Colony-forming bacteria are spread on a nitrocellulose filter laid on an agar plate. The plate is incubated to establish small colonies, which are then replicated to other nitrocellulose filters; these in turn are incubated to establish duplicate colonies. The replicas can then be replicated again. transferred to plates containing different drugs or inducers, or lysed and hybridized. Probing replicas with radioactive nucleic acids can identify a colony of cells carrying sequences homologous to the probe. Keying back from an autoradiogram to the master plate localizes the colony. A few colonies are removed from the region(s) of hybridization and dispersed in medium, and an appropriate dilution is spread on a fresh nitrocellulose filter, to give 100–200 colonies. This enriched population is replicated and probed, allowing isolation of pure clones of the hybridizing species.

Maintaining colonies on filters also provides for long-term storage of large distributions (banks, libraries). A sandwich comprised of two nitrocellulose filters and a colony array can be stored at −55° to −80° for an indefinite period, thawed, and separated to give two viable replicas.

[1] M. Grunstein and D. S. Hogness, *Proc. Natl. Acad. Sci. U.S.A.* **72,** 3961 (1975).
[2] D. Hanahan and M. Meselson, *Gene* **10,** 63 (1980).
[3] M. Grunstein and J. Wallis, this series, Vol. 68, p. 379.

RECOMBINANT DNA
METHODOLOGY

Procedures

Materials

Media and Solutions. Virtually any bacterial growth medium may be used. A typical rich medium is Luria broth supplemented with magnesium (LM): 1% Bacto-tryptone, 0.5% Bacto yeast extract, 10 mM NaCl, 10 mM MgCl$_2$, 1.5% Bacto agar. F plates, used in the preparation of frozen replicas, include 5% glycerol. Chloramphenicol plates are supplemented with 170–250 μg of chloramphenicol per milliliter. Tetracycline-HCl (tet) and sodium ampicillin (amp) are employed at minimum concentrations for the particular strain in use, generally 7–17 μg/ml for tet, and 30–100 μg/ml for amp.

SET is 0.15 M NaCl, 30 mM Tris, pH 8, 1 mM EDTA. Denhardt solution[4] is 0.02% Ficoll, 0.02% polyvinylpyrolidone, and 0.02% bovine serum albumin (BSA). Formamide (MCB FX420) is deionized with Bio-Rad AG501-X8 ion-exchange resin and stored frozen.

Filters. Precut nitrocellulose filters are available from a number of suppliers. All bind DNA quite efficiently. Two important criteria bear on this application of nitrocellulose: (*a*) the probability of establishing a single cell into a growing colony (plating efficiency); and (*b*) dimensional stability and integrity through the various treatments (including melting annealed probes off a filter followed by rehybridization with a different probe). In this context, Millipore HATF (Triton-free) filters have proved to be the most reliable and durable[2,5] and are available in 82-mm and 127-mm diameters for use in 100-mm and 150-mm petri dishes, respectively.

Filters may be prepared in advance for plating and replication and then stored indefinitely. The filters are floated on double-distilled water (dd H$_2$O), submerged when wetted, and sandwiched between dry Whatman 3 MM filters into a stack, which is wrapped in aluminum foil and autoclaved on liquid cycle for 15 min. After sterilization, the pack of filters is sealed in a plastic bag to maintain humidity.

Alternatively, filters may be wetted by placement on an agar plate just prior to use. This is convenient when only a few plates are to be prepared. The filters are sterile as supplied and may be resterilized by ultraviolet irradiation.

Plating and Growth

A sterile nitrocellulose filter is laid upon an agar plate. The filter has a curl to it, and final placement without air bubbles is easier if the curl of the

[4] D. T. Denhardt, *Biochem. Biophys. Res. Commun.* **23,** 641 (1966).
[5] F. G. Grosveld, H. H. M. Dahl, E. de Boer, and R. A. Flavell, *Gene* **13,** 227 (1981).

filter matches the curl up at the edges of the plate. Any remaining bubbles can be nursed out using a bent glass Pasteur pipette.

The cells are applied to the filters (in a 200–400 μl volume for 82-mm, twice this volume for 127-mm filters) and quickly spread with a bent Pasteur pipette (L-shaped) to distribute the cells on the filter. A plating wheel helps to spread the suspension while leaving a blank edge (or rim) on the filter. To assure even distribution, the suspension should not be drying in patches during the spreading process. A uniform film of liquid will be visible (by reflection) immediately upon completion of the spreading. It will dry in a minute or two. If dry patches form during the spreading, use larger volumes of medium.

The plates are then incubated until colonies 0.1 mm in diameter appear. Incubation at 30–32° facilitates control of colony size. The most frequent cause of distortion in colony distributions during replication arises from overgrowth of colonies. For any density, colonies 0.1–0.2 mm in diameter are best. Virtually any visible colony will transfer properly, while colonies greater than 1 mm in diameter can transfer nonuniformly. Heterogeneity in colony size is not a problem if the largest colonies are kept below 0.5 mm in diameter.

An alternative to plating directly on filters is to lift the colonies off an agar plate much in the same manner that phage are lifted off plaques in the method of Benton and Davis.[6] Colonies are established on well dried agar plates, which are then refrigerated for a few hours (to retard smearing). A dry nitrocellulose filter is placed on the colony distribution and allowed to become wet; then the filter is lifted off, carrying the colonies with it. This filter can then either be placed (colonies up) on a fresh plate, replicated immediately, or chloramphenicol amplified, etc.[7,8] Colony lifts work reasonably well at moderate densities, but reliability at high density remains uncertain.

Replication

The template filter is peeled off its agar plate and laid, colonies up, on a bed of sterile Whatman paper. A wetted sterile nitrocellulose filter is held between two flat-bladed forceps and laid upon the template filter. The sandwich is pressed firmly together with a velvet-covered replica plating tool (Fig. 1). The filters are keyed to each other by making a characteristic set of holes in the sandwich with a large needle. The replica is peeled off the template and placed on a fresh agar plate.

---

[6] W. D. Benton and R. W. Davis. *Science* **196,** 180 (1977).
[7] D. Ish-Horowicz and D. F. Burke. *Nucleic Acids Res.* **9,** 2989 (1981).
[8] G. Guild and E. Meyerowitz, personal communication.

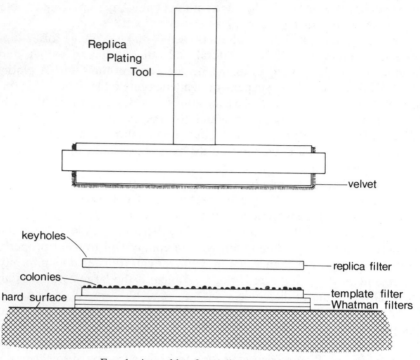

FIG. 1. Assembly of a replication apparatus.

Additional replicas may be made by placing fresh filters on the template and repeating the process. In this case the sandwich is inverted and the existing holes in the template are used to key the two filters.

Four to five replicas may be made off a template without sacrificing complete transfer. If additional replicas are desired, the template filter can be reincubated to replenish its colonies. The replicas are incubated at 30–37° until the colonies develop to the desired size, which is generally 0.5–1 mm in diameter for lysis or amplification, but again only 0.1–0.2 mm in diameter for further replication or frozen storage.

The sterile nitrocellulose filters used in the replication step should be moist and supple but free of visible surface moisture. Overly dry filters can be placed briefly on a fresh agar plate just prior to use, which assures the desired degree of wetness.

Various alternatives to the velvet replica plating device may be used: automobile pistons, a pair of thick glass plates,[5] etc. The velvet seems to help even out the force distribution over the occasionally uneven surface of the filter–colony sandwich.

Lysis and Prehybridization

Replica filters to be probed are lysed (colonies up) on several sheets of Whatman 3 MM paper that have been barely saturated with 0.5 M NaOH. It is convenient to float the layers of Whatman paper in a tray that is partially filled with liquid, which is then poured off, leaving the saturated paper in the tray. After several minutes, the replica filters are peeled off the lysis pad, blotted momentarily on dry paper towels, and then placed on a second pad saturated with 0.5 M NaOH. This is followed by two neutralization steps, first with 1 M Tris, pH 8, and then with 1 M Tris, pH 8, 1.5 M NaCl, each for 3–5 min. The replica filters are then placed (colonies still up) on sheets of dry Whatman 3 MM to blot the surface film of liquid down through the filter. After brief air drying, the filters are sandwiched between Whatman paper and baked under vacuum at 60–80° for 1–3 hr. The filters should still be supple, and not yet bone white when placed in the vacuum oven.

Regardless of the hybridization probe and conditions, prehybridization is important both to occupy nonspecific binding sites and also to clean off cellular debris from the filters. Polypropylene food storage boxes are very useful both in the prehybridization and posthybridization washes. Agitation is very important throughout.

The baked replica filters are floated on double-distilled H_2O, submerging when wetted and then placed in a box containing several hundred milliliters of 5 × Denhardt solution, 0.5% SDS. The filters are incubated with agitation at 68° for at least 2 hr (up to 12 hr). The filters are then removed and placed in the hybridization solution.

Hybridization, Washing, and Visualization

Colony hybridization can be performed under a wide variety of conditions. Those described below have proved to be convenient and give good signal-to-noise ratios. Hybridization in rotating water baths is particularly important for reducing nonspecific background. Filters are placed in heat-sealable freezer bags, the hybridization solution is added, and the bag is sealed and then placed in a water bath of the desired temperature and agitated at 25–100 rpm. A good rule of thumb is to use 7 ml for the first 82-mm filter, and 2 ml for each additional one (2 × for 137-mm filters). Ten large filters can be readily screened in one bag provided that a reasonable volume is used and that the bag is well agitated during the hybridization.

DNA. Standard DNA hybridization conditions are 6 × SET, 1 × Denhardt solution, 0.1% SDS, at 68°. Alternative conditions are 6 × SET, 1 × Denhardt solution, 50% deionized formamide, 0.1% SDS, 42–45°. A number of investigators add single-stranded carrier DNA (50–150 μg/ml) and/or tRNA (100–250 μg/ml) to hybridizations. The usefulness of these additions are dependent on the characteristics of the probe, but are unlikely to be counterproductive even when unnecessary. The hybridization probe is denatured by the addition of 0.1 volume of 1 N NaOH for 2 min at room temperature, neutralized by the addition of 0.1 volume of 1 M Tris, pH 8, and 0.1 volume of 1 N HCl, and then added to the hybridization solution to a final concentration of less than 20 ng/ml.

Several 30-min posthybridization washes are performed with agitation in a few hundred milliliters of 2 × SET, 0.2% SDS, at 68° in large polypropylene boxes. Alternative or additional wash conditions are 0.1 × SET, 0.1% SDS, 53°.

RNA. Standard RNA hybridization conditions are 6 × SET, 1 × Denhardt solution, 50% deionized formamide, 0.1% SDS, 250 μg of tRNA per milliliter, at 42–45° for 10–20 hr. Several washes are performed in 2 × SET, 0.2% SDS, at 68° (with agitation) and/or 0.1 × SET, 0.1% SDS at 53°. The probe is employed in the hybridization at less than 20 ng/ml.

Oligonucleotides.[9-12] Oligonucleotides are hybridized in 6 × SET, 5 × Denhardt solution, 250 μg/ml tRNA, 0.5% Nonidet P-40 (NP-40) (Shell oil) (or 0.1% SDS). Optimal hybridization temperatures depend on the length and nucleotide composition of the oligonucleotide. A good approximation for the hybridization temperature T_H is given by $T_H = T_D - (3°)$, where $T_D = 2° ×$ [the number of A-T base pairs] plus 4°C × [the number of G-C base pairs].[13] For mixed probes of average GC composition, a reasonable approximation is $T_H = 3° ×$ the length of the oligonucleotide. The oligonucleotides are used at a final concentration of 2 ng/ml. The filters are washed with four changes of 6 × SET, 0.5% SDS at 20° over 20 min, followed by a 1-min wash at T_H.[11] Nonspecific hybridization of oligonucleotides to colonies is reduced significantly by prehybridizing for 6–8 hr at 68° in 5 × Denhardt solution, 0.5% SDS, 10 mM EDTA, after which the filters are gently rubbed with a gloved hand, to remove any remaining colonial matter (D. H., unpublished observations).

Mounting and Autoradiography. After posthybridization washes in food storage boxes with gentle agitation, the filters are blotted dry on paper towels, but not allowed to become bone dry and brittle. The filters

[9] J. W. Szostak, J. I. Stiles, B.-K. Tye, D. Chiu, F. Sherman, and R. Wu, this series, Vol. 68, p. 419.

[10] R. B. Wallace, M. J. Johnson, T. Hirose, T. Miyake, E. H. Kawashima, and K. Itakura, *Nucleic Acids Res.* **9,** 879 (1981).

[11] R. B. Wallace, M. Schold, M. J. Johnson, D. Dembek, and K. Itakura, *Nucleic Acids Res.* **9,** 3647 (1981).

[12] M. Smith, *in* "Methods of RNA and DNA Sequencing" (S. M. Weissman, ed.), in press. Praeger, New York, 1983.

[13] S. U. Suggs, T. Hirose, T. Miyake, E. H. Kawashima, M. J. Johnson, K. Itakura, and R. B. Wallace, *ICN–UCLA Symp. Dev. Biol. Using Purified Genes,* VXX, p. 683 (1981).

are sandwiched between two sheets of plastic wrap (which keeps the filters supple) and then mounted on a solid support (e.g., cardboard, old X-ray film). The mounted filters are marked with radioactive ink and subjected to autoradiography, generally using intensifying screens at −70°.

The developed film is aligned to the mounted filters using the radioactive ink marks. The keyholes on the filters can then be marked either directly on the film or on a clear plastic sheet (along with positive hybridization spots). The keyed film can be aligned to the master filter in two ways. The film (or transparent sheet) can be placed on top of the plate. The keyholes are aligned, and the colonies are identified by sighting straight down through the film. An alternative is to put the film on a light box, remove the filter from its plate and lay it on a small square of plastic wrap, and then place the filter (plus plastic wrap) directly on the film, rotating it to align keyholes with their corresponding marks. Circled positives are then readily identified, without parallax problems.

Notes. In contradiction to benefits observed with DNA blots and phage screens, the authors and other investigators have not found dextran sulfate to improve the signal-to-noise ratios in colony hybridization.

Filters may be rehybridized with different probes. The filters are cut out of their plastic wrap mounts (still supple), floated on, and then submerged in several hundred milliliters of 5 × Denhardt solution, 0.5% SDS in a food storage box. The filters are incubated with gentle agitation at 80° for several hours followed by a second wash at 68° with fresh solution. Millipore filters hybridized at 68° can be rehybridized 2–4 times before significant disintegration occurs. Filters hybridized in formamide at 42–45° can be rehybridized considerably more.

High levels of nonspecific hybridization (noise or background) may result from several factors: excessive quantities of probe; an inappropriately low hybridization temperature, which can often be raised several degrees without affecting specific hybridization; or a "dirty" probe, which can be filtered quickly through a 0.2-μm filter prior to its addition to the hybridization. A number of investigators routinely filter all hybridization probes.

Freezing

Distributions of colonies on nitrocellulose may be stored indefinitely at −55 to −80°. Template filters (which may themselves be replicas) are prepared as usual, except that they are either grown on F plates or transferred to F plates and incubated for a few hours (30–37°). Sterile dry nitrocellulose filters are wetted on F plates just prior to use. The replication is carried through the keying step. The two filters are left together and

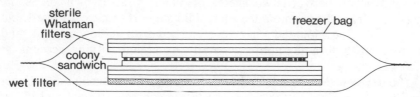

FIG. 2. A colony distribution prepared for frozen storage.

sandwiched between several dry Whatman filters, plus one wet Whatman filter to maintain humidity. The stack is placed inside a freezer bag, sealed, and stored below −55° (Fig. 2). When needed, the bag is removed and brought to room temperature. The filters are peeled apart, laid on fresh agar plates, and incubated until small distinct colonies appear. As with basic replication, the template colonies usually develop more rapidly than those on the replica; incubation at 30–32° allows more control of the colony size.

Chloramphenicol Amplification

A replica filter may be transferred to an agar plate containing chloramphenicol, which upon incubation at 37° for 12–48 hr will effect substantial amplification of appropriate plasmid vectors (e.g., pBR322). For most healthy *E. coli* K12 stains, chloramphenicol is quite effective at 170–250 μg/ml, although levels ranging from 12 μg/ml to 500 μg/ml have been used. Very large plasmids (and cosmids) do not amplify well in some cases, and it is advisable to test the particular host–vector in use to verify the utility of amplification.

Comments

This technique has been employed in the isolation of both cDNA and genomic DNA sequences from a wide variety of organisms. No particular limitations due to fidelity of replication or frozen storage have been observed. Density constraints arise primarily from the quality and specific activity of the hybridization probe. This applies particularly to mixed oligonucleotide probes, where the optimum salt concentration and temperature may be very sensitive to the number and particular sequences of the oligonucleotide mixture, with signal and noise in a delicate balance. It is, however, demonstrably possible to screen a large colony distribution with a mixed oligonucleotide probe.[14]

[14] M. Noda, Y. Furutani, H. Takahashi, M. Toyosato, T. Hirose, S. Inayama, S. Nakanishi, and S. Numa, *Nature (London)* **295,** 202 (1982).

In another variation, this technique has been applied in probing colony arrays on nitrocellulose filters with radioactive antibodies in order to locate sequences through recognition of their gene products. A cDNA clone encoding part of the structural gene for chicken tropomyosin was isolated from a high-density colony distribution, using an antibody to chicken tropomyosin to identify colonies expressing an antigenic portion of the protein.[15] The pilus protein gene of *Nisgeria gonorrhoeae* was cloned into *E. coli,* and the positive colony was identified through its expression and subsequent recognition by a suitable antibody.[16] As with oligonucleotide probes, hybridization conditions and density limitation are likely to vary considerably with the characteristics of the probe and its complement.

This technique has been applied to the construction and storage of cosmid banks and to the isolation of specific sequences from them.[5,17] Densities have been constrained to about 10^4 colonies per 127-mm filter by the low titers of packaged recombinant cosmids obtained when using higher eukaryotic DNA. This does not represent an intrinsic sensitivity limit, as 10^5 cosmids carrying yeast DNA (50 genome equivalents) have been screened on one 127-mm filter using a moderate specific activity DNA probe (10^7 dpm/μg) (D. H. and B. Hohn, unpublished). Steinmetz *et al.*[17] reported that if initial platings of cosmids are incubated first on a low concentration of tet (5 μg/ml) for several hours, and then the filters are transferred to higher tet plates (10 μg/ml), three times the number of colonies form as compared to initial plating on the higher concentration of tet (an amp selection was not similarly tested).

There is no clearly defined maximum colony density for plasmid screening when a pure DNA probe of greater than 10^7 dpm/μg is used, and it is likely that 10^6 colonies can be replicated and screened on a single filter.

Acknowledgments

The authors thank many colleagues for comments and suggestions and Patti Barkley for preparing the manuscript. D. H. is a junior fellow of the Harvard Society of Fellows.

[15] D. M. Helfman, J. R. Feramisco, J. C. Fiddes, G. P. Thomas, and S. H. Hughes, *Proc. Natl. Acad. Sci. U.S.A.* **80,** 31 (1983).

[16] T. F. Meyer, N. Mlawer, and M. So, *Cell* **30,** 489 (1982).

[17] M. Steinmetz, A. Winoto, K. Minard, and L. Hood, *Cell* **48,** 489 (1982).

[15] New Bacteriophage Lambda Vectors with Positive Selection for Cloned Inserts

By JONATHAN KARN, SYDNEY BRENNER, and LESLIE BARNETT

Molecular cloning methods eliminated the necessity for physical fractionation of DNA and permitted, for the first time, the isolation of eukaryotic structural genes.[1-7] In principle, any eukaryotic gene may be isolated from a pool of cloned fragments large enough to give sequence representation of an entire genome. A simple multicellular eukaryote such as *Caenorhabditis elegans* has a haploid DNA content of approximately 8×10^7 bp.[8] Assuming random DNA cleavage and uniform cloning efficiency, a collection of 8×10^4 clones with an average length of 10^4 bp will include any genomic sequence with greater than 99% probability. Similarly, the human genome with 2×10^9 bp will be represented by 10^6 clones of 10^4 bp length.[6] Clones of interest are then identified in these genome "libraries" by hybridization and other assays, and flanking sequences can be obtained in subsequent "walking" steps.

Bacteriophage lambda cloning vectors offer a number of technical advantages that make them attractive vehicles for the construction of genome libraries.[9] DNA fragments of up to 22 kb may be stably maintained, and recombinants in bacteriophage lambda may be efficiently recovered by *in vitro* packaging. The primary pool of clones may be amplified without significant loss of sequences from the population by limited growth of the phage. Subsequently the entire collection may then be stored as bacteriophage lysates for long periods. Finally, bacteriophage plaques from the amplified pools may readily be screened by the rapid and sensitive

[1] P. C. Wensink, D. J. Finnegan, J. E. Donelson, and D. Hogness, *Cell* **3**, 315 (1974).

[2] M. Thomas, J. R. Cameron, and R. W. Davis, *Proc. Natl. Acad. Sci. U.S.A.* **71**, 4579 (1974).

[3] L. Clarke, and J. Carbon, *Cell* **9**, 91 (1976).

[4] S. M. Tilghman, D. C. Tiemeier, F. Polsky, M. H. Edgell, J. G. Seidman, A. Leder, L. W. Enquist, B. Norman, and P. Leder, *Proc. Natl. Acad. Sci. U.S.A.* **75**, 725 (1978).

[5] S. Tonegawa, C. Brach, N. Hozumi, and R. Scholler, *Proc. Natl. Acad. Sci. U.S.A.* **74**, 3518 (1977).

[6] T. Maniatis, R. C. Hardison, E. Lacy, J. Lauer, C. O'Connell, and D. Quon, *Cell* **15**, 687 (1978).

[7] F. R. Blattner, A. E. Blechl, K. Denniston-Thompson, M. E. Faber, J. E. Richards, J. L. Slighton, P. W. Tucker, and O. Smithies, *Science* **202**, 1279 (1978).

[8] J. E. Sulston, and S. Brenner, *Genetics* **77**, 95 (1974).

[9] N. E. Murray, *in* "The Bacteriophage Lambda II," Cold Spring Harbor Laboratory, Cold Spring Harbor, New York.

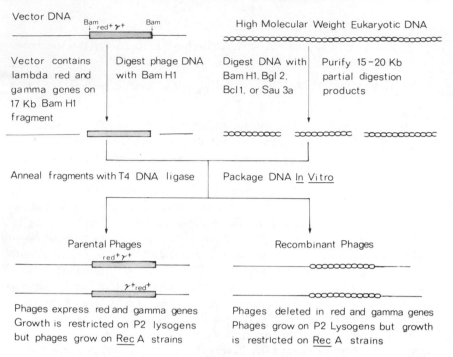

Fig. 1. Schematic diagram outlining the construction of recombinants using the λ1059 vector.

plaque hybridization method of Benton and Davis,[10] genetic selections,[11,12] or immunological assays[13-16] that take advantage of the high levels of transcription that may be achieved with clones in bacteriophage.

Most bacteriophage vectors are substitution vectors that require internal filler fragments to be physically separated from the vector arms before insertions of foreign DNA.[2,6,7,9,17] This step is inefficient and leads to the contamination of the recombinant phage pools with phages harboring one

[10] W. D. Benton, and R. W. Davis, *Science* **196,** 180 (1977).
[11] B. Seed, unpublished results.
[12] M. Goldfarb, K. Shimizu, M. Pervcho, and M. Wigler, *Nature (London)* **296,** 404 (1982).
[13] B. Sanzey, T. Mercereau, T. Ternynck, and P. Kourilsky, *Proc. Natl. Acad. Sci. U.S.A.* **73,** 3394 (1976).
[14] A. Skalka, and L. Shapiro, in "Eucaryotic Genetics Systems" (*ICN-UCLA Symp. Mol. Cell. Biol.* **8**), p. 123. Academic Press, New York, 1977.
[15] S. Broome, and W. Gilbert, *Proc. Natl. Acad. Sci. U.S.A.* **75,** 2746 (1978).
[16] D. J. Kemp, and A. F. Cowman, *Proc. Natl. Acad. Sci. U.S.A.* **78,** 4520 (1981).
[17] N. E. Murray, and K. Murray, *Nature (London)* **251,** 476 (1974).

or more of these fragments.[6,7] Some years ago we developed a bacterio-phage lambda *Bam*HI cloning vector, lambda 1059, with a positive selec-tion for cloned inserts.[18] This feature allows construction of recombinants in lambda without separation of the phage arms. A schematic diagram outlining the strategy we have adopted for cloning in bacteriophage 1059 (and derivative strains) is shown in Fig. 1. Genomic DNA is partially di-gested with restriction endonucleases to produce a population of DNA fragments from which molecules 15–20 kb long are purified by agarose gel electrophoresis. The size-selected fragments are ligated with T4 DNA ligase to the arms of the phage vector cleaved with an appropriate en-zyme. Viable phage particles are recovered by *in vitro* packaging of the ligated DNAs, and a permanent collection of recombinant phages is then established by allowing the phages harboring inserts to amplify through several generations of growth on a strain that restricts the growth of the original vector. Clones of interest are then identified by hybridization with specific probes.

Principle of the Method

Our selection scheme for inserts is based on the spi phenotype of lambda. Spi⁻ derivatives of phage lambda are phages that can form plaques on *Escherichia coli* strains lysogenic for the temperate phage P2. This phenomenon was first described by Zissler *et al.*,[19] who demon-strated that concomitant loss of several lambda early functions at the *red* and *gamma* loci was required for full expression of the phenotype. We reasoned that if the *red* and *gamma* genes were placed on a central frag-ment in a bacteriophage lambda vector, then recombinants that substi-tuted foreign DNA for this fragment should be spi⁻ and distinguished from the parent vector by plating on P2 containing strains. In order to ensure that the *red* and *gamma* genes were expressed in either orientation of the central fragment, we placed these genes under pL control, and specific *chi* mutations[20,21] were introduced into the vector arms in order to assure good growth of the recombinant phages. Selection for the spi phenotype alone does not distinguish between phages that harbor foreign DNA frag-ments and phages that have simply deleted the central fragment. We took advantage of lambda's packaging requirements to complete the selection

[18] J. Karn, S. Brenner, L. Barnett, and G. Cesareni, *Proc. Natl. Acad. Sci. U.S.A.* **77,** 5172 (1980).
[19] J. Zissler, E. R. Signer, and F. Schaefer, *in* "The Bacteriophage Lambda" (A. D. Her-shey, ed.), p. 455. Cold Spring Harbor Laboratory, Cold Spring Harbor, New York, 1971.
[20] F. W. Stahl, J. M. Craseman, and M. M. Stahl, *J. Mol. Biol.* **94,** 203 (1975).
[21] D. Henderson, and J. Weil, *Genetics* **79,** 143 (1975).

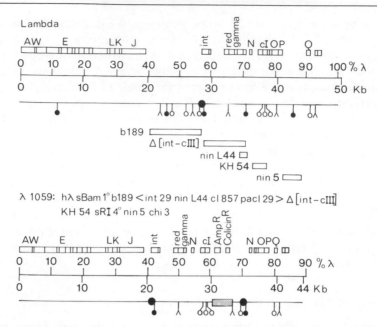

Fig. 2. Structure of λ1059. The top panel shows the *Bam*HI (●), *Eco*RI (∧), and *Hin*-dIII (○) restriction maps of lambda, and the position of many of the known lambda genes. The bars underneath the lambda map indicate the map positions of the deletions used in the construction of λ1059. A restriction map of λ1059 is shown here. The left arm of the phage carries the λ structural genes *A–J*. The *s*baml° mutation and the *b*189 deletion remove the *Bam*HI sites from this arm. The central fragment carries the sequence from the first *att* site (△●P′, shown on the map as a large filled circle) to the *Bgl* site at coordinate 745 in the *cro* gene. At this juncture sequences from the mini ColE1 plasmid pACL29 (stippled region) are introduced. This plasmid introduces the β-lactamase gene (*Amp*ᴿ) and colicin immunity gene (*Colicin*ᴿ). The central fragment terminates in a duplicated λ*att* site (P●P′). This sequence is present in wild-type lambda from the *Eco*RI site at coordinate 543 to the *Bam*HI site at 578. The *Bam*HI site at 714 has been removed from the central fragment by the *nin*L44 deletion. The short right arm carries a deletion △[*int-c*III] originally made *in vitro* by removing DNA from between the two *Bam*HI sites at 580 and 714, the KH54 deletion, which removes the *rex* and *c*I genes, and the *nin*5 deletion. Substitution of the central fragment produces a spi⁻ phage with a *b*189 arm, a single λ*att* site, a 9–22 kb insert cloned between the *Bam*HI sites at 580 and 714, and an immunity arm with the KH54 and *nin*5 deletions. The growth of these phages is enhanced by the *chi* D mutation present on the right arm of the vector.

scheme. Lambdoid phages require genome sizes of between 0.7 and 1.08 of the wild-type DNA properly to fill the phage heads,[22,23] yet all the essential functions required for lambda growth and maturation can be obtained on DNA fragments of approximately 0.6 the genome size. By using

[22] J. Weil, R. Cunningham, R. Martin III, E. Mitchell, and B. Bolling, *Virology* **50,** 373 (1972).
[23] N. Sternberg, and R. Weisberg, *Nature (London)* **256,** 97 (1975).

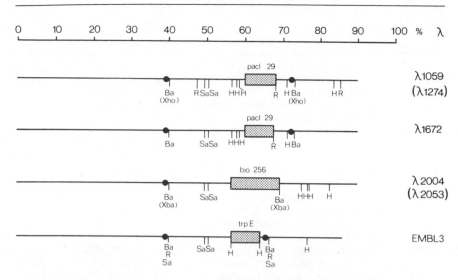

FIG. 3. Restriction endonuclease cleavage maps of λ1059, λ1274, λ1672, λ2004, λ2053, EMBL3. Sites of cleavage for *Bam*HI (Ba), *Hin*dIII (H), *Eco*RI (R), *Sal*I (Sa), *Xba*I (Xba), and *Xho*I (Xho) are indicated. Genotypes are given in Table II. In 2004 and 2053 the pACL29 plasmid has been replaced by an *Eco*RI-*Bam*HI fragment from the *bio*256 substitution in Charon 4a. This removed the *c*I857 gene and the *Hin*dIII sites from the central fragment of the phage. In EMBL3 a *Hin*dIII fragment carrying the *trpE* gene replaces pACL29.

a set of naturally occurring[24-27] and enzymically generated[28] deletions, we were able to construct vectors with appropriately short arms.[29] It should be noted that the packaging requirement places both an upper and a lower limit on the size of DNA fragments to be cloned in the bacteriophage and that this must be taken into account when designing cloning experiments.

Structure of the Bacteriophage Vectors

The restriction endonuclease cleavage maps of our original vector, lambda 1059, is given in Fig. 2, and a number of derivative strains are shown in Fig. 3 and in Table I. The phages are each composed of three fragments, separable by cleavage with an appropriate restriction enzyme: a 19.6 kb left arm carrying the genes for the lambda head and tail proteins, a 12–14 kb central fragment carrying the *red* and *gamma* genes under pL

[24] R. W. Davis, and J. S. Parkinson, *J. Mol. Biol.* **56,** 403 (1971).
[25] J. S. Salstrom, M. Fiandt, and W. Szybalski, *Mol. Gen. Genet.* **168,** 211 (1979).
[26] F. R. Blattner, M. Fiandt, K. K. Hass, P. A. Twose, and W. Szybalski, *Virology* **62,** 458 (1974).
[27] D. Court, and K. Sato, *Virology* **39,** 348 (1969).
[28] L. Enquist, and R. A. Weisberg, *J. Mol. Biol.* **111,** 97 (1979).
[29] S. Brenner, G. Cesareni, and J. Karn, *Gene* **17,** 27 (1982).

TABLE I

LAMBDA CLONING VECTORS WITH POSITIVE SELECTION FOR INSERTS

Strain	Genotype	Chi	Cloning sites	Capacity (kb)
1059	hλsbam1°b189⟨int29ninL44cI857pACL29⟩ Δ[int-cIII]KH54sRI4°nin5	D	BamHI	9–22
1672	hλsbam1°b189⟨int29sRI3°ninL44cI857 pACL29⟩Δ[int-cIII]KH54sRI4°nin5 sRI5°sHindIII6°	C	BamHI	9–22
2004	hλsbam1°b189att int29sRI3°ninL44 Δ[sHindIII3-sHindIII5]Σbio256Δ[int-cIII]cI857sRI4°nin5sRI5°	C	BamHI	7–20
1259	hλsbam1°Eam2001Kam424b189 ⟨int29ninL44pACL29⟩Δ[int-cIII] KH54sRI4°nin5	D	BamHI	9–22
1274	XhoI linker in 1059 BamHI sites	D	BamHI, XhoI	9–22
2053	XbaI linker in 2004 BamHI sites	C	XbaI	7–20
2149	hλsbam1°b189att int (XbaI)ninL44 Δ[sHindIII3-sHindIII5]Σbio256int (XbaI) [int-cIII]cI857	C	XbaI	5–18
EMBL3,4	hλsbam1°b189⟨int(linker)ninL44 Δ[sHindIII3-sHindIII5]ΣtrpE⟩int(linker) Δ[int-cIII]KH54sRI4°nin5sRI5°	D	EcoRI, BamHI, SalI	9–22

control, and a 9–11 kb right arm carrying the lambda replication and lysis genes from which the *red* and *gamma* genes have been deleted. The two arms of the vector contain all the essential functions required for lambda replication and maturation in a DNA sequence less than 65% of the wild-type length. Viable phages are produced when these arms are annealed with internal DNA fragments between 5 and 22 kb; however, the two arms together do not produce viable phages. The left arms of all our phages carry the b189 deletion (17.5%)[24] and the *s*baml° mutation[30] removing the *Bam*HI site in the *D* gene. The right arms are all deleted between the *Bam*HI sites in the lambda *int* gene and the *c*III gene (13.1%)[28] and have defined *chi* sites (either *chi* C or *chi* D) that have been introduced to ensure efficient growth of the recombinant spi phages.[20,21]

Most of the vectors we have constructed are "phasmid" vectors and carry a ColE1 type plasmid (pACL29) on the central fragment.[29] This proved to be a disadvantage in some experiments since commonly used ColE1 plasmid probes such as pBR322[31] will cross-hybridize with these

[30] B. Klein, and K. Murray, *J. Mol. Biol.* **133**, 289 (1979).
[31] F. Bolivar, R. L. Rodriguez, P. J. Greene, M. C. Betlach, H. L. Heyneker, H. W. Boyer, J. H. Crosa, and S. Falkow, *Gene* **2**, 95 (1977).

sequences and detect those parental phages that survive the spi selection procedure. Some derivatives of 1059 have therefore been constructed that substitute other DNA fragments for the plasmid component. We cloned a fragment of biotin operon from a bio256[32,33] phage between the first HindIII site in 1672 (in the cI gene) and the BamHI site on the right arm of lambda 1129.[29] (Note that the fragment is inverted compared with normal bio256 transducing phages and that one lambda att site is deleted.) Lehrach et al.[34] have prepared similar derivatives that substitute a HindIII fragment carrying the E. coli trpE gene for the plasmid component (EMBL 3,4).

Other derivatives of 1059, which introduce defined amber mutations or alter the restriction enzyme sites on the vector, have also been constructed.[35] The XhoI and XbaI vectors were prepared by cloning synthetic oligonucleotide linkers into the BamHI sites of parental phages. These linkers, were decamers composed of G-A-T-C followed by the relevant restriction site (i.e., G-A-T-C-C-T-C-G-A-G and G-A-T-C-T-C-T-A-G-A). These self-anneal to yield double-stranded hexanucleotide sequences with G-A-T-C sticky ends, which may be cloned directly into the BamHI sites. The XhoI linker maintains the BamHI site, whereas the XbaI linker destroys it. The derivatives with amber mutations are of use in genetic selection experiments (see below) as well as providing biological containment.

We now describe the use of these vectors in detail.

Growth of Bacteriophage

Media

CY broth: 10 g of Difco casamino acids, 5 g of Difco Bacto yeast extract, 3 g of NaCl, 2 g of KCl adjusted to pH 7.0. For most experiments this is supplemented with 10 mM Tris-HCl, pH 7.4, and 10 mM MgCl$_2$

Lambda dil: 10 mM Tris-HCl, pH 7.4, 5 mM MgSO$_4$, 0.2 M NaCl, 0.1% gelatin

Lambda agar: 10 g of Difco Bacto-tryptone, 2.5 g of NaCl, 12 g of agar (bottom) or 6 g of agar (top) per plate

[32] E. R. Signer, K. F. Manly, and M. Brunstetter, Virology 39, 137 (1969).

[33] F. R. Blattner, B. G. Williams, A. E. Blechl, K. Denniston-Thompson, H. E. Faber, L. A. Furlong, D. J. Grunwald, D. O. Kiefer, O. D. Moore, J. W. Schumm, E. L. Sheldon, and O. Smithies, Science 196, 161 (1977).

[34] H. Lehrach, and N. Murray, in preparation.

[35] J. Karn, H. Mattes, M. Gait, L. Barnett, and S. Brenner, Gene, in press (1983).

Bacterial Strains

Lambda 1059 and its derivative strains will grow on any lambda-sensitive host. The stringency of the spi selection scheme varies markedly with different strains. In general, *E. coli* C strains harboring P2 are more stringent than the corresponding K strains; however, we routinely work with K strains that are derivatives of C600 (the Q series strains, Table II), since we have found that recombinants grow considerably better on these strains. It is important to use strains that are restriction-deficient in the initial plating of bacteriophage clone collections to prevent loss of recombinants that introduce unmodified restriction sites. Accordingly, we have introduced the $hsr^-_K hsm^+_K$ alleles into our set of isogenic plating strains. Derivatives of 1059 harboring amber mutations must be plated on hosts carrying the appropriate supressor mutations. Table II lists the genotypes and origins of the bacterial strains. The P2 lysogens will segregate on long-term storage in stabs, and it is advisable to keep master stocks as glycerinated cultures at $-70°$.

Phage DNA Preparation

Recombinants in lambda 1059 and related strains grow well, and titers of 10^9 to 10^{10} PFU per milliliter of lysate may be expected. Bacteriophage were grown as liquid lysates on Q358 bacteria using CY medium supplemented with 25 mM Tris-HCl, pH 7.4, and 10 mM MgCl$_2$. Early log-phase cultures were inoculated with the phage from a single purified plaque. Occasionally these starter cultures fail to lyse after 5–7 hr of growth and the bacteria approach saturation. Tenfold dilution of the cul-

TABLE II
BACTERIAL STRAINS

Strain	Relevant features	Source
EQ82	$su_{II}^+ su_{III}^+ hsr_K^- hsm_K^+$	N. Murray
Q276	$recA1 su_{II}^+$	Cambridge
Q342	$recA1 su_{II}^+ su_{III}^+$	Cambridge
Q358	$su_{II}^+ hsr_K^- hsm_K^+$	Cambridge
Q359	$su_{II}^+ hsr_K^- hsm_K^+$ P2	Cambridge
Q360	su_{II}^+ P2	Cambridge
Q364	$su_{II}^+ hsr_K^- hsm_K^+$ P2 $\Delta[lac\text{-}pro]$	Cambridge
CQ6	*E. coli* C, P2	G. Bertani
WR3	$recA1 su^\circ$	M. Gottesman
D91	$\Delta[lac\text{-}pro]$	Cambridge
WX71	su_{III}^+ P2	I. Herskowitz

tures with fresh media allows renewed growth of the bacteria, and lysis usually ensues after 3–4 hr. DNA was prepared from 1-liter cultures inoculated with 2–5 ml of the primary lysate. The phages were recovered from lysates by precipitation with 70 g of polyethylene glycol (PEG-6000) per liter and purified by two cycles of the CsCl density gradient centrifugation.[36] DNA was extracted from concentrated, dialyzed, phage suspensions by phenol extraction and stored at a concentration of 0.5–2.5 mg/ml in 10 mM Tris-HCl, 10 mM NaCl, 0.1 mM EDTA.

Amplifying the Clone Collection

Recombinant phage were plated at a density of approximately 2000 plaques per 10-cm dish of Q359 bacteria. Plate stocks were prepared as follows: 5 ml of lambda dil were added to each dish, and the top agar was scraped off. The agar suspension was vortexed, and bacteria, agar, and debris were removed by centrifugation at 5000 rpm for 10 min in a Sorvall GLC centrifuge. The extracted phage, which typically had titers of 10^9 per milliliter, were stored over chloroform at 4°.

Preparation of DNA Fragments

Random Fragments

Genomic DNA suitable for insertion into the spi vectors (Table I) may be prepared with a variety of enzymes. Vectors with BamHI sites can accommodate fragments prepared with BamHI, BglII, BclI, Sau3a, or MboI. Vectors with XhoI sites can accommodate fragments prepared with either SalI or XhoI. Cleavage of the DNA with a restriction enzyme with a four base-pair recognition sequence, such as Sau3a, produces a nearly random population of fragments, whereas cleavage to completion with restriction enzymes with larger recognition sequences allows purification of particular sequences. Sau3a cleaves at the sequence G-A-T-C and leaves a tetranucleotide extension.[37,38] These fragments may therefore be cloned directly into BamHI sites (G-G-A-T-C-C-) without linker addition.[18,38,39] The Sau3a sites should occur once every 256 bp in DNA with 50% G+C, and only $\frac{1}{80}$th of these sites need to be cleaved to produce

[36] K. R. Yamamoto, B. M. Alberts, R. Benzinger, L. Hawthorne, and C. Treiber, *Virology* **46**, 734 (1970).
[37] J. S. Sussenbach, C. H. Monfoort, R. Schipof, and E. C. Stobberingh, *Nucleic Acids Res.* **3**, 3193 (1976).
[38] R. J. Roberts, *CRC Crit. Rev. Biochem.* **4**, 123 (1976).
[39] G. A. Wilson, and F. E. Young, *J. Mol. Biol.* **97**, 123 (1975).

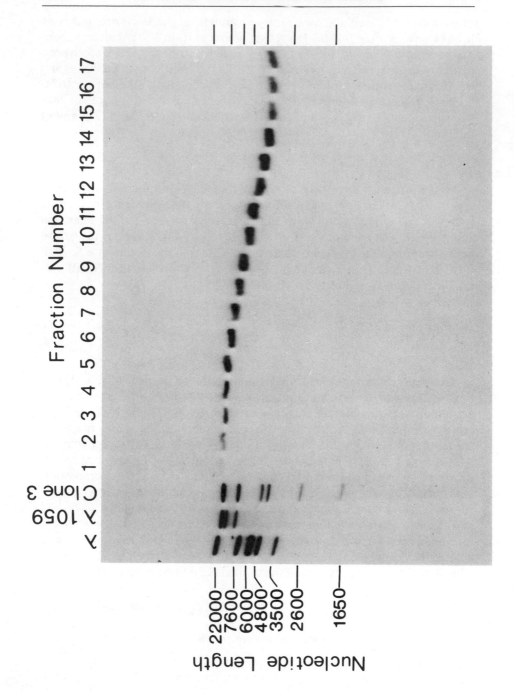

FIG. 4. Fractionation of partially digested nematode DNA. Nematode DNA (N2 DNA) was prepared from frozen animals purified by flotation on sucrose.[8] The worms were pulverized by grinding in a mortar chilled with liquid nitrogen. DNA was released from the disrupted worms by suspending the animals in 1% SDS, 100 mM Tris-HCl, pH 7.4, 1 mM EDTA using 100 ml of buffer per 5 g wet weight of worms. The viscous suspension was extracted with phenol and then phenol–chloroform–isoamyl alcohol (25:24:1), and crude high molecular weight DNA was precipitated by addition of 2 volumes of ethanol. This preparation was further purified by CsCl density gradient centrifugation. Purified DNA was stored at 500 μg/ml in 10 mM Tris-HCl, pH 7.4, 10 mM NaCl, 0.1 mM EDTA at 4°.

Analysis of this material on neutral agarose gels showed the DNA to be greater than 100 kb. N2 DNA was digested with BamHI or Sau3a for 1 hr at 37° in a buffer containing 10 mM Tris-HCl, pH 7.4, 10 mM MgCl$_2$, 10 mM 2-mercaptoethanol, 50 mM NaCl. Aliquots of 20 μg of DNA were digested in 100-μl reactions containing 0.1, 0.2, 0.5, 1.0, and 2.0 units of Sau3a or 1, 2, 5, 10, and 20 units of BamHI. The reaction mixes prepared with each enzyme were pooled, and an aliquot containing 1 μg of DNA was end-labeled by incubation with 0.1 unit of E. coli DNA polymerase I large subunit (Boehringer) in a 10-μl reaction mix containing 10 μCi of [α-^{32}P]dATP (350 mCi/mmol) 500 μM dCTP, 500 μM dGTP, 500 μM dTTP, 10 mM Tris-HCl, pH 7.4, 10 mM MgCl$_2$, 0.1 mM DTT, 50 mM NaCl. After incubation for 20 min at 25°, the reaction was terminated by heat inactivation of the polymerase at 70° for 5 min. The labeled DNA was mixed with the remaining DNA, and the sample was extracted with phenol and then ether. Residual phenol and unincorporated triphosphates were removed by chromatography of the sample on small columns of Sepharose 4B equilibrated with 10 mM Tris-HCl, pH 7.4, 10 mM NaCl, 0.1 mM EDTA.

The excluded peak was concentrated by ethanol precipitation and redissolved at a final DNA concentration of 500 μg/ml. Aliquots containing 50 μg of labeled, digested DNA were fractionated by electrophoresis on columns of 0.5% low melting temperature agarose (BRL). Gels were cast in 1.5 × 20 cm tubes sealed at one end by a piece of dialysis tubing fixed with an elastic band. A flat upper surface was obtained by overlayering the melted agarose with a small layer of butan-2-ol. Samples were applied in 0.3% agarose containing 0.01% bromophenol blue and 0.01% xylene cyanole fast tracking dyes. Electrophoresis was for approximately 18 hr at 150 V, after which time the xylene cyanole dye had moved approximately 15 cm. Both the gel and the electrophoresis buffer contained 40 mM Tris-acetate, pH 8.3, 20 mM sodium acetate, 2 mM EDTA (TAE buffer), and 2 μg of ethidium bromide per milliliter.

After electrophoresis, fractions were cut from the gel with a sterile razor blade. DNA was recovered from the agarose gel slices by melting the agarose at 70° for 5 min. The melted agarose slice was diluted with 10 volumes of H$_2$O and transferred to a 37° water bath. This was loaded on 300-μl columns of phenyl neutral red polyacrylamide affinity absorbent (Boehringer product No. 275, 387) equilibrated with 0.1 × TAE buffer. The columns were washed with 10 ml of 0.1 × TAE, and the DNA was eluted with 2 M NaClO$_4$ in 1.0 × TAE. One-drop fractions were collected, and fractions containing radioactive DNA were pooled. The eluted DNA was concentrated by ethanol precipitation and redissolved at 10 mM Tris-HCl, pH 7.4, 10 mM NaCl, 0.1 mM EDTA. After phenol extraction and subsequent ethanol precipitation, the DNA was redissolved in 10 mM Tris-HCl, pH 7.4, 10 mM NaCl, 0.1 mM EDTA at a final concentration of 500 μg/ml and stored at −20°. Recovery of DNA from agarose gel varied from 50 to 70%.

The autoradiograph depicted in the figure shows fractions of Sau3a-digested nematode DNA prepared as described, analyzed by electrophoresis on a 1% agarose gel. The gel was cast in 0.1 × 1.8 × 20 cm slabs. Electrophoresis was at 100 mA for 4 hr using TAE buffer containing 2.0 μg of ethidium bromide per milliliter. Nick-translated EcoRI-cutλDNA, BamHI-cut 1059 DNA, and a clone of unc54 DNA cut with BamHI were included as size markers. Fractions 3, 4, and 5 contain 15–20 kb Sau3a fragments suitable for insertion into λ1059.

DNA fragments 20 kb long. The frequency of *Sau*3a sites does not vary appreciably with changes in base composition. In DNA with 67% G+C, the sites should occur once every 324 bp. In practice, however, we minimize the possibility of obtaining abnormal distributions of fragments by a single digestion condition and routinely digest genomic DNA in several reactions in which the concentration of enzyme is varied over a 20-fold range (Fig. 4).

Specific Fragments

In some experiments it may be desirable to clone specific restriction fragments produced by limit digestion with a six-base pair enzyme. For example, most of the *unc*54 myosin heavy-chain gene coding sequence is present on a 8.3 kb *Xba* fragment.[40,41] In order to isolate this fragment from strains carrying *unc*54 mutations quickly, we have constructed an *Xba* vector (2149) with slightly extended arms to accommodate this fragment. Since the distribution of *Xba* sites in the nematode genome is nonrandom, a considerable sequence enrichment is obtained simply by purifying a single size fraction from a limit enzyme digest. It should be noted that the use of more than one restriction enzyme in succession would provide additional sequence purification.

Size Fractionation

Rigorous size fractionation of the DNA to be cloned is essential to avoid spurious linkage produced by multiple ligation events. If fragments greater than 14 kb are ligated to the vector arms, any dimers or multimers formed during the ligation reactions will exceed the 22 kb cloning capacity of the phage and will not appear in the recombinant phage population. Fragments less than 12 kb are frequently cloned as multiples. We have found that preparation of DNA fragments by agarose gel electrophoresis is more satisfactory than purification of fragments by velocity sedimentation on sucrose or NaCl density gradients. Any method of recovery of DNA from gels that yields ligatable DNA is satisfactory. In most of our recent experiments we have recovered DNA from low melting temperature agarose gels by phenol extraction. Usually the DNA is sufficiently pure after ethanol precipitation, without additional purification. Figure 4 shows nematode DNA fragments prepared by *Sau*3a partial digestion and size-fractionated by agarose gel electrophoresis. Fractions 2, 3, and 4 contain 15–20 kb DNA fragments suitable for cloning.

[40] A. R. MacLeod, J. Karn, and S. Brenner, *Nature (London)* **291,** 386 (1981).
[41] J. Karn, and L. Barnett, *Proc. Natl. Acad. Sci. U.S.A.,* in press (1983).

Preparation of Recombinants

Enzymes and In Vitro Packaging

Successful and efficient cloning requires highly purified restriction enzymes and active DNA ligase. Commercial preparations have improved markedly in recent years and most are satisfactory; however, we have found it convenient to prepare our own enzymes in order to have large quantities of calibrated materials. T4 DNA ligase was prepared from a lysogen of a lambda-T4 gene 30 recombinant originally prepared by Murray et al.[42] Restriction enzymes were prepared by standard methods. A number of in vitro packaging systems have been developed, and each gives much the same packaging efficiencies. In our experiments we have used extracts prepared from NS 428 supplemented with partially purified protein A, following the method of Sternberg[43] and Becker and Gold[44] as modified by Blattner[7] and ourselves.[18] There should be no incompatibility between our vectors and other packaging extracts.

Yield of Recombinants

We routinely monitor the yield of religated vector molecules and recombinants by plating on Q358 and Q359. Figure 5 shows a cloning experiment in which lambda 1059 DNA was cleaved with BamHI and 2-μg aliquots were religated with T4 DNA ligase in the presence of 0–0.6 μg of 18 kb fragments produced by BamHI or Sau3a cleavage of nematode DNA. Cleavage and religation of the vector DNA in the absence of nematode DNA produces more than 1×10^6 phage particles per microgram of phage DNA. These phages grow on Q358, but fewer than 2×10^3 PFU are detectable on Q359. This background is reduced to less than 2×10^2 PFU per microgram on the more stringent selective strain, CQ6. Cleavage and relegation of 1059 in the presence of nematode DNA fragments produce recombinant phages that are detected by plating on Q359. The yield of recombinants is linear with the amount of nematode DNA added as long as the DNA concentration is low. The ligation reaction is saturated with a greater than 2.0-fold molar excess of insert DNA to vector DNA (0.5 μg insert DNA per 1.0 μg of vector). The yield of recombinants in this experiment ranged from 2.4×10^5 to 5.4×10^5 per microgram of 15–20 kb nematode DNA. This yield is approximately 10-fold higher than the yield reported by Maniatis et al.[6] using Charon 4a vectors. At saturation of the ligation reaction with nematode DNA, approximately

[42] N. E. Murray, S. A. Bruce, and K. Murray, J. Mol. Biol. 132, 493 (1979).
[43] N. Sternberg, D. Tiemeier, and L. Enquist, Gene 1, 255 (1977).
[44] A. Becker, and M. Gold, Proc. Natl. Acad. Sci. U.S.A. 72, 581 (1975).

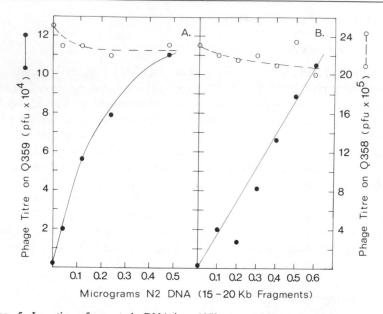

FIG. 5. Insertion of nematode DNA into 1059 arms. 1059 DNA was digested with a threefold excess of *Bam*HI as described in the legend to Fig. 4. The reaction was terminated by incubation at 70° for 5 min. Aliquots of 2.0 μg of cleaved 1059 DNA were ligated in the presence of 0–0.6 μg of 15–20 kb nematode DNA prepared as described in the legend to Fig. 4. In 20-μl reactions containing 0.1 Weiss unit of T4 DNA ligase, 10 m*M* Tris-HCl, pH 7.4, 10 m*M* MgCl$_2$, 50 m*M* NaCl, 0.1 m*M* ATP. Incubation was at 4° for 18 hr. The ligated DNAs were packaged *in vitro* as follows, using extracts of heat-induced NS 428. Extracts were prepared by lysing 10 g of induced cells in 50 ml of 50 m*M* Tris-HCl, pH 8.0, 3 m*M* MgCl$_2$, 10 m*M* 2-mercaptoethanol, 1 m*M* EDTA in a French pressure cell operated at 1000 psi. Cellular debris was removed by centrifugation of the extract for 30 min, 35,000 rpm in a T60 rotor, and aliquots of the supernatant were stored at −70°. Extracts prepared in this manner are active in *in vitro* packaging when supplemented with partially purified protein A prepared as described by Blattner *et al.*[7] Packaging was performed in 150-μl reactions containing 50 μl of extract, 10 μl of protein A, 20 m*M* Tris-HCl, pH 8.0, 3 m*M* MgCl$_2$, 0.05% 2-mercaptoethanol, 1 m*M* EDTA, 6 m*M* spermidine, 6 m*M* putrescene, 1.5 m*M* ATP, and 2.0 μg of cleaved and religated 1059 DNA. After incubation for 60 min at 20°, the extracts were diluted to 1 ml with λdil (10 m*M* Tris-HCl, pH 7.4, 5 m*M* MgSO$_4$, 0.2 *M* NaCl, 0.1% w/v gelatin), and titered on Q358 and Q359 bacteria. Panel A: Yield of total phage (PFU on Q358, O---O) and recombinant phage (PFU on Q359, ●——●) genomes obtained by religation of *Bam*HI-cleaved 1059 in the presence of 0–0.5 μg of 15–20 kb fragments of *Bam*HI-cleaved N2 DNA. Panel B: Yield of total phage (PFU on Q358, O---O) and recombinant phage (PFU on Q359, ●——●) genomes obtained by ligation of *Bam*HI-cloned 1059 in the presence of 0–0.6 μg of 15–20 kb fragments of *Sau*3a-cleaved N2 DNA.

10% of the total phages produced harbor inserts. The total yield of phages decreases somewhat upon addition of nematode DNA to the ligation reaction. This may be due to the addition of trace quantities of inhibitors of the T4 ligase or the result of sequestering of vector arms by broken nematode fragments.

Identification of Specific Clones

Mean Length of DNA Inserts

The distribution of DNA sizes in a lambda phage population can be determined by measuring the density of phages on CsCl density gradients.[45] Since the amount of protein in the phage particles is constant, the buoyant density of a phage is a function of the DNA-to-protein ratio. Changes in the length of lambda DNA of as little as 500 bp may be detected by this method. Figure 6 shows the results of a density gradient analysis of the clone collections prepared in the experiment shown in Fig. 4. An $h434cI857nin5$ phage (46.1 kb) as well as 1059 were used as size markers. The recombinant phages varied in size from 46 to 44 kb with an average of 45 kb. This corresponds to an average insert size of 15 kb. The half-maximal bandwidth of the density distribution of the recombinant phage population was approximately twice that of the marker phage, demonstrating that the recombinants contain DNA inserts with limited heterogeneity.

Plaque Hybridization

In most of our work we have used probes made from the mp series of M13 vectors.[46,47] Originally we used nick-translated RF DNA, but more recently we have been using probes made by priming on M13 single-stranded DNA to the 5' sides of the clone insert and hybridizing with the partially double-stranded material.[48,49] A slight background of hybridization of M13 DNA to *lac* DNA sequences from the host strains was encountered in our early experiments. This can be eliminated by the addition of 20 μg of M13 vector DNA per milliliter as a competitor when

[45] N. Davidson, and W. Szybalski, *in* "The Bacteriophage Lambda" (A. D. Hershey, ed.), pp. 45–82. Cold Spring Harbor Laboratory, Cold Spring Harbor, New York, 1971.

[46] J. Messing, B. Gronenborn, B. Müller-Hill, and P. H. Hofschneider, *Proc. Natl. Acad. Sci. U.S.A.* **74**, 3642 (1977).

[47] J. Messing, R. Crea, and H. Seeburg, *Nucleic Acids Res.* **9**, 309 (1981).

[48] N-t. Hu, and J. Messing, *Gene* **17**, 271 (1982).

[49] D. Brown, J. Frampton, P. Goelet, and J. Karn, *Gene* **20**, 139 (1982).

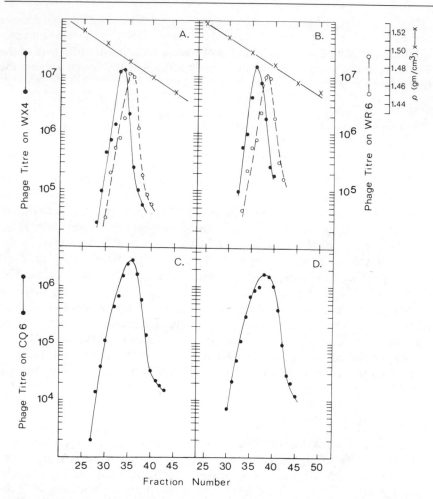

FIG. 6. Analysis of recombinant phage collections by CsCl density gradient centrifugation. 1059 (44 kb) and 308: $h434c$I857sRI 4° $nin5$ sRI 5° $Sam7$ (46.1 kb) were included as density markers. Approximately 10^5 of each marker phage and 10^8 phages from recombinant phage pools were added to 5 ml of 100 mM Tris-HCl, pH 7.4, 10 mM MgCl$_2$. Solid CsCl was added to a final refractive index of 1.3810, and the phage were banded by centrifugation in an SW60 rotor at 40,000 rpm for 24 hr. After centrifugation, one drop fractions were collected by puncturing the bottom of the centrifuge tubes with a needle. The refractive index of every fifth fraction was measured, and 20 μl aliquots of each fraction were added to 1 ml of dil. Each fraction was titered on WX4 (λ^R), ●——●; WR6, ($recA$, $h434^R$), ○---○; and CQ6 ($h434^R$ P2), ●——●, to determine the position of the $h434$, 1059, and recombinant phages, respectively. Panels (A) and (C) plot the distribution of BamHI-generated recombinant phage (panel C) and marker phages (panel A) included in the same gradient. Panels B and D plot the results of a similar analysis of the recombinants generated with Sau3a fragments.

working with the single-stranded DNA probes or by using strains with deletions of the *lac* region for plating (D91 or Q364).

Plasmid probes may also be used to screen libraries, but it is preferable to construct the library in a vector that lacks the plasmid insert (2004, 2139, EMBL4) in order to avoid false positives arising from hybridization to parental phages that escape the spi selection.

Immunological Assay

Recombinants of the spi vectors have a number of features that are of use when immunological detection of cloned sequences is planned.[13-16] Each of the phages is constructed so that DNA is inserted in the *Bam*HI site at 714 in the leftward promoter. This segment is efficiently transcribed from pL and any inserted fragment that contains an intact gene and ribosome binding site will be expressed at a high level when cloned in this site. We cloned the T4 DNA ligase gene 30 into 1059 and found that during lytic infection the protein was produced at the same levels as in the original phages constructed by Murray,[42] which placed this gene under the control of the rightward promoter. Additionally, clones in 1059 and its derivatives retain a single lambda attachment site. All the recombinants may therefore be inserted efficiently into the *E. coli* chromosome with helper phage supplying integrase and repressor.

Genetic Selections

Most genetic selection schemes involve suppression of amber mutations in the vector arms by cloned suppressor tRNA genes. Seed *et al.*[11] have found that plasmids carrying both the tRNA su^+_{III} gene and a cloned insert may be inserted in specific lambda clones through *rec*-mediated "lifting" events. If the vector has amber mutations in essential functions, then only "phasmids"[29] carrying the plasmid and the suppressor gene will grow on su^- strains. A second selection scheme was developed by Goldfarb *et al.*[12] following a suggestion by one of us. DNA is cotransformed into mammalian cells together with su^+_{III} DNA to provide a selective marker. Larger DNA fragments carrying transforming DNA and the su^+_{III} DNA are then selected in a lambda 1059 derivative carrying the *Sam7* mutation. The vector 1259 would also be suitable for this experiment.[50]

---

[50] We have recently constructed another vector, λ2001, which is a derivative of 2053 and carries a Δ[*int-c* III] KH54 *s*RI 4° *nin5* *s*RI 5° *sHin*dIII 6° *chi*C right arm and has the polylinker sequence TCTAGAATTCAAGCTTGGATCCTCGAGCTCTAGA cloned into the *Xba* sites. This phage is a vector for *Eco*RI, *Hin*dIII, *Bam*HI, *Xho*I, *Sac*I, and *Xba*I.

[16] λ Phage Vectors—EMBL Series

By A. M. Frischauf, N. Murray, and H. Lehrach

Introduction

To establish genomic libraries from large genomes and to identify specific genomic clones, either cosmids or λ replacement, vectors have been used. Both systems rely on the packaging of recombinant genomes into phage heads *in vitro* to give infective particles as a means of introducing DNA into the bacterial cell with high efficiency. Packaging requires the cutting of specific DNA sequences, termed *cos*, and infectious particles are formed efficiently only if the length of DNA between two *cos* sequences is between 78 and 105% of the length of wild-type λ (38–55 kb).[1] Hence, the size of the vector DNA determines the capacity remaining for the inserted DNA molecule. Approximately 30 kb of DNA is necessary to encode the functions required for lytic growth, so the capacity of λ vectors has an inherent upper limit of approximately 22 to 24 kb. A vector comprising only the essential DNA regions (i.e., 30 kb in length) would be too small to be packaged; therefore, the theoretical cloning capacity can only be used in replacement vectors in which a nonessential region of the vector, flanked by the restriction sites used for cloning, is replaced by the inserted DNA.

Cosmids, in contrast, require only a few kilobases of DNA to encode the functions essential for their replication and selection, and the *cos* sequence, or sequences necessary for packaging, consequently they allow the cloning of DNA fragments of close to 50 kb in length. This larger capacity makes cosmids very attractive for those experiments in which the aim is either to analyze long regions by overlapping clones, or to clone genes too large to fit into λ vectors. λ replacement vectors, however, offer advantages in the efficiency of library construction and screening. Using the protocol described later, which allows cloning samples of analogous (or even identical) DNA preparations in either λ or cosmid vectors, we find cloning efficiencies of roughly 5×10^5 clones/μg of starting DNA for cloning in EMBL3[2] and approximately 5×10^4 clones/μg DNA for cloning in pcos2 EMBL.[3] Taking the different size of the insert into account,

---

[1] M. Feiss and A. Becker, *in* "Lambda II" (R. Hendrix, J. Roberts, F. Stahl, and R. Weisberg, eds.), p. 305. Cold Spring Harbor Lab., Cold Spring Harbor, New York, 1983.

[2] A.-M. Frischauf, H. Lehrach, A. Poustka, and N. Murray, *J. Mol. Biol.* **170,** 827 (1983).

[3] A. Poustka, H.-R. Rackwitz, A.-M. Frischauf, B. Hohn, and H. Lehrach, *Proc. Natl. Acad. Sci. U.S.A.* **81,** 4129 (1984).

RECOMBINANT DNA
METHODOLOGY

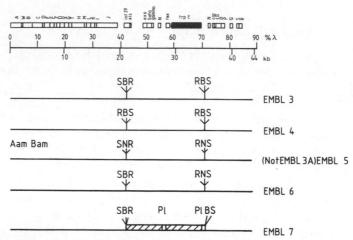

FIG. 1. Schematic structure of the λ vectors EMBL3, EMBL4, EMBL5 (NotEMBL3A), EMBL6, and EMBL7. EMBL7 contains as a middle fragment two head-to-tail copies of pEMBL18 [L. Dente, M. Sollazzo, C. Baldari, G. Cesareni, and R. Cortese, *in* "DNA Cloning" (D. M. Glover, ed.), Vol. 1. IRL, in press] with a modified polylinker sequence (M. Burmeister and F. Michiels, unpublished observations). The sequence of sites in the polylinker (P1) is *Hind*III, *Sph*I, *Pst*I, *Mlu*I, *Bam*HI, *Xma*I (*Sma*I, *Ava*I), *Kpn*I, *Sst*I, *Eco*RI. S, *Sal*I; B, *Bam*HI; R, *Eco*RI; N, *Not*I.

roughly five times more DNA is needed to construct cosmid libraries than λ libraries of equivalent coverage.

In addition, many laboratories find λ libraries somewhat easier to handle, to screen, and to amplify. Cloning in a λ replacement vector may be preferred if there are no special requirements for the larger capacity of cosmid vectors, if only small amounts of DNA are available, or if several libraries have to be constructed and screened.

EMBL3 and EMBL4

EMBL3 and EMBL4[2] are λ replacement vectors derived from λ 1059[4,5] which allow the cloning of DNA fragments with sizes between 8 and 23 kb. Polylinkers flanking the nonessential middle fragment contain symmetrically arranged sites for *Sal*I, *Bam*HI, and *Eco*RI (Fig. 1). Therefore fragments created by these enzymes, or a number of different restriction enzymes generating compatible ends, can be cloned [*Sal*I and *Xho*I in

[4] J. Karn, S. Brenner, L. Barnett, and G. Cesareni, *Proc. Natl. Acad. Sci. U.S.A.* **77,** 5172 (1980).

[5] J. Karn, S. Brenner, and L. Barnett, this volume [15].

the SalI site, BamHI, BglII, BclI, XhoII, and Sau3A (or MboI) in the BamHI site, and EcoRI or EcoRI* in the EcoRI site]. Of these enzymes, Sau3A and its isoschizomer, MboI, are most generally useful, since partial digestion with these enzymes, which recognize a 4-bp sequence, minimizes potential distortion of the sequence representation in the library due to either preferential cutting or irregular distribution of sites. When Sau3A fragments are cloned in a vector cut with BamHI, three out of four ligation events will not recreate the BamHI site. In most cases, therefore, the excision of the cloned fragment from the vector with this enzyme is not possible. However, the flanking sites of the polylinker (SalI in EMBL3 or EMBL3A, EcoRI in EMBL4) can be used to excise the cloned fragment. SalI is especially convenient in this regard since these sites are underrepresented in mammalian DNA so that in many cases, the insert can be recovered in a single SalI fragment.

As in the vectors 1059[4,5] and 2001,[6] the replaceable middle fragment of the EMBL vector carries the red and gam genes of λ, and these can be transcribed, irrespective of their orientation, from a promoter included on this fragment. Expression of red and gam, particularly gam, prevents the growth of the vector phage on Escherichia coli lysogenic for phage P2.[7] This Spi+ (sensitive to P2 interference) phenotype of the vector phage allows strong genetic selection against the vector phage, and therefore enrichment for ligation products in which the middle fragment has been replaced by inserted DNA.

There is, however, a disadvantage associated with this powerful selection system. In the absence of the phage red and gam genes, phage growth requires the host recombination function (for a more detailed discussion, see Refs. 8 and 9). This is a consequence of the fact that multimeric λ DNA, the substrate for packaging, can be produced in two ways, one of which is dependent on recombination, the other on the product of the gam gene. Following infection λ DNA is first replicated in the theta mode to give rise to monomeric circles, which can oligomerize via recombination. Later in the phage infection cycle replication proceeds by a rolling circle mode, giving long concatemeric molecules. This change of mode is dependent on inactivation of ExoV[10] by the phage gam gene product.

[6] J. Karn, H. W. D. Mathes, M. J. Gait, and S. Brenner, Gene 32, 217 (1984).
[7] J. Zissler, E. Signer, and F. Schafer, in "The Bacteriophage Lambda" (A. D. Hershey, ed.), p. 469. Cold Spring Harbor Lab., Cold Spring Harbor, New York, 1971.
[8] G. Smith, in "Lambda II" (R. Hendrix, J. Roberts, F. Stahl, and R. Weisberg, eds.), p. 175. Cold Spring Harbor Lab., Cold Spring Harbor, New York, 1983.
[9] N. Murray, in "Lambda II" (R. Hendrix, J. Roberts, F. Stahl, and R. Weisberg, eds.), p. 395. Cold Spring Harbor Lab., Cold Spring Harbor, New York, 1983.
[10] R. C. Unger and A. J. Clark, J. Mol. Biol. 70, 539 (1972).

Growth of Spi$^-$ (*red*$^-$ *gam*$^-$) phages therefore depends on the host RecA function to generate oligomeric circles by recombination, a process that is only efficient if either the vector or the cloned insert contains a Chi sequence (see Ref. 8). Libraries constructed in EMBL phages therefore are generally propagated on a recA$^+$ strain which might enhance the potential instability of repetitive sequences and might make the occurrence of deletions and rearrangements more likely. Alternatives are the use of recBC$^-$ hosts, the provision of the *gam* gene transcribed from pR' on a helper plasmid in a RecA$^-$ host,[11] or, at the cost of losing the genetic selection for chimeric molecules, the use of one of a number of *gam* replacement vectors of which Charon 34 and Charon 35 are the most versatile.[12]

As an alternative, or in addition to a genetic selection, religation of the original vector molecule can be avoided by cleaving the inner site in the polylinker molecule (*Eco*RI in EMBL3). The *Bam*HI end of the replaceable fragment will be released on a small connector fragment, which can be removed by the selective precipitation of the larger DNA fragments by 2-propanol.

Though the EMBL3 vector itself has not been sequenced, the published sequence of λ and various mutants[13] can be used to predict high-resolution restriction maps. A map based on these data is shown in Fig. 2.

Amber Derivatives of EMBL Vectors

Amber derivatives of EMBL3 have been constructed. These include EMBL3A, carrying amber mutations in the *A* and *B* genes, EMBL3S, carrying a *Sam*7 mutation, and EMBL3AS with the *Aam Sam* mutations. Due to the proximity of the *Sam*7 mutation to the Chi site (ChiD) carried in the EMBL vectors, all *Sam*7 derivatives have lost the Chi mutation, and must therefore rely on Chi sites provided by the cloned sequence.

These vectors can be used to selectively clone sequences carrying suppressor genes, e.g., *supF*.[14] The use of two amber mutations eliminates the background due to reversion of the amber mutations. However, with the use of double amber mutations, the growth of suppressor-independent phages (probably formed by loss of the *Aam Bam* mutations by

[11] G. F. Crouse, *Gene* **40,** 151 (1985).

[12] W. A. M. Loenen and F. R. Blattner, *Gene* **26,** 171 (1983).

[13] D. L. Daniels, J. L. Schroeder, W. Szybalski, F. Sanger, A. R. Coulson, G. F. Hong, D. F. Hill, G. B. Peterson, and F. R. Blattner, *in* "Lambda II" (R. Hendrix, J. Roberts, F. Stahl, and R. Weisberg, eds.), p. 519. Cold Spring Harbor Lab., Cold Spring Harbor, New York, 1983.

[14] M. Goldfarb, K. Shimizu, M. Perucho, and M. Wigler, *Nature (London)* **296,** 404 (1982).

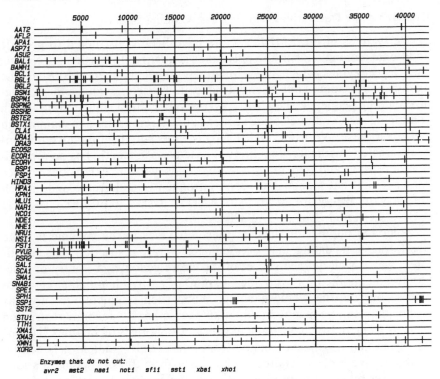

Fig. 2. Restriction map of EMBL3.

recombination with DNA from the *in vitro* packaging extract) has been observed. It is expected that the use of the *Sam* mutation, also present in the prophages used for preparation of packaging extracts, would reduce this background considerably.

An *orf* in the right arm of λ has been shown to be required for efficient phage plasmid recombination.[15] This *orf* is missing in EMBL3A and many other vectors, as the consequence of the *nin5* deletions. EMBL3A, therefore, is not generally useful for the isolation of phage clones by homologous recombination with suppressor-carrying probe plasmids.[16] A derivative of EMBL3 containing the sequence responsible for a high rate of recombination has been described.

[15] H. V. Huang, P. F. R. Little, and B. Seed, *in* "Vectors: A Survey of Molecular Cloning Vectors and Their Uses" (R. Rodriguez, ed.). Butterworth, in press.
[16] B. Seed, *Nucleic Acids Res.* **11**, 2427 (1983).

EMBL Vector Derivatives with Further Cloning Sites

Derivatives of the EMBL vectors have been constructed to provide additional cloning sites in the polylinker sequence. In particular, cloning sites for enzymes which cut a relatively small number of sites in the mammalian genome have been introduced. These enzymes either recognize an 8-bp sequence site (*Not*I, *Sfi*I) or recognition sites of 6-bp length containing the CG dinucleotide sequence underrepresented in mammalian DNA. Of these, *Not*I (GC′GGCCGC) is especially useful, since this enzyme creates fragments of millions of base pairs in length, has no recognition site in λ, and generates cohesive ends with 5′ extensions. Other enzymes that cut mammalian DNA infrequently and create ends which are easy to ligate are *Mlu*I (A′CGCGT), *Bss*HII (G′CGCGC), *Sac*II (CCGC′GG), and *Pvu*I (CGAT′CG).

Since the number of sites for these enzymes in the genome is quite small, fragments ending in or containing such rare restriction sites constitute a small, well-defined subset of the entire genome. A library of junction fragments therefore provides a group of easily identifiable reference points covering the whole genome. When such fragments are used as probes they can be mapped relative to each other by genomic restriction analysis using pulsed field gradient gel electrophoresis techniques.[17] Cloning of purified fragments isolated from pulsed field gradient gels can be used to derive probes hundreds of kilobases from the start point. Various procedures making use of rare cutting enzymes should therefore be very helpful in coping with the large distances in mammalian genomes.

*Not*I-Linking Fragment Cloning

A derivative of EMBL3A in which the *Bam*HI sites of the polylinkers have been replaced by *Not*I sites (NotEMBL3A, Fig. 1) has been constructed. This is used for the selective cloning of fragments containing *Not*I sites (*Not*I-linking fragment clones).

In the cloning procedure *Mbo*I-digested, sized DNA (10–20 kb) is circularized at low concentration in the presence of a *Bam*HI-cleaved plasmid carrying a suppressor sequence.[18] Under appropriate conditions[19] a large fraction of the circles will contain a suppressor plasmid. The circles containing a *Not*I recognition site can be linearized by this enzyme and subsequently cloned in *Not*I-cut NotEMBL3A vector. By plating on a suppressor-free host phages containing *Not*I junction fragments can be

---

[17] D. C. Schwartz and C. R. Cantor, *Cell* **37,** 67 (1984).
[18] A. Levinson, D. Silver, and B. Seed, *J. Mol. Appl. Genet.* **2,** 507 (1984).
[19] F. S. Collins and S. M. Weissman, *Proc. Natl. Acad. Sci. U.S.A.* **81,** 6812 (1984).

selected. If DNA from interspecies hybrid cells has been used, clones from, e.g., the human chromosome(s) can be identified by hybridization with, in this case, human repetitive sequences.[20]

End Fragment Cloning

A derivative of EMBL3 has been constructed in which only one of the two *Bam*HI sites has been substituted by a *Not*I site (EMBL6, Fig. 1). This allows the cloning of the ends of large, gel-purified *Not*I fragments after partial digestion with *Mbo*I.

For a number of enzymes that cut rarely in mammalian DNA, the construction of specific cloning vectors is quite difficult, due to the large number of sites in λ DNA (e.g., *Mlu*I, *Bss*HII). Vectors allowing the cloning of the ends of fragments created by these enzymes have, however, been constructed (Michiels, unpublished) by cloning a *Mlu*I-containing polylinker at the position of the right *Eco*RI site in EMBL3 (EMBL7, Fig. 1). To clone either *Mlu*I or *Bss*HII fragment ends, the right arm from this vector has to be isolated and used in combination with a left vector arm created, e.g., by *Bam*HI digestion of EMBL3.

Additional Derivatives

To provide additional sites for *Xba*I and *Sac*I in the polylinker, Ernst Natt and Gerd Scherer (personal communication) have constructed EMBL3-12, in which the *Sal*I–*Eco*RI fragments of EMBL3 are replaced by the *Sal*I–*Eco*RI polylinker segment of the pUC12 plasmid.[21] The modified polylinker therefore contains sites for *Sal*I, *Xba*I, *Bam*HI, *Sam*I, *Sac*I, and *Eco*RI. This arrangement of sites in the polylinker allows the excision of inserts created by ligation of *Nhe*I and *Spe*I fragments to the ligation-compatible *Xba*I sites by the flanking *Sal*I sites.

DNA Analysis

Analysis of the partial digestion products of λ clones can be facilitated by tagging one *cos* sequence with a complementary radiolabeled oligonucleotide.[22] Since this analysis is complicated by the vector DNA between the labeled cohesive ends and the cloning site, an EMBL4 derivative in which *cos* is adjacent to one of the cloning sites has been constructed.

[20] J. F. Gusella, C. Jones, F. T. Kao, D. Housman, and T. T. Puck, *Proc. Natl. Acad. Sci. U.S.A.* **79,** 7804 (1982).
[21] J. Vieira and J. Messing, *Gene* **19,** 259 (1982).
[22] H. R. Rackwitz, G. Zehetner, A. M. Frischauf, and H. Lehrach, *Gene* **30,** 195 (1984).

Construction of Libraries

Principle

The protocol used by us aims to minimize the manipulations involved in library construction, taking advantage of a combination of genetic and biochemical selection steps to rule out side reactions and to ensure the correct structure of the final clones. In our experience, this leads to high reproducibility, high yields, and the capability to handle very small amounts of material. The protocol described here for the construction of λ libraries is essentially identical to the protocol used by us in cosmid library construction. It is therefore possible, and in many cases advantageous, to construct both λ and cosmid libraries in parallel.

In constructing the library, the DNA to be cloned is first digested partially with Sau3A or MboI to a (number) average size of slightly above 20 kb. This DNA is then dephosphorylated using calf intestine alkaline phosphatase, and ligated to EMBL3 or EMBL4 cleaved with BamHI. Due to the phosphatase treatment of the insert, ligation events between different originally unconnected insert fragments are ruled out, while ligation of inserts to vector can proceed over the 5' phosphates carried by the vector arms.

As mentioned before, religation of the middle fragment to the vector arms can be prevented for EMBL3 by removal of the BamHI-ligatable end as a small fragment by recleavage with EcoRI, followed by selective precipitation of the long DNA with 2-propanol. (Alternatively, or in addition, the Spi selection can be used to remove remaining uncut or religated vector from the library.)

After ligation, the DNA is packaged in vitro. The packaging product can be plated directly. Alternatively, if large amounts of packaged material have to be plated at high density, inhibitory material can be removed from the packaging product by a CsCl step gradient. In general we use either NM538 (SupF, rk⁻, mk⁺, the nonselective host) or NM539 [NM538 (P2cox3), the host used for Spi selection] for plating.

For EMBL3 and EMBL4, many other Rec⁺, restriction strains can, however, be used. EMBL3A is best plated on supF-carrying strains, since it will form very small plaques on hosts providing only supE. To allow recovery of sequences carrying inverted repeats in the library, recB recC sbcB host strains can be used as alternatives to the strains mentioned above.[23,24]

---

[23] D. R. F. Leach and F. W. Stahl, Nature (London) 305, 448 (1983).
[24] A. R. Wyman, L. B. Wolfe, and D. Botstein, Proc. Natl. Acad. Sci. U.S.A. 82, 2880 (1985).

Procedures

Vector Preparation

1. *Plate Stock.* To prepare a plate stock for vector preparation a well-separated single plaque is picked from a fresh BBL plate, and suspended in 300 μl λ diluent. After 1 hr or more for diffusion the phage is plated with DH1 or NM538 bacteria to give a slightly less than confluent plate. Five milliliters λ diluent is left on the plate for 3 hr at room temperature. The plate stock is harvested, treated with 0.2 ml CHCl₃, and titrated on NM538, NM539, and, for amber-containing phages, on a suppressor-free host (NM430, MC1061).

2. *Preparation of Phage DNA.* If the background of Spi⁻ (and suppressor independent) phages is sufficiently low, the plate stock is used to infect an exponentially growing culture of NM538 in L broth containing 5 mM MgSO₄ (multiplicity of infection 1 : 10 for EMBL3 and EMBL4, 1 : 1 for *Aam Bam* derivatives). Phage is precipitated with polyethylene glycol (PEG 6000) and purified by two cycles of CsCl equilibrium density gradient centrifugation. DNA is extracted by standard procedures and stored in TE.

3. *Testing of Vector DNA.* To test the quality of the vector DNA preparations, samples are packaged *in vitro* directly, or cut with *Bam*HI, religated, and packaged. The material is plated on NM538 and NM539 to check for religatability and contamination with nonphage DNA. Religated material should give approximately 10% of the original number of plaques. The ratio of background as the result of the cloning of DNA on NM539/538 should not increase significantly.

4. *Digestion.* EMBL3 DNA (20 μg) is digested with 40 U *Bam*HI in 100 μl 100 mM NaCl, 10 mM MgCl₂, 10 mM Tris · HCl (pH 7.6) for 1 hr at 37°. *Eco*RI (100 U) is added and digestion continued for 30 min at 37°. EDTA is added to a final concentration of 15 mM and the mixture is heated 15 min at 68°. To 120 μl reaction mixture, 18 μl 3 M NaAc and 90 μl 2-propanol are added, the tube is left on ice for 5 min, then centrifuged for 5 min. The pellet is washed once with 100 μl 0.45 M NaAc plus 60 μl 2-propanol, air dried, and dissolved in TE buffer, to a final concentration of 0.25 μg/μl.

Comments

EMBL3 has the option of cutting with *Eco*RI after the *Bam*HI digestion and thus removing the *Bam*HI sticky ends from the middle fragment. (This is not convenient for EMBL4, where the *Sal*I site is very close to the *Eco*RI site and cutting is inefficient.) The small fragment can be removed by selective precipitation of the long DNA with 2-propanol. The efficiency of recutting and small fragment removal is checked by religating *in vitro,* packaging, and plating on NM538. The background should be no more than 1%.

The vector DNA may be phenol extracted and ethanol precipitated before use. Alternatively, heat inactivation of *Bam*HI and *Eco*RI is usually sufficient if the vector is used within a short time.

Preparation of Insert DNA

1. Partial Digestion. The appropriate enzyme concentration is established by partial digestion on an analytical scale. One microgram high-molecular-weight DNA (>100 kb) is digested with 0.1 U *Mbo*I in 10 mM Tris · HCl (pH 7.6), 10 mM MgSO$_4$, 1 mM DTT, 50 mM NaCl in a total volume of 30 μl; 5-μl aliquots are taken after 0, 5, 10, 20, 40, and 80 min. Four microliters of 0.1 M EDTA is added to each aliquot immediately, then the samples are incubated 10 min at 68° and separated overnight on a 0.3% agarose gel at 1 V/cm.

The appropriate enzyme concentration is chosen from the result of the analytical test. The reaction is scaled up 15-fold. Three time points are taken corresponding to 0.5, one, and two times the optimal enzyme concentration. EDTA is added to a final concentration of 15 mM. The samples are heated 15 min at 68°, and aliquots are analyzed on a 0.3% agarose gel. The remainder is ethanol precipitated, centrifuged 1 min in a table top centrifuge, and redissolved overnight in TE.

2. Phosphatase Treatment. To 20 μl *Mbo*I-digested DNA (0.25 μg/ μl), 2.5 μl 500 mM Tris · HCl (pH 9.5), 10 mM spermidine, 1 mM EDTA, and 2 U alkaline phosphatase from calf intestine are added and the samples are incubated 30 min at 37°. After addition of 3 μl 100 mM trinitriloacetic acid (pH 8) the samples are heated 15 min at 68°. The DNA is precipitated with ethanol/NaAc, centrifuged, and dissolved overnight in TE to a concentration of 0.25 μg/μl.

Comments

For partial digest reactions it is important to have a completely dissolved homogeneous DNA solution. Otherwise part of the DNA is degraded rapidly while the rest is not accessible to the enzyme and its molecular weight remains high.

The chosen fractions should show the maximum ethidium bromide staining in the >20 kb range. Digesting too little rather than too much makes it more likely that larger inserts will be obtained. A partially overlapping, slightly larger size cut may be used for the parallel construction of cosmid libraries in a *Bam*HI vector (e.g., pcos2EMBL). If the starting

DNA was somewhat degraded, better results are obtained with more digested samples, because the likelihood of a molecule having Sau3A ends on both sides is increased.

If only very small amounts of starting DNA are available, analytical digests can be carried out on 50 ng DNA. After electrophoresis, DNA is transferred onto nitrocellulose and visualized by hybridization to labeled repetitive DNA.

It is important that during construction of the library, fragments of DNA cannot be ligated to give packageable phage DNA containing inserts not contiguous in the genome. One way to assure this result is to size fractionate. An alternative is the use of phosphatase to prevent ligation of the DNA molecules to each other while leaving them able to be ligated to the unphosphatased vector DNA. To check the phosphatase reaction, aliquots of DNA treated with phosphatase or untreated are ligated to themselves and compared on a gel to unligated samples.

Ligation and Packaging

Eight micrograms of pooled, digested, phosphatase-treated DNA is ligated with 16 μg BamHI- and EcoRI-cut EMBL3 in 80 μl 40 mM Tris $\cdot$ HCl (pH 7.6), 10 mM MgCl$_2$, 1 mM DTT, 0.2 mM ATP, and 400 U ligase for at least 16 hr at 15°.

To 80 μl ligation at room temperature first 80 μl of sonic extract and then 240 μl of freeze-thaw lysate are added, mixed well, and left for 4 hr. λ diluent (800 μl) is added and the mixture is then kept at 4°. Alternatively, the packaged mixture can be applied to a CsCl step gradient [31, 42, 54% CsCl (w/w) in λ diluent].

The gradients are centrifuged in an SW60 rotor for 3 hr at 35,000 at 20°. Fractions are collected from the top with a Pasteur pipet, then tested by plating on NM538. The pooled fractions containing phage are dialyzed against λ diluent before plating.

Comments

A weight ratio of 1 : 2 or 1 : 3 of insert DNA to vector works well. If availability of insert DNA is limiting, a ratio of 1 : 4 or lower may be advisable. Even when vector cut with BamHI only is used, excess of vector does not significantly increase the background.

Ligation is usually carried out at 10–100 μg vector DNA/ml overnight at 15°.

Commercial packaging extracts can be used. We prepare two-component packaging extract essentially as described by Scherer et al.[25]

---

[25] G. Scherer, J. Telford, C. Baldari, and V. Pirotta, Dev. Biol. 86, 438 (1981).

A gradient concentration step is useful if the library construction is relatively inefficient. This can be the case because of unknown technical difficulties, or because there are few clones that should give rise to plaques in a very large background of other ligation products. More pfu's can be added to plating cells after this purification step, facilitating plating of the library at the desired density of plaques per plate.

Plating and Library Amplification

Before amplification the library is titrated on NM538 (both vector and recombinant grow), NM539 (recombinants only grow), and DH1 (vector only grows).

Plating cells are prepared from fresh, saturated overnight cultures by spinning down the cells and resuspending them in 0.5 vol 10 mM MgSO$_4$.

For a 22 × 22 cm plate, 2 ml plating cells is incubated with 1–2 × 10^5 pfu's for 15 min at 37°, 30 ml BBL top agar (or agarose for direct screening) plus 10 mM MgSO$_4$ are added, and the mixture is plated on BBL agar plates.

After incubation at 37° overnight the library is either screened directly or the top layer is scraped off, 30 ml λ diluent and 1 ml CHCl$_3$ are added, and the mixture is stirred for 20 min at room temperature. The agar is then removed by centrifugation and the library is stored over a drop of CHCl$_3$ at 4°.

Comments

In our experience, different preparations of plating cells give different plating efficiencies of *in vitro* packaged phage. It is therefore advisable to test different batches of cells before plating a library. Cells can be kept in 10 mM MgSO$_4$ at 4° for at least 2 weeks without loss of plating efficiency.

If it is not expected that the library will be rescreened a number of times over an extended period, a primary nonamplified library should be screened. Different phage clones grow at quite different rates and the complexity of the library decreases significantly on each amplification step. If the library is amplified on plates, it is advisable not to pool the plate stocks from different plates. This is especially important when screening with probes that occur a few times in the genome (e.g., a gene family with many pseudogenes).

Materials

L broth: 10 g Bacto tryptone (Difco, 0123-01), 5 g yeast extract (Difco, 0127-01), 5 g NaCl, water to 1 liter, adjust to pH 7.2

BBL agar: 10 g Baltimore Biological Laboratories trypticase, peptone (11921), 5 g NaCl, 10 g agar (Difco), add water to 1 liter, adjust to pH 7.2

BBL top layer agar: Like BBL agar, but 6.5 g agar/liter

BBL top layer agarose: Like BBL agar, but 5 g agarose/liter

λ diluent: 10 mM Tris/HCl (pH 7.6), 10 mM MgSO$_4$, 1 mM EDTA

TE: 10 mM Tris (pH 7.6), 1 mM EDTA

Enzymes: Restriction endonucleases were purchased from Bethesda Research Laboratories, New England BioLabs, or Boehringer–Mannheim; T4 DNA ligase was from New England BioLabs; alkaline phosphatase, intestinal, from Boehringer–Mannheim.

[17] λgt 11: Gene Isolation with Antibody Probes and Other Applications

By MICHAEL SNYDER, STEPHEN ELLEDGE, DOUGLAS SWEETSER, RICHARD A. YOUNG, and RONALD W. DAVIS

Genes can be isolated with either nucleic acid probes[1] or antibody probes. For cloning genes that are expressed at a low level or not at all in the organism of study, oligonucleotide probes can be prepared if protein sequence data are available. Alternatively, antibodies can be used as probes to isolate directly recombinant clones producing proteins of interest regardless of whether protein sequence is available. This chapter describes methods for the isolation of eukaryotic and prokaryotic genes by screening *Escherichia coli* expression libraries with antibody probes using the bacteriophage expression vector λgt 11.[2,3] Other uses of λgt 11, including preparation of foreign proteins expressed in *E. coli,* mapping epitope coding regions, and transplason mutagenesis are also described.

Gene Isolation by Immunoscreening

The general scheme for immunoscreening is shown in Fig. 1.[3–5] A recombinant DNA library is first constructed in λgt 11, an *E. coli* expression vector, and foreign antigens are expressed from the DNA inserts in particular *E. coli* host cells. The antigens are transferred onto nitrocellulose filters and then probed with antibodies to detect the desired recombinant.

λgt 11

The features of λgt 11 are shown in Fig. 2.[3] λgt 11 contains the *lacZ* gene of *E. coli* which can express foreign DNA inserts as β-galactosidase fusion proteins. cDNA or genomic DNA fragments are inserted into the unique *Eco*RI site located in the 3' end of the *lacZ* gene, 53 bp upstream of the translation termination codon. When the DNA fragments are intro-

[1] T. Maniatis, E. F. Fritsch, and J. Sambrook, "Molecular Cloning: A Laboratory Manual." Cold Spring Harbor Lab., Cold Spring Harbor, New York, 1982.

[2] R. A. Young and R. W. Davis, *Proc. Natl. Acad. Sci. U.S.A.* **80,** 1194 (1983).

[3] R. A. Young and R. W. Davis, *Science* **222,** 778 (1984).

[4] R. A. Young and R. W. Davis, *Genet. Eng.* **7,** 29 (1985).

[5] M. Snyder and R. W. Davis, *in* "Hybridomas in the Biosciences and Medicine" (T. Springer, ed.), p. 397. Plenum, New York, 1985.

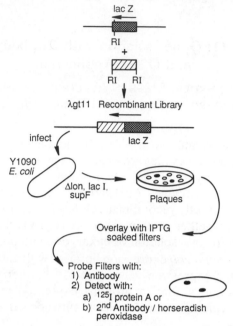

FIG. 1. Screening λgt 11 expression libraries with antibody probes. A recombinant DNA library is first constructed in λgt 11. The library is plated on bacterial cells to form plaques. After initial propagation of the phage, antigen production is induced by overlaying the plates with IPTG-treated filters, and phage growth is continued. The antigen-coated filters are then removed and probed with antibody. The bound antibody is detected with ^{125}I-labeled protein A or second antibody techniques to find the desired recombinant.

duced in the proper orientation and reading frame, the foreign DNA is expressed as a β-galactosidase fusion protein. Fusion of a foreign protein to a stable *E. coli* protein enhances the stability of the foreign protein in at least several cases.[6,7] However, antigens can also be expressed and not fused to *lacZ* (see below).

Since the *lacZ* gene has a strong *E. coli* promoter, and since λgt 11 propagates to high copy number during lytic infection, a significant percentage of total *E. coli* cellular protein is produced from the *lacZ* gene. For *lacZ* alone (unfused to any other sequences), approximately 5% of the total cellular protein is β-galactosidase.[3,5] For the λgt 11 fusions examined thus far, the intact fusion proteins comprise from less than 0.1% up to 4% of the total *E. coli* protein.[3,5]

[6] H. Kupper, W. Keller, C. Kurtz, S. Forss, H. Schaller, R. Franze, K. Strommaier, O. Marquardt, V. Zaslavsky, and P. H. Hofschneider, *Nature (London)* **289**, 555 (1981).
[7] K. Stanley, *Nucleic Acids Res.* **11**, 4077 (1983).

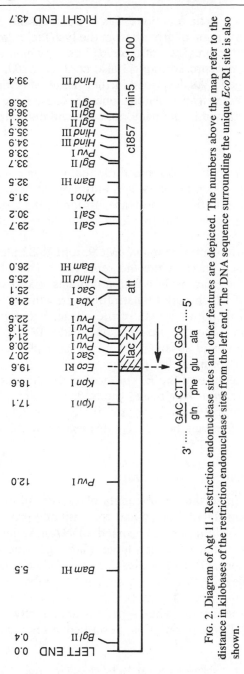

FIG. 2. Diagram of λgt 11. Restriction endonuclease sites and other features are depicted. The numbers above the map refer to the distance in kilobases of the restriction endonuclease sites from the left end. The DNA sequence surrounding the unique *Eco*RI site is also shown.

In addition to lytic functions, λgt 11 contains the genes necessary for lysogeny. Conversion of growth from the lysogenic state to lytic is readily controlled by the presence of the *cI857* gene, which encodes a temperature-sensitive *cI* repressor. λgt 11 also contains *S100*, an amber mutation in the *S* lysis gene. As described further below, this mutation is useful for producing proteins from isolated clones, because large amounts of protein can be accumulated in the bacterial cells without extensive cell lysis.

Bacterial Host

The *E. coli* host, Y1090, has three features that are useful for screening λgt 11 expression libraries with antibody probes. The bacterial strain used is deficient in the *lon* protease. In lon⁻ cells β-galactosidase fusion proteins often accumulate to much higher levels relative to wild-type cells.[3]

The bacterial strain contains pMC9, a pBR322 plasmid harboring the *lacI* gene which encodes the *lac* repressor.[8] This allows the regulated expression of β-galactosidase fusion proteins. Expression of foreign proteins, some of which may be harmful to growth of the bacterial host, is repressed during initial growth of the phage, but is expressed later after induction with isopropyl-β-D-thiogalactoside (IPTG). In the presence of the *lon* mutation and absence of *lac* repressor plasmids, λgt 11 recombinants typically form smaller plaques.

The third feature of the *E. coli* host is the presence of the *supF* suppressor tRNA. *supF* suppresses the *S100* mutation of λgt 11 and allows plaque formation. Plaques are preferable for immunoscreening because they provide a much better signal/noise ratio than colony screening methods.

λgt 11 Recombinant DNA Libraries

The most important components of successful immunoscreening are the recombinant DNA library and the antibody probe. Two types of recombinant DNA libraries can be used: cDNA libraries and genomic DNA libraries. For organisms which have a large genome size, such as mammals (3×10^9 bp), cDNA libraries are screened because currently it is not technically feasible to screen a sufficient number of genomic recombinants to find the desired gene (see below). cDNA libraries may also be preferable for screening when a gene is abundantly expressed. The construction of cDNA libraries is described elsewhere.[9] Near full-length

---

[8] M. P. Calos, T. S. Lebkowski, and M. R. Botchan, *Proc. Natl. Acad. Sci. U.S.A.* **80,** 3015 (1983).

[9] T. V. Huynh, R. A. Young, and R. W. Davis, *in* "DNA Cloning Techniques: A Practical Approach" (D. Glover, ed.), p. 49. IRL Press, Oxford, England, 1984.

cDNAs provide the most antigenic determinants and are preferable for immunoscreening.

However, λgt 11 cDNA expression libraries suffer from the fact that nonabundant mRNAs are represented very rarely, if at all, and often encode only a 3′-terminal portion of the gene. It is useful to be able to clone any gene, independent of its level of expression. This can be achieved by using a genomic DNA library, constructed by randomly shearing and inserting genomic DNA into the λgt 11 *lacZ* gene. For organisms with a genome size equal to *Drosophila* (1.8×10^8 bp) or smaller, the construction and screening of genomic libraries is possible.

In genomic DNA libraries, the probability of having a particular gene fused to *lacZ* depends on the length of the gene and the size of the organism's genome. For instance, fusion of any portion of a 1.5-kb yeast gene (yeast genome = 14,000 kb) to *lacZ* will occur once in every 9,333 times, and one out of six fusions will reside in the proper orientation and reading frame. Therefore, an inframe fusion will occur an average of once in every 5.6×10^4 recombinants. Fusions nearest the amino-terminal protein coding sequences will encode more antigenic determinants than carboxy-terminal protein fusions. In addition, the frequency of detecting a gene within a genomic library may be affected by intervening sequences. Organisms of small genome size, such as *Drosophila* and yeast, usually contain few if any introns within their coding sequences.

One interesting feature of λgt 11 libraries is that not all of the recombinants isolated encode β-galactosidase fusion proteins, a feature often encountered when using genomic libraries (Fig. 3).[10-13] For yeast, greater than 50% of the genes isolated from a genomic library by immunoscreening are expressed but not fused to β-galactosidase. In some instances, the expression of these genes may still be *lacZ* dependent and presumably use the *lacZ* promoter for expression (Fig. 3, Case B). Alternatively, expression can be *lacZ* independent, and insert sequences serve for transcription and translation in *E. coli* (Fig. 3, Case C). The mode of expression will also affect the relative frequency with which a gene is isolated from a genomic DNA library.

A general method for constructing a random shear genomic DNA library is diagramed in Fig. 4. DNA is randomly sheared to an average size of 5 kb, and the ends made flush with T4 DNA polymerase I. The DNA is subsequently methylated, *Eco*RI linkers ligated and cleaved, and

[10] J. L. Kelly, A. L. Greenleaf, and I. R. Lehman, *J. Biol. Chem.* **261**, 10348 (1986).
[11] T. Goto and J. C. Wang, *Cell* **36**, 1073 (1984).
[12] E. Ozkaynak, D. Finley, and A. Varshavsky, *Nature (London)* **312**, 663 (1984).
[13] M. Snyder, S. Elledge, and R. W. Davis, *Proc. Natl. Acad. Sci. U.S.A.* **83**, 730 (1986).

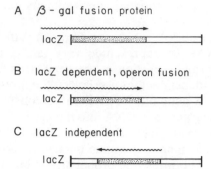

A   β - gal fusion protein

B   lacZ dependent, operon fusion

C   lacZ independent

Fig. 3. Modes of expression from λgt 11 genomic clones. Expression can be (A) as β-galactosidase fusion proteins or (B and C) not fused to β-galactosidase. In Case B antigen production is still *lacZ* dependent and expression occurs via an operon-type structure. In Case C expression is *lacZ* independent and insert sequences serve for transcription and translation in *E. coli*.

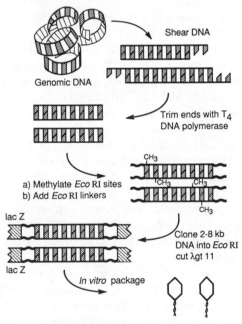

Fig. 4. Construction of a genomic DNA library. Genomic DNA is randomly sheared through a syringe needle to an average size of 5 kb. The ends are trimmed with T4 DNA polymerase and the DNA methylated with *Eco*RI methylase. *Eco*RI linkers are ligated on the ends, cleaved with *Eco*RI, and the DNA size fractionated on a 0.7% agarose gel. The 2- to 8-kb DNA is collected and cloned into *Eco*RI-cleaved λgt 11 that has been treated with calf alkaline phosphatase.

the DNA is size fractionated on an agarose gel. The 2- to 8-kb fraction is then cloned into the *lacZ* gene of λgt 11. A protocol is as follows:

1. Genomic DNA preparation: Shear the genomic DNA (15–20 μg at 100–200 μg/ml) to an average size of 5 kb by passing DNA vigorously through a 27-gauge syringe needle with a 1-ml syringe. The sheared DNA size is monitored by gel electrophoresis using a 0.7% agarose gel.[1]
2. Trim the DNA ends in 100 μl of 50 mM Tris–HCl (pH 7.8), 5 mM MgCl$_2$, 50 μg/ml BSA, 10 mM 2-mercaptoethanol, 20 μM of each dNTP, and 5 units of T4 DNA polymerase. Incubate for 30 min at room temperature.
3. Methylate the DNA in 50 μl of 50 mM Tris–HCl (pH 7.5), 1 mM EDTA, 5 mM DTT, 10 μM S-adenosyl-L-methionine plus 2 μg *Eco*RI methylase. Incubate at 37° for 15 min, then inactivate the enzyme by heating for 10 min at 70°.
4. (Optional step) To ensure blunt DNA ends, add MgCl$_2$ to 10 mM and dNTP to a final concentration of 20 μM. Add 2 units of *E. coli* DNA polymerase I large (Klenow) fragment. Incubate for 30 min at room temperature. Extract once with phenol:CHCl$_3$ (50:50), then once with CHCl$_3$ alone. Ethanol precipitate DNA.
5. Add phosphorylated *Eco*RI linkers on DNA using 50 μl of 70 mM Tris–HCl (pH 7.5), 10 mM MgCl$_2$, 200 μg/ml gelatin, 1 mM DTT, 5% polyethylene glycol, 1 mM rATP, 100 μg/ml phosphorylated *Eco*RI linkers plus 0.5 μl of 1.5 mg/ml T4 DNA ligase. Incubate at room temperature for 6–12 hr. Ethanol precipitate DNA.
6. Dissolve pellet in 100 μl *Eco*RI restriction digest buffer [50 mM NaCl, 100 mM Tris–HCl (pH 7.5), 7 mM MgCl$_2$, 1 mM DTT]. Add 50 units *Eco*RI and incubate 1.5 hr at 37° then 10 min at 70°. Ethanol precipitate the DNA by adding 2.5 volumes of ethanol and chilling at −20° for 1 hr. Repeat ethanol precipitation.
7. Purification of DNA inserts from *Eco*RI linkers and size fractionation: The DNA is completely dissolved in electrophoresis buffer and size-fractionated on a 0.8% agarose gel. The 2- to 8-kb DNA is collected from the gel. Two convenient methods are (1) Cut an empty well in front of the DNA to be recovered. As the gel is run at high current, remove aliquots from the well every 45–60 sec. Recover the DNA by adding sodium acetate (pH 6) to 0.3 M and precipitating with ethanol. (2) Place Schleicher and Schuell NA45 membrane in a gel slot cut just in front of the DNA to be collected. Run the DNA into the membrane at high current. Rinse the NA45 membrane in electrophoresis buffer and elute DNA in 1 M NaCl,

50 mM arginine (free base) at 70° for 2–3 hr. Extract the DNA with phenol : CHCl$_3$ as in step 4 and recover DNA by ethanol precipitation. *Note:* Removing the linkers is a crucial step. If linker contamination presents a problem, repeat the gel isolation step or fractionate the DNA on a P60 column prior to the gel isolation step.

8. λgt 11 vector preparation: Ligate the λgt 11 DNA to form concatamers according to the buffer conditions in step 5 (without linkers) for 2 hr at room temperature. Incubate at 70° for 15 min and ethanol precipitate DNA. Cleave DNA with *Eco*RI. Treat with calf intestine alkaline phosphatase (CAP) in 50 mM Tris–HCl (pH 9.0) using high quality enzyme that has been titered to yield less than 10% vector in the presence of insert DNA. Do not overtreat. Inactivate the CAP by heating at 70° for 10 min. Extract DNA with phenol : CHCl$_3$ as in step 4 and ethanol precipitate DNA.

9. Library construction: Ligate the genomic insert DNA into λgt 11 using a ratio of insert DNA : vector DNA of 0.5–1 : 1 by mass using buffer conditions in step 5. A final DNA concentration of 4–800 µg/ml is used. Incubate at room temperature for 8–20 hr.

10. *In vitro* package DNA and titer library according to Huynh *et al.*[9] Yield is approximately 5 × 10^5 phage/µg λ DNA.

11. Amplify library on *E. coli* strain Y1088 on LB plates containing 50 µg/ml ampicillin [Y1088 = Δ*lacU169 supE supF hsdR$^-$ hsdM$^+$ metB trpR tonA21 proC*::Tn5 + pMC9]. This bacterial strain is *hsdR$^-$* and *hsdM$^+$* and contains the *lacI* plasmid in order to prevent the expression of proteins that are harmful to *E. coli*.

Using these methods λgt 11 genomic DNA libraries containing 10^6–10^7 recombinants from yeast[5] and mycobacteria[14] have been constructed.

Antibody Probes

Successful immunoscreening depends on the quality of the antibody. Antibodies that produce good signals on immunoblots or "Westerns" usually work well for antibody screening. Both monoclonal and polyclonal antibodies have been used successfully. Polyclonal antisera have the advantage that they can recognize multiple epitopes on any given protein. The ability to detect a variety of epitopes on a protein is important, since genomic and cDNA inserts often will not be full length and hence only a portion of the polypeptide will be expressed. Since a single epitope can be shared by multiple proteins, clones other than the desired

[14] R. A. Young, B. R. Bloom, C. M. Grosskinsky, J. Ivanyi, D. Thomas, and R. W. Davis, *Proc. Natl. Acad. Sci. U.S.A.* **82,** 2583 (1985).

recombinants can also be detected. Therefore, immunoscreening with polyclonal antisera or with several monoclonal antibodies recognizing different determinants on the same protein is recommended.

For a polyclonal serum it is usually not necessary to use antibodies that have been affinity purified for the protein of interest. However, this can lead to the isolation of clones other than the recombinant of interest, particularly if the original serum is made against an impure protein.

Most rabbit and human sera contain significant amounts of anti-*E. coli* antibodies. These are easily removed by "pseudoscreening" *E. coli* protein lysates. λgt 11 phage (without inserts) are plated and protein-coated filters prepared according to the screening conditions described in the next section. The filters are treated and incubated with antibodies to absorb the anti-*E. coli* antibodies from the serum. It is usually necessary to repeat the pseudoscreening 2 or more times to remove *all* anti-*E. coli* antibodies.[15] The same λgt 11 plates can be reused: after the first filter is removed the plates are overlayed with a second filter and incubated for an additional 2 hr at 37°.

Screening λgt 11 Libraries with Antibody Probes

1. Plate out library: Grow a culture of Y1090 cells at 37° to saturation in LB media[16] containing 50 μg/ml ampicillin and 0.2% maltose. Per 150-mm LB plate infect 0.1 ml of the saturated culture with 1×10^5 phage. Allow phage to absorb for 25 min at room temperature or 15 min at 37°. Add 6.5 ml LB top agar and plate immediately. Neither top agar nor plates contain ampicillin. Plates work best 3–5 days after preparation. Incubate plates for 3.0 hr at 42° until lawn just becomes visible.

2. Overlay plates with IPTG-treated nitrocellulose filters prepared as described below. Incubate the plates at 37° for 8–10 hr (overnight). At the end of the incubation period, remove the lid from the plates and incubate at 37° for an additional 10–15 min. This step helps prevent top agar from sticking to the filter.

3. Mark the position of the filters with a syringe needle containing ink. Carefully remove filters from plates. For duplicate screening, overlay the plates with a second IPTG-treated filter and return to 37° for an additional 2–4 hr.

   For subsequent steps, filters are treated in such a manner that they are well exposed to the washing and probing solutions. One convenient method is to use two filters per Petri dish.

---

[15] T. St. John, personal communication.
[16] R. W. Davis, D. Botstein, and J. R. Roth, "Advanced Bacterial Genetics." Cold Spring Harbor Lab., Cold Spring Harbor, New York, 1980.

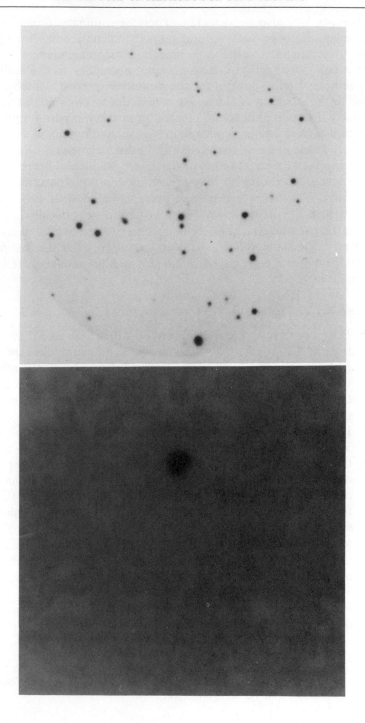

4. Rinse filters 1–2 times in TBS (15 ml/filter) (TBS = 150 mM NaCl, 50 mM Tris–HCl, pH 8.1).
5. Incubate filters for 30 min or longer in TBS plus 0.05% Tween 20 plus 0.5% BSA or TBS plus 20% fetal calf serum (FCS) (15 ml/filter). (These solutions may be saved and reused several times.)
6. Probe filters in the antibody solution diluted in TBS plus 0.05% Tween 20 plus 0.1% BSA or TBS plus 20% FCS (10 ml/filter) for 1–8 hr.
7. Wash filters 10 min each with (1) TBS plus 0.1% BSA, (2) TBS plus 0.1% BSA plus 0.1% NP40, (3) TBS plus 0.1% BSA (Tween 20 optional). The antibody probes can then be detected by either methods 8A or 8B.

8A. ^{125}I-Labeled protein A probe:
    a. Treat filters with ^{125}I-labeled protein A in TBS plus 0.1% BSA for 1.5–2.5 hr. Use 1 μCi of >30 mCi/mg specific activity ^{125}I-labeled protein A (ICN or Amersham) per 132 mm filter.
    b. Wash filters 10 min each with (1) TBS plus 0.1% BSA, (2) TBS plus 0.1% BSA plus 0.1% NP40, repeat once, (3) TBS plus 0.1% BSA.

8B. Horseradish peroxidase/alkaline phosphatase conjugated probes:
    a. Incubate filters with second antibody according to the manufacturer's recommendations (see below).
    b. Wash filters and incubate with substrate solution according to manufacturer's instructions.

Sample positives of these screens using ^{125}I-labeled protein A or horseradish peroxidase probes are shown in Fig. 5.

Comments

1. General methods for plating and handling bacteriophage λ can be found in Maniatis *et al.*[1] and Davis *et al.*[16]
2. IPTG-treated filters are prepared by wetting nitrocellulose filters in solution of 10 mM IPTG dissolved in distilled water and allowed to air dry on plastic wrap.

FIG. 5. Exemplary positives from immunoscreening. Top: A positive yeast DNA polymerase–λgt 11 clone (approximately 30/plate) was plated with 10^4 nonreactive phage on a 90-mm plate. The antigens were immunoscreened with an anti-DNA polymerase I polyclonal serum and detected with ^{125}I-labeled protein A. Bottom: High magnification of a positive signal from a *Mycobacterium leprae* clone immunoscreened with a monoclonal antibody and detected with horseradish peroxidase conjugated probes (Vectastain probes from Vector Laboratories).

3. Incubation times for most of the steps can be varied greatly. Incubation of IPTG-treated nitrocellulose filters on the plates for 8–10 hr produces signals 5–10 times stronger than for a 2-hr incubation. Overnight incubation with antibody yields a 3- to 5-fold stronger signal relative to a 2-hr incubation.

4. Most antibodies produce good signals at room temperature. Signals generally increase 3- to 5-fold if the antibody treatments and subsequent incubations are performed at 4°. This is the recommended temperature for low-affinity antibodies.

5. The antibody solution can be reused many times and is stored with 0.01% sodium azide after each use. For anti-DNA polymerase I antibodies, the serum was reused 12 times with little reduction in intensity.[17] Every time the serum is used, the background is reduced, presumably because of the further removal of anti-*E. coli* antibodies from the serum.

6. For primary screens it is recommended that plates be screened in duplicate to avoid false positives. For ^{125}I-labeled protein A these usually occur at a frequency of one or two per 132-mm filter. The number of phage to be used should be low enough such that the lawn is not completely lysed prior to the application of the second filter.

7. Filters can be stored for several weeks in TBS plus 20% FCS prior to probing with antibody.

8. Control experiments where known amounts of protein antigen have been spotted onto filters and screened with antibody as above indicate that the limit of detection for one high-titer serum is approximately 100 pg of protein.[18] Actual amounts of antigen transferred from a single plaque vary up to approximately 100 times higher than this for the positives thus far isolated.

9. Ampicillin, which is used to select for cells containing the *lacI* plasmid, is omitted in the screening plates because it slows cell growth.

10. High specific activity ^{125}I-labeled protein A (>30 mCi/mg) is often necessary for detecting positive clones.

11. Numerous methods exist for enzymatic detection of the antibody using second antibodies. These include horseradish peroxidase probes (Bio-Rad, Vector Laboratories) or alkaline phosphatase probes (Promega Biotec, Vector Laboratories).

[17] L. M. Johnson, M. Snyder, L. M. S. Chang, R. W. Davis, and J. L. Campbell, *Cell* **43**, 369 (1985).
[18] C. Glover and M. Snyder, unpublished results.

12. Common substitutes for 0.05% Tween 20/0.5% BSA or 20% FCS protein blocks (step 5) are 3% BSA or 5% solution of powdered milk. Substrates that work well for immunoblots also work well for immunoscreening.

Verification of Gene Identity

The most difficult part of isolating genes by antibody screening is to determine whether the immunoreactive clones encode the gene of interest. Several useful methods for determining the identity of clones are as follows.

A. *Affinity Purification of Antibodies Using λgt 11 Clones.* Proteins produced from the λgt 11 clones can be used to affinity purify antibody from the screening serum. The affinity-purified antibody is then used to probe immunoblots of a purified or crude preparation of the protein of interest. This will determine if the isolated clone is antigenically related to the protein of interest, and is particularly useful if the original antiserum was not made against a homogeneous protein. A protocol is as follows[5]:

1. Plate the positive recombinants at $2-5 \times 10^4$ phage/90-mm LB plate. Use 2.5 ml of LB top agar and 50 μl of a saturated culture of Y1090 cells. Follow steps 1–7 of the immunoscreening protocol for preparing filters, and probing them with antibody and washing. One plate per clone produces sufficient antigen for most sera.
2. Place the antibody-treated filter in a capped tube. Add 3.5 ml of 0.2 M glycine–HCl (pH 2.5). Mix the tube for 2 min and no longer. Remove filter.
3. *Immediately* add 1.75 ml of 1.0 M KPO$_4$ (pH 9.0) plus 5% FCS and mix.
4. Dilute to lower ionic strength by adding 6 ml of distilled water plus 2 ml of TBS. Add FCS to 20% and use directly for immunoblots.

Immunoblot protocols are described by Burnette.[19] Alternative antibody elution protocols are detailed by Earnshaw and Rothfield[20] and Smith and Fisher.[21]

B. *Gene Expression.* The insert from the λgt 11 recombinant can be used to recover an intact gene by hybridization techniques. The intact gene can be overexpressed in *E. coli* or other organisms and assayed for gene activity. Overexpression in *E. coli* has been used for yeast topoisomerase II[11] and yeast DNA polymerase I.[17] Overexpression in yeast has

[19] W. N. Burnette, *Anal. Biochem.* **112**, 195 (1981).
[20] W. C. Earnshaw and N. Rothfield, *Chromosoma* **91**, 313 (1985).
[21] D. E. Smith and P. A. Fisher, *J. Cell Biol.* **99**, 20 (1984).

also been performed for yeast DNA polymerase I.[17]

C. *DNA Sequencing.* The DNA sequence of the λgt 11 recombinant insert can be determined, and the deduced amino acid sequence compared with the known protein sequence. Primers are available (New England Biolabs) that are complementary to the *lacZ* portion of λgt 11 adjacent to the *Eco*RI cloning site. These primers can be used to direct dideoxy sequencing by the method of Sanger[22] (see below).

Protein Preparation from λgt 11 Clones

Proteins can be prepared from Y1089 cells [Δ*lacU169 proA*⁺ Δ*lon araD139 strA hflA* [chr::Tn*10*] (pMC9)] or CAG456 cells [*lac(am) trp(am) pho(am) supC*ᵗˢ *rpsL mol(am) htpR165*]. The relevant features of Y1089 are that it contains the *lacI* plasmid, Δ*lon* mutations, and the *hflA* mutations for forming lysogens at a high frequency, and the strain contains no suppressor—this latter feature allows λgt 11 protein products to accumulate to high levels without cell lysis. CAG456 cells also will not be lysed by λgt 11 and contains the *htpR* mutation which is a mutation in the heat shock sigma factor. The *htpR* mutation causes temperature-sensitive growth of the cells, and prevents the synthesis of heat shock proteins, many of which are proteases. This protease deficiency is observed at both 30 and 37°.[23] The level of protein production from two distinct yeast λgt 11 recombinant clones were compared in the two strains. Both fusion proteins were found to accumulate 10- to 100-fold higher in CAG456 cells relative to Y1089 cells.[24] However, the stability of different fusion proteins varies greatly among different strains of *E. coli* so it is useful to test several bacterial strains.

A. Preparation of Lysogens in Y1089

1. Infect a saturated culture of Y1089 cells grown in LB plus 0.2% maltose plus 50 μg/ml ampicillin with λgt 11 clones using an m.o.i. (multiplicity of infection) of 10. Absorb 20 min at room temperature.
2. Plate 200 cells per LB plate and incubate at 30°.
3. Pick individual colonies with a toothpick and test for growth at 42 and 30°. λgt 11 lysogens will not grow at 42°. Usually 30–50% of the colonies are lysogens.

[22] F. Sanger, A. R. Coulson, B. G. Barrell, A. J. H. Smith, and B. Rose, *J. Mol. Biol.* **143,** 161 (1980).
[23] T. A. Baker, A. D. Grossman, and C. A. Gross, *Proc. Natl. Acad. Sci. U.S.A.* **81,** 6779 (1984).
[24] M. Snyder, M. Cai, and R. W. Davis, unpublished results.

B. Small-Scale Preparation of Protein Lysates from λgt 11 Lysogens

1. Four milliliters of lysogens are grown with vigorous shaking at 30° to $OD_{600} = 0.4$.
2. Lysogens are induced by shifting the temperature to 44° for 15 min with vigorous shaking. A rapid temperature shift is important for maximal induction.
3. IPTG is added, and the culture is incubated for 1 hr at 37°. The time of incubation varies with the particular recombinant; some clones promote rapid cell lysis and so must be harvested earlier.
4. Cells are harvested as quickly as possible by sedimentation at 10,000 g for 30 sec and freezing immediately at $-70°$. For SDS gel analysis, SDS sample loading buffer is added just prior to freezing. Four milliliters of cells yields sufficient protein for 8–10 gel lanes.

C. *Preparation of Protein Lysates by Infection.* This procedure is similar to that above. Cells (Y1089 or CAG456) are grown to $OD_{600} = 0.40$ and infected at a multiplicity of infection of 5 instead of inducing the culture. The culture is then incubated at 37° for the appropriate length of time: Y1089 cells, about 1 hr as above; CAG456 cells, 1–4 hr. (Determine the appropriate length of time.) Cells are then harvested as above.

The lysogen induction and the infection procedure produce comparable protein yields. However, the infection procedure requires much more phage, which can present a problem if the procedure is performed on a larger scale.

Mapping Epitopes on λgt 11 Clones

λgt 11 can be used to locate the boundaries of the antigenic determinants of a protein.[25] The strategy involves (1) isolating a DNA clone that encodes the entire antigen of interest and determining its nucleotide sequence, and consequently, the encoded amino acid sequence; (2) constructing a λgt 11 sublibrary containing fragments of the gene; (3) detecting the expression of epitope-coding sequences with monoclonal antibody probes; and (4) isolating and determining the precise nucleotide sequences of the cloned DNA fragments via primer-directed DNA sequence analysis. A comparison of shared sequences among the antibody-positive clones locates the epitope-coding region.

A λgt 11 clone that is capable of expressing the antigen of interest is isolated by immunoscreening or specifically constructed. It is important to confirm that the antigenic determinants of interest can be expressed.

---

[25] V. Mehra, D. Sweetser, and R. A. Young, *Proc. Natl. Acad. Sci. U.S.A.* **83**, 7013 (1986).

The foreign DNA is then sequenced, and the protein amino acid sequence is deduced.

The next step is to construct a λgt 11 gene sublibrary that contains small random DNA fragments from the gene of interest. The aim is to produce recombinant phage in sufficient numbers to obtain DNA insert end points at each base pair in the gene, such that all possible overlapping segments of the coding sequence are expressed. A protocol is as follows:

1. DNA fragments with random end points are generated by digesting with DNase I (1 ng DNase I/10 μg DNA/ml) in a buffer containing 20 mM Tris–HCl (pH 7.5), 1.5 mM MgCl$_2$, and 100 μg/ml BSA at 24° for 10–30 min to produce short random fragments.
2. The DNA is fractionated on a 1% agarose gel, and fragments of 200–1000 bp are isolated and purified.
3. These DNA fragments are end repaired by treatment with T4 polymerase in the presence of dNTPs and then ligated to phosphorylated EcoRI linkers (Collaborative Research).
4. This material is then digested with EcoRI, heat inactivated at 70° for 5 min, and fractionated on a P60 column (Bio-Rad) to remove unligated linkers.
5. The linkered DNA fragments are further purified on an agarose gel, from which they are eluted, phenol extracted, and ethanol precipitated.
6. The EcoRI-linkered DNA fragments are ligated onto phosphatase-treated λgt 11 arms (Promega Biotec), and the ligated DNA is packaged in vitro.
7. The resultant recombinant phage are amplified on E. coli Y1088.

These steps are described in detail above.

The gene sublibrary is screened with monoclonal antibodies to isolate DNA clones that express the epitope-coding sequences. The limits of the sequences that encode an epitope are located by subjecting the recombinant clones to two types of analysis. First, the sequences of the DNA insert end points are determined for each clone. Second, the lengths of the DNA insert fragments are ascertained by restriction analysis.

The sequence of DNA insert end points in λgt 11 is determined as follows:

1. The recombinant DNA is first isolated from phage purified by CsCl block gradient centrifugation.[16]
2. DNA (1–5 μg) is digested with the restriction endonuclease KpnI and SacI, then phenol extracted, ethanol precipitated, and resuspended in 20 μl H$_2$O.
3. The DNA is denatured by adding 2 μl of 2 M NaOH, 2 mM EDTA, and the solution is incubated for 10 min at 37°. The solution is

    neutralized with 6.5 μl of 3 M sodium acetate (pH 5.2), 6.5 μl H_2O
    is added, and the DNA is ethanol precipitated, washed twice with
    70% ethanol, and resuspended in 10 μl H_2O.

4. To the DNA (optimum amount should be determined empirically),
    1 μl of 10 μg/ml DNA primer and 1.5 μl sequencing buffer (75 mM
    Tris–HCl, pH 7.5, 75 mM DTT, 50 mM $MgCl_2$) are added and
    incubated at 55° for 15 min. The two primers used (New England
    Biolabs) are complementary to $lacZ$ sequences adjacent the EcoRI
    site in λgt 11; the sequence of the "forward primer" is GGTG-
    GCGACGACTCCTGGAGCCCG, that of the "reverse" primer is
    TTGACACCAGACCAACTGGTAATG.

5. Primer extension and dideoxy termination reactions are performed
    as described by Sanger et al.,[22] and the products are subjected to
    electrophoresis on an 8% polyacrylamide–8 M urea gel.

    The recombinant subclones are characterized further to ascertain
whether they contain multiple inserts or rearranged insert DNA, either of
which could complicate the interpretation of the data. DNA from each of
the clones is digested with the restriction endonuclease EcoRI and is
subjected to agarose gel electrophoresis to determine the number and
sizes of inserted DNA fragments. Subclones that contain insert DNAs
whose sequenced end points predict a DNA fragment length that agrees
with the size determined by agarose gel electrophoresis of the EcoRI
DNA fragments can be used for subsequent analysis.

    The amino acid sequences that contain an epitope are those that are
shared by all of the subclones that produce positive signals with a particular antibody (Fig. 6). Only clones that produce positive signals with an
antibody can be used to deduce the position of the determinants. Clones
that produce no signal could contain and even express the appropriate
amino acids; however, the antigenic determinant might not be detectable
because it is susceptible to proteolysis or it lacks the ability to form the
correct antigenic structure. Since the epitope lies within the amino acid
sequences shared by signal-producing clones, the resolution of the boundaries of an epitope should improve as larger numbers of recombinant
clones are analyzed.

    An example of this technique for mapping mycobacterial antigenic
determinants is presented in Fig. 6. It is striking that all of the recombinant clones studied thus far that contain coding sequences for the myco-
bacterial antigenic determinants express detectable levels of that determinant (Fig. 6). As discussed above, clones containing DNA encoding an
epitope might not express that epitope at detectable levels, reflecting
different stability or structural constraints. The λgt 11 system, coupling
the expression of fusion proteins with the use of lon protease-deficient

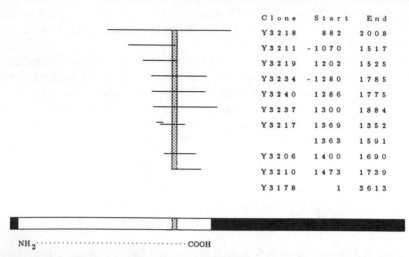

Clone	Start	End
Y3218	882	2008
Y3211	-1070	1517
Y3219	1202	1525
Y3234	-1280	1785
Y3240	1286	1775
Y3237	1300	1884
Y3217	1369	1352
	1363	1591
Y3206	1400	1690
Y3210	1473	1739
Y3178	1	3613

NH$_2$ ·································· COOH

FIG. 6. Deducing an epitope in the *M. leprae* 65-kDa antigen. The heavy horizontal line at bottom depicts the Y3178 insert DNA in which the open box represents the 65-kDa antigen open reading frame. The thin horizontal lines illustrate the extent of the insert DNA fragments from Y3178 subclones. The vertical stippled region indicates the extent of the epitope-coding sequence (15 amino acids) as defined by the minimum overlap among clones that produce a positive signal with the anti-*M. leprae* monoclonal antibody MLIIIE9. To the right is tabulated the precise insert end points for each DNA clone.

host cells, may help express all encoded epitopes at detectable levels. The particular monoclonal antibodies that are used and whether they recognize segmental or assembled topographic determinants may also influence this result.

One application of this method is that recombinant clones from these sublibraries can also be used to elucidate determinants to which T cells respond. *Escherichia coli* lysates containing antigen expressed by λgt 11 recombinants can be used to assay antigen-specific T-cell stimulation *in vitro*.[26]

Transplason Mutagenesis of λgt 11 Clones

Another method for mapping antigenic determinants uses transposon mutagenesis, which was developed in collaboration with Elledge.[13] Mini-Tn*10* transposons were constructed containing *E. coli* selectable markers, *tet*^{R}, *kan*^{R}, and *supF*. They are present on a high copy number pBR322-type plasmid and in a strain, BNN114, that overproduces Tn*10* transpo-

---

[26] A. S. Mustafa, H. K. Gill, A. Nerland, W. J. Britton, V. Mehra, B. R. Bloom, R. A. Young, and T. Godal, *Nature (London)* **319**, 63 (1986).

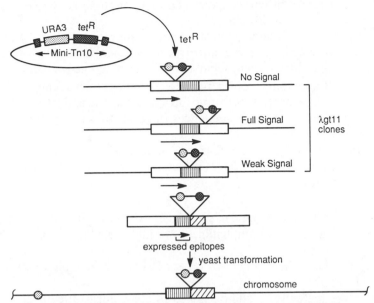

FIG. 7. Transplason mutagenesis of λgt 11 clones. Mini-Tn*10* transposon containing 70 bp of Tn*10* ends, *E. coli* selectable markers, and yeast selectable markers are carried on high copy number plasmids and in an *E. coli* strain that overproduces Tn*10* transposase. λgt 11 clones are grown on these strains, and transposition events into the phage are selected. A correlation of the position of the insertion with the effect on the immunoscreening signal can be used to locate the antigenic coding region. For a polyclonal serum, insertions between the promoter and antigenic coding region result in no immunoscreening signal, insertions outside the promoter and antigen coding region yield full level signals, and insertions within a coding region produce reduced signal. Reduced signals are probably the result of failure to express downstream epitopes, but other explanations are also possible.[13] For yeast *all* relevant insertions can be substituted for the chromosomal sequences by one-step gene transplacement[26] to inactivate the gene.

sase.[13,27] A library of transposon insertions is easily constructed for a λgt 11 clone by growing the phage on the mutagenesis strain and selecting for the transposition events. After mutagenesis the clones are screened with polyclonal or monoclonal antibody probes to determine the effect of the insertion on the level of antigen production. Tn*10* transposons when inserted in the middle of a coding segment obliterate downstream antigen expression. Correlation of the position of the insertion and the effect on the immunoscreening signal can be used to locate the antigen-coding region (Fig. 7). This procedure is useful for mapping antigenic coding seg-

[27] D. J. Foster, M. A. Davis, D. E. Roberts, K. Takeshita, and N. Kleckner, *Cell* **23,** 201 (1981).

ments, regardless of whether the recombinant clone is synthesizing a β-galactosidase fusion protein or not.

Transposons containing yeast selectable markers (*URA3*, *TRP1*) have been constructed.[13] These elements can be used to rapidly mutagenize yeast clones to map coding segments. The mutated genes can be directly substituted for genomic copies by one-step gene transplacement[28] (Fig. 7). Because of their use for both transpo*son* mutagenesis in *E. coli* and for *transpla*cement of genomic sequences in yeast, these elements are called transplasons.

A protocol for mTn*10*/*URA3*/*tet*[R] mutagenesis—a mini-Tn*10* transplason containing the *E. coli* *tet*[R] selectable marker and the *URA3* gene of yeast—is as follows:

1. Grow 0.5 ml of the Tn*10* mutagenesis strain in LB medium containing 50 μg/ml ampicillin, 0.2% maltose plus 10 mM MgSO$_4$ to OD$_{600}$ = 0.4.
2. Infect cells at m.o.i. 2–4 with λgt 11 recombinant phage. Incubate 2 hr at 37°.
3. Lyse the cells by adding 150 μl of chloroform. Incubate 15 min at 37° with shaking.
4. Clear the lysate of unlysed bacteria and debris by centrifugation for 5 min in an Eppendorf centrifuge. Repeat once. Store the lysate over 100 μl of chloroform.
5. Mix 100 μl of lysate (2–10 × 10^9 phage) with 0.4 ml of a saturated culture of BNN91 grown in LB medium plus 0.2% maltose plus 10 mM MgSO$_4$. Absorb 20 min at room temperature [BNN91 = Δ*lacZ hflA150 strA*].
6. Plate a serial dilution of cells on LB plates containing 15 μg/ml tetracycline to select for lysogens. For insert sizes of 3–4 kb, 25–70% of the transposition events occur in the insert DNA.
7. Prepare phage by growing a 5 ml culture of lysogens in LB plus 10 mM MgSO$_4$ (without drug) to OD$_{600}$ = 0.50. Induce phage by heating culture at 44° for 15 min with vigorous shaking. Incubate at 37° for 1.5–2 hr. Add 150 μl of chloroform to lyse cells and pellet bacterial debris as above.
8. Minipreparations of DNA are prepared using polyethylene glycol precipitation according to Snyder *et al.*[13] From a 5 ml culture approximately 5 μg of DNA is obtained.
9. Yeast transformation: Cut 2 μg of DNA with a restriction enzyme that cuts on one or both sides of the yeast insert flanking the trans-

---

[28] R. J. Rothstein, this volume [18].

plason. Ethanol precipitate the DNA and transform yeast colonies using the lithium acetate procedure.[29] Approximately 10–60 transformants/2 μg DNA are obtained.

Examples

Using these procedures, antibodies directed against pure proteins have been used to isolate a large number of genes. These include the yeast genes for RNA polymerases I, II, and III,[3,30] clatherin,[31] topoisomerase II,[10] and ubiquitin.[12] Examples from other organisms include *Plasmodium* antigens,[32] *Mycobacterium tuberculosis* antigens,[14] human terminal deoxynucleotidyltransferase,[33,34] rat fibronectin,[35] and human factor X.[36]

Knowledge of the function a protein is not obligatory for gene isolation. Use of a combination of the methods described above may help determine the function of that gene. For instance, Kelly, Greenleaf, and Lehman purified a yeast protein whose function was unknown *in vivo*.[37,38] The gene was cloned using the above techniques. Inactivation of that gene by transplason mutagenesis led to the discovery that the protein was a mitochondrial RNA polymerase subunit.

Antibodies made against impure proteins can also be used to isolate genes. For example, the gene for yeast DNA polymerase I was isolated using an antiserum directed against an impure preparation of that enzyme (estimated to be 50% homogeneous).[17] This was made possible by (1) using a genomic DNA library and isolating *all* sequences of the yeast genome that would react with the antibody and (2) identifying the correct DNA polymerase clones among a collection of other reactive phage using the independent assays described above.

[29] H. Ito, Y. Fukuda, K. Murata, and A. Kimura, *J. Bacteriol.* **153,** 163 (1983).

[30] J. M. Buhler and A. Sentenac, unpublished results.

[31] G. S. Payne and R. Schekman, *Science* **230,** 1009 (1985).

[32] D. J. Kemp, R. L. Coppel, A. F. Cowman, R. B. Saint, G. V. Brown, and R. F. Anders, *Proc. Natl. Acad. Sci. U.S.A.* **80,** 3787 (1983).

[33] R. C. Peterson, L. C. Cheung, R. V. Mattaliano, L. M. S. Chang, and F. J. Bollum, *Proc. Natl. Acad. Sci. U.S.A.* **81,** 4363 (1984).

[34] N. R. Landau, T. P. St. John, I. L. Weissman, S. C. Wolf, A. E. Silverstone, and D. Baltimore, *Proc. Natl. Acad. Sci. U.S.A.* **81,** 4363 (1984).

[35] J. E. Schwartzbauer, J. W. Tamkun, I. R. Lemischka, and R. O. Hynes, *Cell* **35,** 421 (1983).

[36] S. P. Leytus, D. W. Chung, W. Kisiel, K. Kurachi, and E. W. Davie, *Proc. Natl. Acad. Sci. U.S.A.* **81,** 3699 (1984).

[37] J. Kelly, A. Greenleaf, and I. R. Lehman, submitted for publication.

[38] A. Greenleaf, J. Kelly, and I. R. Lehman, submitted for publication.

Acknowledgments

We thank Tom St. John, Andrew Buchman, Stewart Scherer, and Carl Mann for their contributions to the development of these methods. We are grateful to C. Glover, J. Reichardt, S. Cotterill, and others for their contributions. This work was supported by grants from the National Institutes of Health (GM34365, AI23545, and GM21891) and the World Health Organization/World Bank/UNDP Special Program for Research and Training in Tropical Diseases.

[18] One-Step Gene Disruption in Yeast

By RODNEY J. ROTHSTEIN

The accessibility of the yeast genome to genetic manipulation using plasmid technologies is reviewed in this volume.[1] This chapter describes a relatively simple technique for gene disruption or replacement in yeast that requires a single transformation. The method can be used (*a*) to determine whether a cloned fragment contains a specific gene; (*b*) to determine whether a cloned gene is essential; and (*c*) to alter or completely delete a specific region. The method takes advantage of our previous observations that during yeast transformation free DNA ends are recombinogenic, stimulating recombination by interacting directly with homologous sequences in the genome.[2]

Figure 1 outlines the principle of the procedure. A cloned DNA fragment containing *GENE Z*⁺ is digested with a restriction enzyme that cleaves within the *GENE Z* DNA sequence. Another DNA fragment containing a selectable yeast gene (*HIS3*⁺ in this example) is cloned into the cleaved *GENE Z*. The outcome of this cloning strategy is to disrupt *GENE Z*. As illustrated in Fig. 1, the cloned *gene Z'* and *'gene Z* (we have adopted the nomenclature for protein fusions from *E. coli*) are separated by the selectable yeast fragment containing the *HIS3*⁺ gene. The *in vitro* disrupted gene is liberated from the bacterial plasmid sequences by restriction enzyme digestion. It is important that the linear fragment contain sequences homologous to the chromosomal *GENE Z* region on both sides of the inserted fragment to direct the integration into the *GENE Z* region. The digested DNA is transformed into a *his3*⁻, *GENE Z*⁺ yeast cell and *HIS3*⁺ transformants are selected. Among the transformants are strains that simultaneously become *HIS3*⁺ and *gene Z*⁻. Thus in these transformants *GENE Z* has been replaced with a disrupted *gene Z*⁻ in one step.

The method outlined above may be modified to delete an entire gene or a gene fragment simply by using the appropriate cloning strategy (Fig. 2, type II). It can be used to determine whether or not a gene is essential by first disrupting the gene in a diploid (Fig. 2). Subsequent genetic analysis of a disrupted essential gene will reveal linkage of the lethal function to the selectable genetic marker used to disrupt the gene. As has been

[1] T. L. Orr-Weaver, J. W. Szostak, and R. J. Rothstein, this series, Vol. 101, p. 228.
[2] T. L. Orr-Weaver, J. W. Szostak, and R. J. Rothstein, *Proc. Natl. Acad. Sci. U.S.A.* **78**, 6354 (1981).

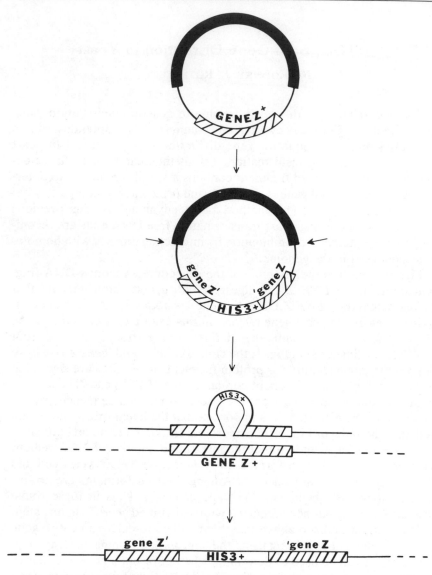

Fɪɢ. 1. One-step gene disruption. The cloned fragment containing *GENE Z*[+] is digested with a restriction enzyme that cleaves within the *GENE Z* sequence. A fragment containing a selectable yeast gene (*HIS3*[+] in this example) is cloned into the site. The fragment containing the disrupted *gene Z* is liberated from plasmid sequences, making certain that homology to the *GENE Z* region remains on both sides of the insert. Transformation of yeast cells with the linear fragment results in the substitution of the linear disrupted sequence for the resident chromosomal sequence.

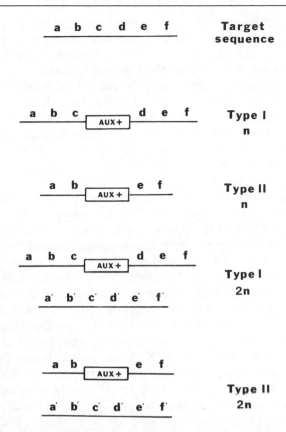

FIG. 2. Various types of one-step gene disruptions. A type I disruption is created after cloning a selectable marker (AUX^+) into a single restriction site within the target sequence. A type II disruption is created when two restriction sites are present and AUX^+ is inserted while c-d is deleted. For diploid type I and II disruptions, a', b', etc., represent the undisrupted homologous chromosome. In the case of the disruption of an essential gene in a diploid, spores containing AUX^+ would be inviable.

pointed out before,[3] gene disruption techniques in yeast offer similar advantages as the use of transposon insertions in prokaryote genetic studies.

Results

The methods for the manipulation of plasmids, yeast transformation and genomic blot analysis are described in this volume.[1] In this section, three examples of successful one-step gene replacement experiments will be described (Fig. 2, type II). The first involves the *put2* gene, which

[3] D. Shortle, J. E. Haber, and D. Botstein, *Science* **217**, 371 (1982).

codes for an enzyme needed for proline utilization,[4] cloned by Brandriss.[5] Brandriss identified a 5.6-kb *Sac*I fragment containing *put2* complementing activity in a yeast shuttle vector and undertook a one-step gene disruption experiment to prove that the fragment coded for the *PUT2*+ gene and not for a phenotypic suppressor. Preliminary restriction mapping and subcloning experiments indicated that two *Bgl*II restriction sites flanked a 0.3-kb region necessary but not sufficient for *PUT2*+ function. Standard cloning methods were used to insert a 1.7-kb *Bam*HI fragment containing the *HIS3*+ region[6] into these *Bgl*II sites. Restriction digestion with *Sac*I liberated a linear fragment that contained the *put2* genetic region that had been disrupted by a 1.7-kb *HIS3*+ insert. The *HIS3*+ insert was flanked by 3.1 kb and 2.2 kb of homology to the *put2* region. Approximately 120 ng of fragment were used to transform 2×10^7 cells of W301-18A (*MATα ade2-1 leu2-3,112 his3-11,15 trp1-1 ura3-1*), a *his3*⁻, *PUT2*+ strain. Among the 14 *HIS3*+ transformants were 3 that simultaneously became *put2*⁻. Yeast colony hybridization was performed on these colonies using vector sequences as probe.[7] Two of the colonies also contained the vector sequences and were presumably due to contaminating linear molecules that integrated and converted the *put2* region during integration (for discussion of conversion associated with integration, see Orr-Weaver *et al.*[1]) The other transformant did not contain vector sequences, but did contain a *put2* region in which the *HIS3* DNA substituted for part of the *put2* gene. This result was confirmed by a genomic blot[8] of a *Sac*I digest of both the parent strain and the transformant probed with labeled *PUT2* DNA (Fig. 3). Tetrad analysis confirmed tight linkage of *put2*⁻ and *HIS3*+.

In a second example of gene disruption, Orr-Weaver and I[9] inserted one selectable genetic function into a region and simultaneously deleted another from the same region (Fig. 2, type II). The recipient strain contained, integrated at the *HIS3* genetic region, a plasmid into which the *SUP3*-**a** suppressor fragment was cloned. In a one-step gene replacement experiment, we successfully substituted a plasmid containing the *LEU2*+ genetic region for the *SUP3*-**a** plasmid sequence. A *LEU2*+, *HIS3*+ plasmid was linearized by a restriction cut within the *HIS3* sequence. The free ends of the incoming plasmid molecule stimulate recombination with the resident *HIS3* sequences. Since in this case the recipient strain contained two *HIS3* regions separated by the *SUP3*-**a** plasmid sequence, homolo-

[4] M. C. Brandriss and B. Magasanik, *J. Bacteriol.* **140,** 498 (1979).
[5] M. C. Brandriss, personal communication, 1982.
[6] K. Struhl and R. W. Davis, *J. Mol. Biol.* **136,** 309 (1980).
[7] A. Hinnen, J. B. Hicks, and G. R. Fink, *Proc. Natl. Acad. Sci. U.S.A.* **75,** 1929 (1978).
[8] E. M. Southern, *J. Mol. Biol.* **98,** 503 (1975).
[9] T. L. Orr-Weaver and R. J. Rothstein, unpublished observations, 1982.

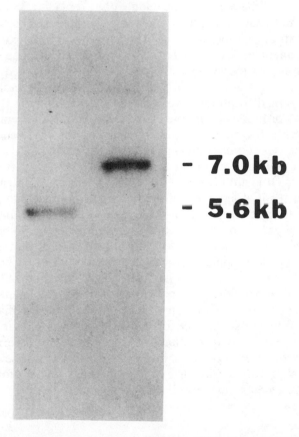

FIG. 3. Genomic blot of the *put2* gene disruption. W301-18A, a *his3⁻ PUT2⁺* strain was transformed in one step to a *HIS3⁺ put2⁻* strain using a *PUT2* fragment which was disrupted within the *put2* region with *HIS3⁺* DNA sequence. DNA was isolated from both the parent W301-18A and the transformant MB1426; 0.5 μg of DNA was digested with *Sac*I, electrophoresed on a 0.7%agarose gel, and transferred to nitrocellulose. Hybridization with radioactively labeled *Sac*I–*PUT2* DNA revealed that the *PUT2* region contained the *HIS3⁺* insert.

gous pairing could result in either targeted integration (see Orr-Weaver *et al.*[1]) or a one-step gene replacement. For the pairing configuration that resulted in the one-step replacement, there was 0.3 kb of homology at one end of the *LEU2*+ insert and 1.7 kb of homology at the other. After transforming with 500 ng of DNA, at least two transformants simultaneously inserted the *LEU2*+ plasmid and deleted the *SUP3*-**a** plasmid as confirmed by genomic blots.

In the third example, Tollervey and Guthrie[10] mutated a small structural RNA transcript (~200 bp) by a one-step gene disruption experiment. They cloned and identified a 3.4-kb fragment from yeast that encodes the RNA transcript. They disrupted the gene by cloning a 2.2-kb *Sal*I-*Xho*I *LEU2*+ fragment into the middle of it, replacing 36 bp of DNA. A linear fragment containing the disrupted gene was used to transform a diploid yeast strain. A diploid was used because it was anticipated that the replacement might be lethal. A *LEU2*+ transformant lacking vector sequences was shown by genomic blots to contain a disrupted gene fragment as well as an unaltered wild-type fragment (Fig. 2, type II diploid).

One-step gene disruptions have been successful when the DNA has as little as 0.3 kb of homology in one of the two regions adjacent to the inserted selectable fragment. It is recommended that the length of homology to the desired chromosomal region be maximized to lower the probability of gene conversion of the selectable marker at its own chromosomal locus. It is not necessary to purify the linear fragment from the vector sequences if at least one of the two restriction sites that liberates the fragment from the vector is at the border of the vector and the fragment. It is desirable to purify the linear fragment when homologous chromosomal sequences are left on both ends of the plasmid vector. Without purification, such a plasmid may interact directly with the homologous chromosomal region, resulting in gap repair and integration.[1,2]

As outlined above, the one-step gene disruption technique requires cloning of a selectable gene fragment into the region to be disrupted. A partial list of fragments available for this procedure is given in the table. Included are the flanking restriction sites that can serve to permit the insertion of the functional gene fragment. The genotypes of several commonly used laboratory strains that contain markers suitable for selecting one-step replacements have also been listed in the table.

Discussion

Cloning techniques have permitted the isolation of numerous yeast genes. Many have been recovered from clone banks of total yeast DNA

[10] D. Tollervey and C. Guthrie, personal communication, 1982.

CONVENIENT DNA FRAGMENTS AND YEAST STRAINS FOR
USE IN ONE-STEP GENE DISRUPTIONS[a]

Gene	Plasmid	Flanking restriction sites	Size of fragment (kb)	Reference[b]
LEU2[+]	YEp13	BglII	3.0	(1)
		SalI, XhoI	2.2	
		HpaI, SalI	1.9	
HIS3[+]	YIp1	SalI, EcoRI	6.0	(2)
		BamHI	1.7	
		BamHI, XhoI	1.4	
URA3[+]	YEp24	HindIII	1.2	(3)
TRP1[+]	YRp7	EcoRI	1.4	(2)
		EcoRI, BglII	0.9	

Strains	Genotype	Reference
LL20	MATα leu2-3,112 his3-11,15	L. Lau, (4)
DBY745	MATα leu2-3,112 ade1-101 ura3-52	(3)
SR25-1A	MATa his4-912 ura3-52	S. Roeder (5)
RH218	MATα trp1 gal2	(6)
W301-18A	MATα ade2-1 trp1-1 leu2-3,112 his3-11,15 ura3-1	R. Rothstein (7)

[a] Gene disruptions with the SalI–XhoI LEU2[+] fragment, the BamHI HIS3[+] fragment, and the HindIII URA3[+] fragment have been successful. The other fragment has not been tried.

[b] Key to references:

(1) J. R. Broach, J. N. Strathern and J. B. Hicks, *Gene* **8**, 121 (1979).

(2) K. Struhl, D. T. Stinchcomb, S. Scherer, and R. W. Davis, *Proc. Natl. Acad. Sci. U.S.A.* **76**, 1035 (1979).

(3) D. Botstein, S. C. Falco, S. E. Stewart, M. Breenan, S. Scherer, D. T. Stinchcomb, K. Struhl, and R. W. Davis, *Gene* **8**, 17 (1979).

(4) T. L. Orr-Weaver, J. W. Szostak, and R. J. Rothstein, *Proc. Natl. Acad. Sci. U.S.A.* **78**, 6354 (1981).

(5) Unpublished strain.

(6) G. Miozzari, P. Niederberger, and R. Hütter, *J. Bacteriol.* **134**, 48 (1978); S. Scherer and R. W. Davis, *Proc. Natl. Acad. Sci. U.S.A.* **76**, 4951 (1979).

(7) Unpublished strain.

by selecting for complementation of mutations in yeast.[11-18] Others have been isolated from clone banks by hybridization with pure or partially purified RNA molecules,[19-23] cDNA probes,[24] or probes from highly conserved genes previously cloned in divergent species.[25-27]

For genes cloned by complementation, it is necessary to demonstrate that a fragment codes for the wild-type gene, not for a phenotypic suppressor. Proof that a fragment actually codes for the complemented gene is generally demonstrated by showing that a plasmid containing the fragment integrates at the genetic locus and shows genetic linkage to the gene. This evidence does not rule out the possibility that the fragment codes for a tightly linked phenotypic suppressor. A gene disruption experiment proves in one step that a cloned fragment codes for the gene function in question by (a) using homology on the fragment to integrate into the corresponding chromosomal region; (b) simultaneously creating a mutation; and (c) showing genetic linkage of the insert by tetrad analysis.

Gene disruption experiments may also be used to examine the phenotype of fragments cloned by hybridization procedures. The two existing methods for gene disruption, transplacement[28] and integration disruption,[3] require several manipulations or detailed information about the cloned fragment. In the Scherer and Davis method[28] an altered sequence cloned on a plasmid is introduced by homologous integration into the de-

[11] A. Hinnen, P. J. Farabaugh, C. Ilgen, G. R. Fink, and J. Friesen, *Eucaryotic Gene Regul. ICN–UCLA Symp. Mol. Cell. Biol.* **14,** 43 (1979).

[12] J. R. Broach, J. N. Strathern, and J. B. Hicks, *Gene* **8,** 121 (1979).

[13] V. M. Williamson, J. Bennetzen, E. T. Young, K. Nasmyth, and B. D. Hall, *Nature (London)* **283,** 214 (1980).

[14] J. Hicks, J. Strathern, and A. Klar, *Nature (London)* **282,** 478 (1979).

[15] K. Nasmyth and K. Tatchell, *Cell* **19,** 753 (1979).

[16] K. A. Nasmyth and S. I. Reed, *Proc. Natl. Acad. Sci. U.S.A.* **77,** 2119 (1980).

[17] M. Crabeel, F. Messenguy, F. Lacroute, and N. Glansdorff, *Proc. Natl. Acad. Sci. U.S.A.* **78,** 5026 (1981).

[18] C. L. Dieckmann, L. K. Pape, and A. Tzagoloff, *Proc. Natl. Acad. Sci. U.S.A.* **79,** 1805 (1982).

[19] R. A. Kramer, J. F. Cameron, and R. W. Davis, *Cell* **8,** 227 (1976).

[20] M. V. Olson, B. D. Hall, J. R. Cameron, and R. W. Davis, *J. Mol. Biol.* **127,** 285 (1979).

[21] J. L. Woolford, Jr., L. M. Hereford, and M. Rosbach, *Cell* **18,** 1247 (1979).

[22] T. P. St. John and R. W. Davis, *Cell* **16,** 443 (1979).

[23] R. A. Kramer and N. Andersen, *Proc. Natl. Acad. Sci. U.S.A.* **77,** 6541 (1980).

[24] M. J. Holland, J. P. Holland, and K. A. Jackson, this series, Vol. 68, p. 408.

[25] L. Hereford, K. Fahvner, J. Woolford, Jr., M. Rosbash, and D. B. Kaback, *Cell* **18,** 1261 (1979).

[26] D. Gallwitz and R. Seidel, *Nucleic Acids Res.* **8,** 1043 (1980).

[27] R. Ng and J. Abelson, *Proc. Natl. Acad. Sci. U.S.A.* **77,** 3912 (1980).

[28] S. Scherer and R. W. Davis, *Proc. Natl. Acad. Sci. U.S.A.* **76,** 3912 (1979).

sired region of the chromosome. The integrated plasmid is flanked by a mutant copy and a wild-type copy of the gene. The next step requires that a recombination event between the duplicated regions flanking the plasmid occur such that the wild-type sequence is excised and the mutant sequence is left in the chromosome. The major disadvantage of this method is that several recombinants must be examined in order to detect the correct replacement. A positive selection for loss of a marker on the plasmid enriches for, but still does not ensure, the successful replacement of the mutant gene. The method of Shortle *et al.*[3] utilizes plasmid integration to create a mutant phenotype. A plasmid containing a cloned internal fragment of a gene is integrated into its homologous chromosomal sequence. Since the internal fragment lacks the 5' and 3' sequences of the gene, the integrated plasmid results in two mutated copies of the gene—a 5' deletion as well as a 3' deletion. The limitations of this technique are that (*a*) it demands the knowledge of restriction sites with respect to the 5' and 3' ends of the desired gene; (*b*) the mutated gene can revert by deletion of the plasmid sequences, since the disrupted gene region is flanked by direct repeats as a consequence of the integration event; and (*c*) the technique may not be applicable to all genes, since it has been observed that some small fragments (less than 200 nucleotides in length) fail to integrate into their homologous region.[1] The methods outlined in this chapter permit simple disruptions (Fig. 2, type I) or deletion-substitution events (Fig. 2, type II) in one step. A detailed restriction map of the fragment is not necessary, since a single insertion (Fig. 2, type I) is sufficient to disrupt a gene. The position, within a fragment, of a gene cloned by complementation may be determined by independently disrupting the fragment at several different restriction sites and testing each for its phenotype. Identification of transformants containing the correct configuration is aided by the fact that the insertion simultaneously results in a mutation whose phenotype can be directly scored. For disruptions of essential genes, the recessive lethal phenotype becomes tightly linked to the insertion, which is nonreverting and can be used as a selectable marker for further genetic manipulation.

To date, successful disruptions have been reported with the *Bam*HI *HIS3*+ fragment, the *Sal*I-*Xho*I *LEU2*+ fragment,[29] and the *Hin*dIII *URA3*+ fragment.[30] The other fragment listed in the table has not been tried in one-step disruption experiments. As with any gene disruption

---

[29] Amar Klar (personal communication) has replaced yeast mating-type information with the *Sal*I–*Xho*I *LEU2*+ fragment using the techniques described in this paper.

[30] N. Abovich and M. Rosbash (personal communication) have disrupted a ribosomal protein gene with the *Hin*dIII *URA3*+ fragment.

method, a successful experiment must be verified by Southern blots on the DNA of the transformants (e.g., Fig. 3). Finally, it is important to consider the possibility that moving a gene into a new environment may inadvertently create a position effect[31] leading to nonexpression of the selectable marker.

Summary

The one-step gene disruption techniques described here are versatile in that a disruption can be made simply by the appropriate cloning experiment. The resultant chromosomal insertion is nonreverting and contains a genetically linked marker. Detailed knowledge of the restriction map of a fragment is not necessary. It is even possible to "probe" a fragment that is unmapped for genetic functions by constructing a series of insertions and testing each one for its phenotype.

Acknowledgments

The author wishes to thank Nadja Abovich, Christine Guthrie, Amar Klar, Terry Orr-Weaver, Michael Rosbash, David Tollervey, and especially Marjorie Brandriss for permission to quote unpublished data. Thanks are also due to Cindy Helms for technical assistance. This work was supported by NIH Grant GM27916, and NSF Grant PCM 8003805, and Foundation of UMDNJ Grant 24-81.

[31] K. Struhl, *J. Mol. Biol.* **152**, 569 (1981).

[19] Cloning Regulated Yeast Genes from a Pool of *lacZ* Fusions

By STEPHANIE W. RUBY, JACK W. SZOSTAK, and ANDREW W. MURRAY

Gene fusions have been used during the last decade to study such diverse problems as transcriptional regulation[1,2] translational regulation,[3] and protein localization.[4,5] Their use has been particularly valuable for the study of genes that are autogenously regulated, or whose products are not easily measured. The *Escherichia coli lacZ* gene has been utilized most often in the generation of these fusions for several reasons: the *lac* operon is well characterized[6]; the *lacZ* gene product, β-galactosidase, is easily assayed[7]; the first 27 amino acids of the N terminus of the protein may be replaced by other peptides with little or no effect on enzyme activity[8]; there are several substrates for the selection of Lac⁻ and Lac⁺ mutants, and there are several nonselective, indicator substrates.[6]

In prokaryotes, various genetic methods[1,9,10] have been employed to create operon and protein fusions. Recombinant DNA techniques are now commonly applied to engineer fusions between eukaryotic and prokaryotic genes. Rose *et al.*[11] isolated many different fusions between the yeast *URA3* and the *lacZ* genes by placing the two genes near each other on a plasmid and then selecting for deletions between them by selecting

[1] N. C. Franklin, *Annu. Rev. Genet.* **12**, 193 (1978).

[2] J. R. Beckwith, *Cell* **23**, 307 (1981).

[3] A. Miura, J. H. Krueger, S. Itoh, H. A. de Boer, and M. Nomura, *Cell* **25**, 773 (1981).

[4] T. Silhavy, H. Shuman, J. Beckwith, and M. Schwartz, *Proc. Natl. Acad. Sci. U.S.A.* **74**, 5411 (1977).

[5] K. Ito, P.H. Bassford, and J. Beckwith, *Cell* **24**, 707 (1981).

[6] J. Beckwith, *in* "The Operon" (J. H. Miller and W. S. Reznikoff, eds.), p. 11. Cold Spring Harbor Laboratory, Cold Spring Harbor, New York, 1972.

[7] J. H. Miller, "Experiments in Molecular Genetics," p. 466. Cold Spring Harbor Laboratory, Cold Spring Harbor, New York, 1972.

[8] E. Brickman, T. J. Silhavy, P. J. Bassford, H. A. Shuman, and J. R. Beckwith, *J. Bacteriol.* **139**, 13 (1979).

[9] P. Bassford, J. Beckwith, M. Berman, E. Brickman, M. Casadaban, L. Guarente, I. Saint-Girons, A. Sarthy, M. Schwartz, H. Shuman, and T. Silhavy, *in* "The Operon" (J. H. Miller and W. S. Reznikoff, eds.), p. 245, Cold Spring Harbor Laboratory, Cold Spring Harbor, New York, 1980.

[10] M. J. Casadaban and S. N. Cohen, *Proc. Natl. Acad. Sci. U.S.A.* **76**, 4530 (1979).

[11] M. Rose, M. J. Casadaban, and D. Botstein, *Proc. Natl. Acad. Sci. U.S.A.* **78**, 2460 (1981).

RECOMBINANT DNA
METHODOLOGY

for a lac$^+$ phenotype in *E. coli*. Guarente and Ptashne[12] joined the yeast *CYCl* gene to the *lacZ* gene directly *in vitro*. Both groups subsequently showed that the gene fusions could be expressed in yeast as protein fusions, and that the levels of β-galactosidase in yeast were regulated in the appropriate manner. Furthermore, they found that yeast cells, which are lac$^-$, can hydrolyze the chromogenic substrate Xgal when they contain an active fusion, and that the resulting blue color of the colony varied with the cellular enzyme levels.

We have utilized *lacZ* gene fusions in the yeast *Saccharomyces cerevisiae* to clone regulated yeast genes. These genes were identified as regulated protein fusions and detected by screening a library of random yeast gene:*lacZ* fusions. This method allows the cloning of yeast genes based solely on the way in which they are regulated and thus is applicable to genes whose products are not easily assayed or whose functions are unknown. It is particularly useful for cloning genes that are similarly or coordinately regulated, and it complements other cloning techniques such as complementation[13,14] and differential plaque hybridization.[15] In this chapter we describe the procedures that we have used to isolate DNA *d*amage *in*duced (*DIN*) genes as an illustration of the usefulness of the fusion cloning method.

General Outline of the Method

The active, regulated gene fusions constructed by the procedures described below contain a yeast DNA fragment with sequences for the control and initiation of transcription and translation joined to a *lacZ* fragment from which the 5' portion including the first few N-terminal amino acid codons of the gene have been deleted. Expression of such fusions in yeast results in hybrid proteins with β-galactosidase activity.

The first step of the method is the creation of a library of random yeast gene fusions. A fusion shuttle vector is described that has a unique restriction endonuclease site at the 5' end of the *lacZ* fragment. Random yeast DNA fragments are inserted at this site to create a pool of ligated DNA molecules, which are then used to transform yeast cells.

The second step is to screen the library of yeast transformants for colonies with regulated β-galactosidase activity. Yeast colonies are replicaplated onto media containing Xgal, and the colonies are monitored for differential expression by differences in color intensities.

[12] L. Guarente and M. Ptashne, *Proc. Natl. Acad. Sci. U.S.A.* **78**, 2199 (1981).
[13] B. Ratzkin and J. Carbon, *Proc. Natl. Acad. Sci. U.S.A.* **74**, 487 (1977).
[14] K. A. Nasmyth and S. I. Reed, *Proc. Natl. Acad. Sci. U.S.A.* **77**, 2119 (1980).
[15] T. P. St. John and R. Davis, *Cell* **16**, 225 (1979).

The third step is to recover the fusion plasmids from the yeast strains of interest and to confirm that the fusion plasmids have the expected regulated expression when reintroduced into yeast.

The final step is to demonstrate by an alternative assay such as Northern blot analysis that the identified genes are regulated in the expected manner.

Materials and Reagents

Strains

Yeast strain A2 (α *his3-11,15 leu2-3,112 canl*) was originally LL20 from L. Lau. Strain A15 (α *gal2, mal*) was originally S288C. The strain DA151 is α *his3-11,15 leu2-3,112, ura3, trpl, tcml*,[*LEU2, his3 : lacZ*] × a *leu2-3,112, met10*, which was derived from a transformant of strain D234-3B (α *his3-11,15, leu2-3,112, ura3, trpl, tcml*) from P. Brown. *Escherichia coli* strain 5346 (C600 *leuB6, thr-, thi-, r-, m-*) was obtained from J. Calvo. The plasmid pMC1403 was obtained from M. Rose. The plasmids pSZ62 and pSZ93 were given by T. L. Orr-Weaver.

Media

Yeast synthetic selective medium (lacking leucine), YPD, and regeneration agar were as described.[16] Yeast calcium-free, Xgal medium (lacking leucine) contained amino acids, adenine, and uracil as in synthetic medium, as well as 13.6 g of KH_2PO_4, 2 g of $(NH_4)_2SO_4$, 4.2 g of KOH, 0.2 g of $MgSO_4 \cdot 7\ H_2O$, 0.5 mg of $FeCl_3 \cdot 6\ H_2O$, 400 μg each of pantothenic acid, biotin, and pyridoxine monohydrochloride, 2 μg of biotin, 2 mg of myoinositol, 20 g of dextrose, 20 g of agar, and 40 mg of Xgal per liter. The KH_2PO_4, $(NH_4)_2SO_4$, and KOH were made up as a 10 × stock, the $MgSO_4$ and $FeCl_3$ were made up as a 1000 × stock, and both unsterile stocks were stored at room temperature. The vitamins were kept as a 100 × frozen, sterile stock, and the amino acids were stored as sterile stock solutions as described.[16] Xgal was prepared as 20 mg/ml in dimethylformamide and stored frozen. To prepare the medium, the amino acids (minus leucine, threonine, aspartic acid, and tryptophan) and salts were brought to a volume of 500 ml with H_2O, the agar and sugar were put into 500 ml of H_2O, and the two solutions were autoclaved separately. After cooling to 55°, the two solutions were mixed, and the vitamins, remaining amino acids (minus leucine), and Xgal were added. Bacterial media were as described.[7]

---

[16] F. Sherman, G. R. Fink, and C. W. Lawrence, "Methods in Yeast Genetics," p. 98, Cold Spring Harbor Laboratory, Cold Spring Harbor, New York, 1979.

Solutions

Z buffer,[7] pH 7.0, contained, per liter, 16.1 g of $Na_2HPO_4 \cdot 7\ H_2O$, 5.5 g of $NaH_2PO_4 \cdot H_2O$, 0.75 g of KCl, 0.25 g of $MgSO_4 \cdot 7\ H_2O$, and 2.7 ml of 2-mercaptoethanol.

$1 \times$ RNA gel buffer, pH 7.3, was 20 mM morpholinopropanesulfonic acid (MOPS), pH 7.3, 50 mM sodium acetate, and 1 mM EDTA.

$20 \times$ Denhardt's solution contained 0.4 g of poly(vinylpyrollidone), 0.4 g of bovine serum albumin, and 0.4 g of Ficoll per 100 ml of H_2O.

$4 \times$ RNA hybridization buffer, pH 7.5, contained 175.3 g of NaCl, 73.1 g of Tris base, 22.1 g of NaH_2PO_4, 64.3 g of $Na_2HPO_4 \cdot 7\ H_2O$, 4 g of sodium pyrophosphate, and 35 ml of 12 N HCl per liter.

$20 \times$ SSC was 3.0 M NaCl and 0.3 M sodium citrate.

Reagents

5-Bromo-4-chloroindolyl-β-D-galactoside (Xgal) and O-nitrophenyl-β-D-galactopyranosyl (ONPG) were obtained from Bachem, Marina del Ray, California, and Sigma, respectively. Formaldehyde and formamide were from Mallinckrodt. Restriction endonucleases and T4 DNA ligase were from New England BioLabs, and all enzyme units were those of the supplier.

Procedures and Results

Fusion Vector

The plasmid pSZ211 was used as a fusion shuttle vector (Fig. 1). It is composed of (*a*) the 3.7 kb *Eco*RI–*Sal* fragment of pBR322, encoding ampicillin resistance and a replication origin for selection and replication in *E. coli;* (*b*) the yeast 2.2 kb *Sal*I–*Xho*I *LEU2*[14] fragment and the 0.84 kb *Eco*RI–*Hin*dIII *ars1*[17] fragment for selection and replication in yeast; and (*c*) the *E. coli* 6.3 kb *Bam*HI–*Sal*I *lac* fragment, which contains the *lacZ* gene deleted for the N-terminal codons of β-galactosidase, the *lacY* gene, and a portion of the *lacA* gene. The *lac* fragment was obtained from plasmid pMC1403, which was constructed by M. Casadaban *et al.*[18]; pSZ211 was constructed by inserting the *Bam*HI–*Sal*I *lac* fragment into *Bam*HI, *Sal*I-digested pSZ93 (not shown).[19]

The plasmid pSZ211 does not produce detectable β-galactosidase activity in yeast, but it does produce easily detectable enzyme activity in

---

[17] D. T. Stinchcomb, K. Struhl, and R. W. Davis, *Nature (London)* **282,** 39 (1979).
[18] M. J. Casadaban, J. Chou, and S. N. Cohen, *J. Bacteriol.* **143,** 971 (1980).
[19] T. L. Orr-Weaver, J. W. Szostak, and R. J. Rothstein, this series, Vol. 101, p. 228.

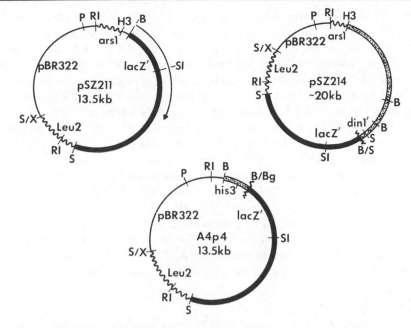

FIG. 1. Restriction maps of the cloning vector (pSZ211), *din1 : lacZ* plasmid (pSZ214), and *his3 : lacZ* plasmid (A4p4). An arrow indicates the direction of transcription of the *lacZ* gene. Its 3′ end indicates the approximate end of the *lacZ* coding sequences, but it does not indicate the 3′ end of a transcript. Plasmids are not drawn to scale. Restriction endonuclease sites are indicated as follows: B, *Bam*HI; Bg, *Bgl*II; H3, *Hin*dIII; P, *Pst*I; RI, *Eco*RI; SI, *Sac*I; S, *Sal*I; S3, *Sau*3A; and X, *Xho*I.

E. coli. It contains an *ars* sequence and therefore transforms yeast at high frequency[20] and is usually maintained extrachromosomally. The plasmid has a unique *Bam*HI site at the 5′ end of the *lacZ* gene into which random yeast DNA fragments can be inserted to create gene fusions.

Construction of Random Fusions

A DNA pool of random fusions was made by ligating yeast genomic DNA that had been partially cut with *Sau*3A, to *Bam*HI-cleaved vector. Yeast DNA from haploid strain A15 was prepared by the method of Davis *et al.*[21] The amount of *Sau*3A required to achieve the desired partial digestion was estimated from a set of small-scale reactions. One of these was

[20] K. Struhl, D. T. Stinchcomb, S. Scherer, and R. W. Davis, *Proc. Natl. Acad. Sci. U.S.A.* **76**, 1035 (1979).
[21] R. W. Davis, M. Thomas, J. Cameron, T. P. St. John, S. Scherer, and R. Padgett, this series, Vol. 65, p. 404.

subsequently scaled up. About 200 μg of A15 DNA were cleaved in 6 mM Tris HCl, pH 7.5, 50 mM NaCl, 6 mM MgCl$_2$, and 6 mM 2-mercaptocth- anol in a volume of 4.5 ml with 3.6 units of Sau3A for 1 hr at 37°, after which the reaction was stopped by heating at 65° for 10 min. A 40-μl ali- quot of this DNA was fractionated by agarose gel electrophoresis and vis- ualized by ethidium bromide staining; the average molecular weight was estimated to be 6 kb. The remainder of the DNA was added to 30 μg of BamHI-cut pSZ211 DNA, and the mixture was brought to a volume of 6 ml and a final concentration of 66 mM Tris-HCl, pH 7.5, 0.4 mM ATP, 10 mM MgCl$_2$, and 10 mM dithiothreitol. Three hundred units of T4 DNA ligase were added, and the mixture was incubated overnight at 4°. The DNA was EtOH precipitated, washed once with 70% EtOH, dried, and resuspended in 200 μl of 10 mM Tris-HCl, pH 7.5, and 1 mM EDTA (TE). The pool of ligated DNA molecules was used to transform yeast directly without prior amplification in $E.\ coli$. Ligated DNA (150 μl) was added to 1.8 ml of competent yeast cells derived from 90 ml of a log-phase culture of A2 cells.[19] Approximately 10,000 colonies were obtained.

Screening

The number of independent colonies that must be screened to ensure isolation of at least one fusion that is regulated in the desired manner was estimated. Assuming that (a) the average yeast gene is 1 kb long; (b) the yeast haploid genome is 2×10^7 bp; (c) the number of yeast genes is 10^4; and (d) one out of six fusions contains yeast and $lacZ$ coding sequences in the same reading frame and orientation, then 8% of the colonies will yield productive fusions. From the Poisson distribution, the number of colonies that must be screened to isolate at least one regulated fusion with a 95% probability is 360,000 divided by the number of regulated genes. The fu- sion cloning method then, is more convenient for identifying fusions to members of a group of genes that are similarly or coordinately regulated than a fusion to a particular gene.

Yeast colonies containing productive fusions were detected by replica plating onto Xgal medium. Since the yeast transformants were embedded in an overlay of regeneration agar, they could not be replica plated directly from the transformation plates. We were not able to detect en- zyme activity consistently in colonies within the agar overlay by including Xgal in buffered regeneration agar. We used two simple procedures for extracting cells from the overlay. One method was to use toothpicks or a wire inoculating needle to transfer each independent colony manually onto a master plate. Alternatively, all the colonies from each transforma- tion plate were pooled by breaking up the overlay in water in a blender and filtering out the agarose with cheesecloth. An aliquot of cells from each pool were plated onto selective medium, and the remainder of each cell pool was stored in 50% glycerol at −70°. Each 100-mm master plate was routinely inoculated with 200 colonies or cells.

The screening for productive fusions, or for regulated fusions, worked best when the colonies were actively growing on the master plate just prior to replica plating. Under these conditions a few yeast colonies were blue on Xgal plates within 4 hr of replica plating. Most colonies, however, required 1 to 2 days to develop a detectable blue color, and a few were blue only after 1–2 weeks of incubation at 30°. When master plates were inoculated with cells extracted from the overlay by either method, 6% of the colonies contained productive fusions.

The transformants were screened for the presence of regulated fusions by two protocols. In the first protocol, all transformants were prescreened to detect productive fusions, and those containing productive fusions were screened for regulated β-galactosidase activity. In the second protocol, the prescreen for productive fusions was omitted, and all transformants were screened for regulated *lacZ* expression. About 1% of all the colonies were pale blue on experimental plates but white on control plates in a screen for DNA *d*amage *in*duced (*din*) fusions, and therefore, would have been missed by the first screening protocol.

The results of the screening for *din* fusions illustrate the plate screening technique. Approximately 20,000 colonies from 20 independent yeast transformant pools were replica plated onto Xgal medium. Nearly 6% of the colonies were blue after 2 weeks of incubation at 30°. The blue colonies were transferred to selective medium, grown for 2 days, and replica plated onto 3 sets of Xgal plates that were treated in the following manner: (*a*) UV irradiated with two doses of 8.4 J/m² separated by 2 hr; (*b*) given nitroquinoline oxide (NQO),[22] a DNA damaging agent, in the medium at a concentration of 0.05 μg/ml; and (*c*) untreated control. After 1 week of incubation, 4 colonies were detected as darker blue under treatment than under control conditions. The increased color of the colonies was not due to growth of mutants generated by the treatments, because whole colonies, rather than sectors, were darker than control colonies. Subsequent quantitative assays (see below) showed that one of the colonies contained a *din : lacZ* fusion. The haploid strain T378 (see below) containing this fusion (*din1 : lacZ*) was used to further optimize the screening conditions for identifying other *din* fusions. Under these conditions,[23] the induction of the *din1 : lacZ* fusion can be detected on a screening plate between 16 and 24 hr after the initial exposure of the colony to the DNA damaging agent.

[22] L. Prakash, *Genetics*, **83**, 285 (1975).
[23] S. Ruby and J. Szostak, in preparation.

Quantitative β-Galactosidase Assays

Once potentially interesting colonies were identified, they were streaked on selective medium to obtain pure clones, which were then analyzed quantitatively during growth in liquid medium. Some fusions appeared to be regulated in the desired manner on Xgal plates, but were not so regulated when assayed quantitatively. In one screen for *din* fusions, only $\frac{4}{20}$ of the candidates identified on Xgal plates behaved as *din* fusions in subsequent quantitative assays.

β-Galactosidase assays using permeabilized cells gave good reproductibility and speed. Guarente and Ptashne[12] have shown for a *cyc1 : lacZ* fusion that the same changes in β-galactosidase specific activity were detected whether permeabilized cells or cell extracts were assayed. If the enzyme activity was high enough, we assayed cells directly in liquid, calcium-free medium (minus Xgal). It was necessary to use a calcium-free medium as a precipitate formed when Z buffer was added to selective medium. Cells having low enzyme activity were concentrated by centrifugation and resuspended in 50 mM potassium phosphate buffer, pH 7.0. The enzyme assay was similar to that used for bacteria.[7] Z buffer (0.4 ml) containing 0.025% SDS and 1 mM phenylmethylsulfonyl fluoride (PMSF) was added to 0.4 ml of cell suspension, and the mixture was vortexed vigorously for 5 sec. After exactly 10 min at 23°, 2 drops of CHCl$_3$ were added to the mixture, which was immediately vortexed vigorously for 5–8 sec and placed in a 28° water bath. After exactly 10 min again, 0.2 ml of ONPG (4 mg/ml in 0.1 M potassium phosphate, pH 7.0) was added, and the sample was vortexed and incubated at 28°. The reaction was terminated by the addition of 0.5 ml of 1 M Na$_2$CO$_3$. The samples were spun briefly at 3000 rpm in a Sorvall GLC-4 centrifuge before measuring the OD$_{420}$ of the supernatants. The units were calculated as follows: units = OD$_{420}$ × 1000/t(min) × V(ml) × OD$_{600}$; where v was the volume of the cell sample assayed and OD$_{600}$ was the cell density of the culture as measured spectrophotometrically.

An example of the type of induced response that is seen for the haploid yeast strain T378, which contains the *din1 : lacZ* fusion stably integrated in genomic DNA, is seen in Fig. 2. Cells were grown at 30° to mid-log phase in selective medium. At 0 hr, 0.05 μg of NQO per milliliter was added to half of the culture. During subsequent incubation at 30°, cell samples were removed at the times indicated and assayed for cell density and β-galactosidase activity. The NQO-treated culture reached 86% of the control culture cell density at stationary phase, which they both entered at the same time, between 4 and 6 hr. Enzyme activity increased 100-fold above that of control levels at 3–4 hr after NQO addition. After 4 hr the induced levels declined. The reason for this decline is not known,

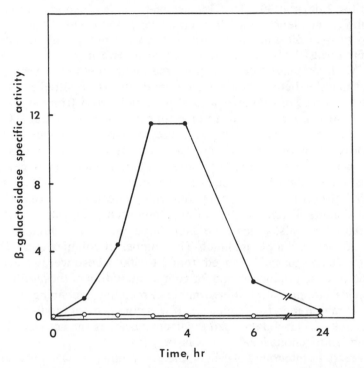

FIG. 2. Kinetics of NQO induction of β-galactosidase in strain T378 containing the *din1 : lacZ* fusion. Mid-log phase cultures received 0.05 μg of NQO per milliliter (●) or no treatment (○) at 0 hr. Cells were assayed for enzyme activity during subsequent growth at 30° at the times indicated as described in text. All samples were assayed in duplicate.

although it is not due to inactivation of NQO. A similar response is observed in T378 cells treated with other DNA damaging agents and in A2 cells containing other *din : lacZ* fusions and other fusions.[24] The decline in β-galactosidase activity occurs when the culture is entering stationary phase. It is also noteworthy that increased levels of β-galactosidase are detectable much more rapidly in quantitative assays than on plates.

Recovery of Plasmids from Yeast Transformants

All *lacZ*-fusion strains of interest were tested for the mitotic stability of the *LEU2* marker of their fusion plasmids. In general, a stable strain contains an integrated plasmid, and an unstable strain contains a free plasmid. The strains were streaked onto nonselective YPD medium and incu-

[24] S. Ruby, unpublished data, 1982.

bated for 2 days at 30°. The colonies were replica plated onto selective medium (minus leucine) and monitored for growth after 24 hr.

To recover plasmids from strains that were mitotically unstable, the extracted yeast DNA was used directly to transform competent bacterial cells. Miniprep yeast DNA was prepared by the method of Davis et al.[21] except that the diethylpyrocarbonate treatment was omitted and the total supernatant was precipitated with isopropanol. DNA from 50 ml of cells was resuspended in 0.5 ml of TE; the concentration was about 100 µg/ml. Competent E. coli 5346 cells were prepared by a modification of the method of Morrison.[25] Cells in 1 liter of LB medium were grown to $OD_{550} = 0.3$, after which they were chilled, pelleted in a centrifuge, washed once with 0.1 M $CaCl_2$, and resuspended in 15 ml of 0.1 M $CaCl_2$ and 15% glycerol. After 16 hr at 0°, the cells were slowly frozen at −70° in 1.0 ml aliquots. From 5 to 20 µl of yeast miniprep DNA was used to transform 0.5 ml of cells, which were then plated onto LB medium containing 100 µg of ampicillin per milliliter. The number of colonies obtained from different fusion plasmids ranged from 0 to 500. These were screened for β-galactosidase production and for complementation of the leuB mutation by replica plating onto buffered minimal medium containing 0.04 mg of Xgal per milliliter and 0.5% glucose. It was not necessary to use Lac⁻ bacteria to detect plasmid encoded β-galactosidase, as the activity of the repressed, endogenous lacZ gene was very low.

To recover integrated plasmids, the yeast miniprep DNA was cut by a restriction endocnuclease and then ligated before being used for bacterial transformation. As there are no SmaI, XbaI, and XhoI sites within the vector sequences, and the HindIII site is between the ars1 and pBR322 sequences, these four endonucleases can be used to cleave genomic DNA while leaving the vector sequences intact. DNA (80 µl) was brought to a volume of 100 µl by the addition of 20 µl of 5 × restriction enzyme buffer, 5 units of restriction enzyme were added, and the sample was incubated at 37° for 2 hr. The reaction was terminated by heating at 65° for 10 min, and a 20-µl aliquot was removed for analysis of the completeness of the reaction by agarose gel electrophoresis. The remaining 80 µl were brought to a volume of 0.8 ml and a final concentration of 66 mM Tris HCl, pH 7.5, 10 mM dithiothreitol, 6.6 mM $MgCl_2$, and 0.4 mM ATP. Thirty units of T4 DNA ligase were added, and the reaction was incubated from 4 hr to overnight at 4°. The solution was brought to 0.2 M NaCl, and the DNA was EtOH precipitated, washed once with 70% EtOH, and dried. The DNA was resuspended in 60 µl of 0.01 M Tris HCl, pH 7.4, 0.1 M $CaCl_2$, and 0.01 M $MgCl_2$, and 10–40 µl of the solution was added to 0.5 ml of competent 5346 cells. The transformants were plated and tested as above.

[25] D. Morrison, J. Bacteriol. **132**, 349 (1977).

DNA of each fusion plasmid was prepared[26] and then analyzed by restriction enzyme mapping and Southern blots[27] to determine the insert size and to detect any insert rearrangements that would result in spurious expression. The *din1 : lacZ* fusion plasmid, pSZ214, (Fig. 1) for example, was recovered after the yeast genomic DNA was cut with *Hin*dIII, ligated to circularize the plasmid, and then used to transform bacteria. It has a 7 kb insert that is larger than the average insert size (4 kb) of the eight other fusion plasmids examined. The 1.3 kb fragment between the *Bam*HI/*Sau*3A junction and the adjacent *Bam*HI site on pSZ214 is found in genomic DNA of wild-type yeast strain A2. Yeast sequences upstream of the *Bam*HI site, however, are not contiguous in A2 DNA.[24] These sequences were probably joined together in pSZ214 during the ligation of genomic DNA fragments to the fusion vector.

To confirm that the isolated fusion plasmids contained the regulated gene sequences, the circular fusion plasmids were used to transform A2 yeast cells.[19] Strain T378 was obtained by transformation with the plasmid, pSZ214. The regulation of β-galactosidase in T378 is not detectably different from that of the original *din1* fusion transformant, T283.

Detection of Regulated Wild-Type Gene Activity by Northern Blot Analysis

The changes in β-galactosidase specific activity in NQO-induced T378 cells reflect changes in wild-type *DIN1* gene activity. Levels of *DIN1* gene transcripts in NQO-induced and uninduced A2 (wild type, untransformed) cells were examined by Northern blot analysis.[28] A2 cells were grown in YPD at 30° to mid-log phase. At 0 hr, 0.05 μg of NQO per milliliter was added to one half of the culture. During subsequent incubation at 30°, cell samples were removed at the times indicated in Fig. 3, and the total RNA was extracted according to Hereford *et al.*[29] Total cell RNA (20 μg per sample in 50% formamide, 2.2 M formaldehyde, and 1 × RNA gel buffer) was heated at 60° for 10 min, cooled rapidly, and then size-fractionated on a 1% agarose gel in 1 × RNA gel buffer containing 2.2 M formaldehyde.[30] The gel was treated with 50 mM NaOH[31] and 150 mM NaCl for 30 min, neutralized with 1 M Tris-HCl, pH 7.6, and 3 M NaCl for 30 min, and washed for 10 min in 10 × SSC. The RNA was transferred

[26] D. S. Holmes and M. Quigley, *Anal. Biochem.* **114**, 193 (1981).
[27] E. M. Southern, *J. Mol. Biol.* **98**, 503 (1975).
[28] P. S. Thomas, *Proc. Natl. Acad. Sci. U.S.A.* **77**, 5201 (1980).
[29] L. M. Hereford, M. A. Osley, J. R. Ludwig, and C. S. McLaughlin, *Cell* **24**, 367 (1981).
[30] H. Lehrach, D. Diamond, J. M. Wozney, and H. Boedtker, *Biochemistry* **16**, 4743 (1977).
[31] B. Seed and D. Goldberg, in preparation, 1982.

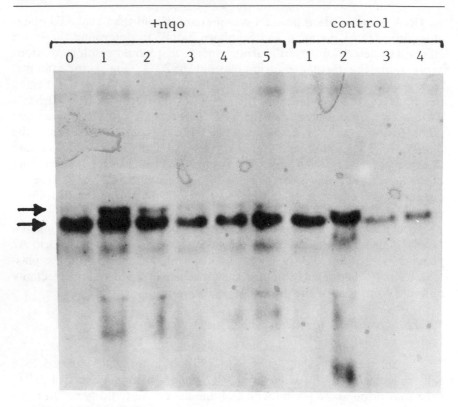

FIG. 3. Levels of *DIN1* transcripts in NQO-induced and uninduced A2 cells. Mid-log A2 cultures received 0.05 μg of NQO per milliliter or no treatment (control) at 0 hr. During subsequent growth, cell samples were removed at the hours indicated, and their RNA was extracted and assayed by Northern blot analysis as described in text. The autoradiogram of a Northern blot probed for *DIN1* (upper arrow) and control gene (lower arrow) transcripts is shown.

to nitrocellulose in $10 \times$ SSC for 4 hr. The blot was washed once briefly in $5 \times$ SSC, and then baked under vacuum at 80° for 2 hr.

The blot was prehybridized in 20 ml of 50% formamide, 0.1% SDS, $4 \times$ Denhardt's solution, $1 \times$ RNA hybridization buffer, 4 mM vanadyl adenosine complex,[32] and 100 μg of boiled calf thymus DNA per milliliter for 30 min at 42°. It was then hybridized in 10 ml of fresh solution to which 10^7 cpm of boiled ^{32}P-labeled DNA probe were added. Incubation of the blot was at 42° overnight. The blot was washed four times at 23° with $2 \times$ SSC and 0.1% SDS for 15 min each, followed by four times with

[32] S. L. Bezger and C. S. Birkenmeir, *Biochemistry* **18**, 5143 (1979).

3 mM Tris base for 5 min each at 23°, and exposed to X-ray film in the presence of an intensifying screen for 4 days. Two ^{32}P-labeled probes specific for DIN1 and control transcripts were prepared by nick translation.[33] The DIN1 probe (7 × 10^7 cpm/μg) was the 2.3 kb SalI–SacI fragment of pSZ214, which contained 0.5 kb of din1 and 1.8 kb of lacZ sequences (Fig. 1). It was obtained by cutting pSZ214 DNA with SalI and SacI, fractionating the DNA by agarose gel electrophoresis, electroeluting the fragment from the gel onto DEAE paper, and eluting it from the paper with 1 M NaCl, 10 mM Tris HCl, pH 7.5, and 1 mM EDTA. The control probe (10^8 cpm/μg) was plasmid pSZ62[34] (not shown), which contains pBR322 and the 1.7 kb BamHI-HIS3 fragment.[35]

As shown in Fig. 3, there was a single size class of DIN1 transcript, which increased within 1 hr to detectable levels in NQO-treated cells, remained present for an additional hour, and then declined. These changes correlate with changes in β-galactosidase levels in T378 cells. There were no detectable DIN1 transcripts in untreated cells. The 2.3 kb control transcript[35] from a gene adjacent to the HIS3 gene reflects the amount of RNA present in each lane. The lacZ sequence alone does not hybridize detectably to yeast RNA under these conditions.

Sensitivity of the Fusion Cloning Method

The lacZ fusion method can be used to detect regulated genes whose products are at very low concentrations within the cell because the development of a blue color depends on cumulative β-galactosidase activity. An estimation of the sensitivity of fusion screening was made by examining the β-galactosidase levels of a yeast strain containing a fusion to a gene whose product levels are known. The diploid yeast strain, DA151, contains a single copy of a his3 : lacZ fusion integrated at the HIS3 locus.[36] The his3 : lacZ fusion plasmid A4p4 is illustrated in Fig. 1. During log-phase growth in histidine-free medium, the β-galactosidase specific activity was 0.67 unit as measured in permeabilized cells. The level of HIS3 mRNA in wild-type, haploid cells growing in liquid under similar conditions has been estimated to be 1 or 2 copies per cell.[35] When this diploid was grown on histidine-free medium and replica plated onto Xgal medium lacking histidine, a light blue color was detectable after 2 days. Fusions expressed at one-tenth of this level can be reproducibly detected.

[33] P. Rigby, D. Deichmann, C. Rhodes, and P. Berg, J. Mol. Biol. **113**, 237 (1977).
[34] T. L. Orr-Weaver, J. W. Szostak, and R. J. Rothstein, Proc. Natl. Acad. Sci. U.S.A. **78**, 6354 (1981).
[35] K. Struhl and R. Davis, J. Mol. Biol. **152**, 535 (1981).
[36] A. Murray, unpublished data, 1981.

Discussion

Genes Not Detectable by This Method

There are probably regulated yeast genes that are not identifiable by the fusion cloning method. A gene fusion product could disrupt normal cellular functions and cause lethality either because of its abundance or location. A fusion on a multicopy replicating plasmid might be improperly regulated. One solution to this problem might be to include a centromere fragment[37] on the fusion vector. Finally, the fusion itself might disrupt regulation if sequences adjacent to the 3' end of the gene are required for proper control.

Not all yeast genes will be represented in a fusion library as constructed here. The *lac* fragment in pMC1403[18] has three restriction endonuclease sites (*Eco*RI, *Sma*I, and *Bam*HI) adjacent to each other and occurring just within the N-terminus codons of the *lacZ* gene. Any one of these sites can be used for making fusions. We chose the *Bam*HI site because *Sau*3A-cut DNA can be ligated into the unique *Bam*HI site of pSZ211. The *Sau*3A recognition sequence occurs frequently enough to be within most yeast genes. A partial *Sau*3A digest would have DNA fragments large enough to contain all the necessary regulatory, and transcription and translation initiation sequences. At least some of the fusions made with *Sau*3A fragments would have a *Bam*HI site regenerated at the fusion junction. Clearly, fusions will not be obtained to genes lacking *Sau*3A sites in the correct reading frame in their coding regions. A similar strategy could be used to obtain complementary fusion libraries by using the other two restriction sites on the *lac* fragment from pMC1403. A single, complete fusion library could be achieved by using sheared DNA fragments since the breakpoints would be sequence independent; however, this is not necessary for the isolation of a few representatives of a large class of genes.

A library of fusions could also be constructed and amplified in *E. coli*.[38] This would enrich the DNA pool for intact replicating plasmids, and would be useful if it was necessary to screen for fusions in several different strains. It would though, reduce the probability that an independent yeast transformant contained a unique species of fusion plasmid. Using the ligated pool of DNA to transform yeast directly as described here saves time and possibly increases the proportion of integrated plasmids among the transformants. Two out of 6 *din* fusions for example, were obtained as mitotically stable transformants.[23] In contrast, for

[37] L. Clarke and J. Carbon, *Nature (London)* **287**, 504 (1980).
[38] L. Clarke and J. Carbon, this series, Vol. 68, p. 396.

closed, circular, replicating plasmids containing the *ars1* sequence and 1.7 kb to 2.3 kb yeast DNA inserts, less than 1% of the yeast transformants are mitotically stable.[19]

Problems in Screening lacZ Fusions

We have encountered certain problems in screening *lacZ* fusions and in subsequently working with specific fusions. To obtain consistent results among screenings of a pool of fusions, the colonies should be of the same size and at the same stages of growth in different experiments. It is important to use optimal time intervals consistently between (*a*) inoculation of the master plate and replica plating and (*b*) replica plating and exposure of the colonies to treatment conditions.

The genetic background of a strain may affect the expression of a fusion, and the development of blue color on Xgal media. The blue color varied greatly among nonisogenic strains containing the same fusion integrated at the same locus. When a diploid formed from two nonisogenic haploid strains was sporulated, the colonies from the four spores of a tetrad had different induced and uninduced color intensities. A significant reduction in this variability often required more than three backcrosses.[24] This variation may give rise to a large number of false positives when screening for differential fusion expression in haploid and diploid cells unless isogenic strains are used. It may also complicate the use of Xgal plates in the study of the effects of mutations on fusion expression, when the mutation is crossed in from a nonisogenic strain.

The biochemical and physiological characteristics of chimeric transcripts and proteins in yeast are possible sources of artifacts in the use of *lacZ* fusions to isolate and study regulated yeast genes. Active β-galactosidase is a tetramer,[39] so that any foreign peptide in the amino terminus of the monomer that would interfere with subunit interactions could reduce or destroy β-galactosidase activity. Oliver and Beckwith[40] have found a significant reduction in β-galactosidase activity of a hybrid protein in *E. coli* when it is membrane-associated compared with when it is in the cytoplasm as the result of a signal sequence mutation in the *malE:lacZ* gene fusion. It is interesting that for different yeast *cyc1:lacZ*[41] and *ura3:lacZ* fusions,[42] in general, the more of the yeast peptide in a fusion, the lower the specific activity of β-galactosidase. We know little about the

---

[39] I. Zabin and A. V. Fowler, *in* "The Operon" (J. H. Miller and W. S. Reznikoff, eds.), p. 89. Cold Spring Harbor Laboratory, Cold Spring Harbor, New York, 1980.

[40] D. B. Oliver and J. Beckwith, *Cell* **25,** 765 (1981).

[41] L. Guarente, personal communication.

[42] M. Rose, personal communication.

processing and degradation of chimeric molecules or their distribution within the yeast cell. It is conceivable that in some cases the blue color developed on Xgal plates might not reflect enzyme levels measured in permeabilized cells or in cell extracts. A formal possibility for some gene fusions is that changes in protein–protein interactions rather than gene activity could lead to changes in β-galactosidase activity. Therefore, the identification of a regulated gene by the use of a gene fusion should be confirmed by independent evidence such as Northern blot analysis of mRNA levels.

Rearrangements in Fusion Plasmids

In the isolation of fusion plasmids we routinely selected *E. coli* transformants that were blue on minimal medium with Xgal in the cases where there were both blue and white colonies on the transformation plates. The restriction map analyses of four different fusions examined thus far have shown that these plasmids had not undergone rearrangements to allow expression in *E. coli*. Since *E. coli* is not capable of expressing all eukaryotic genes, it is expected that not all fusions isolated will be expressed in bacteria. In fact, a *lacZ* fusion to the yeast histone H2A gene is not productive in *E. coli* but is in yeast, while a fusion to the H2B gene is expressed in both organisms.[43] Thus, it is important to examine plasmids for rearrangements if blue bacterial transformants are used for plasmid isolations.

Conclusions

We have described the use of *lacZ* gene fusions for the identification and isolation of regulated yeast genes. A library of random yeast gene:*lacZ* fusions in yeast transformants was constructed and screened for regulated gene fusions. The screen detected differential β-galactosidase activity in fusion-containing yeast colonies on different Xgal media. We have used this method to clone *DIN1* sequences as a *din1*:*lacZ* fusion that is induced in yeast by DNA damaging agents. The method is particularly useful for the cloning of coordinately regulated genes.

Once isolated, gene fusions can be applied to the study of gene structure, function, and regulation. The *lacZ* fusion would be useful for fine-structure mapping and sequencing since the 5' end of the *lacZ* gene has been sequenced.[44] Gene products could be identified by mRNA hybrid selection and *in vitro* translation.[45] Techniques already exist by which fu-

---

[43] M. A. Osley and L. Hereford, personal communication.
[44] A. M. Maxam, personal communication.
[45] M. L. Goldberg, R. P. Lifton, G. R. Stark, and J. G. Williams, this series, Vol. 68, p. 206.

sions can be mutagenized *in vitro*,[46] and the effects of the mutations examined *in vivo*.[47] *lacZ* fusions could be used to create insertion and deletion mutations in the wild-type gene.[47] Gene fusions could also be employed to isolate unlinked, regulatory mutations. This last application would be aided by the development of yeast able to take up lactose and of selectable substrates for fusion-containing yeast.

Acknowledgments

We thank Burt Beames for excellent technical assistance at the beginning of this work and Sandra Phillips for typing. We also thank T. L. Orr-Weaver and C. Potrickus for helpful comments on the manuscript. This work was supported by NIH Grant GM 27862 to J. W. S. S. W. R. was supported by NIH Training Grant 5T32GM07196, and A. W. M. by NIH Training Grant CA 09361.

[46] D. Shortle and D. Nathans, *Proc. Natl. Acad. Sci. U.S.A.* **75,** 2170 (1978).
[47] S. Scherer and R. W. Davis, *Proc. Natl. Acad. Sci. U.S.A.* **76,** 4951 (1979).

[20] Selection Procedure for Isolation of Centromere DNAs from *Saccharomyces cerevisiae*

By LOUISE CLARKE, CHU-LAI HSIAO, and JOHN CARBON

Hybrid plasmid derivatives of pBR322 that contain yeast (*Saccharomyces cerevisiae*) centromere DNA in combination with a yeast DNA replicator (*ars1* or *ars2*[1,2]) and a suitable genetic marker are stably maintained in yeast through both mitotic and meiotic cell divisions and exhibit typical Mendelian segregation (2+ : 2−) through meiosis.

The centromere DNA from yeast chromosome III (*CEN3*) was originally isolated by cloning the centromere III-linked genes *LEU2*[3] and *CDC10*,[4] obtaining flanking regions by overlap hybridization,[5] and testing individual DNA segments from the *LEU2-CDC10* region that were carried on hybrid *ars*-containing plasmids for proper mitotic and meiotic segregation in yeast.[6] Similarly, the centromere from yeast chromosome XI (CEN11) has been isolated by cloning the centromere XI-linked gene *MET14* and assaying flanking regions for centromere function *in vivo*.[7] Obviously these methods are most successful for isolation of those centromeres that are closely adjacent to a known genetic marker.

Taking advantage of the unique mitotic stability of *CEN* hybrid plasmids (minichromosomes) in yeast, however, and the characteristic instability of other *ars* plasmids lacking centromere sequences,[6,8] we have developed[9] and present here a procedure that should permit the isolation of centromere DNA sequences of any yeast chromosome directly from a yeast recombinant DNA library.

Principle of the Method

Escherichia coli–S. cerevisiae shuttle vectors, such as YRp7, that contain a presumptive chromosomal replicator (*ars1*[1]) are unstably main-

[1] D. T. Stinchcomb, K. Struhl, and R. W. Davis, *Nature (London)* **282**, 39 (1979).

[2] C.-L. Hsiao and J. Carbon, *Proc. Natl. Acad. Sci. U.S.A.* **76**, 3829 (1979).

[3] B. Ratzkin and J. Carbon, *Proc. Natl. Acad. Sci. U.S.A.* **74**, 487 (1977).

[4] L. Clarke and J. Carbon, *Proc. Natl. Acad. Sci. U.S.A.* **77**, 2173 (1980).

[5] A. C. Chinault and J. Carbon, *Gene* **5**, 111 (1979).

[6] L. Clarke and J. Carbon, *Nature (London)* **287**, 504 (1980).

[7] M. Fitzgerald-Hayes, J.-M. Buhler, T. G. Cooper, and J. Carbon, *Mol. Cell. Biol.* **2**, 82 (1982).

[8] A. J. Kingsman, L. Clarke, R. K. Mortimer, and J. Carbon, *Gene* **7**, 141 (1979).

[9] C.-L. Hsiao and J. Carbon, *Proc. Natl. Acad. Sci. U.S.A.* **78**, 3760 (1981).

RECOMBINANT DNA
METHODOLOGY

tained in budding yeast cultures[1,8] and are almost completely lost after 10–20 generations of cell growth in nonselective media. However, when the vector contains functional yeast centromere DNA (*CEN3* or *CEN11*), the hybrid plasmid is maintained in up to 90% of the cells after 20 generations of nonselective growth.[6,7] The mitotic stabilization of plasmids in yeast induced by centromere DNAs is sufficient to permit their direct selection from yeast cultures transformed with genomic library DNA constructed in the vector YRp7.

Three alternative routes can lead to mitotic stabilization of plasmids in yeast. The first is integration of the plasmid into host chromosomal DNA; the second is enhanced stability of *ars* plasmids in diploid versus haploid yeast; and the third is the presence of cloned 2 μm (sometimes termed "2 μ") plasmid DNA sequences. These may readily be distinguished, however, by the methods described here.

Materials and Techniques

Commonly employed techniques for construction of genomic libraries, isolation and purification of DNAs, yeast and *E. coli* transformations, genetic manipulations of yeast, and preparation of culture media have all been described in Vols. 68 and 101 of *Methods in Enzymology* and in references cited in footnotes 2–11.

Method

Selection of Mitotic Stabilizing Sequences

Five micrograms of hybrid plasmid DNA from a yeast genomic library consisting of partial *Sau*3A restriction fragments of total yeast DNA ligated into the single *Bam*H1 site of the *E. coli*–yeast shuttle vector Yrp7 (*Trp1 ars1 ampicillin*[R])[10] are used to transform yeast strain Z136-1-13C (a *trp1* strain) to Trp[+], selecting for the *TRP1*[+] marker on the vector. With sterile toothpicks, 500 of the fastest growing transformants are transferred to YEP (1% yeast extract, 2% peptone, 2% glucose; nonselective) agar plates in a grid array and grown overnight at 32°. The colonies are replica-plated onto fresh YEP plates and grown overnight for a total of at least five successive transfers. Finally the transformants are replica-plated onto yeast minimal (0.65% Difco yeast nitrogen base, 2% glucose) agar plates lacking tryptophan, and those colonies that are still strongly Trp[+] after prolonged nonselective growth (5–10% of the original 500 clones) are colony purified and tested for mitotic stability of the *TRP1* marker.

[10] K. A. Nasmyth and S. I. Reed, *Proc. Natl. Acad. Sci. U.S.A.* **77**, 2119 (1980).

[11] F. Sherman, G. R. Fink, and J. B. Hicks, "Methods in Yeast Genetics;" (Laboratory Manual). Cold Spring Harbor Laboratory, Cold Spring Harbor, New York, 1979.

Assay for Mitotic Stabilization

Individual putative stably transformed clones are grown nonselectively on YEP agar or in 10 ml of liquid YEP medium for 24 hr at 32°. Each culture is then streaked or spread for single colonies on YEP agar plates, and the resulting colonies are replica plated onto minimal plates with and without tryptophan. The relative numbers of Trp[+] and Trp[−] colonies are scored after overnight incubation at 32°.

For nearly all the cultures selected from the original 500, the fraction of cells remaining Trp[+] after overnight growth on nonselective media should be roughly 60–100%, whereas the parent vector YRp7 is usually completely segregated from strain Z136-1-13C under identical conditions.

Recovery of Mitotically Stable Hybrid Plasmids from Individual Yeast Clones

A crude DNA preparation (containing both chromosomal and plasmid DNA)[9] is isolated from each stably transformed yeast clone and used to transform *E. coli* strain JA228(W3110 $hsdM_K^+$ $hsdR_K^-$ $argH$ Str^R), selecting for ampicillin resistance, a marker carried on the pBR322-derived Yrp7 shuttle vector. With approximately 85% of these preparations, large numbers of ampicillin-resistant transformants are obtained from which recombinant plasmids may be isolated and purified. The yeast clones whose DNAs yield no bacterial transformants presumably either contain an integrated or recombined *TRP1* gene or are Trp[+] revertants.

The set of hybrid plasmids now contained in *E. coli* may be screened directly by colony hybridization[5] for the presence of all or a portion of the 2 μm yeast plasmid sequences. Thus those plasmids stabilized in yeast by 2 μm DNA are identified and may be eliminated from the set.

Purified plasmid DNAs from the remainder readily transform Trp[−] yeast to Trp[+] with high frequency ($\sim 10^4$ transformants per microgram), and all the transformants should be mitotically stable. These plasmids, therefore, contain cloned DNA segments capable in some way of stabilizing plasmid replication and maintenance in yeast during mitotic cell division.

Identification of CEN-Containing Plasmids by their Meiotic Segregation Pattern

Hybrid plasmids carrying cloned yeast centromere DNA sequences (minichromosomes) segregate as ordinary yeast chromosomes (2+:2−) through meiosis.[6] On the other hand, multiple copy number plasmids, specifically, 2 μm-derived vectors, would be expected to segregate 4+:0−. Therefore, plasmids carrying functional centromere sequences can be distinguished from 2 μm-derived vectors, which segregate predominately 4+:0−, and *ars* vectors lacking centromere sequences, which are unstable and are normally lost in the process of meiosis, by the following genetic manipulations.

The plasmid to be tested is contained in a yeast strain with a mutation in the gene for which there is a wild-type copy on the plasmid (for example, *trp1*⁻ in the host genome and *TRP1*⁺ on the plasmid). The strain is crossed with another strain of the opposite mating type that also contains a mutation in the gene whose wild-type allele is carried by the plasmid. In this way the plasmid may be followed easily through the cross by the presence of the wild-type marker. Either one or both strains in the cross should also contain at least one centromere-linked marker so that sister spores, the products of the second meiotic division, may be identified.

After mating, diploids are purified and sporulated; resulting asci are dissected on YEP agar plates for genetic analysis.[11] At least 10–15 tetrads having four viable spores are scored on appropriate minimal plates for all pertinent markers in the cross. If a plasmid carries functional centromere sequences, it will segregate as a chromosome in the first meiotic division and, as a consequence, is found predominantly in the two sister spores, which are the products of the second meiotic division. Centromere-containing plasmids are not completely stable, however, and segregate 0+:4− in 10–15% of the asci. The marker on the plasmid should behave as a centromere-linked gene, so that when the plasmid marker is scored versus another centromere-linked gene (such as *met14* on chromosome XI), principally parental ditype and nonparental ditype (but very few tetratype) asci are obtained in roughly equal numbers in the cross.

The table lists a number of stable plasmids isolated by the direct selection procedure.[9] Plasmids pCH4 and pCH25 have been shown by hybridization to contain *CEN3* DNA.[9] Plasmids pCH3, pCH9, and pCH10 are also presumably centromere-containing plasmids, since their behavior in genetic crosses is typical of that observed with minichromosomes carrying the previously identified centromere DNAs, *CEN3*[6] and *CEN11*.[7]

A hybrid plasmid would also exhibit 2+:2− segregation if it were integrated or recombined into one of the host chromosomes. If this were the case, however, the small proportion of 4+:0− and 0+:4− tetrads would not be obtained in crosses, nor would all integration events be expected to occur near a centromere. Although the reason for the small percentage of 4+:0− and 0+:4− tetrads seen in crosses involving CEN plasmids is still

MEIOTIC SEGREGATION OF STABLE YEAST PLASMIDS ISOLATED BY THE DIRECT-SELECTION PROCEDURE[a]

Plasmid in cross	Stabilizing DNA	Distribution in tetrads of *TRP1*+ marker on plasmid[b]					Test for centromere linkage of marker on plasmid			Reference centromere marker
		4+:0-	3+:1-	2+:2-	1+:3-	0+:4-	PD	NPD	T	
pCH3	(CEN)	2 (13)	0	13 (37)	0	0	6	7	0	*met14*
pCH4	(CEN3)	7 (27)	0	17 (55)	0	2 (8)	4	13	0	*met14*
pCH9	(CEN)	2 (9)	1 (4)	13 (57)	0	7 (30)	4	9	0	*met14*
pCH10	(CEN)	2 (10)	0	17 (85)	0	1 (5)	8	7	2	*ura3*
pCH25	(CEN3)	0	1 (7)	12 (86)	0	1 (7)	5	7	0	*met14*
pCH21	(2 μm)	16 (100)	0	0	0	0				
pCH27	(2 μm)	19 (76)	0	1 (4)	0	5 (20)				

[a] Transformants in strain Z136-1-13C [*a trp1 ade1 leu1 gal1 a-g4lpCH(TRP1+)*] were crossed with strain X2928-3D-1C (*α trp1 leu1 his2 ura3 ade1 met14 gal1*). PD, parental ditype; NPD, nonparental ditype; T, tetratype.

[b] Values in parentheses represent percentage of total.

unclear, the pattern is characteristic of plasmids carrying *CEN3*,[6] *CEN11*,[7] and presumably other yeast centromeres.[9]

The table gives two examples of the behavior of 2 µm-derived plasmids (pCH21 and pCH27; both isolated by the direct selection procedure) in genetic crosses. As expected, these high copy number plasmids are predominantly distributed to all four progeny of the tetrads analyzed.

Biochemical Confirmation of Autonomous Replication of CEN Plasmids

Confirmation that prospective *CEN*-containing plasmids are replicating autonomously in yeast may be obtained by a routine Southern blot hybridization experiment.[6,9,12] The presence of unintegrated plasmid is shown by hybridization of ^{32}P-labeled pBR322 DNA to Southern blots of fractionated restriction digests of total cell DNA isolated from plasmid-bearing progeny of selected crosses. For example, the cell DNA is predigested with a restriction enzyme that cleaves the plasmid once in the pBR322 vector region. A single radioactive band on the Southern autoradiogram corresponding in mobility to control plasmid DNA digested with the same enzyme indicates the presence of unintegrated plasmid in the cell and the absence of integrated copies of pBR322 DNA.

Similar experiments with 2 µm-derived plasmids yield a single band of much greater intensity, reflecting the high copy number of these autonomously replicating plasmids.[9]

Screening Clones for Haploidy Using Resistance to Canavanine

Standard procedures for the transformation of yeast in the presence of polyethylene glycol result in a significant proportion of diploid (a/a or α/α) transformants. Plasmid YRp7 (*TRP1 ars1*) and its derivatives are mitotically appreciably more stable in these diploid cells[8] and may be confused with stable *CEN* or 2µm-containing plasmids in the original selection. In order to avoid this problem, a simple screening procedure may be carried out on the yeast transformants to distinguish haploid from diploid colonies in the collection. For example, the fast growing clones from the original transformation are distributed in small patches on YEP plates, grown to confluency, and replica-plated onto minimal plates containing 60 µg of the amino acid antagonist canavanine per milliliter. (The plates cannot contain arginine, an inhibitor of canavanine action.) Mutations to canavanine resistance in yeast are recessive, thus haploid cells may readily be distinguished from diploids by a brief exposure to ultraviolet light, resulting in canavanine resistance at a much higher frequency in haploids. Using control haploid and diploid strains, UV exposure conditions are

[12] E. M. Southern, *J. Mol. Biol.* **98,** 503 (1975).

sought that yield good growth in patches of haploid clones and little or none in diploid patches. This test is also useful for identification of haploid (versus diploid) transformants prior to performing genetic crosses or studies of plasmid stability and copy number.

Although centromere-carrying plasmids have been identified with relative ease by the direct selection procedure without introducing the canavanine- resistance screen, the addition of this step does eliminate at the onset further analysis of apparently stable plasmids that turn out to be mitotically unstable when introduced into haploid cell.

Comments

The direct-selection procedure for isolation of functional centromere DNA by mitotic stabilization of *ars* plasmids has been used successfully in *S. cerevisiae* with cloned yeast DNA segments. The applicability of the method to other eukaryotic DNAs or host cells is untested and ultimately depends on the proper functioning and the mitotic behavior of *ars*-carrying plasmids in these systems.

[21] Construction of High Copy Yeast Vectors Using 2-μm Circle Sequences

By JAMES R. BROACH

This chapter addresses one aspect of the problem of expressing any cloned gene at very high levels in yeast. Since, as a first approximation, it is reasonable to assume a gene dosage effect for the expression of cloned sequences in yeast, maximizing expression will involve maximizing the number of copies of that gene in the cell. For reasons elaborated below, this in turn will undoubtably require propagating that gene on a vector derived from the yeast plasmid 2-μm circle. Thus in this chapter I discuss various aspects of the construction and use of 2-μm circle vectors for propagating cloned genes in yeast.

The yeast plasmid 2-μm circle is a 6318 bp double-stranded DNA species present at 60–100 copies per cell in most *Saccharomyces cerevisiae* strains.[1–3] Although the replication of this plasmid during normal exponential growth is strictly under cell cycle control, it can escape from this control in certain situations to increase its copy number from as low as a single copy per cell to its normal high level.[4,5] As discussed below, this capability is a property of the 2-μm circle origin of replication functioning in conjunction with several proteins encoded by the plasmid itself. As a consequence of this amplification potential, many hybrid plasmids constructed from 2-μm circle sequences can establish and maintain high copy number in yeast strains, following their introduction by transformation.[5–7]

[1] J. L. Hartley and J. E. Donelson, *Nature (London)* **286**, 860 (1980).

[2] C. P. Hollenberg, P. Borst, and E. F. J. Van Bruggen, *Biochim. Biophys. Acta* **209**, 1 (1970).

[3] G. D. Clark-Walker and G. L. G. Miklos, *Eur. J. Biochem.* **41**, 359 (1974).

[4] D. C. Sigurdson, M. E. Gaarder, and D. M. Livingston, *Mol. Gen. Genet.* **183**, 59 (1981).

[5] J. B. Hicks, A. H. Hinnen, and G. R. Fink, *Cold Spring Harbor Symp. Quant. Biol.* **43**, 1305 (1978).

[6] K. Struhl, D. T. Stinchcomb, S. Scherer, and R. W. Davis, *Proc. Natl. Acad. Sci. U.S.A.* **76**, 1035 (1979).

[7] C. Gerbaud, P. Fournier, H. Blanc, M. Aigle, H. Heslot, and M. Guerineau, *Gene* **5**, 233 (1979).

RECOMBINANT DNA
METHODOLOGY

In addition, these hybrid plasmids containing 2-μm circle sequences are relatively stable during mitotic growth and meiosis. Although a number of specific yeast chromosomal sequences are also capable of promoting autonomous replication in yeast of plasmids in which they are incorporated[8-10]—a property that is presumably a reflection of their normal function in the cell as chromosomal origins of replication—the ability to establish and maintain high copy number and to propagate stably during mitotic growth is apparently unique to plasmids derived from 2-μm circle sequences.

At this point an unequivocal recommendation of the best possible 2-μm circle vector for achieving high copy number in yeast is not possible. First, our knowledge of the mechanism of 2-μm circle copy control is not complete. In addition, the influence on plasmid copy number—whether that plasmid is 2-μm circle or a hybrid cloning vector—of such factors as genetic background of the host yeast strain, the size of the vector, the presence on the plasmid of extra yeast or bacterial sequences, or the presence of other plasmids in the cell, has not been rigorously examined. Nonetheless, the components of the copy control system of 2-μm circle are reasonably well defined. On the basis of this information we can develop various strategies for using 2-μm circle sequences to maximize the copy number of a cloned gene in yeast. Thus, in this chapter, I discuss various approaches to the construction and use of 2-μm circle vectors that possess the potential for stable, high-copy propagation in yeast. By pursuing several of these avenues and assessing the copy number attained with each, one can optimize the synthesis of a cloned gene in yeast. Other aspects of the problem of maximizing expression of a particular gene in yeast—such as attachment to appropriate promoter and terminator sequences—are addressed in other chapters in this volume.

In the first section of this chapter I describe the structure of 2-μm circle and the features of the plasmid replication apparatus salient to a discussion of vector construction. I then describe hybrid 2-μm circle vectors suitable for use with yeast strains containing endogenous 2-μm circles (designated [cir$^+$] strains). There are some reasons to suspect, however, that one can achieve higher copy numbers by propagating appropriate hybrid 2-μm circle vectors in strains that lack 2-μm circles ([cir^0] strains). Thus, in the third section I provide a list of available [cir^0] strains, present a procedure for isolating [cir^0] derivatives of [cir$^+$] strains, and describe 2-μm circle vectors suitable for use in [cir^0] strains. Then I describe sev-

[8] D. T. Stinchcomb, K. Struhl, and R. W. Davis, *Nature (London)* **282**, 39 (1979).
[9] C. Chan and B.-K. Tye, *Proc. Natl. Acad. Sci. U.S.A.* **77**, 6329 (1980).
[10] D. Beach, M. Piper, and S. Shall, *Nature (London)* **284**, 188 (1980).

eral unusual vectors and suggest several strategies for possible additional improvements in copy levels. Finally, in the last section I describe several rapid procedures for determining copy number of hybrid plasmids in yeast.

In addition to using 2-μm circle vectors to optimize the copy number of cloned sequences, there are other contexts in the molecular biology of yeast in which hybrid 2-μm circle vectors play or can play a significant role. These uses of 2-μm circle vectors are described briefly in the following, and, for each procedure, a reference to the chapter in Vol. 101 of *Methods in Enzymology* in which it is discussed more fully is provided. First, 2-μm circle vectors have been useful for cloning yeast genes by complementation. That is, several hybrid plasmids that will be described in this chapter have been used as vectors for constructing libraries of random genomic sequences from yeast as well as from other organisms. Specific genes have been readily recovered from these libraries by transforming an appropriate yeast strain with the pooled plasmids and selecting for complementation of specific markers.[11,12] This technique was made possible by the high transformation frequencies obtainable with 2-μm circle vectors, the relative stabilities of such plasmids in yeast, and the ease with which such plasmids can be recovered from yeast into *E. coli*. This procedure for cloning genes by complementation is mentioned in the chapter by MacKay [22] in Vol. 101. Second, various 2-μm circle vectors are useful in recovering specific alleles of a cloned gene from the yeast genome.[13] This procedure—which involves transformation by a 2-μm circle vector carrying sequences spanning the gene of interest, after linearizing the plasmid by restriction enzyme digestion at a site within the gene— is addressed in the chapters by Orr-Weaver *et al.* [14] and Stiles [19] in Vol. 101. Finally, 2-μm circle plasmids can be used in a quite novel fashion to determine the chromosomal location of a cloned yeast fragment. This involves integrating 2-μm circle sequences at the chromosomal site homologous to the cloned fragment. High-frequency chromosomal breakage at this site, which results from the specialized recombination system encoded by 2-μm circle, permits a rapid genetic determination of the site of integration.[14]

The Copy Control System of 2-μm Circle

A diagram of the structure of 2-μm circle is presented in Fig. 1. The entire nucleotide sequence of the plasmid has been determined[1] and the

[11] J. R. Broach, J. N. Strathern, and J. B. Hicks, *Gene* **8,** 121 (1979).
[12] K. A. Nasmyth and S. I. Reed, *Proc. Natl. Acad. Sci. U.S.A.* **77,** 2119 (1980).
[13] J. N. Strathern, personal communication.
[14] C. Falco, Y.-Y. Li, J. R. Broach, and D. Botstein, *Cell* **29,** 573 (1982).

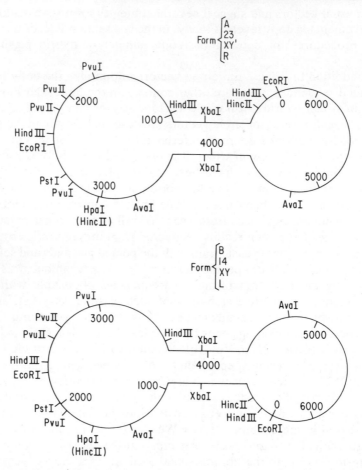

FIG. 1. Structure and restriction map of 2-μm circle. The location of a subset of restriction sites as determined from the published sequence of the plasmid are indicated on schematic diagrams of the two forms of 2-μm circle. Those terms variously used to designate the two forms are listed above each. The circular portion of each figure represents unique sequences, whereas the linear portion of the figures represents the repeated regions. The base-pair numbering system, shown inside each figure, is consistent with that used by Hartley and Donelson.[1]

cleavage sites for a subset of restriction enzymes as determined from the sequence are indicated in the figure. In confirmation of earlier observations on the structure of the plasmid, analysis of the sequence demonstrated that the plasmid contains two regions of 599 bp each, which are precise inverted repeats of each other and divide the molecule approximately into halves. Since recombination readily occurs between these repeated sequences, 2-μm circles isolated from yeast actually consists of a mixed population of two plasmids that differ only in the orientation of one unique region with respect to the other.[15] Both forms of the plasmid— which have been variously designated as A and B, R and L, XY' and XY, or 23 and 14—are diagrammed in Fig. 1.

The salient features of the 2-μm circle replication system are indicated in Fig. 2. The origin of replication spans a region extending from the middle of one of the inverted repeats 100 bp into the adjacent unique sequences (approximately nucleotides 3650 to 4000 in the A form or nucleo-

FORM A

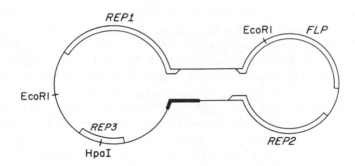

FORM B

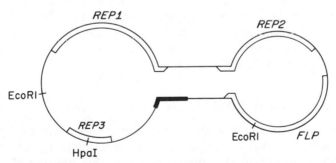

FIG. 2. Components of the 2-μm circle replication system. The locations of the origin of replication (filled line), the three known 2-μm circle genes (open double lines), and the cis-acting replication region (*REP3*) are shown on schematic diagrams of the two forms of the plasmid.

[15] J. D. Beggs, *Nature* (*London*) **275**, 104 (1978).

tides 650 to 1000 in the B form). This site was initially identified as the sole sequence within the 2-μm circle that would promote autonomous replication in yeast of hybrid plasmids into which it was inserted.[16] That this region actually functions *in vivo* as the primary 2-μm circle origin of replication was subsequently confirmed.[17,18]

Although autonomous replication is conferred by the 2-μm circle origin of replication, propagation at high copy number requires, in addition, two proteins encoded by the plasmid itself. Hybrid plasmids containing 2-μm circle sequences spanning only the origin of replication are present at low copy number in [cir^0] yeast strains and are lost rapidly during non-selective growth. However, in a [cir$^+$] strain, the same plasmids are present stably and at high copy number.[16] Similarly, hybrid plasmids containing the entire 2-μm circle genome are present stably and at high copy number even in [cir^0] strains. Thus, some trans-acting function encoded in a site or sites away from the origin of replication is necessary for high copy propagation of the plasmid in yeast. The genes encoding this function have been identified by genetic analysis of hybrid plasmids containing the entire 2-μm circle genome. Mutations that interrupt either of two plasmid genes, designated *REP1* and *REP2* in Fig. 2, abolish efficient, high copy maintenance of the plasmid.[19,20] We have demonstrated that high copy number and stability also require sequences between the *Pst*I and *Ava*I sites in the large unique region. This region is active only in cis with respect to the 2-μm circle origin and probably defines the site through which *REP* proteins act to promote amplification. We have designated this region *REP3*.[20] Thus, the requisite components for high-copy, stable propagation of a hybrid plasmid in yeast are the presence on the plasmid of a fragment from 2-μm circle spanning both the origin of replication and *REP3* and the presence in the cell of the *REP1* and *REP2* gene products. These proteins can be provided either by the endogenous 2-μm circles in the cell or by the hybrid plasmid itself.

Vectors for Use with [cir$^+$] Strains

Since [cir$^+$] strains contain a sufficient complement of 2-μm circle *REP* proteins, all that is required for high copy number propagation of a particular hybrid plasmid is the presence on that plasmid of sequences

[16] J. R. Broach and J. B. Hicks, *Cell* **21**, 501 (1980).
[17] C. S. Newlon, R. J. Devenish, P. A. Suci, and C. J. Roffis, *Initiation of DNA Replication, ICN–UCLA Symp. Mol. Cell. Biol.* **22**, 501, (1981).
[18] H. Kojo, B. D. Greenberg, and A. Sugino, *Proc. Natl. Acad. Sci. U.S.A.* **78**, 7261 (1981).
[19] J. R. Broach, V. R. Guarascio, M. H. Misiewicz, and J. L. Campbell, *Alfred Benzon Symp.* **16**, 227 (1981).
[20] M. Jayaram, Y.-Y. Li, and J. R. Broach, *Cell,* in press (1983).

spanning the 2-μm circle origin of replication and the *REP3* locus. In order to introduce the plasmid into yeast by transformation and assure its maintenance, the plasmid must also contain a gene that can be selected in an appropriate yeast strain. There are essentially two strategies for constructing such a high copy, selectable plasmid that also contains a particular gene of interest. Either the gene can be inserted into a preexistent plasmid vector, or sequences for selection and high copy number propagation can be inserted into a plasmid containing the gene of interest.

Preexisting Vectors. A number of plasmid vectors currently available are capable of selection and high copy number propagation in yeast.[6,11] The relevant properties of a number of these plasmids are identified in Table I, and their structures are diagrammed in Fig. 3. All these plasmids are constructed from the bacterial plasmid pBR322 and thus can be maintained in and purified, in reasonable quantities, from appropriate *E. coli* strains. Each of these vectors carries the 2-μm circle origin of replication and *REP3*, most often in the form of the 2240 bp *Eco* RI fragment from the B form of the plasmid. The vectors differ though in the particular yeast gene used as the selectable marker for transformation and maintenance in yeast and in the availability of restriction sites for inserting a gene of interest. Nonetheless, each of these plasmids is maintained at approximately 30 copies per cell in [cir⁺] yeast strains. In addition, all these plasmids are

TABLE I
USEFUL 2-μm CIRCLE VECTORS

	Selectable markers			
Plasmid	Yeast	*E. coli*	Available cloning sites	Comments
CV7	*LEU2ª*	*Amp^R, LeuB*	*Bam*HI, *Sal*I, *Hind*III	—
YEp13	*LEU2ª*	*Amp^R, Tet^R, LeuB*	*Bam*HI, *Hind*III	Inserts in *Bam*HI can be detected as *Tet^S* clones
YEp13S	*LEU2ª*	*Amp^R, Tet^R, LeuB*	*Bam*HI, *Sal*I, *Hind*III	Inserts in *Bam*HI or *Sal*I can be detected as *Tet^S* clones
YEp6	*HIS3ª*	*Amp^R, HisB*	*Bam*HI	—
YEp24	*URA3ª*	*Amp^R, PyrF*	*Eco*RI, *Bam*HI	—
YEp16	*URA3ª*	*Amp^R, Tet^R, PyrF*	*Bam*HI, *Sal*I, *Hind*III	Inserts into any of these sites can be detected as *Tet^S* clones
CV19	*LEU2ᵇ*	*Amp^R, LeuB*	*Bam*HI, *Sal*I	—
CV21	*LEU2ᵇ*	*Amp^R, LeuB*	*Bam*HI, *Sal*I	—
pSI4	*LEU2ᵇ*	*Amp^R, Tet^R, LeuB*	*Bam*HI, *Sal*I	Derived from pJDB219; insertions detectable as *Tet^S* clones

ª For use with [cir⁺] strains only.
ᵇ Can be used with both [cir⁰] and [cir⁺] strains.

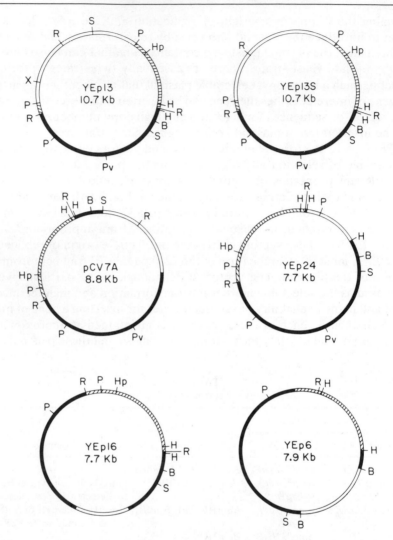

FIG. 3. Hybrid 2-μm circle vectors. In each vector diagram pBR322 sequences are represented by filled double lines, 2-μm circle sequences by hatched lines, and yeast chromosomal sequences by open double lines. For YEp13, YEp13S, and CV7, the chromosomal sequences span the *LEU2* gene; for YEp24 and YEp16, these sequences span *URA3;* and for YEp6 these sequences span *HIS3*. The locations of a number of restriction enzyme sites are indicated on the diagrams. These sites are abbreviated as follows: B, *Bam*HI; H, *Hin*dIII; Hp, *Hpa*I; P, *Pst*I; Pv, *Pvu*II; R, *Eco*RI; S, *Sal*I; and X, *Xho*I.

reasonably stable during mitotic growth, being lost in less than 20% of the cells after 10 generations of growth in nonselective media.[6,16] Since there is apparently little difference among the plasmids in terms of stability or copy number, the choice of a particular plasmid would be dictated by the genetic markers of the yeast strain to be used and by the particular restriction sites bracketing the gene to be inserted.

A hybrid plasmid lacking the 2-μm circle origin of replication, but containing an inverted repeat, can also propagate at moderately high copy number in yeast.[5] This is possible since such a plasmid can recombine very efficiently with an endogenous 2-μm circle in the cell and thus replicate by virtue of its physical attachment to an autonomously replicating molecule.[16] Since the plasmid by itself is not able to replicate, and since the reverse recombination event also occurs frequently, the plasmid is lost at a reasonably high rate. Thus such a plasmid is of little use in the context of maximizing expression, but can be of value in other situations.

Contructing New Vectors. If a gene of interest resides on a plasmid containing a selectable yeast marker or if the gene itself can be selected in yeast, then propagating this gene at high copy number can be accomplished by inserting a DNA fragment spanning the 2-μm circle origin of replication and *REP3* into the plasmid and transforming the resultant plasmid into a [cir⁺] strain. From the B form of 2-μm circle one can recover several different restriction fragments spanning the origin and *REP3* which could be inserted readily into a plasmid at a single restriction site. These include the 2242 bp *Eco*RI fragment, the 2214 bp *Hin*dIII fragment, and the 1576 *Sau*3A fragment (which can be inserted into either a *Bam*HI, a *Bgl*II, or a *Bcl*I site). A ready source of any of these fragments is plasmid pMJ5 (cf Fig. 4), which consists of the entire B form of 2-μm circle cloned at the *Eco*RI site in the small unique region into the *Eco*RI site of pBR322.

For those situations in which the gene of interest resides on a plasmid lacking a suitable selectable marker, there are several plasmids available from which a single restriction fragment can be obtained that carries the 2-μm circle origin of replication as well as a gene that can be selected in yeast. Thus, by inserting such a fragment into a plasmid containing a gene of interest, it is possible to generate in one step a vector by which the gene can be introduced into yeast and propagated at high copy number. For instance, the 3.2 kb *Sal*I to *Hin*dIII fragment from plasmid pC4 [or the equivalent *Bam*HI to *Hin*dIII fragment of the derivative plasmid pC4(B); see Fig. 4] carries both the *LEU2* gene of yeast and the 2-μm circle origin.[21] Similarly, the 3.6 kb *Hin*dIII fragment from pJDB219 carries the origin and the *LEU2* gene[15] (see Fig. 6; the special properties of plasmid

---

[21] J. R. Broach, unpublished results.

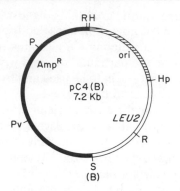

FIG. 4. Plasmid pC4. The representation of the sources of the various portions of the plasmid and the abbreviations for restriction sites on the plasmid are the same as those used in Fig. 3. In plasmid pC4B the *Sal*I site has been converted into a *Bam*HI site. As described in the text, the right-hand *Hin*dIII to *Sal*I restriction fragment of pC4 (or the equivalent *Hin*dIII to *Bam*HI fragment of pC4B) contains all the sequences required for selection and high copy propagation of a plasmid in a *leu2* [cir⁺] yeast strain.

pJDB219 are discussed below). Insertion of any of these fragments into a bacterial plasmid is readily monitered since the *LEU2* gene complements the *E. coli leu*B mutation (present in the commonly used *E. coli* strain C600). There are certainly other plasmids from which one can obtain such dual-function fragments, but an exhaustive list is not provided here. As above, the primary criteria in choosing a particular fragment are the constraints of available restriction sites in the recipient plasmid and of available genetic markers in the recipient yeast strain.

Propagating Cloned Genes in [cir⁰] Strains

If two distinct bacterial plasmids share a common copy control system, then they can exhibit a phenomenon termed incompatibility. One manifestation of incompatibility is that the number of copies of either plasmid in a cell when both plasmids cohabit the cell is less than the number of copies of either plasmid when either is the sole occupant of the cell. There is some indication that 2-μm circles and hybrid 2-μm circle plasmids display incompatibility. That is, the presence of endogenous 2-μm circle in a yeast strain may reduce the potential copy number of a hybrid 2-μm circle plasmid introduced into the strain.[22] Thus, in seeking to optimize copy number of a cloned gene it is worth considering using strains devoid of endogenous circle ([cir⁰] strains) in conjunction with appropriate 2-μm circle-derived vectors.

[22] C. Gerbaud and M. Guerineau, *Curr. Genet.* **1**, 219 (1980).

[*cir⁰*] *Strains.* There are a number of [cir⁰] strains currently available that contain nonreverting alleles (or alleles that revert at low frequency) of loci complemented by the cloned yeast genes most commonly used as selectable markers for transformation. A partial list of such strains is given in Table II. These strains either arose as spontaneous [cir⁰] derivatives of [cir⁺] strains or were induced by the technique described below.[4, 7, 23–25] None of the strains, though, displays any obvious barrier to propagation of 2-μm circles when they are reintroduced by cytoduction or of hybrid 2-μm circle plasmids introduced by transformation.

In addition to these preexisting [cir⁰] strains, [cir⁰] derivatives can be recovered from almost any *leu2* [cir⁺] strain after transformation with the hybrid 2-μm circle-*LEU2*-pMB9 plasmid, pJDB219.[24,25] Since pJDB219 propagates in yeast transformants at unusually high levels (discussed below), it is possible that elimination of endogenous 2-μm circles from such transformants is a manifestation of incompatibility; that is, pJDB219 molecules outcompete for a replication apparatus specific for 2-μm circle origins to the exclusion of the endogenous circles. However, the fact that this incompatibility is observed following introduction of pJDB219 by transformation, but is not observed if pJDB219 is introduced by cytoduction or mating, suggests that the actual explanation of the phenomenon is more complex.[25] Nonetheless, by using the following procedure one can obtain [cir⁰] derivatives of almost any *leu2* strain.

The *leu2* strain is transformed to Leu⁺ with pJDB219 DNA using the standard transformation protocol.[15] Cells from a single transformant are resuspended in selective media (ca. 2 ml of synthetic complete medium minus leucine[11]) and grown overnight at 30° with agitation. A sample of this culture is then diluted 1 : 1000 in YEPD (1% yeast extract, 2% Bacto peptone, and 2% glucose) and grown overnight at 30°. An appropriate dilution of the culture is spread on solid YEPD medium to give ca. 200 colonies per plate, and colonies auxotrophic for leucine are identified by replica plating to synthetic complete minus leucine medium. Approximately

TABLE II
AVAILABLE [cir⁰] STRAINS

Designation	Genotype	Reference
AH22	**a** *leu2-3,112 his4-519 can1-11*	5, 25
SC3	α *trp1 ura3-52 his3* ∇ *gal2 gal10*	4
DC04	**a** *leu2-04 ade1*	16
SB1-3B	α *leu2-04 ade1 ade6*	14

[23] D. M. Livingston, *Genetics* **86**, 73 (1977).
[24] M. J. Dobson, A. B. Futcher, and B. S. Cox, *Curr. Genet.* **2**, 193 (1980).
[25] A. Toh-e and R. I. Wickner, *J. Bacteriol.* **145**, 1421 (1981).

20 Leu⁻ colonies are recovered and tested for the presence of 2-μm circle sequences by colony hybridization[5,7] using ^{32}P-labeled pMJ5 DNA (or any readily available DNA containing 2-μm circle sequences). Leu⁻ colonies which lack 2-μm circle sequences by this criterion can be further tested by Southern analysis of genomic DNA isolated from a small culture of the strain.[26]

There is, in addition, a genetic test for the absence of 2-μm circles that one can use as an alternative procedure for identifying the [cir⁰] clones arising from the pJBD219-transformed strain. This procedure is based upon the observation that the presence of 2-μm circles in a diploid strain in which an inverted repeat from 2-μm circle has been inserted into a chromosome causes a marked instability of that chromosome.[14] The instability can be tested by assessing whether there is an unusually high number of cells, in a culture of the diploid strain, that possess a phenotype associated with the loss of a dominant marker on the insertion-containing chromosome for which the equivalent locus on the homolog chromosome carries a recessive allele. Specifically, for identifying the [cir⁰] clones obtained by the above procedure, the Leu⁻ segregants are mated to strain DC-04::LCV7 (a *leu2-04*::CV7 *ade1 ade6* [cir⁰]) or strain SB1-3B (α *leu2-04*::CV7 *ade1 ade6* [cir⁰]) and diploids are obtained by selecting for Ade⁺ Leu⁺ colonies (both DC-04::CV7 and SB1-3B are Ade⁻ but Leu⁺, since the CV7 plasmid integrated at *leu2* contains an intact *LEU2* gene).

After the diploid has been colony purified, it is streaked on nonselective media (YEPD, for example) for single colonies so that at least 100 individual colonies are distinguishable. (Alternatively, the colony-purified diploid can be grown in liquid YEPD and appropriate dilutions can be spread on YEPD plants to give 100 or so colonies.) These colonies are then replica-plated onto SD minus leucine media. One of two possible outcomes should be obtained: either all the colonies will be Leu⁺ or a reasonable proportion (5% to 25%) of the colonies will be Leu⁻. If all the colonies are Leu⁺, then the strain tested was [cir⁰]; if a reasonable proportion are Leu⁻, then the strain tested was [cir⁺]. This is explained in the following manner. If the diploid strain does not contain 2-μm circles—that is, if the haploid strain being tested is [cir⁰]—then the chromosome in which CV7 is inserted will be stable, the *LEU2* will be retained in all the cells, and all the colonies will be Leu⁺. If, on the other hand, the strain does contain 2-μm circles, then frequently the chromosome in which CV7 is inserted will be lost. Since this chromosome contains the only functional copy of *LEU2*, cells that lose this chromosome will give rise to Leu⁻ colo-

---

[26] R. W. Davis, M. Thomas, J. R. Cameron, T. P. St. John, S. Scherer, and R. A. Padgett, this series, Vol. 65, p. 404.

nies. In this protocol this chromosome loss occurs in approximately 5% to 25% of the cells.

Vectors for Use with [cir⁰] Strains. Since [cir⁰] strains lack the *REP* gene products, both *REP* genes must be carried by a hybrid 2-μm circle plasmid if that plasmid is to propagate at high copy number in the strain. Owing to the genetic organization of 2-μm circles, such vectors have been most readily constructed by inclusion of the entire 2-μm circle genome. Two vectors appropriate for use in [cir⁰] strains are diagrammed in Fig. 5 and listed in Table I. They contain pBR322 plus *LEU2* sequences inserted in the *Eco*RI site either in the small unique region (pCV20) or in the large unique region (pCV21). For pCV20 this insertion interrupts the 2-μm circle *FLP* gene. However, this insertion does not appear to impair significantly the replication potential of the plasmid. Similarly, although the pBR322-*LEU2* sequences in pCV21 interrupt an open coding region within the 2-μm circle genome, the replication potential of the plasmid is apparently unimpaired. Both of these plasmids are maintained at approximately 40 copies per cell in [cir⁰] strains and are lost from less than 20% of the cells following growth in nonselective media for 10 generations. Plasmids pMJ5 and pMJ6 are equivalent to pCV20 and pCV21, respectively, except that they lack *LEU2* sequences and therefore possess intact *Tet*ᴿ and *Amp*ᴿ genes within the pBR322 moiety. Thus sequences containing other selectable yeast genes can be easily cloned onto these plasmids to construct vectors suitable for use with other [cir⁰] strains.

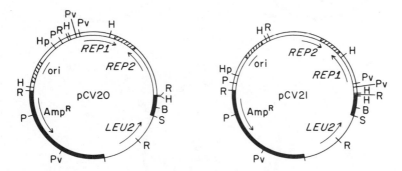

FIG. 5. Hybrid 2-μm circle vectors suitable for use with [cir⁰] yeast strains. pBR322 sequences are shown as filled double lines and 2-μm circle sequences as open double lines with the inverted repeat regions indicated by hatching. The location of chromosomal sequences spanning the *LEU2* gene is indicated by the single line. The directions of transcription of *REP1*, *REP2*, *Bla*, and *LEU2* are shown, as is the approximate location of the 2-μm circle origin of replication (ori). Restriction site abbreviations are the same as those used in Fig. 3.

Other Approaches to Vector Construction

pJDB219 and Derivative Plasmids. The structure of plasmid pJDB219 is diagrammed in Figure 6. It consists of the entire 2-μm circle genome cloned, at the *Eco*RI site in the small unique region, into the *Eco*RI site of the bacterial plasmid pMB9. The plasmid contains a fragment of yeast DNA spanning the *LEU2* gene—isolated from randomly sheared, total genomic DNA—inserted at the *Pst*I site in the 2-μm circle moiety through complementary homopolymer extensions[15] (since the homopolymers in this case are dA and dT, *Pst*I sites on either side of the insert are not reconstructed). This fragment of yeast DNA is 1400 bp or so in length, which is not much larger than the *LEU2* gene itself. Several derivative plasmids constructed from pJDB219 are also diagrammed in Fig. 6. Plasmid pSI4 differs from pJDB219 by the substitution of pBR322 sequences for the pMB9 moiety. Plasmid pJDB207 consists of that region of pJDB219 spanning the *LEU2* gene and the origin of replication cloned

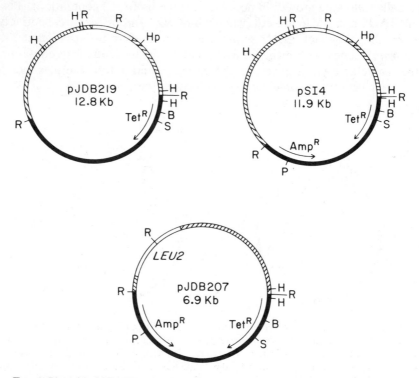

FIG. 6. Plasmid pJDB219 are its derivatives. Sequence source and restriction site abbreviations are the same as those in Fig. 3.

onto the bacterial vector pAT153, a smaller derivative of pBR322.[27] All three of these plasmids can be propagated in *E. coli* and can transform *leu2⁻* yeast strains to Leu⁺ at high frequency. In addition, the latter two plasmids contain intact Amp^R and Tet^R genes, with the *Bam*HI and *Sal*I sites in the Tet^R gene available for insertion of exogenous restriction fragments.

The property of pJDB219 and its derivatives that warrants their separate discussion in this chapter is that they display unusually high copy number in yeast.[27] All three of the plasmids shown in Fig. 6 propagate at an average of 200 copies per cell in yeast (cf. Fig. 7). The reason for this high copy number is unknown. A plasmid essentially identical to pJDB219, except that the *LEU2* insert is somewhat larger, does not display an elevated copy number. At this point the high copy number is best ascribed to an as yet unexplained activation of the 2-μm circle replication origin fortuitously induced by the singular nature of the *LEU2* insertion in the plasmid. As is evident in Fig. 7, the high copy number and stability of pJDB219 derivatives is dependent in part on 2-μm circle *REP* functions. That is, plasmid pSI5, a plasmid essentially equivalent to pJDB207, is present at 200 copies per cell in a [cir⁺] strain, but is present at only 15 copies per cell in a [cir⁰] strain. The *REP* functions can also be provided by the hybrid plasmid itself. This is evident since pJDB219 and pSI4 display high copy number in [cir⁰] as well as [cir⁺] strains.[21]

Vectors Free of Bacterial Sequences. The presence of certain bacterial sequences on hybrid plasmids has been shown to inhibit the replication of these hybrid plasmids in cultured mammalian cells.[28] Whether the presence of extraneous bacterial sequences on 2-μm circle vectors inhibit their replication potential as well has not been established. It is certainly true that, except for pJDB219 and its derivatives, all hybrid vectors— even those containing the entire 2-μm circle genome—are present in the cell at lower copy number than authentic 2-μm circles. Therefore, in attempting to maximize the copy number of a gene, it may be worth considering the use of vectors free of bacterial sequences.

This approach has not been appreciably explored and is presented here primarily as an indication of a potential future direction for the development of yeast expression vectors. In the following, a specific construction is described as an illustration of the way in which this technique can be applied.

A plasmid consisting of pBR322 and the entire 2-μm circle genome, joined through their respective *Pst*I sites (cf. Fig. 1), is digested with an

[27] J. D. Beggs, *Alfred Benzon Symp.* **16**, 383 (1981).
[28] M. Lusky and M. Botchan, *Nature (London)* **293**, 79 (1981).

enzyme that makes a single cut in the plasmid, at a site in the 2-μm circle moiety that is nonessential for replication. The *Bcl*I site, which lies at the 3' end of the *FLP* gene, should be one such site (after isolation of the plasmid from a *dam⁻ E. coli* strain) as may be the *Hpa*I site in the large unique region. Sequences to be used for selection in yeast as well as sequences that one wishes to propagate in yeast are then inserted by ligation into this site and the resultant plasmid is recovered in *E. coli*. After purification of this plasmid from *E. coli,* it is digested with *Pst*I and used to transform an appropriate [cir⁺] yeast strain, or it is digested, religated at low DNA concentration, and used to transform an appropriate [cir⁰] yeast strain. In the former case, the recombinogenic nature of the ends of the 2-μm circle fragment leads to restoration of a covalently closed circle through recombination with endogenous plasmids.[29,30] Thus by this technique one can readily insert and propagate sequences of interest on a 2-μm circle vector that is free of bacterial DNA.

Dobson *et al.* have described a high-copy 2-μm circle hybrid plasmid that should be useful as a yeast vector free of bacterial sequences.[24] The plasmid, designated pYX, arose as a recombinant in yeast between pJDB219 and an endogenous 2-μm circle and consists of the entire 2-μm circle genome containing the pJDB219 *LEU2* insertion in the large unique region. Since this species is the sole extrachromosomal plasmid in their strain MC16 L⁺1 (α *ade2-1 leu2 his4-712 SUF2* [pYX⁺, cir⁰]) and appears to be present at approximately 150–200 copies per cell in this strain (thus constituting ca 5% of the total DNA), it should be readily purifiable in reasonable amounts by isopycnic centrifugation of total genomic DNA in CsCl plus ethidium bromide. Sequences of interest can then be inserted at any one of a number of sites within the purified plasmid (as mentioned above, either the *Bcl*I or the *Hpa*I site should be suitable) and then introduced into yeast by transformation of an appropriate *leu2⁻* [cir⁰] strain.

Determination of Plasmid Copy Number

Since there is presently no definitive vector of choice for maximizing copy number of a sequence of interest, the approach to achieving high-level propagation that I have recommended in this chapter is one of trial and error. That is, several different strategies have been outlined with the anticipation that one or more of these can be pursued for any particular gene of interest. The plasmid that provides the highest copy number is

---

[29] T. L. Orr-Weaver, J. W. Szostak, and R. J. Rothstein, *Proc. Natl. Acad. Sci. U.S.A.* **78,** 6354 (1981).

[30] M. Jayaram, S. Sumida, and J. R. Broach, unpublished results.

then retained. It is obvious, then, that the ability to assess readily the copy numbers of the various constructions must play in integral part of this approach.

Measurements of the cellular copy number of 2-μm circles or of various hybrid yeast vectors have been accomplished through a variety of techniques. All require isolating total DNA from the appropriate yeast strain and then determining the ratio of plasmid-specific DNA to total genomic DNA. This has been done either by direct measurement of circular versus linear DNA in electron micrographs.[2,3] or by determining the amount of plasmid DNA in a given amount of total DNA by various hybridization or reassociation protocols.[22,31,32] None of these techniques includes a suitable control for possible errors arising from differential extraction of chromosomal versus plasmid DNA. In addition, each of these methods possesses one or more specific potential sources of error, which could lead to an inaccurate estimate of the plasmid copy number. Nonetheless, the similarity of the values obtained by each of these procedures for the copy number of 2-μm circles lends greater credence to each procedure than theoretical considerations might warrant.

Although any of the above referenced methods can be used, the easiest method for determining the copy number in yeast of a high copy number plasmid is to compare the intensity of ethidium bromide staining of a plasmid specific restriction fragment with respect to a fragment corresponding to the repeated rDNA sequences in restriction digests of total DNA isolated from an appropriate transformed strain. As can be seen in Fig. 7, both rDNA restriction fragments and plasmid specific restriction fragments stand out sharply against the background of staining of random genomic DNA fragments, after fractionation of the digested DNA by electrophoresis on agarose gels. The relative amount of plasmid DNA versus ribosomal DNA can be quantitiated from densitometer scans of photographic negatives of such a gel, as shown in Fig. 7. Assuming 100 repeats of rDNA per haploid genome[33] and normalizing the staining intensities for the sizes of the fragments (a restriction map of the ribosomal DNA repeat unit is given by Bell et al.[34]), one can determine an absolute value for the copy number of the hybrid plasmid. With lower copy number, pBR322-containing plasmids for which the plasmid specific restriction fragments are not so prominent, the relative copy number can be determined by

[31] D. T. Stinchcomb, C. Mann, E. Selker, and R. W. Davis, *Initiation of DNA Replication*, *ICN–UCLA Symp. Mol. Biol.* **22**, p. 473 (1981).
[32] V. L. Seligy, D. Y. Thomas, and B. L. A. Miki, *Nucleic Acids Res.* **8**, 3371 (1980).
[33] T. J. Zamb and T. D. Petes, *Cell* **28**, 355 (1982).
[34] G. I. Bell, L. J. DeGennaro, D. H. Gelfand, R. J. Bishop, P. Valenzuela, and W. J. Rutter, *J. Biol. Chem.* **252**, 8118 (1977).

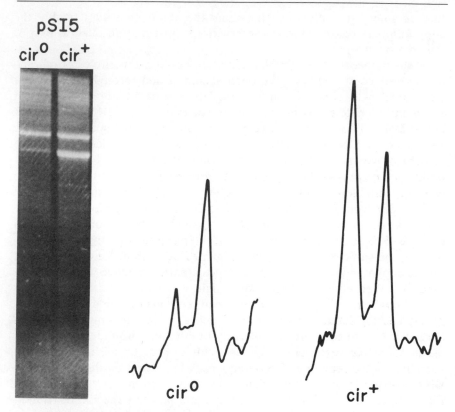

pSI5

cir^0 cir$^+$

cir^0 cir$^+$

FIG. 7. Determination of plasmid copy number. Plasmid pSI5 was transformed into strains DCO4 [cir^0] and DCO4 [cir$^+$], and total DNA was isolated from a representative transformant from each. Plasmid pSI5 consists of the *Hin*dIII B fragment of pJDB219, which spans *LEU2, REP3,* and the 2-μm circle origin of replication, inserted into the *Hin*dIII site of pBR322. Samples (2 μg) of each DNA preparation were digested with *Kpn*I and fractionated by electrophoresis on a 0.8% agarose. The gel was photographed on a UV transilluminator with Polaroid type 665 Pos/Neg film. The positive photograph of the two tracks is shown on the left. The upper band visible against the background is the 9060 bp rDNA repeat fragment. The lower band is the 8080 bp plasmid fragment. Each of these tracks was scanned on the photographic negative using a Joyce-Loebl microdensitometer, the tracings of the relevant portions of which are shown on the right. The areas under the peaks corresponding to plasmid and rDNA bands were determined using a Numonics Corporation digitizer. The copy number of the plasmid in each of the two strains could then be determined using the formula

$$\text{Copy number} = \frac{(\text{area under plasmid peak})(\text{length of rDNA fragment})}{(\text{area under rDNA peak})(\text{length of plasmid fragment})} \times 100$$

The final term is the rDNA repeat value. The data shown in the figure yield values of 15 and 190 for the copy number of pSI5 in the [cir^0] and [cir$^+$] strain, respectively.

Southern analysis, hybridizing the transferred genomic DNA with a pBR322-rDNA probe. As with ethidium staining, an absolute value for the copy number can be determined from densitometer tracings of the autoradiogram, with due consideration given to the relative lengths of the fragments and the extent to which the fragments are covered by the labeled probe. Care should be taken in selecting a plasmid specific fragment upon which to make the measurements. Any 2-μm circle plasmid containing two inverted repeat sequences is capable of undergoing inversion and thus may exist in the yeast in two forms. Thus some of the restriction fragments from such a plasmid may represent only half of the total cellular population of that plasmid. Care should also be taken in assuring oneself that autoradiographic exposures yield a linear response with respect to amount of label hybridized. Using preflashed film usually avoids any nonlinearity of response.

This procedure is an unsophisticated but rapid technique for determining copy number. Besides speed, the principal advantage of the technique is that it incorporates an internal standard to permit ready extraction of absolute values for copy number. The method obviously goes no further in addressing the problem of differential extraction of plasmid versus chromosomal DNA than those previously mentioned. Indeed, this procedure adds the additional complication of potential differential extraction of rDNA, which serves as the normalizing component. Nonetheless, this procedure has provided reproducible values for copy levels of various hybrid plasmids we have examined and has given values of 2-μm circle that are certainly comparable to those obtained by more rigorous methods.

Acknowledgments

I would like to thank M. Jayaram and D. Shortle for critical evaluation of this manuscript.

[22] Improved Vectors for Plant Transformation: Expression Cassette Vectors and New Selectable Markers

By S. G. ROGERS, H. J. KLEE, R. B. HORSCH, and R. T. FRALEY

Introduction

Much of the recent progress in plant molecular biology, particularly in the area of gene function, is due to the availability of facile systems for plant transformation. Improvements in the basic vectors and selectable markers for identification of transformants are appearing at an increasing pace. These are leading to increased efficiency and ease of producing transgenic plants as well as to an extension of the range of plant species which may be transformed. This chapter will review the vectors and methods based on the *Agrobacterium tumefaciens* Ti plasmid and will describe, in detail, the construction and use of several new cassette plasmids for expression of gene-coding sequences in plants.

Agrobacterium-Derived Plant Transformation Systems

By far, the *Agrobacterium tumefaciens* Ti plasmid-derived vectors are the easiest and most utilized of the various schemes for introduction of DNAs into plants. In nature, *A. tumefaciens* infects most dicotyledonous and some monocotyledonous plants[1] by entry through wound sites. The bacteria bind to cells in the wound and are stimulated by phenolic compounds released from these cells[2] to transfer a portion of its endogenous, 200-kilobase (kb) Ti (tumor inducing) plasmid into the plant cell. The transferred portion of the Ti plasmid (T-DNA) becomes covalently integrated into the plant genome where it directs the biosynthesis of phytohormones from enzymes which it encodes. These phytohormones stimulate the undifferentiated growth of the infected and surrounding tissue, leading to the formation of a tumor or gall.

Two regions of the Ti plasmid are essential for the mobilization and integration of the transferred DNA into the plant cell. The *vir* region encodes a group of bacterial genes whose products are required for the excision of the T-DNA from the Ti plasmid, and for its transfer and

---

[1] M. De Cleene and J. De Ley, *Bot. Rev.* **42**, 389 (1976).
[2] S. Stachel, E. Messens, M. Van Montagu, and P. Zambryski, *Nature (London)* **318**, 624 (1985).

RECOMBINANT DNA
METHODOLOGY

integration into the plant genome.[3,4] The expression of these genes has been shown to be induced by compounds released from damaged plant cells at the wound site. The second regions essential for transfer are the T-DNA border sequences; these are almost perfect, direct repeats of 25 bp that flank the T-DNA. The border sequences apparently function as sites both for nuclease cleavage[5] and for integration of the T-DNA.

The Ti plasmid-based vectors all adapt these two essential features for the introduction of new DNAs into plant cells. Since no portion of the T-DNA other than the borders is required for transfer of the DNA into plant cells, vectors could be developed that retain the border sequences but selectable markers that function in plants replace the normal complement of T-DNA genes. The T-DNA functions that are removed encode phytohormone biosynthetic genes that specify the synthesis of cytokinin and auxin, which prevent the morphogenesis of transformed tissue and regeneration of plants. The altered Ti plasmids that have had the phytohormone genes deleted or replaced are referred to as "disarmed." An example is the disarmed pTiB6S3-SE[6] plasmid which is an octopine-type Ti plasmid in which most of the T_L T-DNA, including the phytohormone genes and octopine synthase gene and entire T_R DNA, have been replaced with a bacterial kanamycin resistance marker. Only the left border and approximately 2 kb of the T_L DNA remains. The T-DNA segment that remains is referred to as the left inside homology (LIH) and is essential for formation of the cointegrate T-DNAs. Another approach to disarm the Ti plasmid was taken by Zambryski and co-workers.[7] This group created pGV3850 in which the entire T-DNA, save the border sequences, was replaced with pBR322. In this case, the pBR322 sequences provide homology for cointegrate formation as described below.

Cointegrating Intermediate Vectors

The great size of the disarmed Ti plasmid and lack of unique restriction endonuclease sites prohibit direct cloning into the T-DNA. Instead, intermediate vectors such as pMON200 (Fig. 1) were developed for introducing genes into the Ti plasmid. This method was adapted from one published by Ruvkun and Ausubel[8] for mutagenesis of the large plasmids

[3] G. Ooms, P. Klapwijk, J. Poulis, and R. Schilperoort, *J. Bacteriol.* **144**, 82 (1980).

[4] H. Klee, F. White, V. Iyer, M. Gordon, and E. Nester, *J. Bacteriol.* **153**, 878 (1983).

[5] Z. Koukolikova-Nicola, R. Shillito, B. Hohn, K. Wang, M. Van Montagu, and P. Zambryski, *Nature (London)* **313**, 191 (1985).

[6] R. Fraley, S. Rogers, D. Eichholtz, J. Flick, C. Fink, N. Hoffmann, and P. Sanders, *Bio/Technology* **3**, 629 (1985).

[7] P. Zambryski, H. Joos, C. Genetello, J. Leemans, M. Van Montagu, and J. Schell, *EMBO J.* **2**, 2143 (1983).

[8] G. Ruvkun and F. Ausubel, *Nature (London)* **289**, 85 (1981).

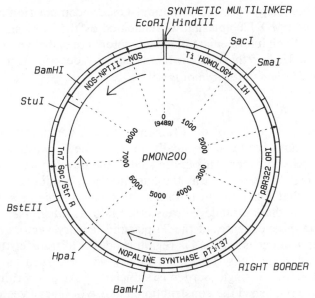

FIG. 1. Map of pMON200: A cointegrating intermediate transformation vector. The pMON200 plasmid has been described in Fraley et al.[6] and in the Appendix to this chapter. The plasmid is made up of a 1.6-kb LIH (left inside homology) segment derived from the octopine-type pTiA6 plasmid which provides a region of homology for recombination with a resident Ti plasmid in *A. tumefaciens,* a 1.6-kb segment carrying the pBR322 origin of replication, a 2.4-kb segment of the nopaline-type pTiT37 plasmid that carries the right border of the nopaline T-DNA and intact nopaline synthase (NOS) gene, a 2.2-kb segment of Tn7 carrying the spectinomycin/streptomycin resistance determinant, a 1.6-kb segment encoding a chimeric NOS-NPTII'-NOS gene that provides selectable kanamycin resistance in transformed plant cells, and a synthetic multilinker containing unique restriction sites for insertion of other DNA segments. This and the following maps were generated using the UWGCG program PLASMID [J. Devereux, P. Haeberli, and O. Smithies, *Nucleic Acids Res.* **12,** 387 (1984)].

of *Rhizobium.* The intermediate vectors are smaller plasmids, usually derivatives of pBR322, that are easily manipulated in *Escherichia coli* and then transferred to *A. tumefaciens* using standard bacterial genetic techniques. Intermediate plasmids that are derivatives of pBR322 cannot replicate in *A. tumefaciens,* therefore these vectors carry DNA segments homologous to the disarmed T-DNA to permit recombination to form a cointegrate T-DNA structure.[9,10]

The pMON200 vector carries a 1.6-kb fragment of the disarmed T-DNA called the left inside homology (LIH) fragment which provides a

[9] L. Comai, C. Schilling-Cordaro, A. Mergia, and C. Houck, *Plasmid* **10,** 21 (1983).
[10] L. Van Haute, H. Joos, M. Maes, G. Warren, M. Van Montagu, and J. Schell, *EMBO J.* **2,** 411 (1983).

region of homology for recombination. Once recombination occurs, the cointegrated pMON200 plasmid is replicated by the Ti plasmid origin of replication. Such intermediate plasmid vectors are called cointegrating or cis vectors. Use of this system has been fully described in a previous volume of *Methods in Enzymology*.[11]

Binary Vectors

Hoekema *et al.*[12] and de Framond *et al.*[13] demonstrated that the T-DNA did not have to be physically joined to the Ti plasmid for transfer into plant cells to occur. This finding led to the development of binary or trans vectors. In these systems the intermediate plasmid has an origin of replication that functions in both *E. coli* and *A. tumefaciens*. When a binary intermediate plasmid is transferred into *A. tumefaciens,* it is able to replicate independently of the Ti plasmid. Binary vectors must carry the border sequences for the *vir* region products to transfer the DNA of interest into plant cells. For small binary vectors, such as pMON505 described below, a single border is sufficient.[14] At this writing, several groups have published the construction and use of binary vector systems for plant transformation.[15–18]

The binary vectors offer the advantage of higher frequency transfer from *E. coli* to *A. tumefaciens* since cointegration is no longer required for their maintenance. The frequency of transfer for the binary vectors is routinely 10% since this percentage of the exconjugant *A. tumefaciens* contains the binary vector. This can be as high as 100% and may be four to six orders of magnitude higher than the rate of cointegrate formation using the pMON200 plasmid.

Design of Expression Cassette Vectors

The expression cassettes have been constructed with the promoter region, transcription initiation site, and a portion of the 5′ nontranslated leader of the promoter of interest joined to a synthetic multilinker and followed by a polyadenylation signal carried in the 3′ nontranslated portion of a plant gene. This design permits the efficient initiation of tran-

[11] S. Rogers, R. Horsch, and R. Fraley, this series, Vol. 118, p. 627.
[12] A. Hoekema, P. Hirsch, P. Hooykaas, and R. Schilperoort, *Nature (London)* **303,** 179 (1983).
[13] A. de Framond, K. Barton, and M.-D. Chilton, *Bio/Technology* **1,** 262 (1983).
[14] R. Horsch and H. Klee, *Proc. Natl. Acad. Sci. U.S.A.* **83,** 4428 (1986).
[15] M. Bevan, *Nucleic Acids Res.* **12,** 8711 (1984).
[16] G. An, B. Watson, S. Stachel, M. Gordon, and E. Nester, *EMBO J.* **4,** 277 (1984).
[17] H. Klee, M. Yanofsky, and E. Nester, *Bio/Technology* **3,** 637 (1985).
[18] P. van den Elzen, K. Lee, J. Townsend, and J. Bedbrook, *Plant Mol. Biol.* **5,** 149 (1985).

scription at the promoter's natural mRNA start site and maintains any features of the 5' end of the messenger RNA required for efficient ribosome interactions. No additional, upstream AUG translational initiator signals are present in the 5' leader sequence to ensure that translation will begin on the AUG of the inserted coding sequence and therefore to maximize translation of the coding sequence.[19,20]

The promoter–leader segment is followed by a synthetic multilinker with unique restriction endonuclease cleavage sites for the insertion of coding sequences derived from prokaryotic or eukaryotic genes, including genomic sequences with introns or cDNAs. Although cDNA clones often contain the signal and site of polyadenylation, including a portion of the poly(A) tail, it is not clear if these signals or transcribed runs of poly(A) are efficient in providing correct message stability and transport from the nucleus.[21] Therefore, an additional, functional polyadenylation signal from the Ti plasmid T-DNA nopaline synthase gene is present downstream of the multilinker.

Two different expression cassettes have been developed using the two cauliflower mosaic virus promoters: the 19 S (P66 or gene VI) promoter and the 35 S (full-length transcript) promoter. Both the 19 and 35 S promoters are constitutively expressed in transformed plants.[22,23] The relative strengths of these promoters have been determined in our laboratory by comparison to the nopaline synthase promoter. The 19 S promoter is at most 5-fold stronger than the nopaline synthase promoter. In contrast, the 35 S promoter is approximately 50 times stronger than the nopaline synthase promoter and is the strongest constitutive promoter identified at this time for expression of foreign coding sequences in transformed plants.[24] It has the additional advantage of functioning in both dicots and monocots.[25,26] The availability of both of these cassettes allows construction of chimeric genes that differ in expression levels over a 10-fold range. Details of the construction of the cassettes and the vectors in which they are carried follow.

[19] M. Kozak, *Cell* **15**, 1109 (1978).
[20] S. Rogers, R. Fraley, R. Horsch, A. Levine, J. Flick, L. Brand, C. Fink, T. Mozer, K. O'Connell and P. Sanders, *Plant Mol. Biol. Rep.* **3**, 111 (1985).
[21] H. Okayama and P. Berg, *Mol. Cell. Biol.* **2**, 161 (1982).
[22] J. Paszkowski, R. Shillito, M. Saul, V. Mandak, T. Hohn, B. Hohn, and I. Potrykus, *EMBO J.* **3**, 2717 (1984).
[23] J. Odell, F. Nagy, and N.-H. Chua, *Nature (London)* **313**, 810 (1985).
[24] S. Rogers, K. O'Connell, R. Horsch, and R. Fraley, *in* "Biotechnology in Plant Science" (M. Zaitlin, P. Day, and A. Hollander, eds.), p. 219. Academic Press, Orlando, Florida, 1986.
[25] M. Fromm, L. Taylor, and V. Walbot, *Proc. Natl. Acad. Sci. U.S.A.* **82**, 5824 (1985).
[26] M. Fromm, L. Taylor, and V. Walbot, *Nature (London)* **319**, 791 (1986).

pMON237: The 19 S-NOS Cassette Vector

The CaMV 19 S promoter fragment was isolated from plasmid pOS-1, a derivative of pBR322 carrying the entire genome of CM4-184 as a SalI insert.[27] This plasmid was kindly provided by R. J. Shepherd. The CM4-184 strain is a naturally occurring deletion mutant of strain CM1841. The nucleotide sequences of the CM1841[28] and Cabb-S[29] strains of CaMV have been published as well as some partial sequences for a different CM4-184 clone.[30] The references to nucleotide numbers (n) in the following discussion are those for the sequence of CM1841.[28] The nucleotide sequences of both 19 and 35 S promoter regions of these three isolates are essentially identical. A 476-bp fragment extending from the HindIII site at bp 5372 to the HindIII site at bp 5848 was cloned into M13 mp8 for site-directed mutagenesis[31] to introduce an XbaI site immediately 5' of the first ATG translational initiation signal in the 19 S transcript.[30] The resulting 400-bp HindIII–XbaI fragment was isolated and joined to a synthetic linker containing BglII and BamHI sites and then cloned adjacent to the nopaline synthase 3' nontranslated region. This cassette was inserted into pMON200 (Fraley et al.[6]) between the EcoRI and HindIII sites to give pMON237 (Fig. 2). The complete sequence of the 19 S promoter-NOS cassette is given in Fig. 3.

Plasmid pMON237 is a cointegrating type intermediate vector with unique XbaI and BglII sites for the insertion of coding sequences carrying their own translational initiation signals immediately adjacent to the 19 S transcript leader sequence. The pMON237 plasmid retains all of the properties of pMON200, including spectinomycin resistance for selection in E. coli and A. tumefaciens, as well as a chimeric kanamycin gene (NOS-NPTII'-NOS) for selection of transformed plant tissue and the nopaline synthase gene for ready scoring of transformants and inheritance in progeny. The pMON237 plasmid is used exactly as is pMON200.[6,11]

pMON316 and pMON530: The 35 S-NOS Vectors

The 35 S promoter was isolated from the pOS-1 clone of CM4-184 as an AluI (n = 7143)–EcoRI* (n = 7517) fragment which was inserted first into pBR322 cleaved with BamHI, treated with the Klenow fragment of

[27] A. Howarth, R. Gardner, J. Messing, and R. Shepherd, Virology 112, 678 (1981).
[28] R. Gardner, A. Howarth, P. Hahn, M. Brown-Luedi, R. Shepherd, and J. Messing, Nucleic Acids Res. 9, 2871 (1981).
[29] A. Franck, H. Guilley, G. Jonard, K. Richards, and L. Hirth, Cell 21, 285 (1980).
[30] R. Dudley, J. Odell, and S. Howell, Virology 117, 19 (1982).
[31] M. Zoller and M. Smith, Nucleic Acids Res. 10, 6487 (1982).

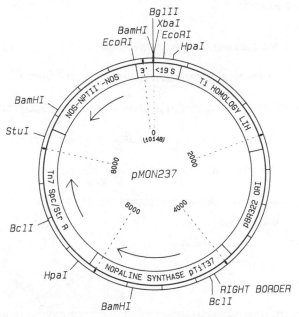

FIG. 2. Map of pMON237: A cointegrating 19 S-NOS cassette vector. The construction of this expression vector is described in the text. <19 S, The CaMV 19 S promoter segment; 3′, the NOS 3′ nontranslated segment. The additional segments are described in the legend to Fig. 1. The complete sequence of the cassette appears in Fig. 3.

DNA polymerase I, and then cleaved with *Eco*RI. The promoter fragment was then excised from pBR322 with *Bam*HI and *Eco*RI, treated with Klenow polymerase, and inserted into the *Sma*I site of M13 mp8 so that the *Eco*RI site of the mp8 multilinker was at the 5′ end of the promoter fragment. The nucleotide numbers refer to the sequence of CM1841.[28] Site-directed mutagenesis was then used to introduce a G at nucleotide 7464 to create a *Bgl*II site. The 35 S promoter fragment was then excised from the M13 as a 330-bp *Eco*RI–*Bgl*II fragment which contains the 35 S promoter, transcription initiation site, and 30 nucleotides of the 5′ non-translated leader but contains none of the CaMV translational initiators nor the 35 S transcript polyadenylation signal that is located 180 nucleotides downstream from the start of transcription.[32,33] The 35 S promoter fragment was joined to a synthetic multilinker and the NOS 3′ nontranslated region and inserted into pMON200 to give pMON316 (Fig. 4). Plasmid 316 contains unique cleavage sites for *Bgl*II, *Cla*I, *Kpn*I, *Xho*I, and *Eco*RI located between the 5′ leader and the NOS polyadenylation signals. The

[32] S. Covey, G. Lomonosoff, and R. Hull, *Nucleic Acids Res.* **9,** 6735 (1981).
[33] H. Guilley, R. Dudley, G. Jonard, E. Balaz, and K. Richards, *Cell* **30,** 763 (1982).

```
CaMV 19 S PROMOTER

HindIII
1|        .          .          .          .          .          70
AAGCTTTAAAGCTGCAGAAAGGAATTACCACAGCAATGACAAAGAGACATTGGCGGTAATAAATACTATA

71       .          .          .          .          .          140
AAGAAATTCAGTATTTATCTAACTCCTGTTCATTTTCTGATTAGGACAGATAATACTCATTTCAAGAGTT

141      .          .          .          .          .          210
TTGTTAACCTTAATTACAAAGGAGATTCAAAACTTGGAAGAAACATCAGATGGCAAGCATGGCTTAGCCA

211      .          .          .          .          .          280
CTATTCGTTTGATGTTGAACATATTAAAGGAACCGACAACCACTTTGCGGACTTCCTTTCAAGAGAATTC

281      .          .          .          .          .          350
AATAAGGTTAATTCCTAATTGAAATCCGAAGATAAGATTCCCACACACTTGTGGCTGATATCAAAAAGGC
         TATA                            5' mRNA
351      |          .          .        .|||       402
TACTACCTATATAAACACATCTCTGGAGACTGAGAAAATCAGACCTCCAAGC

SYNTHETIC MULTI-LINKER

    XbaI          BglII BamHI
    |             |     |
TCTAGACTCCTTACAACAGATCTGGATCCCC

NOS 3'

  434      .          .          .          .          .          500
  GATCGTTCAAACATTTGGCAATAAAGTTTCTTAAGATTGAATCCTGTTGCCGGTCTTGCGATGATTA

  501      .          .          .          .          .          570
  TCATATAATTTCTGTTGAATTACGTTAAGCATGTAATAATTAACATGTAATGCATGACGTTATTTATGAG

          3' End mRNA
  571     . |  | .||||||   .          .          .          640
  ATGGGTTTTTATGATTAGAGTCCCGCAATTATACATTTAATACGCGATAGAAAACAAAATATAGCGCGCA

                                                    EcoRI
  641     .          .          .          .        . | 698
  AACTAGGATAAATTATCGCGCGCGGTGTCATCTATGTTACTAGATCGATCgggaattc
```

FIG. 3. Nucleotide sequence of the CaMV 19 S promoter–NOS 3' cassette. The 5' end of the natural 19 S mRNA is from Dudley *et al.*[30] The sequence of the NOS 3' segment and the 3' end of the NOS mRNA were determined by Bevan *et al.* [M. Bevan, W. M. Barnes, and M.-D. Chilton, *Nucleic Acids Res.* **11,** 370 (1983)].

316 plasmid retains all of the properties of pMON200. The complete sequence of the 35 S promoter, multilinker, and NOS 3' segment is given in Fig. 5. This sequence begins with the *Xmn*I site created by Klenow polymerase treatment to remove the *Eco*RI site located at the 5' end of the 35 S promoter segment.

The 35 S-NOS cassette was also inserted into our binary vector, pMON505. Plasmid pMON505 is a derivative of pMON200 in which the Ti plasmid homology region, LIH, has been replaced by a 3.8-kb *Hind*III to *Sma*I segment of the mini RK2 plasmid, pTJS75.[34] This segment con-

[34] T. Schmidhauser and D. Helinski, *J. Bacteriol.* **164,** 446 (1985).

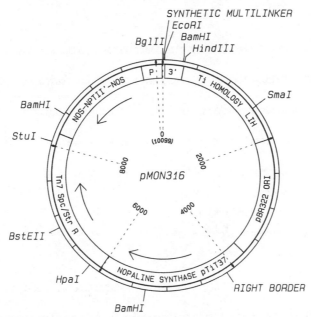

FIG. 4. Map of pMON316: A cointegrating 35 S-NOS cassette vector. The construction of this expression vector is described in the text. P, The CaMV 35 S promoter segment; 3', the NOS 3' nontranslated sequences. The additional segments are described in the legend to Fig. 1. The complete sequence of the cassette appears in Fig. 5.

tains the RK2 origin of replication, *ori*V, and the origin of transfer, *ori*T, for conjugation into *Agrobacterium* using the triparental mating procedure. The detailed construction of pMON505 has been published elsewhere.[14]

Plasmid pMON505 (Fig. 6) retains the important features of pMON200 including the synthetic multilinker for insertion of desired DNA fragments, the chimeric NOS-NPTII'-NOS gene for kanamycin resistance in plant cells, the spectinomycin/streptomycin resistance determinant for selection in *E. coli* and *A. tumefaciens,* an intact nopaline synthase gene for facile scoring of transformants and inheritance in progeny, and a pBR322 origin of replication for ease in making large amounts of the vector in *E. coli.* The 505 plasmid contains a single T-DNA border derived from the right end of the pTiT37 nopaline-type T-DNA. Experiments in our laboratory[14] have shown that this single border sequence is both necessary and sufficient for high-frequency transfer of the pMON505 into plant cells and its stable integration into the plant genome. Southern analyses have shown that the 505 plasmid and any DNA that it carries are integrated into the plant genome. The entire plasmid acts as a T-DNA and is inserted into the plant genome. One end of the integrated DNA is

```
CaMV 35 S PROMOTER

    Filled EcoRI
1   |          .          .          .          .          .          70
GAATTAATTCCCGATCcTATCTGTCACTTCATCAAAAGGACAGTAGAAAAGGAAGGTGGCACTACAAATG

71        .          .          .          .          .          .          140
CCATCATTGCGATAAAGGAAAGGCTATCGTTCAAGATGCCTCTGCCGACAGTGGTCCCAAAGATGGACCC

141       .          .          .          .          .          .          210
CCACCCACGAGGAGCATCGTGGAAAAAGAAGACGTTCCAACCACGTCTTCAAAGCAAGTGGATTGATGTG
                                                                   TATA
211       .          .          .          .          .          |  280
ATATCTCCACTGACGTAAGGGATGACGCACAATCCCACTATCCTTCGCAAGACCCTTCCTCTATATAAGG
                         5' mRNA
281       .          |          .          .          332
AAGTTCATTTCATTTGGAGAGGACACGCTGAAATCACCAGTCTCTCTCTACA

SYNTHETIC MULTI-LINKER

  BglII   ClaI SmaI KpnI SalI EcoRI
  |       |    |    |    |    |
AGATCTATCGATTCCCGGGTACCTCGAGAATTCCC

NOS 3'

        368       .          .          .          .          430
        GATCGTTCAAACATTTGGCAATAAAGTTTCTTAAGATTGAATCCTGTTGCCGGTCTTGCGATG

431       .          .          .          .          .          500
ATTATCATATAATTTCTGTTGAATTACGTTAAGCATGTAATAATTAACATGTAATGCATGACGTTATTTA
                    3' End mRNA
501       .        | |   ||||||          .          .          570
TGAGATGGGTTTTTATGATTAGAGTCCCGCAATTATACATTTAATACGCGATAGAAAACAAAATATAGCG

571       .          .          .          .          .          640
CGCAAACTAGGATAAATTATCGCGCGCGGTGTCATCTATGTTACTAGATCgggatccgtcgacctgcag
    HindIII
641| 648
ccaagctt
```

FIG. 5. Nucleotide sequence of the CaMV 35 S promoter-NOS 3' cassette. The 5' end of the 35 S mRNA is from Odell et al.[23] The sequence of the NOS 3' segment and the 3' end of the NOS mRNA were determined by Bevan et al. [M. Bevan, W. M. Barnes, and M.-D. Chilton, *Nucleic Acids Res.* **11,** 370 (1983)].

located between the right border sequence and the nopaline synthase gene and the other end is between the border sequence and the pBR322 sequences.

A pMON505 derivative carrying the 35 S-NOS cassette was created by transferring the 2.3-kb *StuI–HindIII* fragment of pMON316 into pMON526. Plasmid pMON526 is a simple derivative of pMON505 in which the *SmaI* site was removed by digestion with *XmaI*, treatment with Klenow polymerase, and ligation. The resultant plasmid, pMON530 (Fig. 7), retains the properties of pMON505 and the 35 S-NOS expression cassette now contains a unique cleavage site for *SmaI* between the promoter and polyadenylation signals. Tables listing the major restriction

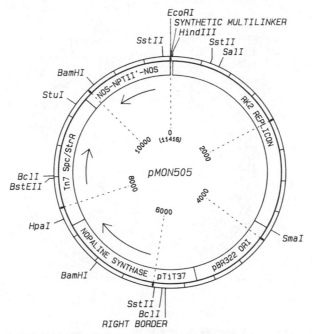

FIG. 6. Map of pMON505: A binary transformation vector. Details of the construction of pMON505 have been described in Horsch and Klee.[14] The 3.8-kb RK2 replicon fragment permits independent replication in *A. tumefaciens*. The additional segments are described in the legend to Fig. 1. More complete information on pMON505 appears in the Appendix.

endonuclease cleavage sites useful for cloning and analysis of clones and transformants are included in the Appendix to this chapter. The nucleotide sequences of most of the segments that comprise the pMON200 and pMON505 vectors have been determined and published by other workers. As an aid to mapping inserts and analysis of transformants, the appropriate references and directions for the assembly of these sequences to give pMON200 and pMON505 are provided in the Appendix.

Development of New Selectable Markers

Chimeric genes for the expression of the Tn5 neomycin phosphotransferase II (NPTII) coding sequence have become the standard markers for plant transformation vectors. The kanamycin resistance provides for direct selection of transformants in a wide variety of plant species, including tobacco, petunia, tomato, lettuce, canola, wheat, and maize.

There are species such as *Arabidopsis thaliana* and certain legumes where the selectability of kanamycin resistance is marginal. New dominant selectable markers are required for these species. Additional select-

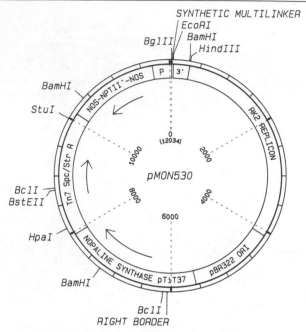

FIG. 7. Map of pMON530: A binary 35 S-NOS cassette vector. The construction of this expression vector is described in the text. P, The CaMV 35 S promoter segment; 3', the NOS 3' nontranslated sequences. The additional segments are described in the legend to Fig. 1. The complete sequence of the cassette appears in Fig. 5.

able markers would also permit retransformation of previously transformed tissues or plants to introduce additional copies of the same gene of interest, or several different genes for enzymes of a biosynthetic pathway, or several different members of the same gene family for the study of their relative expression levels. For genetic studies, additional selectable markers provide the means to follow the segregation of several introduced genes in the progeny of transformants. All of these reasons led us to develop other dominant selectable markers for selection of transformed plants.

Plant tissues are sensitive to many antibiotics that are toxic to bacteria and antimetabolites that are toxic to mammalian cells. This property provides a wide range of potential selective agents for the development of new markers for plants. In addition, many bacterial genes that encode enzymes for the detoxification of antibiotics and eukaryotic genes for target proteins that are insensitive to antimetabolites have been identified. The expression cassette vectors described above and current methods of DNA modification make the construction of chimeric genes and their testing as selectable markers a straightforward process. Specific use of these vectors and methods for the development of two marker genes is described below.

pMON410: pMON530 Carrying a Coding Sequence for Hygromycin Resistance

Hygromycin B is an aminocyclitol antibiotic that inhibits protein synthesis in prokaryotic and eukaryotic cells, including those of plants.[35] Two groups[36,37] have isolated and described a gene from a bacterial R-factor that encodes a hygromycin phosphotransferase (HPH)[38] which inactivates the antibiotic. Chimeric genes that act as dominant selectable markers for yeast[36,39] and mammalian cells[40] have been constructed. These results led us to construct and test a HPH marker for transformation of plants.

Gritz and Davies[36] created a set of fragments carrying the HPH gene by S1 digestion and addition of BamHI linkers. J. Davies provided us with two of their plasmids, pLG89 and 83, which carry synthetic BamHI linkers near the 5' and 3' ends of the HPH coding sequence, respectively. A full-length HPH coding sequence was assembled by joining the 0.26-kb BamHI–EcoRI fragment from pLG89 to a 0.9-kb EcoRI–BamHI fragment from pLG83 and by inserting the 1.2-kb fragment into the BamHI site of M13 mp8. A spurious ATG translational initiator signal, located 5 bp 5' to the HPH initiator, was removed by site-directed mutagenesis. The sequence of the reassembled, altered HPH coding region is shown in Fig. 8. The resulting 1.2-kb BamHI fragment was then transferred to the BglII site of pMON316 to create pMON408, and into the BglII site of pMON530 to create pMON410 (Fig. 9).

Two other groups have assembled chimeric HPH genes that provide hygromycin resistance in plant tissues using either the octopine promoter[41] or the nopaline synthase promoter.[42] Neither of these groups has removed the spurious ATG initiator codon found in the bacterial leader to increase translational efficiency. Deletion of the extra ATG can be expected to increase the levels of HPH protein by 5-fold based on a similar alteration of the neomycin phosphotransferase II leader to remove a spurious ATG codon.[20] This prediction has been confirmed in our laboratory

[35] A. Gonzalez, A. Jimenez, D. Vasquez, J. Davies, and D. Schindler, Biochim. Biophys. Acta 521, 459 (1978).

[36] L. Gritz and J. Davies, Gene 25, 179 (1983).

[37] K. Kaster, S. Burgett, R. Rao, and T. Ingolia, Nucleic Acids Res. 11, 6895 (1983).

[38] R. Rao, N. Allen, J. Hobbs, W. Alborn, H. Kirst, and J. Paschal, Antimicrob. Agents Chemother. 24, 689 (1983).

[39] K. Kaster, S. Burgett, and T. Ingolia, Curr. Genet. 8, 353 (1984).

[40] R. Santerre, N. Allen, J. Hobbs, R. Rao, and R. Schmidt, Gene 30, 147 (1984).

[41] C. Waldron, E. Murphy, J. Roberts, G. Gustafson, S. Armour, and S. Malcolm, Plant Mol. Biol. 5, 103 (1985).

[42] P. van den Elzen, J. Townsend, K. Lee, and J. Bedbrook, Plant Mol. Biol. 5, 299 (1985).

```
HPH SEQUENCE
      10          .          30          .          50          .
ggatccCGGGGGGCAATGAGATATGAAAAAGCCTGAACTCACCGCGACGTCTGTCGAG..
                  MetLysLysProGluLeuThrAlaThrSerValGlu..

HPH' SEQUENCE
      10          .          30          .          50          .
ggatccCGGGGGGCAA___GATATGAAAAAGCCTGAACTCACCGCGACGTCTGTCGAG..
                  MetLysLysProGluLeuThrAlaThrSerValGlu..
```

```
        1220          .          1240          .          1260
....GATTCAGGCCCTTCTGGATTGTGTTGGTCCCCAGGGCACGATTGTCggatcc
```

Fig. 8. Partial sequence of the wild-type (HPH) and modified (HPH') hygromycin phosphotransferase coding sequences. The underlined extra ATG signal has been deleted in the HPH' sequence. The nucleotides in lower case are the *Bam*HI linkers added to the coding sequence clone. The nucleotide numbers are from Gritz and Davies.[36]

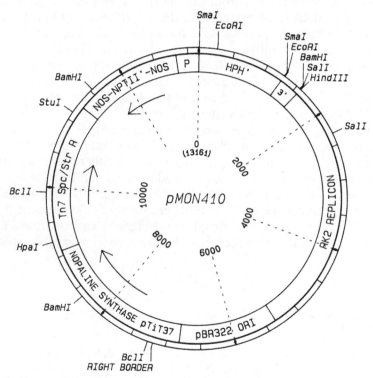

Fig. 9. Map of pMON410: A binary vector carrying the chimeric 35 S-HPH'-NOS selectable marker gene. The insertion of the HPH' coding sequence into pMON530 is described in the text.

```
dhfr' SEQUENCE
        10                    30                    50
         .                     .                     .
ggatccccTGCCATCATGGTTCGACCATTGAACTGCATCGTCGCCGTGTCCCAAAATATG
        MetValArgProLeuAsnCysIleValAlaValSerGlnAsnMet

        70                    90                    110
         .                     .                     .
GGGATTGGCAAGAACGGAGACagaCCCTGGCCTCCGCTCAGGAACGAGTTCAAGTAC...
GlyIleGlyLysAsnGlyAspARGProTrpProProLeuArgAsnGluPheLysTyr...


dhfr SEQUENCE
        70                    90                    110
         .                     .                     .
GGGATTGGCAAGAACGGAGACCTACCCTGGCCTCCGCTCAGGAACGAGTTCAAGTAC...
GlyIleGlyLysAsnGlyAspLEUProTrpProProLeuArgAsnGluPheLysTyr...
```

FIG. 10. Partial sequence of the modified and wild-type mouse dihydrofolate reductase (dhfr') coding regions.[44] The positions of the Leu to Arg substitution are underscored.

by comparison of the transformation efficiency and growth characteristics of transformed plant tissue carrying either the single or two ATG chimeric genes.

pMON321: pMON237 Carrying a Coding Sequence for Methotrexate Resistance

Methotrexate (mtx) is an antimetabolite that inhibits eukaryotic dihydrofolate reductase (dhfr), ultimately preventing biosynthesis of glycine, thymine, and purines. Plant cells and tissues are quite sensitive to mtx. Certain mammalian cell lines selected for resistance to mtx show amplification of the *dhfr* gene[43] while others produce an altered *dhfr* enzyme that does not bind mtx. Simonsen and Levinson[44] cloned the cDNA for an altered mouse *dhfr* from cell line 3T6-R400. The altered *dhfr* produced by this cell line shows a 270-fold decrease in mtx binding compared to the wild-type enzyme. These authors demonstrated that the resistance was due to a single base change that caused an arginine to be substituted for a leucine at position 22 in the protein. We tested the ability of this altered *dhfr* coding sequence to confer mtx resistance on plant cells and tissues in the following manner.

The wild-type mouse *dhfr* coding sequence was obtained on a 1.1-kb *Fnu*4HI fragment from pDHFR-11, a cDNA clone in pBR322[45] provided by R. Schimke. The *Fnu*4HI fragment was treated with Klenow poly-

[43] F. Alt, R. Kellems, J. Bertino, and R. Schimke, *J. Biol. Chem.* **253**, 1357 (1976).
[44] C. Simonsen and A. Levinson, *Proc. Natl. Acad. Sci. U.S.A.* **80**, 2495 (1983).
[45] J. Nunberg, R. Kaufman, A. Chang, S. Cohen, and R. Schimke, *Cell* **19**, 355 (1980).

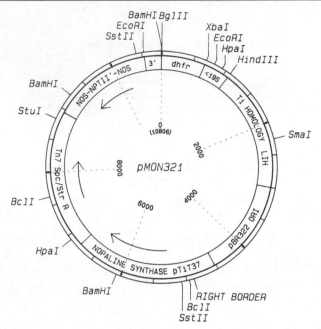

FIG. 11. Map of pMON321: A cointegrating vector carrying the chimeric 19 S-dhfr'-NOS selectable marker gene. The insertion of the *dhfr'* coding sequence into pMON237 is described in the text.

merase and inserted into the *Sma*I site of M13 mp8 for site-directed mutagenesis. Mutagenesis was performed to replace the wild-type 5'-CTA sequence with 5'-AGA resulting in an arginine substitution for leucine at amino acid 22. A partial sequence of the *Bam*HI–*Bgl*II fragment showing the alteration that results in an mtx-resistant *dhfr* enzyme is given in Fig. 10. A 660-bp *Bam*HI to *Bgl*II fragment was then isolated from the mp8 and inserted into the *Bgl*II site of pMON237 to give pMON321 (Fig. 11). A plasmid in which the modified *dhfr* coding sequence is expressed from the CaMV 35 S promoter-NOS 3' cassette was also made. In this plasmid, pMON806, the NOS-NPTII'-NOS selectable marker gene of pMON505, was replaced with the methotrexate-resistant *dhfr* gene. Plasmid pMON806 will be described in a subsequent publication.[46]

Introduction of the pMON Plasmids into *A. tumefaciens*

These pMON plasmids were transferred into *A. tumefaciens* using a simplified triparental mating procedure. Aliquots (0.1 ml) of fresh over-

[46] H. Klee, manuscript in preparation.

night cultures of *E. coli* cells carrying the pMON plasmids were spotted with 0.1 ml of *E. coli* cells carrying the pRK2013 helper plasmid[47] and 0.1 ml of *Agrobacterium tumefaciens* strain 3111 carrying the disarmed octopine-type pTiB6S3-SE plasmid[6] or strain A208 carrying the disarmed nopaline-type pTiT37-SE plasmid[48] on a fresh Luria agar plate and the cells are mixed with a sterile loop. As a control the pRK2013 cells are spotted on a plate with the 3111-SE or A208-SE cells and mixed. After overnight incubation at 28°, aliquots of these mating mixes are transferred to selection plates. For the binary vectors (pMON505 and derivatives) this is simply done by streaking a loopful of the cells on selective plates for single colonies. For the cointegrating vectors (pMON200 and derivatives), the entire mating mix is resuspended in 3 ml of 10 mM MgSO$_4$ with a glass rod, transferred to a culture tube, washed with an additional 3 ml of 10 mM MgSO$_4$, and then a 0.2-ml aliquot is spread on a selection plate. Selection plates contain 25 μg/ml chloramphenicol, 50 μg/ml kanamycin, and 100 μg/ml spectinomycin. After 2–3 days at 28° single colonies are visible. These are picked into LB containing the same antibiotics at the same concentration as used in the selection plates and grown at 28°. We use the 3111-SE or A208-SE *Agrobacterium* strains for both the cointegrating and binary vectors. We have found that these disarmed strains are more efficient for transfer of our binary vectors in petunia and tobacco transformation than is the commonly used LBA4404 disarmed, virulence plasmid strain.[12]

Use of the Hygromycin and Methotrexate Markers for Transformation

A. *tumefaciens* 3111-SE or A208-SE cells containing a pMON vector can efficiently transform plant cells using the leaf disk technique.[11,49] Petunia or tobacco leaf disks are cultured for 2 days on medium containing 4.3 g/liter MS salts (Gibco, Grand Island, N.Y.), 30 g/liter sucrose, B$_5$ vitamins, 1 mg/liter BA, 0.1 mg/liter NAA, and 0.8% agar before infection. Following brief immersion in a liquid culture of the bacteria, the leaf disks are blotted dry and cocultured for 2 days on the same medium before being transferred to medium supplemented with 500 μg/ml carbenicillin to kill residual bacteria, and the appropriate selective agent to inhibit growth of nontransformed plant cells.

The efficiency of selection for antibiotic resistance is influenced by

[47] G. Ditta, S. Stanfield, D. Corbin, and D. Helinski, *Proc. Natl. Acad. Sci. U.S.A.* **77**, 7347 (1980).

[48] S. Rogers and D. Fischhoff, manuscript in preparation.

[49] R. Horsch, J. Fry, N. Hoffmann, M. Wallroth, D. Eichholtz, S. Rogers, and R. Fraley, *Science* **227**, 1229 (1985).

two separate factors: the degree of resistance conferred by the gene construct and the effects of the dying cells that comprise the great mass of tissue from which the resistant cells must grow. The degree of resistance is itself determined by the inherent toxicity of the selective agent compared to the level and specificity of the resistance conferred by the gene construct. Selectability is a combination of this degree of resistance and the toxicity of materials released from the surrounding, dying cells. In extreme cases, the resistant cells may all be poisoned by their dying neighbors despite high-level resistance to the selective agent.[50]

While transformed cells containing pMON410 are resistant to up to 400 μg/ml of hygromycin B, the selectability is poor: many fewer colonies grow from leaf disks on 40 μg/ml hygromycin than on 300 μg/ml kanamycin. The rapid bleaching and browning of the disks on hygromycin may indicate a hypertoxic effect on wild-type cells compared to the effect of kanamycin.

Excellent resistance to high levels (30 μM) of methotrexate is conferred by pMON321, but initially the selectability was poor. Delaying selection for 2 to 3 days after completion of transformation greatly improved the recovery of resistant colonies from leaf disks. This improved selectability could have been due to allowing growth of single transformed cells into colonies of several cells before selection or due to accumulation of more of the resistant *dhfr* protein before selection. The latter was demonstrated to be the case by using the stronger CaMV 35 S to drive expression in plasmid pMON806, resulting in efficient selection immediately after transformation. Thus the cells did not have time to divide but only to accumulate sufficient quantities of the resistant enzyme before selection.

In routine practice, we use 50 μg/ml hygromycin B (CalBehring) or 1 μM methotrexate (Sigma, St. Louis, MO) to select transformed cells and shoots with tobacco and petunia. Hygromycin-selected shoots are transferred to medium without added phytohormones for rooting. In contrast, methotrexate-resistant shoots are transferred to solid medium without added phytohormones containing 1 μM mtx for rooting so that selection may be maintained throughout the regeneration procedure. This represents an improvement over the previously reported chimeric mtx-resistant bacterial *dhfr* marker gene[51] which could only be selected at low levels of mtx as callus and not during morphogenesis.

Expression of hygromycin or mtx resistance may also be assayed by

[50] R. Horsch and G. Jones, *Mutat. Res.* **72,** 91 (1980).

[51] M. De Block, L. Herrera-Estrella, M. Van Montagu, J. Schell, and P. Zambryski, *EMBO J.* **3,** 1681 (1984).

the leaf callus assay by culturing sterile leaves on medium with phytohormones containing 50 μg/ml hygromycin or 1 μM methotrexate.

Hygromycin resistance is the best marker we have found for use in transformation of *Arabidopsis thaliana,* where 20 to 50 μg/ml is optimal for efficient selection.[52] The kanamycin resistance conferred by pMON200 does not function well in *Arabidopsis:* the degree of resistance is low and overlaps with spontaneous resistance that readily develops in control tissues. Here too the hygromycin selection is removed during the regeneration step.

Other Uses of the Expression Cassette Vectors

The expression cassette vectors described here have been used to express coding sequences for a wide range of bacterial, mammalian, and plant genes. These include a plant coding sequence for 5-enolpyruvylshikimate-3-phosphate (EPSP) synthase, the target of glyphosate, the active ingredient of Roundup® herbicide.[53] Transgenic plants containing the chimeric EPSP synthase gene show greatly enhanced tolerance to Roundup®.

A mammalian hormone cDNA for the α-subunit of human chorionic gonadotropin has been inserted into the pMON316 vector. Transformed plant tissues produce the hormone as determined by radioimmunoassay and protein antibody blots.[54]

Recently, the tobacco mosaic virus (TMV) coat protein has been expressed from a cDNA inserted into pMON316. Tobacco and tomato plants expressing this normally cytoplasmic mRNA from a nuclear gene show resistance to superinfecting TMV.[55]

The availability of these simple-to-use expression cassette vectors will permit continued rapid progress in the introduction and testing of many additional prokaryotic and eukaryotic coding sequences in plants.

Appendix

The information given below is to assist researchers in analyzing recombinants of pMON200 or pMON505 and derivatives carrying a new

[52] A. Lloyd, A. Barnason, S. Rogers, M. Byrne, R. Fraley, and R. Horsch, *Science* 234, 464 (1986).
[53] D. Shah, R. Horsch, H. Klee, G. Kishore, J. Winter, N. Tumer, C. Hironaka, P. Sanders, C. Gasser, S. Aykent, N. Siegel, S. Rogers, and R. Fraley, *Science* 233, 478 (1986).
[54] S. Rogers *et al.,* manuscript in preparation.
[55] P. Abel, R. Nelson, B. De, N. Hoffmann, S. Rogers, R. Fraley, and R. Beachy, *Science* 232, 738 (1986).

insert and for mapping the inserts of these vectors in transformed plant tissue and plants. For those who wish to assemble the pMON200 and pMON505 plasmid sequences in their own computer systems, we have provided directions, sequences, coordinates, and references for the various segments that comprise the vectors. It must be noted that neither of the sequences are complete. Regions for which sequence is lacking are represented by inserts of spacer null nucleotides of the appropriate size based on extensive restriction endonuclease mapping.

Table I gives the endpoints of the various segments comprising pMON200 as well as the coordinates of the cleavage sites for the 6-bp and larger recognition site restriction endonucleases. A list of endonucleases that do not cleave the vector are also included.

Assembly of pMON200 begins with the *Eco*RI site of the synthetic multilinker which has the sequence

5′-GAATTCATCGATATCTAGATCTCGAGCTCGCGAAAGCTT

The multilinker ends with a *Hin*dIII site. The next fragment is the LIH or left inside homology segment derived from pTiA6. This sequence is the reverse of bp 1618 to 3395 of the pTi15955 octopine-type T-DNA.[56] This segment begins with a *Hin*dIII site and ends with a *Bgl*II site that was made flush ended by treatment with Klenow polymerase and the four nucleotide triphosphates. This fragment was ligated to a fragment of pBR322[57] from the *Pvu*II site (bp 2069) to the *Pvu*I (bp 3740). This *Pvu*I site was joined to a *Pvu*I site located in the pTiT37 Ti plasmid approximately 150 bp from the end of the published sequence. At this point in the assembled sequence, 150 N's were inserted. Then bp 1 to bp 2102 of the nopaline T-DNA right border flanking sequence, the right border, and entire nopaline synthase gene ending at a *Cla*I site, were added.[58] This segment was joined to the Tn7 dihydrofolate reductase sequence[59] beginning at the *Cla*I site at bp 560 and ending at the end of the published sequence at bp 883. The next segment consists of the other unknown segment. Here 250 N's were inserted. The next segment of sequence comes from the Tn7 spectinomycin/streptomycin resistance determinant[60] starting at bp 1 and ending after the second T of the *Eco*RI site at bp 1614. This *Eco*RI site was treated with Klenow polymerase and the four nucleo-

---

[56] R. F. Barker, K. B. Idler, D. V. Thompson, and J. D. Kemp, *Plant Mol. Biol.* **2,** 335 (1983).

[57] J. G. Sutcliffe, *Proc. Natl. Acad. Sci. U.S.A.* **75,** 3737 (1978).

[58] A. Depicker, S. Staechel, P. Dhaese, P. Zambryski, and H. M. Goodman, *J. Mol. Appl. Genet.* **1,** 561 (1982).

[59] M. Fling and C. Richards, *Nucleic Acids Res.* **11,** 5147 (1983).

[60] M. Fling, J. Kopf, and C. Richards, *Nucleic Acids Res.* **13,** 7095 (1985)

TABLE I
MAJOR REGIONS AND RESTRICTION ENDONUCLEASE CLEAVAGE SITES OF pMON200

Segment name	Coordinates
Synthetic mutilinker	1–36
pTiA6 fragment LIH	36–1700
pBR322 origin fragment	1700–3476
Right border	3842
pTiT37 fragment	3485–5730
Nopaline synthase coding region	4217–5455
Tn7 *Spc/StrR* fragment	5731–7911
Spc/StrR coding region	6390–7144
NOS-NPTII'-NOS kanamycin^R fragment	7934–9489
NPTII' coding region	8366–9159

Endonuclease[a]	Cleavage site	Endonuclease[a]	Cleavage site
*Acc*I	739, 1990	*Nae*I	7040, 7551, 8251, 8534
*Afl*II	5509, 8154		
*Asu*II	8352	*Nar*I	9034
*Ava*I	21, 1214, 3745	*Nco*I	504, 4607, 8600
*Bal*I	7694, 8954	*Nde*I	1252, 2041, 5978
*Bam*HI	5028, 8191	*Nhe*I	523, 4115, 9284
*Bbe*I	9037	*Nru*I	30, 4768
*Bcl*Id	3917, 6587	*Pst*I	3356, 8987, 9190
*Bgl*I	3231	*Pvu*I	3481, 6770, 7934, 8187, 9175
*Bgl*II	17		
*Bst*EII	6549	*Pvu*II	367, 594, 4637, 7897, 8930
*Bvu*I	28, 822, 8675		
*Cla*I	8	*Rsr*II	8517
*Cla*Id	510, 4389, 5731, 7934	*Sac*I	28, 822
		*Sac*II	4033, 9372
*Cvn*I	4100, 9300	*Sma*I	1216
*Dra*I	629, 702, 749, 2977, 2996, 3814, 3891	*Sna*BI	1023
		*Spe*I	7809
		*Sph*I	191, 3771, 5212, 5316, 8635
*Dra*III	6526, 7143		
*Eco*RI	1	*Stu*I	7776
*Eco*RV	12, 723	*Tth*I	1965, 8917
*Fsp*I	3333, 8934	*Xba*Id	14
*Hinc*II	4547, 4715, 5883	*Xho*I	21
*Hind*III	34	*Xma*III	9126
*Hpa*I	5883	*Xmn*I	617, 6034, 7133, 7924
*Mst*I	3333, 8934		

Enzymes that do not cut
   *Aat*II   *Apa*I   *Avr*II   *Bst*XI   *Dra*II   *Kpn*I
   *Mlu*I   *Not*I   *Ppu*MI   *Sal*I   *Sca*I   *Sfi*I

[a] d, The cleavage site is protected by *dam* methylation.

tide triphosphates and joined to a similarly treated *Eco*RI site at the end of a synthetic linker joined to the 3' end of the NOS 3' nontranslated region. The resultant sequence is

5'-GAATTAATTCCCGATCGATC

The ATCGAT is the *Cla*I site located at bp 2102 of the nopaline synthase sequence.[58] The NOS 3' sequence ends at the *Sau*3A site at bp 1847 of the nopaline synthase sequence and adjoins the following linker sequence

5'-GGGGATCCGGGGG

The last three G's of this sequence are from one-half of the *Sma*I site located at bp 1118 of the Tn5 sequence.[61] The Tn5 neomycin phosphotransferase II (NPTII) segment extends from the *Sma*I site at bp 1118 to the *Sau*3A site at bp 140. This *Sau*3A site is immediately adjacent to the following linker

5'-GTCTAGGATCTGCAG

The T of the *Pst*I site (CTGCAG) is the 3' end of the nopaline synthase promoter segment which begins at bp 584 and ends with the *Bcl*I site at bp 284 of the nopaline synthase sequence.[58] The *Bcl*I site was cleaved with *Sau*3A and joined to a linker to give the following sequence which includes the half *Bcl*I site and the *Eco*RI site which is the origin of the pMON200 plasmid

5'-TGATCCGGGGAATTC

When assembled following the above instructions, the total size of the complete pMON200 plasmid will be 9489 bp.

The coordinates of the various segments comprising pMON505 and selected restriction endonuclease cleavage sites are given in Table II. Directions for assembly of the pMON505 plasmid are as follows. Starting with pMON200 the LIH region beginning after the unique *Hin*dIII site is replaced with the following segments. First, 74 N's are inserted followed by bp 617 to bp 1 of the published *ori*V sequence.[62] Next is a run of 54 N's and a *Sal*I site (5'-GTCGAC) and then 694 N's. This is followed by the bp 400 to bp 1618 sequence of the *trfA** gene.[63] Next are 155 N's followed by an *Nco*I site (CCATGG) followed by 500 N's and a second *Nco*I site (CCATGG). This is followed by another run of 435 N's and then the *Sma*I site that is the end of the original *Hin*dIII to *Sma*I fragment from

[61] E. Beck, G. Ludwig, E.-A. Auerswald, B. Reiss, and H. Schaller, *Gene* **19,** 327 (1982).
[62] D. Stalker, C. Thomas, and D. Helinski, *Mol. Gen. Genet.* **181,** 8 (1981).
[63] C. Smith and C. Thomas, *J. Mol. Biol.* **175,** 251 (1984).

TABLE II

Major Regions and Restriction Endonuclease Cleavage Sites of pMON505

Segment name	Coordinates
Synthetic multilinker	1–36
RK2 origin	36–3801
pBR322 origin fragment	3802–5273
Right border	5778
pTiT37 fragment	5274–7665
Nopaline synthase coding region	6153–7391
Tn7 *Spc/StrR* fragment	7666–9838
Spc/StrR coding region	8317–9071
NOS-NPTII'-NOS kanamycin^R fragment	9839–11412
NPTII' coding region	10293–11086

Endonuclease[a]	Cleavage site	Endonuclease[a]	Cleavage site
*Acc*I	785	*Nco*I	2860, 3360, 6543,
*Afl*II	7445, 10081		10527
*Asu*II	10279	*Nde*I	2423, 3838, 7914
*Ava*I	21, 3800, 5681	*Nhe*I	6051, 11211
*Bal*I	491, 9621, 10881	*Not*I	634
*Bam*HI	6964, 10118	*Nsi*I	7146, 7254, 7536,
*Bbe*I	10964		9998
*Bcl*Id	5853, 8514	*Ppu*MI	1835, 2192, 2365
*Bgl*II	17	*Pst*I	5153, 10914, 11117
*Bsp*MI	9843	*Pvu*II	6573, 9824, 10857
*Bst*EI	8476	*Rsr*II	10444
*Cla*I	8	*Sal*I	784
*Cla*Id	6325, 7667, 9861	*Sac*I	28
*Cvn*I	6036, 11227	*Sst*II	640, 5969, 11299
*Dra*I	212, 4774, 4793,	*Sfi*I	1866
	5750, 5827	*Sma*I	3802
*Eco*RI	1	*Spe*I	9736
*Eco*RV	12	*Sph*I	5707, 7148, 7252,
*Fsp*I	5130, 10861		10562
*Hinc*II	537, 572, 786,	*Ssp*I	204, 6060, 6590,
	2051, 6483, 6651,		7263, 11206
	7819	*Stu*I	9703
*Hind*III	34	*Tth*I	10844
*Hpa*I	7819	*Xba*Id	14
*Mst*I	5130, 10861	*Xho*I	21
*Nar*I	10961	*Xma*II	11053
		*Xmn*I	7970, 9060, 9851

Enzymes that do not cut

  *Aat*II *Aap*I *Avr*II *Bst*XI *Eco*K *Kpn*I *Mlu*I *Sca*I *Sna*BI

[a] d, The cleavage site is protected by *dam* methylation.

pTJS75[34] used to construct pMON505. The next segment is a small 43-bp *Sma*I to *Nde*I fragment derived from the pMON200 LIH segment and has the following sequence

5′-CCCGGGATGG CGCTAAGAAG CTATTGCCGC CGATCTTCAT ATG

This is joined to the *Nde*I (bp 2297) to *Pvu*I (bp 3740) fragment of pBR322.[57] This is followed by the pMON200 sequences described above. The restriction endonuclease sites shown on the maps have been verified by restriction endonuclease cleavage. However, since the sequences are incomplete, certain of the sites listed in the tables must be regarded as preliminary.

Section IV

Vectors or Methods for Expression of Cloned Genes

[23] β-Galactosidase Gene Fusions for Analyzing Gene Expression in *Escherichia coli* and Yeast

By Malcolm J. Casadaban, Alfonso Martinez-Arias, Stuart K. Shapira, and Joany Chou

The β-galactosidase structural gene *lacZ* can be fused to the promoter and controlling elements of other genes as a way to provide an enzymic marker for gene expression.[1] With such fusions, the biochemical and genetic techniques available for β-galactosidase can be used to study gene expression.[2] Gene fusions can be constructed either *in vivo,* using spontaneous nonhomologous recombination or semi-site specific transposon recombination,[3] or *in vitro,* with recombinant DNA technology.[4,5] Here we describe *in vitro* methods and list some recently developed β-galactosidase gene fusion vectors. With these *in vitro* methods, gene controlling elements from any source can be fused to the β-galactosidase structural gene and examined in the prokaryote bacterium *Escherichia coli* or the lower eukaryote yeast *Saccharomyces cerevisiae* (see this volume, Ruby *et al.* [19] and Guarente [27], and *Methods in Enzymology*, Vol. 101, Rose and Botstein [9] for additional descriptions of β-galactosidase gene fusion systems in yeast).

β-Galactosidase gene fusions can be constructed both with transcription initiation control signals and with transcription plus translation initiation control signals. This is done by removing either just the promoter region from the β-galactosidase gene or both the promoter and translation initiation regions. The β-galactosidase gene is convenient for making translational fusions because it is possible to remove its translation initiation region along with up to at least the first 27 amino acid codons without affecting β-galactosidase enzymic activity. In a gene fusion this initial part of the β-galactosidase gene can be replaced by the promoter, translation initiation site, and apparently any number of amino-terminal codons from another gene. Here we focus on these transcription–translation fusions because they provide all the gene initiation signals from the other gene.

[1] P. Bassford, J. Beckwith, M. Berman, E. Brickman, M. Casadaban, L. Guarante, I. Saint-Girons, A. Sarthy, M. Schwartz, and T. Silhavy, *in* "The Operon" (J. H. Miller and W. S. Reznikoff, eds.), p. 245. Cold Spring Harbor Laboratory, Cold Spring Harbor, New York, 1978.

[2] J. Beckwith, *in* "The Operon" (J. H. Miller and W. S. Reznikoff, eds.), p. 11. Cold Spring Harbor Laboratory, Cold Spring Harbor, New York, 1978.

[3] M. Casadaban and S. Cohen, *Proc. Natl. Acad. Sci. U.S.A.* **76,** 4530 (1979).

[4] M. Casadaban and S. Cohen, *J. Mol. Biol.* **138,** 179 (1980).

[5] M. Casadaban, J. Chou, and S. Cohen. *J. Bacteriol.* **143,** 971 (1980).

For a description of *in vitro* β-galactosidase transcriptional fusions, see Casadaban and Cohen.[4]

β-Galactosidase expression from a gene fusion can be used not only to measure gene expression and regulation, but also to isolate mutations and additional gene fusions. Mutants with altered gene regulation, for example, can be used to identify regulatory genes and to help decipher regulatory mechanisms. Mutants with increased gene expression, or constructions that join the fused gene to a stronger gene initiation region, can be used to isolate large amounts of the original gene product following removal of the β-galactosidase fused gene segment.[6]

An additional result of translational β-galactosidase gene fusions is the formation of hybrid proteins that can be used in protein studies. The hybrid protein can readily be identified by its β-galactosidase enzymatic activity and purified for studies of protein functions located in the amino terminus, such as DNA or membrane-binding functions.[7,8] These hybrid proteins can also be used to determine amino terminal sequences[9,10] and to elicit antibody formation.[11]

Principle of the Methods

A cloning vector, containing the β-galactosidase structural gene segment without its initiation region, is used to clone a DNA fragment that contains a gene's transcription–translation initiation region and amino terminal codons to form the gene fusion. The fragment either can be purified or can be selected from among the resulting clones. Alternatively, fusions can be made by inserting a DNA fragment (cartridge) containing the β-galactosidase structural gene into a gene already on a vector by cloning (inserting) the β-galactosidase gene into a site within the other gene. For a functional fusion, the gene segments must be aligned in their direction of transcription; and, for a translational fusion, their amino acid codons must be in phase. Gene control regions that cannot conveniently be fused to β-galactosidase by cloning in a single step can be cloned in a nearby position followed by an additional event to form the fusion, such as a deletion to remove transcription termination or to align the codons. These additional events can be obtained either *in vitro,* as by nuclease

---

[6] M. Casadaban, J. Chou, and S. Cohen, *Cell* **28,** 345 (1982).

[7] B. Müller-Hill and J. Kania, *Nature (London)* **249,** 561 (1974).

[8] T. Silhavy, M. Casadaban, H. Shuman, and J. Beckwith, *Proc. Natl. Acad. Sci. U.S.A.* **73,** 3423 (1976).

[9] A. Fowler and I. Zabin, *Proc. Natl. Acad. Sci. U.S.A.* **74,** 1507 (1977).

[10] M. Ditto, J. Chou, M. Hunkapiller, M. Fennewald, S. Gerrard, L. Hood, S. Cohen, and M. Casadaban, *J. Bacteriol.* **149,** 407 (1982).

[11] H. Schuman, T. Silhavy, and J. Beckwith, *J. Biol. Chem.* **255,** 168 (1980).

digestion, or *in vivo* using genetic screens or selections for lactose metabolism.

Once fusions have been made, they can be used to measure gene expression under different regulatory conditions, such as with or without an inducer or repressor, or in different growth media or physical conditions of growth (such as temperature[12]). Self-regulation can be investigated by providing (or removing) in *trans* an expressed wild-type copy of the gene (such as on another plasmid, episome, lysogenic prophage or in the chromosome[13]) and checking the effect on β-galactosidase expression.[14] Linked or unlinked, *cis* or *trans* acting mutations can be sought with appropriate mutagenesis, selection, and screenings using the lactose phenotype.[2,13]

Materials

Escherichia coli strain M182 and its derivatives (see the table) are currently used because they are deleted for the *lac* operon, can be efficiently transformed, and yield good plasmid DNA preparations. Their *gal* mutations may contribute to their transformation properties, since they are missing polysaccharides containing galactose. Strains M182 and MC1060 have no nutritional requirements and so can be used to select for growth on minimal media with a single carbon and energy source (such as lactose) without the addition of any nutritional factor that could be used as an alternative carbon or energy source. Several of these strains have the *hsdR*⁻ host restriction mutation so that DNA unmodified by *E. coli* K12 is not degraded. All strains are *hsdS*⁺ M⁺ so that DNA in them becomes K12 modified and can be transferred efficiently to other *E. coli* K12 strains. Some MC strains have a deletion of the *ara* operon, so they can be used with plasmids containing the *ara* promoter.[4] Strain MC1050 has an amber mutation in the *trpB* gene (for tryptophan biosynthesis) so that amber mutation suppressors can be introduced, a procedure that is useful in studies of regulatory genes.[13] MC1116 is a *recA*⁻ and streptomycin-sensitive derivative. Casadaban and Cohen[4] describe additional derivative strains with features such as the early *lacZ* M15 deletion, which can be intercistronically complemented by the α segment of β-galactosidase.

---

[12] M. Casadaban, T. Silhavy, M. Berman, H. Schuman, A. Sarthy, and J. Beckwith, *in* "DNA Insertion Elements, Plasmids and Episomes" (M. Bukhari, J. Shapiro, and S. Adhya, eds.), p. 531. Cold Springs Harbor Laboratory, Cold Spring Harbor, New York, 1977.

[13] J. Chou, M. Casadaban, P. Lemaux, and S. Cohen, *Proc. Natl. Acad. Sci. U.S.A.* **76**, 4020 (1979).

[14] F. Sherman, G. Fink, and C. Lawrence, "Methods in Yeast Genetics." Cold Spring Harbor Laboratory, Cold Spring Harbor, New York, 1979.

STRAINS USED FOR ANALYZING GENE EXPRESSION[a]

	Strain	Genotype	Reference
E. coli	M182	Δ(lacIPOZYA)X74,galU,galK,strA^r	4
	MC1000	M182 with Δ (ara,leu)	4
	MC1050	MC1000 with trpB9604 amber	4
	MC1060	M182 with hsdR$^-$	4
	MC1061	MC1060 with Δ (ara,leu)	4
	MC1116	MC1000 with strA^s,spcA^r and recA56	
	MC1064	MC1061 with trpC9830	
	MC1065	MC1064 with leuB6, ara$^+$	
	MC1066	MC1065 with pyrF74::Tn5(Km^r)	
	AMA 1004	MC1065 with Δ (lacIPOZ)C29,lacY$^+$	
Yeast	M1-2B	α, trp1, ura3-52	R. Davis
	M1-2BA2	M1-2B with an increased permeability to β-galactosidase substrates	
	JJ4	a, trp1, lys1, ura3-52, Δ (gal7) 102, gal2$^+$	J. Jaehning

[a] ara, gal, and lac refer to mutations that abolish the ability of the cell to use arabinose, galactose, and lactose as carbon and energy source for growth; leu, trp, and pyr (ura) refer to mutations creating a requirement for leucine, tryptophan, or pyrimidine (uracil) to be added to minimal media for growth. For further references, see text.

Strains MC1064, MC1065, and MC1066 are derivatives of M182 with hsdR$^-$ and Δ (lac) that have auxotropic markers that can be complemented by the yeast genes leu2 (leuB), ura3 (pyrF), and trp1 (trpC). The leuB6 and trpC9830 are point mutations that revert infrequently, and the pyrF is a Tn5 kanamycin-resistance transposon insertion that reverts at a higher frequency. AMA1004 is an MC1065 derivative with a deletion of the lacZ gene on the chromosome, which leaves the lacY gene expressed from an unknown chromosomal promoter. This strain is useful for complementing β-galactosidase plasmids without the lacY gene. The yeast strains in the table can be efficiently transformed with plasmid DNA. The ura3 mutation does not revert, and the trp1 reverts at a low rate. Strain M1-2B has a heterologous pedigree, so that diploids derived from it do not sporulate efficiently.

Escherichia coli medium is described by Miller.[15] β-Galactosidase expression *in vivo* can be detected with three types of agar media: medium with colorimetric β-galactosidase substrates, lactose (or other β-galactoside) minimal medium, and fermentation indicator medium. The colorimetric substrate medium is the most sensitive and can be used to detect β-galactosidase levels below those necessary for growth on lactose minimal

[15] J. Miller, "Experiments in Molecular Biology." Cold Spring Harbor Laboratory. Cold Spring Harbor, New York, 1972.

medium. The XG (5-bromo-4-chloro-3-indolyl-β-D-galactoside) substrate is commonly used since its β-galactosidase hydrolysis product dimerizes and forms a blue precipitate that does not diffuse through the agar.[15] The blue color is most easily seen on a clear medium such as M63. XG dissolved in dimethylformamide solution at 20 mg/ml can be conveniently stored for months at −20° and used by spreading on the agar surface just before use, or by adding to the liquid agar just before pouring. Acid-hydrolyzed casamino acids can also be added (to 250 μg/ml) to minimal *E. coli* or yeast media to speed colony growth without affecting the detection of low levels of blue color. Note that the acid-hydrolyzed casamino acids can serve as a source for leucine, but not for pyrimidine or tryptophan, for growth on minimal medium.

β-Galactosidase detection with XG in yeast colonies can be done with the standard yeast SD minimal medium[14] if it is buffered to pH 7 with phosphate buffer to allow β-galactosidase enzymic activity. The phosphate buffer can be prepared as described for M63 medium[15] but must be added separately (usually as a 10-fold concentrate) to the cooled agar medium just prior to pouring in order to avoid a phosphate precipitate. The carbon energy source sugar also is usually prepared and added separately to avoid caramelization. Less efficient carbon sources for yeast than glucose, such as glycerol plus ethanol,[14] result in bluer yeast colonies on SD-XG medium, presumably because less acidity is formed in the growing colony.

Yeast cells are not very permeable to β-galactosidase, and so lower levels of β-galactosidase can be better detected by first permeabilizing the yeast cells. This can conveniently be done by replica plating or growing yeast colonies on filters (such as Millipore No. 1) and permeabilizing the yeast by immersion in liquid nitrogen. The filters can then be placed in β-galactosidase assay Z buffer[15] with the XG substrate. This process also overcomes the problem of the inhibition of β-galactosidase activity by the acid formed within the yeast colony.

Higher levels of β-galactosidase are indicated if lactose (or another β-galactoside) can be utilized for growth on minimal media (such as M63 medium[15]). For detecting even higher levels of β-galactosidase, the fermentation indicator plates, such as lactose MacConkey (Difco), tetrazolium, or EMB[15] can be used. On fermentation indicator plates, both lactose utilizing (Lac⁺) and nonutilizing (Lac⁻) cells form colonies, but the colonies are colored differently. Lactose utilization also requires expression of the *lacY* permease gene unless very high levels of β-galactosidase are made. Yeast *S. cerevisiae* does not have a permease for lactose, and so media of these types cannot be used. There is a Lac⁺ yeast species *Kluveromyces lactis,* with a lactose permease, but no transformation system has yet been devised for it.[16] The M1-2B A2 *S. cerevisiae* mutant we have isolated (see the table, unpublished results), however, does allow a patch of yeast cells expressing β-galactosidase, but not a single yeast cell, to grow on lactose minimal medium. This mutant seems to result in a lethal permeabilization of about 80% of the cells, which can then take up lactose and cross-feed their neighbors with lactose hydrolysis products (unpublished results).

β-Galactosidase assays are as described.[15] For yeast, cells are broken either by freeze-thawing three times with ethanol–Dry Ice, by vortexing with glass beads, or by sonicating. Procedures involving glusulase to form spheroplasts cannot be used, since glusulase contains a β-galactosidase activity. However, the more purified spheroplasting enzyme zymolase can be used.[14] It is important not to centrifuge away broken yeast particulate matter, since much of the β-galactosidase activity can be associated with it. After incubation with the colorimetric β-galactosidase substrate ONPG[15] and stopping the reaction by addition of 1 M NaCO$_3$, the particulate matter can then be removed by centrifugation before reading the optical density of the yellow product at 420 nm. Units are expressed as nonomoles of ONPG cleaved per minute per milligram of protein in the extract, where 1 nmol is equivalent to 0.0045 OD$_{420}$ unit with a 10-mm light path.[15]

Methods

β-Galactosidase Vectors. The prototype vector pMC1403 (Fig. 1) has been described.[5] pMC1403 contains the entire *lac* operon but is missing the promoter, operator, and translation initiation sites as well as the first eight nonessential codons of the *lacZ* gene for β-galactosidase. It contains three unique restriction endonuclease cleavage sites, for *Eco*RI, *Sma*I, and *Bam*HI, into which DNA fragments containing promoter and translation initiation sites can be inserted to form gene fusions (Fig. 1). Figure 2 lists derivatives of pMC1403 with additional restriction sites. Figure 3 lists vectors with the *lac* promoter and translation initiation sites before a series of restriction sites aligned in various translation phases with the β-galactosidase codons. These vectors (unpublished results) are derived from the M13mp7, 8, and 9 cloning vectors.[17] Additional vectors with the p15a plasmid replicon are described by Casadaban *et al.*[5] p15a is a high copy number plasmid that is compatible with pBR322 and ColE1.

[16] R. Dickson, *Gene* **10,** 347 (1980).
[17] J. Messing, R. Crea, and P. Seeburg, *Nucleic Acids Res.* **9,** 309 (1981).

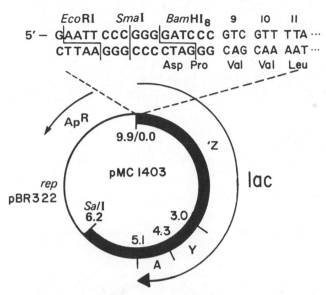

FIG. 1. The prototype β-galactosidase hybrid protein gene fusion vector pMC1403 with three unique cloning sites EcoRI, SmaI (XmaI), and BamHI.[5] It contains the lac operon with the lacZ gene for β-galactosidase, the lacY gene for lactose permease in *Escherichia coli,* and the nonessential lacA gene for transacetylase. It is missing the lac promoter, operator and, translation initiation site, as well as the first seven and one-third nonessential amino acid codons; codons are labeled from eight onward. The non-lac segment is from the EcoRI to SalI sites of pBR322, which has been entirely sequenced [G. Sutcliffe, *Cold Spring Harbor Symp. Quant. Biol.* **43,** 77 (1979)]. Fragments inserted into the unique restriction sites form gene fusions if they contain promoters and translation initiation sites with amino terminal codons that align with the β-galactosidase gene codons. Note that the SmaI site yields blunt ends into which any blunt-ended DNA fragment can be integrated, and that any such inserted fragment will become flanked by the adjacent EcoRI and BamHI sites contained in the synthetic EcoRI-SmaI-BamHI linker sequence on the plasmid. These adjacent sites can be used separately to open the plasmid on either side of the inserted fragment for making exonuclease-generated deletions into either side of the fragment. There is a unique SacI (SstI) site in lacZ and a unique BalI site (blunt ended) in the pBR322 sequence just after the lac segment. This plasmid has no BglII, HindIII, KpnI, XbaI, or XhoI sites. There are two BalI and HpaI sites and several AccI, AvaI, AvaII, BglI, BstEII, EcoRII, HincII, PvuI, and PvuII sites.

Figure 4 describes vectors containing yeast replicons *ars1* and *2* with the selectable yeast genes *trp1, ura3,* and *leu2.* All these vectors are derived from the pBR322 cloning vector. Unique sites are present for EcoRI, BamHI, and SmaI. pMC1790 and 2010 have all these three unique cloning sites, are relatively small, and are deleted for the lacY and lacA genes.

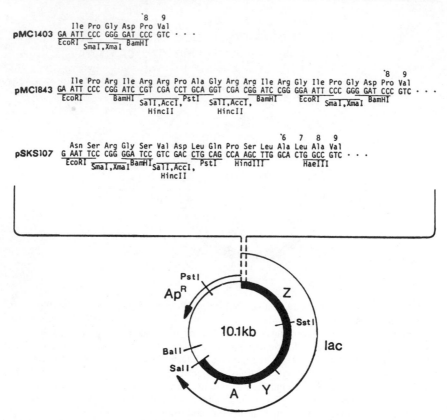

```
                              ˙8  9
                   Ile Pro Gly Asp Pro Val
pMC1403  GA ATT CCC GGG GAT CCC GTC · · ·
         ‾‾‾‾‾‾‾ ‾‾‾‾‾‾‾ ‾‾‾‾‾
         EcoRI  SmaI,XmaI BamHI
```

```
                                                                                                    ˙8  9
                   Ile Pro Arg Ile Arg Arg Pro Ala Gly Arg Arg Ile Arg Gly Ile Pro Gly Asp Pro Val
pMC1843  GA ATT CCC CGG ATC CGT CGA CCT GCA GGT CGA CGG ATC CGG GGA ATT CCC GGG GAT CCC GTC · · ·
         ‾‾‾‾‾‾‾         ‾‾‾‾‾‾‾ ‾‾‾‾‾‾‾     ‾‾‾‾‾ ‾‾‾‾‾‾‾ ‾‾‾‾‾‾‾ ‾‾‾‾‾‾‾ ‾‾‾‾‾‾‾ ‾‾‾‾‾‾‾
         EcoRI          BamHI  SalI,AccI, PstI SalI,AccI, BamHI   EcoRI  SmaI,XmaI BamHI
                               HincII          HincII
```

```
                                                          ˙6  7  8  9
                   Asn Ser Arg Gly Ser Val Asp Leu Gln Pro Ser Leu Ala Leu Ala Val
pSKS107  G AAT TCC CGG GGA TCC GTC GAC CTG CAG CCA AGC TTG GCA CTG GCC GTC · · ·
         ‾‾‾‾‾‾ ‾‾‾‾‾‾‾ ‾‾‾‾‾ ‾‾‾‾‾‾‾    ‾‾‾‾‾ ‾‾‾‾‾‾‾    ‾‾‾‾‾‾‾
         EcoRI  SmaI,XmaI BamHI SalI,AccI, PstI HindIII   HaeIII
                               HincII
```

FIG. 2. Translation fusion vectors without a promoter. The numbers above refer to the original *lacZ* codons. pMC1403 is from Fig. 1. pMC1843 is pMC1403 with the polylinkers from M13mp7[17] inserted into the *Eco*RI site of pMC1403. pSKS107 is from pSKS105 (Fig. 3) with the removal of the *Eco*RI fragment containing the *lac* promoter. Additional *Acc*I, *Hinc*II, and *Hae*III sites are not shown.

Alternative β-Galactosidase Insertion Fragments. Instead of forming β-galactosidase fusions by inserting gene control fragments into β-galactosidase vectors, it is also possible to insert β-galactosidase fragments without replicons ("cartridges") into sites within genes already on replicons. Figure 5 describes plasmids from which β-galactosidase gene fragments can be excised with restriction sites on both sides of *lacZ*. pMC931[5] has in addition the *lacY* gene, whereas pMC1871 has only the *lacZ* gene with no part of *lacY*. The *lac* fragment from pMC931 can be obtained with *Bam*HI or *Bam*HI plus *Bgl*II, which yields identical sticky ends. The *lac* fragment from pMC1871 can be obtained with *Bam*HI, *Sal*I, or *Pst*I.

```
              1   2   3   4   5   6   7  8                    8   9
            Thr Met Ile Thr Asn Ser Leu                          Val
pMC1513  ATG ACC ATG ATT ACG AAT TCA CTG G(AATTCCCGGGGATC)CC GTC · · ·
                                         EcoRI SmaI,XmaI BamHI
```

```
              1   2   3   4   5   6                                        4   5   6   7   8
            Thr Met Ile Thr Asn Ser Pro Asp Pro Ser Thr Cys Arg Ser Thr Asp Pro Gly Asn Ser Leu Ala
pSKS104  ATG ACC ATG ATT ACG AAT TCC CCG GAT CCG TCG ACC TGC AGG TCG ACG GAT CCG GGG AAT TCA CTG GCC · · ·
                            EcoRI          BamHI SalI,AccI, PstI SalI,AccI, BamHI      EcoRI       HaeIII
                                                HincII           HincII
```

```
              1   2   3   4   5   6                          6   7   8
            Thr Met Ile Thr Asn Ser Arg Gly Ser Val Asp Leu Gln Pro Ser Leu Ala Leu Ala
pSKS105  ATG ACC ATG ATT ACG AAT TCC CGG GGA TCC GTC GAC CTG CAG CCA AGC TTG GCA CTG GCC · · ·
                            EcoRI SmaI,XmaI BamHI SalI,AccI, PstI HindIII        HaeIII
                                                 HincII
```

```
              1   2   3   4
            Thr Met Ile Thr Pro Ser Leu Ala Ala Gly Arg Arg Ile Pro Gly Asn Ser Leu Ala
pSKS106  ATG ACC ATG ATT ACG CCA AGC TTG GCT GCA GGT CGA CGG ATC CCC GGG AAT TCA CTG GCC · · ·
                            HindIII    PstI SalI,AccI, BamHI SmaI,XmaI EcoRI  HaeIII
                                            HincII
```

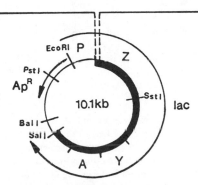

FIG. 3. Translation fusion vectors that have an upstream *lac* promoter, operator, and initial *lacZ* codons. For pMC1513 these codons are out of phase with the rest of *lacZ*. In the other vectors the codons are in phase and result in full β-galactosidase expression. pMC1513 was formed by cloning the *Eco*RI (*lacP*)UV5 promoter fragment from pKB252 [K. Backman and M. Ptashne, *Cell* **13,** 65 (1978)] into pMC1403. (The (*lacP*)UV5 results in catabolite independence of the *lac* promoter). pSKS104, 105, and 106 were made by inserting *Pvu*II fragments containing the *lac* promoters and translation initiation regions from M13mp7, M13mp8, and M13mp9[17] into the *Sma*I site of pMC1403, followed by homologous recombination between the *lacZ* segments to result in full β-galactosidase expression. In all cases an *Eco*RI site remains to the left (upstream) of *lacP*.

Alternatively, the 1871 fragment can be obtained with a combination of restriction enzymes (see Fig. 5). Fragments can also be obtained with different restriction ends from the plasmids of Figs. 1–4. Note that a fragment with blunt ends on both sides can be obtained using *Sma*I plus *Bal*I from many of these plasmids. In addition, *lac* fragments can easily

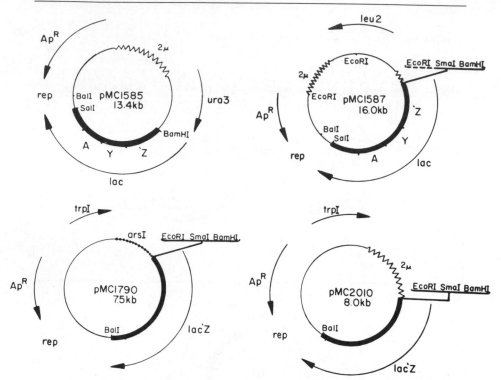

FIG. 4. Yeast replicon β-galactosidase vectors. All are derivatives of pMC1403 with the yeast 2μ or arsI replicons and the yeast leu2, ura3, or trp1 gene.[24] These genes can be selected in either *Escherichia coli* or yeast as complementing appropriate auxotrophs (see the table and the Materials section). Unique cloning sites are underlined. pMC1585 and 1587 have the same lac segment as on pMC1403 of Fig. 1. pMC1790 and 2010 are similar but have a deletion of lacY and A gene formed by removing an AvaI fragment extending from lacY into pBR322, thereby also removing the SalI site but keeping the BalI site. This AvaI deletion was first isolated by M. Berman (Fredrick Cancer Research Center, Fredrick, Maryland).

be combined using a unique SstI (SacI) site in lacZ so that any sites before lacZ can be combined with any sites after the lac segment.

Choice of Fragment. To form a gene fusion with a β-galactosidase fusion vector it is necessary to obtain a fragment with the control elements of the gene of interest. This is most conveniently done if the sequence or restriction map is known so that appropriate sites can be identified. The fragment can have ends that match the ends of cleaved β-galactosidase fusion vectors, or the ends can be altered to match (as by filling in, chewing back, or adding linkers). One cleavage site should be well upstream to all the controlling sequences, and the other should be within the coding part of the gene such that the codons of the gene are in phase with the β-galactosidase codons. If the codons are not aligned, they can be made to align in a second step by obtaining a frameshift mutation (see below). Also, if the second site is beyond the gene, a second step deletion can be formed to fuse the gene.

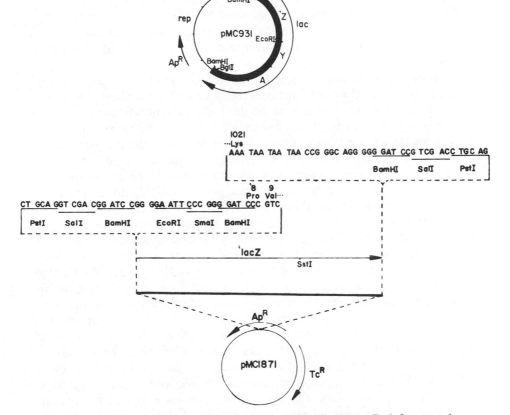

FIG. 5. β-Galactosidase hybrid protein fusion cartridge fragments. Both fragments have the *lacZ* gene without the control region of the promoter, operator, translation initiation region, and first eight nonessential *lacZ* codons, as in pMC1403 of Fig. 1. The *lac* fragment on pMC931[5] has the end of the *lac* operon (*lacY* and *A*) on the pACYC177 vector. The *lac* fragment on pMC1871 has only *lacZ*, without any part of *lacY* inserted in the *PstI* site of pBR322. *lacY* was removed by cleaving at an *HaeIII* site located between *lacZ* and *Y* [D. Buchel, B. Gronenborn, and B. Müller-Hill, *Nature (London)* **283**, 541 (1980)] and joining to a polylinker sequence on a plasmid derived from M13mp7.[17] The beginning of the pMC1871 *lac* fragment is from pMC1403 (Fig. 1) with additional sites from the M13mp7 polylinker.

The vectors of Fig. 1–5 have several different cloning sites. The *Sma*I site can be used with blunt-ended fragments. The *Bam*HI site can be used with *Bgl*II, *Bcl*I, or *Sau*3A (*Mbo*I) digested DNA since all these enzymes yield identical sticky ends. *Sau*3A recognizes only four base pairs and so should cleave within an average gene, whereas the other enzymes recognize six nucleotides and so should not cleave in an average gene. The *Eco*RI site can be used with fragments formed by *Eco*RI* digestion. The *Sal*I site can be used with *Xho*I-cleaved DNA. The *Eco*RI, *Bam*HI, *Xma*I, *Sal*I, and *Hin*dIII sites can also be used with blunt-ended DNA fragments either by making them blunt by filling in their ends with a polymerase or by adding synthetic linker DNA fragments onto the blunt fragment ends. Each of these manipulations can yield different fusion frames.

Selection of Fusion Clones. Fusions can be made with purified fragments or with mixtures of fragments followed by a screening of the resulting clones to identify those with the desired fragment containing the gene control region. Fusion clones can be identified by their resulting *lac* expression (Lac phenotype), or they can be selected by their physical structure using standard DNA cloning techniques without regard to their Lac phenotype. Fragments containing promoters which are expressed in *E. coli* and for which translation is aligned with the β-galactosidase codons can be detected as a resulting clone that forms red colonies on lactose MacConkey agar, dark blue colonies on XG agar, or colonies on lactose minimal medium. These clones can easily be distinguished from the original vector if a vector is used for which the *lac* genes were not expressed, as is the use for the vectors of Fig. 1, 2, and 4. The lactose MacConkey red colony phenotype is the most diagnostic since only efficient *lac* fusions have enough expression for this lactose fermentation indicator test.

For promoters that are expressed in yeast, only the XG indicator can be used, since there is no lactose permease in yeast. (Note that some but not all yeast promoters are expressed in *E. coli*[18]). Since yeast cells are also relatively impermeable to colorimetric β-galactosidase substrates, high levels of β-galactosidase expression are needed for yeast colonies to turn blue with XG. Lower levels of β-galactosidase formed by less efficient yeast fusions can be detected by permeablizing the yeast cells or by placing them into pH 7 buffered medium (see Materials).

In *E. coli*, inefficient fusions can be detected with XG agar. Fusions with good promoters but with codons out of phase or without any translation initiation site, or fusions without good promoters but with codons in phase, can usually be detected as blue colonies on XG but white colonies on lactose MacConkey.

[18] L. Guarante and M. Ptashne, *Proc. Natl. Acad. Sci. U.S.A.* **78,** 2199 (1981).

Some β-galactosidase vectors such as those of Fig. 3 already have promoters and translation start sites for expression of β-galactosidase. These can be used to detect clones of fragments that do not result in efficient β-galactosidase expression. Such a system has been used for transcriptional fusions with the pMC81 vector,[5] which has the arabinose *ara* operon promoter; fragments that do not express β-galactosidase well can be detected, as they result in less expression of β-galactosidase than is the case for the original vector when the *ara* promoter is induced by L-arabinose. This vector can also be used in the standard way in the absence of the *ara* inducer L-arabinose to detect fragments that promote β-galactosidase expression.

Each DNA fragment in a particular orientation yields a specific level of β-galactosidase expression. Thus, clones from a mixture of fragments can be sorted by their levels of *lac* expression, i.e., by their colors on *lac* indicator plates. This facilitates the identification of clones of different fragments or of a fragment in different orientations.[4]

Fusions can also be made by cloning DNA fragments without regard to their *lac* expression using standard cloning techniques: cloning can be done with an excess of fragments followed by screening of candidates for their plasmid DNA size and structure or by colony hybridization. Alternatively, the ends of a cleaved vector can be treated with phosphatase to prevent their ligation without an inserted fragment.

Frame-Shifting. For a DNA fragment with a promoter and translation initiation signals there are three possible phases with which it can be aligned with the *lacZ* codons. Fusions that are out of phase can be detected by their low levels of β-galactosidase but with high levels of the downstream *lacY* gene product. Expression of the *lacY* lactose permease can be detected using melibiose MacConkey agar at 37° or above. This classical *lac* procedure relies on the fact that the sugar melibiose can be transported by the *lacY* gene permease and that the original melibiose permease in *E. coli* K12 is temperature sensitive.[2]

An out of phase fusion can still be used to study regulation merely by assaying the low levels of β-galactosidase expression. The fusion can be made in phase by *in vitro* or *in vivo* manipulation. *In vitro* the fragment can be recloned into another β-galactosidase vector with the same cloning site in a different phase, or its ends can be altered and inserted into a different site to achieve a different phase. Alternatively, an out of phase fusion can be opened at or near its β-galactosidase joint and the phase changed by removing (as with the exonuclease *Bal*31) or by adding a few nucleotides (as with a linker).

Frames can be shifted *in vivo* by selecting for mutants with increased *lac* expression. This is most conveniently done by selecting for mutants that allow growth on lactose minimal media or yield red "papilla" growth from a colony incubated several days on lactose MacConkey agar. Normally we select spontaneous mutations formed without the use of a mutagen. These frame-shifts can be small changes or large deletions. Frameshift mutations can even be selected from fusions without a good promoter or translation initiation: the Lac⁺ selection is powerful enough to select double genetic changes, one to frame shift and one to improve a promoter or translation initiation site.[6]

An Example. An example of a fusion construction that employs some of the points we have discussed is our fusion of the yeast *leu2* gene promoter to the β-galactosidase gene. At first we did not know the sequence of this gene. It had been cloned on a plasmid, and its major restriction sites were mapped.[19] We noted that there was an *Eco*RI site within the gene. From the location of sites on the restriction map we assumed that the *Eco*RI site was within the coding region of *leu2*, not in a control element. The *leu2* gene was known to be expressed in *E. coli*, but its direction of transcription was not known. We chose a clone of *leu2*, YEp13,[19] which had two *Eco*RI sites in addition to the one in *leu2*, such that when digested with *Eco*RI, one of the three fragments would contain the promoter of the *leu2* gene. We chose the vector pMC1403 (Fig. 1) because it had a unique *Eco*RI cloning site. (At this point the yeast replicon vectors pMC1790 and pMC2010 had not yet been constructed.) The mixture of the three *Eco*RI digested fragments of YEp13 was ligated with *Eco*RI-cleaved pMC1403 vector DNA, and the resulting ligation mixture was used to transform the efficiently transformable, *lac* deletion, *E. coli* strain M182 (see the table). Transformants were selected on both M63-glucose–XG-ampicillin and on lactose MacConkey ampicillin. About 10% of the transformants made blue colonies on the XG plates, and all of these were equally blue. On lactose MacConkey all the colonies were initially white, but after prolonged incubation for 2 days about 10^{-3} to 10^{-4} of the colonies had a small red papilla. Several of the blue colonies and the red papilla colonies were picked and purified by streaking to single colonies. Plasmid DNA from all of them was indistinguishable on 0.7% agarose gels. They each had two *Eco*RI bands, one the size of the pMC1403 vector and the other the size of one of the three *Eco*RI fragments from the *leu2* plasmid. Furthermore, digestion with other enzymes revealed that all clones had the same fragment orientation, and that this orientation placed the end of the *Eco*RI fragment that was within *leu2* nearest to β-galactosidase. From this we concluded that the *leu2* promoter was within the cloned *Eco*RI fragment and that the *Eco*RI site was probably out of phase

[19] B. Ratzkin and J. Carbon, *Proc. Natl. Acad. Sci. U.S.A.* **74,** 487 (1977).

with the *lacZ* codons from the *Eco*RI site in pMC1403. The plasmids from the papilla cells probably had a small frame-shift, since they did not have a DNA structure that was altered as judged on the agarose gel. The fusion was then joined to a yeast replicon and placed into yeast, where it was found that β-galactosidase expression was repressed by leucine and threonine as expected for the *leu2* gene. This showed that the *Eco*RI fragment cloned contained the gene control signals for regulation in yeast. These observations were verified when the sequence was obtained (P. Schimmel and A. Andreadis, personal communication).

The sequence predicted new restriction sites that would yield a smaller *leu2* gene control fragment of approximately 650 base pairs between an *Xho*I and a *Bst*EII site. These enzymes yielded a fragment with sticky ends that were filled in with *E. coli* DNA polymerase I. This fragment was ligated into the blunt *Sma*I site of pMC1790 (Fig. 4), which yielded an in-frame fusion that was detected as a red colony clone on lactose MacConkey ampicillin agar using strain MC1066. Strain MC1066 is deleted for *lac,* transforms well, is *hsdR*⁻ restriction minus so as not to cleave unmodified DNA, and has the *E. coli trpC9830* mutation, which is complemented to Trp⁺ by the *trp1* gene from yeast on pMC1790. Plasmid DNA of this clone was used to transform yeast strain M1-2B to Trp⁺, (SD-glucose plates). These transformants were assayed for β-galactosidase after being grown with and without leucine and threonine (1 m*M* each). As expected β-galactosidase expression was repressible. Yeast cells containing this plasmid did not make enough β-galactosidase to form blue colonies on SD-glucose-XG plates, therefore they were assayed by replica plating onto filters and permeabilizing by freeze-thawing (see above).

The *leu2* promoter inserted into the *Sma*I site of pMC1790 did not have the *Sma*I site regenerated, but it could be excised at the adjacent *Eco*RI and *Bam*HI sites. These sites are the same as on pMC1403 (Fig. 1). Sequences upstream to the *leu2* promoter on this fragment could readily be deleted on this plasmid by using the unique *Eco*RI site on the vector. The plasmid clone was opened in the upstream site with *Eco*RI, partially degraded with the exonuclease *Bal*31, and the remaining fragment was excised by cutting on the other side with *Bam*HI. These deleted fragments, with a blunt end on one side formed by the *Bal*31 and a *Bam*HI end on the other side, were then ligated into the original pMC1790 vector, which had been cleaved with the blunt *Sma*I and with *Bam*HI. This resulted in a series of deletions of sequences on the upstream side of the *leu2* promoter, without any removal of vector sequences. These deletions are now being examined for their size and their *leu2* promoter regulation function in yeast.

Comments

We note that it may not be possible to fuse some gene control elements to the β-galactosidase gene with these plasmids. Some fragments may be lethal to the cell on a high copy plasmid. A promoter may be so strong that too much of a *lac* gene product is made, or transcription may proceed into the pBR322 replicon to repress replication (Remaut *et al.*[20] discuss this problem). Some hybrid protein β-galactosidases are lethal to the cell when synthesized at a high level, as is known for some membrane-bound or membrane-transported protein fusions.[21] In addition, the regulation of a gene may be altered when it is present in high copy number.

Some of these problems can be reduced by incorporating the fusion into a chromosome or a stable low copy number plasmid. In *E. coli,* a fusion can be incorporated onto a λ *lac* phage and integrated into the chromosome by lysogenization.[22] Alternatively, a fusion on a ColE1 replicon plasmid (as are the plasmids in this chapter) can be directly integrated by homologous recombination following introduction into a *pol* A⁻ cell that cannot replicate ColE1 replicons.[23] In yeast, plasmids without a yeast replicon can be integrated by homologous recombination after transformation into a strain with selection for a gene on the plasmid.[24]

Constructing a β-galactosidase gene fusion is more than a way to measure gene expression. Once it is made, it can be used as a source for further studies such as to isolate mutants with altered gene expression or regulation.[25] A translational fusion can also be used as a source of hybrid protein (see above).

β-Galactosidase fusions can be used in more species for which DNA transformation is possible, such as in gram-positive bacteria *Bacillus subtilis* and *Streptomyces* and in higher eukaryotes, such as plants and mammalian tissue culture cells.

Acknowledgments

We would like to acknowledge suggestions, help, and comments from our colleagues M. Berman, M. Ditto, M. Malamy, D. Nielsen, S. J. Suh, and H. Tu. This work was supported by NIH Grant GM 29067. S.K.S. is a Medical Scientist Trainee supported by NIH grant PHS 5T32 GM07281.

[20] E. Remaut, P. Stanssens, and W. Fiers, *Gene* **15,** 81 (1981).
[21] M. Hall and T. Silhavy, *Annu. Rev. Genet.* **15,** 91 (1981).
[22] M. Casadaban, *J. Mol. Biol.* **104,** 591 (1976).
[23] D. Kingsbury and D. Helinski, *Biochem. Biophys. Res. Commun.* **41,** 1538 (1970).
[24] D. Botstein, C. Falco, S. Stewart, M. Brennan, S. Scherer, D. Stinchcomb, K. Struhl, and R. Davis, *Gene* **8,** 17 (1979).
[25] J. Beckwith, *Cell* **23,** 307 (1981).

[24] The Use of pKC30 and Its Derivatives for Controlled Expression of Genes

By MARTIN ROSENBERG, YEN-SEN HO, and ALLAN SHATZMAN

There are numerous gene products of biological interest that cannot be obtained in quantities sufficient for detailed study of their structure and function. Over the past few years recombinant technology has offered new approaches to this problem. One such approach has been the development of vector systems designed to achieve efficient expression of cloned genes in bacteria.

In general, the rationale used in the design of these systems involves insertion of the gene of interest into a multicopy vector system (e.g., a plasmid) such that the gene is transcribed from a "strong" bacterial promoter. This usually ensures efficient transcription of the gene, but does not necessarily guarantee its expression. In particular, for those genes that do not naturally contain the proper signals for ribosome recognition and translation initiation in *Escherichia coli,* special procedures must be devised to supply this information. This is done either by fusing the gene to a bacterial ribosome binding site or to the N-terminal coding region of a bacterial gene. In the first case, some difficulties in obtaining efficient translation have been encountered owing to sequence alterations made in the ribosome recognition region prior to or as a result of gene insertion. In the latter case, the gene product is a fusion protein carrying additional peptide information at its N terminus. The fusion products may have physical and functional properties that differ from the normal protein, thereby limiting their value for biological study. In addition to factors such as promoter strength, gene copy number, and translational efficiency, a number of other factors may influence the expression of a cloned gene in bacteria. These include (*a*) reduction of transcription resulting from polarity effects; (*b*) the stability of the mRNA; (*c*) the stability of the gene product; and (*d*) the potential lethality of the product to the growth of the host.

This chapter describes a set of plasmid cloning vehicles that were constructed to achieve efficient expression of cloned genes in *E. coli.* These vectors contain transcriptional and translational signals that derive from the bacteriophage lambda genome. The design of the vectors and the rationale for using these particular regulatory sequences are discussed in relation to each of the above considerations. Procedures for using these systems are described in detail.

RECOMBINANT DNA
METHODOLOGY

Rationale of the Method

The system to be described utilizes a plasmid vehicle (a pBR322 derivative) carrying regulatory signals derived from the bacteriophage λ genome. Phage regulatory information was chosen because, in general, these signals tend to be more efficient than their host-derived counterparts. For example, with a vector system designed specifically for studying transcriptional regulatory signals,[1] the phage λ promoter P_L was shown to be 8–10 times more efficient than the bacterial promoter of the lactose operon (P_{lac}). In fact, P_L was as efficient, or more so, as all other bacterial promoters tested.[2,3]

Plasmids carrying P_L are often unstable, presumably owing to the high level of P_L-directed transcription.[4] This problem of instability was overcome by repressing P_L transcription, using bacterial hosts that contain an integrated copy of the λ genome (i.e., bacterial lysogens). In these cells, P_L transcription is controlled by the phage λ repressor protein (cI), a product that is synthesized continuously and regulated autogenously in the lysogen.[5] It was demonstrated that certain lysogens synthesize sufficient repressor to inhibit P_L expression completely on the multicopy vector.[6] Thus, the cells can be stably transformed and the vector maintained in these lysogenic hosts. Moreover, by using a lysogen carrying a temperature-sensitive mutation in the λ cI gene (cI857),[7] P_L-directed transcription can be activated at any time. Induction is accomplished by simply raising the temperature of the cell culture from 32° to 42°. Thus, cells carrying the vector can be grown initially to high density without expression of the cloned gene (at 32°), and subsequently induced to synthesize the product (at 42°). The ability to control gene expression on the vector, coupled with the rapidity of the induction procedure and the efficiency of P_L, ensures high-level expression of the product in a relatively short time period. These features are particularly useful for the expression of gene products that may be lethal and/or rapidly turned-over in bacteria.

In addition to providing a strong, regulatable promoter, the system

[1] K. McKenney, H. Shimatake, D. Court, U. Schmeissner, C. Brady, and M. Rosenberg, in "Gene Amplification and Analysis," Vol. 2: Analysis of Nucleic Acids by Enzymatic Methods" (J. G. Chirikjian and T. S. Papas, eds.), p. 383. Elsevier/North-Holland, Amsterdam, 1981.

[2] M. Rosenberg, K. McKenney, and D. Schümperli, in "Promoters: Structure and Function" (M. Chamberlin and R. L. Rodriguez, eds.) p. 387. Praeger, New York, 1982.

[3] A. Shatzman and M. Rosenberg, unpublished data.

[4] R. N. Rao, unpublished data.

[5] M. Ptashne, K. Backman, M. Z. Humayun, A. Jeffrey, R. Maurer, B. Meyer, and R. T. Sauer, Science 194, 156 (1976).

[6] H. Shimatake and M. Rosenberg, Nature (London) 292, 128 (1981).

[7] R. Sussman and F. Jacob, C. R. Hebd. Seances Acad. Sci. 254, 1517 (1962).

also ensures that P_L-directed transcription efficiently traverses any gene insert. This is accomplished by providing both the phage λ anti-termination function, N, and a site on the P_L transcription unit necessary for N utilization (*Nut* site). N expression from the host lysogen removes transcriptional polarity, thereby inhibiting termination within the P_L transcription unit.[8,9] Hence, any transcriptional polarity caused by sequences that occur before or within the coding sequence is eliminated by the N plus *Nut* system. As demonstrated below, this system leads to a dramatic increase in product yield and also allows much greater flexibility in inserting genes into the vector.

In order to extend this system to the expression of genes lacking *E. coli* translational regulatory information, efficient ribosome recognition and translation initiation sites were engineered into the P_L transcription unit. The site chosen was that of an efficiently translated λ phage gene, *cII*. The entire coding region of this gene was removed, leaving only its initiator fMet codon and regulatory sequences upstream. Neither the sequence nor the position of any nucleotides in the ribosome binding region was altered. Instead, a restriction site for insertion of the desired gene was introduced immediately downstream from the ATG initiation codon. As described below, this system allows direct fusion of any coding sequence to this translational regulatory signal and, furthermore, allows any gene to be adapted for insertion into the vector.

Cloning Prokaryotic Genes into pKC30

The plasmid pKC30 (Fig. 1A) is used to overexpress bacterial genes that contain their own translational regulatory information. This vector contains a unique *Hpa*I restriction site positioned 321 bp downstream from the P_L promoter. Blunt-ended DNA fragments are inserted into this blunt-ended restriction site. The appropriate fragments can be generated: (*a*) directly by restriction; (*b*) subsequent to the removal of single-strand overhanging ends by the action of S_1 or mung bean nuclease; or (*c*) subsequent to the "fill-in" of single-strand overhanging ends by the use of DNA polymerase (Klenow fragment). The *Hpa*I site of pKC30 also can be adapted for the insertion of other DNA fragments by first introducing various synthetic linkers into the site. In addition, the site can be used in combination with other unique restriction sites positioned downstream

---

[8] S. Heinemann and W. Spiegelman, *Cold Spring Harbor Symp. Quant. Biol.* **35,** 315 (1971).

[9] M. Rosenberg, D. Court, D. L. Wulff, H. Shimatake, and C. Brady, *in* "The Operon" (J. Miller, ed.), p. 345. Cold Spring Harbor Laboratory, Cold Spring Harbor, New York, 1978.

A

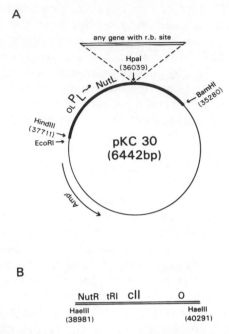

B

			NutR	tRI	cII		O	
Haelll							Haelll	
(38981)							(40291)	

FIG. 1. (A) Schematic diagram of plasmid pKC30. This plasmid is a derivative of pBR322 and contains a 2.4-kb *Hind*III-*Bam*HI restriction fragment derived from phage λ inserted between the *Hind*III and *Bam*HI restriction sites within the tetracycline gene of pBR322.[4] The λ insert contains the operator (O_L), the promotor signal (P_L), and a site for N recognition (*Nut L*). [N. C. Franklin and G. N. Bennett, *Gene* **8**, 197, (1979)]. Fragments are cloned into the *Hpa*I restriction site that occurs within the N gene coding region. The numbering system of the λ DNA segment is that of F. Blattner (personal communication). (B) The 1.3 kb *Hae*III restriction fragment of phage λ, which was inserted into the *Hpa*I site of pKC30. This fragment contains the entire *cII* coding region, a site for N recognition (*Nut R*), the rho-dependent transcription termination site (*tRI*), and most of the O gene region.

from the *Hpa*I site on pKC30 (Fig. 1A). Note that the *Hpa*I cloning site interrupts the coding region of the λ N gene, which occurs on pKC30. Hence, insertion into this site also gives rise to N protein fusion products that result from N gene translation entering the promoter proximal part of the DNA insert. The effect of this translation on the expression of genes cloned into pKC30 has not been examined.

The pKC30 vector has been used to express efficiently several bacterial gene products. The application of the system is probably best exemplified by its use in production of the phage λ transcriptional activator protein *cII*,[6,10] and eight *cII* protein variants that differ by only single amino

[10] Y. S. Ho, M. Lewis, and M. Rosenberg, *J. Biol. Chem.* **257**, 9128 (1982).

acid substitutions.[11,12] Although *cII* is rapidly turned over in *E. coli* and is lethal to cell growth, its insertion and expression in pKC30 allowed levels of synthesis approaching 3–5% of total cellular protein.[6] The following describes in detail the cloning of such a gene fragment into pKC30.

Preparation of HpaI-Restricted pKC30. pKC30 DNA was transformed[13] into an *E. coli* N99λ*cI*+ lysogen, and plasmid DNA was prepared. The yield was approximately 1 μg of DNA per milliliter of cell culture. Plasmid DNA (5–20 μg) was restricted with *Hpa*I (all restriction conditions and enzymes are those of New England BioLabs). The reaction was stopped, and the DNA was recovered by phenol extraction and ethanol precipitation. The DNA was resuspended in 10 mM Tris, pH 8.0 (10–50 μl).

Preparation of the Gene Insert. A 1.3-kb *Hae*III fragment from phage λ DNA that carries the *cII* gene (Fig. 1B) was inserted into the *Hpa*I site. Phage DNA (250 μg) was restricted with *Hae*III (200 units) and, after phenol extraction and ethanol precipitation, was dissolved in 600 μl of sample buffer (40 mM Tris-HAc, pH 7.9, 2.0 mM EDTA, 2.5% glycerol, 0.012% bromophenol blue and xylene cyanol). The sample was loaded into a 5.2 cm wide slot of a preparative 5% polyacrylamide gel (0.15 × 14 × 40 cm³), and the products were resolved by electrophoresis at 200 V for 18 hr. After electrophoresis, the gel was stained with methylene blue (0.02% for 20 min). The 1.3 kb DNA fragment was cut out of the gel and eluted electrophoretically into 1 ml of 20 mM Tris-HAc, pH 8.0. The DNA sample was extracted with phenol and precipitated twice with cthanol. Fragment recoveries vary between 50 and 75%.

Ligation. Fragments are inserted into the *Hpa*I site by blunt-end ligation[14] using T4 DNA ligase (P-L Biochemicals). It is important that reaction volumes be kept small (~25 μl) in order to increase the efficiency of fragment insertion. The ratio of fragment insert to pKC30 vector used in the ligation reaction is ~2:1 (e.g., 0.4 pmol fragment: 0.2 pmol pKC30). The ligation reaction is carried out for 14 hr at 12–15°, a somewhat lower temperature than is usually recommended. This is done to increase the efficiency of blunt-end ligation at A·T rich restriction sites (e.g., *Hpa*I). The ligation reaction is stopped by heating at 70° for 5 min. If the inserted fragment neither recreates nor contains an *Hpa*I restriction site, *Hpa*I (5 units in 50 μl of *Hpa*I restriction buffer) is added to the ligation mixture to

---

[11] Y. S. Ho, M. Mahony, and M. Rosenberg, in preparation.
[12] D. Wulff, M. Mahony, A Shatzman, and M. Rosenberg, in preparation.
[13] R. W. Davis, D. Botstein, and J. R. Roth, *in* "Advanced Bacterial Genetics." Cold Spring Harbor Laboratory, Cold Spring Harbor, New York, 1981.
[14] A. Ullrich, J. Shine, J. Chirguin, R. Pictet, E. Tischer, W. Rutter, and H. Goodman, *Science* **196**, 1313 (1977).

recut those pKC30 molecules that rejoined without an insert. The reaction mixture is then phenol extracted, ethanol precipitated, dried, and redissolved in 50 μl of H_2O.

Transformation and Clone Analysis. The ligated DNA (25 μl) is then used to transform a λ lysogen carrying a wild-type repressor gene (e.g., *E. coli* N100λcI^+). Ampicillin-resistant recombinants are obtained and screened by size and restriction analysis for the presence and orientation of the insert. This is done by preparing rapid-plasmid DNA isolates[13] of individual clones and examining this DNA on 1% agarose gels using appropriate size markers. Since blunt-end ligation can result in the fragment being inserted in either orientation, it is important to select an appropriate restriction analysis to distinguish the different orientations.

Expression of the Gene Product. In order to express the cloned gene, the pKC30 derivative (e.g., pKC30cII) is first transformed into a λ lysogen carrying a temperature-sensitive mutation in its repressor gene. The transformed cells are grown overnight in LB broth containing ampicillin (50 μg/ml) at 32°; 2 ml of this cell culture are then inoculated into 200 ml of LB containing ampicillin (50 μg/ml). The culture is incubated at 32° until $A_{650} = 1$. At this time, an equal volume of LB, prewarmed to 65° is added with swirling to elevate the temperature rapidly to 42°, and the culture is incubated at 42° for another 60–90 min.[15,16] The cells are harvested

---

[15] The temperature induction also has been carried out in 50–100-liter volumes using a standard fermentor. Raising the temperature of the culture from 32° to 42° requires approximately 3 min.

[16] The time period of induction resulting in the best yield varies (45–120 min) and appears to depend upon the nature of the gene product, the host cell, and type of phage lysogen. For each cloned gene, expression should be monitored and compared in a number of lysogenic hosts.

FIG. 2. An SDS–polyacrylamide gel analysis of proteins made in a $\lambda cI857$ lysogen carrying pKC30cII before and after heat induction. (A) Autoradiogram of pulse-labeled proteins. Cells were grown at 32° in modified minimal media M56 [M. E. Gottesman and M. B. Yarmolinsky, *J. Mol. Biol.* **31**, 487 (1968)] with ampicillin (50 μg/ml) until $A_{650} = 0.6$. The cells were harvested by centrifugation and resuspended in minimal media M56, supplemented with 0.2% glucose, 0.2 m*M* amino acid mixture lacking methionine, 1 μg of thiamine and 50 μg of ampicillin per milliliter. Aliquots (200 μl) of these cells were pulse-labeled for 1 min with 20 μCi of [^{35}S]methionine (Amersham, >600 Ci/mmol) at 32° and at 42°, 20 min after incubation at these temperatures. After pulse labeling, 30 μl of 10% SDS and 35% 2-mercaptoethanol were added, and the samples were heated at 95° for 1 min, followed by quick freezing in Dry Ice–ethanol bath, and precipitated with trichloroacetic acid (final concentration 10%). The precipitates were washed twice with cold ethanol, resuspended in sample buffer, heated at 95° for 5 min, and subjected to electrophoresis on a 15% polyacrylamide gel. The gels were finally dried and fluorographed [W. M. Bonner and R. A. Laskey, *Eur. J. Biochem.* **46**, 83 (1974)]. (B) Coomassie brilliant blue-stained SDS–polyacrylamide gel analysis of total cellular protein prepared before and after a 40-min heat induction as described previously.[6]

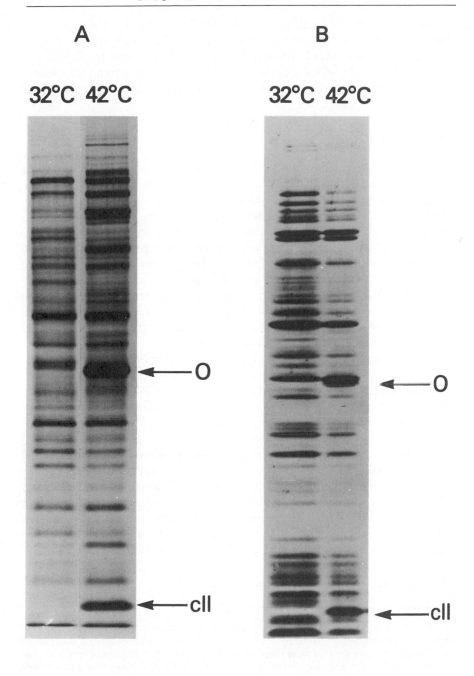

by centrifugation, washed once with TMG buffer (Tris–MgSO$_4$–gelatin buffer, pH 7.5) to remove the ampicillin, and then immediately frozen. Using this procedure, cII protein (subunit molecular weight 10,500) was produced as 3–5% of total cellular protein and purified to a final yield of 1 mg of homogeneous protein per gram wet weight of cell culture (Fig. 2).[6,10]

The Use of Antipolarity: $N + Nut$

When a gene is cloned into pKC30, sequences upstream of the coding region, which are untranslated or carry transcription termination signals, may be inserted as well. The polar effects on transcription caused by these sequences will reduce gene expression. For example, in the construction of pKC30cII a termination site, *tR1*, positioned immediately upstream of *cII* was inserted along with the gene. The presence of *tR1* could have drastically reduced *cII* expression. To circumvent these problems, we use the phage λ antitermination factor, N, which funtions on P_L-directed transcription. N is provided in single copy from the host lysogen and its expression is induced by temperature concomitant with P_L transcription on the vector. The regulatory site required for N utilization (*Nut* site) is provided on all pKC30 derivatives.

The importance of N was first demonstrated with the vector pKC30cII. *cII* production was found to be ~8 times higher in lysogens that provided N (N^+) as opposed to those that did not (N^-).[17] In more recent experiments,[3] it was shown that the single-copy lysogen provides sufficient N to antiterminate completely all transcription through a terminator on a multicopy plasmid. Moreover, because N generally relieves all transcriptional polarity, it can even increase expression of genes cloned into pKC30 that do not have specific terminators preceding them. For example, the *E. coli* galactokinase gene (*galK*) was cloned into the *Hpa*I site of pKC30. Expression was monitored before and after induction of both an N^+ and N^- lysogen. The results (Fig. 3) indicate that, after induction, 3–4 times more *galK* is produced in the N^+ cells than in the N^- cells. No *galK* is detected in either background before induction.

---

[17] H. Shimatake and M. Rosenberg, unpublished data.

FIG. 3. A Coomassie Brilliant Blue-stained SDS–polyacrylamide gel analysis of total cellular protein obtained from an N^+ and an N^- defective lysogen carrying the plasmid pKC30galK. Protein samples were prepared from cells before (at 32°) and 60 min after thermal induction (at 42°) and analyzed as described in Fig. 2. The position of the galactokinase gene product is indicated by an arrow.

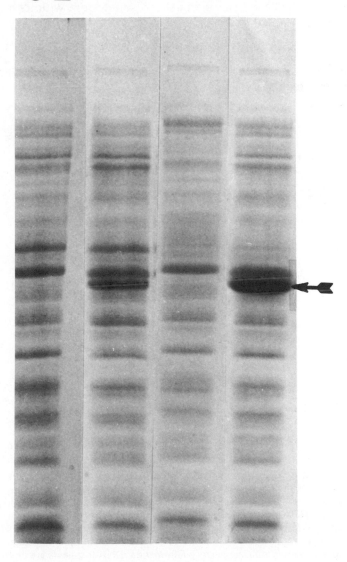

The effect of N on *galK* expression can be attributed primarily to its ability to relieve polarity occurring before or within the *galK* gene. However, it is also possible that N may be directly affecting the translation of genes that come under its transcriptional control. Whatever the case, the effect is significant (*galK* is produced as >20% of total cellular protein in the N^+ condition) and eliminates the necessity for manipulating sequences upstream of the gene that might otherwise interfere with its expression.

Adapting pKC30 for the Expression of Eukaryotic Genes

The plasmid constructed for the expression of genes that do not normally carry regulatory signals for their translation in bacteria is shown in

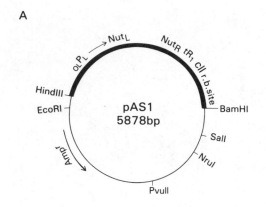

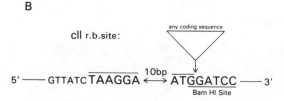

FIG. 4. (A) Schematic diagram of plasmid pAS$_1$. This vector is a derivative of pKC30 and is made of phage λ sequences (thickened line) inserted between the *Hin*dIII and *Bam*HI restriction sites in pBR322. The region of pKC30 between the *Hpa*I (36060) and *Bam*HI (35301) site has been deleted, and a portion of the fragment shown in Fig. 1B, extending from the *Hae*III site (38981) to and including the ribosome binding site (r.b. site) and ATG initiation codon of *c*II, has been inserted. See text for other details.   (B) The region of pAS$_1$ that surrounds the *cII* translational regulatory information. The ribosome binding site and ATG initiation codon (overscored) of *cII* are indicated, as is the unique *Bam*HI site (underscored), which provides access to this regulatory information. DNA fragments are inserted into this site as detailed in the text.

Fig. 4A. This vector, pAS$_1$, is essentially identical to pKC30cII except that all λ sequences downstream of the *cII* initiation codon have been deleted. The *Bam*HI site of pBR322 is now fused directly to the *cII* ATG (Fig. 4B). This fusion retains the *Bam*HI site and positions one side of the staggered cut immediately adjacent to the ATG codon permitting ready access to the *cII* translational regulatory information. Eukaryotic and/or synthetic genes, can be adapted and fused to this translation initiation signal. It is most important that all fusions between the gene coding sequence and the *cII* initiation codon maintain the correct translation reading frame. Below, various procedures are described for inserting genes into the pAS$_1$ vector. Note that all cloning experiments with pAS$_1$, like those for pKC30, are carried out in a *cI*$^+$ lysogen in order to maximize stability of the vector. Expression of the cloned gene takes place in the *cI*^{ts} lysogen using procedures identical to those described above for pKC30.

Cloning and Expression of Genes in pAS$_1$

Direct Insertion at the BamHI Site. The only genes that can be fused directly to the *cII* initiation codon are those that contain a *Bam*HI, *Bgl*II, *Sau*3A, or *Bcl*1 restriction site[18] at or near their own initiation codon. The necessary restriction site may occur naturally within the gene or be engineered into the gene by recombinant or synthetic techniques. Standard procedures are used for the cloning and for clone analysis and expression.

Two genes have been cloned and expressed in pAS$_1$ using this technique, the β-galactosidase gene (*lacZ*) of *E. coli* and the metallothionein II gene from monkey. The *lacZ* gene was engineered to contain a unique *Bam*HI site near its 5′ end,[19] whereas the metallothionein gene naturally contained a *Bam*HI site at its second amino acid codon.[20] In both cases direct *Bam*HI ligation of the gene into pAS$_1$ appropriately positioned the coding sequence in frame with the *cII* ATG codon. Expression of both genes was controlled entirely by transcriptional and translational signals provided by pAS$_1$. As shown in Fig. 5A, the pAS$_1$*lacZ* construction results in high level expression of β-galactosidase. Similar results were obtained with the monkey metallothionein gene (not shown).

Adapting pAS$_1$ for Other Genes. Most genes do not contain the restriction information necessary for their direct insertion into the *Bam*HI site of pAS$_1$. Thus, it was necessary to provide greater versatility for inserting

---

[18] These sites all share the common four-base 5′-overhanging end pGATC...

[19] This construction was made and kindly provided by M. Casadaban, J. Chou, and S. Cohen, *J. Bacteriol.* **143**, 971 (1980).

[20] Kindly provided by D. Hamer.

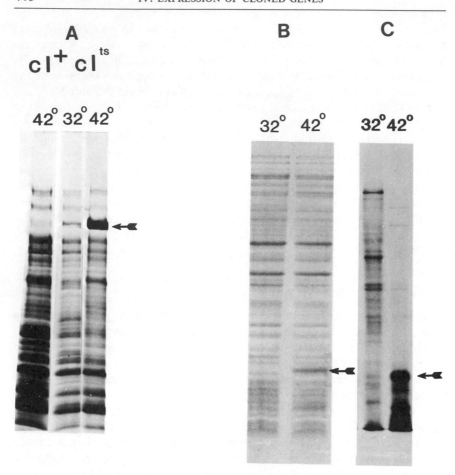

FIG. 5. (A) A Coomassie Blue stained SDS–polyacrylamide gel analysis of total cellular protein obtained from a cI^+ and a cI^{ts} host lysogen carrying the plasmid pAS$_1$βgal. Temperature inductions, sample preparations, and gel electrophoresis are the same as described in Figs. 2 and 3. The position of the β-galactosidase gene product is indicated (arrow).   (B) Analysis (as in A) of total cellular protein obtained from a cI^{ts} lysogen carrying the plasmid pAS$_1$t. The position of the SV40 small t antigen is indicated (arrow).   (C) An autoradiogram of an SDS–polyacrylamide gel analysis of ^{35}S pulse-labeled proteins synthesized in a cI^{ts} lysogen carrying pAS$_1$t before and 60 min after heat induction. All procedures are similar to those described in Fig. 2A.

DNA fragments into the vector. This was accomplished by converting the *Bam*HI site of pAS$_1$ into a blunt-ended cloning site. The four-base 5′-overhanging end of the *Bam*HI cleavage site can be removed using mung bean nuclease, thereby creating a blunt-end cloning site immediately adjacent to the *cII* initiation codon. Mung bean nuclease is used because this enzyme has a high exonucleolytic specificity for single-stranded DNA. We have found this enzyme to be reproducibly better than S1 nuclease for selectively removing the four-base overhang sequence. The following conditions are used for mung bean nuclease removal of the four base overhang generated by BamHI cleavage of pAS$_1$.

pAS$_1$ DNA (1.5 pmol; 5 μg) is linearized by BamHI restriction and then treated with mung bean nuclease (25–30 units, P-L Biochemicals) in 50 μl of 30 mM NaOAc (pH 4.6), 250 mM NaCl, 1 mM ZnCl$_2$, 5% glycerol for 30 min at 30°. The reaction is stopped by the addition of SDS (to 0.2%), extracted with phenol, and precipitated twice with ethanol. Approximately 50% of those molecules that have lost their BamHI site are blunt-ended precisely to the correct site.

Any gene containing any restriction site properly positioned at or near its 5' end can now be inserted into this vehicle. Blunt-ended fragments can be inserted directly, whereas other restriction fragments must first be made blunt-ended. This is accomplished by either removing the 5'- and 3'-overhanging ends with mung bean nuclease (as above) or "filling-in" the 5'-overhanging ends with DNA polymerase.[21] The fill-in reaction is carried out as follows. The DNA (0.3 pmol) is dissolved in 100 μl of 6 mM Tris-HCl (pH 7.4), 6 mM MgCl$_2$, 60 mM NaCl containing 20 μM of each of the four deoxynucleotide triphosphates. DNA polymerase (1 unit; large fragment after Klenow, Boehringer-Mannheim) is added and the reaction incubated at 15° for 3 hr. The reaction is stopped by heat inactivating the polymerase at 65° for 5 min. The reaction mix is extracted with phenol, and the DNA is recovered by ethanol precipitation.

This procedure was used to adapt, insert, and express the metallothionein I gene of mouse in the pAS$_1$ system. The mouse gene contains a unique AvaII restriction site at its second codon (5' $\overline{\text{ATG GAC CCC}}$ 3'). Cleavage with AvaII, followed by fill-in of the three base 5'-overhanging end with DNA polymerase creates a blunt-end before the first base pair of the second codon. This blunt-end fragment was inserted into the mung bean nuclease-treated pAS$_1$ vector. The resulting fusion places the second codon of metallothionein in-frame with the initiation codon of cII. This construction results in high levels of synthesis of the authentic metallothionein gene product in $E.\ coli$.

Adapting Any Gene for Expression in pAS$_1$. The procedures described above still limit the use of pAS$_1$ to those genes that contain appropriate restriction information near their 5' termini. In order to make the pAS$_1$ system generally applicable to the expression of any gene, a procedure was developed that allows precise placement of a new restriction site at the second codon (or any other codon) of any gene. Creation of this site

---

[21] F. Rougeon, *Nucleic Acids Res.* **2**, 2365 (1975).

permits fusion of the gene in-frame to the *cII* initiation codon as pAS_1. The general procedure is outlined below.[22]

1. The gene of interest is cloned into a plasmid vector such that the vector can be opened uniquely at a restriction site just upstream of the gene (e.g., RS1, Fig. 6A).

2. The plasmid, after cleavage at RS1 is digested with the double-stranded exonuclease *Bal*31.[23] Conditions are selected which ensure digestion into the region surrounding the second codon of the gene. For example, a typical reaction will contain 100 μg of plasmid DNA dissolved in 500 μl of 20 mM Tris-HCl (pH 8.0), 12 mM CaCl$_2$, 12 mM MgCl$_2$, 200 mM NaCl, 1 mM EDTA, and *Bal*31 exonuclease (2–5 units, Bethesda Research Laboratories). The reaction is carried out at 29° and the time of incubation varies depending upon the number of base pairs to be removed. Although the rate of degradation depends upon the particular

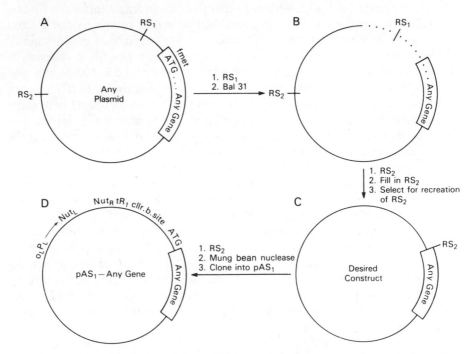

FIG. 6. Schematic diagram of the procedure used to adapt any gene for precise insertion into the pAS_1 expression system. See text for details.

[22] A similar procedure has been used to introduce a *Sal*I site at the initiation codon of the bacteriophage T7 gene *1.1*; N. Panayotatos and K. Truong, *Nucleic Acids Res.* **9**, 5679 (1981).

[23] H. Gray, D. Ostrander, J. Hodnett, R. Legerski, and D. Robberson, *Nucleic Acids Res.* **2**, 1459 (1975).

DNA sequence being digested, a time estimate can be calculated according to Gray *et al.*[23] In order to monitor more precisely the extent of digestion, aliquots (~20 μg) of the reaction are removed at various times and subjected to detailed restriction analysis. In each case the reaction is stopped by addition of EDTA (to 20 mM) and phenol. Two successive phenol extractions are carried out, and the DNA is recovered by ethanol precipitation.

3. After digestion by *Bal*31 to the proper extent, the DNA sample is then restricted at a second unique site (RS2, Fig. 6B), positioned well upstream of RS1. This second site should be far enough upstream from RS1 so as not to have been affected by the *Bal*31 digestion. Most important, RS2 should have a six-base pair recognition sequence that gives rise to a four-base 5'-overhanging end and in addition, have as its sixth base pair the same base as the first bp of the second codon of the gene to be expressed (e.g., *Eco*RI, GAATTC or *Bam*HI, GGATCC for codons starting with CXX; *Bgl*II, AGATCT or *Hind*III, AAGCTT for codons starting with TXX; *Xho*I, CTCGAG or *Xma*I, CCCGGG for codons starting with GXX; and *Xba*I, TCTAGA or *Bcl*I, TGATCA for codons starting with AXX).

4. The four-base 5'-overhanging end of RS2 is filled-in with DNA polymerae (to recreate five-sixths of the RS2 site) and then blunt-end ligated to the heterogeneous ends created by *Bal*31 digestion (using procedures detailed above).

5. Recombinant plasmids are selected and screened by restriction for those that have recreated RS2 (Fig. 6C). Only those molecules that end in the appropriate base pair will re-form the RS2 site.

6. Those clones containing RS2 are examined by restriction analysis and/or direct sequence analysis. The correct construction has recreated the RS2 site by fusing it to the first base of the second codon of the gene.

7. Restriction of this new construction at RS2, followed by treatment with mung bean nuclease (as above), provides a blunt-ended cloning site adjacent to the first base of the second codon. This blunt end can be fused directly in-frame to the *cII* initiation codon, similarly made blunt-ended by mung bean treatment of the *Bam*HI restricted pAS$_1$ vector (Fig. 6D).

The above procedure has been used to adapt several genes for insertion and expression into pAS$_1$. For example, the small t antigen gene of SV40 does not contain an appropriate restriction site at its 5' end. Using *Bal*31 digestion from an upstream site, the first base pair of the second codon of the small *t* gene (ATG GAT···) was fused to an upstream, filled-in *Ava*I restriction site (···CCCGA).[24] The fusion, (···CCCGAGAT···) recreates the *Ava*I site precisely at the second codon of the small *t* gene.

[24] This construction was made and kindly provided by C. Queen.

Restriction of this vector with AvaI followed by mung bean nuclease digestion produces a blunt end that was fused in-frame to the blunt-ended BamHI site of pAS$_1$. The resulting vector, pAS$_1$t, expresses authentic SV40 small t antigen from phage regulatory signals. After only a 60-min induction period, small t antigen represents some 10% of the total cellular protein (Fig. 5B). ^{35}S-pulse labeling experiments indicate that small t is the major product synthesized in these bacteria (Fig. 5C). Apparently, the pKC30-pAS$_1$ vector system offers the potential for efficiently expressing essentially any gene in $E.\ coli$.

Acknowledgments

We thank Gail Taff for typing and editing the manuscript.

[25] Expression of Heterologous Unfused Protein in Escherichia coli

By Erik Remaut, Anne Marmenout, Guus Simons, and Walter Fiers

Introduction

One of the major achievements of recombinant DNA technology is the high-level expression of adventitious genes in *Escherichia coli*. In particular, a steadily increasing number of eukaryotic proteins that are difficult to obtain from their natural sources are now being produced in massive amounts by "engineered" bacteria. The cloning of a eukaryotic gene onto a bacterial plasmid and its introduction into *E. coli* does not, in general, lead to an efficient synthesis of the corresponding protein, because the eukaryotic DNA lacks the specific signals necessary for it to be recognized by the host's transcription–translation machinery. To remedy this problem, expression vectors have been developed that incorporate the essential control elements to ensure efficient transcription and translation of essentially any coding region in *E. coli*. The salient features of such an expression vector are (1) the presence of a strong and preferably regulatable promoter and (2) the presence of an efficient ribosome-binding site and initiation codon which are easily accessible for insertion of adventitious coding sequences. The most widely used strategy to clone eukaryotic genes is reverse transcription of (semi-) purified mRNA. The cDNA clones obtained in this way contain sequences upstream from the coding region that are not suitable for expression in bacterial systems. In cases where the mRNA codes for a secreted protein the coding region of the mature protein is preceded by the signal peptide which is removed during the process of secretion. In order to obtain efficient expression of the mature protein in *E. coli,* the cDNA needs to be tailored in such a way that the codon for the first amino acid of the mature protein can be precisely fused to the initiation codon of the bacterial expression vector.

This chapter describes the use of plasmid expression vectors based on the inducible leftward promoter of coliphage λ. These vectors have been engineered to incorporate bacterial ribosome binding sites designed for easy insertion of adventitious coding sequences. Examples of tailoring cloned cDNA to obtain precise fusion of coding sequences to the prokaryotic translation signals are discussed.

RECOMBINANT DNA
METHODOLOGY

Principle of the Method

   There are two aspects to be considered in the overall design of an expression system for unfused proteins: (1) construction of an expression vector suitable for easy insertion of adventitious coding sequences, and (2) methods to tailor coding sequences such that these can be precisely joined to the expression signals of the vector.

1. Design of an Expression Vector

   The expression vectors used in this work are derivatives of pBR322[1] containing a 247-bp DNA fragment carrying the leftward operators and leftward promoter (P_L) of phage λ.[2] This strong promoter is tightly regulated by a repressor protein, product of the phage gene *cI*. Mutants are available that synthesize a thermolabile repressor.[3] This property allows the control of the activity of the P_L promoter by a simple temperature shift. The P_L expression plasmids[2] are propagated at 28° in *E. coli* strains which synthesize a temperature-sensitive repressor. While repression of the P_L promoter is virtually complete at 28°, full activity is obtained by raising the temperature to 42°. The ability to regulate expression is an important feature of an expression system since continuous high-level expression of an adventitious protein may be lethal to the cell.[4] The *cI* repressor product can be supplied to the system in a variety of ways. Strains of *E. coli* are available which synthesize a thermolabile repressor from a defective prophage, e.g., strain K12ΔH1Δ*trp*.[5] Alternatively, the mutant *cI* gene can be present on a compatible plasmid such as pRK248cIts,[5] pcI857,[6] or on an F' episome.[7] The dual plasmid system allows the use of essentially any *E. coli* strain as a host for P_L expression vectors. This versatility is of special interest, as it may be that proper choice of the host results in elevated levels of accumulation of a given protein. While the presence of a strong regulatable promoter on an expression plasmid generally ensures efficient transcription of cloned DNA sequences, efficient translation into protein depends on the presence on the cloned sequence of a ribosome-binding site that is recognized by the *E. coli* host. As outlined in the Introduction, eukaryotic genes are unlikely

---

[1] F. Bolivar, R. L. Rodriguez, P. Y. Greene, M. C. Betlach, H. L. Heynecker, H. W. Boyer, Y. H. Crosa, and S. Falkow, *Gene* **2,** 95 (1977).

[2] E. Remaut, P. Stanssens, and W. Fiers, *Gene* **15,** 81 (1981).

[3] M. Lieb, *J. Mol. Biol.* **16,** 149 (1966).

[4] E. Remaut, P. Stanssens, and W. Fiers, *Nucleic Acids Res.* **11,** 4677 (1983).

[5] H.-U. Bernard, E. Remaut, M. V. Herschfield, H. K. Das, D. R. Helinski, C. Yanovsky, and N. Franklin, *Gene* **5,** 59 (1979).

[6] E. Remaut, H. Tsao, and W. Fiers, *Gene* **22,** 103 (1983).

[7] M. Mieschendal and B. Müller-Hill, *J. Bacteriol.* **164,** 1366 (1985).

to display this feature. It is therefore necessary to develop expression vectors which incorporate a strong ribosome-binding site of bacterial origin. Moreover, the nucleotide sequence at the initiation codon has to be engineered in such a way that it becomes easily accessible for precise fusion of a eukaryotic coding region to the ATG codon. This is most frequently accomplished by "creating" a unique restriction site which partly overlaps with the initiation codon. Ways to achieve this end, some of which are detailed in the Methods and Results section, are outlined below (see Table I).

a. *Removal of 5' Protruding Ends with Single-Strand-Specific Nucleases.* The initiation codon ATG is engineered in such a way that it is part of a restriction site recognized by a restriction enzyme generating 5' protruding ends. Following cleavage with the restriction enzyme and removal of the 5' protruding end with a single-strand-specific nuclease, the ATG codon can be exposed as a blunted end (see Table IA). Single-strand-specific nucleases currently available are *Aspergillus oryzae* S1 nuclease[8] and mung bean nuclease.[9] It should be noted that these nucleases are often not very reliable in their action, i.e., base-paired nucleotides adjacent to the single strand may be removed during the reaction.

b. *Filling in of 5' Protruding Ends.* The enzyme *Nco*I cleaves the sequence 5'-CCATGG-3', between the two C residues, generating 5' protruding ends. These ends can be filled in using the 5' to 3' polymerizing reaction of DNA polymerases, thus generating a blunted ATG codon (see Table IB). The large fragment of *E. coli* DNA polymerase 1 (Klenow fragment)[10] or T4 DNA polymerase are frequently used for this purpose. Both enzymes lack the 5' to 3' exonuclease activity. In the presence of deoxyribonucleoside triphosphates, the 3' to 5' exonuclease activity of these enzymes is inhibited in favor of the polymerizing reaction. The fidelity of this reaction is very high.

c. *Resection of 3' Protruding Ends.* A number of restriction enzymes recognize a palindromic sequence starting with a 5' G residue and cleave that sequence leaving 3' protruding ends. These 3' ends can be resected by the 3' to 5' exonuclease activity of, e.g., Klenow fragment or T4 DNA polymerase. When the reaction is performed in the presence of dGTP, 3' to 5' resection into the duplex DNA is inhibited, because the much more active 5' to 3' polymerizing activity takes over by exchanging the G residue opposite the C. Restriction sites of the type mentioned here can be engineered to partly overlap with an ATG codon so that the enzymatic

[8] V. M. Vogt, *Eur. J. Biochem.* **33,** 192 (1973).

[9] W. D. Kroeker, D. Kowalski, and M. Laskowski, *Biochemistry* **15,** 4463 (1976).

[10] H. Jacobsen, H. Klenow, and K. Overgaard-Hansen, *Eur. J. Biochem.* **45,** 623 (1974).

TABLE I
METHODS TO EXPOSE THE ATG CODON OF AN EXPRESSION VECTOR AS A BLUNTED END[a]

	Specific example	Alternatives

A. Removal of 5'-protruding ends with single-strand-specific nucleases

Specific example:

↓
5'-ATGAATTC-
3'-TACTTAAG-

↓ EcoRI ↑

5'-ATG
3'-TACTTAA

↓ S1

5'-ATG
3'-TAC

Alternatives:

↓
5'-ATGTGCAC   ApaLI
↓
5'-ATGGTACC   Asp718
↓
5'-ATGGATCC   BamHI
↓
5'-ATGCGCGC   BssHII
↓
5'-ATGCTAGC   NheI
↓
5'-ATGTCGAC   SalI

B. Filling in of 5'- protruding ends

Specific example:

↓
5'-CCATGG-
3'-GGTACC-

↓ NcoI ↑

5'-C
3'-GGTAC

↓ Klenow PolI or T4 Pol

5'-CCATG
3'-GGTAC

Alternatives:

Not known to date

C. Resection of 3'-protruding ends

Specific example:

↓
5'-ATGGTACC
3'-TACCATGG

↑
↓ KpnI

5'-ATGGTAC
3'-TAC

dGTP ↓ Klenow PolI or T4 Pol

5'-ATG
3'-TAC

Alternatives:

↓
5'-ATGGGCCC   ApaI
↓
5'-ATGACGTC   AatII
↓
5'-ATGGCGCC   BbeI
↓
5'-ATGAGCTC   SacI
↓
5'-ATGCATGC   SphI

[a] Examples of sequences in which an ATG codon partly overlaps with a restriction site are illustrated. Methods to expose the ATG codon as a blunted end are worked out for a specific example. Alternative possibilities are indicated with only one strand of the duplex DNA being displayed. Only hexameric recognition sequences were considered. The small arrows indicate cleavage points of the restriction enzymes.

reactions will result in blunt-end exposure of the ATG codon (see Table IC). As in the previous example the reaction can be easily controlled.

2. Tailoring of Coding Regions for Precise Fusion to the Initiation Signals of an Expression Vector

Only in rare cases will a cloned cDNA possess a restriction site positioned thus that the coding region can be precisely fused to the ATG codon of an expression vector by simple cleavage and joining reactions. Far more commonly one has to rely on a restriction site present some distance downstream from the desired fusion point. In this approach the gene is cleaved at the internal restriction site and synthetic DNA fragments are used to establish the link between the ATG codon of an expression vector and the restriction site of the gene. The nucleotide sequence of the DNA linkers is designed to restore the missing coding information between the desired point of fusion and the restriction site used (see Methods and Results section for specific examples). A prerequisite of this general approach is, of course, that the nucleotide sequence of the gene has been determined.

Materials and Reagents

Enzymes and Chemicals

1. Restriction enzymes: New England BioLabs, Beverly, Massachusetts, or Boehringer–Mannheim, West Germany
2. T4 DNA ligase: Purified from the overproducing strain C600 (pcI857) (pPLc28lig8)[6]
3. DNA polymerase I (Klenow fragment): Boehringer–Mannheim, West Germany
4. T4 DNA polymerase: New England BioLabs, Beverly, Massachusetts
5. S1 nuclease: Boehringer–Mannheim, West Germany
6. T4 polynucleotide kinase: New England Nuclear Corp., Boston, Massachusetts
7. Serva Blue R: Serva, Heidelberg, West Germany
8. LB medium: 1% Bacto tryptone, 0.5% Bacto yeast extract, 0.5% NaCl (Difco Corp., Detroit, MI)
9. Synthetic DNA linker fragments: Oligonucleotide linker fragments were synthesized following the phosphoramidite method.[11] The fragments were purified by HPLC chromatography

[11] A. Chollet, E. Ayala, and E. Kawashima, *Helv. Chim. Acta* **67**, 1356 (1984).

10. Gene Screen: New England Nuclear Corp., Boston, Massachusetts
11. EN³HANCE: New England Nuclear Corp., Boston, Massachusetts

Bacterial Strains

All bacterial strains used are derivatives of *E. coli* K-12.

K12ΔH1Δ*trp*[5]: Sm^R, *lacZ*am, Δ*bio-uvrB*, Δ*trpEA2*, (λ*Nam7Nam-53cI857*ΔH1)

MC1061[pcI857][12,13]: Sm^R, *araD*139, Δ(*ara leu*)7697, Δ*lacX*74, *galU*, *galK*, r$_K^-$m$_K^+$, [pcI857]

C600[pcI857][13,14]: *thr*-1, *leuB6*, *thi*-1, *sup*E44, *lac*Y1, *tonA*21, [pcI857]

GM119(λ)[15]: *dam*-3, *dcm*-6, *metB*1, *galK*2, *galT*22, *lac*Y1, *tsx*-78, *sup*E44(λwt)

Plasmids

pcI857: A derivative of pACYC177 that contains the *cI*857 allele. This plasmid specifies resistance to kanamycin (50 μg/ml) and is compatible with ColE1-derived replicons[6]

pPLc236*trp*: A ColE1-type plasmid derived from pBR322; specifies resistance to carbenicillin (100 μg/ml); carries the P$_L$ promoter and a ribosome-binding site derived from the tryptophan attenuator region[13]

pPLc245: As above, except that the ribosome-binding site is derived from the replicase gene of the RNA phage MS2[4]

pPLcmu299: As above, except that the ribosome-binding site is derived from the *ner* gene of phage Mu[16]

pAT153: A general ColE1-type cloning vector, specifying resistance to carbenicillin and tetracycline[17]

Plasmid constructions engineered to give optimal expression of a specific cloned gene are detailed in the Methods and Results section.

---

[12] M. J. Casadaban and S. N. Cohen, *J. Mol. Biol.* **138,** 179 (1980).
[13] G. Simons, E. Remaut, B. Allet, R. Devos, and W. Fiers, *Gene* **28,** 55 (1984).
[14] R. K. Appleyard, *Genetics* **39,** 440 (1954).
[15] This strain was kindly provided by M. G. Marinus.
[16] G. Buell, J. Delamarter, G. Simons, E. Remaut, and W. Fiers, *Basic Life Sci.* **30,** 949 (1985).
[17] A. J. Twigg and D. Sheratt, *Nature (London)* **283,** 216 (1980).

Methods and Results

General DNA Methodology

Procedures for preparation of plasmid DNA, restriction and ligation of DNA fragments, as well as conditions for bacterial transformation have been described.[18]

S1 nuclease was used to remove 5' protruding ends in a restriction buffer consisting of 25 mM sodium acetate, pH 4.4, 250 mM NaCl, 4.5 mM ZnSO$_4$. About 5 μg of CsCl-purified plasmid DNA was incubated in 100 μl of buffer with 100 U of enzyme at 18° for 30 min.

E. coli DNA polymerase I, Klenow fragment was used to fill in 5' sticky ends. A typical reaction mixture contained 1 pmol of DNA termini and 1 U of enzyme in 30 μl of a buffer consisting of 25 mM Tris · HCl, pH 7.4, 10 mM MgCl$_2$, 10 mM dithiothreitol, and 0.1 mM of all four deoxyribonucleoside triphosphates. The reaction was performed at 18° for 45 min.

Resection of 3' protruding ends was carried out with T4 DNA polymerase. The reaction was performed at 15° for 3 hr in a 30-μl reaction mixture containing 3 pmol of DNA termini in 65 mM Tris · HCl, pH 7.9, 20 mM KCl, 10 mM MgCl$_2$, 5 mM dithiothreitol, and 0.1 mM of any one of the four deoxynucleoside triphosphates.

Enzymatic reactions were stopped by phenol extraction. The aqueous layer was extracted twice with 5 vol of ethyl ether. The DNA was precipitated from the aqueous phase by addition of 2 vol of ethanol, 0.1 vol of 3 M potassium acetate, pH 4.5, and standing at −20° for at least 2 hr. The precipitated DNA was pelleted in an Eppendorf centrifuge (5 min at maximum speed). The pellet was washed with 70% ethanol and after careful draining resuspended in the desired buffer.

Specific DNA fragments were purified by electrophoresis in 0.8–2% agarose gels and recovered by the squeeze–freeze method[19] or in more recent experiments by filtration through Gene Screen membranes.[20] The latter procedure combines high yields of recovery and excellent quality of the DNA with respect to secondary restriction and ligation.

Construction of Expression Plasmids

1. *Human Immune Interferon.* cDNA clones containing the coding sequence for human immune interferon (IFN-γ) were obtained by reverse

[18] E. Remaut, P. Stanssens, G. Simons, and W. Fiers, this series, Vol. 119, p. 366.
[19] R. W. J. Thuring, J. P. H. Sanders, and P. Borst, *Anal. Biochem.* **66**, 213 (1975).
[20] J. Zhu, W. Kempenaers, D. Van der Straeten, R. Contreras, and W. Fiers, *Biotechnology* **3**, 1014 (1985).

transcription of an enriched mRNA fraction prepared from splenocyte cultures stimulated with staphylococcal enterotoxin A.[21,22] The IFN-γ cDNA was originally obtained via oligo-dG tailing and annealing with a filled in BamHI site which had been tailed with oligo-dC. In this way, the BamHI site is restored and the fragment can be cleaved out with BamHI.[22] The nucleotide sequence was determined using the chemical degradation method of Maxam and Gilbert.[23] As IFN-γ is a secreted protein, the coding sequence for mature protein is preceded by a signal peptide which is cleaved off during the process of secretion. At the time these studies were performed the exact N-terminus of the mature protein had not been determined. By analogy with the known N-terminal amino acid sequence of human IFN-α, the cleavage point was postulated to occur before a cysteine residue at position 21 in the IFN-γ coding sequence.[24,25] The constructions detailed below aimed at precise fusion of this cysteine codon to the ATG codon of an expression vector.[26]

The coding sequence for mature IFN-γ contains a unique AvaII site located at a position corresponding to the fifth amino acid of the mature protein. This site was used as a starting point for plasmid constructions. The AvaII site overlaps with an EcoRII site and is consequently refractory to cleavage because of C-methylation. To alleviate this problem, plasmids were propagated in a dam strain [GM119(λ)] which is defective in C-methylation. AvaII cleavage of IFN-γ DNA removes the first four amino acid residues of the mature protein. The missing coding information was restored using synthetic linker fragments as schematically shown in Fig. 1. As a result of the enzymatic reactions the TGT codon, corresponding to the presumptive N-terminal cysteine residue of the mature protein, was exposed as a blunt end. This end was then ligated to the ATG codon of an expression vector. In this approach we used a vector which carries the ATG codon as part of an EcoRI site. Following EcoRI cleavage, this site was blunted with S1 nuclease and ligated to the treated IFN-γ sequence. Full details of the constructions are given in the legend

[21] R. Devos, H. Cheroutre, Y. Taya, and W. Fiers, J. Interferon Res. 2, 409 (1982).

[22] R. Devos, H. Cheroutre, Y. Taya, W. Degrave, H. Van Heuverswyn, and W. Fiers, Nucleic Acids Res. 10, 2487 (1982).

[23] A. M. Maxam and W. Gilbert, Proc. Natl. Acad. Sci. U.S.A. 74, 560 (1977).

[24] N. Mantei, M. Schwarzstein, M. Streuli, S. Panem, S. Nagata, and C. Weissman, Gene 10, 1 (1980).

[25] P. W. Gray, D. W. Leung, D. Pennica, E. Yelverton, R. Najarian, C. C. Simonsen, R. Derynck, P. J. Sherwood, D. M. Wallace, S. L. Berger, A. D. Levinson, and P. V. Goedell, Nature (London) 295, 503 (1982).

[26] It has since been established that mature IFN-γ starts with a GIN residue at position 24 of the coding region [E. Rinderknecht, B. H. O'Connor, and H. Rodriguez, J. Biol. Chem. 259, 6790 (1984)].

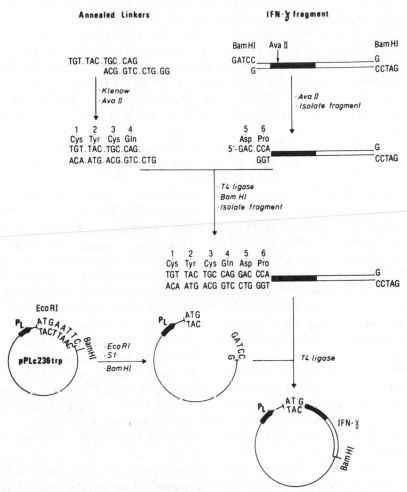

FIG. 1. Construction of a vector that expresses mature human immune interferon. As a consequence of the cDNA cloning procedure,[22] the IFN-γ fragment can be obtained as a *Bam*HI fragment. This fragment contains an *Ava*II site located at the fifth amino acid residue of mature IFN-γ. Following cleavage with *Ava*II, a 1100-bp fragment was isolated by gel electrophoresis and elution. Two synthetic linker molecules were phosphorylated using T4 polynucleotide kinase, annealed to each other, and filled in with Klenow fragment of DNA polymerase I. Following cleavage with *Ava*II, a duplex is obtained containing (1) a blunt end, (2) codons for the first four amino acids of mature IFN-γ, and (3) a sticky *Ava*II end. Following ligation with the isolated IFN-γ fragment, the mixture was digested with *Bam*HI to resolve dimers originating from ligated *Bam*-HI ends. The fragment was purified by gel electrophoresis and ligated to a vector molecule having an ATG codon as a blunt end and a sticky *Bam*HI end downstream. The plasmid used, pPLc236trp, contains a unique *Eco*RI site partially overlapping with the ATG codon. To expose the ATG codon and the *Bam*HI site, the plasmid was sequentially cleaved with *Eco*RI, treated with S1 nuclease, and cleaved with *Bam*HI.

to Fig. 1. In the example discussed the vector contains the P_L promoter and a manipulated ribosome-binding site derived from the *E. coli* tryptophan attenuator region.[13] In a similar way IFN-γ expression plasmids were constructed which contain the ribosome-binding site of the phage MS2 replicase gene.[4,13] The construction of further derivatives of the expression modules which differ in the vector part (e.g., presence of a transcription terminator) will not be detailed here. The characteristics of these vectors are schematically shown in Table II. Full details on their construction can be found in Simons *et al.*[13]

2. *Human Tumor Necrosis Factor.* The construction and identification of cDNA clones containing the coding sequence for human tumor necrosis factor (TNF) have been described.[27]

The TNF cDNA sequence was engineered for expression of mature protein according to the procedure detailed in Fig. 2. A unique *Ava*I site is present in the coding region of TNF and overlaps with the eighth amino acid residue of the mature protein. Following cleavage with *Ava*I, the sequence coding for the first seven amino acids is removed and the triplet of the eighth amino acid is exposed as a 5' protruding *Ava*I end. Two synthetic DNA linker fragments were synthesized which, after annealing, form a double strand that codes for the first seven amino acids and has one 5' protruding *Ava*I end, and another 5' protruding *Nco*I end. The *Ava*I site was used to link the fragment to the body of TNF coding region while the *Nco*I site was used to join the restored coding region to an expression vector. For this purpose we used a vector in which the ATG codon overlaps with an *Nco*I site. Plasmid pPLcmu299 contains the ribosome-binding site of the *ner* gene of phage Mu.[17] Similar constructions were obtained in pPLctrp321 which carries likewise a *Nco*I site, in this case part of the ribosome-binding site derived from the *E. coli* tryptophan attenuator region.

3. *Human Interleukin 2.* cDNA clones coding for human interleukin 2 (IL-2) were obtained by reverse transcription of an enriched mRNA fraction prepared from splenocyte cultures[28] using the method of oligo-dG tailed cDNA annealed into a filled in and oligo-dC tailed *Bam*HI site as described above.[22]

[27] A. Marmenout, L. Fransen, J. Tavernier, J. Van der Heyden, R. Tizard, E. Kawashima, A. Shaw, M.-J. Johnson, D. Semon, R. Müller, M.-R. Ruysschaert, A. Van Vliet, and W. Fiers, *Eur. J. Biochem.* **152**, 515 (1985).

[28] R. Devos, G. Plaetinck, H. Cheroutre, G. Simons, W. Degrave, J. Tavernier, E. Remaut, and W. Fiers, *Nucleic Acids Res.* **11**, 4307 (1983).

TABLE II
LEVELS OF ACCUMULATED PROTEIN

Expressed gene	Ribosome-binding site[a]	Transcription termination[b]	Percentage of total protein[c]
IFN-γ	MS2	—	3.5
		T4	10
	trp	—	15
		T4	24
		fd	24
TNF	*ner*	—	9
	MS2	—	0.5
	trp	—	0.1
IL-2	*ner*	—	5
	trp	—	10

[a] Origin of ribosome binding site. MS2, Derived from the replicase gene of the RNA phage MS2 [E. Remaut, P. Stanssens, and W. Fiers, *Nucleic Acids Res.* **11**, 4677 (1983)]; *trp*, derived from the *E. coli* tryptophan attenuator [G. Simons, E. Remaut, B. Allet, R. Devos, and W. Fiers, *Gene* **28**, 55 (1984)]; *ner*, derived from the *ner* gene of phage Mu [G. Buell, J. Delamarter, G. Simons, E. Remaut, and W. Fiers, *Basic Life Sci.* **30**, 949 (1985)].

[b] Origin of the terminator cloned downstream from the expressed gene. T4, Derived from gene 32 of phage T4 [H. M. Krisch and B. Allet, *Proc. Natl. Acad. Sci. U.S.A.* **79**, 4937 (1982)]; fd, the central terminator of phage fd.

[c] This value was determined as described in the text.

The several construction steps performed to obtain expression of mature IL-2 are outlined in Fig. 3. In this case all steps were carried out using cutting and joining reactions without the need for synthetic DNA linker fragments. Mature IL-2 starts with the sequence Ala-Pro. A unique *Hgi*AI site is present at this region. Following *Hgi*AI cleavage and resection of the 3′ protruding ends with T4 DNA polymerase, the CCT codon of the proline residue is exposed as a blunt end. To add an alanine codon, the CCT codon was blunt-end ligated to a filled in *Nar*I site of an acceptor plasmid. The resulting sequence GGCGCC constitutes a *Ban*I site. The latter enzyme cleaves between the G residues, leaving 5′ protruding ends. After filling in, a GCG codon (Ala) is added to the proline codon. The GCG codon is then ligated to a filled in *Nco*I site of the expression vector pPLcmu299.

4. *Detection of Expressed Protein.* Bacteria harboring one of the expression plasmids discussed above were inoculated at a density of 2 ×

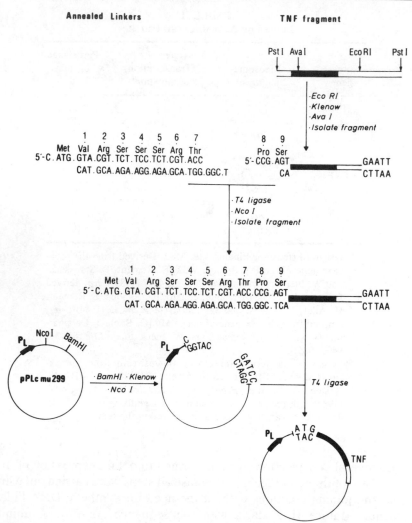

FIG. 2. Construction of a vector that expresses mature human tumor necrosis factor. A *Pst*I fragment obtained from the TNF cDNA clone[27] was sequentially digested with *Eco*RI, treated with Klenow fragment of DNA polymerase I, and cleaved with *Ava*I. The resulting 669-bp fragment was purified by gel electrophoresis and ligated to a double-stranded synthetic DNA fragment containing (1) a sticky *Nco*I end, (2) codons corresponding to the first seven amino acids of mature TNF, and (3) a sticky *Ava*I end overlapping with the eight amino acid residue. Following ligation, the reaction mixture was cut with *Nco*I to resolve dimers and the fragment was purified on an agarose gel. This fragment was ligated into the expression vector pPLcmu299. Plasmid DNA was first blunted at the *Bam*HI site by cleavage with *Bam*HI and treatment with Klenow fragment of DNA polymerase I, and subsequently opened at the unique *Nco*I site. The isolated TNF fragment having an *Nco*I end and a blunted *Eco*RI end was ligated between the *Nco*I end and the blunted *Bam*HI end of the vector.

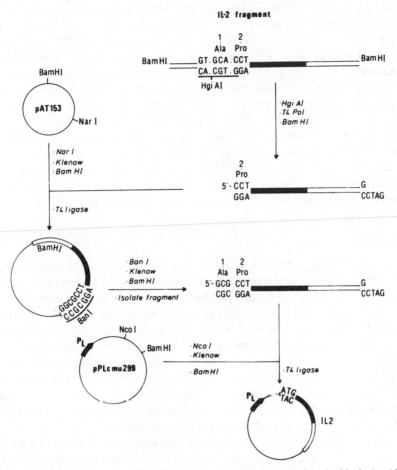

Fig. 3. Construction of a vector that expresses mature human interleukin 2. A pAT153 derivative containing the IL-2 cDNA as a *Bam*HI fragment was sequentially cleaved with *Hgi*AI, treated with T4 DNA polymerase in the presence of dGTP, and cleaved with *Bam*HI. The resulting 700-bp fragment was ligated between the *Bam*HI site and a filled in *Nar*I site of pAT153. The reconstructed *Nar*I site was cleaved by the enzyme *Ban*I. The 5' protruding ends were filled in with Klenow polymerase, and following *Bam*HI restriction a 700-bp fragment was isolated. This fragment was cloned between a filled in *Nco*I site and a *Bam*HI site of pPLcmu299.

10^6/ml in LB medium containing 5 μCi/ml U-^{14}C-labeled protein hydrolysate. The cultures were incubated at 28° with vigorous agitation to a density of 2 × 10^8/ml. Half of the culture was then shifted to 42° and incubation was continued for up to 6 hr. At various time points after induction, aliquots were collected by centrifugation. The pellet was dissolved in sample buffer and boiled for 5 min before electrophoresis in SDS–polyacrylamide gels (15%). The gels were fixed in 10% TCA and stained with 0.05% Serva Blue R in 30% methanol and 7% acetic acid. Autoradiographs were obtained after treatment with EN³HANCE and exposure of the dried gel to X-ray film at −70°. To determine the percentage of accumulated protein, the relevant band was excised from the dried gel and its radioactivity compared to the total radioactivity present in the same track. Under the conditions used, the cellular proteins have been uniformly labeled so that the incorporated radioactivity is an accurate measure of the amount of protein synthesized.

Examples of expressed protein using different expression plasmids and different host cells are shown in Fig. 4. A survey of the efficiency of expression obtained with different plasmids is given in Table II.

Comments

In the present communication we have emphasized the mechanics of precisely joining a coding region to the exposed ATG codon of an expression vector. Of the methods discussed, the one using single-strand-specific nucleases is the less reliable one. Indeed, it has been observed that removal of the single strand is not always accurate, so that the ATG codon may be damaged in the process. The methods of filling in 5' protruding ends or resecting 3' protruding ends (see Nishi et al.[29]), on the other hand, have proved to be very reliable. When choosing an expression vector for optimal expression of an unfused protein, other considerations regarding the type of ribosome-binding site used, as well as other features of the host–plasmid sytem, become important.

The nature of the ribosome-binding site may greatly influence expression levels. For evident reasons, expression vectors were provided with ribosome-binding sites known to be very effective in their natural context. It soon became clear, however, that the efficiency of a ribosome-binding site is not an absolute parameter but depends to a large extent on the nature of the downstream coding region. It is believed that the secondary structure of the mRNA in the vicinity of the ATG codon plays a major role in determining efficiency of expression.[30,31] It should be noted, however, that current models to predict secondary structure are frequently inadequate to foresee the effect on expression levels.[32] It is therefore

[29] T. Nishi, M. Sato, A. Saito, S. Itoh, C. Takaoka, and T. Taniguchi, *DNA* **2,** 265 (1983).
[30] D. Iserentant and W. Fiers, *Gene* **9,** 1 (1980).
[31] D. Gheysen, D. Iserentant, C. Derom, and W. Fiers, *Gene* **17,** 55 (1982).
[32] P. Stanssens, E. Remaut, and W. Fiers, *Gene* **36,** 211 (1985).

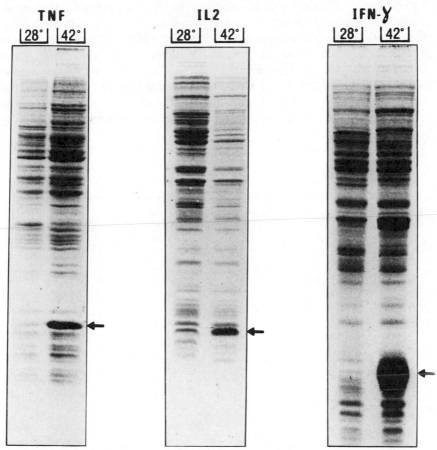

FIG. 4. Stained protein profiles obtained following SDS–PAGE of extracts from induced (42°) and uninduced (28°) cells containing an expression plasmid for IFN-γ, TNF, and IL-2, respectively. The induction period was 4 hr.

worthwhile to try out a number of ribosome-binding sites in combination with the gene to be expressed. Table II lists some examples of the large differences that can be observed.

Another parameter (probably of less consequence) in optimizing expression is the presence of a strong transcription terminator downstream from the cloned gene. In the case of clones expressing immune interferon, this feature increased the yield of the product by a factor of 1.5–3 (Table II). This is not primarily related to an increased efficiency of synthesis, but is rather due to a prolonged maintenance of the level of synthesis.[33]

[33] G. Simons, unpublished results (1984).

The bacterial strain used is another, poorly documented, element affecting accumulation of the expressed protein. For instance, human fibroblast interferon (not discussed in this chapter) was found to accumulate to levels of 0.4 and 2%, respectively, in *E. coli* strains MC1061[pcI857] and K12ΔH1Δ*trp*.[4] Human interleukin 2, on the other hand, was recovered in higher yields from the former strain than from the latter.[34] The phenomenon has not been studied in any detail. Variations between strains in mRNA and/or protein stability might conceivably contribute to the observed differences.

Acknowledgments

This research was supported by Biogen, SA, Geneva, and by a grant from the Gekoncerteerde Onderzoeksakties of the Belgian Ministry of Science.

[34] R. Leemans, unpublished results (1985).

[26] Expression and Secretion of Foreign Proteins in *Escherichia coli*

By Guy D. Duffaud, Paul E. March, and Masayori Inouye

Whether the gene for a given protein has been cloned or a DNA fragment is found to contain an open reading frame, the expression of protein, preferably in large amounts, often becomes a main objective. Our laboratory has developed a set of expression and secretion vectors designed to overcome many of the problems encountered when expressing a given gene in *Escherichia coli*. These vectors have been designed for use at specific steps during the cloning and expression of proteins as well as for the particular needs that might come with the nature of the protein to be produced, such as the necessity to secrete products that are toxic for cell growth. We discuss here the different vectors we have constructed and how they can best be utilized for cloning, expression, and secretion.

Cloning and Expression in *E. coli*

Expression Vectors

The first step in achieving expression of a protein is to clone the DNA fragment in generalized expression vectors. These vectors must possess a certain number of essential properties: a multiple restriction cloning site, an upstream promoter region which will promote the expression of the foreign gene, and some way of controlling that expression, particularly if the foreign protein is suspected of being lethal to the cell. Other properties are also important: a high copy number of the vector, a suitable marker for selection, a capacity for cloning large fragments. We have constructed vectors that encompass all of these properties: the pIN-II-A and pIN-III-A series (Table I). The pIN vectors are derived from pBR322, a plasmid produced in roughly 30 copies per bacterial cell.[1] To promote expression, the pIN vectors employ the efficient *lpp* gene promoter (*lpp*P).[2] This gene is responsible for the constitutive expression of lipoprotein, a major outer membrane protein, and has been integrated into these vectors. Downstream of *lpp*P, the *lac*UV5 promoter operator (*lac*PO) fragment has been inserted. Thus, expression will occur only in the presence of a *lac* in-

---

[1] D. Stueber and H. Bijard, *EMBO J.* **1,** 1399 (1982).
[2] K. Nakamura and M. Inouye, *EMBO J.* **1,** 771 (1982).

TABLE I

EXPRESSION VECTORS BASED ON THE *lpp–lac* PROMOTER SYSTEMS AND CORRESPONDING HOST STRAINS

Plasmid	*E. coli* host	Comments	Reference
pIN-I-A1 pIN-I-A2 pIN-I-A3	W620	Constitutive expression, product in the cytoplasm	3
pIN-II-A1 pIN-II-A2 pIN-II-A3 pIN-II-B1 pIN-II-B2 pIN-II-B3 pIN-II-C1 pIN-II-C2 pIN-II-C3	SB221	Inducible promoter; lpp^P-lac^{PO}. Transformation should be carried out in an *E. coli* strain containing *lacI*^{q}. The pIN-II-B2 plasmid is unstable due to the creation of an open reading frame during construction. Its use is not recommended	3
pIN-III-A1 pIN-III-A2 pIN-III-A3 pIN-III-B1 pIN-III-B2 pIN-III-B3 pIN-III-C1 pIN-III-C2 pIN-III-C3	W620 or SB221	Inducible promoter; lpp^P-lac^{PO}. Plasmid contains *lacI*. The pIN-III-B2 plasmid is unstable as discussed above for the pIN-II-B2 plasmid	3
pIN-III-ompA1 pIN-III-ompA2 pIN-III-ompA3	W620 or SB221	Secretion expression vector. For use when extracytoplasmic localization of the product is desired	10
pIN-III (lpp^{P-5})	W620 or SB221	Improved *lpp* promoter; 3- to 4-fold higher production	4
pIC-III	SB4288	Hybrid protein expression vector; cloning sites are *Eco*RI, *Sma*I, *Bam*HI in the *lacZ* coding region	3

Strain	Description of *E. coli* strains, phenotype	
W620	K-12 F⁻ *thi-I pyrD36 gltA6 galK30 strA*129 *supE44 relA recA*	21
SB221	K-12 F⁻ *hsdR leuB6 lacY thi ΔtrpE5 lpp*/F' *lacI*^{q} *proAG lacZYA*	15
SB4288	K-12 F⁻ *recA thi-1 relA mal-24 spc*12 *supE-50* DE5(Δ*lac-proB*)	23

ducer. Downstream of this tandem promoter region a multiple restriction site linker has been inserted (EcoRI, HindIII, BamHI) at three positions determined by the properties of lpp: the A, B, and C sites. At each site, the linker has been inserted in such a way that cloning at any restriction site can be accomplished in all three possible reading frames (Fig. 1). This allows the choice of a vector with a reading frame compatible with the fragment to be cloned. The A sites are located right after the initiation codon of lipoprotein, the B sites right after the cleavage site of the lipo-protein precursor, and the C sites nine amino acid into the mature portion of lipoprotein. Constructions employing the B and C sites give rise to hybrid proteins which can use the properties of the lipoprotein signal peptide for localization in the different E. coli compartments, as discussed below. Cloning into any of these sites results in the usage of the lipopro-tein Shine–Dalgarno sequence, initiation codon, and termination codon, which is convenient if the fragment to be cloned lacks one or more of these essential features. In addition there is a termination codon in each of the reading frames and a ρ-independent efficient transcription termination signal. Detailed reviews on the construction of all these vectors can be found elsewhere.[2,3]

Utilization of the pIN-II and pIN-III Expression Vector

Cloning and Maintenance. Successful utilization of expression vec-tors requires certain considerations. For cloning it is desirable that the gene have compatible ends with the available restriction sites (six differ-ent types of restriction fragments can be cloned). However, if one or two of the sites are not compatible, it is always possible to clone together with commercially available gene linkers in order to make the sites compatible. It should be pointed out that if the gene contains its own ribosome-binding region and initiation codon, a unique XbaI site after lac^{PO} can be used as an additional cloning site.

Another important consideration is the choice of host. For the expres-sion to be properly regulated it is necessary to have a host in which the lac repressor (lacI) is expressed. The pIN-II vectors require a host containing the lacI gene, such as E. coli strain SB221 (Table I). This will suppress expression of the product during the cloning procedure. If such a strain is not available or cannot be used, the vector of choice becomes one of the pIN-III series. These vectors have the same characteristics as pIN-II except that they have the lacI gene inserted at a unique SalI site within the expression vector itself. In this way expression of the gene is blocked

[3] Y. Masui, J. Coleman, and M. Inouye, in "Experimental Manipulations of Gene Expres-sion" (M. Inouye, ed.), p. 15. Academic Press, New York, 1983.

A

—GAGGGTATTAATAATGAAACGTACTAAACTGGTACTGGGCGCGGTAATCCTGGGTTGCTCCAGCAAGCGTAAATCGATCAGCTGTCT—
MetLysAlaThrLysLeuValLeuGlyAlaVallleLeuGlySerThrLeuLeuLeuAlaGlyCysSerSerAsnAlaLysIleAspGlnLeuSer
↑ 20

A1
EcoRI HindIII BamHI
—GGTATTAATAATGAAACGGAATTCCAAGCTTGGATCCGGCTGAGC—
MetLysGlyIleAsnSerLysLeuGlySer

A2
EcoRI HindIII BamHI
—GGTATTAATAATGAAACGGAAGGAATTCCAAGCTTGGATCCGGCTGAGC—
MetLysGlyLysGluPheGlnAlaTrpIle

A3
EcoRI HindIII BamHI
—GGTATTAATAATGAAACGGGGAATTCCAAGCTTGGATCCGGCTGAGC—
MetLysGlyGlyIleProSerLeuAspPro

B1
EcoRI HindIII BamHI
—CTGGCTGGAATTCCAAGCTTGGATCCGGCTGAGC—
LeuAlaGlyAsnSerLysLeuGluSer

B2
EcoRI HindIII BamHI
—CTGGCTGGAAGGAATTCCAAGCTTGGATCCGGCTGAGC—
LeuAlaGlyLysGluPheGlnAlaTrpIle

B3
EcoRI HindIII BamHI
—CTGGCTGGGGAATTCCAAGCTTGGATCCGGCTGAGC—
LeuAlaGlyGlyIleProSerLeuAspPro

C1
EcoRI HindIII BamHI
—CTGGCAGGTTGCTCCAGCAACGCTAAATCGATGGAATTCCAAGCTTGGATCCGGCTGAGC—
LeuAlaGlyCysSerSerAsnAlaLysIleAspGlyIleAspArgAsnSerLysLeuGlySer

C2
EcoRI HindIII BamHI
—CTGGCAGGTTGCTCCAGCAACGCTAAATCGATCGAAGGAATTCCAAGCTTGGATCCGGCTGAGC—
LeuAlaGlyCysSerSerAsnAlaLysIleAspArgLysGluPheGlnAlaTrpIle

C3
EcoRI HindIII BamHI
—CTGGCAGGTTGCTCCAGCAACGCTAAATCGATCGGGGAATTCCAAGCTTGGATCCGGCTGAGC—
LeuAlaGlyCysSerSerAsnAlaLysIleAspArgArgGlyIleProSerLeuAspPro

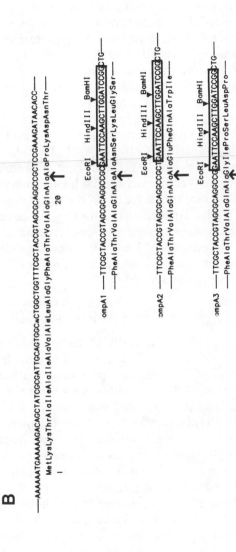

FIG. 1. (A) DNA sequence of the three reading frames of the A, B, and C sites of the pIN vectors. The natural lipoprotein gene is shown at the top and the positions of the A, B, and C sites are aligned below. (B) DNA sequence of the three reading frames of the ompA secretion vectors. The entire ompA signal sequence and cleavage site region is shown at the top and the ompA1, ompA2, and ompA3 reading frames are shown below. The solid arrows denote signal peptide cleavage sites. In the pIN-B series the signal peptide cleavage site is unknown, as indicated by the dotted arrow. The linker DNA is boxed and the positions of restriction enzyme cleavage sites are indicated by arrow heads.

until a *lac* inducer such as isopropyl-β-D-thiogalactopyranoside (IPTG) is added to the growth medium. It is important to note that we have found that certain culture media such as yeast extract can contain forms of *lac* activator that induce expression of a cloned gene to significant levels even in the absence of IPTG. Levels of background induction can be high enough to be lethal to the cell if the gene product is toxic per se or if high amounts are toxic. To avoid these problems, it is preferable to use a minimal medium supplemented with casamino acid. For the same reasons, it is also advisable to store the cloned gene in the purified plasmid form rather than to maintain the plasmid in the host cell in a growth medium where background induction may result in mutations in the cloned protein and/or its expression system. If such a mutant plasmid is used, the protein might be altered in its function and/or its level of expression might be lowered in comparison to the original plasmid.

Expression. Upon induction with IPTG, the level of protein production of a cloned gene in a pIN vector might reach higher than 20% of the total cellular protein. However, this level of production depends on the particular protein under study. To compensate for this, we have modified *lpp*[P] by means of oligonucleotide-directed site-specific mutagenesis. The *lpp*[P] was systematically altered at the -35 and -10 regions in order to bring them closer to the consensus sequence and create a more effective promoter.[4] The *lpp*[P] (wild type) sequence is

-35 region   -10 region
ATCAAAAAAAAT<u>ATTCTCAAC</u>ATAAAAAACTTTGTGT<u>AATACTT</u>GTAAC

and it was changed to *lpp*[P-5], the most effective of the designed promoters, by two substitutions in the -35 region (TT<u>GACA</u>) and one single substitution in the -10 region (<u>T</u>ATACT). This new promoter was found to increase the level of expression of β-galactosidase by approximately 4-fold over the already efficient wild-type *lpp*[P] and can easily be substituted in any pIN vector by simply replacing the *Pst*I–*Xba*I fragment from the vector with the cloned gene with the equivalent fragment from pIN-III-(*lpp*[P-5]). When this is done, the level of expression is expected to increase at least 3- to 4-fold. A word of caution is necessary when using this powerful promoter: overproduction per se might be lethal. If a protein is already expressed at a level of 10–20% of the total cellular protein, attempting to increase this percentage may be counterproductive.

The size of the protein to be expressed is another factor to be considered. It would seem that the protein size would be a limiting factor when using *E. coli* as the host cell. However, a recent report suggests that this

---

[4] S. Inouye and M. Inouye, *Nucleic Acids Res.* **13**, 3101 (1985).

might not be critical. Two to four *lacZ* genes have been fused head to tail, in phase, and have been shown to express the corresponding dimers, trimers, and tetramers of β-galactosidase (up to 460 kDa).[5] All products showed enzymatic activity, which was, however, strongly reduced for the trimer and tetramer forms. Also, the latter were largely degraded to small products (such as a dimer) as could be seen on SDS–PAGE gels. We have also recently been able to express a fused pentameric unit of β-galactosidase using a different strategy: gene fusion was done in such a way that only the tail unit could be active since only this unit possessed the entire *lacZ* coding sequence. This pentamer was expressed with a pIN-III-A3-*lpp*[P-5] vector, showed activity, and was clearly identifiable, with little degradation, upon SDS–PAGE stained with Coomassie blue (K. Tsung and M. Inouye, unpublished results). As was the case for the tetrameric β-galactosidase, foreign proteins might be proteolitically degraded in *E. coli*. In such instances it is advisable to transfer the expression vector with the cloned gene to an *E. coli* strain deficient in proteolytic activity such as a lon⁻ strain.[6]

Other Expression Vectors and Their Usage

The pIN-I vectors are like the pIN-II vectors except that the expression of a cloned gene is not under *lac*[PO] control.[3] Thus they can be used only for constitutive expression of proteins that are known to be at least tolerated by the host cell (or even needed). Their advantage is that they do not require induction and thus have more flexibility in the choise of a host cell and media. Constitutive expression is advantageous because of the potential to make large amounts of product while at the same time minimizing manipulations during cell growth. These vectors could also be suited, for example, to produce proteins meant to be used as antigens for antisera production or even for large-scale production of the antigenic fragment when properly cloned.

pIC-III vectors are different from pIN-III vectors in that they generate a hybrid of a desired protein and β-galactosidase. A linker has been inserted into a pIN-III-*lacZ* vector immediately after the initiation codon in such a way that *lacZ* is out of phase with the initiation codon. A DNA fragment inserted within this linker, in the appropriate reading frame, results in the expression of a β-galactosidase hybrid protein.[3] Since such a hybrid is expected to have a β-galactosidase activity selection of the hybrid is made easy by using agar plates containing β-galactosidase indicator. This vector is best used in the production of hybrid protein for

[5] W. Kuchinke and B. Muller-Hill, *EMBO J.* **4**, 1067 (1985).
[6] S. A. Goff and A. L. Goldbert, *Cell* **41**, 587 (1985).

rapid preparation of antisera, since any *lacZ* hybrid protein is very large and quite easy to purify in a single step via preparative SDS–PAGE.

Secretion Vectors in *E. coli*

There are several advantages in using a secretion vector: (1) the amino terminus can be changed to other than a methionine by adjusting the cleavage site in such a way that, after cleavage of the signal peptide, the desired amino acid is at the amino terminus; (2) the usage of secretion vector is essential for proteins such as nucleases or proteases which are harmful if kept in the cytoplasm after production; (3) proteins secreted into the periplasmic space (the space between the outer membrane and the cytoplasmic or inner membrane) are considered to be more stable because there is a lower proteolytic activity in the periplasmic space; (4) the reducing environment of the cytoplasm does not allow the formation of disulfide bonds which can easily be formed in the oxidizing environment of the periplasm; (5) proteins secreted to the periplasmic space can easily be released from the cell by osmotic shock and readily purified. Based on these considerations we have developed secretion vectors that can accomplish the processes discussed above. We describe below their constructions and methods that can be used with them.

Secretion in E. coli: Requirements for a Secretion Vector

Escherichia coli contains four major subcellular compartments: the cytoplasm, the cytoplasmic or inner membrane, the outer membrane, and the periplasmic space between the inner and outer membranes. Secretory proteins are synthesized first in the cytoplasm as precursor proteins which have a peptide extension of 20–25 amino acids, termed the signal peptide, at their amino termini. This signal peptide is an essential element for directing the precursor protein to the inner membrane and subsequent translocation across the inner membrane. In *E. coli*, outer membrane proteins, periplasmic proteins, and some inner membrane proteins have been shown to be produced as precursors containing a signal peptide. Signal peptides have been found to have unique features that account for their functions. Upon translocation of the precursor protein across the inner membrane, the signal peptide is cleaved by a signal peptidase located in the inner membrane. The mature protein then reaches its final location (outer membrane, periplasmic space, or inner membrane) by information contained within its amino acid sequence.[7] In general, pro-

[7] G. D. Duffaud, S. K. Lehnhardt, P. E. March, and M. Inouye, *Curr. Top. Membr. Transp.* **24,** 65 (1985).

teins normally destined for export, such as hormones and serum proteins, will be easily secreted by *E. coli*. Whether other type of proteins can be secreted across the membrane is not easily predicted.

The signal peptide has been shown to guide translocation of the amino terminus of the mature protein to the outside surface of the inner membrane. The amino terminus of any fused protein should also be translocated in a similar fashion. Whether the rest of the protein is translocated across the membrane depends on the structure of the protein. For example, a protein containing a long stretch of hydrophobic amino acids could likely remain in the inner membrane, since a hydrophobic stretch is not easily translocated across the membrane. Such a protein might still be translocated across the membrane by engineering the hydrophobic stretch as discussed later. Conversely, proteins normally located in the cytoplasm can be guided to localize to the inner membrane by fusing them to a noncleavable signal peptide (pIN-II-B or pIN-III-B). Or a protein might be guided to the outer membrane by fusing it to a cleavable signal peptide of an outer membrane protein (pIN-II-C or pIN-III-C). These vectors have been created in our laboratory and are listed in Table I. The construction and function of these vectors has been reviewed elsewhere and will only be discussed briefly in this chapter.[8,9]

pIN-III-OmpA Secretion Vectors

The pIN-III-OmpA vector has been constructed by replacing the fragment between the unique *XhoI* site and the unique *EcoRI* site of the pIN-III-A3 vector with the DNA fragment carrying the coding region for the signal peptide of the OmpA protein, a major outer membrane protein.[10] The signal peptide of OmpA was chosen for two reasons: (1) OmpA is a major protein of the *E. coli* outer membrane and thus its signal peptide is thought to be efficient in the translocation process and (2) the OmpA signal peptide does not need to be modified before processing, as is the case for the lipoprotein signal peptide, and it is processed efficiently by the major processing pathway that uses signal peptidase I. Thus this vector is the system of choice when the primary objective is the translocation of a protein outside the cytoplasm: it provides all of the advantages of the pIN-III expression vectors and at the same time the necessary information for secretion.

[8] C. A. Lunn, M. Takahara, and M. Inouye, this series, in press.

[9] C. A. Lunn, M. Takahara, and M. Inouye, *Curr. Top. Microbiol. Immunol.* **125,** 59 (1986).

[10] J. Ghrayeb, H. Kimura, M. Takahara, H. Hsiung, Y. Masui, and M. Inouye, *EMBO J.* **3,** 2437 (1984).

Utilization of Secretion Vectors

General cloning strategies and expression protocols follow the same requirements addressed above. The question now is which vector to choose and what to expect. Some precautions must be taken before a secretion vector is utilized. First, as discussed earlier, a protein might not be secreted even if it is fused to a signal peptide: β-galactosidase, a cytoplasmic protein, was not exported to the outer membrane when fused to the signal peptide of *lam*B which codes for an outer membrane protein.[11] Second, it is also important to consider that the fusion of a signal peptide with a foreign protein might not yield a functional cleavage site, thus impairing secretion. This particular problem and ways to solve it are discussed below. Finally, the ultimate localization of the protein will depend on information intrinsic to the sequence itself, not on the origin of the signal peptide.

pIN-III-OmpA has been utilized to secrete both bacterial and eukaryotic proteins to the periplasmic compartment of *E. coli*.[10,12,13] The first protein to be expressed using this vector was β-lactamase.[10] The mature portion of this protein was fused to the OmpA signal peptide by insertion at the *Eco*RI site of pIN-III-OmpA3 (see Fig. 1). Upon induction of gene expression, fully proceed β-lactamase was secreted into the periplasmic space and after a 3-hr induction accumulated to 20% of total cellular protein. The OmpA signal peptide was correctly removed, resulting in the production of β-lactamase with four extra amino acid residues at its amino terminus due to the linker sequence in the vector. As a next step, this linker sequence was deleted by site-specific mutagenesis and the new product had the same amino terminus as authentic β-lactamase. No significant accumulation of precursor was observed.

Since β-lactamase is an *E. coli* protein normally secreted into the periplasmic space, pIN-III-OmpA was tested for the production and secretion of an extracellular protein from a gram-positive bacteria: staphylococcal nuclease A from *Staphylococcus aureus*. This protein was cloned from an *E. coli* vector in which it is poorly expressed to the *Bam*HI site of pIN-III-OmpA.[13] The nuclease was now produced to an extent of 3% of the total cellular protein, was accurately processed at the cleavage site of the OmpA signal peptide (i.e., with a peptide extension originating

[11] F. Moreno, A. V. Fowler, M. Hall, T. J. Silhavy, I. Zabin, and M. Schwartz, *Nature* (*London*) **286,** 356 (1980).

[12] J. Ghrayeb and M. Inouye, *J. Biol. Chem.* **259,** 463 (1984).

[13] M. Takahara, D. W. Hibler, P. J. Barr, J. A. Gerlt, and M. Inouye, *J. Biol. Chem.* **260,** 2670 (1985).

from the linker and an N-terminal portion of the nuclease gene), and was translocated to the periplasmic space of *E. coli*. The nuclease accounted for 10% of the protein in the periplasm. When the linker portion was deleted via site-specific mutagenesis, nuclease A was produced in its authentic form and accounted for 12% of the periplasmic protein of which one-sixth was in the unprocessed form. To increase the level of expression of the nuclease, the *lpp* promoter was replaced with *lpp*[P-5] as described above. With this vector the production of nuclease was increased almost 4-fold after a 3-hr induction. However, overproduction of this protein resulted in a higher accumulation of the unprocessed precursor (about one-third of the total nuclease protein).[4]

Recently, human growth hormone (hGH) has been successfully expressed and secreted using pIN-III-OmpA3.[14] A DNA fragment carrying the hGH gene was first cloned into the *Eco*RI site out of frame. The linker portion was then removed by site-specific mutagenesis to adjust the reading frame and this new clone was able to express authentic hGH to an amount of 6% of the total cellular protein. This protein was exported to the periplasmic space, where it constituted approximately 30% of the total protein. The protein was purified to homogeneity in a two-step procedure and characterized by trypsin digestion, CD spectroscopy, and amino acid sequencing of the N-terminus. The expressed protein had the same structural characteristics as authentic hGH, including the correct formation of two disulfide bridges. It is interesting to note that in this case expression was obtained without the addition of IPTG and that the addition of IPTG resulted in the accumulation of precursor without a significant increase in the total level of hGH.

These examples underline some of the advantages of the pIN-III-OmpA secretion vector. They can be employed to secrete proteins of different origin (β-lactamase, nuclease A, and hGH). These proteins are identical to their authentic counterpart in terms of activity, conformation, amino terminal residue, and disulfide bond formation. They are secreted in large amounts (even if toxic for the cell, like nuclease A) to the periplasmic space from which they can easily be purified.

Problems Inherent to Secretion Vectors

Secretion of Cytoplasmic Proteins. Cytoplasmic proteins do not contain a signal peptide and remain in the cytoplasm as soluble proteins. Whether they can be secreted across the membrane or not depends on their primary structure, as discussed earlier. Complete translocation

---

[14] H. M. Hsiung, N. G. Mayne, and G. W. Becker, *Bio/Technology*, in press.

might be inhibited by the presence of hydrophobic sequences, as discussed earlier. The existence of such sequences results in a stop-transfer
sequence which blocks the translocation process across the inner membrane or will promote insertion within the inner membrane. It has been
shown that a secretory protein can be engineered to contain a stop-transfer signal, making this protein an inner membrane protein.[15] The function
of the stop-transfer signal can be destroyed by altering the hydrophobic
region to a more polar or hydrophilic sequence, allowing translocation.
When designing such mutations care should be taken not to alter the
structure or function of the protein.

   *Structural Requirements for Processing and Translocation of Secre-
tory Proteins; Structural Compatibility between the Signal Peptide and
the Secretory Protein.* There is a possibility that the signal peptide of the
hybrid protein might not be cleaved by an *E. coli* signal peptidase. The
properties of the signal peptides should be considered, when creating a
hybrid protein, to ensure that it will be processed. Several features are
conserved among all signal peptides and are thought to account for their
function. The 20- to 25-amino acid prokaryotic signal peptide can be
divided into three parts. The amino terminal portion (usually four to five
amino residues) is always positively charged due to the presence of up to
three basic amino acids. The positive charges arising from these residues
are thought to help the initial binding to the nascent peptide chain to
negative charges of the inner membranes and the putative secretory components. The amino terminal portion of the signal peptide is followed by a
highly hydrophobic segment of about 10 residues, usually punctuated by
glycine and/or proline residue. According to the loop model this hydrophobic segment is able to insert in the lipid bilayer by forming a loop
structure.[7,16] The cleavage site region follows the hydrophobic core of the
signal peptide. It is usually joined to the hydrophobic core by serine and/
or threonine residues and it is separated from the cleavage site itself by a
short, rather hydrophobic, segment. The C-terminus of the signal peptide
is always a small side chain amino acid: glycine, alanine, or serine. The
first amino acids of the mature protein are not as well conserved. However, by using the Chou and Fasman parameters to predict the secondary
structure at the cleavage site, one observes that precursor proteins might
have a conserved β-turn structure in this area.[17] It is possible that the
small side chain amino acid at the cleavage site of the signal peptide is a

---

[15] J. Coleman, M. Inukai, and M. Inouye, *Cell* **43,** 351 (1985).
[16] M. Inouye, R. Pirtle, I. Pirtle, J. Sekizawa, K. Nakamura, J. Di Rienzo, S. Inouye, S.
   Wang, and S. Halegoua, *Microbiology,* 34 (1979).
[17] P. Y. Chou and G. D. Fasman, *Biochemistry* **47,** 251 (1977).

specific requirement of the signal peptidase and that the conserved β-turn structure at the cleavage site is a main requirement for accurate processing, thus explaining why it is not necessary to conserve the amino acids at the N-terminus of the mature sequence.[18,19] Evidence for the function of the different regions of the signal peptide described above have been provided by a whole set of mutations naturally found or created by site-specific mutagenesis and other methods.[7]

It has been shown that mutation on either side of the cleavage site could hamper processing, and that a combination of mutations could make processing totally defective.[19] Thus when cloning a protein into a secretion vector care should be taken that the newly created cleavage site has the required structural elements, as well as a proper amino acid at the C-terminus of the signal peptide. If these conditions are not met defective processing might result. At best the defect might result in a slow processing rate and at worst in an unexpected cleavage site or a totally defective processing. However, when such a problem arises it should be possible to at least partially remedy it. For example, a linker could be inserted into the construction in such a way that new amino acids are encoded to create the desired structure. More simply, site-specific mutagenesis could be performed on either side of the cleavage site in order to change one or two amino acids, preferably in the signal peptide, so as not to change the amino acid sequence of the secretory protein. We have also found that the signal peptide can have a direct effect on the rate of synthesis of the secretory protein. Pollitt et al. have found that the deletion of the glycine residues at positions 9 and 14 of the lipoprotein signal peptide resulted in a 4-fold increase of lipoprotein synthesis in vivo as compared to the rate of synthesis of the wild-type lipoprotein (when expressed with the same vector).[20] A significant increase was also observed in the in vitro rate of synthesis. However, when the mutant lipoprotein signal peptide was fused to β-lactamase no effect was observed on the rate of lipo-β-lactamase synthesis (C. A. Lunn and M. Inouye, unpublished results). It seems that a certain amount of compatibility between the signal peptide and the secretory protein is required if the signal peptide is to have an effect on the rate of protein synthesis.

Other Secretion Vectors: pIN-II-B, pIN-II-C, pIN-III-B, and pIN-III-C

These vectors have cloning sites right after the lipoprotein signal peptide (B site) or nine amino acid residues after the cleavage site (C site). A

[18] S. Pollitt, S. Inouye, and M. Inouye, *J. Biol. Chem.* **261**, 1835 (1986).
[19] S. Inouye, G. D. Duffaud, and M. Inouye, *J. Biol. Chem.* **261**, 10970 (1986).
[20] S. Pollitt, S. Inouye, and M. Inouye, *J. Biol. Chem.* **260**, 7965 (1985).

protein cloned at one of these sites should be translocated across the cytoplasmic membrane. However, localization of the protein to the outer membrane is possible when cloning into the C sites. The hybrid protein expressed in a pIN-II-C or pIN-III-C vector contains the first nine amino acids of lipoprotein, including the lipid-modified cysteine at position +1. These amino acid residues have been shown to direct the secretion of β-lactamase, a periplasmic protein,[12] and staphyloccoccal nuclease A (M. Takahara and M. Inouye, unpublished results) to the outer membrane.

When a protein is fused to the B site, the hybrid is expected to be exported to the periplasmic space. However, the hybrid may still be bound to the inner membrane by the uncleaved lipoprotein signal peptide. This is the case for staphyloccoccal nuclease A which was found to be localized in the inner membrane fraction when it was cloned into the B site (M. Takahara and M. Inouye, unpublished results).

General Considerations

Cloning, expression, and secretion of any protein is greatly simplified by using specifically designed vectors such as the ones that are described here. Regardless of the strategy used, there is a set of common considerations that apply to expression and/or secretion of a protein when these vectors are utilized.

Expression Levels

The pIN vectors offer a ready solution to many of the problems that might be encountered when one seeks to express a protein. However, in some cases the expression of a given protein might not be as high as one would desire. Any of the vectors described above can be changed to a higher level of expression just by changing its promoter (lpp^P) by a still more efficient promoter: lpp^{P-5}. This procedure is made easy by the existence of two unique restriction sites on each side of the promoter region: *Xba*I and *Pst*I.

If the expression is still not high enough there are other factors that should be considered, such as the codon usage, the secondary structure of the mRNA, the protein stability, and the toxicity of the product to the cell. If the cloned fragment contains codons rarely used by *E. coli* then

[21] E. T. Wurtzel, R. N. Movva, F. Ross, and M. Inouye, *J. Mol. Appl. Genet.* **1,** 61 (1981).
[22] P. J. Green and M. Inouye, *J. Mol. Biol.* **176,** 431 (1984).

low levels of expression might result. Alteration of some of the codons by site-specific mutagenesis could improve the overall expression.

Host Strains

The choice of a host strain is also critical. The nature of the vector used determines what type of host have to be used, e.g., a strain containing the *lacI* gene or not. Also it is important to use a lon⁻ strain or strains having low proteolytic activity if the protein is particularly sensitive to proteolytic activity.

Growth Conditions

Establishing satisfactory growth conditions is an important first step in the successful expression of a protein. Expression of a particular protein might be more successful at the exponential phase than at the stationary phase. With the pIN vector system expression can be induced at any point of the growth curve by addition of IPTG. Another important factor is the medium used for growth. Care should be taken that no *lac* inducer is present when using an inducible vector. The nature of the medium might also have a more indirect role. A growth medium can be altered in its osmolarity, reducing agents, nutrient composition, etc., any of which may affect overall expression levels. Selection of the right components of the growth medium could further improve expression of the protein either by ensuring a more stable secreted product, or by giving the cell optimum growth conditions to produce the protein. Growth temperature can also be an important factor.

Some problems such as the glycosylation of certain eukaryotic proteins are out of the reach of the *E. coli* sysem at present. However, it should be noted that the *E. coli* system is able to provide an environment adequate for the formation of the correct disulfide bonds of hGH in the periplasmic space.

Toxicity

There are several ways to deal with products that are toxic for the cell. One is to have the protein secreted by using a secretion vector. Another way is to induce only tolerable levels of the protein. If the product is particularly stable the cells could even be left to die and the protein harvested later.

Maintenance of Expression Vectors

Whether proteins are toxic to the *E. coli* cell or not, the cloned genes are always subjected to spontaneous mutations. In particular toxic products will be under more pressure to have either their toxicity diminished by a mutation in the structural gene or the expression level lowered by mutations in the promoter region. To avoid such problems the newly obtained clones should be stored in the form of purified DNA form as soon as constructed.

Acknowledgments

This work has been supported by Grants GM11145 and GM19043D from NI General Medical Sciences, Grant NP387M from the American Cancer Society, and a Grant from Eli Lilly Company. Guy Duffaud is recipient of a Fujii Fellowship.

[27] Yeast Promoters and *lacZ* Fusions Designed to Study Expression of Cloned Genes in Yeast

By Leonard Guarente

Experiments probing gene regulation in bacteria have yielded two pertinent kinds of results. First, they have shed light on what bacterial signals regulate gene expression. Second, they have provided systematic methods to monitor expression, the most notable of which is the *lac* fusion technique (for review, see Bassford *et al.*[1]). This technique involves fusing genes of the *lac* operon (e.g., *lacZ*) to the promoter under study so that regulation can be monitored by changes in levels of *lac* enzymes (e.g., β-galactosidase), which can be assayed with ease. The above results have led to the development of methods that employ bacterial regulatory signals to direct expression in *Escherichia coli* of cloned genes from any source.[2]

It is now possible to apply most of the above methodology to the yeast *Saccharomyces cerevisiae*. Laying the groundwork for this extension was the development of a yeast transformation procedure[3] and shuttle plasmids that can be moved back and forth between *E. coli* and yeast.[4] More recently, the *lacZ* fusion technique has been adapted to yeast.[5,6] In this report, I describe plasmids and methods for constructing fusions of any cloned gene to *lacZ* for study in yeast. One application of this technique to be discussed is in the study of regulation of expression of yeast genes. Our studies to date in this area have given rise to plasmids containing yeast promoter regions and conveniently located restriction enzyme sites that facilitate fusion of the promoters to cloned genes. The potential use of these plasmids and *lacZ* fusions to obtain expression in yeast of cloned (heterologous) genes is discussed.

[1] P. J. Bassford, J. Beckwith, M. Berman, E. Brickman, M. Casadaban, L. Guarente, I. Saint-Girons, A. Sarthy, M. Schwartz, H. A. Shuman, and T. Silhavy, *in* "The Operon" (J. H. Miller and W. S. Reznikoff, eds.), p. 245. Cold Spring Harbor Laboratory, Cold Spring Harbor, New York, 1978.

[2] L. Guarente, G. Lauer, T. M. Roberts, and M. Ptashne, *Cell* **20**, 543 (1980).

[3] A. Hinnen, J. B. Hicks, and G. R. Fink, *Proc. Natl. Acad. Sci. U.S.A.* **75**, 1929 (1978).

[4] D. Botstein, C. Falco, S. Stewart, M. Brennan S. Scherer, D. Stinchcomb, K. Struhl, and R. W. Davis, *Gene* **8**, 17 (1979).

[5] L. Guarente and M. Ptashne, *Proc. Natl. Acad. Sci. U.S.A.* **78**, 2199 (1981).

[6] M. Rose, M. J. Casadaben, and D. Botstein, *Proc. Natl. Acad. Sci. U.S.A.* **78**, 2460 (1981).

RECOMBINANT DNA
METHODOLOGY

Materials, Reagents, and Assays

DNA Enzymology

Restriction Digests. These are performed in prescribed buffers (New England BioLabs).

Conversion of Cohesive Ends into Flush Ends. 5' or 3' extensions are converted into flush ends in a 50 μl reaction containing about 1 unit[6a] of DNA polymerase I large fragment (New England BioLabs), 6.6 mM MgCl$_2$, 6.6 mM Tris-HCl, pH 7.4, 50 mM NaCl, 6.6 mM 2-mercaptoethanol, 500 μM of each of the four deoxynucleoside triphosphates, and 1–10 pmol of ends. The reaction is carried out at 23° for 1 hr and terminated by heating to 70° for 20 min, or by phenol extraction of the DNA.

Ligations. DNA ligations are performed in a 25-μl reaction volume containing 5×10^3 units[6a] of T4 DNA ligase (New England BioLabs), 50 mM Tris-HCl pH 7.8, 10 mM MgCl$_2$, 20 mM dithiothreitol, 600 mM ATP, and 1–10 pmol of ends to be joined. The reaction is performed at 15° for 2 hr. Linkers to be attached are added to 0.01 A_{260} unit. After ligation, the DNA either is used directly for transformation or gel electrophoresis (see below) or is extracted by phenol if additional enzymic reactions are desired.

Isolation of Component DNA Fragments for Use in Constructions. DNA fragments are typically separated on a 3.5% acrylamide gel (30, 0.8, acrylamide–bis), and the gel is stained in H$_2$O containing 0.02% methylene blue and destained in H$_2$O until the DNA bands are visible. Bands are excised from the gel and pulverized in a round-bottom plastic tube with a tight-fitting Teflon plunger; the DNA is eluted overnight at room temperature in a buffer containing 50% saturated phenol, 0.25 M NaCl, 5 mM Tris-HCl, pH 8.0, and 5 mM EDTA. DNA recovered from the aqueous phase is then precipitated in ethanol, dried down, and subsequently used in a ligation reaction.

Transformation

Escherichia coli. One hundred milliliters of a log-phase culture are chilled on ice for 10 min; the cells are spun down, resuspended in 50 ml of

---

[6a] Units. DNA polymerase large fragment: the amount of enzyme that will convert 10 nmol of deoxyribonucleotides to an acid-insoluble form in 30 min at 37°. T4 DNA ligase: the amount of enzyme required to give 50% ligation of *Hae*III fragments of phage λ DNA in 30 min at 16° with a fragment end concentration of 0.12 μM. β-galactosidase: 1000 $\times$ OD$_{420}$/time (min) $\times$ volume assayed (in ml) $\times$ OD$_{600}$.[7]

0.1 *M* CaCl$_2$, chilled for 20 min on ice, spun again, and resuspended in 2 ml of 0.1 *M* CaCl$_2$. Cells are then chilled on ice overnight and used in 0.1-ml aliquots per transformation. Cells prepared in this way may be used for 3–4 days.

Yeast. The method of Hinnen *et al.*[3] is used with the following modifications: 10 μg of calf thymus DNA are added along with the DNA to be used, and spheroplasts are incubated at 30° for 20 min in 150 μl of 1 *M* sorbitol, 33% yeast extract peptone dextrose media, 6.5 m*M* CaCl$_2$, 6 μg or uracil per milliliter (for URA$^+$ selection), and then added to regeneration agar for plating.

The *E. coli* strain LG90 (F$^-$, Δ*lacpro XIII*)[2] and the yeast strain BWG2-9A (α, *URA3-52, Ade$^-$, gal4$^-$, his4-519*) (B. Weiffenbach, unpublished observations) are routinely used because they can be transformed at high efficiency by the above methods.

β-Galactosidase Assays

Plates. These contain an M63 salt base (per liter, 13.6 g of KH$_2$PO$_4$, 2 g of (NH$_4$)$_2$SO$_4$, 0.5 mg of FeSO$_4$·7H$_2$O, pH adjusted to 7.0),[7] and, per milliliter, 4 μg of thiamin, 200 μg of biotin, 4 μg of pantothenic acid, 20 μg of inositol, 4 μg of pyridoxine, 40 μg of required amino acids, 2% carbon source, and 40 μg of 5-bromo-4-chloro-3-indolyl-β-D-galactoside (XG). It is important that the pH of the plates be about 7.0.

Liquid Cultures. Cells are grown in minimal medium. Then 10^6 to 10^7 cells are spun down and resuspended in 1 ml of Z buffer (per liter, 16.1 g of Na$_2$HPO$_4$·7 H$_2$O, 5.5 g of NaH$_2$PO$_4$·II$_2$O, 0.75 g of KCl, 0.246 g of MgSO$_4$·7 H$_2$O, and 2.7 ml of 2-mercaptoethanol; pH adjusted to 7.0).[7] Three drops (Pasteur pipette) of CHCl$_3$ and 2 drops of 0.1% SDS are added, and cells are vortexed at high speed for 10 sec. *o*-Nitrophenyl-β-D-galactoside (ONPG) hydrolysis is measured as described.[7] Briefly, 0.2 ml of a 4 mg/ml solution of ONPG (dissolved in H$_2$O) is added to samples preincubated at 28°. The reaction is stopped by adding 0.5 ml of 1 *M* Na$_2$CO$_3$, and the cell debris is spun out. The OD$_{420}$ is measured. Assays are normalized to the OD$_{600}$ of the culture and to the assay time. Alternatively, cells in Z buffer may be broken by adding glass beads and vortexing. The ONPG hydrolyzing activity of the extract is then determined, and activity is normalized to the protein concentration of the extract. Both methods are satisfactory.

[7] J. H. Miller, ed., "Experiments in Molecular Genetics." Cold Spring Harbor Laboratory, Cold Spring Harbor, New York, 1972.

Methods

Construction of lacZ Fusions to Cloned Genes

DNA CONSTRUCTION

Plasmid pLG670-Z, shown in Fig. 1, has been developed for the construction of *lacZ* fusions for study in yeast.[5] It contains markers selectable in *E. coli* and yeast (*Amp*[R] and *URA3*, respectively) and origins of replication that function in the respective organisms (from ColE1 and the yeast 2 μm plasmid). Also carried on the plasmid is a large 3' fragment of *lacZ* immediately preceded by a *Bam* site. Insertion of DNA encoding the amino terminus of a protein into that site will result in a gene fusion to *lacZ*, as long as the translational reading frame is preserved across the junction. The restriction sites *Xho*, *Sal*, and *Sma* lie upstream of *Bam* and facilitate the insertion of DNA fragments into the plasmid as described below. All these sites occur but once on the plasmid.

Several possible approaches for constructing fusions include insertion of *Sau* 3A fragments, fusion by joining flush ends, and use of *Bam* linkers.

Insertion of Sau3A Fragments The enzyme *Sau*3A recognizes a 4-base sequence and cleaves to generate cohesive ends with the same se-

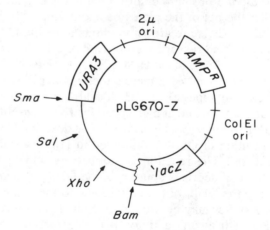

FIG. 1. Depicted is the shuttle vector pLG670-Z used for constructing *lacZ* gene fusions for study in yeast. It contains the markers *Amp*[R] (*Escherichia coli*) and *URA3* (yeast) and origins of replication from ColE1 (*E. coli*) and the 2 μm circle (yeast). Insertion of gene fragments into the *Bam* site can result in fusion to *lacZ*. The *Xho*, *Sal*, and *Sma* sites are also of use in constructing fusions (see text). The *Xho* and *Sal* sites actually lie in a segment of DNA from the *CYC1* leader. This region, however, does not drive significant synthesis of β-galactosidase in yeast because the *CYC1* promoter has been inactivated by a deletion of the UAS.[5]

quence as those generated by *Bam* (GATC). Thus, if a gene contains an internal *Sau*3A site, its fusion to *lacZ* is facilitated. A DNA fragment extending from an internal *Sau*3A site to a *Sau*3A site upstream of the gene can readily be inserted into a pLG670-Z recipient molecule that has been cleaved with *Bam*. If one uses 1 μg of the recipient and insert DNA, about 10^3 to 10^4 transformants of *E. coli* may be obtained. Treatment of the recipient DNA with alkaline phosphatase prior to ligation will increase the percentage of transformants that contain the *Sau*3A insert. *A priori,* the insert will have a 1:2 probability of being correctly oriented; of those correctly oriented, the inserted coding sequence will have a 1:3 probability of being in frame with *lacZ*.

In some cases, for reasons that are not clear, recovered plasmids may show a strong bias in the orientation of their inserts. This problem may complicate the task of constructing in-frame fusions. If *lacZ* is being fused to an *E. coli* or yeast gene, the in-frame fusion may be identified by its encoded β-galactosidase activity in the homologous host organism. However, if *lacZ* is being fused to a gene from some other source, enzyme activity in *E. coli* or yeast resulting from the fused gene is not assured. In this case, it is desirable to insert the gene fragment in a directed orientation.

To direct the gene fragment to insert in the desired orientation only, one must employ a fragment extending from the internal *Sau*3A site to a site recognized by a different enzyme that lies upstream of the gene. If restriction at the upstream site leaves a flush end, or if the end has been rendered flush-ended by DNA polymerase (see Materials, Reagents, and Assays), then the fragment can be inserted into a pLG670-Z backbone that extends from the *Sma* (flush) site to the *Bam* site. Such inserts will all be in the desired orientation. (In addition, the number of background products of ligations, particularly plasmid backbones that self-ligate, is minimized) Alternatively, DNA fragments extending from a *Xho* (or *Sal*) upstream site to an internal *Sau*3A site may be cloned into a pLG670-Z backbone extending from the *Xho* (or *Sal*) site to the *Bam* site.

Fusion by Joining Flush Ends. If the gene to be fused to *lacZ* contains no appropriate *Sau*3A sites, alternative strategies to the above must be employed. One such strategy is to generated a gene fragment that terminates in a flush end. This may be done by cleaving the gene with an enzyme that gives flush ends, or by use of DNA polymerase following cleavage by an enzyme that leaves cohesive ends. Alternatively, exonucleases such as *Bal*31[8] or exonuclease III plus S1,[9] can be used to generate flush

[8] H. B. Gray, D. Ostrander, H. Hodne, R. J. Legerski, and D. Robberson, *Nucleic Acids Res.* **2**, 1459 (1975).

[9] T. M. Roberts and G. D. Lauer this series, Vol. 68, p. 473.

ends after cleavage. In this case, a population of fragments will be generated with different (flush) ends internal to the gene.

If the cloned gene is preceded by a *Sal* (or *Xho*) site, a DNA fragment may be prepared extending from that site to the site internal to the gene that has been rendered flush. This fragment could be inserted in directed fashion into a pLG670-Z recipient molecule extending from the *Sal* (or *Xho*) site to the *Bam* site preceding *lacZ*, which itself has been rendered flush ended. Joining of the flush ends will result in gene fusion if the translational reading frame is preserved across the junction. Alternatively, a DNA fragment bearing two flush ends may be inserted into a pLG670-Z molecule that has been cleaved with *Bam* and rendered flush ended.

Use of Bam Linkers. Oligonucleotide molecules containing the sequence cleaved by *Bam* (*Bam* linkers) may be attached by DNA ligase to gene fragments that terminate with flush ends. Cleavage of the ligated molecules with *Bam* will generate gene fragments that can be fused to the *lacZ* gene of pLG670-Z that itself has been digested with *Bam*. The orientation of such inserts may be directed as described for insertion of *Sau*3A fragments. Methods of constructing fusions using flush ends or linkers provide sufficient flexibility so that virtually any cloned gene can be fused in frame to *lacZ*.

DETECTION OF IN-FRAME FUSIONS

If the DNA sequence of the gene to be fused to *lacZ* is known, in-frame fusions can be made simply by choosing the appropriate gene fragment. In the absence of sequence information, confirming that a fusion is in frame must proceed on a case by case basis. In constructing gene fusions between the yeast gene encoding the iso-1-cytochrome *c* (*CYC1*) and *lacZ*, it was noticed that the in-frame fusion displayed significant β-galactosidase activity in *E. coli*.[5] This expression is likely to originate from transcription that is initiated in the *CYC1* leader DNA and translation that is initiated at the ATG at the start of *CYC1* coding DNA (Guarente, unpublished results). Thus, the in-frame fusion could be detected by the appearance of β-galactosidase activity after transformation of a Lac⁻ *E. coli* indicator strain with DNA containing the fused gene.

Fusions to Yeast Genes

In general, in-frame fusions of yeast genes to *lacZ* should result in β-galactosidase expression in yeast. This expression can be monitored by a plate assay that employs the chromogenic β-galactosidase substrate, XG, or by a quantitative determination of ONPG hydrolyzing activity in a liquid-grown culture (see Materials, Reagents, and Assays). Enzyme expressed from such fusions should respond to the regulatory signals of the yeast genes to which they are fused.

Use of 2 μm Plasmids. We have studied fusions of *lacZ* to regulatory signals of each of two different yeast genes, *CYC1*[9] and *GAL10*.[10] In each case, the fusion resides on 2 μm-containing plasmids and β-galactosidase expression follows the regulation of the promoter to which *lacZ* is fused (see Fusion of Yeast Promoters, below).

In some cases, the high copy number (about 20/cell) of the 2 μm plasmids might interfere with normal gene regulation. Here, the use of vectors that integrate into the yeast genome or that replicate autonomously in low copy is called for.

Effects of Vector Sequences. The vector sequences that precede the yeast DNA is a yeast gene–*lacZ* fusion may affect gene regulation. For example, we have found that a sequence of pBR322 DNA (between the *Hin*dIII and *Sal* sites) is homologous to a portion of the *CYC1* promoter region (unpublished observation). This region of the wild-type *CYC1* promoter lies about 250 base pairs upstream from the transcriptional start (−250) and activates gene expression. If this upstream activation site (UAS) is deleted, expression is reduced about 500-fold. However, if it is replaced with the pBR322 sequence, expression is only slightly reduced.

Fusion of Yeast Promoters to Cloned Genes

The CYC1 Promoter. Transcription of a wild-type *CYC1* gene has been shown to be glucose repressible.[11] The *CYC1*–*lacZ* fused gene directs synthesis of 100 units (see Materials, Reagents, and Assays) of β-galactosidase in minimal glucose medium, and 1000 units of enzyme in medium supplemented with glycerol and ethanol as carbon sources.[10] The *CYC1* promoter contains two important regions, the UAS mentioned above, and the region around the TATA box and sites of transcription initiation.[10] (*CYC1* mRNA, has been found to contain multiple 5′ ends.[12])

Two shuttle vectors have been constructed that carry the *CYC1* promoter followed by a *Bam* site (Fig. 2). In vector C1 the *Bam* site immediately follows the ATG at the start of *CYC1* coding DNA.[10] In vector C2, this ATG and three nucleotides that precede *CYC1* coding DNA have

[9] L. Guarente, R. Yocum, and P. Gifford, *Proc. Natl. Acad. Sci. U.S.A.* **79,** 7410 (1982).
[10] L. Guarente and T. Mason, *Cell,* in press.
[11] R. Zitomer, D. Montgomery, D. Nichols, and B. Hall, *Proc. Natl. Acad. Sci. U.S.A* **76,** 3627 (1979).
[12] G. Faye, D. Leung, K. Tatchell, B. Hall, and B. Smith (1981). *Proc. Natl. Acad. Sci. U.S.A.* **78,** 2258 (1981).

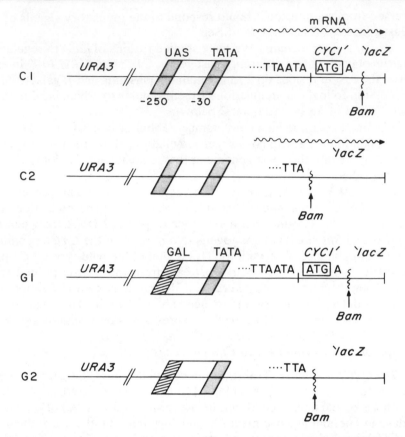

FIG. 2. Yeast expression vectors are shown. These vectors have the same markers and origins of replication as pLG670-Z (Fig. 1). C1 bears the *CYC1* promoter which contains a TATA box and an upstream activation site (UAS) at -250 required for activation of the promoter. The terminal nucleotides of *CYC1* leader DNA and the ATG at the start of *CYC1* coding DNA (boxed) are shown. C2 is identical to C1 except that the sequence ATA[ATG]A (adjacent to the *Bam* site of C1) has been deleted. G1 and G2 are analogous to C1 and C2 except that the UAS has been substituted with a segment of DNA that lies upstream of *GAL10* (hatched box). This segment confers galactose inducible regulation on the promoter. The use of these plasmids for the expression of cloned genes in yeast is discussed in the text.

been deleted (Guarente, unpublished observation). (In both cases, the *Bam* site is immediately followed by the 3' *lacZ* fragment.) Thus, the C1 DNA upstream of the *Bam* site encodes a mRNA leader and AUG initiator codon, whereas the C2 DNA upstream of the *Bam* site encodes the leader without the initiator codon.

Expression of a cloned gene, X, in yeast may be achieved using either

C1 or C2. If the X coding sequence is inserted into the *Bam* site of C1, it may be expressed as a result of transcription that is directed by the *CYC1* promoter, and translation that initiates at the *CYC1*-encoded AUG. Translation would proceed through any X leader RNA into the X coding sequence. Expression of *lacZ* has been achieved using C1 and analogous plasmids.[5] If the X coding sequence is inserted into the *Bam* site of C2, expression of the unfused X protein may be achieved. In this case, translation would initiate at the start of the X coding sequence. Although we have not yet probed the expression potential of C2, genetic experiments in yeast[13] and biochemical experiments in higher eukaryotic systems[14] suggest that translation initiates at the AUG closest to the 5' end of the mRNA. If the inserted X leader DNA does not encode any AUG triplets, then the AUG encoded at the start of X coding DNA would be the first AUG of the *CYC1*–X chimeric mRNA. Translation of such an mRNA, therefore, might be expected to initiate at this codon. This point in greater detail in the next section.

A set of vectors, G1 and G2, that are analogous to C1 and C2 have also been constructed[9] (Fig. 2). G1 and G2 differ from C1 and C2 only in the DNA in the −250 region of the promoter. In these plasmids, the UAS has been replaced by a DNA segment that lies upstream of the *GAL10* gene of *S. cerevisiae*. Although this segment does not encode a transcriptional start site, it does encode galactose-inducible expression.[9] Thus, the chimeric promoter, consisting of the *GAL10* segment and the *CYC1* TATA box and initiator region directs synthesis of <1 unit of β-galactosidase in medium with glucose and about 1000 units in galactose medium. Translation of sequences encoded by a gene X cloned into G1 and G2 will proceed in a manner identical to that of C1 and C2. In this case, however, because of the strong galactose inducibility of the promoter, levels of expression of the X product may be varied over a 1000-fold range.

Will Codons at the Start of Cloned Genes Function as Initiators in Yeast?

More detailed experiments designed to troubleshoot problems encountered in expressing cloned genes in yeast are described here. For example, if one does not observe expression of a protein product after cloning a gene X into C2 or G2 as described above, one may wish to pinpoint where the block occurs. The block may be due to an inability of yeast ribosomes to initiate translation at the AUG triplet encoded at the start of

[13] F. Sherman, J. Stewart, and A. Schweingruber, *Cell* **20**, 215 (1980).
[14] M. Kozak, *Cell* **22**, 7 (1980).

X, or to other factors, such as instability of the protein in yeast. Therefore, to determine directly whether the AUG encoded by the start of X
functions as an initiator in yeast, I have devised the following scheme.
First, a DNA fragment encoding an amino-terminal portion of X is cloned
into C2 (or G2) so that an X–lacZ fused gene is formed (see Fig. 3B). Detection of β-galactosidase expression from such a fusion would demonstrate that the AUG encoded at the start of X functions as an initiator in
yeast. In this case, insertion into C2 of the intact X gene should result in
synthesis of the unfused protein product. If the product is not observed, it
is likely that it is unstable in yeast.

　　　If no β-galactosidase expression derives from the fused gene in the
above case, it is likely that the AUG encoded at the start of X does not

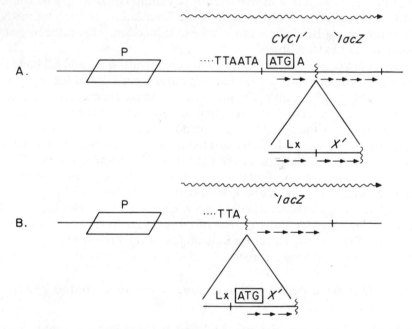

Fɪɢ. 3. Illustrated in the utilization of C1 and C2 (Fig. 2) (or G1 and G2) to probe whether
AUG triplets encoded at the starts of cloned genes are recognized as initiator codons in
yeast. (A) A segment of DNA encoding the leader (L_x) and a portion of the coding sequence
of a cloned gene, X, has been inserted into C1 (or G1). Translation (arrows) will initiate at
the AUG corresponding to the CYC1 ATG (boxed), proceed through the leader and gene X
sequences, and finally through lacZ encoded sequences, providing that the translational
reading frame is in register across each junction. (B) The X gene segment has been inserted
into C2 (or G2). In this case, translation may initiate at the triplet corresponding to the gene
X ATG (boxed) and proceed through gene X and lacZ sequences, if the translational reading
frame is in register across the X–lacZ junction. The use of these manipulations to infer
whether the gene X encoded AUG functions as an initiator in yeast is discussed in the text.

function as an initiator in yeast, but other possibilities exist to explain this result. For example, it is possible that internal codons of the X gene are not utilized in yeast or that the X–β-galactosidase hybrid protein is inactive. To distinguish between these possibilities, the X gene fragment may be fused to the *lacZ* fragment in C1 (or G1) (Fig. 3A). In this case, care must be taken so that the AUG initiation codon of *CYC1* is in frame with the inserted X codons and that these in turn are in frame with *lacZ*. Translation, thus, will initiate at the AUG encoded by *CYC1*, proceed through the X leader and coding sequences, and finally into *lacZ* sequences. Failure of such a construct to give rise to β-galactosidase expression would suggest that at least one X codon is not utilized in yeast, or possibly that the hybrid protein is inactive. Moreover, if β-galactosidase is expressed upon insertion of the X gene fragment into C1, but *not* C2, then it is likely that the AUG encoded at the start of X does not function as an initiator in yeast.

With this information, site-specific mutagenesis around the start of X could be performed, and mutations isolated that render the AUG encoded at the start of X as an active initiator. These mutations could be easily recognized, since they would result in β-galactosidase expression, which can be detected on plates containing the dye XG (see Reagents section). After the isolation of such mutations, the X gene could be regenerated intact by recombination *in vitro*. A procedure analogous to this has been developed to express cloned genes in *E. coli*.[2]

Comments

Plasmids described in this chapter may be used to probe the signals that govern the initiation of transcription and translation in *S. cerevisiae*. These methods provide powerful tools in the analysis of the expression of yeast genes. Further, the plasmids should facilitate the expression in yeast of any cloned gene to produce the native, unfused product. If there is a block to the expression of a particular gene, the methods outlined herein should pinpoint the stage at which the block occurs. With this information, it should be possible to isolate mutations that overcome the block, as discussed.

[28] The Use of *Xenopus* Oocytes for the Expression of Cloned Genes

By J. B. GURDON and M. P. WICKENS

General Properties of Oocytes Compared to Other Recipient Cells

The injection of amphibian oocytes was one of the first systems in which purified DNA was correctly transcribed and expressed as protein.[1] Since then, cell-free systems have been developed which initiate transcription accurately, and the expression of eukaryotic genes has been obtained by infecting cells with genetically manipulated viruses, by transfecting Ca-precipitated DNA into cultured cells, and by the direct injection of DNA into cultured cells or mouse eggs. Compared to these other systems, amphibian oocytes have three characteristics. First, a very small amount of DNA needs to be injected into a single oocyte to obtain recognizable transcription and expression (see sections Transcription and Expression as Protein). Second, the expression of DNA can be monitored within a few hours of injection, during which time it is not replicated or integrated into host cell chromosomes, but is assembled with nucleosomes into an apparently normal chromatin structure. In most other expression systems, DNA is integrated and replicated, as a result of which it may undergo genetic changes and its expression may be influenced by the properties of adjacent host DNA. Third, oocyte injection makes it possible to introduce any cell components, such as RNA, chromatin, or nuclear proteins; this is likely to be particularly valuable for analyzing the regulation of gene expression.

An amphibian oocyte is a single large cell, surrounded by several thousand small follicle cells. It is in meiotic prophase, and active in RNA and protein synthesis, but totally inactive in DNA synthesis. Its composition and general properties are summarized in Table I.

---

[1] Many of the conclusions stated in this chapter have been documented in recent reviews which should be consulted for all unreferenced statements [see J. B. Gurdon and D. A. Melton, *Annu. Rev. Genet.* **15,** 189 (1981); A. Kressman and M. L. Birnstiel, *in* "Transfer of Cell Constituents into Eukaryotic Cells" (J. Celis, ed.) pp. 383–407 Plenum, New York, 1980; M. P. Wickens and R. A. Laskey, *in* "Genetic Engineering" (R. Williamson, ed.), Vol. 1, pp. 103–167. Academic Press, New York, 1981]. We supply original references in this article only when they provide a source of technical information.

RECOMBINANT DNA
METHODOLOGY

TABLE I

Composition and General Properties of a Full Sized (1250 μm Diameter) Oocyte of *Xenopus laevis*[a]

	Cytoplasm	Nucleus	Follicle cells
Volume (% of total)[b]	0.5 μl (90%)	40 nl (~10%)	—
DNA content			
Chromosomal[c]	None	12 pg	30,000 pg
Nucleolar[c]	None	25 pg	30 pg
Mitochondrial[c]	4000 pg	None	150 pg
DNA synthesis per day[d]	None	None	Significant
RNA content			
Ribosomal (28 + 18 S)[e]	5 μg (10^{12} ribosomes)		
5 S[f]	60 ng	—	
4 S[f]	60 ng	—	
Poly(A)-containing[g]	70[h] ng (10% polysomal)	10 ng	
RNA accumulation[i] (total per day)	Whole oocyte: 20 ng (1 ng mitochondrial)		
Protein content			
Yolk	250 μg	None	
Nonyolk	25 μg	2.5 μg	
Histones[j]	70 ng	70 ng	
Nucleoplasmin[k]	5 ng	250 ng	
RNA polymerase[l]	—	10^5 × somatic cell	
Protein synthesis (total accumulated/day)[m]	400 ng	None	

Content per oocyte, ignoring follicle cells (pmol)

rATP[m]	1700
rUTP	1200
rCTP	500

rGTP	250
dTTP	7
Methionine[o]	44
Glutamic, aspartic acids	2900, 1600
Other amino acids	30–300

[a] Much of the information in this table on synthesis is discussed in detail in Ref. 28. —, not known.

[b] Total volume is 1 µl, of which half is yolk and is therefore metabolically unavailable space.

[c] Tetraploid nucleus with 1000-fold amplified ribosomal genes. The haploid genome consists of about 3×10^9 base pairs or 3 pg DNA. See Refs. 34–37 for genomic and mitochondrial DNA values. About 5000 follicle cells surround each oocyte.

[d] No replication of double-stranded DNA; single-stranded DNA becomes double-stranded (see Ref. 16). About 5% of follicle cells take up [³H]thymidine.

[e] See Refs. 38 and 39.

[f] See Ref. 40.

[g] See Refs. 41, 44, 49, and 51; 90% is untranslated, not polysome-associated.

[h] Partly mitochondrial, see Refs. 36, 37, and 46.

[i] The typical transcription rate for an oocyte is 15 nucleotides per second, with RNA polymerases spaced 100–200 bp apart (Ref. 28). See also Refs. 11, 36, 42, and 45. At this rate of RNA accumulation (<20 ng/day), it would take 250 days for a full-sized oocyte (5 µg RNA) to be formed.

[j] See Ref. 43.

[k] See Ref. 32.

[l] An oocyte has equal activities of polymerases I, II, and III. Its activity exceeds that of a cultured cell by 60,000 for polymerases I and II, and by 500,000 for polymerase III. Almost all oocyte activity except for IIA is in its nucleus. See Ref. 33.

[m] See Ref. 47, p. 136 and Ref. 48. The rate given may increase by a few fold under some physiological conditions. The typical translation rate for an oocyte is one codon (3 nucleotides) per second, with ribosomes spaced 100 nucleotides apart. Protein synthesis does not include yolk, which is synthesized in the liver and transported to growing oocytes.

[n] For measurement of nucleotide pools, see Ref. 50.

[o] For measurement of amino acid pools, see Ref. 29.

Methods

Oocytes

The collection and culture of oocytes, as well as the instruments and techniques needed for injection, have all been described.[2] This source supplements the information given below.

In a typical experiment, 25 nl of DNA at 200 μg/ml (i.e., 5 ng) is injected into an oocyte. The best medium for culturing oocytes is modified Barth solution (MBS) (see Table II).

Labeling of RNA and Protein

For labeling RNA it is generally best to inject 1 μCi of [^{32}P]GTP or [^{32}P]CTP per oocyte (mixed with DNA if desired). ATP or UTP may be used, but the pool sizes are larger (see Table I) and UTP is efficiently converted in oocytes into non-RNA molecules. Nucleosides and amino acids, but not nucleotides, are taken up from the medium. Thus RNA may also be labeled by adding [^{3}H]guanosine to the medium. In this case, transcripts synthesized in the large number of follicle cells account for more than 90% of the total ^{3}H-labeled RNA (see below for follicle cell removal).

The most generally useful procedure for labeling proteins is to incubate an oocyte in 10 μl MBS containing 5 μCi of [^{35}S]methionine for 18–24 hr. The amount of incorporation is not greatly affected by the volume of medium in which the oocytes are incubated, at least between 3 and 15 μl per oocyte.

With isotopes of high specific activity, the amount of incorporation is directly proportional to the amount of radioactive precursor used, at least up to 10 μCi per oocyte.

Injection Technique

The transcription of injected DNA takes place only in the nucleus or germinal vesicle of an oocyte, which is not normally visible. DNA can be deposited in the germinal vesicle with about an 80% success rate by penetrating the oocyte in the center of the black pigmented hemisphere, until the pipette tip is judged to be one-third of the way from this point to the opposite pole of the oocyte.[3] This technique may be readily learned by practising the injection of a concentrated trypan blue solution. Opening the oocyte with forceps just after injection will show whether the dye was

[2] J. B. Gurdon, *Methods Cell Biol.* **16,** 125, 139 (1977).
[3] J. B. Gurdon, *J. Embryol. Exp. Morphol.* **36,** 523, 540 (1976).

TABLE II
MODIFIED BARTH SOLUTION (MBS) AND
ITS PREPARATION

	Concentration in medium (mM)	10 × stock[a] (g/liter)
NaCl	88	51.3
KCl	1.0	0.75
NaHCO$_3$	2.4	2.0
Hepes, pH 7.5	10.0	23.8
MgSO$_4$·7H$_2$O	0.82	2.0
Ca(NO$_3$)$_2$·4H$_2$O	0.33	0.78
CaCl$_2$·6H$_2$O	0.41	0.90

[a] The 10 × stock solution should be prepared by adding reagents in the above order to 900 ml H$_2$O, finally making up to 1 liter. After filter sterilization, this stock may be stored for months at +4°. After dilution for use, the pH should not need adjustment, but penicillin and streptomycin may be added to give a final concentration of 10 mg/liter.

deposited in the oocyte's nucleus. Some workers prefer to centrifuge oocytes lightly so as to bring the germinal vesicle to the surface where it can be seen,[4] but this can reduce viability. Usually a manually controlled syringe is used to control the volume of fluid injected,[2] but a more sophisticated apparatus has been described[5] which avoids problems of the pipette becoming blocked by backflow.

A single injected oocyte is often adequate for the detection of RNA or protein. However, individual oocytes vary somewhat in the amount of RNA or protein which they synthesize from an injected template. Such individual differences are easily averaged out by injecting 10 or more oocytes from the same ovary with each sample.

Follicle Cell Removal

Each oocyte is closely surrounded by about 5000 follicle cells, which greatly affect the composition and synthesis of ovarian material (Table I) unless removed. This can be done by gently swirling small clusters of oocytes (20 per cluster) overnight in MBS containing 2 mg/ml collagenase.[6] This procedure removes all follicle cells as judged by scanning electron

[4] A. Kressman, S. G. Clarkson, V. Pirrotta, and M. L. Birnstiel, *Proc. Natl. Acad. Sci. U.S.A.* **75**, 1176 (1978).

[5] D. L. Stephens *et al.*, *Anal. Biochem.* **114**, 299 (1981).

[6] T. J. Mohun, C. D. Lane, A. Colman, and C. C. Wylie, *J. Embryol. Exp. Morphol.* **61**, 367 (1981).

microscopy, and leaves oocytes with unimpaired viability.[6] It seems preferable to the use of pronase.[7] In many cases, the presence of follicle cells can be ignored and not removed, e.g., after labeling RNA by the injection of ^{32}P-labeled nucleotides which do not penetrate the follicle cells from oocyte cytoplasm.

RNA Extraction

For the analysis of transcripts, RNA may be extracted from oocytes with good recovery and minimal risk of degradation by the following procedure (other procedures have also been reported[8-11] and may be used successfully). Homogenize 5–15 oocytes in 0.5 ml of 0.3 M NaCl, 2% SDS, 50 mM Tris, pH 7.5, 1 mM EDTA (room temperature). This is conveniently done in a 1.0-ml glass homogenizer. Quickly transfer the homogenate to a 1.5-ml microfuge tube containing 0.5 ml of phenol: chloroform (1:1), and vortex immediately. Centrifuge in a microfuge for 5–10 min. Remove the aqueous phase. Add 0.5 ml of the same homogenization buffer to the phenol: chloroform phase, vortex, and centrifuge again. Remove the aqueous phase, combine with the first, and add 2 volumes of ethanol. A large flocculent white precipitate of carbohydrate (see below for removal) will form immediately upon addition of the alcohol. Ethanol precipitate as desired (− 20° for 15 min usually is adequate); then recover the precipitate by centrifugation and wash it with 70% ethanol. Again drain off the alcohol, and dry briefly. If the precipitate is contaminated with brown material, as is often the case when extracting 15 or more oocytes per 0.5 ml, redissolve the precipitate in 50 mM Tris, pH 7.5 and extract once more with an equal volume of phenol: chloroform (1:1). Remove the aqueous phase, adjusting to 0.3 M NaCl, and precipitate with 3 volumes of ethanol.

Proteinase K may also be included in the homogenization buffer (at 1 mg/ml) when isolated germinal vesicles and cytoplasms are being analyzed.[12] It is simplest to transfer the nucleus or cytoplasm directly into homogenization buffer. Using proteinase K-supplemented buffer, cytoplasm and nuclei may be accumulated for at least 15 min without the RNA suffering any degradation.

When preparing RNA from single oocytes, homogenize each oocyte in 0.2–0.4 ml as above. It is not necessary to add carrier, since each oocyte

[7] L. D. Smith and R. E. Ecker, *Dev. Biol.* **19**, 281 (1969).
[8] J. E. Mertz and J. B. Gurdon, *Proc. Natl. Acad. Sci. U.S.A.* **74**, 1502 (1977).
[9] D. D. Brown and E. Littna, *J. Mol. Biol.* **8**, 669 (1964).
[10] A. Kressmann *et al.*, *Cold Spring Harbor Symp. Quant. Biol.* **42**, 1077 (1978).
[11] D. M. Anderson and L. D. Smith, *Cell* **11**, 663 (1977).
[12] E. Probst *et al.*, *J. Mol. Biol.* **135**, 709 (1979).

contains 5 μg of rRNA and a considerable amount of carbohydrate. How-ever, in preparing RNA from single germinal vesicles which contain little nucleic acid or carbohydrate, it is prudent to add exogenous RNA or DNA (e.g., tRNA, to a final concentration of 50 μg/ml) to the above ho-mogenization buffer, so as to minimize losses during extraction and pre-cipitation.

Analysis of Labeled RNA

Any standard method used to analyze labeled or unlabeled RNA can, in principle, be applied to transcripts from oocytes. Each of the tech-niques which follow is discussed in detail elsewhere in *Methods in Enzy-mology*. Here we discuss only those points which are particularly impor-tant for work with oocytes.

1. Direct gel electrophoresis. For injected genes transcribed by RNA polymerase III (tRNA and 5 S RNA), this clearly is the method of choice. Enough radioactive RNA is synthesized in a few hours to be detected by gel electrophoresis of total RNA from less than one oocyte. Although oo-cytes synthesize tRNA and 5 S RNA from their own genes, this endoge-nous background is inconsequential in most experiments, since the tran-scripts of injected genes are usually at least 20 times more abundant. Genes encoding mRNA, transcribed by polymerase II, cannot always be assayed by direct gel electrophoresis of total labeled RNA. This is be-cause (a) such genes initiate 100 to 1000 times less frequently than those transcribed by polymerase III,[1,12] (b) precise initiation and termination are required for the production of a detectable RNA band, and neither pro-cess may be efficient on a particular DNA template, and (c) polyadenyla-tion and multiple splicing events may complicate the pattern observed. In addition, several endogenous transcripts are prominent in those regions of an agarose gel in which mRNAs generally lie, and so contribute to the background (see Ref. 46 for a detailed description). In spite of these diffi-culties, mRNA transcripts from some genes, notably the sea urchin his-tone genes and the SV40 late genes, are of discrete size, and can be read-ily detected in total RNA preparations from less than one oocyte.

2. Purification of specific transcripts by hybridization to DNA bound to paper. DNA immobilized on DBM-paper filters may be used to purify template-specific transcripts which are only a small fraction of total oo-cyte RNA. This reduces the background by two to three orders of magni-tude relative to direct gel analysis and provides a reasonable recovery of specific transcripts (10–50% of template-specific transcripts in a 4-hr hy-bridization). In such experiments as much as 30 oocytes' worth of RNA is redissolved in 100 μl of 50% formamide (deionized), 0.4 M NaCl, 0.2% SDS, 20 mM Pipes, pH 6.4, 5 mM EDTA. Redissolving the RNA at this concentration may require repeated pipetting and some patience. This so-

lution is then hybridized to paper-bound DNA which is then washed and eluted. The large amount of carbohydrate does not increase the background or interfere with the hybridization.

3. Southern filter hybridization. To determine which regions of injected DNA direct the synthesis of labeled transcripts, 50,000–500,000 cpm of ^{32}P-labeled RNA from injected oocytes can be hybridized to a filter bearing DNA restriction fragments transferred after gel electrophoresis.

Analysis of Unlabeled RNA

1. Northern analysis. Sufficient material is generally synthesized from injected genes (Table III) to permit rapid detection. However, it is essential to remove the DNA which was injected from the RNA sample prior to electrophoresis, since it can otherwise contribute a very high background. This may be achieved by standard biochemical techniques, e.g., digestion with RNase-free DNase. It must also be noted that each oocyte contains 5 μg of ribosomal RNA which may prevent the detection of transcripts of similar size.

2. S1 nuclease digestion of RNA:DNA hybrids. The versatility of this method and the large amount of RNA synthesized in oocytes make this technique very useful for analyzing transcripts synthesized by polymerase II. Again, standard techniques may be used without modification. Less than one oocyte's worth of RNA is often adequate for an overnight exposure using an end-labeled DNA fragment probe of only 10^6 cpm/μg specific activity.

A contaminant of oocyte RNA (possibly the carbohydrate) can distort the migration of protected DNA fragments on thin sequencing gels, such that lanes narrow toward the bottom. Results are interpretable, though it may be difficult to deduce precise lengths by comparison with undistorted marker lanes. This problem can be circumvented either by using only a little oocyte RNA in each lane (less than one-tenth oocyte per 5-mm-wide slot), by mixing markers with the protected fragments, or by purifying the RNA free of carbohydrate prior to hybridization [see DBM-paper technique (2) above].

Methods of Protein Analysis

The methods used to prepare homogenates for protein analysis must, of course, vary with the conditions required for the stability of the particular protein examined. A generalized technique useful for direct gel analysis of radioactive proteins is as follows. It includes extraction of the homogenate with freon (see below); this selectively removes yolk proteins, which distort the high-molecular-weight region of SDS polyacrylamide gels, but does not appear to selectively remove any other proteins from an oocyte homogenate (as judged by SDS–polyacrylamide gel electrophoresis). In contrast, the removal of yolk by direct centrifugation results in highly specific losses of basic proteins (R.A. Laskey, personal communication).

Homogenize 10–30 oocytes in 1 ml of ice-cold 15 mM Tris, pH 6.8, and 150 μg/ml PMSF. Add an equal or greater volume of freon (1,1,2-trichlorotrifluoroethane) and vortex. Separate the upper, aqueous phase from the lower, freon phase by centrifugation for 10 min in a microfuge. A large dark interphase containing yolk protein and pigment granules will be obvious. Remove the aqueous phase and, if necessary, centrifuge for 10 min to clarify. This homogenate may then be analyzed directly by SDS–polyacrylamide gel electrophoresis.

Immunological techniques may also be used to recognize proteins synthesized in the oocyte from injected DNA templates. It is advantageous that the oocyte extensively and accurately modifies protein posttranslationally (see section Posttranslational Events). Furthermore the selective secretion by oocytes of only those proteins which are normally secreted results in a considerable purification, since all the endogenous nonsecreted proteins are effectively removed.

To detect materials secreted into the medium,[6,13] oocytes should be isolated individually from the loosely attached follicular tissue. It is essential to remove any dead oocytes, which may release proteolytic enzymes. Incubations should be carried out with one or two oocytes in a 5 or 10 μl drop of MBS (Table II) in a water-saturated atmosphere. Increasing the volume in which the oocyte is incubated does not much decrease the amount of [^{35}S]methionine which it takes up from the medium. The medium may be collected after 1–2 days.

Isolation of Germinal Vesicles and Cytoplasms

For some experiments, it is useful to separate the germinal vesicle and cytoplasm of injected oocytes. When this is anticipated, it is desirable to inject only about 15 nl into the germinal vesicle, so that it is not damaged by inflation. The germinal vesicle and cytoplasm can be separated by opening the oocyte with forceps in MBS (Table II), and removing adhering yolk by passing the germinal vesicle into and out of a pipette. Alternatively, an incision can be made with a syringe needle (26 G) in the oocyte's animal pole (center of pigmented region) and the germinal vesicle gently squeezed out with forceps.[2] Since the germinal vesicles are small, they are best collected in homogenization medium containing proteinase K and carrier DNA, RNA, or protein (see above). Isolated germinal

[13] A. Colman and J. Morser, *Cell* **17**, 517 (1979).

vesicles of DNA-injected oocytes may be used for the electron micro-scope examination of active transcription complexes. To avoid spending time on the analysis of germinal vesicles which happened to miss an injec-tion of DNA, a trace of any iodinated large protein such as serum albu-min, may be added to the DNA which is injected.[14] Individual germinal vesicles are then counted in a drop in a gamma counter before making nu-clear spreads.

Transcription

The configuration of DNA injected into oocytes has a substantial ef-fect on the efficiency of its transcription. A small linear molecule (5000 base pairs long) yields 10–20 times less RNA in oocytes than the same kind of DNA in circular form.[1,12] On the other hand all forms of circular molecule are equally well transcribed. Single-stranded molecules, such as M13 phage DNA, are copied into a double-stranded form,[15] and nicked circles are ligated. These are then converted into double-stranded super-coiled molecules and assembled with nucleosomes.[16]

A generally appropriate amount of DNA to inject is 5 ng (25 nl at 200 μg/ml). This amount seems to saturate the transcriptional capacity of an oocyte, at least with genes transcribed by polymerase II and III,[8,17] though an oocyte has sufficient histones to assemble much more than 5 ng DNA (Table I). The efficiency with which DNA is transcribed (tran-scripts/gene/hour) increases when less DNA is injected, but the absolute amount of gene product is less.

The amount of RNA typically obtained from injected DNA, and its specific activity, can be calculated as follows, assuming circular mole-cules containing one copy of a gene are injected: 1 μCi (2×10^6 dpm) of [^{32}P]nucleoside triphosphate (either GTP or CTP) is injected. Since the oocyte contains about 250 pmol of GTP (Table I), this corresponds to 8000 dpm/pmol GTP, or 5 dpm/pg RNA, which therefore has a specific activ-ity of about 5×10^6 dpm/μg.

Oocytes not injected with DNA generally incorporate about 5% of the GTP pool in 24 hr, corresponding to 12.5 pmol GTP, or 17 ng RNA. Most of the stable RNA seen after a 24-hr incorporation period is rRNA; in shorter labeling periods, pre-rRNAs predominate. In addition to their en-dogenous transcription, oocytes injected with DNA transcribed by poly-merase II generally incorporate 1% of the GTP pool into RNA com-

[14] M. F. Trendelenburg and J. B. Gurdon, *Nature* (London) **276**, 292 (1978).
[15] R. Cortese, R. Harland, and D. A. Melton, *Proc. Natl. Acad. Sci. U.S.A.* **77**, 4147 (1980).
[16] A. H. Wyllie, *et al., Dev. Biol.* **64**, 178 (1978).
[17] J. B. Gurdon and D. D. Brown, *Dev. Biol.* **67**, 346 (1978).

plementary to injected DNA (or about 5% with DNA transcribed by polymerase III). These values correspond to 3 ng (polymerase II) and 8 ng (polymerase III) of transcript accumulated per day in each oocyte (Table III).

The amounts of accumulated transcript per day given above relate to *stable* RNAs and are presented as an aid in experimental design. They are not equivalent to *rates* of synthesis, since as much as 90% of the newly synthesized RNA from some templates—SV40, for instance—may be rapidly degraded.[17a]

α-Amanitin can be used to determine the type of oocyte polymerase which transcribes injected genes (Fig. 1). The injection of α-amanitin (mixed with DNA) at 5 pg per oocyte eliminates polymerase II transcription without decreasing the activity of polymerases III or I; 500 pg α-amanitin per oocyte greatly reduces polymerase III transcription, but even 5 ng per oocyte has no substantial effect on transcription by polymerase I (Fig. 1). It has been found so far that eukaryotic genes are transcribed in injected oocytes by the same type of polymerase as is used in the cells where these genes are normally expressed.

TABLE III

RNA SYNTHESIS IN DNA-INJECTED *Xenopus* OOCYTES[a]

DNA injected (No. of genes)[b]	Synthesis of complementary (5 S or SV40) RNA[b]		
	Total RNA dpm (% of [^{32}P]GTP injected)	dpm in RNA (% of total labeled RNA)	Amount of 5 S or SV40 RNA[c]
None	100,000 (5%)	[5 S: 0.25% (endogenous); SV40: none]	(5 S: 40 pg)
Xenopus 5 S in plasmid[b] (5 ng; 5 × 10^8 mol)	200,000 (10%)	50,000[d] (25%)	8 ng; 10^{11} mol
SV40 DNA[b] (5 ng; 5 × 10^8 mol)	120,000 (6%)	20,000 (16%)	3 ng; 10^9 mol[e]

[a] Labeled for 24 hr with 1 μCi [^{32}P]GTP per oocyte. These results apply only to newly synthesized RNA which is stable for several hours.

[b] See Ref. 17 for quantitation of results with 5 S genes in a plasmid, and Refs. 8 and 30 for SV40 transcription.

[c] Values calculated assuming that an uninjected oocyte synthesizes 20 ng RNA per day (Table I).

[d] The remaining 50,000 dpm RNA synthesized from injected DNA is complementary to the plasmid region of the injected DNA.

[e] Assuming an average transcript length of 3000 nucleotides.

[17a] A. A. Miller *et al. Molec. Cell Biol.* (1982) (in press).

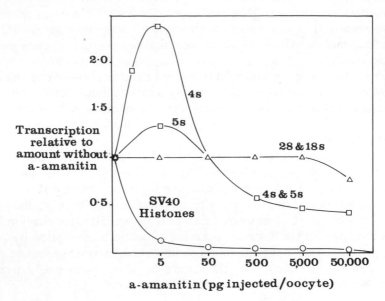

FIG. 1. α-Amanitin sensitivity of transcription in DNA-injected oocytes.

Injected prokaryotic plasmids (pBR322, ColE1, etc.) are transcribed by polymerase II, and produce approximately the same amount of transcript per day per oocyte as do eukaryotic genes which direct accurate initiation and termination, like SV40. Not surprisingly then, read-through transcription from vector DNA into eukaryotic genes inserted into recombinant clones can complicate studies of transcription initiation. Consequently, eukaryotic DNAs are often excised from the vector using a restriction enzyme and are then religated into a circle free of vector DNA prior to injection.[12]

The fidelity with which injected genes are transcribed by oocytes varies considerably according to the type of gene used.[1] Genes transcribed by polymerase III (e.g., 5 S genes and tRNA genes) generally show good strand selectivity, accurate initiation, and good but not perfect termination. Some genes transcribed by polymerase II, such as herpes virus thymidine kinase, certain sea urchin histone genes, and the SV40 late genes also show accurate initiation and termination, while others, such as ovalbumin (in a plasmid or on its own), do not. The cause of this difference in the transcription of these genes is not yet clear.

In those cases in which the oocyte has been used to map promoter regions, it appears that oocytes provide more *in vivo*-like conditions than cell-free systems, in that regions of DNA other than the TATA box are required. In this context the fact that injected DNA is assembled into nucleosomes may be relevant.

Posttranscriptional Processing

RNA polymerase III transcribes tRNA genes injected into oocytes into a primary transcript which is then matured in the nucleus into tRNA by a series of modifications. These include the removal of 5' and 3' sequences, base modification, and the precise excision of an intervening sequence. Fully matured tRNA is produced with high efficiency and is transported to the cytoplasm.

Oocytes also carry out several posttranscriptional modifications of mRNA precursors synthesized by polymerase II from injected genes. For example, the proper 3' end of SV40 late transcripts is synthesized efficiently in oocytes; in infected cells, it has been shown that this requires posttranscriptional cleavage of an RNA precursor.[18] Furthermore, roughly 80% of those transcripts which terminate at the proper position are polyadenylated, as judged by their binding to poly(U)-Sepharose and their electrophoretic mobility.[18] Other genes tested in these respects include histones and herpes virus thymidine kinase. Some of the histone genes direct efficient and accurate termination, while other histone genes[19] and the thymidine kinase gene do not.[20] Neither histone nor thymidine kinase transcripts are efficiently polyadenylated. The presence of a 5' cap has not been investigated.

The splicing of mRNA precursors has been inferred when proteins are synthesized from genes in which intervening sequences interrupt the protein-coding region. This is true for T antigen of SV40 and for ovalbumin (see the following section). The direct demonstration of RNA splicing has so far been presented only for tRNA genes transcribed by RNA polymerase III and for SV40 late genes transcribed by polymerase II.[18] At present it is difficult to predict whether a particular type of injected gene will produce abundant correctly spliced transcripts.

The passage of transcripts from an occyte nucleus to the cytoplasm has been followed in some detail because it is easy to separate manually the nucleus from the cytoplasm. The most significant result of such investigations is that incorrect or incomplete transcripts generally fail to reach the cytoplasm whereas mature correctly processed transcripts clearly do so. This has been documented for tRNA genes,[1] histone genes,[12] and SV40[18]

[18] M. P. Wickens and J. B. Gurdon, *J. Mol. Biol.* **163,** 1 (1983); J. P. Ford and M.-T. Hsu, *J. Virol.* **28,** 795 (1978).
[19] C. C. Hentschel and M. L. Birnstiel, *Cell* **25,** 301 (1981).
[20] S. L. McKnight and E. R. Gavis, 1980. *Nucleic Acids Res.* **8,** 5931 (1980), and personal communication.

TABLE IV

PROTEIN SYNTHESIS IN INJECTED OOCYTES[a]

Material injected[b]	Complementary RNA synthesized[c]	Functional mRNA	Protein synthesis due to injection	
			dpm in specific protein (% of total)[d]	Amount[e]
SV40 DNA[b] (1 ng; 10^8 mol)	3 ng; 10^9 mol	10^6 mol[f]	5000 dpm VP1 (0.5%)	1.25 ng; 3×10^{10} mol
Chick ovalbumin[b] DNA in plasmid (5 ng; 3×10^8 mol)	3 ng; 3×10^8 mol	10^4 mol[f]	100 dpm (0.01%)	25 pg; 5×10^8 mol
Rabbit β-globin[b] mRNA (1 ng)	—	3×10^9 mol	500,000 dpm (50%)	120 ng; 5×10^{12} mol

[a] Labeled for 24 hr, with 5 μCi [^{35}S]methionine per oocyte.

[b] Results based on Ref. 31 (for SV40 DNA), Ref. 30 (for chick ovalbumin DNA), and Ref. 21 (for rabbit globin mRNA).

[c] Based on values in Table III for SV40, and in Ref. 30 for ovalbumin.

[d] One oocyte incubated in 5 μCi [^{35}S]methionine in 10 μl MBS for 24 hr synthesizes a total of 10^6 dpm protein.

[e] Calculated from percentage values in last column assuming that an oocyte synthesizes 250 ng protein per day (Table I).

[f] Estimated assuming that each molecule of mRNA makes 30 proteins per minute (Ref. 21).

Expression as Protein

DNA injected into oocytes has been shown to be expressed as protein for SV40 and polyoma virus (T antigens and virion proteins), *Drosophila* and sea urchin histones, chick ovalbumin, and thymidine kinase.[1] Protein will possibly be found whenever sufficiently sensitive methods of detection are used. The amount of protein synthesized can be estimated as follows. Under standard conditions (see Methods section) an oocyte incorporates about 10^6 dpm of [^{35}S]methionine into protein per day, during which time it synthesizes 250 ng of total protein. Virion protein 1 of SV40 is synthesized in DNA-injected oocytes in greater amounts than most other proteins coded for by injected genes (Table IV). It constitutes 0.5% of total synthesis, corresponding to 5000 cpm (1 ng) after standard labeling conditions for 1 day. A similar value is obtained if a calculation is based on the specific activity of an oocyte's methionine pool (44 pmol) and the methionine content of the protein.

There is considerable variation in the yield of protein synthesized by different types of injected DNA. For example 50 times less ovalbumin is synthesized than SV40 VP1 from similar amounts of DNA (Table IV). The efficiency with which oocytes translate mRNA is known from the injection of many different kinds of purified mRNA. A pure preparation of rabbit β-globin mRNA is translated 30 times per minute.[21] If we assume that fully processed cytoplasmic mRNA for SV40 proteins and ovalbumin are all translated at the same rate, we can deduce the amounts of these mRNAs in DNA-injected oocytes. It is clear from Table IV that only a small fraction of the total RNA transcribed from injected DNA becomes translatable mRNA.

Posttranslational Events

Many proteins are accurately modified in oocytes after translation. These posttranslational steps include the modification of amino acids, proteolytic cleavage of a primary protein chain, and the transport and secretion of selected proteins within and from a cell.[22] The ability of oocytes to carry out these activities has been tested in mRNA rather than DNA-injection experiments. In general oocytes perform correctly most posttranslational events characteristic of wholly unrelated cells of different types and different species. For example,[23] the N-terminal methionine is

[21] J. B. Gurdon, "The Control of Gene Expression," p. 59. Harvard Univ. Press, Cambridge, 1974.

[22] C. D. Lane, *Cell* **24**, 281 (1981).

[23] C. D. Lane and J. S. Knowland, *in* "Biochemistry of Development" (R. Weber, ed.), Vol. 3. Academic Press, New York, 1974.

acetylated on calf lens crystalline, a terminal 15 amino acids are removed
from mouse light chain immunoglobulin, ovalbumin is glycosylated, and
the primary polypeptide of mouse encephalomyocarditis virus is cleaved
into virion proteins.

The selective secretion of proteins also takes place in injected oocytes.[13] For example, mammalian interferon and the milk protein casein
are secreted from mRNA-injected oocytes, as they are from cells in which
they are normally synthesized. In contrast, globin and other proteins not
normally secreted are also not secreted from oocytes. The recovery of
materials secreted by oocytes is described in the Methods Section.

It seems, from these examples, that the cellular mechanisms responsible for posttranslational events are fairly universal in their occurrence in
cells, and particularly in oocytes. Proteins synthesized from injected eukaryotic genes of nonamphibian species are therefore very likely to undergo their normal modifications even though synthesized in oocytes. The
selective secretion of proteins can greatly help their identification. The recovery of proteins from the oocyte culture medium rather than from a
crude oocyte homogenate already eliminates most of the high background
of normal oocyte proteins, and antibody precipitations can be further
used to recognize minute amounts of material.

Other Uses of Oocytes

Assay for mRNA Purification. Gene isolation usually requires at least
partially pure mRNA for the preparation of cDNA or for screening a genomic DNA library directly. Since oocytes were first used for translating
mRNA, cell-free systems have been greatly improved and generally have
a lower background than oocytes. However, the efficiency of mRNA
translation in oocytes is much greater than cell-free systems. Thus if only
small amounts of mRNA are available, oocytes provide an especially sensitive assay.[24] As far as is known, all types of eukaryotic mRNA [includ-
ing some that do not normally carry a poly(A) tail] are translated in oocytes, whether they come from mammals, nonamphibian vertebrates, or
invertebrates. Large amounts of contaminating rRNA or tRNA can be tolerated (up to 1 mg/ml or 50 ng per oocyte) but, since all mRNAs are in
competition for a limited translational capacity, contaminating mRNAs
reduce the efficiency of translation of the one being assayed.

DNA Expression in Somatic Cells. DNA can be conveniently injected
into the cytoplasm of fertilized eggs. More than 250 pg per egg usually
causes abnormal development,[25] and least damage is sustained if about

---

[24] S. Nagata *et al. Nature (London)* **284,** 316 (1980).
[25] J. B. Gurdon, *Nature (London)* **248,** 772 (1974).

30 nl containing 200 pg (or 2×10^7 molecules of 5 kb DNA) of DNA is injected into the vegetal pole of eggs undergoing cleavage into the two-cell stage. The injected DNA replicates, and by the late blastula stage has increased the injected amount by 10 times or more.[26,27] Some of the injected DNA is expressed and is probably integrated into the host chromosomes. The large amount of yolk in amphibian eggs makes it impossible to see the egg pronuclei (as can be done in mouse eggs), but DNA deposited in the cytoplasm has a good chance of becoming included in nuclei as they undergo over 10 rounds of rapid division during the 12 hr that follow fertilization.

Acknowledgments

We gratefully acknowledge the comments of L. Dennis Smith, E. J. Ackerman and P. Farrell.

[26] M. M. Bendig, *Nature (London)* **292**, 65 (1981).
[27] S. Rusconi and W. Schaffner, *Proc. Natl. Acad. Sci. U.S.A.* **78**, 5051 (1981).
[28] E. H. Davidson, "Gene Activity in Early Development." Academic Press, New York, 1976 (contains original references).
[29] J. J. Eppig, Jr., and J. N. Dumont, *Dev. Biol.* **28**, 531 (1972).
[30] M. P. Wickens *et al.*, *Nature (London)* **285**, 628 (1980).
[31] D. Rungger and H. Turler, *Proc. Natl. Acad. Sci. U.S.A.* **75**, 6073 (1978).
[32] A. D. Mills *et al.*, *J. Mol. Biol.* **139**, 561 (1980).
[33] R. G. Roeder, *J. Biol. Chem.* **249**, 249 (1974).
[34] I. B. Dawid, *J. Mol. Biol.* **12**, 581 (1965).
[35] D. D. Brown and I. B. Dawid, *Science* **160**, 272 (1968).
[36] J. W. Chase and I. B. Dawid, *Dev. Biol.* **27**, 504 (1972).
[37] A. Webb and L. D. Smith *Dev. Biol.* **56**, 219 (1977).
[38] D. D. Brown and E. Littna, *J. Mol. Biol.* **20**, 95 (1966).
[39] L. Golden, U. Schafer, and M. Rosbash, *Cell* **22**, 835 (1980).
[40] M. Mairy and H. Denis, *Dev. Biol.* **24**, 143 (1971).
[41] M. Rosbash and P. J. Ford, *J. Mol. Biol.* **85**, 87 (1974).
[42] I. B. Dawid, *J. Mol. Biol.* **63**, 201 (1972).
[43] H. R. Woodland and E. D. Adamson, *Dev. Biol.* **57**, 118 (1977).
[44] M. O. Cabada *et al.*, *Dev. Biol.* **57**, 427 (1977).
[45] A. C. Webb, M. J. La Marca, and L. D. Smith, *Dev. Biol.*, **45**, 44 (1975).
[46] E. Rastl and I. B. Dawid, *Cell* **18**, 501 (1979).
[47] E. D. Adamson and H. R. Woodland, *Dev. Biol.* **57**, 136 (1977).
[48] W. J. Wasserman, J. D. Richter, and L. D. Smith, *Dev. Biol.* **89**, 152 (1982).
[49] G. J. Dolecki and L. D. Smith, *Dev. Biol.* **69**, 217 (1979).
[50] J. Maller, M. Wu, and J. C. Gerhart, *Dev. Biol.* **58**, 295 (1977).
[51] D. G. Capco and W. R. Jeffrey, *Dev. Biol.* **89**, 1 (1982).

[29] Expression and Secretion Vectors for Yeast

By Grant A. Bitter, Kevin M. Egan, Raymond A. Koski,
Matthew O. Jones, Steven G. Elliott, *and* James C. Giffin

It is now common practice to express heterologous genes in the yeast *Saccharomyces cerevisiae*. This is in large part due to the utility of this organism as a host for the production of commercially relevant proteins. Vectors designed for efficient expression of heterologous genes have been developed which employ promoter elements derived from the alcohol dehydrogenase,[1] phosphoglycerate kinase,[2] acid phosphatase,[3] glyceraldehyde-3-phosphate dehydrogenase,[4] galactokinase,[5] and mating factor-α[6] genes. These promoters, in general, result in efficient transcription of the heterologous gene. However, the level of accumulation of the heterologous protein varies widely depending on the foreign gene expressed. This is presumably due to different translational efficiencies determined by mRNA secondary structure, untranslated leader sequences, codon utilization, or protein stability. Considerable basic research is being conducted by expressing certain genes in yeast where high-level expression of the given protein is not required or attempted. These studies employ various other promoters and expression strategies. This chapter, however, will focus on methods used to efficiently express and secrete biologically active proteins from *Saccharomyces cerevisiae*.

Standard recombinant DNA methods will not be discussed in this chapter since they have been reviewed in detail elsewhere.[7-9] Methods of introducing recombinant DNA into yeast cells have also been described in

[1] R. A. Hitzeman, F. E. Hagie, H. L. Levine, D. V. Goeddel, G. Amerer, and B. D. Hall, *Nature (London)* **293,** 717 (1981).

[2] M. R. Tuite, M. J. Dobson, N. A. Roberts, R. M. King, D. C. Burke, S. M. Kingsman, and A. J. Kingsman, *EMBO J.* **1,** 603 (1982).

[3] A. Miyanohora, A. Toh-E, C. Nozaki, F. Hamada, N. Ohtomo, and K. Matsubara, *Proc. Natl. Acad. Sci. U.S.A.* **80,** 1 (1983).

[4] G. A. Bitter and K. M. Egan, *Gene* **32,** 263 (1984).

[5] C. G. Goff, D. T. Moir, T. Kohro, T. Gravins, R. A. Smith, E. Yamasaki, and A. Taunton-Rigby, *Gene* **27,** 35 (1984).

[6] G. A. Bitter, K. K. Chen, A. R. Banks, and P.-H. Lai, *Proc. Natl. Acad. Sci. U.S.A.* **81,** 5330 (1984).

[7] R. Wu, L. Grossman, and K. Moldave (eds.), this series, Vol. 100.

[8] R. Wu, L. Grossman, and K. Moldave (eds.), this series, Vol. 101.

[9] T. Maniatis, E. F. Fritsch, and J. Sambrook, *in* "Molecular Cloning: A Laboratory Manual." Cold Spring Harbor Lab., Cold Spring Harbor, New York, 1982.

RECOMBINANT DNA
METHODOLOGY

detail.[10,11] This chapter will describe yeast expression vectors utilizing episomal vectors. The expression cassettes from these vectors can also be integrated into the yeast chromosome where they will be present at controlled copy number and exhibit a high degree of mitotic stability. Methods of integrating yeast DNA into the chromosome, which will not be discussed in this chapter, have been described in detail.[12,13] Finally, there are several other recent reviews of yeast cloning and secretion methods which are directly relevant to this chapter.[14–19]

Extrachromosomal Replication Vectors

Selectable Markers

The first yeast genes to be molecularly cloned were those encoding biosynthetic enzymes and these genes are now routinely utilized as selectable markers on recombinant yeast plasmids. The cloned *TRP1, LEU2, HIS3,* and *URA3* genes are commonly used as markers and numerous auxotrophic strains exist with mutations in the appropriate chromosomal gene. Medium lacking tryptophan, leucine, histidine, or uracil is used to select cells containing the plasmid. In addition, several dominant selectable markers for yeast have been developed. These include selection for aminoglycoside G418 resistance[20] [encoded by the *Escherichia coli Tn601(903)* aminoglycoside phosphotransferase gene], copper resistance[21] (encoded by the yeast *CUP1* gene), and methotrexate resistance[22,23] (*E. coli* plasmid R388 or mouse *DHFR* gene). Use of dominant selectable markers greatly increases the range of host strains which may be used with a particular plasmid, including those strains which lack any auxotrophic markers. Additionally, increasing the selective pressure may allow establishment of strains with higher plasmid copy numbers.

[10] A. Hinnen, J. B. Hicks, and G. R. Fink, *Proc. Natl. Acad. Sci. U.S.A.* **75,** 1929 (1978).
[11] H. Ito, Y. Fukuda, and A. Kimara, *J. Bacteriol.* **153,** 163 (1983).
[12] R. J. Rothstein, this volume [18].
[13] F. Winston, F. Chumley, and G. R. Fink, this series, Vol. 101, p. 211.
[14] A. Hinnen and B. Meyhack, *Curr. Top. Microbiol. Immunol.* **96,** 101 (1982).
[15] C. P. Hollenberg, *Curr. Top. Microbiol. Immunol.* **96,** 119 (1982).
[16] N. Gunge, *Annu. Rev. Microbiol.* **37,** 253 (1983).
[17] K. Struhl, *Nature (London)* **305,** 391 (1983).
[18] R. A. Hitzeman, C. Y. Chen, F. E. Hagie, J. M. Lugovoy, and A. Singh, "Recombinant DNA Products: Insulin, Interferon, and Growth Hormone," p. 47. CRC Press, Boca Raton, Florida, 1984.
[19] G. A. Bitter, *Microbiology,* 330–334 (1986).
[20] A. Jimenez and J. Davies, *Nature (London)* **287,** 869 (1980).
[21] S. Fogel and J. W. Welch, *Proc. Natl. Acad. Sci. U.S.A.* **79,** 5342 (1982).
[22] A. Miyajima, I. Miyajama, K.-I. Arai, and N. Arai, *Mol. Cell. Biol.* **4,** 407 (1984).
[23] J. Zhu, R. Contreras, D. Gheysen, J. Ernst, and W. Fiers, *Bio/Technology* **3,** 451 (1985).

DNA Replicators

A number of DNA sequence elements have been isolated which confer the capability of autonomous replication of colinear DNA in yeast. These elements have been termed ARS (for autonomously replicating sequence) and those isolated from yeast DNA presumably are chromosomal origins of replication.[24] ARS-containing vectors replicate once per cell cycle[25] but accumulate to multiple copies per cell due to segregational bias. They are mitotically unstable, exhibiting a segregation frequency of 10–30% per generation under nonselective conditions.[26] In a typical culture grown under selective conditions, only 20–30% of the cells contain plasmid and cells lacking plasmid are capable of undergoing a limited number of additional cell divisions. Such behavior of plasmids is clearly suboptimal for efficient heterologous gene expression.

An alternative to ARS elements is the origin of DNA replication of the yeast 2μ plasmid. This 6318-bp plasmid[27] has no known function and is present at 50–100 copies/cell in most laboratory strains of *Saccharomyces cerevisiae*. The origin of DNA replication of the yeast 2μ plasmid has been localized to a 222-bp segment.[28] A DNA sequence, termed the *REP3* or *STB* locus, has been localized to a 294-bp segment[29] and directs mitotic equipartitioning of the 2μ plasmid. Thus, a plasmid which includes the 2μ origin of replication and *REP3* locus will be stably maintained at high copy numbers when the appropriate replication proteins are present in the cell. These may be supplied *in trans* by the endogenous yeast 2μ plasmid.

Figure 1 is a restriction endonuclease map of plasmid pYE,[6] which has been utilized for construction of both expression and secretion vectors. The plasmid includes the *E. coli* plasmid pBR322 with an intact origin of replication and β-lactamase gene (expression of which confers resistance to ampicillin). These features allow propagation in *E. coli* for further recombinant DNA manipulations. The yeast *TRP1* gene allows selection of yeast cells containing the plasmid. Tryptophan prototrophy is a convenient selection for yeast since the selective medium is easily prepared (0.67% yeast nitrogen base without amino acids, 2% dextrose, 0.5% casamino acids). Plasmid pYE has a unique restriction endonuclease recognition site for *Bam*HI located in the pBR322 portion of the plasmid. The 2μ DNA segment contains a *Cla*I restriction site. However, in this sequence context (GATCGATC) it is methylated in *dam*+ strains of *E. coli* (such as

[24] D. H. Williamson, *Yeast* **1**, 1 (1985).
[25] W. L. Fangman, R. H. Hice, and E. Chlebowicz-Slediewska, *Cell* **32**, 831 (1983).
[26] A. W. Murray and J. W. Szostak, *Cell* **34**, 961 (1983).
[27] J. L. Hartley and J. E. Donelson, *Nature (London)* **286**, 860 (1980).
[28] J. R. Broach, Y.-Y. Li, J. Feldman, M. Jayaram, J. Abraham, K. A. Nasmsyth, and J. B. Hicks, *Cold Spring Harbor Symp. Quant. Biol.* **47**, 1165 (1982).
[29] M. Jayaram, A. Sutton, and J. R. Broach, *Mol. Cell. Biol.* **5**, 2466 (1985).

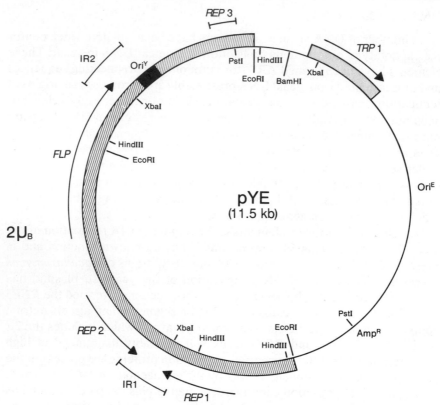

FIG. 1. *Saccharomyces cerevisiae–E. coli* shuttle vector pYE. DNA sequences derived from pBR322 are depicted as a single line. The yeast *TRP1* gene, cloned by blunt-end ligation into the *Sal*I site of pBR322, is depicted by the stippled segment. The yeast 2μ plasmid is indicated by the hatched segment. The yeast and *E. coli* origins of DNA replication are indicated. The location of yeast genes and loci are depicted and the direction of transcription of the yeast genes is indicated by arrows.

HB101). The methylated sequence is not cleaved by *Cla*I. Thus, by propagating the shuttle vector in these hosts, only the *Cla*I site in pBR322 is subject to cleavage and may be used as a unique cloning site.

The entire yeast 2μ plasmid, linearized at the *Eco*RI site in the large unique region, is cloned in pYE. In this construction, the origin of replication, the *REP3* locus, and the *FLP, REP1,* and *REP2* genes are all intact. The *FLP, REP1,* and *REP2* proteins, expressed from their endogenous 2μ promoters on pYE, are involved in amplification of the plasmid to high copy numbers. This has been demonstrated to occur even in strains lacking the endogenous 2μ plasmid (*cir°* strains; see below). It is advantageous to utilize *cir°* hosts for several reasons. Recombination between the endogenous 2μ plasmid and the recombinant plasmid may restructure the expression vector. Because of the expression of the 2μ genes present on pYE, it is capable of self-amplification in the absence of endogenous 2μ plasmid. By propagating the vector in $cir°$ strains, high copy numbers may be achieved without the potential complications due to recombination with the endogenous 2μ plasmid. Furthermore, by utilizing $cir°$ hosts, the copy number of the recombinant expression vector may be increased since it represents the only 2μ plasmid in the cell.

The 2μ plasmid has two inverted repeats (IR1 and IR2; Fig. 1). Recombination between these two sites is catalyzed by the product of the *FLP* gene.[30] Since the *FLP* gene is expressed on the plasmid pYE in yeast, rapid recombination between IR1 and IR2 will result in two forms of plasmid. This phenomenon is illustrated in Fig. 2. Expression vector pGPD-1(HBs),[4] which includes the entire yeast 2μ plasmid cloned as in pYE, was propagated in a $cir°$ yeast strain and then shuttled back into *E. coli*.[31] More than 600 *E. coli* clones were pooled (in order to be representative of the plasmid population present in yeast) and the plasmid amplified and purified. By digesting with a restriction enzyme which cleaves the 2μ DNA on each side of an inverted repeated (e.g., *Hind*III), two new restriction fragments are observed in the plasmid preparation recovered from yeast which were not present in the parent plasmid constructed and propagated in *E. coli* (Fig. 2, lanes 1 and 2). The new fragments are due to the second form of the 2μ plasmid generated by recombination in yeast. If individual *E. coli* clones are analyzed, only one form of the plasmid is observed (Fig. 2, lanes 3 and 4). The staining intensity of the restriction fragments in the plasmid population recovered from yeast (Fig. 2, lane 2) indicates an equivalent abundance of each form of the plasmid, which is corroborated by analysis of a large number of individual *E. coli* clones (data not shown).

The above data demonstrate that plasmids such as pYE, which self-amplify in $cir°$ yeast hosts, recombine with each other to generate a steady state population of plasmids with two forms of the 2μ insert. We have observed no other types of restructuring of plasmids using this system. However, this theoretical possibility should be borne in mind when evaluating heterologous gene expression systems.

Centromere Plasmids

High-level expression of heterologous genes is generally facilitated by increasing the copy number of the expression vectors. The previous section described vectors capable of stable maintenance at high copy numbers. There may be instances, however, where it is preferable to maintain

[30] J. R. Broach, V. R. Guarasio, and M. Jayaram, *Cell* **29**, 227 (1982).
[31] G. Miozzari, P. Neiderberger, and R. Huffer, *J. Bacteriol.* **134**, 48 (1978).

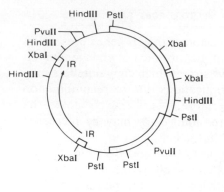
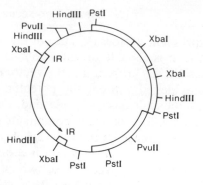

FORM A   FORM B

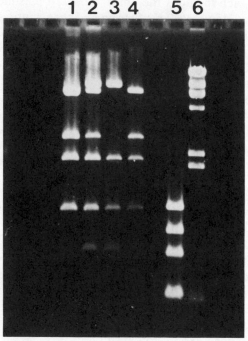

FIG. 2. Interconversion of two forms of 2μ plasmid in yeast. The expression vector pGPD-1(HBs) has been described elsewhere.[4] This plasmid was constructed and propagated as form B in *E. coli* and was expected to interconvert to form A when introduced into yeast. Restriction endonuclease maps of both forms of the plasmid are presented. The plasmid was introduced into *Saccharomyces cerevisiae* RH218 (*Mata gal2 trp*1 *cir*°31) by transformation and shuttled back into *E. coli*. The plamsid was purified from a pool of more than 600 *E. coli* clones as well as from 2 individual *E. coli* clones. The purified DNA samples were restricted with *Hind*III and analyzed by agarose gel electrophoresis. Lane 1, pGPD-1(HBs) form B constructed and propagated in *E. coli*. Lane 2, plasmid population recovered from yeast (pool of *E. coli* clones). Lane 3, pGPD-1(HBs) form A recovered from a single *E. coli* clone. Lane 4, pGPD-1(HBs) form B recovered from a single *E. coil* clone. Lane 5, phage ϕX 174 DNA restricted with *Hae*III. Lane 6, phage λ DNA restricted with *Hind*III.

expression vectors at low copy number. These include cases where expression of the heterologous gene is toxic to yeast and plasmid stability is low. It has been demonstrated that addition of centromere (CEN) sequences to unstable plasmids can reduce segregational loss to 1–3% per generation under nonselective conditions.[32] While the copy number of CEN-containing plasmids (even those including 2μ replicons) is controlled at one to two per haploid cell,[33] the increased stability may be an advantage in particular situations (i.e., in cases where there is strong selection against maintenance of the plasmid). Additionally, it appears that heterologous protein secretion may be more efficient in certain cases when the expression system is stably maintained at low copy numbers[34] (S. Elliott, unpublished results).

Measurement of Plasmid Stability. Plasmid stability of a culture is determined by plating appropriate culture dilutions on nonselective medium, then replica plating the resulting colonies onto medium which is selective for cells containing the plasmid. Plasmid stability is expressed as the percentage of cells containing plasmid, as indicated by the percentage of total colonies on the nonselective plate that grow on the selective plate.

Recombinant yeast plasmids containing the entire 2μ plasmid are lost at a rate of 0.08–4% per generation during nonselective growth.[35] In selective medium, recombinant plasmids reach steady state plasmid stability levels. The steady state level is determined by a number of factors including the plasmid selectable marker, replication efficiency, segregation efficiency, and possible deleterious effects on the cell due to maintaining the plasmid (such as expression of a toxic heterologous protein). For example, we have determined that a recombinant 2μ plasmid which is lost at a rate of 0.3% per generation under nonselective conditions is maintained in 96% of the cells when the *TRP1* gene[36] is used as a selectable marker. In contrast, if a gene encoding human immune interferon (IFN-γ) is constitutively expressed from the 2μ plasmid, the rate of plasmid loss increases to 3% per generation under nonselective conditions and the steady state plasmid stability is 70% under *TRP1* selection (data not shown). Plasmid stability thus varies widely depending on the particular 2μ recombinant vector. As mentioned above, heterologous gene stability may be increased by incorporating CEN sequences into the plasmid, or by integrating the gene into the chromosome. Since these manipulations result in lower gene dosage, such an approach may not yield optimal heterologous gene expression levels.

[32] L. Clarke and J. Carbon, *Nature (London)* **287,** 504 (1980).
[33] G. Tschumper and J. Carbon *Gene* **23,** 221 (1983).
[34] R. A. Smith, M. J. Duncan, and D. T. Moir, *Science* **229,** 1219 (1985).
[35] A. B. Futcher and B. S. Cox, *J. Bacteriol.* **157,** 283 (1984).
[36] G. Tschumper and J. Carbon, *Gene* **10,** 157 (1980).

Measurement of Plasmid Copy Number. It is widely assumed that recombinant plasmids incorporating the 2μ replication/amplification system are maintained at high copy numbers, and a rule of thumb has developed in the literature such that 2μ-based vectors are assumed to be present at an average 20 copies per cell. In actual fact, there is a wide variability in average plasmid copy number per cell for different recombinant vectors. This fact was anticipated from the observed differences in plasmid stability (above). In characterizing a yeast expression system, therefore, it is imperative that plasmid copy number be measured.

High copy number plasmids may be visualized by ethidium bromide staining of restriction enzyme-digested whole-cell DNA which has been size fractionated by agarose gel electrophoresis. This method may be quantitative[37] by comparison of the plasmid band to the staining intensity of the rRNA genes (of defined copy number) but does not detect low copy number plasmids. An attractive alternative, therefore, is hybridization of radioactively labeled specific DNA probes to size-fractionated DNA. In order to accurately quantitate plasmid copy number, a hybridization probe is utilized which is complementary to both the plasmid and a native chromosomal yeast gene. Such DNA probes may derive from either the promoter, transcription terminator, or selectable marker of the expression vector (see below) since these sequences are all also represented in the yeast genome. Whole-cell DNA is isolated as described in Procedures, digested with an appropriate restriction enzyme, and fractionated by size in an agarose gel. The DNA is transferred to nitrocellulose and hybridized to the radioactive probe.[9] By utilizing a restriction enzyme that does not cleave within the DNA fragment used as probe, two of the size-fractionated restriction fragments will hybridize to the probe. Knowledge of the genomic and plasmid restriction map will allow identification of the fragment representing either the chromosomal gene or the recombinant plasmid. The chromosomal fragment in each sample is an internal standard for quantitation and generally represents one (haploid strains) or two (diploid strains) genes per cell. Comparison of the amount of probe hybridized to the plasmid fragment to that hybridized to the chromosomal fragment (of defined copy number) allows quantitation of the plasmid copy number per cell. It has been noted that the transfer/binding efficiency of DNA fragments in Southern blot analyses is size dependent. Therefore, if the sizes of the genomic and plasmid restriction fragments are very different, the copy number should be confirmed by utilizing a different restriction enzyme which generates genomic and plasmid fragments of similar size. Alternatively, a method which employs *in situ* hybridization of the separated DNA fragments in the gel may be

---

[37] J. R. Broach, this volume [21].

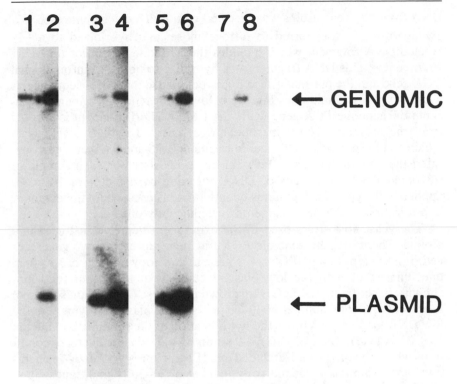

FIG. 3. Determination of plasmid copy numbers. *Saccharomyces cerevisiae* containing expression vectors pGPD(s)γ4 (lanes 1 and 2) or pGPD(G)γ4–9 (lanes 3, 4, 5, 6, 7, and 8) were cultured as described in the text and whole-cell DNA isolated as described in Procedures. The DNA was digested with *Bgl*II, fractionated by agarose gel electrophoresis, transferred to nitrocellulose, and hybridized to a nick-translated yeast *PGK* terminator (*Bgl*II–*Hin*dIII 380-bp segment[39]).

employed.[37a] This "unblot" method is independent of transfer or binding efficiencies associated with the Southern blot procedure.

An example of this method of plasmid copy number determination is depicted in Fig. 3. Expression vector pGPD-2 has been described[4] (see below). Expression vector pGPD(s)-2 is identical except for the introduction of a *Sal*I site 240 bp 5′ to the TATA box of the GPD promoter.[38] Expression vector pGPD(G)-2 contains a DNA sequence from the yeast *GAL*1, 10 intergenic region cloned in the *Sal*I site of pGPD(s)-2 which renders the GPD promoter inactive in glucose and induced by galactose.[38] A gene encoding human immune interferon (IFN-γ) was cloned into each vector to generate pGPD(s)γ4 and pGPD(G)γ4-9.[38] Whole-cell

[37a] M. Purrello and I. Balazs, *Anal. Biochem.* **128,** 393 (1983).
[38] G. A. Bitter and K. M. Egan, manuscript in preparation (1987).

DNA from different strains was digested with BglII, electrophoresed in a 1% agarose gel, transferred to nitrocellulose, and hybridized to a ^{32}P-labeled DNA fragment which includes the 3' coding region of the yeast PGK gene.[39] This DNA fragment will hybridize to both the chromosomal PGK gene and the plasmids since both vectors use this DNA sequence as a transcription terminator (Fig. 5, below). The 4-kbp fragment represents the genomic PGK gene while the 1.6-kbp BglII fragment is derived from the plasmid. (This experiment was repeated with EcoRI digests which yielded genomic and plasmid fragments of 5 and 6 kb, and corroborated the quantitation obtained with BglII digests.) For each sample of restricted DNA, two loads of DNA, differing by a factor of 10, were applied to the gel. This facilitates quantitation in cases where the extent of hybridization to the two fragments is significantly different.

The yeast host strain used in the experiment depicted in Fig. 3 is a diploid. Therefore, the genomic PGK fragment represents two genes per cell. It is evident that pGPD(s)γ4 is present at a copy number of less than one, consistent with the low plasmid stability of the strain (data not shown). In contrast, pGPD(G)γ4-9 grown in glucose (Fig. 3, lanes 3 and 4) is present at a copy number of 40–60 per cell (the plasmid fragment in the low DNA load has 2–3 times the hybridization of the genomic band in the 10× DNA load). The pGPD(G)γ4-9 strain was grown in galactose (conditions where the promoter is induced and IFN-γ expressed) for 3–4 generations and exhibited a plasmid copy number of 20–30 (Fig. 3, lanes 5 and 6). If the same strain was serially cultured for more than 50 generations in galactose (conditions in which IFN-γ is expressed), the plasmid copy number dropped to less than 1 per cell (Fig. 3, lanes 7 and 8).

The yeast host strain used for the experiment in Fig. 3 was cir°. Both expression vectors analyzed contain the entire 2μ plasmid cloned at the EcoRI site in the large unique region (as in plasmid pYE). The results demonstrate, therefore, that plasmid copy number may vary between 1 and 60 per cell depending on the vector, growth conditions, and heterologous gene insert (and its expression).

It should be noted that plasmid copy number measured in this way represents average copy number for the cell population. There is likely to be some variability in plasmid copy number per cell but direct methods for measuring this distribution in the population are not currently available. Since some percentage of the cells in the culture do not contain plasmid, however, a more accurate plasmid copy number is obtained by dividing the measured copy number by the fraction of cells containing plasmid (as measured by the plasmid stability test, above).

[39] R. A. Hitzeman, F. E. Hagie, J. S. Hayflick, C. Y. Chen, P. H. Seeburg, and R. Derynck, Nucleic Acids Res. **10**, 7791 (1982).

Direct Expression Vectors

In addition to an appropriate shuttle vector, three additional functional components are required for heterologous gene expression: (1) An efficient promoter element is required to support high-level transcription initiation. Thus far, only homologous yeast promoters have been shown to be functional for this purpose; (2) for optimal mRNA accumulation, DNA sequences which impart transcription termination and/or polyadenylation should be included. These sequences can be of either yeast origin or from heterologous organisms[18] and there does not appear to be a quantitative effect on expression level due to use of different yeast terminators[4]; (3) the gene encoding the heterologous protein must be appropriately positioned between the transcription promoter and terminator elements in the shuttle vector. The heterologous gene must include a translation termination codon as well as the ATG initiator codon. The mRNA sequence encoded by the heterologous gene determines the intrinsic translational efficiency. Additional factors, such as mRNA abundance and protein half-life, contribute to the steady state levels of the heterologous protein.

A generic yeast expression cassette is depicted in Fig. 4. Most yeast promoter elements have been isolated as DNA fragments truncated within the untranslated leader of the native yeast gene and include at least 300 bp of upstream DNA. The amount of 5′ DNA required for full promoter activity has not been precisely defined for all yeast promoter elements thus far studied. However, functional yeast promoters, as in higher eukaryotes, are considerably larger than those of prokaryotes. The isolated promoter elements include the yeast transcription start site but not the ATG translation initiation codon. Translation initiation in yeast, as in higher eukaryotes, generally starts at the first AUG in the mRNA. Since no essential consensus sequences have been identified in yeast untranslated leaders, there appears to be no analogy to the bacterial ribosome-binding site. Assembling a yeast expression cassette thus involves posi-

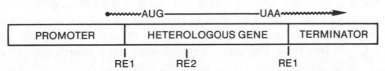

FIG. 4. Generic yeast expression cassette. The three functional components (promoter, heterologous gene, transcription terminator) are indicated. RE1 and RE2 represent restriction enzyme recognition sites in the expression vector. Several of the methods used to assemble expression cassettes (see text) result in loss of the restriction site (RE1) which separates the promoter from transcription terminator of the generalized expression vector (e.g., the *Bam*HI site of pGPD-1 and pGPD-2, Fig. 5).

tioning the promoter element, which includes the transcription start site, upstream of the heterologous gene in such a manner that the ATG translation start codon of the heterologous gene represents the first AUG in the mRNA. If other ATG triplets are present before the intended translation start, translation initiation at the first AUG in the mRNA will either represent a frame shift mutation with no heterologous protein synthesis or, in the case of an in-frame upstream ATG, may result in a heterologous protein fusion. The DNA sequence between the promoter and intended translation initiation codon, including sequences generated in the cloning manipulations, must therefore be carefully examined in constructing expression vectors.

Although there are no consensus sequences in yeast untranslated leader regions, genes which encode highly expressed yeast proteins are markedly A rich and G deficient in this region. Certain heterologous genes will encode untranslated leaders which are clearly suboptimal for translation in yeast. In such instances, it is advisable to resynthesize the 5′ end of the heterologous gene. Thus, using the example in Fig. 4, a DNA segment may be chemically synthesized[40] and used to replace the natural sequence between the leftmost RE1 and RE2. The synthetic segment should incorporate an optimized untranslated leader (e.g., the native sequence from a highly expressed yeast gene) and a sequence downstream of the ATG which encodes the desired amino acid sequence of the foreign protein. It may be preferable to use optimal yeast codons[41] (i.e., those preferentially utilized in highly expressed yeast genes) in this region. Employing such a strategy for expression of the hepatitis B virus surface antigen gene in yeast resulted in more than a 10-fold increase in protein expression level.[4] In designing such synthetic segments, mRNA secondary structure, which may decrease translation efficiency, should be minimized. Additionally, thought should be given to codon selection in the synthetic gene. By chemically synthesizing a complete gene encoding human immune interferon (IFN-γ) which incorporated an optimized untranslated leader and optimal yeast codons, IFN-γ was expressed in yeast as 10% of the total cell protein.[42]

A number of yeast genes contain introns and the splicing mechanism in yeast has been studied in detail. However, when genes containing heterologous introns are expressed in yeast, the splicing appears to be aberrant.[18] At this time, therefore, heterologous gene expression in yeast

[40] M. H. Caruthers, in "Chemical and Enzymatic Synthesis of Gene Fragments, a Lab Manual" (H. G. Gassen and A. Lang, eds.), pp. 71–79. Verlag Chemie, Weinheim, Federal Republic of Germany, 1982.

[41] J. L. Bennetzen and B. D. Hall, J. Biol. Chem. 257, 3026 (1982).

[42] J. L. Fieschko, K. M. Egan, T. Ritch, R. A. Koski, M. D. Jones, and G. A. Bitter, Biotechnol. Bioeng. 29, 1113 (1987).

is limited to cDNA, genomic clones which lack introns, or synthetic genes.

As examples of generalized yeast expression vectors, restriction endonuclease maps of two vectors employing the glyceraldehyde-3-phosphate dehydrogenase gene (*GPD*) promoter are depicted in Fig. 5. Both vectors incorporate the entire 2μ plasmid linearized at the *Eco*RI site in the large unique region (see above). The large size of yeast expression vectors limits the available unique restriction sites which may be engineered between the promoter and terminator elements. The expression vectors pGPD-1 and pGPD-2 (Fig. 5) have been constructed[4] with unique *Bam*HI sites between the *GPD* promoter and transcription terminator. In these constructions, cloning strategies were developed which prevented the generation of any additional inverted repeats which might lead to plasmid structural instability.

Both expression vectors employ the yeast *TRP*1 gene for selection. In pGPD-1, the *TRP*1 gene also supplies transcription termination/polyadenylation signals for RNA polymerase II molecules which have initiated transcription at the *GPD* promoter. The transcription terminator region from the yeast *PGK* gene is employed in pGPD-2. No difference in levels of heterologous gene expression were observed with the two transcription terminators.[4]

Assembly of Expression Cassettes

The paucity of unique restriction sites in yeast expression vectors, as well as constraints on sequences surrounding the translation initiation codon, do not allow for inclusion of polylinkers between the promoter and terminator. Since it is unlikely that the heterologous gene has the appro-

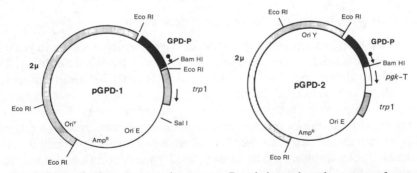

FIG. 5. Generalized yeast expression vectors. Restriction endonuclease maps of expression vectors pGPD-1 and pGPD-2 are presented. DNA derived from the yeast *TRP1* gene, 2μ plasmid, and pBR322 are presented as in Fig. 1. The GPD portable promoter is represented by the black segment and the *PGK* terminator by the open segment. Reproduced from Bitter and Egan[4] with permission of the publisher.

priate flanking restriction sites, the cloning of these genes into the expression vector employs one of three manipulations. Synthetic linkers may be added[9] to the heterologous gene such that the termini are complementary with those of the linearized expression vector (e.g., *Bam*HI cleaved pGPD-1 or pGPD-2). These manipulations will add DNA sequences to the untranslated leader region which must be evaluated (above) for potentially deleterious effects on translation. Alternatively, certain noncomplementary restriction ends may be rendered cohesive by *in vitro* manipulations. These methods have been described in detail elsewhere.[43] For example, a *Bam*HI-restricted vector can be ligated to a *Sal*I restriction fragment by partial end filling of each terminus. This cloning strategy has the advantage that the partially end-filled termini of the vector are no longer complementary. Thus, the frequency of clones containing the inserted gene is greatly increased.

The third method of expression cassette assembly involves blunt-end ligation. Nonhomologous 5' overhang restriction ends may be end filled with Klenow fragment. This introduces additional DNA sequences in the untranslated leader which should be evaluated. Alternatively, blunt ends may be generated by removal of cohesive termini with either mungbean or S1 nuclease. While blunt-end ligation is the most generally applicable method, it suffers the disadvantage of low ligation frequency. Plasmids containing the heterologous gene insert often need to be screened for by colony hybridization. In order to facilitate screening for such clones, we have used synthetic oligonucleotide probes, the sequence of which spans the junction between the promoter and heterologous gene insert. Under the appropriate hybridization conditions, such probes will detect only *E. coli* colonies which contain the heterologous gene cloned in the correct orientation.

Analysis of Heterologous Gene Transcription

A first step in analyzing expression of the heterologous gene in yeast is analysis of its mRNA. By analyzing the transcript by Northern blots, it is possible to determine whether transcription is initiating and terminating near the expected sites in the promoter and terminator. Precise transcription start sites may be determined by S1 nuclease[44] or primer extension mapping.[45] In addition, mRNA abundance may be measured in a manner similar to plasmid copy number measurement (above). Total cell RNA is extracted, electrophoresed in agarose under denaturing conditions, and subjected to hybridization (with or without transfer to nitrocellulose) as described in Procedures. A hybridization probe is utilized which will de-

[43] M. C. Hung and P. C. Wensink, *Nucleic Acids Res.* **12,** 1863 (1984).
[44] A. J. Berk, and P. A. Sharp, *Cell* **12,** 721 (1977).
[45] S. L. McKnight, E. R. Gravis, R. Kingsbury, and R. Axel, *Cell* **25,** 385 (1981).

tect both the heterologous transcript and a native yeast transcript. For example, for expression vectors derived from pGPD-2 (Fig. 5), the *PGK* terminator DNA fragment will detect both the heterologous transcript and the yeast *PGK* transcript since both mRNAs are homologous to portions of this DNA. An example of such an analysis is presented in Fig. 6. The

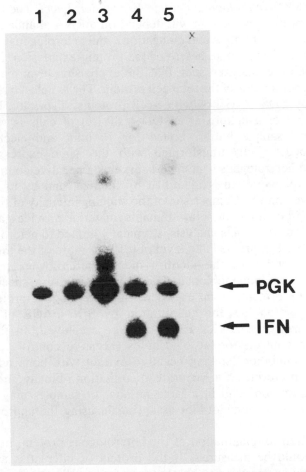

FIG. 6. Measurement of relative transcript abundance. Total RNA was extracted from various *Saccharomyces cerevisiae* strains as described in Procedures, electrophoresed in an agarose gel, and the gel dried. A *PGK* gene terminator fragment (legend to Fig. 3) was hybridized directly to the gel. Lane 1: nontransformed strain. Lane 2: cells containing expression vector pPG73 (M. O. Jones and R. A. Koski, unpublished results) containing the yeast *PGK* promoter and terminator but no heterologous gene insert. Lane 3: cells containing the intact yeast *PGK* gene on a multicopy plasmid. Lanes 4 and 5: cells containing plasmids expressing human immune interferon analogs from vectors employing the *PGK* gene promoter and terminator.

yeast *PGK* transcript serves as an internal standard in each lane for measuring heterologous transcript abundance. In contrast, however, to plasmid copy number determinations (above), the Northern analysis yields a relative transcript abundance. An absolute measurement of mRNA abundance is not obtained. Thus, this method allows determination of relative transcription efficiency of different expression vectors by comparison to a given native yeast mRNA.

Analysis of Heterologous Protein Expression Level. The level of heterologous protein expression will depend on mRNA abundance, translational efficiency (initiation and elongation), and heterologous protein half-life. In order to quantitate heterologous protein expression, a crude cell lysate should be analyzed such that the expression level may be determined as a percentage of the total cell protein. The simplest quantitation is obtained by SDS–PAGE of whole-cell proteins. Typically, 0.5 OD · ml (e.g., 0.5 ml of a culture at an OD_{600} of 1.0) of cells is resuspended in Laemmli sample buffer, boiled for 10 min, and electrophoresed as described.[46] Cells transformed with the same expression vector lacking the heterologous gene insert serve as a negative control. The expressed foreign protein is identified by its size, and by its absence in the negative control. Comparison of the staining intensity of the heterologous protein relative to total stainable material provides an estimate of expression level. This analysis is typically limited to heterologous proteins which are expressed at a level of 0.5% or more of the total cell protein. Confirmation of the identity of the heterologous protein may be obtained by immunoblot analysis[46a] using antisera specific for the heterologous protein. In cases where the heterologous protein expression level is too low for quantitation by SDS–PAGE of whole-cell lysates, sensitive RIA's or EIA's may be employed. In these cases, appropriate controls for false positives (negative control yeast lysate) as well as inhibition (spiking negative extract with bona fide product) should be performed for accurate quantitation. Finally, heterologous protein expression level in a large number of samples may be readily quantitated by immunodot-blot assay, again using the appropriate controls.

In addition to quantitation of the heterologous protein, its biological activity should be measured. If the protein of interest is an enzyme, standard enzyme assays can be used with yeast cell extracts. Hormones, lymphokines, and other biological response modifiers expressed in yeast require *in vitro* biological assays with appropriate target cells. The potency of potential vaccines produced in yeast can be tested by measuring

[46] U. K. Laemmli, *Nature (London)* **227,** 680 (1970).
[46a] W. N. Burnette, *Anal. Biochem.* **112,** 195 (1981).

antibody response of appropriately purified and formulated product in animal systems.

Gentle cell lysis methods are required to prepare samples for biological assays. Small samples can be prepared by hypotonic lysis of spheroplasts, or by vortexing cell suspensions in glass test tubes with 0.5-mm glass beads. French press lysis is convenient for disrupting 1–5 g of cells. Cells should be lysed in buffers that are compatible with the biological activity of the protein of interest. In some cases, addition of urea or a detergent such as Triton X-100 releases the protein of interest from cell debris.

All of the above assays require recovery of intact proteins in yeast cell extracts. Since proteases have historically been a problem with protein purifications from yeast, care should be taken to minimize their activity. Cell extracts should be prepared at 0–4° in the presence of protease inhibitors and assayed as soon as possible. Protease-deficient yeast strains such as 20B-12,[47] available from the Yeast Genetic Stock Center (University of California, Berkeley), can be used to construct appropriate host strains with reduced protease activity. Methods for reducing proteolysis during purification of proteins from yeast have been reviewed.[48] Intracellular turnover of heterologous proteins can be monitored with [^{35}S]methionine pulse–chase experiments. If the foreign protein is much less stable than native yeast proteins, then other expression strategies should be considered. These include regulated (inducible) promoters and secretion systems.

Secretion Vectors

Saccharomyces cerevisiae produces certain of its own proteins at very high levels and efficient expression of heterologous proteins has been obtained using the approaches outlined above. In addition, this organism secretes a restricted number (0.5% of total cell protein) of its own proteins into the culture medium. Initial attempts at secretion of heterologous proteins from *Saccharomyces cerevisiae* were motivated by the anticipation that recovery and purification of a secreted protein would be more efficient. A variety of heterologous signal peptides have been demonstrated to direct protein secretion in *Saccharomyces cerevisiae,* indicating conservation among different organisms in the mechanism of protein translocation.[19] Recent work has focused on homologous yeast leader peptides to program protein secretion. Heterologous protein secretion has been obtained using secretion signals from the yeast

[47] E. W. Jones, *Genetics* **85,** 23 (1977).
[48] J. R. Pringle, *Methods Cell Biol.* **12,** 149 (1975).

α-factor pheromone precursor,[6,49-51] invertase,[52] and acid phosphatase[52] precursors.

There are a number of situations in which it is desirable to produce heterologous proteins in a secretion system. These include foreign proteins that are toxic to yeast or unstable when produced by direct expression in the cytoplasm. Many proteins being considered for use as human therapeutics are themselves secreted proteins which are processed from a larger precursor. These proteins thus have specific NH_2 termini (rarely methionine). Direct expression of these genes results in proteins with NH_2 terminal methionine. Methionine aminopeptidases will remove this residue in certain sequence contexts.[53] However, the methionine will not be removed in other sequence contexts or may be removed from only a fraction of the molecules when the protein is produced at high levels. Under these conditions, the expressed protein represents an analog of the natural material which may have reduced activity or other undesirable properties. Production of such proteins in a secretion system, however, provides a method for the generation of native NH_2 termini. Finally, many proteins of commercial import contain intramolecular disulfide bonds. The cytoplasm of a cell is a reducing environment and few, if any, cytoplasmic proteins contain disulfide bonds. Direct expression of many heterologous proteins thus results in improperly folded (and possibly insoluble) products which must be extracted and subjected to an *in vitro* refolding process. This refolding occurs with variable efficiency and may be impossible for certain complex structures. The formation of accurate disulfide bonds in secreted proteins appears to be catalyzed by protein disulfide isomerase, an enzyme localized to the luminal side of the endoplasmic reticulum in eukaryotic secretory cells.[54] It seems likely, therefore, that heterologous proteins produced in a yeast secretion system will assume the correct tertiary structure. It has recently been demonstrated, by tryptic peptide mapping, that an α-interferon secreted from yeast has the identical disulfide bond structure as the natural human protein.[55]

[49] A. J. Brake, J. P. Merryweather, D. G. Coit, U. A. Heberlein, F. R. Masiarz, G. T. Mullenbach, M. S. Urdea, P. Valenzuela, and P. J. Barr *Proc. Natl. Acad. Sci. U.S.A.* **81,** 4642 (1984).

[50] A. Singh, E. Y. Chen, J. M. Lugovoy, C. N. Chang, R. A. Hitzeman, and P. H. Seeburg, *Nucleic Acids Res.* **11,** 4049 (1983).

[51] A. Miyajima, M. W. Bond, K. Otsu, K. Arai, and N. Arai, *Gene* **37,** 155 (1985).

[52] R. A. Smith, M. J. Ducan, and D. T. Moir, *Science* **229,** 1219 (1985).

[53] F. Sherman and J. W. Stewart, *in* "The Molecular Biology of the Yeast Saccharomyces: Metabolism and Gene Expression" (J. N. Strathern, E. W. Jones, and J. R. Broach, eds.), pp. 301–333. Cold Spring Harbor Lab., Cold Spring Harbor, New York, 1982.

[54] R. B. Freedman, *Trends Biochem. Sci.* **106,** 438 (1984).

[55] K. M. Zsebo, H. S. Lu, J. C. Fieschko, L. Goldstein, J. Davis, K. Duker, S. V. Suggs, P. H. Lai, and G. A. Bitter, *J. Biol. Chem.* **261,** 5858 (1986).

Use of the prepro leader region of the yeast α-factor pheromone precursor, which has proved to be generally useful for secretion of heterologous proteins from yeast, will be described in this section. Although use of other signal peptides[52] will necessitate different secretion vector constructions, the analysis and characterization of the secreted product will be similar. The methods and considerations for direct expression vector construction described above also apply to the development of secretion vectors.

Construction of Prepro-α-Factor Gene Fusions

We have developed two plasmids for construction of prepro-α-factor gene fusions[6,55] (Fig. 7). Plasmids pαC2 and pαC3 contain a 1.7-kb yeast genomic fragment which includes the *MFα*1 structural gene as well as its promoter and transcription termination sequences. Each plasmid contains the yeast DNA segment cloned as a *Bam*HI fragment in pBRΔHS (a derivative of pBR322 in which the *Hind*III and *Sal*I sites were each deleted by separate restriction, end filling, and blunt-end ligation). Plasmid pαC3 was derived from plasmid pαC2 by site-directed mutagenesis.[55] The serine codon at position 81 of the prepro-α-factor leader was converted to an AGC serine codon which introduces a *Hind*III site beginning 10 bp

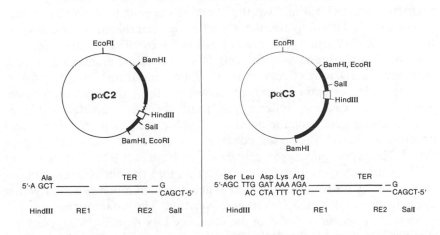

FIG. 7. Plasmids for construction of α-factor gene fusions. Restriction endonuclease maps for plasmids pαC2 and pαC3 are depicted. DNA sequences required at the 5' termini of the heterologous gene for creating the in-frame gene fusion are depicted. The first codon of the heterologous protein-coding region occurs immediately after the GCT[Ala] codon in pαC2 and immediately after the AGA[Arg] codon in pαC3. RE1 and RE2 refer to restriction enzyme sites within the heterologous gene to which synthetic linkers may be fused for assembly of the gene fusion. TER refers to the in-frame termination codon required in the heterologous gene coding region. Refer to the text for a description of construction strategies.

upstream from the start of the Lys codon of the KR endoprotease recognition sequence (see below). In both plasmids, the internal HindIII repeats of the MFα1 structural gene have been deleted.

The strategy for assembly of heterologous gene fusions, as well as the required DNA sequence at the MFα1 gene fusion site, are also depicted in Fig. 7. The heterologous gene encoding the protein to be secreted is cloned into pαC2 or pαC3 as a HindIII to SalI fragment. In the case of constructions using pαC2, the heterologous gene may actually contain a "pseudo-HindIII" site at the 5' end. The GCT of the HindIII recognition site represents the last alanine codon of the spacer peptide. Thus, the first nucleotide of the first codon of the heterologous gene occupies the position of the last T in the HindIII site of pαC2. For codons which begin with T, the HindIII site is regenerated. In other cases, however, a HindIII cohesive terminus is used without regenerating the HindIII site. Since, for both pαC2 and pαC3, the resulting gene encodes a protein fusion, the cloning manipulations must be very precise. This is most readily accomplished by chemically synthesizing a gene as a HindIII–SalI fragment. If this is not possible, a synthetic linker from the HindIII site in pαC2 or pαC3 to a restriction site (RE1; Fig. 7) near the 5' end of the heterologous gene may be synthesized which encodes the proper protein fusion sequence. If the heterologous gene does not include a SalI site on the 3' side and close to the translation stop codon, then a synthetic linker fusing a restriction site (RE2; Fig. 7) at the 3' end of the gene to the SalI site may be employed. Alternatively, the 3' end of the heterologous gene may be fused to the SalI site by blunt-end ligation. In isolating the heterologous gene fragment, excessive 3' flanking DNA should be eliminated since it may have deleterious effects on transcription and translation in yeast.

The native α-factor precursor as well as the hybrid precursors encoded by constructions utilizing pαC2 and pαC3 are depicted in Fig. 8. The 13-amino acid α-factor peptide is synthesized as a 165-amino acid prepro-polyprotein precursor containing 4 copies of the α-factor peptide.[56] The native precursor contains a hydrophic 20- to 22-amino acid NH_2 terminus which initiates translocation into the endoplasmic reticulum. There is a 61-amino acid pro segment of unknown function which contains three sites of N-linked glycosylation. The α-factor peptides are separated by spacer peptides of the sequence Lys-Arg(Glu-Ala)$_{2-3}$ or Lys-Arg-Glu-Ala-Asp-Ala-Glu-Ala. The repeating peptide units are excised from the precursor by cleavage on the carboxyl side of arginine by an endoprotease encoded by the yeast KEX2 gene and termed the KR endoprotease.[57] The Lys-Arg on the COOH terminus of the first three units

[56] J. Kurjan and I. Herskowitz, Cell 30, 933 (1982).
[57] D. Julius, A. Brake, L. Blair, R. Kunbawa, and J. Thorner, Cell 37, 1075 (1984).

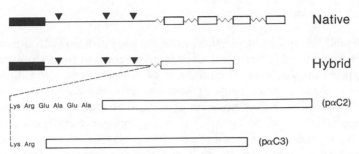

FIG. 8. Structure of native and hybrid α-factor precursors. The hydrophobic NH₂ terminus of the prepro leader region is depicted as a black box while the N-linked glycosylation
sites on the pro segment are represented by inverted triangles and the spacer peptides are
indicated by squiggles. The α-factor peptides in the native precursor and the heterologous
protein in the hybrid precursor are depicted by open boxes. The amino acid sequence of the
spacer peptide is indicated for gene fusions constructed in pαC2 (top) or pαC3 (bottom).

excised is presumably removed by a carboxypeptidase B-like activity
which may be the YscE enzyme identified *in vitro* in yeast membrane
functions.[58] The Glu-Ala or Asp-Ala dipeptides are excised from the NH₂
terminus by dipeptidyl aminopeptidase A which is encoded by the yeast
*STE*13 gene.[59] These three proteolytic events yield the fully processed,
mature α-factor peptide.

The hybrid precursors (Fig. 8) encoded by gene fusions assembled in
pαC2 and pαC3 each include the first 83 amino acids of the native α-factor
precursor. This segment includes signals which direct secretion of the
hybrid precursor. The fusion constructed in pαC2 includes the spacer
peptide Lys-Arg(Glu-Ala)₂ while that constructed in pαC3 has the spacer
peptide Lys-Arg. The first fusion protein thus contains processing sites
for the *KEX*2 and *STE*13 gene-encoded proteases. The gene fusion constructed in pαC3 contains the KR endoprotease recognition sequence but
is independent of processing by dipeptidyl aminopeptidase A since the
Glu-Ala dipeptides have been eliminated from the precursor. Neither hybrid precursor requires COOH-terminal processing since termination codons are included at the end of the coding region of the heterologous gene.

The gene fusions constructed in pαC2 and pαC3 may be subcloned as
a *Bam*HI fragment into a shuttle vector (such as pYE, above) for expression in yeast. The *Bam*III fragment encoding the gene fusion includes the
native yeast *MFα*1 promoter and transcription terminator sequences and
is efficiently expressed in yeast.[55] However, it should be noted that alternative promoters and terminators may be employed for expression according to the strategies outlined in Direct Expression Vectors.

[58] J. Achstetter and D. H. Wolf, *EMBO J.* **4**, 173 (1985).
[59] D. Julius, L. Blair, A. Brake, G. Sprague, and J. Thorner, *Cell* **32**, 839 (1983).

Analysis of Heterologous Protein Secretion

The methods utilized to analyze heterologous protein secretion are the same as those used for proteins expressed directly in the cytoplasm (Direct Expression Vectors). The methods of sample preparation, however, differ considerably since proteins secreted into the culture medium are generally more dilute than proteins present in concentrated cell extracts. Secretion offers an advantage over direct cytoplasmic expression in that, within limits, the culture medium conditions may be adjusted to be compatible with retention of biological activity of the secreted product. Thus, the pH, ionic strength, temperature, etc., of the cell culture medium, may be specified appropriately if the optimal parameters have been determined for the specific secreted protein. It should be noted, however, that conditions optimal for protein stability and activity may be suboptimal for efficient fermentation of the yeast strain. Thus, culture conditions should be formulated which optimize both aspects of the production process.

Measurement of Secretion Efficiency. It has been demonstrated that different heterologous proteins are secreted with different efficiencies using the prepro-α-factor leader region.[55] Thus, certain proteins are secreted entirely into the culture medium while others accumulate primarily intracellularly with only a small percentage secreted into the medium. The mechanisms which determine this partitioning between the cell and medium have not been clearly defined and no simple rules exist which are predictive of the secretion efficiency of a particular protein. In addition to differential partitioning, the total amount of heterologous protein synthesized may vary significantly for different α-factor gene fusions.

The heterologous protein secreted into the culture medium may be quantitated by the methods described in Direct Expression Vectors. If the product concentration is too low, culture supernatants can be rapidly concentrated by ultrafiltration, dialysis against polyethylene glycol, lyophilization, or acetone precipitation. Depending upon the method chosen, samples may be readily desalted or buffer conditions changed during the process of concentration. Alternatively, proteins can be concentrated specifically by immunoprecipitation, lectin columns, or other purification methods specific to the particular protein. The degree to which a sample should be concentrated will depend upon the sensitivity of the technique used for analysis. Nonspecific methods, such as SDS–PAGE, usually require a higher degree of concentration than do immunological methods. If the gel is to be stained with Coomassie blue, it may be necessary to concentrate a sample as much as 100- to 1000-fold, whereas with most silver staining methods the sample may require only 10-fold concentration. For most immunological methods such as RIA, ELISA, and immunoblot analysis, sample concentration is usually unnecessary.

The amount of heterologous protein secreted does not always increase in proportion to the culture cell mass, particularly in high cell density fermentations. Therefore, the product concentration in the culture medium should be correlated to cell mass as a function of growth of the culture. In addition, the amount of heterologous protein which accumulates intracellularly should be measured. In examining secretion efficiency, it is convenient to separate total cell-associated material into intracellular and wall-associated or periplasmic fractions. This can be readily accomplished by harvesting the cells and digesting the cell wall with glusulase or zymolyase (in the presence of osmotic stabilizing agents such as 1.0 M sorbitol) and separating the postenzyme supernatant from the spheroplasts. The material present in the spheroplasts is operationally defined as intracellular protein, and that in the supernatants as wall-associated or periplasmic protein. Boiling each fraction in the presence of a detergent, such as SDS, is usually sufficient to solubilize the protein for quantitation by SDS–PAGE or immunological methods. Less severe extraction conditions are generally required for retention of biological activity of the product.

By determining the total amount of heterologous protein produced, the efficiency of the expression system may be evaluated. If these levels are low, then total protein production should be increased by optimizing transcription and/or translation efficiencies (Direct Expression Vectors). Analysis of the heterologous protein partitioning between the cytoplasm, cell wall, and culture medium defines the secretion efficiency. It is highly desirable to achieve 100% secretion of the heterologous protein into the culture medium. Optimization of this process is currently an area of intense research.

Analysis of Hybrid Precursor Processing

The yeast secretory process is a complex series of events associated with membrane-bound subcellular compartments. In addition to the carbohydrate modifications of the pro segment, the native α-factor precursor is processed by three different proteolytic enzymes to generate the mature α-factor peptides. It is important, therefore, to characterize the processing of the hybrid precursor and secreted heterologous protein.

KR Endoprotease Processing. It has been demonstrated that hybrid protein precursors which have not been cleaved by KR endoprotease are secreted into the culture medium.[55] These fusions proteins may have reduced or no biological activity. Their presence, therefore, should be quantitated prior to measuring the specific activity (biological activity units per milligram protein) of the secreted heterologous protein. This quantitation is readily performed by immunoblot analysis[46] using antisera raised against the heterologous protein.

The hybrid precursor will contain N-linked carbohydrate on the three glycosylation sites within the pro segment (Fig. 8). Because of the large size and heterogeneous nature of yeast carbohydrate modifications,[60] the precursors are not readily visualized after SDS–PAGE. However, by treating the secreted proteins with endoglycosidases which remove the N-linked carbohydrate (see Procedures) the heterogeneously sized precursors are reduced to a single band. An example of this analysis is depicted in the Western blot shown in Fig. 9. Lane 1 contains untreated medium proteins from a yeast strain secreting IFN-Con$_1$.[55] A single major band of apparent molecular weight 18,000 is present. The sample in lane 2 was treated with Endo H prior to gel electrophoresis. A new major band with apparent molecular weight of 28,000 appears which is reactive with the antisera. This protein is the deglycosylated hybrid protein precursor and, in this preparation, represents approximately 90% of the total secreted IFN-Con$_1$ immunoreactive material.

The secretion of unproteolyzed hybrid precursor may be due to limiting KR endoprotease activity when the precursor is produced at high levels from multicopy vectors. This deficiency might be corrected by incorporating the cloned KEX2 gene[57] into multicopy plasmids. This strategy, however, may result in internal cleavages at dibasic residues within the heterologous protein. For at least one hybrid precursor, the KR endoprotease processing efficiency appeared greater when Glu-Ala dipeptides were present in the spacer peptide.[55] This indicates that amino acid sequence context may affect the efficiency of cleavage by KR endoprotease.

If conditions can be developed in which complete cleavage by KR endoprotease occurs, then the yield of correctly processed secreted heterologous protein will be optimized. If this is not possible, then the precursor must be separated from the processed heterologous protein. This separation is readily accomplished for heterologous proteins which are not glycosylated since the large mass of carbohydrate on the pro segment of the precursor allows resolution of the two forms by gel exclusion chromatography. In cases where the heterologous protein is glycosylated, however, different procedures may be required to separate the precursor and processed forms of the heterologous protein.

---

[60] C. E. Ballou, in "The Molecular Biology of the Yeast Saccharomyces: Metabolism and Gene Expression" (J. N. Strathern, E. W. Jones, and J. R. Broach, eds.), pp. 335–360. Cold Spring Harbor Lab., Cold Spring Harbor, New York, 1982.

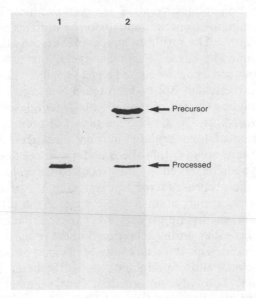

FIG. 9. Detection of secreted unproteolyzed hybrid precursor. Cell-free culture supernatants from a yeast strain secreting IFN-Con$_1$ were subjected to immunoblot analysis as
described.[55] Lane 1, untreated medium proteins. Lane 2, proteins were digested with Endo
H as described in Procedures prior to SDS–PAGE.

NH$_2$ Terminal Analysis of Secreted Heterologous Proteins. One of the
advantages of a secretion system is production of heterologous proteins
with authentic NH$_2$ termini. Although size analysis by SDS–PAGE may
indicate accurate processing of the heterologous protein, this may be
demonstrated only by direct NH$_2$ terminal amino acid sequence analysis.
These methods have been reviewed in detail elsewhere.[61–63] The secreted
heterologous protein (processed form) may be purified by conventional
means for this purpose. Alternatively, if the secreted product is abundant
and well resolved from other secreted yeast proteins, it may be purified
by preparative SDS–PAGE and directly subjected to NH$_2$ terminal
analysis.[64]

It has been demonstrated that secreted proteins programmed by gene
fusions constructed in pαC2 contain authentic (fully processed) NH$_2$ termini.[6] However, when these precursors are overproduced, the secreted
product contains NH$_2$ terminal Glu-Ala extensions due to rate-limiting
proteolysis by dipeptidyl aminopeptidase A.[6,55] Two strategies exist for

[61] P. Edman and G. Begg, *Eur. J. Biochem.* **1,** 80 (1967).
[62] R. M. Hewick, M. W. Hunkapiller, L. E. Hood, and W. J. Dreyer, *J. Biol. Chem.* **256,**
7990 (1981).
[63] P. H. Lai, *Anal. Chim. Acta* **163,** 243 (1984).
[64] M. W. Hunkapiller, E. Lujan, F. Ostrander, and L. E. Hood, this series, Vol. 91, p. 227.

eliminating these dipeptides from the secreted product. The Glu-Ala codons may be eliminated from the gene fusion by utilizing plasmid pαC3 for
construction (Fig. 7). The resulting hybrid precursor is thus independent
of dipeptidyl aminopeptidase A. Alternatively, the processing enzyme
may be overproduced using the cloned *STE*13 gene.[59] This latter approach
might allow more efficient KR endoprotease cleavage[55] as well as complete exoproteolytic removal of the Glu-Ala dipeptides.

Aberrant Proteolysis. In our original report of prepro-α-factor leader-
directed secretion of β-endorphin,[6] internal cleavages on the carboxyl
side of lysines were noted within the β-endorphin peptide. Such internal
cleavages have been observed in other secreted heterologous proteins as
well as within the prepro-α-factor leader region.[55] Aberrant proteolysis
occurs, in general, in only a small percentage of the total heterologous
protein. However, the hybrid precursor is exposed to a number of different proteolytic activities during secretion and the integrity of the product
should be monitored. This is most readily performed by immunoblot analysis using polyclonal antisera directed against the heterologous protein.

Glycosylation. The α-factor precursor pro segment contains three
sites of N-linked glycosylation (Fig. 8). It has been demonstrated that
heterologous proteins containing potential sites of N-linked glycosylation
are, in fact, glycosylated during prepro-α-factor-directed secretion from
yeast (S. G. Elliott and G. A. Bitter, unpublished results). Carbohydrate
addition to glycoproteins is dependent on the accessibility of the glycosylation sites, which is affected by protein folding (both rate and final conformation). Therefore, it cannot be assumed, a priori, that all potential
sites of glycosylation in heterologous proteins will be modified. *Saccharomyces cerevisiae* also effects O-linked glycosylation, and this modification may occur on mammalian proteins secreted using the α-factor
system.

The methods used for analysis of carbohydrate addition are described
in Procedures. It should be noted that the structure of yeast oligosac-
charides is markedly different than the structures present on mammalian
glycoproteins (reviewed by Ballou[60]). Thus, the effect of yeast carbohydrate addition on the biological activity of these products should be tested
in appropriate *in vivo* systems.

Procedures

Mung Bean Nuclease Digestion. This reaction should be done in glass
tubes rather than plastic. Suspend the phenol-extracted and ethanol-precipitated restricted DNA in TNE buffer (6 mM Tris–HCl, pH 8.0, 6 mM
NaCl, 0.2 mM EDTA) to a concentration of 100 ng/μl. Add one-tenth
volume of 10× mung bean buffer (300 mM sodium acetate, pH 4.7, 500

mM NaCl, 10 mM ZnCl$_2$). Add 0.5–1.0 U of mung bean nuclease/1 µg of DNA. (If the nuclease is to be diluted, dilute it into 1× mung bean buffer). Incubate the reaction for 10 min at 37°. To stop the reaction, add 1 µl of 20% SDS/100 µl of reaction mix, and heat at 65° for 5 min. Add one-tenth volume of 500 mM Tris, pH 9.5, and one-tenth volume of 8 M LiCl; then extract the solution with an equal volume of phenol/chloroform (1 : 1), followed by an extraction with ether. Add 2 vol of cold 100% ethanol to precipitate the DNA.

S1 Nuclease Digestion of Cohesive Termini. Suspend the phenol-extracted and ethanol-precipitated restricted DNA in double-distilled H$_2$O to a concentration of 100 ng/µl. To this solution, add one-tenth volume of 10× S1 buffer (2.5 M NaCl, 10 mM ZnSO$_4$, 300 mM sodium acetate, pH 4.5). Chill the reaction mix to 14° (in order to minimize strand separation and therefore to prevent excess single-strand cutting by S1 nuclease). Add the S1 nuclease to a concentration of 200 U/pmol DNA. (If the S1 nuclease must be diluted, dilute it into S1 buffer.) Incubate the reaction for 60 min at 14°. Stop the reaction with 25 mM EDTA, and bring to pH 8.0 with one-tenth volume of 500 mM Tris, pH 9.5. Then extract the solution with an equal volume of phenol/chloroform (1 : 1), followed by an extraction with ether. Add 2 vol of cold 100% ethanol to precipitate the DNA.

End Filling with Klenow. The Klenow end-filling reaction on 5' overhang termini works well in a variety of buffers including standard restriction enzyme buffers and ligation buffers. To the restricted DNA in the appropriate buffer at a concentration of 100 ng/µl, add one-tenth volume of 1 mM deoxynucleoside triphosphates (only those necessary to fill in the recessed 3' end need be added). Add the Klenow fragment to a concentration of about 1 U/pmol DNA. Incubate the reaction for 10 min at room temperature. Stop the reaction by heating for 10 min at 65°. The solution can now be precipitated with 2 vol of ethanol, or used directly for subsequent ligation or restriction reactions.

Yeast RNA Preparation. Cells from approximately 50 OD · ml of late log phase cultures (e.g., 50 ml of culture at OD$_{600}$ = 1.0) are harvested by centrifugation, resuspended in 10 ml ice-cold H$_2$O, and centrifuged again. Resuspend the cell pellets in 3 ml RNA extraction buffer (0.15 M NaCl, 5 mM EDTA; 4% SDS, 50 mM Tris–HCl, pH 7.5). Add 3 ml phenol–chloroform–isoamyl alcohol (50 : 50 : 1) and, after 2 min at room temperature, vortex vigorously for 5 min with glass beads. Centrifuge and collect the aqueous phase. Reextract the organic phase with 3 ml of extraction buffer. Pool the aqueous phases and extract once with phenol/chloroform/isoamyl alcohol and once with chloroform. Precipitate the RNA by the addition of 2.5 vol of ethanol, rinse with 70% ethanol, and dry *in vacuo*.

Yeast DNA Preparation. Cells from approximately 100 OD · ml of late log phase cultures are harvested by centrifugation, resuspended in 10 ml H_2O, and centrifuged again. Prepare spheroplasts rapidly by resuspending the cells in 3 ml SCE (1 M sorbitol, 0.1 M sodium citrate, 0.01 M EDTA, pH 5.8) containing 2.5 mg/ml zymolyase 60,000 (Miles), and incubating 3–5 min at 37°. Dilute the spheroplast suspension with 3 vol of ice-cold SCE, and collect the spheroplasts by centrifugation at 4000 g for 5 min. The spheroplasts are lysed in 5 ml 25 mM EDTA, 0.15 M NaCl and extracted with 0.5 vol of equilibrated phenol, and once with 0.5 vol of chloroform. DNA is precipitated from the aqueous phase with 1.5 vol of 2-propanol. The final DNA pellets are washed with 70% 2-propanol and dried *in vacuo*.

Analysis of Carbohydrate Additions to Heterologous Proteins. The presence of oligosaccharide additions to the heterologous protein may be demonstrated by sensitivity to a variety of enzymes. Three commercially available enzymes are effective at removing N-linked carbohydrate from yeast glycoproteins. These include endoglycosidase H (Endo H; Miles, New England Nuclear and Genzyme), endoglycosidase F (Endo F; New England Nuclear), and N-glycanase (Genzyme). Endo H and Endo F treatment of glycoproteins leaves the sugar N-acetyl-galactosamine (GlcNac) bound to the asparagine at the glycosylation site. N-glycanase treatment, on the other hand, results in complete removal of protein carbohydate. Endo F does have an N-glycanase activity and exhaustive digestion can result in removal of the asparagine bound GlcNac. The following reaction conditions will meet most needs but may need to be adjusted for each particular glycoprotein.

Endo H. Boiling the sample in 0.05% SDS prior to deglycosylation results in faster and more complete deglycosylation. SDS concentrations higher than 0.05% may inhibit the enzyme. To the SDS-treated protein sample (100 μg/ml–2 mg/ml) is added a one-tenth volume of 1 M sodium citrate, pH 5.5, followed by 10–50 mU/ml final concentration of enzyme. Incubate at 37° until digestion is complete (4–16 hr).

Endo F. Some preparations of endo F have a contaminating protease activity. EDTA is required to minimize protease activity. Reaction conditions are 100 mM sodium phosphate, pH 6.1, 50 mM EDTA, 1% Nonidet P-40, 0.1% SDS, and 1% 2-mercaptoethanol. The sample containing 0.1–2 mg/ml protein is boiled for 10 min followed by the addition of 10–50 mU/ml Endo F. Incubate at 37° until digestion is complete (4–16 hr).

N-Glycanase. The glycoprotein sample (0.1–2 mg/ml) is boiled in 0.5% SDS and 0.1 M 2-mercaptoethanol for 10 min. The sample is then diluted into sodium phosphate buffer. The final concentrations are 0.17% SDS, 33 mM 2-mercaptoethanol, 0.2 M sodium phosphate, pH 8.6, 1.25% Nonidet P-40. N-Glycanase is added to 2–10 U/ml and incubated 4–16 hr at 37°.

Section V

Methods for Oligonucleotide-Directed Mutagenesis

[30] Oligonucleotide-Directed Mutagenesis of DNA Fragments Cloned into M13 Vectors

By Mark J. Zoller and Michael Smith

The isolation and sequencing of genes has been a major focus of biological research for almost a decade. The emphasis has now turned toward identification of functional regions encoded within DNA sequences. A variety of *in vivo*[1] and *in vitro*[2-4] mutagenesis methods have been developed to identify regions of functional importance. These techniques provide useful information pertaining to the location and boundaries of a particular function. Once this has been established, a method is required to define the specific sequences involved by precisely directing mutations within a target site. Oligonucleotide-directed *in vitro* mutagenesis provides a means to alter a defined site within a region of cloned DNA.[5] This powerful technique has many far-reaching applications, from the definition of functional DNA sequences to the construction of new proteins.

The method stemmed from the combination of a number of recent discoveries and observations about nucleic acids: (*a*) marker rescue of mutations in ϕX174 by restriction fragments[6,7]; (*b*) the stability of DNA duplexes containing mismatches[8-11]; and (*c*) the ability of *Escherichia coli* DNA polymerase to extend oligonucleotide primers hybridized to single-stranded DNA templates.[12,13] The basic principle involves the enzymic extension by *E. coli* DNA polymerase of an oligonucleotide primer hybridized to a single-stranded circular template. The oligonucleotide is completely complementary to a region of the template except for a mis-

[1] J. W. Drake and R. H. Baltz, *Annu. Rev. Biochem.* **45,** 11 (1976).

[2] W. Muller, H. Weber, F. Meyer, and C. Weissmann, *J. Mol. Biol.* **124,** 343 (1978).

[3] D. Shortle, D. DiMaio, and D. Nathans, *Annu. Rev. Genet.* **15,** 265 (1981).

[4] C. Weissmann, S. Nagata, T. Taniguichi, T. Weber, and F. Meyer, in "Genetic Engineering," (J. K. Setlow and A. Hollaender, eds.), Vol. 1, p. 133. 1979.

[5] M. Smith and S. Gillam, in "Genetic Engineering," (J. K. Setlow and A. Hollaender, eds.), Vol. 3, p. 1. 1981.

[6] P. J. Weisbeek and J. H. van de Pol, *Biochim. Biophys. Acta* **224,** 328 (1970).

[7] C. A. Hutchison III and M. H. Edgell, *J. Virol.* **8,** 181 (1971).

[8] C. R. Astell and M. Smith, *J. Biol. Chem.* **246,** 1944 (1971).

[9] C. R. Astell and M. Smith, *Biochemistry* **11,** 4114 (1972).

[10] C. R. Astell, M. T. Doel, P. A. Jahnke, and M. Smith, *Biochemistry* **12,** 5068 (1973).

[11] S. Gillam, K. Waterman, and M. Smith, *Nucleic Acids Res.* **2,** 625 (1975).

[12] M. Goulian, A. Kornberg, and R. L. Sinsheimer, *Proc. Natl. Acad. Sci. U.S.A.* **58,** 2321 (1967).

[13] M. Goulian, S. H. Goulian, E. E. Codd, and A. Z. Blumenfield, *Biochemistry* **12,** 2893 (1973).

RECOMBINANT DNA
METHODOLOGY

match that directs the mutation. Closed circular double-stranded molecules are formed by ligation of the newly synthesized strand with T4 DNA ligase. Upon transformation of cells with the *in vitro* synthesized closed circular DNA (CC-DNA), a population of mutant and wild-type molecules are obtained. Mutant molecules are distinguished from wild type by one of a number of screening procedures.

In their initial studies, Hutchison *et al.*[14] and Gillam and Smith[15] created a number of mutations in the single-stranded phage φX174. Razin *et al.* reported a similar experiment using φX174.[16] Wallace and co-workers applied the basic principle to create mutations in genes cloned into pBR322.[17,18] Wasylyk *et al.* demonstrated the first example of oligonucleotide mutagenesis of a cloned fragment in a vector derived from a single-stranded phage, fd.[19] More recently, phage M13 and M13 derived vectors have been used in a number of oligonucleotide mutagenesis experiments.[20–25] Vectors derived from single-stranded phage, such as M13 and fd, are more convenient for oligonucleotide-directed mutagenesis than double-stranded vectors, because isolation of pure single-stranded circular template DNA in these systems is quite simple.[26–28] However, the use of a double-stranded vector may be required if cloning of a particular fragment into M13 proves to be difficult or in the creation or destruction of a particular restriction site in the vector itself. Table I summarizes the

[14] C. A. Hutchison III, S. Phillips, M. H. Edgell, S. Gillam, P. A. Jahnke, and M. Smith, *J. Biol. Chem.* **253,** 6551 (1978).

[15] S. Gillam and M. Smith, *Gene* **8,** 81 (1979).

[16] A. Razin, T. Hirose, K. Itakura, and A. Riggs, *Proc. Natl. Acad. Sci. U.S.A.* **75,** 4268 (1978).

[17] R. B. Wallace, M. Schold, M. J. Johnson, P. Dembek, and K. Itakura, *Nucleic Acids Res.* **9,** 3647 (1981).

[18] R. B. Wallace, P. F. Johnson, S. Tanaka, M. Schold, K. Itakura, and J. Abelson, *Science* **209,** 1396 (1980).

[19] B. Wasylyk, R. Derbyshire, A. Guy, D. Molko, A. Roget, R. Teoule, and P. Chambon, *Proc. Natl. Acad. Sci. U.S.A.* **77,** 7024 (1980).

[20] P. D. Baas, W. R. Teertstra, A. D. M. van Mansfield, H. S. Janz, G. A. van der Marel, G. H. Veeneman, and J. H. van Boom, *J. Mol. Biol.* **152,** 615 (1981).

[21] G. F. M. Simons, G. H. Veeneman, R. N. H. Konigs, J. H. van Boom, and J. G. G. Schoenmakers, *Nucleic Acids Res.* **10,** 821 (1982).

[22] I. Kudo, M. Leineweber, and U. RajBhandary, *Proc. Natl. Acad. Sci. U.S.A.* **78,** 4753 (1981).

[23] C. G. Miyada, X. Soberon, K. Itakura, and G. Wilcox, *Gene* **17,** 167 (1982).

[24] C. Montell, E. F. Fisher, M. H. Caruthers, and A. J. Berk, *Nature (London)* **295,** 380 (1982).

[25] G. F. Temple, A. M. Dozy, K. L. Roy, and Y. W. Kan, *Nature (London)* **296,** 537 (1982).

[26] J. Messing, this series, Vol. 101 [2].

[27] G. Winter and S. Fields, *Nucleic Acids Res.* **8,** 1965 (1980).

[28] P. H. Schreier and R. A. Cortese, *J. Mol. Biol.* **129,** 169 (1979).

TABLE I

EXAMPLES OF OLIGONUCLEOTIDE-DIRECTED MUTAGENESIS EXPERIMENTS

Target	Alteration	Oligonucleotide length	Vehicle	Selection or screening procedure	% Mutants
ϕX174 gene E^a	A → G	12	ϕX174	Biological	15
	G → A	12	ϕX174	Biological	15
ϕX174 gene E^b	A → G	14	ϕX174	Biological	1.9
	A → G	17	ϕX174	Biological	13.9
ϕX174 gene E^c	A → G	11	ϕX174	Biological	39
	G → A	10	ϕX174	Biological	23
gene A	T → G	10	ϕX174	Biological	22
	G → T	11	ϕX174	Biological	13
ϕX174 ribosome binding site[d]	del T	10	ϕX174	*In vitro* selection and DNA sequencing	100[n]
Yeast tRNA[Tyr] intervening sequence[e]	del 14 bases	21	pBR322	Hybridization	4
Chicken con-albumin pro-moter (TATA)[f]	T → G	11	fd103	New restriction site	4
Human β-glo-bin[g]	T → A	19	pBR322	Hybridization	0.5
E. coli tRNA[Tyr h]	T → A	10	M13mp3	Biological	7–10
ϕX174 origin[i]	T → C	16	ϕX174	Biological	—
M13 gene IX^j	T → G	13	M13	Biological	1.9
E. coli araBAD operon[k]	del 3 bases	18	M13mp2	Hybridization	7.5
E. coli CRP-cAMP binding site[k]	del 3 bases	19	M13mp2	Hybridization	2.4
Human adenovirus splice junc-tion[l]	T → G	12	M13Goril	*In vitro* selection and DNA sequencing	13[n]
Human tRNA[Lys m]	AAA → TAG	15	M13mp7	*In vitro* selection and DNA sequencing	100[n]

[a] Hutchison *et al.*[14] [b] Razin *et al.*[16] [c] Gillam and Smith.[15] [d] Gillam *et al.*[40] [e] Wallace *et al.*[18] [f] Wasylyk *et al.*[19] [g] Wallace *et al.*[17] [h] Kudo *et al.*[22] [i] Baas *et al.*[20] [j] Simons *et al.*[21] [k] Miyada *et al.*[23] [l] C. Montell *et al.*[24] [m] Temple *et al.*[25] [n] Enrichment by multiple rounds of mutagenesis. See Gillam *et al.*[40]

targets and vehicles utilized in a number of recent mutagenesis experiments.

This chapter describes a mutagenesis procedure that is simple, versatile, and efficient. Three criteria guided the development of the methodology: (a) that a mutation could be produced at any position in a cloned fragment of known sequence; (b) that the efficiency of mutagenesis be sufficiently high to facilitate screening; and (c) that identification of a mutant could be made without biological selection or direct sequencing. As an example, a single G to A transition in the *MATal* gene of the yeast *Saccharomyces cerevisiae* has been produced as part of a study into the role of an inframe UGA codon within the coding region of the gene. The chapter is divided into two parts. The first presents a step-by-step description of the mutagenesis experiment, from the design of the mutagenic oligonucleotide to the screening procedures and verification of the mutant by sequence determination. The second part presents a detailed protocol.

Experimental Rationale

The mating type of the yeast *S. cerevisiae* is determined by the specific gene, *a* or *α*, present at the *MAT* locus on chromosome III.[29,30] These genes are thought to code for regulatory proteins that control the expression of mating type specific genes that act in mating and sporulation.[31,32] Genetic analysis has revealed that *MATα* consists of two complementation groups (*α*1 and *α*2), and that *MATa* consists of only one (a1).[32,33] *MATα1* acts to turn on the expression of *α* specific genes. *MATα2* functions in the repression of a-specific genes and is also active in sporulation. *MATal* is required for sporulation and for the maintenance of the a/*α* diploid cell phenotype, but functions only in the presence of *α*2. In the absence of *α*2, a-specific genes are constitutively expressed. Both *MATa* and *MATα* have been cloned[34,35] and sequenced.[36] Each codes for two transcripts, whose position and direction of synthesis have been physically mapped.[37] The putative protein sequences of a1, *α*1, and *α*2 can be

[29] R. K. Mortimer and D. C. Hawthorne, in "The Yeasts" (A. H. Rose and J. S. Harrison, eds.), Vol. 1, p. 385. Academic Press, New York, 1969.

[30] I. Herskowitz and Y. Oshima, in "The Molecular Biology of the Yeast *Saccharomyces*" (J. N. Strathern, E. Jones, and J. R. Broach, eds.), p. 181. Cold Spring Harbor Laboratory, Cold Spring Harbor, New York, 1982.

[31] V. L. Mackay and T. R. Manney, *Genetics* **76,** 273 (1974).

[32] J. N. Strathern, J. Hicks, and I. Herskowitz, *J. Mol. Biol.* **147,** 357 (1981).

[33] G. F. Sprague, J. Rine, and I. Herskowitz, *J. Mol. Biol.* **153,** 323 (1981).

[34] J. B. Hicks, J. N. Strathern, and A. J. S. Klar, *Nature (London)* **282,** 478 (1979).

[35] K. A. Nasmyth and K. Tatchell, *Cell* **19,** 753 (1980).

[36] C. R. Astell, L. Ahlstrom-Jonasson, M. Smith, K. Tatchell, K. A. Nasmyth, and B. D. Hall, *Cell* **27,** 15 (1981).

[37] K. A. Nasmyth, K. Tatchell, B. D. Hall, C. R. Astell, and M. Smith, *Nature (London)* **289,** 244 (1981).

predicted because in each case there is only one extended open reading frame. Inspection of these sequences revealed that the *MATa1* gene codes for a UGA stop codon 43 codons from the AUG initiator codon. The open reading frame continues for another 103 codons, terminating with UAA. *In vitro* constructed mutants on either side of this in-frame UGA codon resulted in the loss of a1 function.[38] This suggested that the UGA must be read-through. In order to understand the role of this in-frame UGA codon, two mutants have been constructed: (*a*) the UGA changed to UAA, a strong stop codon; (*b*) the UGA changed to UGG, coding for tryptophan. A 4.2 kb *Hin*dIII fragment containing the *MATa* gene was cloned into the vector M13mp5, derived from the single-stranded phage M13. Single-stranded circular DNA, containing the + strand of M13mp5 and the noncoding strand of the *MATa* gene, was isolated and used as template for oligonucleotide-directed mutagenesis.

Design of the Mutagenic Oligonucleotide

Once the experimental rationale and the desired changes have been formulated, the next step is the design of the oligonucleotide to direct the mutagenesis. Three factors should be considered in this regard: (*a*) the synthetic capability; (*b*) placement of the mismatch within the oligonucleotide; and (*c*) competing sites within the M13 vector or in the cloned fragment.

Studies by Gillam and Smith demonstrated the efficacy of short oligonucleotides 8–12 long to direct mutagenesis in ϕX174.[15,39,40] For mutagenesis of cloned fragments in M13, we have chosen to utilize oligonucleotides ranging in length from 14–21 nucleotides. Oligomers in this range serve as primers for DNA synthesis at room temperature and above. In addition, they are more likely to recognize the unique target in the M13-recombinant DNA than are shorter oligonucleotides. The synthesis of oligonucleotides 14–21 long is well within the capability of the chemical methods presently available.[41-46]

[38] K. Tatchell, K. A. Nasmyth, B. D. Hall, C. R. Astell, and M. Smith, *Cell* **27**, 25 (1981).
[39] S. Gillam and M. Smith, *Gene* **8**, 99 (1979).
[40] S. Gillam, C. R. Astell, and M. Smith, *Gene* **12**, 129 (1981).
[41] M. Edge, A. M. Greene, G. R. Heathcliffe, P. A. Meacock, W. Schuch, D. B. Scanlon, T. C. Atkinson, C. R. Newton, and A. F. Markham, *Nature (London)* **292**, 756 (1981).
[42] K. Itakura and A. Riggs, *Science* **209**, 1401 (1979).
[43] M. L. Duckworth, M. J. Gait, P. Goelet, G. F. Hong, M. Singh, and R. C. Titmas, *Nucleic Acids Res.* **9**, 1691 (1981).
[44] F. Chow, T. Kempe, and G. Palm, *Nucleic Acids Res.* **9**, 2807 (1981).
[45] G. Alvarado-Urbina, G. M. Sathe, W-C. Liu, M. F. Gillen, P. D. Duck, R. Bender, and K. K. Ogilvie, *Science* **214**, 270 (1981).
[46] M. D. Matteucci and M. H. Caruthers, *J. Am. Chem. Soc.* **103**, 3185 (1981).

```
                    H  F  K  D  S  L  *  I  N
          5'-TTTCATTTCAAGGATAGCCTTTGAATCAATTTA-3'        coding strand
                                  Hinf I

mutagenic            5'-AAGGATAGCCTTTAAATC-3'             MS-1 (UAA stop)

oligonucleotides     5'-AAGGATAGCCTTTGGATC-3'            MS-2 (UGG TRP)
```

FIG. 1. The DNA sequence of the coding strand of *MATa1* in the region containing the in-frame UGA codon. Below are shown the sequences of two oligonucleotides synthesized to create point mutations. Each mutation destroys the *Hin*fI site in this region (underlined).

The oligonucleotide is designed so that the mismatch(es) is located near the middle of the molecule. This is especially important if the oligomer will be used as a probe to screen for the mutant. Placement of the mismatch in the middle yields the greatest binding differential between a perfectly matched duplex and a mismatched duplex.[47] A second consideration concerns protection of the mismatch from exonuclease activity of DNA polymerase. *Escherichia coli* DNA polymerase I contains intrinsic $5' \rightarrow 3'$- and $3' \rightarrow 5'$-exonuclease activities. The "large fragment" (Klenow derivative) lacks the $5' \rightarrow 3'$-exonuclease activity, and therefore is used in oligonucleotide-directed mutagenesis to prevent correction of the mismatch. Placement of the mismatched nucleotide near the $3'$ end might result in repair of the mismatch by the $3' \rightarrow 5'$-exonuclease activity still present in DNA polymerase (large fragment). Gillam and Smith have demonstrated that three nucleotides following the mismatch are enough to protect against this problem.[15] The final step in designing the mutagenic oligonucleotide is to conduct a computer analysis on the DNA sequences of the M13 vector and the cloned fragment for regions of partial complementarity with the mutagenic oligonucleotide. The purpose of this step is to identify competing sites from which priming might also occur. Such analysis might indicate the need to extend the oligomer, or whether one strand of the cloned fragment is a better template than the other. Generally, a competing target site with a predominance of matches with the $3'$ end of the oligonucleotide will be a more efficient priming site than one that exhibits complementarity to the $5'$ end of the oligomer.[47]

Figure 1 shows the DNA sequence of the coding strand of *MATa1* in the region of the UGA codon.[36] The octadecanucleotide 5'-AAGGA-TAGCCTTTAAATC-3' (MS-1) was designed to change the UGA to UAA, a strong stop codon in yeast. The second oligonucleotide (MS-2) was designed to change the UGA to UGG, coding for tryptophan. Table II summarizes the results of a computer analysis for sequences in the +

[47] M. Smith, *in* "Methods of RNA and DNA Sequencing" (S. M. Weissmann, ed.). Praeger, New York, 1983 (in press).

strand of the vector M13mp5 and *MATa* that are complementary to the oligonucleotide. A potential competing site is the sequence in M13 spanning nucleotides 1533–1550, which has nine consecutive matches with the oligonucleotide. Originally we synthesized a 10-long oligonucleotide that spanned this region. Although oligonucleotides 10–12 long have proved to be very efficient mutagens,[15] preliminary experiments indicated that the 10-mer primed from multiple sites in both the *MATa* insert and M13 vector. To avoid priming from these competing sites, the oligonucleotide was extended. The results of the computer analysis shown in Table II suggested that an 18-mer would prime specifically from the desired site in *MATa*. As demonstrated in a subsequent section, the 18-mer specifically primed from the desired site. This example demonstrated the importance of conducting a thorough computer search before synthesizing the oligonucleotide.

TABLE II

SEQUENCES IN *MATa* AND M13mp5 PARTIALLY COMPLEMENTARY TO
OLIGONUCLEOTIDE MS-1[a,b]

Sequence	Position[c]	Matches/18
MATa		
5'-GAATTAAGGGATATATTA-3'	(1355–1338)	12
5'-GGAATTAAGGCTTTGCTT-3'	(1870–1853)	12
M13mp5		
5'-AATTAAAACGCGATATTT-3'	(339–356)	12
5'-CAATTAAAGGCTCCTTTT-3'	(1533–1550)	12
5'-GCTTTAATGAGGATCCAT-3'	(2210–2237)	12
5'-GATGTAAAAGGTACTGTT-3'	(4370–4387)	12
5'-TTTTTAATGGCGATGTTT-3'	(4962–4979)	12
5'-TGTTTTAGGGCTATCAGT-3'	(4975–4992)	12
5'-GCTTTACACTTTATGCTT-3'	(6141–6158)	12
5'-GATTGACATGCTAGTTTT-3'	(6839–6856)	12

[a] Computer search for DNA sequences in *MATa* 4.2 kb *Hin*dIII fragment (noncoding strand) and in M13mp5 (+ strand) that exhibit partial complementarity to oligonucleotide MS-1 (5'-AAGGATAGCCTTTAAATC-3'). Positions that form matched base pairs are underlined.

[b] Computer program for DNA sequence analysis was developed by A. Delaney and described in *Nucleic Acids Res.* **10,** 61 (1982).

[c] The sequences in *MATa* are numbered according to Astell *et al.*[36] The sequences in M13 are numbered according to P. M. G. F. van Wezenbeek, T. J. M. Hulsebos, and J. G. G. Schoenmakers, *Gene* **11,** 129 (1980).

Synthesis of the Oligonucleotide

Short oligonucleotides can be synthesized *in vitro* by either enzymic or chemical means. The enzymic methods provide relatively low yields of final products and are dependent on slow empirical reactions.[48,49] The chemical procedures based on phosphotriester[41-43] or phosphite[44-46] chemistry are the most widely used. The trend has turned toward the development of solid-phase procedures that are amenable to either manual or automatic operation. The amount of oligonucleotide required for one mutagenesis experiment is small (less than 250 pmol) compared to the amount one obtains from the chemical procedures (5–50 nmol).

The octadecanucleotide 5′-AAGGATAGCCTTTAAATC-3′ for mutagenesis of *MATa1* was synthesized by the solid-phase phosphotriester procedure.[41] Purification was accomplished using ion-exchange and C_{18} reverse-phase HPLC.

Determination of the Sequence of the Oligonucleotide

Once the oligonucleotide has been synthesized, it should be sequenced before preceding with the mutagenesis experiment. Two methods are available: (*a*) two-dimensional homochromotography[50]; (*b*) a modified Maxam and Gilbert procedure designed for oligonucleotides.[51,52] We have found the later procedure to be easier; however, the homochromatography method might be used in order to determine whether two different nucleotides were present at the same position, as would be the case in the deliberate synthesis of a mixture of oligonucleotides. The oligonucleotide is phosphorylated at the 5′ end with [γ-^{32}P]ATP and polynucleotide kinase[44] or at the 3′ end with terminal transferase,[53,54] then subjected to a modified Maxam and Gilbert procedure. Figure 2 shows an autoradiogram of a sequencing gel verifying the sequence of oligomer MS-1.

Preparation of Template DNA

There are a number of vectors available that are derived from the single-stranded phage M13[26,55] or fd.[19] The application of these phage

[48] S. Gillam and M. Smith, this series, Vol. 65 [65].
[49] D. A. Hinton, C. A. Brennan, and R. I. Gumport, *Nucleic Acids Res.* **10**, 1877 (1982).
[50] E. Jay, R. Bambara, R. Padmanbhan, and R. Wu, *Nucleic Acids Res.* **1**, 331 (1973).
[51] A. Maxam and W. Gilbert, this series, Vol. 65 [57].
[52] A. M. Banuszuk, K. V. Deugau, and B. R. Glick, unpublished. Biologicals Incorporated, Toronto.
[53] G. Chacoras and J. H. van de Sande, this series, Vol. 65 [10].
[54] R. Roychoudhury and R. Wu, this series, Vol. 65 [7].
[55] J. C. Hines and D. Ray, *Gene* **11**, 207 (1980).

vectors to oligonucleotide-directed site-specific mutagenesis provides for simple and rapid isolation of single-stranded template DNA and makes screening for the mutant clone easy. We have used the M13 vectors developed by Messing and co-workers.[26] These contain a number of useful cloning sites, usually exhibit the "blue to white" plaque coloration upon insertion of a DNA fragment, and can be used in conjunction with a number of commercially available sequencing primers. The size of the insert is for the most part one of convenience. We have conducted successful mutagenesis experiments on inserts ranging from 200 bases to 4.2 kilobases.

The basic strategy for cloning a fragment into an M13 vector (e.g., M13mp8) is depicted in Fig. 3. For mutagenesis of *MATa1*, a 4.2 kb *Hin*dIII fragment containing the entire *MATa* gene was inserted into the *Hin*dIII site of the vector M13mp5. We have also conducted experiments using M13mp7, M13mp8, and M13mp9 (unpublished). For a detailed discussion of M13 cloning see Messing.[26] A number of white plaques were picked, and single-stranded DNA was prepared from 1-ml cultures. The orientation of the insert was determined by sequencing a number of clones into the gene from the *Hin*dIII cloning site using an universal M13 sequencing primer. The sequence of the 4.2 kb *MATa* fragment near each *Hin*dIII site had been previously determined by C. Astell in our laboratory (unpublished results). One clone with the desired orientation was chosen and used as a source of single-stranded template DNA for the mutagenesis experiment.

A single mutagenesis experiment requires only 1–2 μg of single-stranded template DNA. Thus, a 2–3 ml culture, yielding 6–15 μg of single-stranded DNA,[28] should provide enough template DNA to carry out preliminary *in vitro* tests and the actual mutagenesis experiment.

Preliminary *in Vitro* Tests To Determine Specific Targeting by the Mutagenic Oligonucleotide

Once the oligonucleotide has been synthesized and sequenced and the template DNA has been isolated, it is important to carry out a number of preliminary experiments. The purpose of these tests is to demonstrate that the oligomer efficiently primes from the desired site. In addition, priming from this site is optimized by testing several primer:template ratios and priming temperatures. Two experiments can be done:

Primer Extension and Restriction Endonuclease Digestion. In this experiment, $5'$-^{32}P-labeled oligonucleotide is annealed with the template, then extended by addition of deoxyribonucleotide triphosphates and DNA polymerase (large fragment). After extension, duplex DNA is cleaved with a restriction endonuclease that recognizes a site downstream from the desired priming site. The resulting fragments are denatured and electrophoresed on a polyacrylamide gel under denaturing conditions; the

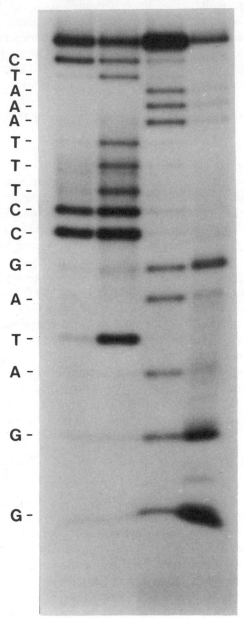
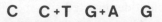

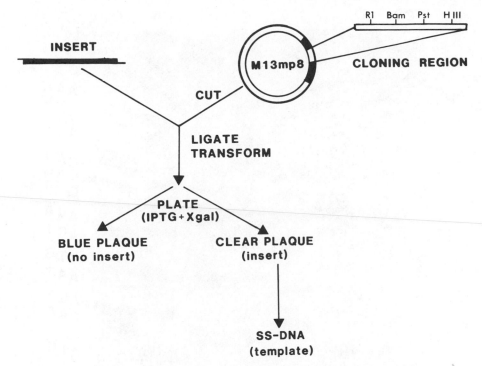

FIG. 3. General strategy for cloning a DNA fragment in the vector M13mp8.[26]

gel is autoradiographed. The resulting autoradiogram will indicate how many priming sites exist and which one is the major site.

Chain Terminator Sequencing. This test utilizes the mutagenic oligonucleotide as a primer in a chain-terminator sequencing reaction.[56,57] The resulting sequence is compared with sequences downstream from the priming site to determine whether the desired site is recognized. The

[56] F. Sanger, A. R. Coulson, B. G. Barrell, A. J. H. Smith, and B. A. Roe, *J. Mol. Biol.* **143**, 161 (1981).
[57] F. Sanger, S. Nicklen, and A. R. Coulson, *Proc. Natl. Acad. Sci. U.S.A.* **74**, 5463 (1977).

FIG. 2. DNA sequence determination of the oligonucleotide MS-1. Autoradiograph of a 20% polyacrylamide gel. A 5′-³²P-labeled oligonucleotide (500,000 cpm, 20 pmol) was sequenced according to the procedure of Maxam and Gilbert, modified for short oligonucleotides.[51,52] 10,000 ³²P cpm from the "C" and "G" reactions and 20,000 ³²P cpm from the "C + T" and "A + G" reactions were loaded into their respective lanes. Electrophoresis was carried out at 1400 V until the bromophenol blue dye migrated half way down the gel.

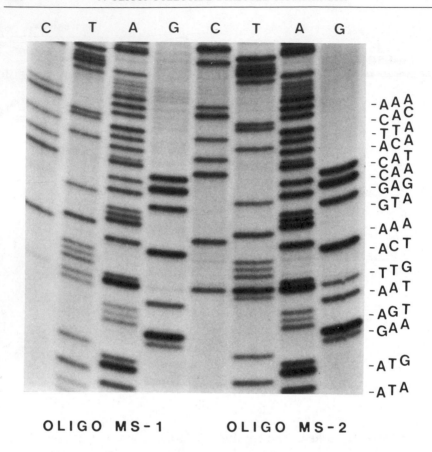

OLIGO MS-1   OLIGO MS-2

FIG. 4. Preliminary *in vitro* test for specific priming by the mutagenic oligonucleotide. Autoradiograph of a chain-terminator sequencing gel. Single-stranded M13 recombinant DNA containing the noncoding strand of the *MATa1* gene was subjected to chain-terminator sequencing using each mutagenic oligonucleotide as a primer.[56,57] In each reaction, the primer:template ratio was 20:1 and enzymic primer extension was carried out at 23°. The sequence corresponds to the region of the *MATa1* gene adjacent to the desired priming site. The clarity of the pattern indicates the high specificity of priming.

specificity of priming can be assessed by the clarity of the sequencing ladder and the absence of a background pattern.

These experiments should be done using several priming temperatures and primer:template ratios. Figure 4 shows an example of the use of MS-1 and MS-2 as sequencing primers. The template:oligomer ratio was 1:20. Annealing was carried out by heating the mixture to 55° for 5 min, then cooling to room temperature. Extension was carried out at room temperature (23°) as described by Sanger *et al.*[56,57] These are the conditions that generally apply to oligonucleotides from 14 to 18 long directing a one- or two-base change. The specific conditions for each experiment should be determined to ensure success.

The sequence in Fig. 4 corresponds to the sequence in *MATa1* approximately 30 bases downstream from the UGA.[36] The clarity of the pattern demonstrates that priming occurred specifically at the desired site.

Procedure for Oligonucleotide-Directed Mutagenesis

Figure 5 shows the general scheme for oligonucleotide-directed site-specific mutagenesis. The mutagenic oligonucleotide and single-stranded template DNA are annealed and extended by DNA polymerase (large fragment) using the conditions determined in the preliminary tests. The newly synthesized strand is radioactively labeled by incorporation of [α-^{32}P]dATP during the initial period of extension. The labeled DNA can thus be followed in subsequent steps. Extension continues at 15° for 12–20 hr producing a population of covalently closed molecules. During this period, the primer is extended around the circular template, and the two ends of the newly synthesized strand are ligated by T4 DNA ligase. We have observed that the conversion of single-stranded circular molecules into closed double-stranded molecules is incomplete. Closed circular DNA (CC-DNA) is separated from unligated and incompletely extended molecules by alkaline sucrose gradient centrifugation. The CC-DNA fraction is pooled and used to transform CaCl$_2$-treated JM101 cells. Phage are prepared from 36 individual plaques and are screened for the desired mutant.

A major problem in oligonucleotide-directed mutagenesis has been the background of wild-type molecules resulting from inefficient conversion of single-stranded template DNA to CC-DNA. This increased the number of recombinants that must be screened to obtain a mutant. The work of Gillam and Smith demonstrated the substantial increase in mutagenesis efficiency by enrichment for closed circular molecules by treatment with S1 nuclease.[39] Other useful methods include binding single-stranded DNA to nitrocellulose,[14] agarose gel electrophoresis,[20] CsCl density gradient centrifugation,[21] and alkaline sucrose gradient centrifugation.[22] Alkaline sucrose gradient centrifugation is utilized in the present method because it effectively enriches for CC-DNA, it provides a diagnostic indication of the efficiency and kinetics of the extension and ligation reactions, and it is an exceedingly simple procedure.

Figure 6 shows the gradient profiles from two experiments. After extension and ligation, the DNA was precipitated by addition of a polyethylene glycol (PEG)–NaCl solution to remove ^{32}P-labeled triphosphate, then

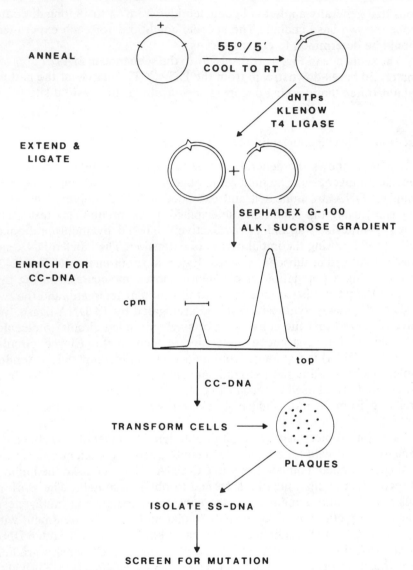

FIG. 5. General scheme for oligonucleotide-directed mutagenesis. CC-DNA, closed circular DNA; SS-DNA, single-stranded DNA.

layered onto a 5 to 20% alkaline sucrose gradient. Centrifugation was carried out at 37,000 rpm for 2 hr using a SW 50.1 rotor. Aliquots were collected by puncturing the bottom of the tube. ^{32}P cpm in each fraction was determined by scintillation counting (Cerenkov). In the first experiment, extension and ligation proceeded for 5 hr at 15°. No CC-DNA was

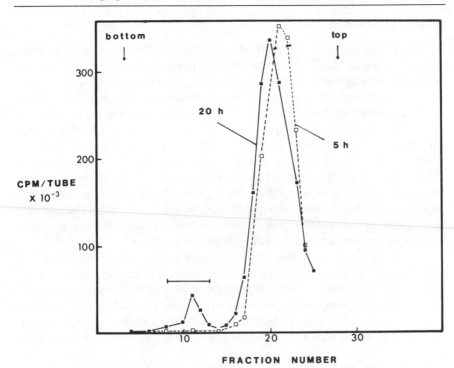

FIG. 6. Profile of ^{32}P-labeled DNA molecules separated by alkaline sucrose gradient centrifugation. After *in vitro* synthesis of closed circular DNA molecules, the sample was denatured with NaOH and centrifuged through a 5 to 20% alkaline sucrose gradient. Centrifugation was carried out at 37,000 rpm for 2 hr at 4° in an SW50.1 rotor. Fractions (175 μl) were collected by puncturing the bottom of the tube, and ^{32}P-cpm in each fraction was measured by scintillation counting. Extension and ligation proceeded at 15° for 5 hr (----) and 20 hr (——).

observed. In the second experiment, the reaction proceeded for 20 hr at 15°. In contrast to the first experiment, a peak of radioactivity representing CC-DNA was observed in the gradient from the second experiment. The CC-DNA fraction (tubes 3–8) was pooled, neutralized with 1 *M* Tris-citrate (pH 5), and used to transform CaCl₂-treated JM101 cells.

Transformation and Phage Isolation

Aliquots of neutralized CC-DNA are used to transform CaCl₂-treated *E. coli* JM101 cells as described by Sanger *et al.*[56] Single-stranded phage are isolated from 36 individual plaques and subjected to the appropriate screening procedure.

Screening for the Mutant

The CC-DNA obtained from the sucrose gradient is a heteroduplex consisting of one wild-type strand and one mutant strand. Transformation with this DNA results in a mixture of wild-type and mutant phage. These are distinguished from each other by a selection or screening procedure. Biological selection has been utilized in a number of reports. However, to create mutations in any fragment of DNA, a screening procedure that does not rely on a biological method is required. Three screening procedures are presented; two of them require isolation of single-stranded DNA, and the third is carried out using an aliquot of phage containing supernatant from a 1-ml culture. The choice of which method to use is dependent on the location of the mutation with respect to the cloning site, the availability of another sequencing primer, or whether the mutation deletes or produces a restriction endonuclease site.

Single-Track Sequencing. Single-stranded DNA from each clone is subjected to single-channel chain-terminator sequencing using the dideoxynucleotide that would differentiate wild-type from mutant clones.[56] A universal M13 sequencing primer is used in the case where the mutation is close to the cloning site. Otherwise, another synthetic oligonucleotide or a restriction fragment that primes adjacent to the mutation site can be used.

Restriction Site Screening. This method can be used if the desired mutation creates or destroys a restriction endonuclease site. Single-stranded DNA from each clone is annealed to either an M13 sequencing primer or a restriction fragment primer. Enzymic extension is carried out by DNA polymerase in the presence of deoxyribonucleotide triphosphates, one of which is α-^{32}P-labeled. The DNA is cleaved with the appropriate restriction enzyme, and the resulting fragments are separated on a polyacrylamide gel, then visualized by autoradiography. A mutant clone will exhibit a different restriction pattern from the wild type.

Hybridization Screening. The simplest and most versatile method is the hybridization method, in which the labeled mutagenic oligonucleotide is used as a probe for mutant clones. The effect of mismatches on the thermal stability of heteroduplexes formed between short oligonucleotides and poly(dT) or poly(dA) cellulose was investigated by Astell *et al.*[10] and by Gillam *et al.*[11] These studies demonstrated that a mismatch dramatically destabilized a heteroduplex compared with a perfectly complementary duplex. Wallace and co-workers showed that this principle could be utilized in a screening procedure to differentiate mutant from wild-type molecules produced by site-specific mutagenesis.[58,59] The

[58] R. B. Wallace, M. J. Johnson, T. Hirose, T. Miyake, E. H. Kawashima, and K. Itakura, *Nucleic Acids Res.* **9,** 879 (1981).
[59] R. B. Wallace, J. Shaffer, R. F. Murphy, J. Bonner, T. Hirose, and K. Itakura, *Nucleic Acids Res.* **6,** 3543 (1979).

method presented here is independent of the position of the mutation within the cloned fragment, is carried out on an aliquot of intact phage, and identifies mutant phage in a single day. The principle behind this procedure is that the mutagenic oligonucleotide will form a more stable duplex with a mutant clone, with perfect match, than with a wild-type clone bearing a mismatch. The mutant is detected by forming a duplex at a low temperature where both mutant and wild-type DNA interact, then carrying out washes at increasingly higher temperatures until only mutant molecules hybridize. Phage from each clone are spotted directly onto a nitrocellulose filter, then baked at 80° for 2 hr. There is no need to carry out the denaturation steps as in other *in situ* hybridization procedures.[60,61] The probe solution is added to the filter after a prehybridization step. Hybridization proceeds for 1 hr at a temperature at which both mutant and wild-type molecules hybridize with the probe. The hybridization solution is removed, and the filter is washed, then autoradiographed. The same filter is washed again at a higher temperature, then autoradiographed. This process is repeated several times, each time increasing the wash temperature, until only mutant molecules exhibit a hybridization signal.

Comments. In the example with *MATa*, the mutation lies approximately 2.7 kb downstream from the cloning site. This position cannot be sequenced using a standard M13 primer. However, the desired change destroys a restriction site for the enzyme *Hin*fI (GAATC to AAATC). Figure 7 shows an example of 6 out of 36 clones screened by this method. Based on the wild-type sequence[37] the loss of the *Hin*fI site around position 1667 would yield a new fragment of 605 bp (320 + 285 bp) and would result in the loss of the two smaller fragments. Clones 7, 32, and 34 exhibited a new fragment at approximately 650 bp (see arrow). Clone 21 (not shown) also exhibited a 650 bp *Hin*fI fragment.

The hybridization method offers a versatile and simple way to identify mutant clones. Figure 8 shows an example of 36 clones screened by hybridization to ^{32}P-labeled MS-1 oligonucleotide. The four clones that exhibited hybridization signals at 47° corresponded directly with the clones that showed the restriction site alteration. Thus, approximately 11% of the isolated phage were mutant.

Plaque Purification of a Suspected Mutant

We have observed that the suspected mutant clones can contain up to 50% wild-type molecules. In a hybridization screen, these clones exhibit signals of lower intensity compared with signals from homogeneous mutant clones. Heterogeneous mutants arise because transformation and

[60] M. Grunstein and D. S. Hogness, *Proc. Natl. Acad. Sci. U.S.A.* **72**, 3961 (1975).
[61] W. D. Benton and R. W. Davis, *Science* **196**, 180 (1977).

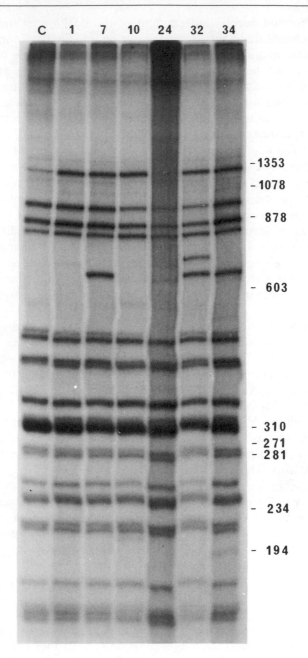

plating were carried out without a cycle of replication and reinfection. Any single infected cell should contain a mixture of wild-type and mutant phage. However, we have observed that phage isolated from a single phage comprise predominantly one type. Nevertheless, it is important to replate the suspected mutant phage and isolate homogeneous mutant phage.

This was done for one of the suspected *MATa1* mutants. Phage from this clone were replated, and single-stranded DNA was isolated from 16 individual plaques, then screened by the restriction site method. In all cases, the new fragment of 650 bp was observed. In other experiments in our laboratory, isolates of phage from a suspected mutant contained 0–50% wild-type molecules determined by dot-blot hybridization analysis (unpublished).

Sequencing the Suspected Mutant

The final step of this procedure is to verify that the desired mutation has been produced. This can be accomplished by chain-terminator sequencing on the single-stranded DNA[56,57] or by the procedure of Maxam and Gilbert using isolated double-stranded DNA.[51]

Double-stranded phage DNA was prepared from one plaque-purified mutant phage[62] and sequenced in the region containing the in-frame UGA codon according to the procedure of Maxam and Gilbert. Figure 9 shows the sequence of mutant compared with the wild-type sequence around this site. The only difference between the two sequences is a C to T transition at the desired site.

MS-2 UGA to UGG Mutant

Construction of this mutant was carried out as described above. The oligonucleotide used to direct the change is shown in Fig. 1 and was synthesized according to the procedure of Edge *et al.*[41] The yield of mutant phage was 16.6%, identified by the restriction enzyme screening method. The DNA sequence within the region of the mutation was deter-

---

[62] D. B. Clewell and D. R. Helinski, *Proc. Natl. Acad. Sci. U.S.A.* **62**, 1159 (1969).

FIG. 7. Screening for mutants by restriction fragment analysis. Autoradiograph of a nondenaturing 5% polyacrylamide gel separating restriction fragments of potential mutants. The autoradiograph shows an example of six recombinants screened by the restriction site method for the loss of a particular *Hinf*I site. Double-stranded ³²P-labeled DNA was synthesized *in vitro* by enzymic primer extension, cleaved with the restriction enzyme *Hinf*I, and electrophoresed. Three mutants are shown that exhibited a new fragment of approximately 650 bp (arrow).

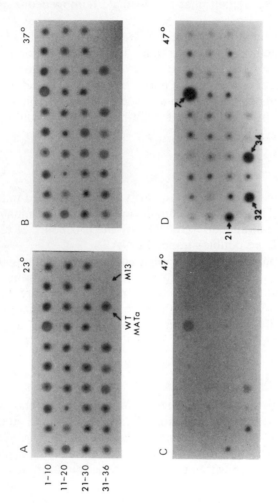

FIG. 8. Dot-blot hybridization screening for mutants using ^{32}P-labeled mutagenic oligonucleotide MS-1. Single-stranded DNA from 36 phage was bound to nitrocellulose and hybridized to ^{32}P-labeled oligomer MS-1. Hybridization was carried out at 23° for 1 hr. The filter was washed in 6 × SSC at 23°, 37°, and 47°, respectively. After each wash, the filter was autoradiographed for 1 hr (panels A–C) or 8 hr (panel D) using Kodak NS-5T film.

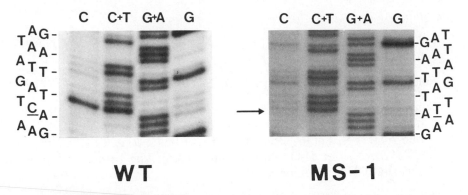

FIG. 9. Comparison of mutant and wild-type (WT) DNA sequences. An autoradiograph of two 12% polyacrylamide gels comparing the sequences of the wild-type and mutant (MS-1) DNA in the region of the *MATa1* gene containing the in-frame UGA codon. A 420 bp *Dde*I–*Bgl*II fragment was 3'-^{32}P-labeled at the *Dde*I site and sequenced by the procedure of Maxam and Gilbert.[51] The only difference between these two sequences is the C to T change in the mutant DNA at the desired position (arrow).

mined from DNA isolated from a plaque-purified phage by the procedure of Maxam and Gilbert.[51] A single T to C transition was observed within this region (data not shown).

Materials and Methods

Escherichia coli JM101 and M13mp5 were the gift of J. Messing (University of Minnesota). *Escherichia coli* DNA polymerase (large fragment) and T4 DNA ligase were obtained from Bethesda Research Laboratories. T4 polynucleotide kinase was purchased from New England BioLabs. [α-^{32}P]ATP were obtained from New England Nuclear. Deoxyribonucleotide triphosphates were purchased from P-L Biochemicals. RiboATP, Trizma-base, and ultrapure sucrose were obtained from Sigma. The octadecanucleotide 5'-AAGGATAGCCTTTAAATC-3' was the gift of A. Markham and T. Atkinson (ICI). Hydrazine (99%) and acrylamide were purchased from Kodak Chemicals. Dimethyl sulfate was obtained from BDH. Formic acid (91%), liquefied phenol, $CaCl_2$, EDTA, and PEG-6000 were purchased from Fisher. Nitrocellulose filters were obtained from Schleicher & Schuell.

The mutagenesis reactions are carried out in 0.5-ml siliconized Eppendorf tubes. Mix by gentle vortex or hand agitation. Additions of 5 μl or less are made directly into the tube using siliconized 5 μl graduated micropipettes. Keep the contents of the tube at the bottom by a short spin (1–2 sec) in an Eppendorf centrifuge. For a convenient siliconizing procedure, see Roychoudhury and Wu.[54]

Solutions

Solution A: 0.2 *M* Tris-HCl (pH 7.5, 23°), 0.1 *M* MgCl$_2$, 0.5 *M* NaCl, 0.01 *M* dithiothreitol (DTT)

Solution B: 0.2 *M* Tris-HCl (pH 7.5, 23°), 0.1 *M* MgCl$_2$, 0.1 *M* DTT

Solution C: 1 μl of solution B, 1 μl of 10 m*M* dCTP, 1 μl of 10 m*M* dTTP, 1 μl of 10 m*M* dGTP, 0.5 μl of 0.1 m*M* dATP, 1 μl of 10 m*M* ribo-ATP, 1.5 μl of [α-^{32}P]dATP (7 μCi/μl), 1.5 μl of T4 DNA ligase (2 U/μl), 2 μl of H$_2$O. Prepare in a 0.5-ml vial before adding to annealed DNA. Keep on ice.

Alkaline sucrose gradient stock solutions: *X*% sucrose, 1 *M* NaCl, 0.2 *M* NaOH, 2 m*M* EDTA; *X* = 20, 17.5, 15, 10, and 5. Autoclave, store at 4°.

Sucrose–dye–EDTA solution (for nondenaturing gels): 60% sucrose, 0.02% (w/v) bromophenol blue, 0.02% (w/v) xylene cyanole FF, 0.025 *M* EDTA

Formamide–dye–EDTA solution (for denaturing gels): 0.02% bromophenol blue, 0.02% xylene cyanole FF, 0.025 *M* EDTA; in 90% formamide (deionized)

20× SSC (concentrated stock solution): 3 *M* NaCl, 0.3 *M* sodium citrate, 0.01 *M* EDTA. Adjust pH to 7.2 with HCl.

100× Denhardt's (concentrated stock solution): 2% bovine serum albumin, 2% poly(vinylpyrrolidone), 2% Ficoll

Media

Supplemented M9: Prepare the following solutions:
1. 10× M9 salts (per liter): Na$_2$HPO$_4$ · 2 H$_2$O, 70 g, KH$_2$PO$_4$, 30 g, NaCl, 5 g, NH$_4$Cl, 10 g.
2. 20% glucose
3. 10 m*M* CaCl$_2$
4. 1 *M* MgSO$_4$
5. 0.1% thiamin (filter sterilize)
Prepare 50 ml of 1× M9 salts in a 200-ml Ehrlenmeyer flask using the 10× stock solution. Autoclave. After solution has cooled add: 0.5 ml of 20% glucose, 0.5 ml of 10 m*M* CaCl$_2$, 50 μl of 1 *M* MgSO$_4$, 50 μl of 0.1% thiamin.

YT medium: Per liter: 8 g of tryptone, 5 g of yeast extract, 5 g of NaCl. For YT plates add 15 g of agar per liter; for YT top agar add 8 g of agar per liter.

2× YT medium: Per liter: 16 g of tryptone, 10 g of yeast extract, 5 g of NaCl

Schedule for a Typical Mutagenesis Experiment

Day 1
  5′ phosphorylate oligonucleotide (procedure IA)
  Start *in vitro* mutagenesis reaction (procedure IV)
  Prepare CaCl$_2$ treated JM101 cells (procedure VA)
  Prepare alkaline sucrose step gradient (procedure IVD)
Day 2
  PEG–NaCl precipitate DNA (procedure IVC)
  Alkaline sucrose gradient centrifugation (procedure IVD)
  Transform JM101 with CC-DNA (procedure VB)
  Start an overnight culture of JM101 (procedure VA,1)
Day 3
  Prepare phage from 36 individual plaques (procedure VI)
Day 4
  Prepare hybridization probe (procedure IB)
  Screen by dot-blot hybridization (procedure VIIC)
  Plate out suspected mutant phage to purify
Day 5
  Prepare phage from 6 individual plaques of suspected mutants (procedure VI)
Day 6
  Screen by dot-blot hybridization (procedure VIIC)

Procedures

Procedure I. 5′ Phosphorylation of the oligonucleotide[53]
  The oligonucleotide should be lyophilized after synthesis, then resuspended in water. A convenient working concentration is 10–20 pmol/μl.
  A. For mutagenesis
    1. Add to a 0.5-ml siliconized Eppendorf tube: 200 pmol oligonucleotide (dry down if necessary), 3 μl of 1 M Tris-HCl (pH 8), 1.5 μl of 0.2 M MgCl$_2$, 1.5 μl of 0.1 M DTT, 3 μl of 1 mM riboATP, H$_2$O to 30 μl total volume.
    2. Mix

3. Add 4.5 units of T4 polynucleotide kinase.
4. Mix and incubate at 37° for 45 min.
5. Stop by heating at 65° for 10 min.

The oligonucleotide must bear a 5'-phosphate group in order to ligate. An aliquot from the reaction vial is used directly for the mutagenesis reaction. To check that the kinase reaction has gone to completion, end label and chromatograph an aliquot as discussed below. This preparation of kinased oligonucleotide can serve as a stock for a number of experiments. Store at −20°.

B. For hybridization screening

1. Dry down 20 μCi of [γ-^{32}P]ATP (2000 Ci/mmol) in a 0.5 ml Eppendorf tube.
2. Add 20 pmol of oligonucleotide, 3 μl of 1 M Tris-HCl (pH 8), 1.5 μl of 0.2 M MgCl$_2$, 1.5 μl of 0.1 M DTT, H$_2$O to 30 μl total volume.
3. Follow steps 2–5 as in A.
4. Chromatograph on Sephadex G-25 to remove unincorporated [α-^{32}P]dATP. (Alternatively, see Wallace et al.[17])

The column (6 × 200 mm) should be made of either plastic or siliconized glass. Chromatograph the sample in 50 mM ammonium bicarbonate (pH 7.8). Collect 100-μl aliquots in 1.5-ml Eppendorf tubes, and measure the radioactivity in each fraction by scintillation counting (Cerenkov). The first radioactive peak contains the labeled oligonucleotide. Pool the major fractions.

The purity of the ^{32}P-labeled oligonucleotide can be checked by chromatographing an aliquot of the reaction mixture on a strip of Whatman DE-81 paper using 0.3 M ammonium formate (pH 8). The oligonucleotide remains at the origin and [γ-^{32}P]ATP migrates near the solvent front. This analysis can be used to assess either reaction A or B. To check reaction A, add 10 μCi of [γ-^{32}P]ATP to the reaction solution, and chromatograph 1 μl.

Incorporation can be measured qualitatively using a hand-held Geiger counter or quantitatively by cutting out the origin and counting it in a scintillation counter. Generally, procedure IB results in the incorporation of 2 to 6 × 10^6 ^{32}P cpm (Cerenkov) per 20 pmol of oligonucleotide. The 5'-^{32}P-labeled oligonucleotide from reaction IB is also used for sequencing (procedure II) and in the preliminary tests (procedure IVA).

Procedure II. Sequence determination of the mutagenic oligonucleotide.

End-labeled oligonucleotide is subjected to a modified Maxam and Gilbert procedure.[51,52]

1. Dry down 5×10^5 cpm ^{32}P-labeled oligonucleotide in a siliconized Eppendorf tube.
2. Resuspend in 32 μl of H_2O.
3. Add 5 μl to the "C" and "G" tubes. Add 10 μl to the "C + T" and "A + G" tubes.
4. Add 2 μl of 1 mg/ml denatured calf thymus DNA to each tube.
5. Add 20 μl of 5 M NaCl to the "C" tube, 15 μl of H_2O to the "C + T" tube, 10 μl of H_2O to the "A + G" tube, 300 μl of cacodylate buffer to the "G" tube.[51]
6. Mix.
7. Add 30 μl of hydrazine to the "C" and "C + T" tubes. Mix and incubate at 45° for 20 min and 5 min, respectively.
8. Add 3 μl of 50% (v/v) formic acid to the "A + G" tube. Mix and incubate at 37° for 20 min.
9. Add 2 μl of dimethyl sulfate to the "G" tube. Mix and incubate at 37° for 3 min.
10. Stop all reactions as described.[51]
11. Carry out subsequent precipitation and piperidine cleavage steps as described.
12. Load 10,000 cpm (Cerenkov) into the "C" and "G" slots, and 20,000 cpm into the "C + T" and "A + G" slots.
13. Electrophorese the samples on a 25% polyacrylamide–7 M urea gel as described[51] until the bromophenol blue is approximately half way down the gel (~2 hr).
14. Autoradiograph for 20 hr using Kodak NS-5T film at −20°.

The standard ethanol precipitation steps following the modification reactions inefficiently precipitate the oligonucleotide, resulting in a significant loss of ^{32}P cpm. This has been taken into account in the present procedure by using more ^{32}P-labeled oligonucleotide than usually is required in a standard sequencing reaction.

Procedure III. Preliminary *in vitro* tests for specific priming
  A. Primer extension
    1. Add to a siliconized 0.5-ml Eppendorf tube: M13 recombinant DNA (0.5 pmol); 5′-end-labeled oligonucleotide (10–30× molar excess over template), specific activity = 2×10^4 cpm/pmol; 1 μl of solution A; H_2O to 10 μl.
    2. Heat in a water bath at 55° for 5 min.
    3. Remove tube from bath and let stand at room temperature (23°) for 5 min.
    4. Add to the annealed DNA: 1 μl of 2.5 mM dCTP, 1 μl of 2.5 mM

dATP, 1 μl of 2.5 m*M* dTTP, 1 μl of 2.5 m*M* dGTP, 0.5 μl of solution A.

5. Mix.
6. Add 1 unit of DNA polymerase (large fragment).
7. Mix and incubate at room temperature for 5 min.
8. Stop reaction by heating at 65° for 10 min.
9. Cool to room temperature.
10. Adjust buffer for appropriate restriction enzyme.
11. Add 1–5 units of enzyme.
12. Incubate at appropriate temperature for 1 hr.
13. Add equal volume of formamide–dye solution.
14. Boil for 3 min; immediately chill on ice.
15. Electrophorese the sample on a 5% polyacrylamide gel under denaturing conditions.[14,40]
16. Autoradiograph.

B. Chain-terminator sequencing. Follow the procedure by Sanger *et al.* for chain-terminator sequencing using M13.[56,57] Substitute the mutagenic oligonucleotide for a universal M13 sequencing primer.

1. Add to a siliconized tube: M13-recombinant DNA (0.1–0.3 pmol) oligonucleotide (10–30× over template DNA), 1 μl of solution A, H_2O to 10 μl.
2. Mix and seal in a glass capillary tube.
3. Heat at 55° for 5 min; cool to room temperature.
4. Follow the procedure of Sanger *et al.*[56]

Use the conditions outlined above for oligonucleotides between 14 and 18 long that bear one or two mismatches. For each experiment, examine a number of different primer : template ratios and priming temperatures.

Procedure IV. Mutagenesis

Steps A–D are carried out in a single siliconized 0.5-ml Eppendorf tube.

A. Annealing

1. Add to a 0.5-ml siliconized Eppendorf tube: M13-recombinant DNA (0.5–1.0 pmol), 5′-phosphorylated oligonucleotide (10–30× molar excess over template DNA), 1 μl of solution A, H_2O to 9 μl total volume.
2. Mix and heat tube in a water bath at 55° for 5 min.
3. Remove tube from bath and let stand at room temperature (23°) for 5 min.

The phosphorylated oligonucleotide is taken directly from the reaction vial in procedure IA. The above conditions generally apply when producing point mutations using oligomers 14–18 long. For shorter oligonucleotides, heat at 55°, then chill to 0–10°. The specific conditions should be determined in the preliminary *in vitro* tests (procedure III).

B. Extension and ligation
1. Prepare solution C in a separate vial. Keep on ice.
2. Add 10 μl of solution C to the annealed DNA.
3. Mix, then add 2.5 units of DNA polymerase (large fragment).
4. Mix and incubate for 5 min at room temperature.
5. Add 1 μl of 10 mM dATP.
6. Mix, then incubate at 15° for 12–20 hr.

For shorter oligonucleotides, carry out the initial extension reaction at 0–10°, then continue at 15° as outlined above.

C. PEG–NaCl precipitation. This step is carried out to remove unincorporated [α-^{32}P]dATP from the sample. This allows subsequent determination of the percentage of CC-DNA produced.
1. Add to the reaction tube: 30 μl of H_2O, 50 μl of 1.6 M NaCl/13% PEG-6000.
2. Mix and incubate on ice for 15 min.
3. Spin in an Eppendorf centrifuge at room temperature for 5 min.
4. Withdraw all of the solution using a 50-μl glass micropipette.
5. Add 100 μl of cold 0.8 M NaCl–6.5% PEG-6000 to rinse the sides of the tube. Do not resuspend pellet at this point.
6. Spin for 30 sec.
7. Withdraw solution as in 4.
8. Resuspend in 180 μl of 10 mM Tris-HCl (pH 8)–1 mM EDTA.
9. Determine total incorporation of [^{32}P]ATP by scintillation counting (Cerenkov).

D. Alkaline sucrose gradient centrifugation. Prepare one 5 to 20% sucrose gradient for each sample in a 0.5-inch × 2-inch centrifugation tube. This is used with a SW 50.1 rotor. Make a step gradient with 1 ml each of the sucrose stock solutions, starting with the 20% solution. Carefully let the solution run down the side of the tube using either a serological or a Pasteur pipette. Hold the tip close to the top of the preceding layer to avoid mixing. After adding all five solutions, ensure that four schlieren bands are sharply defined. Let the tube sit at 4° for 6–16 hr to linearize the gradient. Alternatively, leave the gradient at room temperature for 2 hr, then at 4° for another 2 hr. After the gradient has formed, prepare the sample procedure from IVC.

1. Add 20 μl of 2 N NaOH to the resuspended sample.
2. Mix and incubate at room temperature for 5 min.
3. Chill on ice for 1 min.
4. Place the gradient tube into the SW 50.1 bucket.

5. Apply the sample to the top of the gradient, using a disposable micropipette.

If the top of the gradient solution is too close to the top edge of the centrifugation tube before adding the sample, withdraw 50–100 μl from the top with a Pipetman. The final height of the solution should be within 2–3 mm of the tube edge.

Centrifugation is carried out using a SW 50.1 rotor at 37,000 rpm for 2 hr at 4° with the brake off. After centrifugation, collect aliquots into 1.5-ml Eppendorf tubes by puncturing the bottom of the gradient tube. An apparatus such as the Hoeffer tube holder works well. Collect about 30 fractions (5–7 drops per fraction). Determine the amount of ^{32}P cpm in each fraction by scintillation counting (Cerenkov). The CC-DNA (double stranded) is in the bottom half of the gradient, whereas the single-stranded circular and linear molecules are in the top half. Pool the CC-DNA fraction and neutralize the solution with 1 M Tris-citrate (pH 5). Add about 50 μl of Tris-citrate per 300 μl. Check that the pH is between 7 and 8 by spotting a 2-μl aliquot onto pH paper. Measure the amount of ^{32}P-labeled DNA in the pooled CC-DNA fraction and calculate the percentage of CC-DNA formed.

Procedure V. Transformation
  A. Prepare CaCl$_2$-treated *E. coli* JM101 cells.[63]

1. Inoculate 5 ml of supplemented M9 medium with a single colony of JM101 from a minimal plate (+ glucose). Incubate with shaking at 37° overnight.
2. Add 0.5 ml of overnight culture to 50 ml of supplemented M9. Incubate with shaking at 37° to an $A_{600} = 0.3$.
3. Harvest cells in two 40-ml centrifuge tubes at 5000 rpm for 8 min.
4. Resuspend cells in 25 ml ice cold 50 mM CaCl$_2$. Keep on ice for 20 min.
5. Combine cells into one tube and harvest as in 3.
6. Resuspend in 5 ml cold 50 mM CaCl$_2$.

Cells are stored on ice in the cold room and can be used up to 5 days later. However, transformation efficiency increases during the first 24 hr and thereafter gradually decreases.[63]

  B. Transformation of CaCl$_2$-treated cells with CC-DNA[26,27,56]

1. Pipette 0.2 ml of CaCl$_2$-treated JM101 cells into culture tubes (13 × 100 mm). Keep on ice.
2. Add 1, 2, 5, and 10 μl neutralized CC-DNA to individual tubes.

[63] M. Dagert and S. D. Ehrlich, *Gene* **6**, 23 (1979).

Prepare three additional tubes for controls. To one tube add no DNA, to the other two add 1 and 0.1 ng of double-stranded M13 vector, respectively.

3. Incubate tubes on ice for at least 40 min.
4. Heat tubes in a water bath at 45° for 2 min.
5. Add 2.5 ml of molten YT top agar to each tube, vortex to mix, and pour onto a YT plate.
6. Allow agar to harden at room temperature for 15 min, then incubate plates at 37° overnight. Plaques appear in 6–9 hr.

The yield of phage plaques range from 10–200 plaques per microliter of CC-DNA solution depending on the age of the CaCl$_2$-treated cells and the percentage of CC-DNA obtained. The controls should result in approximately 1000 and 100 plaques, respectively. In the event that no plaques are obtained from the CC-DNA fraction, dialyze the pooled DNA against two changes of 1 liter of 2 mM Tris-HCl (pH 8)–0.01 mM EDTA, and transform again with the same volumes of DNA solution suggested in step 2. If this does not result in the appearance of phage plaques, then concentrate the sample 10-fold by extraction with 1-butanol or lyophilization and transform again.

Procedure VI. Isolation of phage.[26,27,56]

Phage-infected cells exhibit a clear "plaque-like" appearance because they grow much slower than the uninfected lawn cells. Prepare phage cultures from 36 individual plaques by the following procedure.

1. Add 0.1 ml of overnight JM101 cells to 5 ml of 2× YT media.
2. Incubate with shaking for 2 hr at 37°.
3. Add this 5-ml culture to 45 ml of 2× YT.
4. Mix and distribute 1 ml to each of 36 culture tubes (13 × 100 mm).
5. Core each phage plaque from the plate with a 50-μl disposable micropipette and blow the agar plug into the culture tube.
6. Incubate tubes with vigorous shaking at 37° for 5 hr. Longer incubation is not recommended.
7. Pour the contents of each tube into 1.5-ml Eppendorf tubes. It is not necessary to remove all the solution.
8. Spin in an Eppendorf centrifuge for 5 min.
9. Pour phage-containing supernatant into fresh 1.5-ml Eppendorf tubes.
10. Add 200 μl of 20% PEG-6000–2.5 M NaCl to each tube.
11. Vortex, and let stand at room temperature for 15 min.
12. Spin in an Eppendorf centrifuge for 5 min at room temperature.
13. Carefully pour off the supernatant.

14. Spin each tube again for 1–2 sec and remove all residual traces of PEG solution with a disposable micropipette.
15. Resuspend the phage pellet in 100 μl of 10 mM Tris-HCl (pH 8)– 0.1 mM EDTA.
16. For hybridization screening, stop here and proceed to procedure VII. For sequencing or restriction site screening, continue through step 28.
17. Add 50 μl of Tris-saturated phenol.
18. Vortex for 10 sec, then spin in an Eppendorf centrifuge for 1 min.
19. Remove aqueous (top) layer using a Pipetman or disposable micropipette into a fresh 1.5-ml Eppendorf tube.
20. Extract 2× with 500 μl of diethyl ether.
21. Add 10 μl of 3 M sodium acetate (pH 5.5) and 300 μl of ethanol.
22. Precipitate DNA by chilling tubes at −70° for 1 hr or at −20° overnight.
23. Spin tubes in an Eppendorf centrifuge for 5 min.
24. Carefully remove the supernatant.
25. Rinse tube with cold ethanol. Do not disturb the pellet.
26. Spin for 2 min in an Eppendorf centrifuge. Remove the supernatant.
27. Dry under vacuum.
28. Resuspend DNA in 50 μl of 10 mM Tris-HCl (pH 8)–0.1 mM EDTA.

Procedure VII. Screening procedures

Determine which procedure is suitable for the particular mutation. Use method A if the mutation can be detected by sequencing with an M13 sequencing primer, another synthetic oligonucleotide, or a restriction fragment. Use method B in the case where the mutation creates or destroys a restriction site. The hybridization screen, method C, is the simplest and most versatile procedure.

A. Chain-terminator sequencing.[56,57] Use 5 μl M13-recombinant DNA (from procedure VI,28) as template with only one dideoxyribonucleotide in a chain terminator reaction.[57] The primer can be a universal M13 primer, another synthetic oligonucleotide, or a restriction fragment. Suspected mutants are sequenced again using all four dideoxynucleotide mixtures.

B. Restriction site screening

1. Add to a siliconized Eppendorf tube: 5 μl of M13-recombinant DNA (from procedure VI,28); 1 μl of solution A; M13 universal sequencing primer (1.5 pmol); H_2O to 10 μl.

2. Mix, heat at 55° for 5 min, then cool to room temperature. If a restriction fragment is used as a primer, heat the mixture for 3 min at 100°, hybridize at 67° for 30 min, then cool to room temperature.
3. Add: 1 μl of 0.5 mM dCTP, 1 μl of 0.5 mM dTTP, 1 μl of 0.5 mM dGTP, 0.5 μl of 0.1 mM dATP, 5 μCi of [α-^{32}P]dATP, 0.5 μl of solution A.
4. Mix.
5. Add 1 unit of DNA polymerase (large fragment).
6. Mix and incubate for 10 min at room temperature.
7. Add 1 μl of 2.5 mM dNTPs.
8. Mix and incubate for 5 min at room temperature.
9. Heat at 65° for 10 min, then cool to room temperature.
10. Adjust the buffer for desired restriction enzyme.
11. Add 1–5 units of enzyme.
12. Mix and incubate for 1 hr at appropriate temperature.
13. Stop by addition of 5 μl of sucrose–dye–EDTA solution.
14. Load the entire sample onto a 5% nondenaturing polyacrylamide gel with 1.5-mm spacers.[51]
15. Electrophorese at 200 V for 16 hr.
16. Autoradiograph.

C. Dot-blot hybridization

1. Spot 2 μl of phage (from procedure VI,15) onto a dry sheet of nitrocellulose.
2. Let air-dry, then bake *in vacuo* at 80° for 2 hr.
3. Prehybridize with 6× SSC + 10× Denhardt's + 0.2% SDS at 67° for 1 hr. This can be done in a sealable bag or in a closed petri dish. Use 10 ml/100 cm^2.
4. Remove the prehybridization solution and rinse the filter in 50 ml of 6× SSC for 1 min at 23°.
5. Place the filter into a sealable bag, add probe, and seal. The probe is prepared according to procedure IB; 10^6 cpm (Cerenkov) are added to 6× SSC + 10× Denhardt's solution (without SDS). Use 4 ml of probe solution/100 cm^2 filter.
6. Incubate at room temperature (23°) for 1 hr.
7. Remove filter and wash in 6× SSC at room temperature. Use 3 × 50 ml for a total of 10 min.
8. Autoradiograph for 1 hr using Kodak NS-5T film at 23°.
9. Wash the filter at a higher temperature with 50 ml of 6× SSC for 5 min. Preheat the wash solution to the desired temperature in a glass dish in a water bath.

10. Autoradiograph as in step 8.
11. Repeat steps 9 and 10, using several wash temperatures.

This procedure can also be carried out using DNA instead of phage. Spot the DNA (1 μl from procedure VI,28) onto a filter and carry out the procedure as outlined above with the exception of steps 3 and 4. Instead, bake the filter, then prehybridize with 5 ml of 6× SSC + 10× Denhardt's solution for 15 min at room temperature. Remove this solution and proceed with step 5. Another variation of the dot-blot method uses an aliquot of the phage supernatant from a 1 ml of culture prior to PEG precipitation of the phage. This requires a plexiglass "dot-blot" apparatus such as the unit marketed by Bethesda Research Laboratories. The device consists of a vacuum manifold with 96 sample wells through which DNA solutions of relatively large volumes are applied to a single nitrocellulose filter. The size of the resulting "dot" is small despite the large sample volume. Fifty microliters of phage containing supernatant from procedure VI,9 are added to individual wells and washed twice with 100 μl of 10 mM Tris-HCl (pH 8). The filter is removed from the manifold and treated as described above for phage. This procedure eliminates the PEG–NaCl precipitation step. The remainder of each culture is saved and used for isolation of plaque-purified mutant phage.

The choice of temperatures at which to wash is dependent on the length and particular composition of the oligonucleotide. In addition, the specific mutation constructed will contribute to the ability to discriminate between wild-type and mutant molecules. In the construction of single or double point mutations, increment the wash temperature by 5–10° depending on the length of the oligonucleotides. For other examples, see references cited in footnotes 23, 58, and 59.

Acknowledgments

The authors would like to acknowledge helpful discussions with Dr. Shirley Gillam and Dr. Greg Winter during the development of these procedures. We would like to thank Dr. Alex Markham and Tom Atkinson for synthesizing the two oligonucleotides used in this work. This investigation was supported by grants awarded to M. S. by the Medical Research Council of Canada. M. S. is an M.R.C. Career Investigator.

[31] Oligonucleotide-Directed Construction of Mutations via Gapped Duplex DNA

By WILFRIED KRAMER and HANS-JOACHIM FRITZ

Introduction

Oligonucleotide-directed construction of mutations has become the method of choice to introduce predetermined structural changes into DNA. The basic concept of the method, as originally outlined by Hutchison and Edgell,[1] rests on annealing a short (synthetic) oligonucleotide to a target region on a single-stranded replicon that is almost, but not entirely, complementary to the target region. A partial DNA duplex with at least one mispaired or unpaired nucleotide results, and the marker carried by the oligonucleotide can be salvaged into complete progeny replicons by a combination of *in vitro* and *in vivo* reactions on the heteroduplex DNA.

Several laboratories have evolved techniques to improve the versatility and efficiency of the process (see Ref. 2 for a review; also see Chap. [32] in this volume and Chaps. [17] and [20] in Vol. 154 of *Methods in Enzymology*).

cleotide-directed mutation construction.[3-6] The key intermediate in this process is a partial DNA duplex of a recombinant M13 genome (gapped duplex DNA, gdDNA) which has only the target region of mutation construction exposed in single-stranded form and which, furthermore, carries distinguishable genetic markers in the two DNA strands in such a way that a rigorous selection can be applied in favor of phage progeny arising from the shorter strand (i.e., the minus strand of the M13 genome). Since the synthetic oligonucleotide becomes physically integrated into the latter DNA strand, an indirect selection for the synthetic marker is applied without the need of the constructed mutation itself being associated with a discernible phenotype. Marker scrambling is suppressed by the use of a transfection host that is deficient in DNA mismatch repair.

Two alternative variants of the gdDNA method are applicable: the "fill-in" protocol (route B, Fig. 1), which incorporates DNA polymerase/

[1] C. A. Hutchison III and M. H. Edgell, *J. Virol.* **8**, 181 (1971).
[2] M. Smith, *Annu. Rev. Genet.* **19**, 423 (1985).
[3] W. Kramer, K. Schughart, and H.-J. Fritz, *Nucleic Acids Res.* **10**, 6475 (1982).
[4] W. Kramer, V. Drutsa, H.-W. Jansen, B. Kramer, M. Pflugfelder, and H.-J. Fritz, *Nucleic Acids Res.* **12**, 9441 (1984).
[5] W. Kramer, A. Ohmayer, and H.-J. Fritz, *Nucleic Acids Res.,* in press (1986).
[6] H.-J. Fritz, J. Hohlmaier, W. Kramer, A. Ohmayer, and J. Wippler, *Nucleic Acids Res.,* in press (1986).

RECOMBINANT DNA
METHODOLOGY

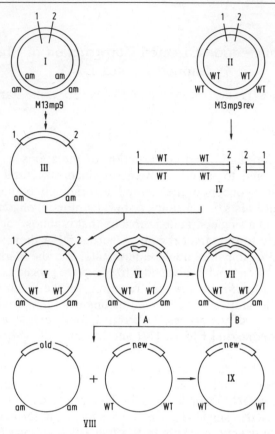

FIG. 1. Schematic representation of the gapped duplex DNA method of mutation construction. For details refer to the section Introduction. Symbols: am, amber codon; WT, corresponding wild-type codon; 1, 2, recognition sites for restriction endonucleases.

DNA ligase reactions *in vitro*, and the simplified "mix–heat–transfect" protocol (route A, Fig. 1), which bypasses these enzymatic manipulations (step 6, see below). The individual steps of the process are as follows (see Fig. 1):

1. Cloning of the target DNA fragment for the mutation construction into phage M13mp9 (I). This phage carries two amber codons in vital phage genes.[7]
2. Preparation of (single-stranded) virion DNA of this recombinant M13mp9 phage (III).
3. Cleavage of (double-stranded) replicative form (RF) DNA of phage M13mp9rev (II) with the same restriction enzyme(s) used for cloning the target DNA into M13mp9 (IV). In M13mp9rev the two

[7] J. Messing, R. Crea, and P. H. Seeburg, *Nucleic Acids Res.* **9,** 309 (1981).

amber codons mentioned under step 1 are replaced by wild-type codons.[4]

4. Construction of gdDNA by *in vitro* DNA/DNA hybridization (V).
5. Annealing of the synthetic oligonucleotide to the gdDNA (VI).
6. Filling in the remaining gaps and sealing the nicks by a simultaneous *in vitro* DNA polymerase/DNA ligase reaction (VII, route B). This step can be omitted without drastic loss of marker yield (route A).
7. Transfection of a host deficient in mismatch repair and production of mixed phage progeny (VIII).
8. Elimination of plus strand progeny by reinfection (at very low multiplicity) of a host unable to suppress amber mutations. The synthetic marker is enriched together with the minus strand progeny (IX).
9. Screening of the clones resulting from the reinfection for the presence of the desired mutation plus DNA sequence verification.

The gapped duplex DNA method combines the following favorable features:

1. Annealing of the synthetic oligonucleotide to inappropriate sites within the recombinant M13 genome is suppressed since only a relatively small window of single-stranded DNA is available for hybridization. This helps to minimize unwanted side reactions such as formation of deletions.
2. Since the entire vector part of the recombinant phage genome is already covered by the complementary minus strand in the gdDNA, the enzymatic fill-in reaction to yield fully double-stranded DNA is greatly facilitated and can be safely left to the natural mechanisms within the transfected cell.
3. As outlined above, an indirect and general selection can be applied in favor of the synthetic marker independent of any phenotype associated with this marker.

Following route B, marker yields of 88% have been achieved with a gap of 120 nucleotides in length, 65% with a larger gap (~1.64 kb).[5] Route A, tested with the larger gap, lead to marker yields of up to 53%.[6] Clearly, these yields are sufficiently high to make screening of a few randomly picked clones by DNA sequence analysis fast and economic. This combines two indispensable steps of analysis into one: identification of the correct mutant and verification of the integrity of the rest of the DNA segment under study. Starting from DNAs II and III (Fig. 1) plus oligonucleotides for mutation construction and DNA sequence analysis (chain termination method[8]), the entire procedure takes about 4–5 days.

[8] A. T. Bankier, K. W. Weston, and B. Barrell, this series, Vol. 155, p. 51.

The marker yields stated above were obtained in model experiments in which large samples of phage progeny were screened to ensure statistically meaningful results. In addition, the method has been used in a sizable number of biological studies in this laboratory and others. Results of these experiments suggest that yields in the range indicated above can generally be expected—possible exceptions being rare cases of special target DNA structure (see Scope, Limitations, and Practical Hints).

Route B, as outlined below, makes use of an enzyme combination[5] consisting of T4 DNA polymerase, *Escherichia coli* DNA ligase, and gp 32, the single-stranded DNA binding protein encoded by gene *32* of phage T4. A similar combination was described by Craik *et al.*[9] Alternatively, *E. coli* DNA polymerase I (large fragment) and T4 DNA ligase can be used as described earlier.[4] The last procedure is slightly less efficient.

Materials

Chemicals and Buffers

Buffer A: 1.5 M KCl, 100 mM Tris–HCl, pH 7.5

Buffer B: 500 mM Tris–HCl, 600 mM ammonium acetate, 50 mM MgCl$_2$, 50 mM dithiothreitol, pH 8.0

100 mM CaCl$_2$, autoclaved

Chloroform/isoamyl alcohol: 24 : 1 (v/v) (analytical grade, Merck)

CsCl (analytical grade), solid

Diethyl ether (analytical grade)

10 (4 × 2.5) mM dNTPs (deoxynucleoside triphosphates): equimolar mixture of dGTP, dATP, dTTP, and dCTP (Boehringer Mannheim), each at a final concentration of 2.5 mM

DOC solution: 50 mM Tris–HCl, pH 8.0, 1% Brij 58 (polyethylene glycol monostearyl ether; Serva), 0.4% sodium deoxycholate, 62.5 mM EDTA

EDTA (ethylenediaminetetraacetate), pH 8, 500 and 250 mM

Ethanol (analytical grade), precooled to −20°

Ethidium bromide solution (10 mg/ml)

2 mM NAD (nicotinamide adenine dinucleotide; Boehringer Mannheim)

Phage buffer: 20 mM sodium phosphate, 0.002% gelatin, pH 7.2; autoclaved

Phenol (analytical grade, Merck), saturated with TE buffer

Phenol/chloroform: 1 : 1 (v/v) mixture of TE-saturated phenol and chloroform/isoamyl alcohol mixture (24 : 1, v/v)

3 M sodium acetate (analytical grade)

---

[9] C. S. Craik, C. Largman, T. Fletcher, S. Roczniak, P. J. Barr, R. Fletterick, and W. J. Rutter, *Science* **228,** 291 (1985).

Solution A: 2.5 M NaCl, 20% polyethylene glycol 6000
Solution B: 25% sucrose (w/v), 50 mM Tris–HCl, pH 8.0
TE buffer: 10 mM Tris–HCl, 0.1 mM EDTA, pH 8.0
TES buffer: 50 mM NaCl, 10 mM Tris–HCl, 1 mM EDTA, pH 8.0
Tetracycline stock solution (2 mg/ml), filter sterilized

Microbiological Growth Media

Antibiotic Medium 3: 17.5 g Antibiotic Medium 3 (Difco; premixed),
1 liter H_2O; autoclaved
EHA plates: 13 g Bactotryptone (Difco), 8 g NaCl, 10 g Bactoagar
(Difco), 2 g sodium citrate·$2H_2O$, 1 liter H_2O; autoclaved; add 7 ml
20% glucose (sterile) after autoclaving; use 25–30 ml per petri dish
(85 mm diameter)
M9 medium: 7 g $Na_2HPO_4 \cdot 2H_2O$, 3 g KH_2PO_4, 1 g NH_4Cl, 1 liter
H_2O; autoclaved; add the following sterile solutions after autoclaving: 25 ml 20% glucose (w/v), 1 ml 100 mM $CaCl_2$, 1 ml 1 M
$MgSO_4$, 5 ml 100 μM $FeCl_3$, and 1 ml thiamin (1 mg/ml)
Top agar: 10 g Bactotryptone (Difco), 5 g NaCl, 6.5 g Bacto-agar
(Difco), 1 liter H_2O; autoclaved

Enzymes

Escherichia coli DNA ligase (New England Biolabs)
T4 DNA polymerase (New England Biolabs)
gp 32 (gene 32 protein from T4-infected E. coli; Pharmacia)
Lysozyme (Sigma)

Bacterial and Phage Strains

BMH 71-18: [Δ(lac–proAB), thi, supE; F', laci^q, ZΔM15, proA$^+$B$^+$]
BMH 71-18 mutS: (BMH 71-18, mutS215::Tn10)
MK 30-3: [Δ(lac–proAB), recA, galE, strA; F', laci^q, ZΔM15,
proA$^+$B$^+$].
M13mp9[10]
M13mp9rev[4]

Procedures

Procedure 1: Preparation of Virion DNA of Recombinant M13mp9 (Adapted from Walker and Gay[11])

1. Dilute a fresh overnight culture of strain BMH 71-18 1:100 with
Antibiotic Medium 3, transfer 2.5 ml of the dilution to a sterile 15

[10] J. Messing, this series, Vol. 101, p. 20.
[11] J. E. Walker and N. J. Gay, this series, Vol. 97, p. 195.

ml culture tube, infect the culture with 3–5 μl of a phage stock (see Procedure 6) of the recombinant M13mp9 clone, and incubate overnight with agitation at 37°.

2. Transfer 1.4 ml of the culture to an Eppendorf tube and centrifuge for 5 min in an Eppendorf centrifuge. (The rest of the culture may be used for preparation of phage stock, see Procedure 6.)

3. Withdraw 1 ml of the supernatant with a Gilson or Eppendorf pipet. To avoid transfer of cells, slide the pipet tip down along the wall of the tube opposite the cell pellet.

4. Centrifuge again for 5 min and transfer 800 μl of the supernatant to another tube as in step 3.

5. Add 200 μl of solution A to the 800 μl of supernatant, mix, and leave for 15 min at room temperature.

6. Centrifuge for 5 min and remove the supernatant carefully with a pasteur pipet; recentrifuge for 30 sec and remove all remaining traces of the supernatant with a drawn-out capillary pipette.

7. Add 110 μl TE buffer to the phage pellet (which should be visible), then 50 μl of TE-saturated phenol, vortex for 15 sec, shake the tube for 10 min at room temperature on an Eppendorf mixer, and vortex again for 15 sec.

8. Centrifuge for 3 min, transfer the aqueous phase to a fresh Eppendorf tube, and reextract with 50 μl phenol/chloroform then with 100 μl chloroform/isoamyl alcohol.

9. Transfer 90 μl of the aqueous phase after the chloroform extraction to a fresh Eppendorf tube, add 10 μl 3 M sodium acetate and 300 μl ethanol (−20°), and keep the mixture at −20° for at least 1 hr.

10. Centrifuge at 4°, 15,000 rpm (rotor type SS34, Sorvall, or equivalent) for 15 min and remove the supernatant with a drawn-out pasteur pipet.

11. Add 1 ml ethanol (−20°) to the tube, vortex for 10 sec, and centrifuge again as in step 10. Remove the supernatant with a drawn-out pasteur pipet and dry the pellet (which is normally not visible) in a Speed Vac concentrator (or vacuum desiccator).

12. Dissolve the pellet in 30 μl TE buffer and store the DNA at −20°.

13. Analyze 3 μl of the virion DNA solution by gel electrophoresis (1% agarose). Use M13 virion DNA of known concentration for comparison of band intensities. The typical yield of this procedure is 2–3 μg of virion DNA.

Procedure 2: Preparation of M13mp9rev RF-DNA[12] (Adapted from Clewell and Helinski[13])

1. Inoculate 2 × 75 ml Antibiotic Medium 3 with 1 ml each of fresh overnight culture of strain BMH 71-18 and infect one of the cultures immediately with 100 μl of M13mp9 phage stock suspension (see Procedure 6). Shake both culture flasks overnight at 37°.
2. Centrifuge the infected culture at 15,000 rpm (rotor type SS34, Sorvall, or equivalent), 4°, for 10 min. Carefully decant the phage suspension and store at 4° until use.
3. Dispense 3 liters Antibiotic Medium 3 into three sterile 2-liter Erlenmeyer flasks, inoculate each flask with 25 ml of the uninfected overnight culture of BMH 71-18, and shake at 37° until the culture has reached an OD_{546} of 0.5–0.6.
4. Infect each culture with 25 ml of the phage suspension (step 2) and continue shaking at 37° for another 3–4 hr.
5. Collect the cells by centrifugation, decant the supernatant, and resuspend the cells at 0° in 22.5 ml solution B using a 10-ml pipet. (If several centrifuge buckets were used for centrifugation, add the 22.5 ml to the first one, resuspend the cells, transfer the suspension to the next bucket, and so on.)
6. After resuspending, add 6 ml lysozyme solution (5 mg/ml in 50 mM Tris–HCl, pH 8.0) and shake gently for 10 min at 0°.
7. Add 7.5 ml 250 mM EDTA, pH 8, and continue incubation at 0° for 10 min, again with gentle shaking.
8. Add 27.5 ml DOC solution and again gently shake at 0° for 10 min.
9. Transfer the mixture (which should now be viscous and cloudy) carefully to screw-cap tubes (for the rotor type 30, Beckman, or equivalent) and centrifuge at 30,000 rpm, 4° for 60 min.
10. Pool the supernatants ("cleared lysate") in a 100-ml graduated cylinder (be careful not to transfer any material from the viscous pellets), add 0.97 g CsCl and 20 μl ethidium bromide solution (10 mg/ml) per milliliter of supernatant, and distribute the solution equally into two 35-ml polyallomer Quick Seal centrifuge tubes (Beckman).

[12] If RF-DNA of M13mp9rev is required for only very few experiments, a scaled-up procedure of the alkaline lysis method as described by T. Maniatis, E. F. Fritsch, and J. Sambrook ("Molecular Cloning: A Laboratory Manual," p. 368. Cold Spring Harbor Lab., Cold Spring Harbor, New York, 1982) may be used.
[13] D. B. Clewell and D. R. Helinski, *Proc. Natl. Acad. Sci. U.S.A.* **62**, 1159 (1969).

11. Fill up the tubes with taring solution [50 mM Tris–HCl, pH 8.0, with 0.97 g CsCl and 20 μl ethidium bromide solution (10 mg/ml) added per milliliter starting buffer], seal the tubes, and centrifuge for at least 14 hr at 45,000 rpm, 15°, in a rotor type VTi50 (Beckman) or equivalent.

12. After centrifugation, two (sometimes three) bands should be visible, corresponding to linear plus "open circle" DNA and to supercoiled DNA (the occasional third band most likely is single-stranded DNA). Puncture the top of the tube for aeration, pierce the wall of the tube with a wide-bore canula mounted to a 10-ml hypodermic syringe just below the upper band(s), and draw off these bands. Without removing the first syringe, repeat the procedure for the lowest band containing the supercoiled DNA (which should be approximately in the middle of the tube). This band should be collected in a total volume of 5 ml per tube.

13. Transfer these two 5-ml samples to two 5-ml Quick Seal tubes (Beckman) and recentrifuge for at least 8 hr at 45,000 rpm, 15°, in a rotor type VTi65 (Beckman) or equivalent.

14. Collect the band containing the supercoiled DNA as in step 12 in a volume of 1–2 ml each. Add 1 volume TE buffer, mix, then add 4 times the starting volume ethanol (room temperature), mix again, and store in the dark for about 1 hr at room temperature.

15. Centrifuge for 30 min at 12,000 rpm, room temperature, in a rotor type HB 4 (Sorvall) or equivalent. Decant the supernatant and wipe the tube with a lint-free paper towel, being careful not to dislodge the pellet. Dry the pellet under reduced pressure and redissolve in approximately the starting volume of TE buffer.

16. Transfer the DNA solution to a 14-ml capped polypropylene tube (Greiner), add an equal volume of TE-saturated phenol, mix by shaking, and centrifuge until clear phase separation is achieved (~10 min). Transfer the aqueous (upper) phase to a new tube and repeat the extraction another 2 times.

17. Extract the aqueous phase 3 times with 2 volumes diethylether, evaporate residual ether by warming to 65° for 15–30 min, and dialyze 2–3 times against 1000 volumes of TES buffer (~6 hr each time).

18. Determine the DNA yield spectrophotometrically ($\varepsilon_{260} = 2 \times 10^{-2}$ cm^2/μg) using the last dialysis buffer as reference. The ratio of A_{260}/A_{280} should be close to 2.0. The yield is typically 100–400 μg RF-DNA per liter of culture.

19. Store DNA at $-20°$.

Procedure 3: Construction of Gapped Duplex DNA (gdDNA)

1. Cleave 0.5 μg (0.1 pmol) RF-DNA of M13mp9rev with the restriction enzyme(s) used for construction of the recombinant M13mp9 clone in a 1.5-ml Eppendorf tube. It is important that the cleavage reaction(s) be driven to completion.
2. Extract the DNA solution with phenol/chloroform and precipitate the DNA with ethanol (analogously to Procedure 1, steps 9 and 10). Wash the pellet with 70% ethanol and dry in a Speed Vac concentrator (or vacuum desiccator).
3. Add 0.5 pmol (~1.3 μg) of the recombinant M13mp9 virion DNA (Procedure 1), 5 μl buffer A, and water to a final volume of 40 μl. Dissolve the pellet by flicking the tube several times.
4. Heat the reaction mixture (in a metal block) to 100° for 3 min, then incubate it at 65° for 5–10 min.
5. Electrophorese an aliquot (about three-fifths) on a 1% agarose gel to check formation of gdDNA. Apply to the same gel as markers 300 ng RF-DNA of M13mp9rev, 100 ng linearized RF-DNA of M13mp9rev, and 300 ng of the recombinant M13mp9 virion DNA (compare Fig. 2).
6. If necessary, the procedure can be interrupted at this point by storing the rest of the hybridization mixture at −20°.

Procedure 4: Hybridization of the Mutagenic Primer to the gdDNA and Optional DNA Polymerase/DNA Ligase Reaction

1. Mix 8 μl of the hybridization mixture and 2 μl of an aqueous solution of the mutagenic primer (2 pmol/μl) in a 1.5-ml Eppendorf reaction tube and heat the mixture to 65° for 5 min. Keep the reaction tube at ambient temperature for approximately 10 min. (In route A, i.e., for mutation construction experiments without polymerase/ligase reactions, 90 μl of 100 mM CaCl₂ is added and the resulting mixture is immediately used for transfection; Procedure 5, step 4.)
2. Add 4 μl buffer B, 4 μl 2 mM NAD, 4 μl of 4 × 2.5 mM dNTPs, and 15 μl water, mix, and add 1 μl *E. coli* DNA ligase (5 U/μl) and 1 μl T4 DNA polymerase (1 U/μl).
3. Incubate for 15 min at 25°, add 2 μl gp 32 (2 mg/ml; if necessary dilute with water immediately prior to use), and incubate again for 90 min at 25°.
4. Stop the reaction by adding 1 μl 500 mM EDTA, pH 8.0, and heating to 65° for 10 min.

1 2 3 4 5 6

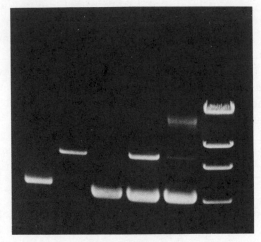

FIG. 2. Gel electrophoretic analysis of different intermediates in the construction of gdDNA (gap size, 120 nucleotides). Lane 1, RF-DNA; lane 2, linear RF-DNA fragment; lane 3, single-stranded virion DNA; lane 4, hybridization mixture (DNA as in lane 2 plus DNA as in lane 3) before denaturation; lane 5, same mixture as in lane 4, but after denaturation/ renaturation; lane 6, λ DNA, cleaved with HindIII. Electrophoresis (field 3.5 V/cm) was carried out on a 1% agarose gel. The DNA was stained with ethidium bromide. The electrophoretic mobility of gdDNA molecules of small gap size as in the example shown here is practically indistinguishable from that of relaxed fully double-stranded DNA (compare uppermost bands of lanes 1 and 5). In lane 5 several minor bands beside remaining singlestranded virion DNA, reformed linear DNA, and gdDNA can be observed. Such bands appear occasionally upon hybridization but do not interfere with the mutation construction experiment.

5. Extract the mixture once with 40 μl phenol/chloroform and 3 times with 200 μl diethyl ether. Evaporate residual ether at 65°, put on ice, add 60 μl 100 mM CaCl$_2$, and use immediately for transfection (Procedure 5, step 4).

Procedure 5: Transfection and Segregation (Transfection Adapted from Cohen et al.[14])

1. Inoculate 50 ml of Antibiotic Medium 3 with 0.5 ml fresh overnight culture of BMH 71-18 mutS grown in M9 medium (optionally containing tetracycline at a concentration of 20 μg/ml).
2. Grow the culture with shaking at 37° to an OD$_{546}$ of 0.4, collect the cells by centrifugation (6,000 rpm at 4°, rotor type SS34, Sorvall, or equivalent), and resuspend the cells in 20 ml ice-cold 100 mM CaCl$_2$.

[14] S. N. Cohen, A. C. Y. Chang, and L. Hsu, *Proc. Natl. Acad. Sci. U.S.A.* **69,** 2110 (1972).

3. Centrifuge as before, resuspend the cells in 10 ml ice-cold 100 mM CaCl$_2$, centrifuge again, resuspend in 2 ml ice-cold 100 mM CaCl$_2$, and keep on ice for at least 30 min.

4. Add 200 μl of this cell suspension to the 100 μl DNA mixture prepared by Procedure 4 and keep the resulting mixture on ice for 90 min.

5. Vortex briefly, heat the suspension for 3 min to 45°, and inoculate 25 ml of Antibiotic Medium 3 with (the bulk of) this suspension (see below). Shake overnight at 37°. (It is advisable to check the transfection efficiency. To this end, remove 20 μl of the 300 μl suspension, dilute 1 : 10 and 1 : 100 in 100 mM CaCl$_2$, and plate 100 μl of the dilutions analogously to steps 9–11.)

6. Inoculate 5 ml Antibiotic Medium 3 with strain MK 30-3 and incubate overnight with agitation at 37°.

7. Transfer a 1.4-ml aliquot of the transfected BMH 71-18 mutS culture (step 5) to a sterile 1.5-ml Eppendorf tube, centrifuge for 5 min, and transfer the supernatant into another sterile Eppendorf tube. (Store at 4°.)

8. Dilute the supernatant 1 : 10^6, 1 : 10^7, 1 : 10^8, and 1 : 10^9 in phage buffer.

9. Place five sterile 12-ml glass tubes in a 45° water bath, dispense 2.5 ml of thoroughly melted top agar into each tube. After the agar is cooled to 45°, add 3 drops of the overnight culture of strain MK 30-3 (step 6) to each tube.

10. Add 100 μl of each of the different phage dilutions (step 8) to four of these tubes, mix by vortexing, and pour the top agar onto EHA plates. The content of the fifth tube without phages is also plated as a control to determine whether the host strain is contaminated by phage.

11. Keep the plates faceup at room temperature for about 20 min until the top agar has solidified and then incubate the plates upside down at 37° for at least 7 hr.

Procedure 6: Preparation of M13 Phage Stocks and Template DNA for Nucleotide Sequence Analysis

1. Dilute a fresh overnight culture of strain BMH 71-18 1 : 100 with Antibiotic Medium 3, transfer 2.5 ml of the diluted cell suspension into a sterile culture tube.

2. Punch out a plaque resulting from a segregation step (Procedure 5) with a sterile glass tube (e.g., a 200-μl capillary pipet) and blow the agar disk from the pipet into the culture tube. Vortex the mixture and incubate with agitation for 4.5–5.5 hr at 37°.

3. The preparation of template DNA for nucleotide sequence analysis

can now be carried out as described in Procedure 1, steps 2–13.

4. For the preparation of a phage stock, transfer 1 ml of the infected culture to a sterile Eppendorf tube. Centrifuge the sample for 5 min in an Eppendorf centrifuge and transfer the supernatant into another (sterile) Eppendorf tube. Carry-over from the bacterial pellet should be avoided.

5. Store at 4°. The tube can be sealed with moldable laboratory film (Parafilm M). Phages from this stock should be viable for at least 1 year.

6. The phage suspension resulting from step 4 may be lyophilized in a Speed Vac concentrator for transport. The lyophilized sample can be redissolved by simply adding the starting volume of water. Prior to lyophilization the phages may be precipitated with polyethylene glycol (see Procedure 1, steps 5 and 6) and redissolved in phage buffer. The phage titer may drop rather drastically upon lyophilization. In case of a low titer (which is sometimes observed after lyophilization), step 1 of Procedure 1 may be repeated.

Procedure 7: Maintenance of Bacterial Strains (Adapted from Miller[15])

For storage of bacterial strains three methods are commonly used: plates, glycerol cultures, and stab cultures.

Agar Plates. Agar plates are good for strains that are used frequently. For strains BMH 71-18, BMH 71-18 mutS, and MK 30-3 minimal plates are used to select for the *pro* marker of the F' episome.

1. For preparation of minimal plates, mix 7 g $Na_2HPO_4 \cdot 2H_2O$, 3 g KH_2PO_4, 1 g NH_4Cl with 400 ml H_2O and, in a second flask, 15 g Bactoagar (Difco) with 600 ml H_2O. Autoclave both mixtures separately, then combine them. Add the following sterile solutions: 25 ml 20% glucose, 1 ml 100 mM $CaCl_2$, 1 ml 1 M $MgSO_4$, 5 ml 100 μM $FeCl_3$, and 5 ml thiamin (1 mg/ml). Use about 35–40 ml per petri dish (85 mm diameter).

2. Prepare an overnight culture of the bacterial strain to be stored, plate 0.2 ml overnight culture per plate, and incubate the plates upside down at 37°, until a bacterial lawn is visible (1–2 days).

3. Seal the plates with moldable laboratory film (Parafilm M) and store at 4°. The strain should be viable for about 1 month.

4. To start a liquid culture, scrape bacteria with a sterile glass rod from the plate (one touch or a short streak) and swirl the rod in culture medium for a few seconds.

[15] J. H. Miller, "Experiments in Molecular Genetics," p. 434. Cold Spring Harbor Lab., Cold Spring Harbor, New York, 1972.

Glycerol Cultures. Glycerol cultures are convenient for less frequent use and for storage of large numbers of different bacterial clones. For preparation of glycerol cultures, buffered growth media are preferred over standard rich media since the low pH prevalent in an overnight culture leads to reduced shelf life.

1. A fresh overnight culture of the strain to be stored is diluted with sterile glycerol to a final concentration of 50% glycerol.
2. Transfer the mixture to a sterile 5-ml screw-cap tube and store the tube at −20°. Avoid repeated warming of the culture. The strain should be viable for at least 1 year.
3. To start a liquid culture, transfer a small aliquot of the glycerol culture into culture medium.

Stab Cultures. Stab cultures are chosen for backup storage and transport of strains. They cannot be used very often.

1. Add 6.5 g Bactoagar (Difco) to 1 liter Antibiotic Medium 3 (Difco) and melt the agar.
2. Fill 3-ml aliquots into 4-ml screw-cap glass vials (autoclavable). Autoclave and allow to cool to room temperature.
3. Dip a sterile glass rod, or (preferably) a platinum loop, into a fresh overnight culture of the strain to be stored and stab into the agar in the vial.
4. Close the vial, seal it with moldable laboratory film (Parafilm M) and keep at room temperature. After 2–3 days bacterial growth should be visible in the stab culture along the path of the loop.
5. Store the culture at room temperature in the dark. The strain should be viable for at least 1 year.
6. To start a liquid culture, remove some bacteria from the agar with a sterile glass rod or platinum loop and transfer to culture medium (swirl rod in medium for a few seconds).

Scope, Limitations, and Practical Hints

A large portion of the accumulated experience on oligonucleotide-directed mutation construction did not originate from systematic model investigations but, rather, was collected in a more or less incidental way while applying the method to a great number of biological problems. Thus, there exists a fair amount of "soft" but (from a practical point of view) useful information. The following section on experimental requirements, possible pitfalls, etc., rests in part on that body of experience with no attempt to make reference to the numerous and diverse individual cases.

Structure of the Target Region

Gap Size. Gap size is a significant, yet not overwhelming factor for the efficiency of the method. Thus, somewhat higher marker yields can be expected if the length of the DNA fragment cloned into M13mp9 is kept as small as possible. However, gaps in the size range of a few thousand bases are still good substrates and the difference in efficiency between the "fill-in" (route B) and the "mix–heat–transfect" protocols (route A) is smaller with larger gaps.

Sequence Redundancies. Sequence redundancies are a more serious concern: Direct repeats will result in equivocal choice of the primer hybridization target or bridging of two remote DNA sites and formation of deletions. Inverted repeats can form snapback structures and prevent primer annealing altogether, if the hybridization target of the mutagenic oligonucleotide is part of the double-stranded stem of such a structure. Hairpin/loop structures outside the hybridization target can block the *in vitro* fill-in reaction catalyzed by purified DNA polymerase. In this situation, route A offers a potential advantage.

In the gdDNA method, such sequence redundancies are rendered insignificant if occurring between insert and vector part because the latter is covered by the complementary DNA strand. If possible, however, the gap sequence should be checked for repetitions. In problematic cases possible countermeasures can be (1) cloning a subfragment which does not contain the repeated DNA sequence, (2) use of longer oligonucleotides for mutation construction to give the correct hybridization product an energetic advantage, or (3) covering the nontarget repeat by an additional DNA segment complementary to that part of the gdDNA (restriction fragment or synthetic oligonucleotide). Circumstantial evidence exists that in addition to the above-mentioned constraints there are *other* (poorly understood) *factors* of target DNA structure that can lower the efficiency of the method. Extremely high G + C content of the target DNA may be one such factor.

Structure of the Synthetic Oligonucleotide

Length of Primer. The mutagenic primer consists of a core sequence to be left unpaired after annealing to the M13 genome and two flanking sequences complementary to the target DNA which provide site-specific and stable hybridization. For a sequence of normal G + C content and an unpaired core of just one nucleotide these flanking sequences should be about 7–9 nucleotides long (route B). The flanking sequences should be longer if they have a high A + T content or if extended unpaired regions must be accommodated in the core. With the "mix–heat–transfect" protocol (route A) mutagenic primers about 10 nucleotides longer should be used to ensure sufficient hybrid stability during transfection.

Chemical Nature of the Primer Near its 5′ Terminus. A fraying 5′ end of the primer annealed to the template seems to invite loss of the mutagenic oligonucleotide by strand displacement after enzymatic elongation of the 3′ end of the gdDNA strand has reached this point. Whenever possible, the primer sequence at the 5′ terminus should therefore consist of one or several G or C residues. In route A, the gap filling and sealing reactions are left to the enzymes of the transfected cell. In this situation, the mutagenic oligonucleotide not only can be lost by strand displacement but also by nick translation. We have found it useful to protect the 5′ end of the primer against 5′ to 3′ exonucleolytic attack by one or several phosphorothioate internucleotidic linkages.[6] Such bonds can be prepared conveniently[16,17] by automated DNA synthesis.[6]

Sequence Redundancies within the Mutagenic Primer. Inverted repeats within the primer sequence (contiguous or interrupted) give the synthetic oligonucleotide (partial) self-complementarity. Because of oligonucleotide duplex formation, this can strongly interfere with annealing to the target site.

Direct repeats near one of the primer ends (as small as three nucleotides in length) can lead to primer–template slippage with loop formation on the primer side. This results in an unwanted insertion. When designing the primer sequence to be synthesized, attention should be paid to such local redundancies. Here too, synthesis of a longer primer can be helpful.

Purity of the Primer. Chemical homogeneity of the primer is essential for specific mutation construction. Two alternative purification schemes have proven successful: (1) two consecutive rounds of high-pressure liquid chromatography (HPLC) exploiting different separation principles in each run, and (2) a combination of HPLC and preparative gel electrophoresis. Current synthetic methods (for a comprehensive treatise, see Ref. 18) yield oligonucleotides with a 5′-terminal dimethoxytrityl group attached. This group provides a convenient handle for purification by HPLC on a reverse stationary phase[19] (bonded hydrocarbon phases with chain lengths ranging from C_4 to C_{18}).

After detritylation,[20] the oligonucleotide is enzymatically phosphorylated[20] before either of the two purification procedures outlined below is

[16] B. A. Connolly, B. V. L. Potter, F. Eckstein, A. Pingoud, and L. Grotjahn, *Biochemistry* **23**, 3443 (1984).

[17] W. J. Stec, G. Zon, W. Egan, and B. Stec, *J. Am. Chem. Soc.* **106**, 6077 (1984).

[18] M. J. Gait (ed.), "Oligonucleotide Synthesis: A Practical Approach." IRL Press, Oxford, England, 1984.

[19] H.-J. Fritz, R. Belagaje, E. L. Brown, R. H. Fritz, R. A. Jones, R. G. Lees, and H. G. Khorana, *Biochemistry* **17**, 1257 (1978).

[20] B. Kramer, W. Kramer, and H.-J. Fritz, *Cell* **38**, 879 (1984).

carried out: (1) rechromatography by HPLC using either an anion-exchange column[21] or a reverse-phase column as above but operated in the paired-ion mode,[22] or (2) preparative polyacrylamide gel electrophoresis as described.[23,24] Both methods separate by chain length *and* state of the 5' terminus (phosphorylated/unphosphorylated) and yield 5'-phosphorylated oligonucleotides of high purity. If, however, the synthetic material, as isolated from the first round of HPLC, is significantly contaminated by products of lower chain length, method (1) is preferable.

Side Reactions

Errors of DNA Polymerase. The "fill-in" protocol (route B), like most techniques of oligonucleotide-directed mutagenesis used to date, requires a DNA polymerase reaction *in vitro*. It is known that a purified DNA polymerase has a high error frequency.[25] Thus, additional mutations outside the mutagenic primer may occasionally be expected. Such events have indeed been observed and, for this reason, it is important to sequence the entire DNA segment under study after mutagenesis. Conceivably, the "mix–heat–transfect" protocol (route A), which leaves all enzymatic reactions to the transfected cell, may be more accurate in that respect. At present, however, the body of experience with route A is not large enough for a meaningful comparison.

Mutator Phenotype of the Transfection Host. Escherichia coli strains deficient in DNA mismatch repair (such as the mutS strain used in this method) have an increased frequency of spontaneous mutations (mutator phenotype). It is possible, in principle, that additional and unwanted mutations may be introduced into the target DNA during propagation of the recombinant M13 phage. This, however, is of no practical significance, since the rate of spontaneous mutation is still lower by several orders of magnitude than the frequency of the constructed mutation.

Hydrolytic Deamination of Cytosine Residues. The construction of gdDNA involves heating of the DNA solution to 100°. Under these conditions, hydrolytic deamination of 2'-deoxycytidine residues to 2'-deoxyuri-

[21] L. W. McLaughlin and N. Pier, *in* "Oligonucleotide Synthesis: A Practical Approach" (M. J. Gait, ed.), p. 117. IRL Press, Oxford, England, 1984.

[22] H.-J. Fritz, D. Eick, and W. Werr, *in* "Chemical and Enzymatic Synthesis of Gene Fragments" (H. G. Gassen and A. Lang, eds.), p. 199. Verlag Chemie, Weinheim, Federal Republic of Germany, 1982.

[23] T. Atkinson and M. Smith, *in* "Oligonucleotide Synthesis: A Practical Approach" (M. J. Gait, ed.), p. 35. IRL Press, Oxford, England, 1984.

[24] R. Wu, N.-H. Wu, Z. Hanna, F. Georges, and S. Narang, *in* "Oligonucleotide Synthesis: A Practical Approach" (M. J. Gait, ed.), p. 135. IRL Press, Oxford, England, 1984.

[25] A. R. Fersht and J. W. Knill-Jones, *J. Mol. Biol.* **165,** 633 (1983).

dine may be a significant reaction. Such reactions would be detrimental because they would provide start points of unwanted repair synthesis when occurring in the double-stranded portion of the gdDNA. More seriously, a C/G to T/A transition would be induced at the site of such a lesion when it was located within the single-stranded gap. In one out of many dozens of cases studied (route B), we have found an additional C/G to T/A transition. It is, of course, not possible to decide whether this mutation was due to the described cytosine deamination or to an error of the DNA polymerase (see above). Again, route A may offer an advantage due to the expected *in vivo* killing of entering gdDNA molecules that carry a 2'-deoxyuridine residue in their single-stranded part.

Flexibility

The applicability of the gdDNA method reaches beyond oligonucleotide-directed mutation construction. Gapped duplex DNA molecules are also ideal substrates for other methods of directed mutagenesis such as forced gap misrepair[26] or attack by single-strand selective chemicals (e.g., bisulfite[27]). Studies in "reversed genetics" can be planned in such a way that a stock of gdDNA is prepared and the last two (or similar) methods are used for saturation of the gap sequence with (single) point mutations in order to identify functionally critical points within the cloned gene or regulatory DNA segment. Once these are known, the same original gdDNA can be used to ask specific questions via oligonucleotide-directed construction of predetermined mutations.

[26] D. Shortle, P. Grisafi, S. J. Benkovic, and D. Botstein, *Proc. Natl. Acad. Sci. U.S.A.* **79**, 1588 (1982).
[27] D. Shortle and D. Botstein, this series, Vol. 100, p. 457.

[32] Rapid and Efficient Site-Specific Mutagenesis without Phenotypic Selection

By Thomas A. Kunkel, John D. Roberts, and Richard A. Zakour

The deliberate alteration of DNA sequences by *in vitro* mutagenesis has become a widely used and invaluable means of probing the structure and function of DNA and the macromolecules for which it codes. In an ideal experiment, alterations are produced at high efficiency and with minimum effort; such features become especially important when sequence changes produce silent, unknown, or nonselectable phenotypes. To overcome some of the limitations that lead to low efficiency, several variations of *in vitro* mutagenesis techniques have been developed.[1-5] Each procedure has its own advantages, but each also requires additional time and/or technical expertise. An alternative method[6] is presented here that is simple, rapid, and efficient. This method takes advantage of a strong biological selection against the original DNA template which is preferentially destroyed on transfection. The use of this special template can be combined with many of the previously described *in vitro* mutagenesis methods. What we describe here is not, therefore, a new procedure for site-directed mutagenesis but is, rather, the use of standard and well-established procedures in conjunction with an unusual template. This combination permits flexibility in the choice of mutagenesis techniques and makes possible the highly efficient recovery of mutants.

Principle

The basis of this method is the performance of site-directed mutagenesis using a DNA template which contains a small number of uracil residues in place of thymine.[6] The uracil-containing DNA is produced within an *Escherichia coli dut⁻ ung⁻* strain. *Escherichia coli dut⁻* mutants lack the enzyme dUTPase[7,8] and therefore contain elevated concentrations of

[1] M. Smith, *Annu. Rev. Genet.* **19**, 423 (1985).
[2] M. Zoller, this series, Vol. 154, p. 329.
[3] W. Kramer and H.-J. Fritz, this volume [31].
[4] P. Carter, this series, Vol. 154, p. 382.
[5] G. Cesarini, C. Traboni, G. Ciliberto, L. Dente, and R. Cortese, *in* "DNA Cloning: A Practical Approach" (D. M. Glover, ed.), Vol. 1, pp. 137–149. IRL Press, Oxford, England, 1985.
[6] T. A. Kunkel, *Proc. Natl. Acad. Sci. U.S.A.* **82**, 488 (1985).
[7] E. B. Konrad and I. R. Lehman, *Proc. Natl. Acad. Sci. U.S.A.* **72**, 2150 (1975).
[8] S. J. Hochhauser and B. Weiss, *J. Bacteriol.* **134**, 157 (1978).

RECOMBINANT DNA
METHODOLOGY

dUTP which effectively competes with TTP for incorporation into DNA. *Escherichia coli ung⁻* mutants lack the enzyme uracil *N*-glycosylase[9] which normally removes uracil from DNA.[10] In the combined *dut⁻ ung⁻* mutant, uracil is incorporated into DNA in place of thymine and is not removed.[11-13] Thus, standard vectors containing the sequence to be changed can be grown in a *dut⁻ ung⁻* host to prepare uracil-containing DNA templates for site-directed mutagenesis.

For the *in vitro* reactions typical of site-directed mutagenesis protocols, uracil-containing DNA templates are indistinguishable from normal templates. Since dUMP in the template has the same coding potential as TMP,[14] the uracil is not mutagenic, either *in vivo* or *in vitro*. Furthermore, the presence of uracil in the template is not inhibitory to *in vitro* DNA synthesis. Thus, this DNA can be used *in vitro* as a template for the production of a complementary strand that contains the desired DNA sequence alteration but contains only TMP and no dUMP residues.

After completing the *in vitro* reactions, uracil can be removed from the template strand by the action of uracil *N*-glycosylase.[10] Glycosylase treatment can be carried out with purified enzyme[6] but is most easily achieved by transfecting the unfractionated products of the *in vitro* incorporation reaction into competent wild-type (i.e., *ung⁺*) *E. coli* cells. Treatment with the glycosylase, either *in vitro* or *in vivo*, releases uracil-producing apyrimidinic (AP) sites only in the template strand.[15] These AP sites are lethal lesions, presumably because they block DNA synthesis,[13,16,17] and are sites for incision by AP endonucleases which produce strand breaks.[15] Thus, the template strand is rendered biologically inactive, and the majority of progeny arise from the infective[6,18-20] complementary strand which contains the desired alteration. The resulting high efficiency of mutant production (typically >50%) allows one to screen for mutants by DNA sequence analysis, thus identifying mutants and confirming the desired alteration in a single step. This feature is particularly advantageous when no selection for the desired mutants is available.

[9] B. K. Duncan, P. A. Rockstroh, and H. R. Warner, *J. Bacteriol.* **134**, 1039 (1978).
[10] T. Lindahl, *Proc. Natl. Acad. Sci. U.S.A.* **71**, 3649 (1974).
[11] B.-K. Tye and I. R. Lehman, *J. Mol. Biol.* **117**, 293 (1977).
[12] B.-K. Tye, J. Chien, I. R. Lehman, B. K. Duncan, and H. R. Warner, *Proc. Natl. Acad. Sci. U.S.A.* **75**, 233 (1978).
[13] D. Sagher and B. Strauss, *Biochemistry* **22**, 4518 (1983).
[14] J. Shlomai and A. Kornberg, *J. Biol. Chem.* **253**, 3305 (1978).
[15] T. Lindahl, *Annu. Rev. Biochem.* **51**, 61 (1982).
[16] R. M. Schaaper and L. A. Loeb, *Proc. Natl. Acad. Sci. U.S.A.* **78**, 1773 (1981).
[17] T. A. Kunkel, *Proc. Natl. Acad. Sci. U.S.A.* **81**, 1494 (1984).
[18] P. Rüst and R. L. Sinsheimer, *J. Mol. Biol.* **23**, 545 (1967).
[19] J. E. D. Siegel and M. Hayashi, *J. Mol. Biol.* **27**, 443 (1967).
[20] E. L. Loechler, C. L. Green, and J. M. Essigmann, *Proc. Natl. Acad. Sci. U.S.A.* **81**, 6271 (1984).

TABLE I
Escherichia coli STRAINS

Strain designation	Genotype	Source
BW313	HfrKL16 PO/45 [*lysA*(61-62)], *dut1, ung1, thi1, relA1*	a
CJ236[b]	*dut1, ung1, thi1, relA1*/pCJ105 (Cm^r)	c
RZ1032	As BW313, but Zbd-279::Tn*10, supE44*	d
NR8051	[Δ(*pro–lac*)], *thi, ara*	a
NR8052	[Δ(*pro–lac*)], *thi, ara, trpE9777, ung1*	a
KT8051	[Δ(*pro–lac*)], *thi, ara*/F′ (*proAB, lacI*$_q^-$*Z*$^-$ ΔM15)	e
KT8052	[Δ(*pro–lac*)], *thi, ara, trpE9777, ung1*/F′ (*proAB, lacI*$_q^-$*Z*$^-$ ΔM15)	e
CSH50	[Δ(*pro–lac*)], *thi, ara, strA*/F′ (*proAB, lacI*$_q$*Z*$^-$ ΔM15, *traD36*)	f
NR9099	[Δ(*pro–lac*)], *thi, ara, recA56*/F′ (*proAB, lacI*$_q^-$*Z*$^-$ ΔM15)	g

[a] As described by T. A. Kunkel, *Proc. Natl. Acad. Sci. U.S.A.* **82**, 488 (1985), but see "Uses, Maintenance, and Characteristics of Bacterial Strains" in this chapter.

[b] The plasmid pCJ105 (Cm^r) was constructed as described by C. M. Joyce and N. D. F. Grindley, *J. Bacteriol.* **158**, 636 (1984).

[c] C. M. Joyce, Yale University, New Haven, Connecticut.

[d] See text.

[e] K. Tindall, National Institute of Environmental Health Sciences, Research Triangle Park, North Carolina.

[f] T. A. Kunkel, *Proc. Natl. Acad. Sci. U.S.A.* **81**, 1494 (1984).

[g] R. M. Schaaper, B. N. Danforth, and B. W. Glickman, *Gene* **39**, 181 (1985).

Materials and Reagents

Bacterial Strains

The *E. coli* strains and their sources are listed in Table I. Their use and maintenance are described below.

Growth Media

YT medium: Bactotryptone, 8 g; Bactoyeast extract, 5 g; NaCl, 5 g. Add to 1 liter of H_2O and sterilize in an autoclave.

2× YT medium: Bactotryptone, 16 g; Bactoyeast extract, 10 g; NaCl, 10 g; pH adjusted to 7.4 with HCl. Add H_2O to 1 liter and sterilize in an autoclave.

Soft agar: NaCl, 9 g; Difco agar, 8 g. Add H_2O to 1 liter and sterilize in an autoclave.

VB salts (50×): $MgSO_4 \cdot 7H_2O$, 10 g; citric acid (anhydrate), 100 g; K_2HPO_4, 500 g; $Na_2HPO_4 \cdot 2H_2O$, 75 g. Dissolve the above in 670 ml dH_2O, bring volume to 1 liter, and sterilize in an autoclave. After dilution, pH is 7.0–7.2.

Minimal plates: Add 16 g of Difco agar to 1 liter of dH$_2$O and sterilize
in an autoclave. When the agar has cooled to 50°, add 0.3 ml of 100
mM IPTG, 20 ml of 50× VB salts, 20 ml of 20% glucose, and 5 ml
of 1 mg/ml thiamine–HCl. [Each of these solutions was sterilized
either by filtration (0.2-μm pore) or in an autoclave prior to their
addition to the 50° agar.] The mixture is mixed well and dispensed
into sterile petri dishes (30 ml/plate).

Enzymes and Reagents

T4 DNA polymerase and T4 polynucleotide kinase were from Pharma-
cia, Molecular Biology Division. T4 DNA ligase was from New England
Biolabs or International Biotechnologies, Inc. Deoxynucleoside triphosphates (HPLC grade, 100 mM solutions) were purchased from Pharmacia,
Molecular Biology Division, and used without further purification. 5-
Bromo-4-chloroindolyl-β-D-galactoside (Xgal) was from Bachem Chemicals. Isopropylthio-β-D-galactoside (IPTG) was from Bethesda Research
Laboratories. Phenol (ultrapure grade) was obtained from Bethesda Research Laboratories or International Biotechnologies, Inc., and used
without further purification. All other chemicals were obtained from standard suppliers of molecular biological reagents.

Stock Solutions

Xgal: 50 mg/ml in N,N-dimethylformamide (DMF), stored at −20°.
   Avoid exposure to light.
IPTG: 24 mg/ml in dH$_2$O, stored at −20°.
PEG/NaCl (5×): Polyethylene glycol 8000, 150 g; NaCl, 146 g. Dis-
   solve in dH$_2$O, adjust volume to 1 liter, and filter sterilize using a
   0.2-μm filter.
Phenol extraction buffer (PEB): 100 mM Tris–HCl (pH 8.0); 300 mM
   NaCl; 1 mM EDTA.
Phenol: equilibrated versus multiple volumes of PEB until the pH of
   the aqueous phase is ~8.0, stored in a brown bottle at 4°.
TE buffer: 10 mM Tris–HCl (pH 8.0); 0.1 mM EDTA.
Kinase buffer (10×): 500 mM Tris–HCl (pH 7.5); 100 mM MgCl$_2$; 50
   mM dithiothreitol.
SSC (20×): 3 M NaCl; 300 mM sodium citrate.
SDS dye mix (10×): 10% sodium dodecyl sulfate; 1% bromophenol
   blue; 50% glycerol.
TAE buffer (50×): Tris base, 242 g; glacial acetic acid, 57.1 ml;
   EDTA, 100 ml of a 500 mM solution (pH 8.0). Dissolve in dH$_2$O,
   adjust volume to 1 liter.

Methods

Uses, Maintenance, and Characteristics of Bacterial Strains

Uracil-containing DNA was first prepared as a template for *in vitro* mutagenesis as described by Sagher and Strauss[13] using *E. coli* strain BW313. This strain was chosen on the basis of three criteria which are crucial for the successful production of uracil-containing viral DNA templates: (1) susceptibility to infection by small filamentous bacteriophages (e.g., M13), which requires the F (sex factor) pilus; (2) the presence of the Dut⁻ and Ung⁻ phenotypes, which are required for the stable incorporation of uracil into phage DNA; and (3) a low rate of spontaneous mutation in the progeny phage, so that unwanted mutations are not introduced into the DNA target.

In the original publication of this method,[6] BW313 was incorrectly described as F'*lysA*. BW313 is actually an Hfr strain with the integrated F factor providing the pilus needed for phage attachment. Since there is no selective pressure that can be used to maintain the Hfr phenotype, we store BW313 frozen (−70°) in multiple aliquots containing 3 ml of a mid-log culture (2×10^8 to 2×10^9 cells/ml) mixed with 0.3 ml of DMSO. When needed, a vial is thawed and cells are streaked on a YT plate (or any rich medium plate) to obtain single colonies. This plate may be used as a source of colonies for over 2 months when stored at 4°. We have not encountered problems in achieving M13 infections using this procedure. However, since others have observed a loss of infectability with this strain, a second strain was produced by introducing a selectable F' that confers resistance to chloramphenicol into a BW313 strain that had lost its competence for M13 infection. This strain (CJ236, produced by Catherine Joyce at Yale) stably retains susceptibility to M13 infection when selective pressure is maintained, and it has been used successfully by us to prepare uracil-containing DNA for several site-directed mutagenesis experiments.

Since BW313 does not contain an amber suppressor, a third *E. coli* *dut⁻ ung⁻* strain was constructed for use with cloning vectors which contain amber mutations in essential genes (e.g., M13 mp8). Strain JM101 (*supE44*) was transduced to tetracycline resistance using P1 grown on strain SK2255 which carries the *tet*ʳ marker on a transposable element, Zbd-279::Tn*10*. A tetracycline-resistant derivative of JM101 was used to grow P1 which were then used to transduce strain BW313 to tetracycline resistance. The BW313 Tetʳ transductants were tested for their ability to support growth of phage M13mp8 (which contains two amber mutations) and for the ability to produce M13mp2 phage containing uracil in their DNA (due to the host *dut1* and *ung1* mutations). The resulting Hfr strain (RZ1032) fulfills these criteria and templates prepared from this strain perform well in subsequent *in vitro* mutagenesis experiments. The *supE44* is maintained with tetracycline selection. This strain grows somewhat more slowly than BW313, and characteristically produces smaller M13 plaques and lower phage yields in liquid cultures.

As with BW313, RZ1032 loses its Hfr phenotype at a low but bothersome frequency and a single-colony isolate may not support M13 infection. (At present, a selectable F' has not been placed into RZ1032.) To overcome this problem with either BW313 or RZ1032, several individual colonies should be picked from a plate and liquid cultures (YT medium), each from a single colony, should be screened by plaque assays on plates to test for M13 infectability. Once a competent culture is identified, aliquots can be stored indefinitely at $-70°$ in 10% DMSO. We have not encountered instability of the *dut⁻* and *ung⁻* markers in BW313, CJ236, or RZ1032 when they are grown in rich medium. Likewise, beyond the usual slight increase in mutation frequency associated with *dut⁻ ung⁻* strains (see below), which is negligible in site-directed mutagenesis protocols, we have observed neither high mutation frequencies nor spurious mutations on growing vectors in these strains.

The other strains listed in Table I have been useful in establishing the utility of uracil-containing DNA, but are not required for most mutagenesis protocols. NR8052 (*ung⁻*) and its wild-type (*ung⁺*) parent, NR8051, can be used to measure the relative survival of uracil-containing DNA upon transfection. Similarly, the newly constructed derivatives of these strains, prepared by Kenneth Tindall of the National Institute of Environmental Health Sciences and designated KT8052 and KT8051, respectively, can be used for a similar analysis of intact phage since they contain an F'. Survival data can be obtained by comparing phage titers on BW313 (or CJ236 or RZ1032, all *ung⁻* strains) with titers on any wild-type (*ung⁺*) strain. However, NR8052 and NR8051 have another advantage in that they can also be used to determine spontaneous mutation rates in the phage-borne *lacZα* gene by following the loss of α-complementation as previously described.[6,17] Such experiments were carried out to establish that growth in the *dut⁻ ung⁻* host is not mutagenic.[6]

Escherichia coli CSH50 and NR9099 are *ung⁺* and are routinely used for α-complementation experiments in our laboratory. These, as well as other *ung⁺ E. coli* strains (for example, the JM series of α-complementation strains), are acceptable hosts for transfection of the products of the *in vitro* mutagenesis reactions with uracil-containing templates.

Growth of Phage

Uracil-containing viral DNA template is isolated from intact M13 phage grown on an *E. coli dut⁻ ung⁻* strain. Phage can be produced as previously described[6] or by a simpler method which we present here.

Using a sterile pipet tip, remove one plaque [usually 10^9–10^{10} plaque-forming units (pfu)/plaque for M13] from a plate and place it in 1 ml of sterile YT medium in a 1.5-ml Eppendorf tube. Incubate the tube for 5 min at 60° to kill cells, vortex vigorously to release the phage from the agar, then pellet cells and agar with a 2-min spin in a microcentrifuge. Place 100 μl of the resulting supernatant (containing 10^8–10^9 pfu) into a 1-liter flask containing 100 ml of YT medium supplemented with 0.25 μg/ml uridine; we have found that neither the thymidine nor the adenosine supplementation originally described[6] is necessary, since omitting these effected neither phage yield nor mutation frequency. Add 5 ml of a mid-log culture of the appropriate *E. coli dut⁻ ung⁻* strain. These proportions result in a multiplicity of infection of ≤1. Most or all phage infect cells and are thus "passaged" through the *dut⁻ ung⁻* strain. Since few uracil-lacking phage remain, a single cycle of growth results in a sufficient survival difference (as measured by titers on *ung⁺* and *ung⁻* hosts) to make the DNA suitable for the *in vitro* mutagenesis protocol.

The flask is incubated with vigorous shaking at 37°. We have prepared phage from cultures incubated for as short as 6 hr or as long as 24 hr. Shorter times are recommended for vectors that contain unstable inserts, since this will help to avoid the growth advantage of phages which have deleted the insert. (A *recA⁻* derivative of a *dut⁻ ung⁻* strain might be useful to stabilize otherwise unstable DNA sequences, but at present we do not have such a strain.)

After incubation at 37°, the culture is centrifuged at 5000 g for 30 min. The clear supernatant contains the phage at about 10^{11} pfu/ml. (This yield may vary depending on the vector and strain used; our experience with RZ1032 suggests that phage titers of 2–5 × 10^{10} pfu/ml are not unusual.) Before preparing viral template DNA, the phage titers should be compared on *ung⁻* and *ung⁺* hosts. Phage which contain uracil in the DNA have normal biological activity in the *ung⁻* host but greater than 100,000-fold lower survival in the *ung⁺* host. Phage produced in the *dut⁻ ung⁻* host show only a slight (~2-fold) increase in mutation frequency when compared to phage produced in wild-type *E. coli*. Loss of α-complementation occurs in about 0.1% of the uracil-containing phage,[6] a negligible background compared to frequencies of 50–90% in site-directed mutagenesis experiments.

Preparation of Template DNA

Phage are precipitated from the clear supernatant by adding 1 volume of 5× PEG/NaCl to 4 volumes of supernatant, mixing, and incubating the phage at 0° for 1 hr. The precipitate is collected by centrifugation at 5000 g for 15 min, and the well-drained pellet is resuspended in 5 ml of PEB in a 15-ml Corex tube. After vigorous vortexing, the resuspended phage solution is placed on ice for 60 min and then centrifuged as above to remove residual debris. (This step has proved useful in reducing the level of endogenous priming in subsequent *in vitro* DNA polymerase reactions.) The supernatant containing the intact phage is extracted twice with phenol (previously equilibrated with PEB) and twice with chloroform : isoamyl alcohol (24 : 1). The DNA is precipitated [by adding 0.1 volume of 3 M sodium acetate (pH 5.0) and 2 volumes of ethanol and chilling the mixture to $-20°$], collected by centrifugation and resuspended in TE buffer. The DNA concentration is determined spectrophotometrically at 260 nm using an extinction coefficient of 27.8 ml/mg cm (i.e., 1 OD_{260} = 36 $\mu g/ml$) for single-stranded DNA. The purity of the DNA is examined by agarose gel electrophoresis as described below, overloading at least one lane to visualize trace contaminants.

We have found no need to further purify the DNA in order to achieve high efficiencies of *in vitro* mutagenesis. If problems related to template purity are encountered, or if mutant production approaching 100% is needed, the DNA can be subjected to any standard purification procedure, since the substitution of a small percentage of thymine residues by uracil should not affect the physical properties of the DNA.

In principle, any cloning vector that can be passaged through an *E. coli dut⁻ ung⁻* strain can be used with the uracil selection technique. Once the uracil-containing DNA is prepared, it can be used as would be any standard template, in a variety of *in vitro* methodologies for altering DNA sequences.[1] We present below a typical oligonucleotide-directed mutagenesis experiment to make several points of interest. Some of these have previously appeared in the literature (see Smith[1] and references therein, as well as several other chapters in this volume and in Vol. 154 of *Methods in Enzymology*[2-4]) and are well known to investigators acquainted with this field. However, these notes may be useful to those less familiar with *in vitro* mutagenesis.

Example of the Method

For reasons to be published elsewhere, we required a mutant of M13mp2 containing an extra T residue in the viral-strand run of four consecutive Ts at positions 70 through 73 (where position 1 represents the first transcribed base) in the coding sequence for the α peptide of β-galactosidase.[21]

Description and Phosphorylation of the Oligonucleotide

A 22-base oligonucleotide, complementary to positions 62–82 and containing an extra (i.e., fifth) A residue, was purchased from the DNA

---

[21] T. A. Kunkel, *J. Biol. Chem.* **260**, 5787 (1985).

Synthesis Service, Dept. of Chemistry, Univ. of Pennsylvania. (The ability of this or any oligonucleotide to prime *in vitro* DNA synthesis at the appropriate position should be examined.[1]) The 5' OH of the 22-mer was phosphorylated (for subsequent ligation) in a 20-μl reaction containing 50 mM Tris–HCl (pH 7.5), 10 mM MgCl$_2$, 5 mM dithiothreitol, 1 mM ATP, 2 units of T4 polynucleotide kinase, and 9.0 ng of the oligonucleotide. The reaction was incubated at 37° for 1 hr and terminated by adding 3 μl of 100 mM EDTA and heating at 65° for 10 min.

Hybridization of the Oligonucleotide to the Uracil-Containing Template

To the phosphorylated oligonucleotide was added 1 μg (in 0.6 μl) of single-stranded, uracil-containing, circular wild-type M13mp2 DNA and 1.2 μl of 20× SSC. After mixing and spinning the sample briefly (5 sec) in a microcentrifuge, the tube was placed in a 500-ml beaker of water at 70° and allowed to cool to room temperature. After another 5-sec centrifugation to spin down condensation, the tube was placed on ice.

We typically perform hybridization at a primer:template ratio between 2:1 and 10:1 since higher ratios do not yield more of the desired product and in some cases inhibit ligation. Hybridization conditions should be chosen to optimize heteroduplex formation with the particular oligonucleotide and template being used, and these conditions are expected to vary widely depending on the resulting heteroduplex (see Smith[1] and references therein for more details).

In Vitro DNA Synthesis and Product Analysis

The sequence contained within the oligonucleotide is converted to a biologically active, covalently closed circular (CCC) DNA molecule by DNA synthesis and ligation. These reactions are performed in a volume of 100 μl containing 20 μl of the above hybridization mixture; 20 mM HEPES (pH 7.8); 2 mM dithiothreitol; 10 mM MgCl$_2$; 500 μM each of dATP, TTP, dGTP, and dCTP; 1 mM ATP; 2.5 units (as defined by the supplier) of T4 DNA polymerase; and 2 units of T4 DNA ligase. All components are mixed at 0° (enzymes being added last), and the reaction is incubated at 0° for 5 min. The tube is placed at room temperature for 5 min, then at 37° for 2 hr. The rationale for this pattern of incubation is as follows.

The reaction is begun at lower temperatures (0°, then room temperature) to polymerize a small number of bases onto the 3' end of the oligonucleotide, thus stabilizing the initial duplex between the template phage DNA and the mutagenic oligonucleotide primer. However, since T4 DNA polymerase does not utilize long stretches of single-stranded DNA template well at low temperature, synthesis is then completed at 37°. We have not encountered significant pausing by T4 DNA polymerase under these conditions. The high concentration of dNTPs (500 mM) serves to optimize DNA synthesis and to reduce the 3'-exonuclease activity of the T4 DNA polymerase.

The reaction is terminated by addition of EDTA to 15 mM (3 μl of a 500 mM stock). The products of the reaction are then examined by subjecting 20 μl (to which is added 2.5 μl of SDS dye mix) to electrophoresis in a 0.8% agarose gel (in 1× TAE buffer containing 0.5 μg/ml ethidium bromide). For comparison, an adjacent lane should contain the appropriate standards: single-stranded, circular viral DNA, and double-stranded replicative form I (supercoiled CCC) and form II (nicked circular) DNAs.

The product of the *in vitro* DNA synthesis reaction should migrate at the same rate as the RF I standard, indicating that the DNA has been converted from primed circles to RF IV (duplex, CCC relaxed DNA) by the combined action of DNA polymerase and ligase. (Note that, in gels containing ethidium bromide, the dye will bind to the CCC relaxed DNA, generating positive supercoils and causing the RF IV to migrate like RF I.) Our experiments typically yield 80% conversion to primarily double-stranded DNA, but only 10–50% of this material is ligated to form covalently closed circles. Lower quality enzymes, unusual inserts, contaminants, or less than optimum reaction conditions may reduce the yield. (For more comments on the DNA synthesis step, see Troubleshooting below.)

Transfection and Plating

After incorporation of the oligonucleotide into double-stranded DNA *in vitro,* the DNA can be used to transfect (or transform) any competent *E. coli* strain (prepared with either CaCl$_2$[22] or by the method of Hanahan[23]). Provided the strain is *ung*[+], one can take advantage of the selection against the uracil-containing template strand. Unless high biological activity is required, the products of the *in vitro* DNA synthesis reaction can be used directly without further manipulation to remove reaction components.

For the reaction described above, 10 μl of the DNA synthesis mixture (containing approximately 80 ng of the double-stranded DNA) was added to 1 ml of competent CSH50 cells (prepared according to Hanahan[23]) in a sterile glass tube. A second transfection with a known amount of RF

[22] A. Taketo, *J. Biochem.* **72,** 973 (1972).
[23] D. Hanahan, *J. Mol. Biol.* **166,** 557 (1983).

DNA was performed to determine the transfection efficiency. The cells were gently mixed and incubated on ice for 30 min, heat shocked at 42° for 2 min, then returned to ice. Small volumes of this mixture (1, 5, 10, 50, or 100 µl) were added to tubes containing 2.5 ml of soft agar (at 50°), 2.5 mg Xgal (previously mixed well with the media to disperse the DMF), 0.24 mg IPTG, and 0.2 ml of mid-log culture of CSH50 cells. The mixtures were poured onto minimal plates and allowed to solidify. Plates were incubated at 37° overnight to allow blue color to develop as a measure of α-complementation. Nonconfluent plates (in this case the 1- and 5-µl plates) were scored for total plaques. Transfection using 10% (~80 ng in 10 µl) of the *in vitro* DNA synthesis reaction produced 193,000 plaques, an efficiency of about 2400 pfu/ng. In this particular instance, the desired mutants are expected to be colorless due to the addition of an extra base in the *lacZα* coding region. Colorless plaques comprised 70% of the total.

To confirm that the colorless mutants contained the expected sequence alteration, DNA from several phage was prepared for sequencing. First, 10 colorless plaques were harvested into 0.9% NaCl, serially diluted, and replated (as above) to obtain plaques derived from single phage particles. This genetic purification eliminates the possibility that the DNA to be sequenced comes from a plaque which contained two genotypes (as is possible in the original plaques) due either to transfection by a heteroduplex DNA molecule or to plaque overlap. This step becomes unnecessary if, following heat shock of the transfection mixture, one adds rich medium and continues incubation at 37° to allow production of progeny phage, which can then be diluted and plated. We have not used this extra incubation in experiments described here; thus, quantitation of the efficiency of mutant production does not require a correction for differences in growth rates of mutant versus wild-type phages. In either instance, once a purely mutant shock is obtained, the DNA can be sequenced by standard techniques. In our example, all 10 colorless plaques contained the expected change, an extra T residue in the viral strand.

Troubleshooting

Unsuccessful experiments are usually characterized by one of two outcomes. The first is low biological activity on transfection. To eliminate the obvious possibility that the *E. coli* cells were not competent, we always perform a parallel transfection with a known quantity of normal DNA to determine the transfection efficiency. Since biological activity depends on a complete complementary strand, inefficient DNA synthesis may also lead to low numbers of plaques. When this problem appears, a careful product analysis is warranted. In our experience, the presence of completely double-stranded DNA in the transfection mix has always been a harbinger of good biological activity. However, even when the yield of fully double-stranded DNA is poor, the strong selection against incompletely copied or uncopied uracil-containing templates which occurs on transfection permits the desired mutant to be recovered. In fact, we have obtained the desired sequence alterations from transfection of reactions that contained no double-stranded DNA as judged by agarose gel electrophoresis. These reactions yielded very few plaques, but 40% of the surviving DNA carried the mutant sequence.

In two instances (out of ~50 separate experiments), however, we observed no biological activity from mixtures which exhibited a band of DNA at the position of RF I on agarose gel electrophoresis in the presence of ethidium bromide. In these two cases, a further analysis was performed to determine the nature of the reaction products. If the product band were indeed the desired species, that is covalently closed circles of RF IV, this DNA would migrate like RF II DNA in the absence of the intercalating dye. Another possibility is that the product was the result of incomplete synthesis of the complementary strand and that its migration at the position of RF I in the initial agarose gel was fortuitous. If this were true, the migration of this DNA would not be affected by ethidium bromide. In both cases of low biological activity, the synthesis products migrated at the position of RF I in both the presence and absence of ethidium bromide, indicating that DNA synthesis was incomplete and explaining the lack of biological activity. (In both cases, the incomplete synthesis was due to the use of an old T4 DNA polymerase preparation that was no longer fully active.)

Incomplete synthesis can result from several factors, including inefficient hybridization of the oligonucleotide primer, inactive (or excess) DNA polymerase, contaminants in the DNA, the polymerase, or the reagents, or a DNA template which contains structures (e.g., hairpin loops) that block polymerization. (For example, with several T4 DNA polymerase preparations we have found as much as 10 units of enzyme may be required to achieve complete synthesis.) Such problems must be dealt with on an individual basis, e.g., by varying hybridization conditions, by ensuring the activity of the DNA polymerase, by repurifying the DNA, or by using alternative incubation temperatures or single-stranded DNA binding protein to assist the polymerase in synthesis on unusually difficult templates.

Low biological activity could also result from dNTP contamination by dUTP (e.g., by deamination of dCTP), which, when incorporated *in vitro,* provides targets for the production of lethal AP sites.[24] For this reason,

---

[24] P. D. Baas, H. A. A. M. van Teeffelen, W. R. Teerstra, H. S. Jansz, G. H. Veeneman, G. A. van der Marel, and J. H. van Boom, *FEBS Lett.* **110,** 15 (1980).

high quality dNTP substrates should be used to eliminate the need for dUTPase treatment of the deoxynucleoside triphosphates.[6,14] (dUTPase is not commercially available.)

The second undesirable outcome is a low percentage of mutants among the progeny. To confirm that there is indeed strong selection for the mutant strand over the template strand due to uracil residues in the latter, a titer of the phage from which the template is purified should be performed on *ung⁻* and *ung⁺* hosts. The difference in titers should be greater than 100,000-fold, although phage that show a smaller difference will still yield mutants at high efficiencies.[6] If a large difference in titer is not obtained, there is insufficient uracil in the viral DNA and new templates should be prepared. (Another but less likely possibility is that the putative *ung⁺* host is genetically and/or phenotypically *ung⁻*.)

A low percentage of mutants can also result from impurities in the template DNA which provide endogenous primers for complementary strand synthesis. The amount of endogenous priming can be determined by performing a DNA synthesis reaction without added oligonucleotide and examining the products by gel electrophoresis. When template is prepared as described above, the amount of RF IV DNA produced *in vitro* should be negligible in the absence of oligonucleotide. With impure template preparations, essentially all the single-stranded circular DNA can be converted to double-stranded product without added primer, in which case biological activity but no mutants will be obtained. The impurities can be removed by standard techniques (e.g., alkaline gradients) or the template can be prepared anew.

Another source of low mutant yield is displacement of the oligonucleotide during *in vitro* DNA synthesis of the strand which carries the mutation (see line 4, Table 2, in Ref. 6). This possibility is not a concern when one uses T4 DNA polymerase, since this enzyme does not perform strand displacement synthesis under the conditions given above. However, such problems may arise with some primers and/or templates that have more, or stronger, polymerase pause sites than do our templates. In these situations, the polymerase can be assisted by its homologous single-stranded DNA binding protein (the T4 gene *32* protein). Alternatively, the Klenow fragment of *E. coli* DNA polymerase I, which efficiently synthesizes complete complementary strands without significant pausing, can be used at low temperature (to prevent displacement of the oligonucleotide). Even this enzyme has difficulty with some templates at reduced temperatures; in these instances *E. coli* single-stranded DNA binding protein may be helpful.

One pathway by which the mutant, complementary strand of DNA may be selected against is the methyl-directed mismatch correction system present in most *E. coli* strains.[25] Repair synthesis which uses the uracil-containing strand as a template before it is destroyed will eliminate the artificially produced mutation from the complementary strand. Although this process is not a major concern (since we routinely achieve high efficiency, 50–70%), two means of reducing mismatch repair and improving mutant production can be employed in those situations where one desires the highest efficiency possible. The first is to transfect the DNA into competent cells made from *E. coli* mutator strains (*mutH*, *mutL*, or *mutS*) which are deficient in mismatch correction.[25] [In a typical experiment, the efficiency of mutant production was improved from 51% (wt) to 57% (*mutH*), 67% (*mutL*), and 59% (*mutS*).] A second strategy is to treat the product DNA with uracil *N*-glycosylase as described[6] to form AP sites in the template strand, followed by alkali treatment to hydrolyze the phosphodiester bonds at AP sites and to disrupt hydrogen bonding. This protocol eliminates the transforming activity of the parental strand and leaves only the covalently closed, single-stranded DNA which contains the desired sequence alteration as a source of transforming DNA. In practice, this treatment improved efficiency from 51 to 89%.[6] We do not routinely utilize either mutator strains or glycosylase, since, for our purposes, 50–70% efficiency is more than sufficient and since uracil *N*-glycosylase is not commercially available.

We originally observed good biological activity but low efficiency of mutagenesis when ligase was omitted from the *in vitro* reaction (see line 4, Table 2, in Ref. 6). This observation was made with DNA products synthesized by the DNA pol I Klenow fragment under conditions that allow strand displacement (37°). More recently, using T4 DNA polymerase as described above, we attempted oligonucleotide-directed mutagenesis of the gene for the *E. coli* cyclic AMP receptor protein. We used a 36-mer which primed DNA synthesis efficiently but could not be ligated to form CCC DNA (i.e., only RF II but no RF IV DNA was produced), perhaps because of abberant chemistry during oligonucleotide synthesis or impurities in the oligonucleotide preparation. Despite the fact that the product DNA was not covalently closed, the mutant yield, with no selection, was 67% (six of the nine clones sequenced having the appropriate mutation). We conclude that, with T4 DNA polymerase (i.e., in the absence of strand displacement), ligation *in vitro* is not required for highly efficient mutant production. Presumably, on transfection, the cell performs the necessary processing at the termini before the template DNA is destroyed.

Lest these comments on troubleshooting discourage use of the technique, we reiterate that, after the preparation of the uracil-containing

[25] B. W. Glickman and M. Radman, *Proc. Natl. Acad. Sci. U.S.A.* **77**, 1063 (1980).

template, oligonucleotide-directed mutagenesis at or above 50% efficiency is a simple procedure, consisting of a polymerization reaction and a transfection.

Variations

We have presented here a simple oligonucleotide-directed mutagenesis protocol to demonstrate the utility of the uracil selection technique for generating mutants with high efficiency (≥50%). Uracil-containing DNA can be prepared for any vector that can be passaged through an *E. coli* *dut⁻ ung⁻* strain. Such templates can be used in conjunction with the wide variety of established procedures (gapped duplexes, double priming, etc.) and vectors (single-stranded phage, pBR derivatives, shuttle vectors, etc.). This procedure may also prove to be useful for investigating the mutagenic potential of specific DNA adducts located at defined positions in genes and for studies of mutational specificity of *in vitro* DNA synthesis.[21] The applications of these techniques for engineering DNA sequences and the proteins for which they code are limited only by the need and the imagination of the investigator.

[33] An Improved Method to Obtain a Large Number of Mutants in a Defined Region of DNA

By RICHARD PINE and P. C. HUANG

Introduction

In recent years production of mutations *in vitro* has supplemented the ability to learn from naturally occurring mutations about specific genes and gene products. Two widely used methods of mutagenesis include oligonucleotide-directed mutagenesis of specific nucleotides and treatment of target regions with base-specific chemical mutagens.[1-36] These

[1] D. Botstein and D. Shortle, *Science* **229**, 1193 (1985).

[2] P. J. Carter, G. Winter, A. J. Wilkinson, and A. R. Fersht, *Cell* **38**, 835 (1984).

[3] P. Carter, H. Bedouelle, and G. Winter, *Nucleic Acids Res.* **13**, 4431 (1985).

[4] P. Carter, this series, Vol. 154, p. 382.

[5] D. J. Der, T. Finkel, and G. M. Cooper, *Cell* **44**, 167 (1986).

[6] R. D. Everett and P. Chambon, *EMBO J.* **1**, 433 (1982).

[7] O. Fasano, T. Aldrich, F. Tamanoi, E. Taparowsky, M. Furth, and M. Wigler, *Proc. Natl. Acad. Sci. U.S.A.* **81**, 4008 (1984).

[8] W. Kramer and H.-J. Fritz, this volume [31].

[9] S. S. Ghosh, S. C. Bock, S. E. Rokita, and E. T. Kaiser, *Science* **231**, 145 (1986).

[10] S. Gutteridge, I. Sigal, B. Thomas, R. Arentzen, A. Cordova, and G. Lorimer, *EMBO J.* **3**, 2737 (1984).

[11] M. Hannink and D. J. Donoghue, *Proc. Natl. Acad. Sci. U.S.A.* **82**, 7894 (1985).

[12] K. Itakura, J. J. Rossi, and R. B. Wallace, *Annu. Rev. Biochem.* **53**, 323 (1984).

[13] D. K. Jemiolo, C. Zwieb, and A. E. Dahlberg, *Nucleic Acids Res.* **13**, 8631 (1985).

[14] J. T. Kadonaga and J. R. Knowles, *Nucleic Acids Res.* **13**, 1733 (1985).

[15] T. A. Kunkel, J. D. Roberts, and R. A. Zakour, this volume [32].

[16] R. J. Leatherbarrow, A. R. Fersht, and G. Winter, *Proc. Natl. Acad. Sci. U.S.A.* **82**, 7840 (1985).

[17] S.-M. Liang, D. R. Thatcher, C.-M. Liang, and B. Allet, *J. Biol. Chem.* **261**, 334 (1986).

[18] D. F. Mark, A. Wang, and C. Levenson, this series, Vol. 154, p. 403.

[19] R. Pine, M. Cismowski, S. W. Liu, and P. C. Huang, *DNA* **4**, 115 (1985).

[20] R. Rohan and G. Ketner, *J. Biol. Chem.* **258**, 11576 (1983).

[21] M. Schold, A. Colombero, A. A. Reyes, and R. B. Wallace, *DNA* **3**, 469 (1984).

[22] D. Shortle and D. Nathans, *Proc. Natl. Acad. Sci. U.S.A.* **75**, 2170 (1978).

[23] D. Shortle, D. DiMaio, and D. Nathans, *Annu. Rev. Genet.* **15**, 265 (1981).

[24] D. Shortle and D. Botstein, this series, Vol. 100, p. 457.

[25] M. Smith, *Annu. Rev. Genet.* **19**, 423 (1985).

[26] J. W. Taylor, J. Ott, and F. Eckstein, *Nucleic Acids Res.* **13**, 8765 (1985).

[27] P. G. Thomas, A. J. Russell, and A. R. Fersht, *Nature (London)* **318**, 375 (1985).

[28] J. A. Tobian, L. Drinkard, and M. Zasloff, *Cell* **43**, 415 (1985).

[29] P. V. Viitanen, D. R. Menick, H. K. Sarkar, W. R. Trumble, and H. R. Kaback, *Biochemistry* **24**, 7628 (1985).

RECOMBINANT DNA
METHODOLOGY

methods embody two rationales for obtaining mutant genes. In the use of random target-directed methods, the genotype must be correlated with the phenotype after the mutagenesis. In contrast, each mutation is exactly specified in advance by oligonucleotide-directed mutagenesis. The desired mutants should be examined by DNA sequencing to determine or confirm the mutations produced by either method.

With oligonucleotide-directed protocols, specific individual mutations can be obtained by targeting relatively few residues of known interest without reliance on the phenotype. This approach has been useful especially for structural gene modifications which have been used to confirm and elaborate X-ray crystallographic data or specific biochemical evidence on the precise amino acid involvement in a catalytic or folding interaction.[2,5,9–11,16–18,27,29–33] Creating various replacements of a single key residue can reveal how the native amino acid functions in the protein.[2,5,11] Although in theory every base can be altered by oligonucleotide-directed mutagenesis, a significant limit is the fact that the entire procedure must be undertaken to obtain a given mutant. Thus this method is best suited to production of a small number of mutant molecules.

Mutations produced randomly at relatively low frequency (0.5–1.6% of mutable sites) can be screened or selected for specific molecular characteristics or desired biological phenotypes. The fraction of target molecules having mutations will be determined by the number of bases and the proportion that are mutable in the target sequence. Systems in which the DNA of interest can be mapped to a particular gene segment but for which more specific information is not available have been used in applications of base specific mutagenesis.

Several examples of this approach have used sodium bisulfite as the mutagen. This chemical can react specifically with cytosine in the single-stranded regions of DNA, causing deamination of the cytosine to produce uracil. On copying of the mutated single-strand DNA by DNA polymerase, adenosine pairs with the uracil, and finally thymidine replaces the original cytosine. The end result is replacement of a CG base pair with a TA base pair; thus each strand in the target has undergone a transition

[30] J. E. Villafranca, E. E. Howell, D. H. Voer, M. S. Strobel, R. C. Ogden, J. N. Abelson, and J. Kraut, *Science* **222,** 782 (1983).
[31] G. Weinmaster, M. J. Zoller, M. Smith, E. Hinze, and T. Pawson, *Cell* **37,** 559 (1984).
[32] A. J. Wilkinson, A. R. Fersht, D. M. Blow, and G. Winter, *Biochemistry* **22,** 3581 (1983).
[33] G. Winter, A. R. Fersht, A. J. Wilkinson, M. Zoller, and M. Smith, *Nature* (*London*) **299,** 756 (1982).
[34] M. J. Zoller and M. Smith, this volume [30].
[35] M. J. Zoller and M. Smith, *DNA* **3,** 479 (1984).
[36] M. J. Zoller and M. Smith, this series, Vol. 154, p. 329.

mutation. Various experiments have identified alterations in properties of viral growth[22] and gene regulation[6,20] or *ras* protein transforming potential.[7] Structure–function studies of the 3' minor domain of *E. coli* 16 S rRNA also have been based on *in vitro* mutagenesis with sodium bisulfite.[13] Analogous procedures which produce CT to GA transitions with hydroxylamine[14] or methoxylamine[28] have been used in studies of protein or RNA processing, respectively.

A significant advantage of random mutagenesis is the ability to obtain mutations at various sites and in different combinations with a single treatment. In addition, conditions can be controlled to alter the extent of mutagenesis and thus the distribution of mutations from one experiment to another.

An approach in which large numbers of altered molecules are produced with random mutations in single or multiple occurrence is required for certain targets. Such a method can be designed by adaptation of low frequency mutagenesis with sodium bisulfite in combination with appropriate cloning and screening. In addition to the criteria on which other uses of random mutagenesis are based, these targets should be sequences with many putatively important sites. Mutants containing varied but specific base alterations throughout a gene would provide a sufficient choice of replacements to be compared and studied. This method is well suited to investigation of conserved proteins or the conserved sequences of gene families in general. The degenerate triplet code ensures that certain conserved codons always will be subject to coding changes that are not found naturally, while some may be mutated both in a degenerate and nondegenerate manner. Others though will never be changed by a given base-specific mutagen. High frequency mutagenesis and production of large numbers of mutants allow a great likelihood of obtaining all possible alterations of each conserved site. Such an approach is especially desirable when many sites of interest occur within the target and when the effects of single, double, or multiple mutations may reveal important interactions between the bases of nucleic acids or the amino acids in proteins.

The metallothionein (MT) system is an excellent model for the above approach to mutagenesis *in vitro*. MT is a ubiquitous, highly conserved protein which binds 7 g atoms of transition class IIB (or IB) metals with tetrahedral coordination via mercaptide bonds.[37–46] Typical MTs have no

[37] J. H. R. Kagi and M. Nordberg, *in* "Metallothionein" (J. H. R. Kagi and M. Nordberg, eds.), p. 41. Birkhauser Verlag, Basel, Switzerland, 1979.

[38] Y. Boulanger, I. M. Armitage, K.-A. Miklossy, and D. R. Winge, *J. Biol. Chem.* **257**, 13717 (1982).

[39] Y. Boulanger, C. M. Goodman, C. P. Forte, S. W. Fesik, and I. M. Armitage, *Proc. Natl. Acad. Sci. U.S.A.* **80**, 1501 (1983).

aromatic amino acids or histidine, and 20 of 61 amino acids are cysteines.[46-65] The protein is comprised of two metal binding domains.[38-41,66,67] It is suggested that at least the 20 cysteines[37,39,40,43-45,68] and probably many other conserved amino acids such as lysines[69] act together in the structure and function of MT. Thus this system is a candidate for production of a large number of point mutations which may provide more

[40] R. W. Briggs and I. M. Armitage, J. Biol. Chem. **257**, 1259 (1982).

[41] K. B. Nielson and D. R. Winge, J. Biol. Chem. **258**, 13063 (1983).

[42] K. B. Nielson, C. L. Atkin, and D. R. Winge, J. Biol. Chem. **260**, 5342 (1985).

[43] M. Vasak and J. H. R. Kagi, Proc. Natl. Acad. Sci. U.S.A. **78**, 6709 (1981).

[44] M. Vasak and J. H. R. Kagi, in "Metal Ions in Biological Systems" (H. Sigel, ed.), Vol. 15, p. 213. Dekker, New York, 1983.

[45] M. Vasak, G. E. Hawkes, J. K. Nicholson, and P. J. Sadler, Biochemistry **24**, 740 (1985).

[46] P. D. Whanger, S. M. Oh, and J. T. Deagan, J. Nutr. **111**, 1207 (1981).

[47] R. D. Andersen, B. W. Birren, T. Ganz, J. E. Piletz, and H. R. Herschman, DNA **2**, 15(1983).

[48] R. D. Andersen, B. W. Birren, S. J. Taplitz, and H. R. Herschman, Mol. Cell. Biol. **6**, 302 (1986).

[49] D. M. Durnam, F. Perrin, F. Gannon, and R. D. Palmiter, Proc. Natl. Acad. Sci. U.S.A. **77**, 6511 (1980).

[50] N. Glanville, D. M. Durnam, and R. D. Palmiter, Nature (London) **292**, 267 (1981).

[51] B. B. Griffith, R. A. Walters, M. D. Enger, C. E. Hildebrand, and J. K. Griffith, Nucleic Acids Res. **11**, 901 (1983).

[52] I.-Y. Huang, A. Yoshida, H. Tsunoo, and H. Nakajima, J. Biol. Chem. **252**, 8217 (1977).

[53] I.-Y. Huang, M. Kimura, A. Hata, H. Tsunoo, and A. Yoshida, J. Biochem. **89**, 1839 (1981).

[54] M. Karin and R. I. Richards, Nature (London) **299**, 797 (1982).

[55] M. M. Kissling and J. H. R. Kagi, in "Metallothionein" (J. H. R. Kagi and M. Nordberg, eds.), p. 145. Birkhauser Verlag, Basel, Switzerland, 1979.

[56] S. Koizumi, N. Otaki, and M. Kimura, J. Biol. Chem. **260**, 3672 (1985).

[57] Y. Kojima, C. Berger, B. L. Vallee, and J. H. R. Kagi, Proc. Natl. Acad. Sci. U.S.A. **73**, 3413 (1976).

[58] Y. Kojima, C. Berger, and J. H. R. Kagi, in "Metallothionein" (J. H. R. Kagi and M. Nordberg, eds.), p. 153. Birkhauser Verlag, Basel, Switzerland, 1979.

[59] K. Munger, U. A. Germann, M. Beltramini, D. Niedermann, G. Baitella-Eberle, J. H. R. Kagi, and K. Lerch, J. Biol. Chem. **260**, 10032 (1985).

[60] R. I. Richards, A. Heguy, and M. Karin, Cell **37**, 263 (1984).

[61] C. J. Schmidt and D. H. Hamer, Gene **24**, 137 (1983).

[62] C. J. Schmidt, M. F. Jubier, and D. H. Hamer, J. Biol. Chem. **260**, 7731 (1985).

[63] P. F. Searle, B. L. Davison, G. W. Stuart, T. M. Wilkie, G. Norstedt, and R. D. Palmiter, Mol. Cell. Biol. **4**, 1221 (1984).

[64] U. Varshney, N. Jahroudi, R. Foster, and L. Gedamu, Mol. Cell. Biol. **6**, 26 (1986).

[65] D. R. Winge, K. B. Nielson, R. D. Zeikus, and W. R. Gray, J. Biol. Chem. **259**, 11419 (1984).

[66] K. B. Nielson and D. R. Winge, J. Biol. Chem. **260**, 8698 (1985).

[67] D. R. Winge and K.-A. Miklossy, J. Biol. Chem. **257**, 3471 (1982).

[68] J. D. Otvos and I. M. Armitage, Proc. Natl. Acad. Sci. U.S.A. **77**, 7094 (1980).

[69] J. Pande, M. Vasak, and J. H. R. Kagi, Biochemistry **24**, 6717 (1985).

detailed information concerning the secondary structural folds and tertiary structural domains of MT, its mechanism of metal binding, as well as the regulation of MT genes and the physiological function of MT proteins.

Figure 1 shows the alterations of the Chinese hamster ovary cell (CHO) MT II coding sequence that can be induced by sodium bisulfite mutagenesis of either its sense or antisense strand. Naturally occurring differences in MT sequences which have evolved in human, monkey, horse, cow, mouse and rat, as well as CHO MT I, are also included for comparison. This representation of natural differences is culled from published reports of protein[46,52,53,55–59,65] and/or nucleic acid sequences.[47–51,54,60–64] As can be seen, the sequence of the target for these studies, CHO MT II cDNA, ensures that mutagenesis with sodium bisulfite can be used on its sense or antisense strand to allow significant alteration of conserved codons. At the same time, mutagenesis of the complementary strand subjects a given codon to only degenerate mutations. Also, the codons of many nonconserved amino acids may be mutated to encode residues which have not been detected at the respective position

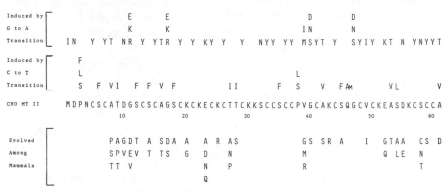

FIG. 1. Comparison of CHO MT II amino acid sequence to mutations that can be induced by sodium bisulfite and to differences evolved among other mammalian MTs. The single letter abbreviations for the amino acids are used; an amber nonsense codon is indicated by Am. Amino acid changes encoded due to induced G to A or C to T transition mutations are shown above the amino acid sequence of CHO MT II. Amino acid differences that have evolved among other mammalian MTs, as determined from protein analysis or deduced from DNA sequences, are shown below the respective positions in the CHO MT II sequence. It should be noted that the occurrence of glutamine at position 52 is reported only for mouse MT II,[53] while glutamate is deduced for that position from DNA sequence data.[63] Also, cysteine at position 58 and serine at position 59 are only reported in the amino acid sequences of MT I and MT II from horse[58] and human.[55] However, analysis of four apparently functional human MT genes has shown serine and cysteine to be encoded at positions 58 and 59, respectively.[54,60,62,64] At this time, human genes known to be functional and which encode valine at positions 10 or 12, threonine at position 53, or aspartate at position 55 have not been cloned.

in any naturally occurring MT. Only an asparagine and the lysine codons cannot be mutated with sodium bisulfite, while the conserved glutamine codon at position 46 can be mutated only to an amber nonsense codon. Thus, determination of the importance and role of conserved residues as well as the leeway for changes at nonconserved positions in MT is made accessible if large numbers of mutants are produced.

Principles

Conditions for mutagenesis of nucleic acids with sodium bisulfite have been developed based on the well-characterized reactions of this reagent with the bases of nucleosides, nucleotides, and nucleic acids.[70] Sulfonation at the 6 position of cytosine, uracil, and thymine occurs readily at 37° between pH 3 and 8. The reaction with uracil and thymine is fully reversible, while the adduct of cytosine can undergo hydrolytic deamination to produce 5,6-dihydrouracil-6-sulfonate. The 5,6-dihydrouracil-6-sulfonate adduct yields uracil and bisulfite irreversibly at pH 9. Thus bisulfite can be used to specifically convert cytosine to uracil. Although the rate for deamination of cytosine is maximal at pH 5, protection of acid-labile components of nucleic acids is effected by the buffering capacity of bisulfite when the reaction is carried out at pH 6–7. Oxidation of bisulfite can produce free radicals which react in various ways with DNA. However, this process can be prevented by using a relatively high concentration of bisulfite (≥ 1 M) and a free radical scavenger such as hydroquinone. In addition, these relatively high concentrations are necessary for treatment of nucleic acids since the susceptible residues in polynucleotides react manyfold more slowly than the corresponding constituents free in solution. Importantly, the reaction of bisulfite with bases that are hydrogen bonded in Watson–Crick pairs is negligible. Thus for all practical purposes, only single-stranded portions of nucleic acids are reactive.

These considerations underlie the protocol for reaction with sodium bisulfite to mutagenize DNA.[22,24] The initial reaction is carried out at pH 6–7 and 37° using 1–3 M bisulfite. During the incubation 5,6-dihydrocytosine-6-sulfonate adduct formation and deamination occur within the single-strand portion of a target molecule. Dialysis at pH 6.8 and 0° stops the reaction as the bisulfite is removed. Further dialysis at pH 9 desulfonates any remaining adducts and thus completes the conversion of cytosine to uracil.

In the method described here, M13 phage have been used as vectors

---

[70] H. Hayatsu, *Prog. Nucleic Acid Res. Mol. Biol.* **16,** 75 (1976).

for the target DNA. Several advantages make this an attractive choice. Phage particles are very stable at −20 or 4°. Single-stranded genomic DNA can be easily isolated from these particles. To obtain material for recombinant DNA manipulations, closed circular replicative form (RF) DNA can be extracted from bacterial cells infected by the phage. A library can be propagated initially and individual elements recovered from the phage first secreted after chimeric M13 DNA is transformed into *Escherichia coli*. However, M13 is not ideal for the amplification of a library of cloned DNA since deletions are more likely than in some other vectors. Strains of M13 have been constructed to allow rapid cloning and sequencing of DNA fragments.[70] The design of these strains includes a multiple cloning site which provides numerous restriction sites to facilitate insertion of DNA into the vector. Additionally, the cloned DNA can later be subcloned via fragments isolated with enzymes other than those used in the first place. Pairs of M13 strains have a given multiple cloning site inserted in opposite orientations. A target sequence can be cloned into each of the pair of strains so that the recombinant fragment will have the same flanking sequence in the double-stranded chimeric DNA derived from either strain. Each strand of the cloned DNA will be part of the single-stranded phage DNA in one strain or the other.

Single-stranded recombinant phage DNA is a ready target for reaction with sodium bisulfite, but the phage genes must be protected from mutagenesis since they are all essential to the phage life cycle. Double-stranded vector DNA, linearized by restriction digestion at the same site or sites used for construction of the chimeric DNA, is annealed to the respective recombinant single-stranded phage DNA to form a gapped duplex molecule. Thus, each strand of the cloned sequence remains available as a target while the viability of the M13 vector is protected.

The final step needed to obtain fixed mutations with this method is propagation of the treated DNA. Since these molecules contain deoxyuridine in a single-strand region they are susceptible to attack by uracil-DNA glycosylase and degradation by further nucleolytic action in wild-type *E. coli* cells. However, cells which are *ung*⁻ lack uracil-DNA glycosylase and will not carry out the first step of this process. Thus gapped duplex DNA containing uracil in the single-stranded region can be directly transformed into such *E. coli* strains. Adenine will pair with uracil and then thymine will pair with adenine; thus a cytosine to thymine transition is fixed during replication of the phage. Phage initially secreted from these cells constitute a mutant library which can be stored safely. Individual elements are obtained by plating the phage with a normal host strain to give single plaques. They can be exactly and efficiently characterized by extracting and sequencing the DNA of phage amplified from the plaques. Subsequently, the RF DNA of chosen mutants can be prepared to provide material for whatever manipulations are needed for further studies. Thus the combination of sodium bisulfite mutagenesis, M13 vectors, and an *ung*⁻ strain of *E. coli* provides a powerful yet simple and direct way to produce and handle large numbers of mutants having the desired distribution of random mutations.

Materials

Bacteria, Phage, and Plasmids

The following bacterial strains were used in this study: *Escherichia coli* BD1528,[71] which lacks uracil-DNA glycosylase, as a recipient for DNA mutagenized by sodium bisulfite and *E. coli* JM101[70] as a host for M13 phage. Phage M13mp8 and M13mp9[71] were used as vectors for DNA to be mutagenized and for DNA sequencing. *Escherichia coli* cultures carrying chimeric plasmids containing CHO MT cDNA clones were obtained from the Genetics group, Los Alamos National Laboratory, courtesy of Drs. Barbara and Jack Griffith.[51] Subcloning of MT sequences into M13mp9 and M13mp8, which produced M13MT2S and M13MT2A, respectively, has been described.[19] These contain the sense and antisense strands of the MT coding sequence in the single-strand genomic phage DNA.

Reagents

Chemicals. Acrylamide and N,N'-methylenebisacrylamide were purchased from Bio-Rad. Deoxynucleotide triphosphates and dideoxynucleotide triphosphates were from P-L Biochemicals. Pentadecamer sequencing primer was from New England Biolabs. Other specialty reagents were obtained from Bethesda Research Laboratories. [α-³²P]dATP (>3000 Ci/mmol) was purchased from Amersham. Hydroquinone (99+ %), from Aldrich, was stored under N_2. All other chemicals were reagent grade from various standard suppliers.

Enzymes. T4 DNA ligase was purchased from either Boehringher Mannheim Biochemicals or Bethesda Research Laboratories. All other enzymes were purchased from Bethesda Research Laboratories.

---

[71] J. Messing, this series, Vol. 101, p. 20.

Methods and Results

Microbiology

Escherichia coli BD1528 was maintained on NZYDT plates[72] or grown in NZYDT broth at 37°. M13 phage, recombinants therein, and the host, *E. coli* JM101, were kept and grown as described,[71] using 2× YT broth or B media for agar plates and top agar. Transformation of DNA into bacterial cells was by a calcium chloride procedure essentially as described,[73] except that cells were 10-fold concentrated from the log-phase culture.

Preparation of DNA

Phage from a single plaque were first amplified. An overnight culture of JM101 was diluted 1 : 100 into 1.5- and 10-ml aliquots of 2× YT broth. The small aliquot was inoculated with phage from a single plaque and both aliquots were grown at 37° with vigorous aeration. After 2.5 hr the small culture was added to the large culture and growth was continued for 5 hr. The culture was centrifuged at 13,000 g for 10 min, and the supernatant, containing phage at 10^{12} pfu/ml, was decanted and stored at $-20°$. M13 RF DNA was obtained as follows. Mid-log cultures of JM101 (400 ml at 2 × 10^8 cells/ml) were infected with 4 ml of phage at 10^{12} pfu/ml and incubation was continued for 4 hr at 37° with aeration. Chloramphenicol at 25 μg/ml was added for the last 30 min, then DNA was extracted according to Birnboim,[74] omitting RNase treatment and adding CsCl banding[72] for 16 hr at 190,000 g (45,000 rpm) in a Beckman VTi65 rotor. M13 genomic DNA was prepared as described.[74]

Sodium Bisulfite Mutagenesis of MT Coding Sequences

A gapped heteroduplex was formed by annealing the chimeric M13 plus strand with the denatured double-strand M13 vector DNA, which had been linearized by restriction digestion at the multilinker site originally used for subcloning the MT sequence.[19] The appropriate vector DNA was mixed with recombinant DNA, M13MT2S (sense strand insert in M13qmp9) or M13MT2A (antisense strand insert in M13mp8), at 2.5 : 1.0 (w/w) in 3 mM sodium citrate, 30 mM sodium chloride (pH 7.0),

[72] T. Maniatis, E. F. Fritsch, and J. Sambrook, "Molecular Cloning: A Laboratory Manual." Cold Spring Harbor Lab., Cold Spring Harbor, New York, 1982.

[73] R. W. Davis, D. Botstein, and J. R. Roth, "Advanced Bacterial Genetics," p. 140. Cold Spring Harbor Lab., Cold Spring Harbor, New York, 1980.

[74] H. C. Birnboim, this volume [8].

10 mM magnesium chloride to give a final total DNA concentration of 15 μg/ml. The DNA was heated to 90° and held for 2.5 min, cooled quickly to 80° and held for 3 min, cooled quickly to 70° and held for 10 min, cooled quickly to 65° and held for 50 min, and then slowly cooled to 25°. The extent of annealing can be assessed by electrophoresis of a small aliquot on a 1% agarose gel in Tris–acetate buffer,[72] which resolves linear duplex, gapped duplex circular, and single-stranded circular molecules. However, samples should not be heated before being loaded, and the gel should be run slowly, at or below 2 V/cm, until the bromophenol blue marker dye is at the end of the gel. Gapped duplex circular molecules have the same mobility as fully duplex relaxed circular molecules. Conversion of single-stranded circular DNA to gapped circular duplex DNA should be observed. This assay has been used to determine that the annealed DNA can be stored for several days at 4°.[75] For routine use, the annealing reaction is so reliable that this analysis is generally not necessary.

Fresh stocks of 50 mM hydroquinone and 4 M sodium bisulfite (pH 6.0) (1.56 g NaHSO$_3$, 0.64 g Na$_2$SO$_3$, 4.3 ml H$_2$O) were used to produce uracil in the single-strand MT portion of the heteroduplex essentially as described.[22,24] The annealed DNA was mixed with bisulfite stock and water in 1.5-ml polypropylene tubes to yield varying final concentrations of mutagen. Addition of 0.04 volumes of hydroquinone completed the reaction mixture, which was then overlaid with paraffin oil and incubated at 37° in the dark. Reactions with 1 M sodium bisulfite were in a final volume of 190 μl or 380 μl with a final DNA concentration of 7.5 μg/ml while reactions with 3 M sodium bisulfite were in 760 μl with a final DNA concentration of 3.75 μg/ml. Thus reactions in 190 μl contained 0.2 μg of chimeric DNA and 0.5 μg of vector DNA while the others contained 0.4 μg of chimeric DNA and 1.0 μg of vector DNA. The final concentration of bisulfite and the length of reaction prior to dialysis were varied to meet the goal of producing a large number of singly or multiply mutated molecules.

Following the incubation samples were recovered from below the paraffin oil with a pasteur pipet and transferred to $\frac{1}{4}$-inch dialysis tubing. Dialysis was performed at 0° versus 5 mM potassium phosphate (pH 6.8), 0.5 mM hydroquinone, twice for 2 hr each time, then once again at 0° for 2 hr versus 5 mM potassium phosphate (pH 6.8). Each change was greater than about 200 volumes relative to the sum of the reaction mixture volumes. Further dialysis against 200 mM Tris (pH 9.2), 50 mM NaCl, 2 mM EDTA was performed for 16–20 hr at room temperature, followed by dialysis against 10 mM Tris, 1 mM EDTA (pH 8.0) for 4 hr at 4°. These dialyses used at least 400 volumes relative to the sum of the reaction

[75] R. Pine, unpublished observations (1985).

mixture volumes. Finally mutagenized DNA was recovered and propagated. The use of dialysis to complete the conversion of cytosine to uracil is especially appropriate for handling many samples at one time, as may be required to produce a range of mutation frequencies for a given target sequence.

The target in this example, CHO MT II cDNA, consists of 62 codons, inclusive of termination,[51] as expected of that protein in mammalian species. It has 55 G and 55 C residues in its coding sequence. The choice of M13 as a vector allows use of sodium bisulfite to obtain exclusively G to A transitions in the codons (GA clones) as the indirect result, via base pairing, of C to T transitions in the original antisense single-stranded DNA. Codons of the sense strand directly reflect C to T mutations (CT clones) made in that single-stranded target. As described for the general case in the section Principles, each strand of the MT coding sequence was obtained with identical vector flanking sequence by cloning the sense and antisense strands of the CHO MT II sequence into the complementary M13 strains mp9 and mp8, respectively. These clones, M13MT2S and M13MT2A, served as a target for *in vitro* mutagenesis with sodium bisulfite. However, different proportions of degenerate and nondegenerate mutations will result from C to T transitions compared to G to A transitions in the coding sequence. Thus it was necessary to use different reaction times or concentrations of sodium bisulfite for the mutagenesis of the sense and antisense single strands to achieve comparable numbers of coding changes due to C to T or G to A mutations in the codons. The conditions were chosen based on earlier reports of mutagenesis with sodium bisulfite.[7,20,22,24] As the data will show, this empirical approach produced the desired results, which now provide an additional guide for other investigators.

Propagation of Mutated DNA and Analysis of the Mutant Library

The library of mutant MT sequences is composed of collections obtained from reactions with 1 M sodium bisulfite carried out for 3, 5, or 7 hr, and from incubations of 20 or 40 min with 3 M sodium bisulfite. *Escherichia coli* BD1528 (ung^-)[76] was transformed with 10-μl aliquots of the mutagenized DNA. Cells were incubated with 1 ml 2× YT broth for 2.5 hr at 37°. DNA replication produced C to T transitions, thus fixing the C to U changes generated *in vitro*. At this time, the beginning of phage secretion, cells were removed by centrifugation. To allow sequencing of individual mutants, single plaques at a density of 50–150 per plate were

[76] B. K. Duncan and J. A. Chambers, *Gene* **28,** 211 (1984).

obtained by infecting JM101 with 0.1-ml portions of the secreted phage. The remainder of the phage in the supernatants, constituting the mutant library, was stored at −20°.

Dideoxy sequencing of single-stranded recombinant phage DNA was carried out as described[77] except that annealing of primer to template was done in 500-μl polypropylene tubes, and sequencing reactions were done at 30° in 400-μl polypropylene tubes. Reaction products were analyzed on 50 cm long by 0.4 mm thick 6% (w/v) polyacrylamide gels (40 : 1 acrylamide : N,N'-methylenebisacrylamide) containing 7 M urea, 100 mM Tris–borate (pH 8.3), 2 mM EDTA. Aliquots of samples cloned in M13mp8 were electrophoresed until the xylene cyanol marker dye had migrated 31 cm, while two aliquots of samples cloned in M13mp9 were electrophoresed so that either the bromophenol blue or the xylene cyanol FF had migrated 47 cm. The gels were dried and autoradiographed for 16–24 hr at room temperature.

For the library as a whole,[19] mutations were obtained at every G residue. Thus, all the possible coding changes caused by G to A transitions indicated in Fig. 1 were produced, either alone or in various combinations. Only 24 of 34 possible degenerate C to T transition mutations were found, but 19 of 21 possible nondegenerate C to T transitions encoding all the amino acid changes shown in Fig. 1 except Pro$_{38}$ to Ser and Ala$_{42}$ to Val were obtained. Overall, only 56 of the 183 GA clones examined were found to have no alterations, and an additional 14 clones had only degenerate mutations. Of the 81 CT mutants sequenced only one had no changes, while 23 more had only degenerate mutations. Extrapolating from the data collected, 45% of the GA elements and 34% of the CT elements are unique within their collections. Since only 264 plaques of a possible 10^5 have been examined to date, the phage stocks are expected to contain still more unique mutated sequences.

The sequence analysis of individual mutated DNA molecules showed that the collections within the library had different desired characteristics. Treatment of the antisense strand with 3 M sodium bisulfite for 20 and 40 min or with 1 M sodium bisulfite for 3 or 5 hr produced G to A mutations in the codons (GA clones). The sense strand was treated for 5 or 7 hr with 1 M sodium bisulfite to produce C to T changes in the codons (CT clones).

Another view of the data obtained from sequence analysis of this library is shown in Table I as the percent of cytosine residues which reacted with bisulfite for the GA clones. The results validate the empirical

---

[77] F. Sanger, A. R. Coulson, B. G. Barrell, A. J. H. Smith, and B. A. Roe, *J. Mol. Biol.* **143,** 161 (1980).

TABLE I
EXTENT OF MUTAGENESIS FROM VARIOUS REACTION CONDITIONS[a]

GA collection	Reaction time (min)	Sodium bisulfite concentration (M)	Number of clones sequenced	Number of target sites	Target sites mutagenized (%)
A	20	3	42	2,310	5.3
B	20	3	37	2,035	10.5
C	180	1	56	3,080	1.6
D	300	1	48	2,640	2.2

[a] GA collections were obtained by treatment of the CHO MT II cDNA antisense strand, which contains 55 dC residues. The number of target sites is calculated from the number of sequenced clones multiplied by 55.

choice of conditions based on earlier reports, and can serve as a guide for future experiments. These clones and other collections in the library essentially show a Poisson distribution of mutation frequency which varies as expected in accord with the average percent of mutations. These distributions reveal important characteristics when various collections are compared. Figure 2A shows that treatment with 1 M sodium bisulfite for 5 hr allowed more total changes per template for the sense strand than was seen from 3 hr of treatment with 1 M sodium bisulfite for the antisense strand, as expected. As intended, the extent of coding changes caused by these treatments was comparable for both collections. Among the mutated sequences, the modal number of coding changes from these reactions is one per template (Fig. 2B). Obtaining additional and more frequent changes desired in the G to A collection required a greater degree of reaction with sodium bisulfite. The occurrence of more multiple mutations and codon changes from 20 and 40 min reactions with 3 M sodium bisulfite is shown in Figs. 3A and 3B, respectively. The former collection again has a modal number of one coding change per template, but includes elements with up to four mutations. The latter collection has a modal number of four to five coding changes per template, and contains elements having up to ten coding changes per template.

Mutants with single coding changes are the most obviously useful. However, sets of mutants with one common and one or more other coding changes may help reveal interactions between amino acids. Since many proteins, including MT, have more than one structural domain, multiple mutations can be very useful if they fall in different domains. Given the appropriate distribution of mutations, highly mutated molecules may eliminate the function of one domain while only slightly altering that of

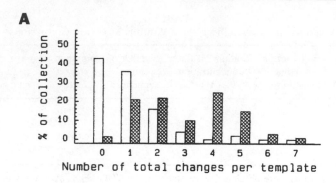

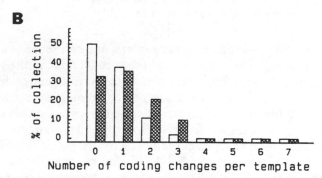

FIG. 2. Transition mutations per template for GA and CT collections in the MT coding sequence point mutant library. The data are shown as histograms (GA clones, open bars; CT clones, crosshatched bars). (A) Total changes in GA clones ($n = 56$) produced by treatment of CHO MT II cDNA antisense strand for 3 hr with 1 M sodium bisulfite; and in CT clones ($n = 67$) produced by treatment of CHO MT II cDNA sense strand for 5 hr with 1 M sodium bisulfite. (B) Coding changes in GA clones analyzed in (A) and in CT clones analyzed in (A).

another. Examples of such DNA sequences have been found in the MT point mutant library.[19]

The differences between the induced and evolved amino acid changes in MT shown by Fig. 1 presumably reflect evolutionary constraints on MT. It is perhaps coincidental, but of the 31 codons that could mutate by a single base change to encode an aromatic amino acid, only 10 are found in CHO MT II cDNA. Five of those encode cysteine or serine. In fact, a nonfunctional human MT I pseudogene encodes Tyr in place of Cys_5 and Phe in place of Ser_{35}, as well as a nonsense codon at position 40.[62]

Although this library is quite extensive, it is also limited. To resolve this paradox, other techniques for point mutagenesis can be employed.[1,4,8,15,18,23,25,36] Alternative libraries can be made with other target-directed protocols, as described in Vol. 154 of *Methods in Enzymology*

A

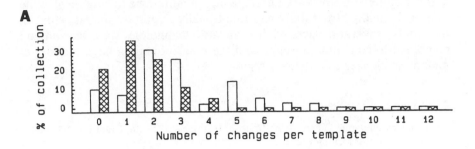

B

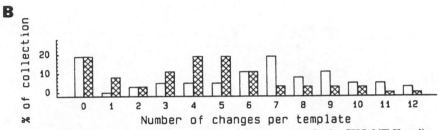

FIG. 3. High frequency collections of GA transition mutants in the CHO MT II coding sequence point mutant library. The data are shown as histograms (total changes, open bars; coding changes, crosshatched bars). (A) Total and coding changes in GA clones ($n = 42$) produced by treatment of the CHO MT II cDNA antisense strand for 20 min with 3 M sodium bisulfite. (B) Total and coding changes in GA clones ($n = 37$) produced by treatment of the CHO MT II cDNA antisense strand for 40 min with 3 M sodium bisulfite.

and elsewhere.[1,14,23,25] However, specific site-directed mutations are likely to be required in the future. While the CT mutants have the advantage of not altering any cysteine codon, and the GA clones include mutations of each cysteine codon (Fig. 1), oligonucleotides can be designed to allow introduction of particular, desired mutations which are rare or unavailable in the bisulfite-generated library.

It is necessary to actually obtain mutant proteins to reap the benefits of available mutated structural gene sequences. Some of these sequences[78] as well as MT deletion mutants[79] have been expressed as hybrid proteins in *E. coli*. This has allowed the use of a common scheme to obtain substantial purification of those proteins. Their structure and metal binding properties are under investigation.

The MT coding sequence provides a broadly applicable model of how the reaction with sodium bisulfite can be controlled and analyzed. MT also is an example of the type of system that initially requires a large

---

[78] M. Cismowski, R. Pine, S. W. Liu, and P. C. Huang, manuscript in preparation (1986).
[79] R. Pine and P. C. Huang, manuscript in preparation (1986).

number of mutants and varying frequency of mutations to help resolve the role of the many structurally and functionally important amino acids. In this study, several aspects of the available technology were chosen and combined to constitute an improved method to obtain a large number of mutant sequences in a defined region of DNA.

Acknowledgment

Work was supported in part by the National Institutes of Health, Grant R01 GM 32606. R.P. is an NIA postdoctoral trainee.

[34] Mutagenesis with Degenerate Oligonucleotides: An Efficient Method for Saturating a Defined DNA Region with Base Pair Substitutions

By DAVID E. HILL, ARNOLD R OLIPHANT, and KEVIN STRUHL

Introduction

Since the advent of recombinant DNA technology, structure–function relationships of genes and genetic elements have been studied primarily by mutating cloned DNA segments and assessing the phenotypic consequences upon introduction into living cells. Although genetic elements can be localized on cloned DNA segments with deletion and insertion mutations, a more detailed description requires single base pair changes. In principle, DNA sequence requirements of a particular genetic element can be determined by making all possible point mutations within the region of interest and then analyzing the phenotypic effects. Such an approach avoids the bias introduced by genetic selections and hence makes it possible to obtain mutations that confer wild-type phenotypes.

Numerous chemical and enzymatic methods for generating point mutations within defined regions of DNA have been described. However, these mutagenesis procedures usually have drawbacks that make them more useful at the earlier stages of an investigation when the region of interest is less defined. Typical problems are (1) severe restrictions in the kinds of mutations that are produced, (2) low frequency of mutagenesis, thus making it necessary to use genetic or physical selections to isolate mutations, and (3) technical difficulties in performing the procedures. As an alternative to these region-specific methods, synthetic oligonucleotides of defined sequence have been used extensively for site-directed mutagenesis of DNA. However, such procedures are extremely expensive as well as time consuming. For example, as there are 30 possible single base pair changes of 10-bp region, saturation would require 30 (or 60) oligonucleotide syntheses, each of which would have to be processed separately. Here, we describe an efficient method to create numerous point mutations within a given region by using the products of a single oligonucleotide synthesis.

Principles of the Method

The procedure utilizes synthetic oligonucleotides that are mutagenized by including low concentrations of the three non-wild-type nucleotide precursors at each step of the synthesis.[1-5] The product of such a DNA synthesis is a degenerate oligonucleotide, i.e., a complex mixture of related molecules, each of which has a defined probability of being altered from the wild-type sequence. The frequencies and types of single, double, and higher order mutations can be set simply by choosing appropriate amounts of non-wild-type precursors at each step of nucleotide addition.

For the cloning procedure shown in Fig. 1, degenerate oligonucleotides are converted to the double-stranded form by mutually primed synthesis.[4,6] In most cases, oligonucleotides are synthesized such that their heterogeneous regions are bounded at their 5' and 3' ends by sequences recognized by restriction endonucleases. Because the 3' ends are palindromic, two oligonucleotide molecules can hybridize such that they will serve as mutual primers for extension with DNA polymerase I. The product of this mutually primed synthesis is a double-stranded molecule containing two oligonucleotide units that are separated by the original 3' restriction site and are flanked by the original 5' restriction sites. The double-stranded molecules are cleaved with restriction endonucleases that recognize the 5' and 3' ends to generate homoduplex versions of the original oligonucleotides. After ligation to an appropriate vector, the resulting products are introduced into *Escherichia coli*. Thus, each transformant represents the cloning of a single oligonucleotide from the original collection. DNA preparations from these transformants are then subjected to nucleotide sequence analysis in order to determine the nature of the mutation(s). In principle, mutations should occur at the frequency that was programmed into the DNA synthesis, and they should be located randomly throughout the region of interest.

[1] M. D. Matteucci and H. L. Heyneker, *Nucleic Acids Res.* **11,** 3113 (1983).
[2] J. B. McNeil and M. Smith, *Mol. Cell. Biol.* **5,** 3545 (1982).
[3] C. A. Hutchison, S. K. Nordeen, K. Vogt, and M. H. Edgell, *Proc. Natl. Acad. Sci. U.S.A.* **83,** 710 (1986).
[4] A. R Oliphant, D. E. Hill, A. L. Nussbaum, and K. Struhl, *Gene* **44,** 177 (1986).
[5] K. M. Derbyshire, J. J. Salvo, and N. D. F. Grindley, *Gene* **46,** 145 (1986).
[6] A. R Oliphant and K. Struhl, this series, Vol. 155, p. 568.

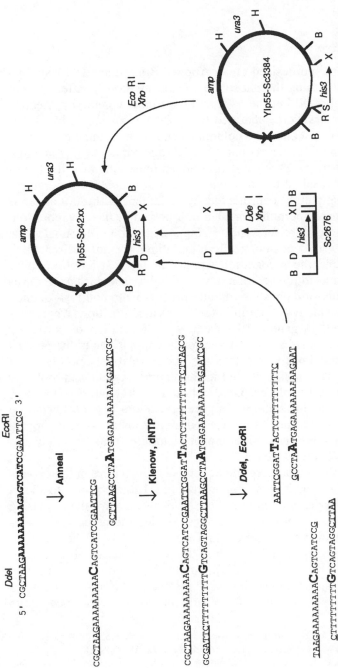

FIG. 1. Saturation mutagenesis using degenerate oligonucleotides. The top line of the left part of the figure indicates the *EcoRI–DdeI*-activated oligonucleotide containing 17 central bases (boldface letters) that were synthesized with 90% of the base indicated and 3.3% of the other three bases. Double-stranded, degenerate oligonucleotides suitable for cloning were produced as described in the text. The large, boldface residues indicate base pair substitution mutations of the central region. The *EcoRI–DdeI* oligonucleotide mixture was combined with the 9-kb *EcoRI–XhoI* fragment of YIp55-Sc3384 and the 0.9-kb *DdeI–XhoI* fragment of Sc2676 (the relevant parts of these molecules are indicated in boldface) to produce the desired molecules (indicated as YIp55-Sc42xx). The locations of the *amp*, *ura3*, and *his3* genes as well as restriction endonuclease sites for *Bam*HI (B), *Eco*RI (R), *Sac*I (S), *Dde*I (D), *Xho*I (X), and *Hin*dIII (H) are indicated. The positions of *Dde*I sites located outside of the 1.8-kb *his3 Bam*HI fragment are not shown. The large X in the circle represents the mutated *Eco*RI site at the junction between vector and *his3* chromosomal sequences. The *EcoRI–SacI* oligonucleotides, as described in the section concerning saturation mutagenesis, were cloned directly between the *Eco*RI and *Sac*I sites of YIp55-Sc3384.

Methods

Oligonucleotide Synthesis

Oligomers were synthesized on an Applied Biosystems DNA synthesizer (Model 380A) using the phosphite triester method.[7,8] After detachment and removal of all but the 5'-dimethoxytriphenylmethyl protecting groups, the oligomers were separated from shorter congeners by HPLC chromatography on a Waters C-8 column using a 40-min linear gradient of 0.1 M triethylammonium bicarbonate (pH 7.1) and from 0 to 25% acetonitrile. The peak containing the trityl chromogen (emerging near the top of the gradient) was desalted by flash evaporation *in vacuo* at temperatures below 30° and completely deprotected by treatment with 80% aqueous acetic acid at room temperature for 20 min, followed by flash evaporation.

The general procedure for synthesizing a degenerate oligonucleotide proceeds as follows. At positions where mutations are not desired, such as those composing the recognition sequences for restriction endonuclease cleavage, nucleotide addition is performed by standard procedures using a single nucleoside phosphoramidite precursor. In contrast, defined mixtures of nucleoside phosphoramidites are used at positions where mutations are desired. In general, the frequency of addition of particular nucleotides is determined simply by the relative molarities of the precursors, although sometimes G residues are added relatively poorly. This problem may be due to instability of the G precursor,[9] and it can probably be avoided by using freshly prepared solutions. Although certain DNA synthesizers can combine solutions of pure precursors, the most accurate and reproducible way to achieve a desired mixture is to combine appropriate amounts of solid nucleoside phosphoramidites prior to solubilization in anhydrous acetonitrile.

The design of the oligonucleotide depends on the nature of the experiment. In many cases, such as diagrammed in Fig. 1, the object is to alter a region of DNA at a defined mutation rate. This is accomplished by using four mixtures, each composed of one major nucleotide precursor (corresponding to the wild-type sequence) and equal amounts of the three remaining precursors (representing each of the possible base pair changes). At each position to be mutagenized, the mixture containing the wild-type precursor as the major component is used instead of the solution containing the pure precursor. The use of premade mixtures as described above

[7] M. D. Matteucci and M. H. Caruthers, *J. Am. Chem. Soc.* **103**, 3185 (1981).

[8] S. L. Beaucage and M. H. Caruthers, *Tetrahedron Lett.* **22**, 1859 (1981).

[9] G. Zon, K. A. Gallo, C. J. Samson, K.-L. Shao, M. F. Summers, and R. A. Byrd, *Nucleic Acids Res.* **13**, 8181 (1985).

ensures that the selected mutation rate is maintained at each step of the DNA synthesis. In other experiments where the purpose is to mutate a particular base to all possible alternatives, an equimolar mixture of the three non-wild-type precursors is used at the relevant addition step. More specialized cases can be accommodated simply by choosing appropriate mixtures for the appropriate positions of the oligonucleotide. The expected results from any particular experiment can be calculated according to the laws of probability.

Conversion of Oligonucleotides to Double-Stranded DNA by Mutually Primed Synthesis[4,6]

A degenerate oligonucleotide (approximately 1 μg) is diluted into 10 μl of 3× buffer [30 mM Tris (pH 7.5), 150 mM NaCl, 30 mM MgCl$_2$, 15 mM dithiothreitol, and 0.1 mg/ml gelatin], hybridized for at least 1 hr at 37°, and then allowed to cool slowly to room temperature. Deoxynucleoside triphosphates (to a final concentration of 250 μM for each of the four) and [^{32}P]dATP (10 μCi) are then added, and the reaction mixture is diluted to a final volume of 30 μl. Then 5 units of the Klenow fragment of $E.$ $coli$ DNA polymerase I are added, and the reaction mixture is incubated at 37° for at least 1 hr. The products of the mutually primed synthesis reaction are extracted with phenol and precipitated with ethanol. A small portion of the resuspended reaction products is then analyzed by electrophoresis in acrylamide gels containing 7 M urea.[10] The desired product, which is visualized by autoradiography, is a homoduplex molecule of length 2A + 2N + B (where A is the length of the 5'-flanking sequences, N is the length of the heterogeneous central region, and B is the length of the 3'-flanking sequences). For calibrating the size of the product, the best markers are oligonucleotides of defined length that have been labeled at their 5' ends with T4 polynucleotide kinase and [γ-^{32}P]ATP.

In order to clone the degenerate oligonucleotides, the double-stranded molecules produced by mutually primed synthesis are cleaved with appropriate restriction endonucleases and then ligated into vector molecules by standard means. The cleavage reactions are monitored by electrophoresis in denaturing acrylamide gels as described above. At some stage prior to the ligation reaction, it is useful to remove the excess, unreacted, single-stranded oligonucleotides as these reduce the ligation efficiency. In most cases, the initial product is cleaved to completion with the restriction endonuclease recognizing the outside sites (originally the 5' site), extracted with phenol, and concentrated by ethanol precipitation. After

[10] F. Sanger, A. R. Coulson, B. G. Barrell, A. J. Smith, and B. A. Roe, $J.$ $Mol.$ $Biol.$ **143**, 161 (1980).

electrophoresis in a native acrylamide gel (6–12% depending on the size of the product), the desired double-stranded molecule containing two oligonucleotide units is eluted in 0.5 M ammonium acetate and 1 mM EDTA for 4–24 hours at 37° and then concentrated by ethanol precipitation. The purified DNA is cleaved with the restriction endonuclease recognizing the central site (the original 3' site) to produce the final product, a double-stranded version of the oligonucleotide mixture with 5' and 3' ends suitable for ligation into standard vector molecules. In our experience, it is better to perform the gel purification step prior to cleavage at the central restriction site; the reasons for this are unknown.

Standard ligation reactions using T4 DNA ligase are carried out at 15° in 20-μl reactions containing 50 mM Tris (pH 7.5), 10 mM MgCl$_2$, 10 mM dithiothreitol, and 500 μM ATP. As the yield of the final oligonucleotide product is somewhat variable, the amount to be added to a given amount of vector is determined empirically in order to optimize the ligation reaction. Typically, ligation reactions containing varying amounts of oligonucleotide are processed in parallel.

Results

Mutagenesis of the his3 Regulatory Region Using a Single Degenerate Oligonucleotide

Extensive deletion analysis of the yeast *his3* promoter defines and localizes a positive regulatory site that is critical for the induction of *his3* transcription in response to conditions of amino acid starvation.[11,12] The regulatory region maps between nucleotides −84 and −102 with respect to the transcriptional initiation site and it contains the sequence TGACTC, which is present in front of and implicated in the expression of coregulated genes.

In order to generate a large number of base pair substitution mutations throughout a 17-bp region containing the essential regulatory sequence, we synthesized an oligonucleotide in which the *his3* region was bounded, respectively, at its 5' and 3' ends by the restriction endonuclease cleavage sites for *Dde*I and *Eco*RI. A 10% mutation rate was achieved by using four mixtures, each composed of a wild-type precursor (90%) and equal amounts of the three mutant precursors (10% total). The DNA synthesizer was programmed to use the appropriate 90/10 mix at each position of the 17-base region and the appropriate pure precursors at the remaining

[11] K. Struhl, *Nature (London)* **300**, 284 (1982).
[12] K. Struhl, W. Chen, D. E. Hill, I. A. Hope, and M. A. Oettinger, *Cold Spring Harbor Symp. Quant. Biol.* **50**, 489 (1985).

positions. The number of mutations per oligonucleotide should be defined by a binomial distribution centered around 1.7, the average value; thus, single and double mutations should predominate. The oligonucleotide mixture was converted into double-stranded DNA by mutually primed synthesis, and then cleaved with EcoRI and DdeI (Fig. 2).

Because of the availability of reliable methods for determining the DNA sequence of double-stranded molecules,[13] we cloned the oligonucleotide mixture such that the resulting molecules could be tested directly for the phenotypes that they conferred in vivo (Fig. 1). A ligation reaction containing the EcoRI–DdeI oligonucleotide mixture, the 9-kb EcoRI–XhoI vector fragment of YIp55-Sc3384,[12] and the 0.9-kb DdeI–XhoI fragment of Sc2676[14] was performed, and the ligation products were introduced into E. coli by selecting for transformants resistant to ampicillin. Because ligations involving three DNA fragments occur at relatively low frequency, we used colony filter hybridization[15] to identify transformants containing oligonucleotide inserts. Using 5' end ^{32}P-labeled oligonucleotide as a probe, 41 out of 97 transformants appeared to contain inserted oligonucleotides. The large number of non-oligonucleotide-containing transformants was probably due to incompletely cleaved vector DNA or to incorrect ligation. DNAs were prepared from the 41 putative, oligonucleotide-containing transformants, and the DNA sequence of the his3 regulatory region was determined by the chain termination method using a his3-specific primer corresponding to positions +26 to +42 of the antisense strand.[14]

All 41 plasmids contained oligonucleotide inserts, as expected from the colony filter hybridization. From the entire collection, 23 unique mutants were obtained, and, as expected, single and double mutants comprise the largest classes (Fig. 3). Six of the mutant sequences were duplicates, a frequency much higher than expected. These probably arose during the transformation procedure as a result of incubating the E. coli cells for 90 min prior to plating on ampicillin-containing medium. Finally, 12 wild-type sequences were obtained, most of which probably represent independent cloning events. The average number of mutations per oligonucleotide was 1.2, a value in fair agreement with the theoretical prediction of 1.7. The pattern of mutations in terms of the positions of alterations was in good accord with expectations. Thus one degenerate oligonucleotide can be used for creating a large number and variety of base pair substitution mutations within a small, defined region.

[13] E. Y. Chen and P. H. Seeburg, DNA 4, 165 (1985).
[14] K. Struhl, Nucleic Acids Res. 13, 8587 (1985).
[15] M. Grunstein and D. S. Hogness, Proc. Natl. Acad. Sci. U.S.A. 72, 3961 (1975).

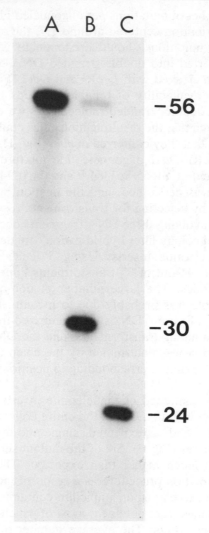

FIG. 2. Conversion of the degenerate oligonucleotide into clonable DNA. The DNAs were electrophoretically separated in a 10% acrylamide gel containing 7 M urea.[10] Lane A corresponds to the initial product of mutually primed synthesis (see Fig. 1). This product was subsequently cleaved with EcoRI (lane B) and EcoRI + DdeI (lane C). The lengths (in nucleotides) of these products are shown at the right side of the autoradiogram.

Saturation Mutagenesis of the TGACTC Core

As essentially all yeast genes subject to general control contain the sequence TGACTC, we decided to saturate this 6-base sequence with single base changes. Six oligonucleotides were synthesized, each of

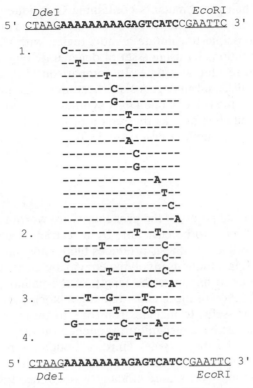

FIG. 3. Mutations of the *his3* regulatory region. The DNA sequences of 23 derivatives containing from one to four substitutions within the 17-bp region that was synthesized at a 10% mutation level (see Fig. 1); only the mutated nucleotides are indicated.

which was mutated at one particular base of the TGACTC sequence. For example, mutagenesis of the G residue was accomplished by using a mixture containing equimolar amounts of the A, C, and T precursors at the appropriate step of the DNA synthesis. Thus, the oligonucleotide product is actually a mixture of three mutant sequences that could be resolved into the individual components by molecular cloning. The degenerate oligonucleotides were cloned as *Eco*RI–*Sac*I fragments between the *Eco*RI and *Sca*I sites of YIp55-Sc3384 (Fig. 1). Unlike the first example, these ligation reactions were more efficient because they contained only two DNA segments with heterologous ends. For this reason, it was unnecessary to screen the transformants by colony filter hybridization prior to DNA sequence analysis. From 37 transformants containing an oligonucleotide insertion (out of 52 that were subject to DNA sequence analysis), 15 out of the 18 possible base pair substitution mutations were obtained.

We did not obtain three mutants containing G residues in place of the wild-type residue, possibly for reasons discussed in the methods section.[9]

Thus, it is possible to saturate a short region with all possible single base pair substitutions by using a set of degenerate oligonucleotides that mutagenize one residue at a time. In comparison to more conventional oligonucleotide-directed mutagenesis procedures, this method produces three mutants for the price of one DNA synthesis. Obviously this method becomes more cumbersome as the region of interest becomes larger; in such situations, the method described in the previous section is more practical.

Comments

Several technical points of the procedures are worth noting. First, the 5′ end does not have to be cleavable by a restriction endonuclease because mutually primed synthesis produces blunt ends that are suitable for cloning. Second, in situations involving enzymatic cleavage at the 5′ end, it is advantageous to minimize the length of the palindrome in order to disfavor hybridization of the 5′ ends that might block complete extension. In addition, it is useful to include from one to three extra nucleotides beyond the endonuclease recognition sequences at the 5′ end in order to facilitate cleavage by the enzyme. Third, although the palindrome at the 3′ end, which is required for mutually primed synthesis, can be as short as 6 bases, the reaction works more efficiently when the region is 8 bases in length. However, in situations using an 8-base palindrome and a restriction endonuclease with a 6-base recognition sequence, each oligonucleotide unit will contain an extra base between the restriction site and the degenerate central region. Fourth, as a palindromic restriction site at the 3′ end of the oligonucleotide is the only requirement for mutually primed synthesis, many different sequences are available. Moreover, as there are almost no limitations on the 5′-end sequences, degenerate oligonucleotides can be cloned into an extremely wide variety of double-stranded DNA molecules. This makes it possible to insert the degenerate oligonucleotides into vectors that can be used directly to examine the phenotypes of the mutant sequences. Fifth, the size of region that one can mutagenize is limited by the length of the oligonucleotide that is synthesized. In other experiments, we have synthesized a degenerate oligonucleotide whose heterogeneous region is 55 bases long.

The mutagenesis procedures described here have several advantages for determining the relationship of structure and function of genetic elements. The cloning efficiency is high because the oligonucleotides are converted to double-stranded restriction fragments, thus making it possible to obtain a large number of mutations. Moreover, in many cases the vast majority of transformants contain an inserted oligonucleotide, thus eliminating the need for a hybridization screen prior to DNA sequence analysis. Unlike other methods that produce mismatches between mutant oligonucleotides and the wild-type sequence, the oligonucleotides described here are cloned as homoduplex molecules. This avoids biases due to differential stability and preferential repair of heteroduplexes, and to screening procedures that depend on mismatch hybridization to distinguish mutants from nonmutants. Most importantly, essentially all possible mutations can be obtained without regard to their phenotypes *in vivo*. Thus, it is possible to determine directly which nucleotides are critical for a particular genetic function and which ones are unimportant.

Section VI

Miscellaneous Methods

[35] Fusogenic Reconstituted Sendai Virus Envelopes as a Vehicle for Introducing DNA into Viable Mammalian Cells

By A. Vainstein, A. Razin, A. Graessmann, and A. Loyter

It is well established that, under certain conditions, eukaryotic cells can take up DNA molecules from the medium and transport them into the cell nucleus.[1,2] The method commonly used for gene transfer in mammalian cells is the addition of DNA molecules to the cell culture in the presence of facilitators, such as polyornithine[3] or DEAE-dextran,[4] or precipitation of the DNA molecules on the cell surface by calcium phosphate.[5] Coprecipitation of DNA with calcium phosphate has proved, so far, to be the most useful method for transformation of eukaryotic cells by selectable genes.[5,6] However, using this method the reported frequencies of gene transfer have been relatively low.[5,7] This has limited the method for those genes for which there exists a good selective system.[6] In addition, the method necessitates, in most cases, the use of specific cell lines, such as the mouse L cell, as recipient.[7] Attempts to transfer specific genes into other cells, such as hamster or human cell lines, either failed or the frequencies of gene transfer obtained were several orders of magnitude lower than with mouse L cells.[7]

Using DNA molecules stained with the fluorescent dye 4, 6-diamino-2-phylindole-dihydrochloride (DAPI), it has been shown that the calcium phosphate–DNA complexes enter mammalian cells by endocyte-like processes.[8] Most of the added DNA was found trapped within phagocytic vacuoles.[8,9] It can be assumed that the low frequencies observed in DNA-mediated gene transfer by the use of calcium phosphate as facilitator could be attributed, at least in part, to the fact that intracellularly these

[1] E. H. Szybalska and W. Szybalski, *Proc. Natl. Acad. Sci. U.S.A.* **48,** 2026 (1962).

[2] P. M. Bhargava and G. Shannugan, *Nucleic. Acids. Res.* **11,** 103 (1971).

[3] F. Farber, J. L. Melnick, and J. S. A. Butel, *Biochim. Biophys. Acta* **390,** 298 (1975).

[4] J. S. Pagano, *Prog. Med. Virol.* **12,** 1 (1970).

[5] F. L. Graham, S. Bacchetti, and R. McKinon, *in* "Introduction of Macromolecules into Viable Mammalian Cells" (R. Baserga, C. Croce, and G. Rovera, eds.), p. 3. Liss, New York, 1979.

[6] M. Wigler, A. Pellicer, S. Silverstein, and R. Axel, *Cell* **14,** 725 (1978).

[7] W. H. Lewis, P. R. Srinivasan, N. Stokes, and L. Siminovitch, *Somatic Cell Genet.* **6,** 333 (1980).

[8] A. Loyter, G. A. Scangos and F. H. Ruddle, *Proc. Natl. Acad. Sci. U.S.A.,* **79,** 422 (1982).

[9] A. Loyter, G. Scangos, D. Juricek, D. Keene, and F. H. Ruddle, *Exp. Cell Res.* **139,** 223 (1982).

RECOMBINANT DNA
METHODOLOGY

DNA molecules are not free but are enclosed within the phagocytic vacuoles.[8] From these observations it is clear why the method is applicable mostly to cells of high endocytic activity, such as mouse L cells.[7-9]

In light of the above observations and from all the information accumulated during the past few years, it appears that, for an efficient method for the introduction of DNA molecules into as many different cell lines as possible,[6-8,10] it may be necessary that (a) the DNA molecules be protected from hydrolysis by serum or cell secreted nucleases[8,9]; (b) the method be based on a principle that can be shared by most of the cell lines; (c) a high number of DNA molecules, namely, of gene copies, be introduced directly into the cytoplasm of the recipient cells.

It seems that resealed, loaded membranous vesicles may serve as an efficient carrier for the introduction of DNA molecules into mammalian cells. DNA molecules can be enclosed within such vesicles, and a strategy for injecting large numbers of cells simultaneously is to fuse them with the DNA-loaded vesicles. Three different approaches have been used to design such a membranous carrier: the use of vesicles made of pure phospholipids (liposomes) as carriers to transfer macromolecules such as protein or DNA into cells[11,12]; fusion between bacterial protoplasts, carrying plasmids and specific genes, and eukaryotic cells[13]; the use of reconstituted envelopes obtained from fusogenic animal viruses,[14] e.g., envelopes of Sendai virus particles.

Infection of cells by Sendai virus, an enveloped virus belonging to the paramyxovirus group, comprises two steps[15]: (a) binding of virus particles to cell surface receptors, mainly sialic acid residues of glycoproteins and glycolipids[15]; and (b) fusion of the viral envelopes with the plasma membrane of recipient cells, with the concomitant injection of the viral nucleocapsid. Sendai virus particles contain six different polypeptides,[15] of which two are located within the viral envelope: the HN protein with hemagglutinin and neuraminidase activity, and the F protein, which is required for vius–cell fusion and virus infection.[15] Evidently, the virus–cell binding and the virus–cell fusion activities of Sendai virus particles are located in its envelope. Therefore, it appears that an ideal vehicle for the delivery of DNA molecules directly to the cytoplasm of living cells should

[10] G. Scangos and F. H. Ruddle, *Gene* **14,** 1 (1981).

[11] G. Poste, D. Papahadjopoulos, and W. J. Vail, Methods Cell Biol. **14,** 33, (1976).

[12] R. Farley, S. Subramani, P. Berg, and D. Papahadjopoulos, *J. Biol. Chem.* **255,** 10431 (1980).

[13] M. Rassoulzadegan, B. Binetruy, and F. Cuzin, *Nature (London)* **295,** 257 (1982).

[14] A. Vainstein, J. Atidia, and Λ. Loyter, *in* "Liposomes in Study of Drug Activity and Immunocompetent Cell Function" (A. Paraf and C. Nicolau, eds.), p. 95. Academic Press, New York, 1981.

[15] G. Poste and A. Pasternak, *in* "Membrane Fusion" (G. Poste and G. L. Nicolson, eds.), p. 305. Elsevier/North-Holland, Amsterdam, 1978.

be a membranous vesicle that will resemble membrane-enclosed viruses or, more specifically, the envelope of Sendai virus. DNA enclosed within such a vesicle will be protected from digestion by any nucleases present during the transformation process. These vesicles, like intact virus particles, should attach efficiently to specific receptors on cell surfaces and then fuse with their plasma membranes.[15]

Principle of the Method

The envelope of Sendai virus particles, like other biological membranes, can be subjected to solubilization by either nonionic or ionic detergents. Since the viral nucleocapsid is not part of the viral membrane, it remains detergent insoluble when intact viral particles are dissolved by detergents.[16,17]

Fusogenic reconstituted Sendai virus envelopes (RSVE) can be obtained after solubilization of Sendai virus particles either by Triton X-100[16] or by Nonidet P-40 (NP-40).[17] Sendai virus particles can be dissolved also by other detergents such as octylglucoside, cholate, deoxycholate, Lubrol, or lysolectin. However, these detergents cause irreversible inactivation of the virus' fusogenic activity.[18]

The main features of the method for preparation of fusogenic reconstituted Sendai virus envelopes are summarized in Fig. 1. The viral particles are dissolved by Triton X-100, and the detergent-insoluble viral nucleocapsids are removed by centrifugation.[16] The clear supernatant contains the viral phospholipids and the two viral envelope glycoproteins (the HN protein, M_r 67,000, and the F protein, M_r 54,000) (see also Fig. 2). After removal of the detergent, resealed membranous vesicles that contain the two viral glycoproteins are formed (Fig. 2). Owing to the presence of these two glycoproteins which, as in intact virus particles, form spikes in the viral envelope, the RSVE are fusogenic[16] (see Fig. 2).

Figure 1 also summarizes schematically the way in which RSVE can be used as carriers of DNA. When a water-soluble macromolecule such as DNA (or protein) is added to the reconstitution system, namely, to the mixture of detergent-soluble viral phospholipids and glycoproteins, it will be trapped within the membrane vesicles formed after removal of the detergent (see also scheme in Fig. 4). It should be noted, however, that if a water-insoluble membrane component is added to the detergent-solubilized mixture, it will be inserted into the reconstituted envelope itself, resulting in the formation of "hybrid" virus vesicles.[19]

[16] D. J. Volsky and A. Loyter, *FEBS Lett.* **92**, 190 (1978).
[17] Y. Hosaka and K. Shimizu, *Virology* **49**, 627 (1972).
[18] D. J. Volsky, A. Loyter, E. Ferber, and H. U. Weltzien, unpublished results.
[19] D. J. Volsky, Z. I. Cabantchik, M. Beigel, and A. Loyter, *Proc. Natl. Acad. Sci. U.S.A.* **76**, 5440 (1979).

A.   Intact Sendai virus particles solubilized

in Triton X-100

(3 mg of viral proteins + 6 mg of Triton X-

-100, 10^{-4} M PMSF in 200 μl of Buffer I)

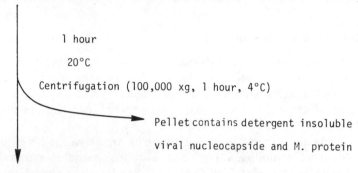

1 hour

20°C

Centrifugation (100,000 xg, 1 hour, 4°C)

Pellet contains detergent insoluble

viral nucleocapside and M. protein

B.   Supernatant contains the viral F and

HN glycoproteins

(0.8 mg in 200 μl of 3% Triton X-100)

+

DNA (50-150 μg in 100 μl of Buffer I)

Dialysis against Buffer II containing

SM2 Bio-Beads to remove Triton X-100

for 48-72 hours at 4°C

Centrifugation 100,000 xg, 1 hour

C.   After washing and DNase treatment, the

DNA-loaded RSVE (200-300 μg protein) pellet

is suspended in 200 μl of Solution Na to give

protein concentration of 1 mg/ml, and stored

either at 4°C or -70°C

FIG. 1. Schematic summary of the method for preparation of DNA-loaded fusogenic Sendai virus envelopes. For details, see text.

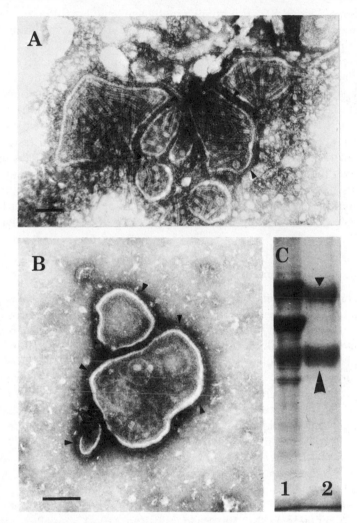

Fig. 2. Electron micrographs (A, B) and gel electrophoresis pattern (C) of intact Sendai virus particles and reconstituted Sendai virus envelopes (RSVE). Intact Sendai virus particles and RSVE were prepared for electron microscopy and observed by the negative staining technique as previously described.[16] (A.) Intact virus particles. (B) RSVE. Arrows show spikes extending from the envelopes of both intact virus and RSVE. (C) Electrophoretic pattern of intact virus (1) or RSVE (2) polypeptides (HN, small arrowhead; F, large arrowhead) (7.5% acrylamide).[16] The major Sendai virus polypeptides were classified as described before.[16] Bars in (A) and (B), 80 nm.

Fusion of loaded RSVE with mammalian cells will result in fusion-mediated injection of the RSVE content into the cytoplasm of the recipient cells. Uchida *et al.*[20] showed that the protein CRM-45, a nontoxic mutant protein related to diphtheria toxin,[20] can be trapped within RSVE. This protein is not toxic to cells, unless it is injected directly into the cell cytoplasm, because it lacks an amino acid sequence necessary for binding to the surface of susceptible cells.[20] RSVE can be used for the introduction of CRM-45 to cells in culture, which are killed after a short exposure to CRM-45-containing RSVE.[20] In our laboratory in Jerusalem (Israel) we have shown that RSVE can be used to introduce ferritin particles or ^{125}I-labeled IgG into cells in culture.[14] In addition, we have shown that DNA molecules, such as SV40-DNA or PTKx1 (*Herpes simplex* virus, type I, thymidine kinase gene cloned in pBR322),[21] can be trapped within RSVE and transferred to the appropriate living cells with relatively high efficiency.[22]

Materials and Reagents

Buffers and Reagents Used for the Preparation of Intact Sendai Virus Particles and RSVE. For injecting into fertilized eggs, Sendai virus particles are suspended in phosphate-buffered saline (PBS) with 50 μg of streptomycin per milliliter and 100 units of penicillin G per milliliter. Intact Sendai virus particles in Solution Na (160 mM NaCl, Tricine-NaOH, pH 7.4) are stored at $-70°$. Buffer I is 100 mM NaCl, 50 mM Tris-HCl, pH 7.4. Buffer II is 10 mM Tris-HCl, pH 7.4, 2 mM MgCl$_2$, 2 mM CaCl$_2$, and 1 mM NaN$_3$. Chicken erythrocytes are obtained from blood that was collected from the necks of decapitated chickens into an Erlenmeyer flask containing heparin (100 units/ml).

Buffers and Reagents Used for Preparation of DNA-Loaded RSVE and Their Fusion with Cultured Cells. Triton X-100 (Kotch & Light) is removed by dialysis in semimicro cellulose dialysis tubing (Spectra/por, M_r cutoff 12,000–14,000, Spectrum Medical Industries, Inc.; 15.9 mm in diameter). SM2 Bio-Bead (Bio-Rad) (10 g) is swollen in methanol (100 ml) overnight and then washed on filter (glass fiber paper, Whatman GF/A) with methanol (200 ml) followed by washing with water (2000 ml), and stored in water solution at 4°. PMSF was dissolved in methanol to a concentration of 10^{-2} M. DNase I (Worthington) was dissolved in 5 mM Tris, pH 7.4.

[20] T. Uchida, M. Yamaizumi, E. Mekada, and Y. Okada, in "Introduction of Macromolecules into Viable Mammalian Cells" (R. Baserga, C. Croce, and G. Rovera, eds.), p. 169. Liss, New York, 1980.

[21] L. W. Enquist, G. F. Van de Woude, M. Wagner, R. Smiley, and W. C. Summers, *Gene* **7**, 342 (1979).

[22] A. Loyter, A. Vainstein, M. Graessman, and A. Graessmann, *Exp. Cell Res.*, in press.

SV40-DNA, unlabeled or labeled with [³H]thymidine, was isolated 48 hr after infection with SV40 strain (777) of TC7 by the method of Hirt.[23] Plasmids pMB9, φX174RFI-DNA, pBR322, and pBR322 containing the 3.2-kilobase TK-DNA fragment of *Herpes simplex* virus, type I (PTKx1) were purified by the method of Clewell.[24] φX174RFI-DNA was nick-translated with [α-³²P]dATP and dCTP, as described before.[25]

Hepatoma tissue culture (HTC) cells were grown in S77 medium supplemented with 10% newborn calf serum (BioLab).

CV1, an African green monkey kidney cell line,[26] TC7, a subline of CV1 cells,[26] F 1' 1-4, an established rat cell line,[26] and mouse LM(TK⁻), also an established cell line, were grown in Dulbecco's modified Eagle medium (DMEM) containing 10% (v/v) calf serum, with the following exceptions: (*a*) dTKinase, deficient mouse LM(TK⁻) cells, which were propagated in medium containing 25 μg of BUdR except for the passage immediately preceding an experiment; and (*b*) a cell line converted to the dTKinase-positive phenotype by fusion-mediated injection of kinase-deficient cells with PTKx1-loaded RSVE, which was transferred to growth medium (36 hr after infection) supplemented with 10^{-4} M, hypoxanthine M aminopterin, 4×10^{-7} and 4×10^{-5} M thymidine (HAT medium), which was changed every 4–5 days for about 3–4 weeks until TK⁺ clones developed.

Methods and Results

Preparation of Intact Sendai Virus Particles and Its Quantitative Estimation

Particles of Sendai virus, one of the parainfluenza group I, are estimated quantitatively by their hemagglutinin titer (Salk's pattern method), namely, by their ability to agglutinate chicken erythrocytes at room temperature.[27] In principle, serial dilutions of Sendai virus preparation are incubated with 0.5% (v/v) of chicken red blood cells in PBS, pH 7.0, at room temperature. Hemagglutinating units (HAU) of the Sendai virus preparation are estimated as described before.[27] Generally about 10^4 HAU are present per milligram of viral protein.

High amounts of intact Sendai virus particles can be obtained by propagation in the allantoic cavity of 10-day-old chicken erythrocytes.[27] For this, about 0.2 ml of sterile Sendai virus preparation, containing 2 HAU,

[23] B. Hirt, *J. Mol. Biol.* **26,** 365 (1967).
[24] D. B. Clewell, *J. Bacteriol.* **110,** 667 (1972).
[25] R. Weinstock, R. Sweet, M. Weiss, H. Cedar, and R. Axel, *Proc. Natl. Acad. Sci. U.S.A.* **75,** 1299 (1978).
[26] A. Graessmann, M. Graessmann, W. C. Topp, and M. Botchan, *J. Virol.* **32,** 989 (1979).
[27] Z. Toister and A. Loyter, *J. Biol. Chem.* **248,** 422 (1973).

is injected into each fertilized egg, after which the injected eggs are incubated in a rotary incubator for 48–72 hr, as previously described.[27] At the end of the incubation period, the allantoic fluid is collected and its hemagglutinating units are estimated. Under optimal growth conditions, about 1500 HAU of Sendai virus are obtained per milliliter of allantoic fluid. About 5 ml of allantoic fluid are obtained from each fertilized egg.

Purification of Viral Particles

The procedure given is described for 100 ml of allantoic fluid, essentially as described before.[27,28] The allantoic fluid (containing about 1.5×10^5 HAU of Sendai virus particles/100 ml) is centrifuged at low speed (1400 g for 10 min); the supernatant is collected and recentrifuged at 37,000 g for 60 min in the cold. The pellet, containing virus particles, is suspended in 5–10 ml of Solution Na, homogenized to obtain a homogeneous suspension, and then centrifuged at 600 g for 5 min to remove large clumps of virus aggregates. The supernatant is loaded on a 15% (w/v) sucrose cushion and centrifuged at 100,000 g for 1 hr. The pellet obtained contains purified viral particles that are suspended in Solution Na to give about 1×10^4 HAU/mg of protein per milliliter. Approximately 1×10^5 HAU of Sendai virus can be obtained from each 100 ml of allantoic fluid. Other purification methods are described elsewhere.[28,29]

Reconstitution of Sendai Virus Envelopes

Fusogenic envelopes of Sendai virus particles are obtained essentially as described before[16] and summarized in Fig. 1. A pellet of Sendai virus particles, usually containing 3–20 mg of viral protein, is solubilized by Triton X-100 (Triton–viral protein, 2:1, w/v), in a final volume of 0.2–1.3 ml (final concentration of Triton X-100, 3%). It is recommended to solubilize the viral particles in the presence of 10^{-4} M PMSF to inhibit degradation of detergent-solubilized proteins by virus-associated proteases.[30]

The turbid suspension obtained contains detergent-soluble viral envelopes and detergent-insoluble viral nucleocapsids. The detergent-insoluble material is removed by centrifugation (100,000 g, 60 min), and the clear supernatant contains the viral envelope phospholipids and glycoproteins (HN and F proteins, see Fig. 2), all dissolved in Triton X-100. The Triton X-100 is removed by dialysis for 48–72 hr in the cold against buffer

[28] Y. Hosaka, H. Kitano, and S. Ikeguchi, *Virology* **29,** 205 (1966).

[29] R. Rott and W. Schaefer, *Virology* **14,** 298 (1961).

[30] A. Loyter, D. Ginzberg, D. J. Volsky, and N. Zakai, *in* "Receptors for Neurotransmitters and Peptide Hormones" (G. Pepeu and M. J. Kuhar, eds.), p. 33. Raven, New York, 1980.

II, which contains SM2 Bio-Beads, (1.2 g) or bovine serum albumin (1 mg/ml).

Efficient removal of Triton is achieved when either Neflex or Spectrophore-2 tubing is used during the dialysis.[16] By the use of ^{3}H-labeled Triton, it was demonstrated that the temperature of the dialysis period greatly affects the rate at which Triton is removed. About 0.01–0.02% of the Triton X-100 is left after 20–30 hr of dialysis at room temperature, whereas 0.015–0.03% remains after 60 hr of dialysis at 4°.[16,31]

The turbid suspension obtained after the end of the dialysis period contains resealed membrane vesicles that are very heterogeneous in size, ranging from 30 to 300 nm. Most of the vesicles resemble viral envelopes, since they reveal a high concentration of spikes extending from their external surface (Fig. 2). In some vesicles the spikes are attached to both sides of the reconstituted envelope.

From 3 mg of protein of intact virus particles, about 0.2–0.4 mg of RSVE are obtained. After centrifugation (100,000 g, 60 min), the RSVE are suspended in Solution Na to give about 1 mg/ml, they are kept at 4° for immediate use, or otherwise stored at −70°. Frequent freezing and thawing promotes fusion between the RSVE themselves, resulting in the formation of huge vesicles.

Enclosure of DNA Molecules in RSVE

For enclosure of DNA molecules within the RSVE (see Fig. 1) about 50–150 μg of DNA (dissolved usually in buffer I) are added to the clear supernatant of the detergent-solubilized virus envelopes, to give about 6–20 μg of DNA and 100 μg of detergent-dissolved viral envelope glycoproteins. Subsequent steps are as described above for the reconstition of "empty" unloaded viral envelopes. For removal of externally adsorbed DNA molecules, the RSVE obtained (1 mg of protein per milliliter) are incubated with pancreatic DNase I (200 μg of viral glycoproteins are incubated with 20 μg DNase in 1 ml of Solution Na that contains 5 mM MgCl$_2$) for 25 min at room temperature. At the end of the incubation period the DNA-loaded RSVE are collected by centrifugation.

"Biological" Assays (Induction of Hemolysis and Membrane Fusion) of Intact Sendai Virus Particles and RSVE

The amount of intact Sendai virus particles and of RSVE can be estimated, as described above, by their hemagglutinin titer. However, agglutination of chicken erythrocytes represents only binding activity of the viral particles, not necessarily their fusogenic ability.

[31] A. Loyter and D. J. Volsky, in "Cell Surface Reviews" (G. Poste and G. L. Nicolson, eds.) Vol. 8, p. 215. Elsevier, Amsterdam, 1982.

The quickest and easiest way to check the fusogenic ability of Sendai virus particles or RSVE is by testing their ability to hemolyse and fuse aged human erythrocytes.[32] Fusion of Sendai virus particles with mammalian cells lead to transient leakage of molecules (lysis), a process from which the cells can recover.[33] Fusion with human erythrocytes, on the other hand, induced permanent lysis, concomitantly with the promotion of cell–cell fusion. Eventually giant polyghosts are formed.[32] Therefore, estimation of the extent of hemolysis—which is subjected to inhibition of antiviral antibody—may give a quantitative measurement of the virus–cell fusion process. Hemolysis is estimated essentially as previously described[27,32] and as follows.

Intact Sendai virus particles or RSVE (400 HAU in 10 ml of Solution Na) are added at 4° to a volume of 0.5 ml of washed, aged human erythrocytes (2.5% v/v in Solution Na). Agglutination in the cold reflects virus–cell binding. No agglutination is observed when virus particles (or RSVE) are added to neuraminidase-treated erythrocytes.[34] Treatment with neuraminidase removes the cell membranes' sialic acid residues, which serve as virus receptor.[34] After 10 min of incubation in the cold, the cell agglutinates are incubated at 37° with gentle shaking. Cell–cell fusion can be followed by phase microscopy. At the end of the incubation period (30 min at 37°), the cells or the erythrocyte ghosts are centrifuged (26,000 g, 10 min), and the amount of hemoglobin in the supernatant is estimated at 540 nm. The extent of the hemolysis (reflected by the amount of hemoglobin in the supernatant) is linearly correlated to the amount of viral particles present and the duration of incubation at 37°. A highly active virus preparation is considered as one whose 150–200 HAU will induce 50% hemolysis during a 10-min incubation at 37°. A preparation of either intact virus particles or RSVE, which is able to promote cell agglutination but does not induce hemolysis of human erythrocytes at 37°, is considered to be an inactive, nonfusogenic preparation.

Fusion-Mediated Microinjection of RSVE Content into Living Animal Cells

Fusion of RSVE with Cells Grown in Monolayers. Most of our experiments were performed with monolayers of cells growing in plates containing 16-mm wells (Nunelon). Thus, the experimental procedure described here will be related to these conditions. However, the described procedure can be adapted, with slight modifications, to cells grown in petri dishes of various diameters. The description given is for mouse L cells.

[32] H. Peretz, Z. Toister, Y. Laster, and A. Loyter, *J. Cell Biol.* **63**, 1 (1974).

[33] M. Wasserman, A. Loyter, and R. G. Kulka, *J. Cell Sci.* **27**, 157 (1977).

[34] R. Rott and H. O. Klenk, *in* "Cell Surface Reviews" (G. Poste and G. L. Nicolson, eds.), Vol. 2, p. 47. North-Holland and Publ. Amsterdam, 1977.

TABLE I
FUSION-MEDIATED INJECTION OF DNA INTO CULTURED CELLS IN MONOLAYERS

Step	Procedure
Binding	DNA-loaded RSVE (10 μg of viral proteins) are added to cells (10^4 to 5×10^4 cells/well) grown in monolayers or to cells in suspension (10^6 to 10^7 cells/ml) for 10–15 min at 4°.
Fusion-mediated injection	Cells are transferred to 37° for 20–30 min of incubation.
Culturing of cells and selection of transformed cells	Cells are washed twice with growth medium and left to grow in a CO_2-incubator.

Mouse L cells were grown in 16-mm wells (DMEM + 10% calf serum) to give a nonconfluent layer of cells about 4 to 5×10^4 cells per 16-mm well). After two washings with cold Solution Na, a volume of 100 μl of Solution Na, containing 10 mM $CaCl_2$ and 10 μg of loaded or unloaded "empty" RSVE (about 500 HAU), is added to each well. The wells are then incubated for 10 min in the cold to allow binding of the RSVE to the surface of cells (Table I). To promote RSVE-cell fusion, the wells are transferred at the end of the cold incubation period to CO_2-incubator and incubated at 37° for 20–30 min. During this period some of the RSVE are fused with the cell plasma membranes, thus injecting their content directly into the cell cytoplasm (Table I). Experiments with radiolabeled RSVE showed that about 15–50% of the RSVE remain associated with the cells at the end of this period.[19] Since the cells in the wells are at subconfluent density, the RSVE promote very little, if any, cell–cell fusion. At the end of the incubation period, unbound RSVE are removed by washing with 2 ml of growth medium (DMEM + 10% calf serum); 2 ml of a warm medium are added, and the cells are left for recovery from the fusion event in the CO_2-incubator (Table I).

Fusion of RSVE with Cells Grown in Suspension. Most of these experiments were performed by fusion of RSVE with hepatoma tissue-cultured (HTC) cells, and the procedure described here is based on experiments with these cells.

Fusion-mediated microinjection into cells grown in suspension is somewhat more complicated than microinjection into cells in monolayer. Obviously, in order to ensure high efficiency of injection with high cell viability, virus-cell fusion should be induced without much promotion of cell–cell fusion. This can be achieved, to a large extent, by gentle vortexing of the large cell agglutinates formed after addition of the RSVE in the cold.

A volume of 20–30 μl (20–30 μg of RSVE, containing 1000–2000 HAU) is added to 10^7 cells suspended in 10 ml of Solution Na, containing

0.15 mM La^{3+};[33] for other cell lines, Mn^{2+} or Ca^{2+} could be added.[33] To ensure RSVE-cell binding, the suspension is incubated for 10 min at 4° (Table I). At the end of the cold incubation period, the agglutinates formed are gently vortexed to receive as many single cells as possible. This prevents massive cell–cell fusion when the suspension is subsequently incubated at 37° (Table I). The vortexed suspension is incubated at 37° for 20–30 minutes. The cells are then washed twice with warm growth medium (DMEM), and the final pellet of the cells is suspended in 10 ml of growth medium and further incubated at 37° in a CO$_2$ incubator (Table I).

In order to follow quantitatively both RSVE-cell binding and RSVE-cell fusion, ^{125}I-radiolabeled RSVE can be prepared from ^{125}I-labeled viral particles. A procedure for the preparation of ^{125}I-radiolabeled fusogenic Sendai virus particles was previously developed in our laboratory.[35] ^{125}I-labeled viral particles will yield RSVE containing ^{125}I-labeled HN and ^{125}I-labeled F glycoproteins with high specific activity.

Trapping of Radiolabeled (^3H or ^{32}P) DNA in RSVE

When DNA molecules are added to the Sendai virus envelopes' reconstitution system, they are adsorbed to and trapped within the RSVE formed upon removal of the detergent. The results (summarized in Table II) are as follows.

1. Regardless of the origin of the DNA, about the same amount remains associated with the viral envelopes after addition of 10–40 μg of DNA to 1 mg of viral protein. As can be seen, about 3–8% of the total DNA remain associated with the RSVE, under the conditions used, at the end of the reconstitution process when four different species of DNA molecules were employed.

2. The amount of the DNA remaining associated with the RSVE was dependent, in a nonlinear fashion, on the amount of the DNA added. With both pBR322 and SV40-DNA (not shown), it was found that there exists an optimal concentration of DNA at which maximum trapping was observed. The reduction in trapping efficiency which was found at higher concentrations of DNA, may indicate that at this concentration of 100–160 μg of pBR322-DNA per milligram of viral protein, for example, the DNA molecules interfere with the reconstitution process.

Table II also shows that about 60–90% of the RSVE-associated DNA are susceptible to hydrolysis by external DNase. DNA molecules associated with RSVE, but not susceptible to hydrolysis by added DNase, are considered to be "trapped" DNA, i.e., DNA molecules that are enclosed within the viral envelope vesicles. Mixing DNA molecules with resealed-reconstituted viral envelopes does not render the DNA molecules resist-

---

[35] D. Walf, I. Kahana, S. Nir, and A. Loyter, *Exp. Cell Res.* **130,** 361 (1980).

TABLE II
ENCLOSURE OF VARIOUS DNA MOLECULES WITHIN RECONSTITUTED
SENDAI VIRUS ENVELOPES (RSVE)[a]

DNA	DNA added to the reconstitution system (μg/mg of viral protein)	DNA associated with RSVE, after washing (μg)[b]	DNA enclosed within RSVE (after DNase treatment)	
			μg[b]	% of total added
³H-pBR322-DNA	2	0.8	0.08	4.0
(M_r 2.7 × 10⁶)	4	2.2	0.26	6.5
	10	5.5	0.55	5.5
	40	5.2	0.40	1.0
	80	1.0	0.30	0.4
	160	1.2	0.30	0.2
³H-pMB9	10	3.0	0.20	2.0
(M_r 3.5 × 10⁶)				
³²P-φX174 RFI-DNA	43	3.0	0.30	0.7
(M_r 3.5 × 10⁶)				
³H-SV40-DNA	40	8.0	1.20	3.0
(M_r 3.5 × 10⁶)				

[a] The different species of radiolabeled DNA were trapped within RSVE and the DNA-loaded RSVE were treated with DNase, as described in Methods and Results and elsewhere.[22] Protein was determined by the method of Lowry et al. [O. H. Lowry, N. J. Rosebrough, A. L. Farr, and R. J. Randall, J. Biol. Chem. 193, 265 (1951)] with 0.1% sodium dodecyl sulfate in the reaction mixture using bovine serum albumin as standard.
[b] The values are expressed as micrograms of DNA associated with or trapped within RSVE obtained from 1 mg of protein of intact virus particles.

ant to hydrolysis, and they remain susceptible to degradation by added DNases, as do free DNA molecules in solution (not shown).

3. Table III shows that about 55% of the RSVE-associated DNase-resistant DNA molecules can be precipitated by antivirus antiserum. (At higher concentrations of antiserum, up to 85% of the RSVE-enclosed ³²P-labeled molecules can be precipitated.) As can be seen in Table III, only negligible amounts of the ³²P-labeled DNA molecules are precipitated under the same conditions when either free DNA molecules or a mixture of free DNA molecules and unloaded RSVE are incubated with the antivirus antiserum. These observations (resistance to DNase and precipitation by antivirus antiserum) strongly indicate that the DNA molecules are indeed trapped within the viral envelopes.

About 15% of ³²P-labeled-DNA-loaded RSVE were found to be adsorbed to HTC cells at 4° when ³²P-labeled-φX174 RFI-DNA was used as

TABLE III
PRESENCE OF ^{32}P-ϕX174RFI-DNA IN RECONSTITUTED SENDAI VIRUS
ENVELOPES (RSVE) PRECIPITATED BY ANTISERUM AGAINST THE VIRUS[a]

System	^{32}P-DNA precipitated (% of total added)
I. ^{32}P-labeled DNA enclosed within RSVE (after DNase treatment)	55.0
II. ^{32}P-labeled DNA	0.3
III. ^{32}P-labeled DNA mixed with RSVE	0.6

[a] DNA was enclosed within RSVE, and the DNA-loaded vesicles were treated with DNase, as described in Methods and Results. Incubation with antiviral antiserum was as follows: 50 μl of antiviral antiserum were incubated with DNA-containing vesicles [50 μg of protein (I)], ^{32}P-labeled DNA [1μg, 3 × 10^5 cpm/μg (II)], or a mixture of ^{32}P-labeled DNA (1 μg) and unloaded RSVE (50 μg of protein) (III) for 1 hr at room temperature and then overnight at 4°. After precipitation (700 g), the pellet obtained was washed twice in Solution Na, and the radioactivity was determined after solubilization of the pellet in scintillation liquid (Insta-Gel II, Packard).

a marker (Table IV). In other experiments, up to 50% of either intact virus particles or added RSVE remained attached to the recipient cells at 4°.[19,36] When either free ^{32}P-labeled DNA or a mixture of free ^{32}P-labeled DNA and unloaded RSVE were incubated with the cells under the same conditions, only 1% of the radioactivity remained associated with the cells (Table IV). After incubation at 37° (fusion-mediated microinjection step) and cultivation for 2.5 hr, about 14% and 5% of the added DNA-loaded RSVE remained associated with the cells, respectively (Table IV). This represents about a 10 to 15-fold increase over the amount of DNA that remained associated with the cells after addition of free DNA (Table IV).

Gel electrophoresis studies, both with SV40-DNA[22] and with pTKx1 (Fig. 3), revealed that (a) during the reconstitution process, the trapped DNA molecules remained intact and were not hydrolyzed or substantially sheared during the long dialysis period required for efficient removal of the detergent; and (b) with pTKx1 only the supercoiled molecules were trapped, but not hydrolyzed by external DNases (Fig. 3). The same was observed with SV40-DNA.[22]

Use of RSVE for Microinjection of SV40-DNA and HSV-TK Genes into Living Cells

SV40-DNA-loaded RSVE can be used as carrier to introduce the trapped DNA either to SV40-susceptible or SV40-resistant cell lines.[22]

---

[36] M. Beigel, G. Eytan, and A. Loyter, in "Targeting of Drugs" (G. Gregoriadis, J. Senior, and A. Trowet, eds.), p. 125. Plenum, New York, 1982.

TABLE IV

ASSOCIATION OF ^{32}P-LABELED ϕX174 RFI-DNA-CONTAINING RECONSTITUTED
SENDAI VIRUS ENVELOPES (RSVE) WITH HTC CELLS AT 4° AND AT 37°[a]

	^{32}P-DNA associated with HTC cells (% of total label)		
System	4° (I)	37° (II)	Cultivation in growth medium (III)
^{32}P-DNA-loaded RSVE	15.0	14.4	5.5
^{32}P-DNA	1.0	1.0	0.6
^{32}P-DNA mixed with RSVE	1.0	1.0	0.3

[a] ^{32}P-labeled ϕX174 RFI-DNA (130 μg of 3×10^5 cpm/μg) was added to the viral envelope reconstitution system (3 mg of viral glycoprotein in 3% Triton X-100). The ^{32}P-labeled DNA-loaded RSVE obtained were treated with DNase, as described in Methods and Results. DNA-loaded RSVE (15 μg of protein), ^{32}P-labeled DNA (1 μg), and a mixture of ^{32}P-labeled DNA (1 μg) and RSVE (15 μg of protein) were incubated with hepatoma tissue culture (HTC) cells. Briefly, various samples of HTC cells, each containing 2×10^6 cells per milliliter of Solution Na with 0.15 mM La^{3+33} were incubated with RSVE for 15 min at 4°. At the end of each incubation, a sample was withdrawn and, after washing in Solution Na (100 g, 5 min), solubilized in solubilization liquid (Insta-Gel II, Packard), and the radioactivity associated with the cells was determined by scintillation counter (Packard) (I). The same was performed after incubation at 37° for 20 min (fusion-mediated injection step II; see Table I) and after culturing in growth medium for 150 min (III).

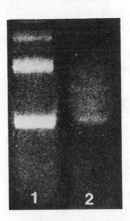

FIG. 3. Gel electrophoresis analysis of pTKx1 trapped within the reconstituted Sendai virus envelopes (RSVE). pTKx1 (150 μg) was trapped within Sendai virus envelopes, as described in Methods and Results. The pellet containing the RSVE was washed twice with Solution Na, suspended in 2 ml of Solution Na (of which 20 μl were withdrawn), incubated with 4 μg of DNase, in a final volume of 100 μl as described before[22] and above. DNase-treated RSVE were washed once with 2 ml of Solution Na containing 2 mM EDTA, and the final pellet was dissolved in 1.5% Triton X-100 and loaded on agarose gel (2). In (1) the gel pattern of the plasmid pTKx-1 (3 μg) used for trapping in RSVE is shown.

TABLE V

APPEARANCE OF SV40-T ANTIGEN AND LTK[+] CELLS AFTER FUSION-MEDIATED
INJECTION OF SV40-DNA AND pTKx1, RESPECTIVELY[a]

System	T antigen-positive cells (% of total)		
	F 1' 1-4	TC7	CV1
Experiment I			
SV40-DNA-loaded RSVE	1	3	20
Naked SV40-DNA mixed with RSVE	0.15	0.20	0.25
Unloaded RSVE	0.10	0.10	0.10
Naked SV40-DNA	0.10	0.10	0.10
Intact SV40	0	Not determined	Not determined
Experiment II	LTK[+] colonies (% of total cells)		
pTKx1-loaded RSVE	0.08–0.1		
pTKx1 mixed with RSVE	0		
Unloaded RSVE	0		
pTKx1	0		

[a] Fusion of cells with loaded RSVE was performed as described in Methods and Results. Briefly, 10–20 μl of RSVE (0.1–0.3 μg of DNA) were incubated with 10[4] to 5 × 10[4] cells for 15 min at 4° and then for 20–30 min at 37°. After this incubation period, cells were washed twice with DMEM supplemented with 10% newborn calf serum or fetal calf serum, and allowed to grow for different periods of time. In case of SV40-DNA loaded RSVE, after growth for 24–48 hr, cells were air dried, fixed with acetone–methanol for 10 min at −20°, and stained with rhodamine B-conjugated immunoglobulin from hamster anti-SV40 tumor serum, as described by Graessmann et al.[26] When pTKx1-loaded RSVE were used, cells were first grown for 36 hr in DMEM medium supplemented with 10% fetal calf serum and then transferred to HAT selective medium. TK[+] colonies appeared after 3–4 weeks of growth.

Epstein-Barr virus (EBV)-DNA was also transferred into various cell lines by using EBV-DNA-loaded RSVE.[37]

RSVE were loaded with naked SV40-DNA as described above in Fig. 1 and Table I. The RSVE were fused with monolayers of CV1 cells (an African mouse kidney cell line permissive to infection by intact SV40 virus) or TC7 cells (subline of CV1). Expression of the inserted SV40-DNA was studied by the appearance of SV40-T antigen in the nucleus of the recipient cells (Table V).[22] Detection of nuclei-associated specific SV40-T antigen is done by the use of fluorescently labeled anti-T antigen antiserum, as previously described.[26] Table V shows that 3% and 20% of TC7 and CV1 cells, respectively expressed SV40-T antigen after fusion-mediated microinjection of SV40-DNA. This high efficiency of transfor-

[37] I. Shapiro, G. Klein, and D. J. Volsky, submitted for publication (1982).

mation was obtained only when SV40-DNA-loaded RSVE were used (Table V). Incubation of the above SV40-permissive cell lines either with naked SV40-DNA or with a mixture of naked SV40-DNA and resealed-unloaded RSVE resulted in the appearance of specific SV40-T antigen in none or only a few of the recipient cells (Table V). This low fluorescence may represent an unspecific background seen always when cells are stained with fluorescently labeled anti-T antigen antiserum.

As sialoglycoproteins and sialoglycolipids, serving as receptors of Sendai virus particles, are present in a wide variety of animal cell lines, RSVE may be used almost as a "universal" fusogenic syringe. Indeed, RSVE can be used to transfer SV40-DNA to cells that are resistant to infection by the intact SV40. Table V shows that F 1' 1-4 cells, an established rat cell line, that do not synthesize SV40-T antigen after incubation with intact SV40 (Table V),[26] are able to synthesize this protein after incubation with SV40-DNA-associated RSVE. About 1% of the "infected" (fused with RSVE) F 1' 1-4 cells show the appearance of nuclei-associated T antigen, as was assayed with fluorescently labeled anti-T antigen antiserum. It should be mentioned that attempts to transform these cells by coprecipitation of DNA with calcium phosphate have failed.[26] F 1' 1-4 cells are resistant to infection by intact SV40, since they probably lack receptors of these viruses. This was established by experiments showing that RSVE can be used to transfer membrane components from TC7 cells to the F 1' 1-4.[22] After such fusion-mediated implanation of membrane fragments from SV40-susceptible cells into F 1' 1-4, the latter became susceptible to infection by intact SV40. These experiments demonstrate, probably for the first time, transplantation of functional receptors for SV40.

Table V also shows that RSVE can be used to transfer the *HSV-TK* gene (*Herpes simplex* thymidine kinase gene) to LTK⁻ cells with relatively high efficiency. This was demonstrated by the ability of pTKx1-loaded RSVE to transform LTK⁻ to LTK⁺ cells. The LTK⁺ cells, after cultivation in HAT medium, showed an eightfold increase in the activity of the enzyme thymidine kinase, as compared with the activity of the enzyme in control untransformed cells (LTK⁻) (not shown).

Conclusions and Comments

Figure 4 illustrates schematically the main steps of the method of using RSVE as a carrier for the introduction of functional DNA molecules and specific genes into cells in culture. From the results described above and from previous work using RSVE as a carrier of macromolecules,[20,37] it appears that the method allows (*a*) simultaneous injection of macromolecules, including DNA, into large numbers of cells; (*b*) introduction of

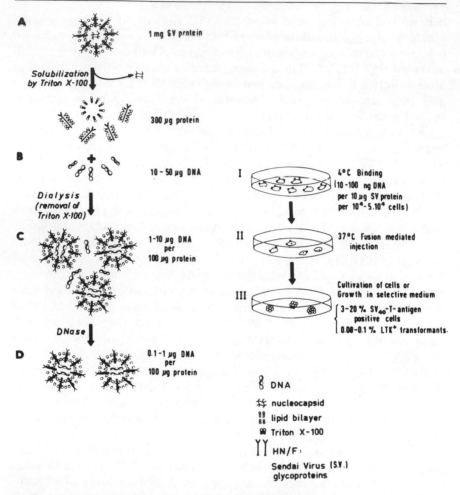

FIG. 4. A schematic illustration of trapping of DNA molecules within reconstituted Sendai virus envelopes (RSVE) and the use of DNA-loaded RSVE for transfer of DNA into cells.

DNA molecules into a broad range of cell lines that have previously been shown to be resistant to transformation by DNA-calcium phosphate complexes or by intact viruses. The only requirement for fusion-mediated injection by RSVE is the presence of sialic acid residues on cell surface glycoproteins or glycolipids.[34]

It is anticipated that in the near future various aspects of the method will be studied and modified, such as the phospholipids' composition of the RSVE or a better characterization of their internal volume, thus allowing a better trapping efficiency. However, in all cases the method will be based on two steps: (*a*) trapping of DNA molecules within viral envelopes during their reconstitution; and (*b*) fusion of the loaded reconstituted vesicles with plasma membranes of recipient cells. The presence of the virus F protein (fusion factor)[15,20,31] in the reconstituted vesicles promotes their fusion with cell plasma membranes without the need for an external fusogenic agent such as polyethylene glycol. This may, on the one hand, render this method more useful than other vesicle-mediated microinjection techniques;[11–13] on the other hand, it may allow, eventually, the use of such vesicles for the delivery of macromolecules, especially DNA molecules, into specific cells in the living organism.

A schematic summary of the method is also given in Fig. 4. The calculations made are based on the use of four different species of radiolabeled DNA molecules and of two specific genes (*SV40-DNA* and *HSV-TK*), whose transformation of cells was followed quantitatively. It can be seen that about 10 μg of viral vesicles, containing 0.1–1 μg of DNA, are required to promote synthesis of T antigen in 3% and 20% of TC7 and CV1 cells, respectively. About the same amount of vesicles (10 μg of protein) and of DNA (0.1–1 μg) were required to transform about 0.1% of LTK⁻ into LTK⁺ by fusion-mediated injection of pTKx1. It appears that the same amount of DNA used by the calcium phosphate coprecipitation method produced 0.5–5%[13] T antigen-positive cells and 0.001–0.01% LTK⁺ cells.[5,7] Based on these numbers, it appears that frequency of transformation obtained by the method described here, namely, using RSVE as a vehicle for the introduction of DNA, is 5 to 15-fold higher than that obtained with the calcium phosphate method. However, we believe that the main advantage of this system is the possibility that RSVE will allow transfer of functional DNA molecules into cells that were not transformed by previous methods. So far, this was demonstrated with two different cell lines and with two genes. DNA-loaded RSVE have been able to induce SV40-T antigen synthesis in F 1′ 1-4 with SV40-DNA (Table V) and to transform hamster TK⁻ cells (E36TK⁻ derivative of V79) to TK⁺ cells with *HSV-TK* (unpublished results). Both cell lines are resistant to transformation by the calcium phosphate coprecipitation technique (Table V).[26,38]

Theoretically, it can be assumed that transformation of cells or transfer of functional DNA molecules by DNA-loaded RSVE should be as efficient as transformation by intact DNA animal viruses. Reaching this goal, however, will necessitate significant improvement of the current method, especially modification of the steps involved in trapping the functional DNA. This might be achieved by using reconstitution systems that will

---

[38] L. Siminovitch, *in* "Genes, Chromosomes and Neoplasia" (F. E. Arrighi, P. N. Rao, and E. Stubblefield, eds.), p. 157. Raven, New York, 1981.

permit better trapping of macromolecules such as the one used for obtaining large unilamellar liposomes[39] or by addition of specific phospholipids[40] or proteins that will interact specifically with the DNA and force it to concentrate within the RSVE formed. The possibility of diluting the functional DNA with carrier-nonfunctional DNA should not be excluded. Carrier-nonfunctional DNA has been shown to be useful in transforming cells by the calcium phosphate coprecipitation methods.[5] Using carrier DNA may significantly increase the efficiency of cell transformation by DNA-loaded RSVE. This may be due to the introduction of DNA sequences present in the carrier-DNA that are required for the efficient reconstitution-insertion of the added DNA into the cell chromosomal DNA[10] or by decreasing the amount of the functional DNA required.

Acknowledgments

This work was supported by grants (to A. L.) from the March of Dimes Birth Defects Foundation; from the National Council for Research and Development, Israel; and from the G.S.F., Munich, Germany; and by a grant (to A.G.) from the Deutsche Forschungsgemeinschaft, No. GR 384-8.

[39] F. J. Szoka and D. Papahadjopoulos, *Proc. Natl. Acad. Sci. U.S.A.* **75**, 4194 (1978).
[40] R. M. Hoffman, L. B. Margolis, and L. D. Bergelson, *FEBS Lett.* **39**, 365 (1978).

[36] *In Vitro* Transcription: Whole-Cell Extract

By JAMES L. MANLEY, ANDREW FIRE, MARK SAMUELS, and
PHILLIP A. SHARP

Three classes of nuclear DNA-dependent RNA polymerases have been identified in eukaryotic cells.[1] RNA polymerase I catalyzes the synthesis of rRNA precursors, RNA polymerase II transcribes primarily the genes that give rise to mRNA, and RNA polymerase III transcription results in the production of tRNAs, 5 S RNA, and several other RNAs of unknown function. It has been clear for many years that in order to study the mechanisms of transcription as well as to identify the factors and nucleotide sequences that control and regulate gene expression, cell free systems that accurately and specifically transcribe exogenously added DNA are required. Early attempts at achieving this aim, which utilized purified RNA polymerases, were unsuccessful. In the last several years, however, *in vitro* systems have been developed in which accurate transcription by all three types of RNA polymerases can be obtained. Two basic approaches have been successful. In one, purified RNA polymerase is supplemented with cell extracts that contain factors required for accurate transcription.[2] This method has been used primarily for RNA polymerase II-mediated transcription and is described in Vol. 101 [36] of *Methods in Enzymology*. The other approach is to prepare concentrated cell lysates that contain not only the factors required for transcription, but also sufficient amounts of RNA polymerase so that addition of purified enzyme is not required. For RNA polymerase I[3] and III[4] such systems can be simply prepared from cytoplasmic extracts, because sufficient amounts of these polymerases and their required factors leak out of the nucleus at isotonic salt concentrations. RNA polymerase II, on the other hand, remains almost entirely within the nucleus. Thus, to obtain extracts containing this activity, a whole cell extract must be prepared.[5] We describe here the preparation and properties of such an extract, which contains all the factors and enzymic activities necessary for accurate and specific transcription, not only by RNA polymerase II, but also by RNA polymerases I and III.

[1] R. G. Roeder, *in* "RNA Polymerase" (R. Losick and M. Chamberlin, eds.), p. 285. Cold Spring Harbor Laboratory, Cold Spring Harbor, New York, 1976.
[2] P. A. Weil, D. S. Luse, J. Segall, and R. G. Roeder, *Cell* **18**, 469 (1979).
[3] I. Grummt, *Proc. Natl. Acad. Sci. U.S.A.* **78**, 727 (1981).
[4] G. J. Wu, *Proc. Natl. Acad. Sci. U.S.A.* **75**, 2175 (1978).
[5] J. L. Manley, A. Fire, A. Cano, P. A. Sharp, and M. L. Gefter, *Proc. Natl. Acad. Sci. U.S.A.* **77**, 3855 (1980).

RECOMBINANT DNA
METHODOLOGY

Most experiments to date have utilized extracts prepared from human cells that grow in suspension culture (HeLa or KB). Such extracts show quite broad species specificities for RNA polymerases II and III. Polymerase III genes from virtually all higher eukaryotes that have been tested are accurately transcribed in HeLa lysates. Whole-cell extracts do not seem able to transcribe yeast polymerase II genes accurately, but have been shown to be capable of transcribing *Drosophila*[6] and chicken[7] polymerase II genes as well as many such genes from higher eukaryotes and their viruses.

Synthesis of mature RNA molecules requires additional enzymes and factors other than those needed to bring about accurate transcriptional initiation (e.g., processing enzymes). Soluble HeLa lysates appear to contain virtually all the enzymes required for tRNA processing, including the splicing enzymes.[8,9] Although work with RNA polymerase I systems is just beginning, soluble extracts appear to contain at least one processing enzyme.[10,11] RNA polymerase II transcripts synthesized *in vitro* are efficiently capped and methylated at their 5′ ends.[2,5] Although one report in the literature claims that an extract efficiently spliced RNA,[12] we have not observed in a variety of experiments either splicing or creation of polyadenylated 3′ termini in the cell-free system described here.

Methods

Preparation of Extract

Extracts are prepared by modification of a procedure originally described by Sugden and Keller,[13] who used the method as a first step in RNA polymerase purification. We have used HeLa cells almost exclusively. These cells are easy to obtain in large quantities, and the resultant extracts are relatively free of nuclease activity at 30°. Lysates with transcriptional activity have been prepared from a few other cells lines; most other cell lines and tissues, however, have yielded extracts without detectable levels of transcription or with high levels of nuclease. Cells are

[6] R. Morimoto, unpublished observations.
[7] B. Wasylyk, C. Kedinger, J. Corden, O. Brison, and P. Chambon, *Nature* (*London*) **285**, 366 (1980).
[8] D. N. Standring. A. Venegas, and W. J. Rutter, *Proc. Natl. Acad. Sci. U.S.A.* **78**, 5963 (1981).
[9] F. Laski, A. Fire, U. L. RájBhandary, and P. A. Sharp, unpublished observations.
[10] I. Grummt, E. Roth, and M. R. Paule, *Nature* (*London*) **296**, 173 (1982).
[11] K. G. Miller and B. Sollner-Webb, *Cell* **27**, 165 (1981).
[12] B. Weingartner and W. Keller, *Proc. Natl. Acad. Sci. U.S.A.* **78**, 4092 (1981).
[13] B. Sugden and W. Keller, *J. Biol. Chem.* **248**, 3777 (1973).

grown in suspension culture in Eagle's minimal essential medium supplemented with 5% horse serum to a density of 4 to 8 × 10^5 cells/ml. The cell density appears not to be crucial, although we have obtained slightly more active extracts with cells harvested at the lower end of the range indicated.

The following operations are carried out at 0–4°.

1. Cells are harvested by centrifugation and washed twice with phosphate-buffered saline.

2. The volume of the resultant cell pellet is determined, and the cells are resuspended in four packed-cell volumes (PCV) of 0.01 M Tris-HCl (pH 7.9), 0.001 M EDTA, and 0.005 M dithiothreitol (DTT) (6 × 10^8 cells yield approximately 2 ml of packed cells). At this point, the cells should visibly swell.

3. After 20 min, the cells are lysed by homogenization in a Dounce homogenizer with eight strokes using a "B" pestle.

4. Four PCV of 0.05 M Tris-HCl (ph 7.9), 0.01 M MgCl$_2$, 0.002 M DTT, 25% sucrose, and 50% glycerol are added, and the suspension is gently mixed. With continued gentle stirring, one PCV of saturated (NH$_4$)$_2$SO$_4$ is added dropwise. After this addition, the highly viscous lysate is gently stirred for an additional 30 min. Stirring must be very gentle to prevent shearing of the DNA, which would interfere with its removal in the next step. Nuclear lysis can be detected by increased viscosity after approximately half the (NH$_4$)$_2$SO$_4$ has been added. Occasionally, lysates appear clumpy and only slightly viscous, rather than extremely viscous and uniform as usually observed. We have obtained active extracts from both types of lysates, although more reproducibly from the latter.

5. The extract is carefully poured into polycarbonate tubes and centrifuged at 45,000 rpm in a SW 50.2 rotor for 3 hr.

6. The supernatant is decanted so as not to disrupt the pellet (the last 1 or 2 ml are left behind), and protein and nucleic acid are precipitated by addition of solid (NH$_4$)$_2$SO$_4$ (0.33 g/ml of solution). After the (NH$_4$)$_2$SO$_4$ is dissolved, 1 N NaOH [0.1 ml/10 g solid (NH$_4$)$_2$SO$_4$] is added and the suspension stirred for an additional 30 min.

7. The precipitate is collected by centrifugation at 15,000 g for 20 min (the supernatant should be completely drained off), and resuspended with 5% of the volume of the high-speed supernatant with 0.025 M HEPES (adjusted to pH 7.9 with NaOH), 0.1 M KCl, 0.012 M MgCl$_2$, 0.5 mM EDTA, 2 mM DTT, and 17% glycerol.

8. The suspension is dialyzed against two changes of 50–100 volumes each of the resuspension buffer for a total of 8–12 hr. The volume of the solution increases 30–50% during dialysis. The conductivity of a 1:1000 dilution of dialyzed extract into distilled H$_2$O (23°) should be 12–14 μmho.

9. The dialyzate is centrifuged at 10,000 g for 10 min to remove insol-

uble material. The supernatant is divided into small aliquots (0.2–0.5 ml), quick frozen in liquid nitrogen or powdered dry ice, and stored at $-80°$. Extract can be thawed and quick frozen several times without loss of activity and retains full activity at $-80°$ for at least a year.

10. Lysates contain between 15 and 30 mg of protein per milliliter and up to 2 mg of nucleic acid per milliliter. We routinely start with 1–50 liters of cells. One liter of cells should yield about 2 ml of whole-cell extract (WCE), or enough for 100–400 assays. More concentrated extracts are desirable because with these the same optimal protein concentration can be obtained in reaction mixtures with a smaller volume of lysate. In this manner, the salt concentration in the *in vitro* reaction mixture can be lowered (high salt severely inhibits transcription; see below). Attempts to obtain more concentrated extracts by resuspending the pellet in a smaller volume after precipitation, or by tying the dialysis bag tightly (to reduce expansion during dialysis), have not been reproducibly successful owing to increased protein precipitation during dialysis. Likewise, dialysis against buffer containing lower salt concentrations results in less active lysates, again as a result of increased protein precipitation.

The Transcription Reaction

Reactions can be done in volumes of a few microliters or more. Analytical reactions are conveniently performed in 20 μl. A typical reaction mix might contain the following: 30–60% whole-cell extract in its dialysis buffer, 0.2–1.5 μg of template DNA (see below for effects of DNA and extract concentrations), 50 μM ATP, 50 μM GTP, 50 μM CTP, 5 μM UTP, 5 mM creatine phosphate, and 10 μCi of [α-^{32}P]UTP [commercial preparations of aqueous nucleotides can be obtained at a high enough concentration ($\sim$10 mCi/ml) and a sufficient specific activity ($>$200 Ci/mmol) to be added directly to the transcription]. After incubation for 30–120 min, the reactions can be extracted directly or placed at $-80°$ for up to a week before extraction.

Extraction of RNA and Resolution of Products by Gel Electrophoresis

To terminate transcription, 200 μl of stop buffer [7 M urea, 100 mM LiCl, 0.5% sodium dodecyl sulfate (SDS), 10 mM EDTA, 250 μg/ml tRNA, 10 mM Tris-HCl (pH 7.9)] and 300 μl of PCIA (phenol–chloroform–isoamyl alchohol, 1:1:0.05, water saturated and buffered with 20 mM Tris, pH 7.9) are added, the tubes are blended in a vortex mixer and centrifuged at 12,000 g for 15 min. The aqueous phase (discarding interface) is extracted once more with PCIA and once with chloroform and then pooled with 200 μl of 1.0 M ammonium acetate and precipitated with 900 μl of ethanol. The pellet is washed with ethanol and resuspended in

20 μl of 10 mM Na$_2$HPO$_4$ (pH 6.8)–1 mM EDTA; to this is added 50 μl of 1.4 M deionized glyoxal–70% dimethyl sulfoxide–10 mM Na$_2$HPO$_4$ (pH 6.8)–1 mM EDTA–0.04% bromophenol blue. After 1 hr at 50°, 25 μl of the samples are loaded on 1.4% agarose gels (run in 10 mM PO$_4$–1 mM EDTA).[14] This extraction procedure removes most of the free nucleotides from the RNA preparation.

For some techniques, larger reaction volumes are necessary. The above extraction protocol can be scaled up with modifications as follows: After removal of the first aqueous phase, three reaction volumes of stop buffer are added to the first organic phase, and the mixture is again homogenized and spun at 12,000 g (for 1 min). The organic phase is removed, and an equal volume of chloroform is added. After brief homogenization and centrifugation, the aqueous phase can be easily removed. The two aqueous phases are then pooled and reextracted once with PCIA, and twice with chloroform. Any precipitate at the interface of these extractions should be discarded. After the first ethanol precipitation, the pellet is resuspended in 200 μl of 0.2% SDS and 1 mM EDTA. An equal volume of 2 M ammonium acetate is added, and nucleic acid is reprecipitated with ethanol. The pellet is washed with ethanol and can be resuspended in the buffer of choice.

For analysis by hybridization and S1 nuclease digestion,[15] it is important first to remove the template DNA. The pellet is resuspended in 0.3 M sodium acetate (pH 5.2) and reprecipitated and washed with ethanol. The dried pellet is resuspended in 100 μl of 10 mM Tris-HCl (pH 7.5), and 100 mM NaCl. RNase-free DNase (treated with iodoacetate[16]) to 50 μg/ml, and MgCl$_2$ to 10 mM are added. After 5 min at 37°, 100 μl of 10 mM EDTA, 0.2% SDS, 150 mM NaCl are added, and the mixture is extracted with PCIA and chloroform. The final aqueous is reprecipitated with 0.25 ml of ethanol as above, and the pellet is resuspended in 60 μl of 0.2% Sarkosyl–1 mM EDTA (pH 8.0) and stored at −20°.

Sizing and Mapping in Vitro RNA

In general, RNA polymerase II does not terminate transcription *in vitro*. However, distinct length RNA products can be generated by the "runoff" assay. This method uses, as template, DNA molecules that have been cleaved by a restriction enzyme that cuts downstream from a putative transcription start site. RNA polymerases that transcribe this DNA will stop or fall off when they reach the end of the DNA. If a sub-

---

[14] G. K. McMaster and G. C. Carmichael, *Proc. Natl. Acad. Sci. U.S.A.* **74,** 4835 (1977).
[15] P. A. Sharp, A. J. Berk, and S. M. Berget, this series, Vol. 65, p. 750.
[16] S. B. Zimmerman and G. Sandeen, *Anal. Biochem.* **14,** 269 (1966).

stantial number of enzymes initiate transcription at the same site, then a population of molecules of a discrete size will be produced; such a population will migrate as a band on gel electrophoresis. DNA segments that have been cleaved by different restriction enzymes are used as templates in separate reaction mixtures; the transcription start site can be deduced by comparison of the sizes of the RNAs produced. This technique has been widely used for promoter mapping with *in vitro* transcription systems.

An example of the technique is shown in Fig. 1. The RNAs were transcribed from recombinant plasmids containing the adenovirus late promoter and various segments of the long (30 kb) late transcription unit. Several points are exemplified by this experiment. The *in vitro* system is capable of synthesizing very long RNAs, up to 7–8 kb, and hence contains little nuclease activity. Also, the glyoxal method of analyzing RNA is sensitive over a wide range of sizes. Plots of log molecular weight vs migration are linear for transcripts from 0.2 kb to over 5 kb.

Analysis of runoff transcripts is a simple, sensitive, and accurate method for determining the structure of *in vitro* synthesized RNA. However, it does have some limitations. The WCE contains relatively high levels of nucleic acid. Since most of this is 18 S and 28 S rRNA, it is impossible to load more than 25–50% of the RNA obtained from a 20-μl reaction mix onto a standard size gel slot (6 mm $\times$ 3 mm) without producing severe overloading of the gel in the regions occupied by these RNAs. An additional problem is that specific transcripts produced by very weak promoters can sometimes be obscured by nonspecific initiation or termination, or by exogenous nucleic acids labeled by end labeling activity in the extract (particularly rRNA and its breakdown products). This latter activity is insensitive to amanitin and actinomycin D and can thus be distinguished from *de novo* RNA synthesis. Use of radioactive GTP as tracer produces the least end labeling. Use of CTP or ATP produces high levels of tRNA labeling by enzymes exchanging the 3′-terminal CCA. A third limitation of analyzing runoff RNAs by denaturation and agarose gel electrophoresis is that end points of transcripts can be mapped only to within ~20 nucleotides at best. To map the 5′ ends of RNAs more precisely, short runoff transcripts can be analyzed on polyacrylamide sequencing gels.

The above analysis can be extended, and some of the problems circumvented, by using several variants on the technique of hybridization and Sl nuclease digestion. By using labeled RNA and a nonradioactive DNA probe, the problem of spurious RNA labeling can be eliminated, since RNA not complementary to the DNA probe is destroyed by the nuclease. Use of nonradioactive RNA and a DNA probe 5′ end-labeled 50–200 nucleotides downstream from the promoter allows resolution of ±1

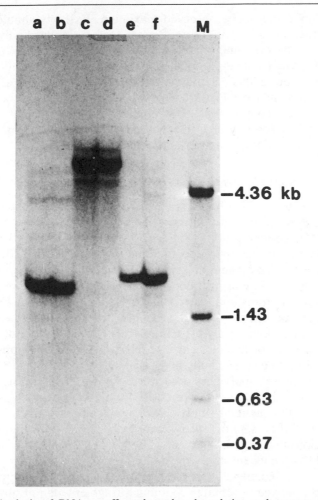

FIG. 1. Analysis of RNA runoff products by glyoxalation and agarose gel electrophoresis. Recombinant plasmids containing the adenovirus late promoter and various segments of the late transcription unit were constructed (R. Jove and J. L. Manley, unpublished), cleaved with restriction enzymes, and used as templates for *in vitro* transcription. DNAs were cleaved so that runoff transcripts of 1.8 kb (lanes a and b), 7.0 kb (lanes c and d), and 1.95 kb (lanes e and f) would be produced. Size markers (lane M) were also produced by *in vitro* transcription (see text).

nucleotide. The structure of the 5′ end of *in vitro* synthesized RNA can also be studied by classical RNA fingerprinting techniques.[2,5]

Transcriptional Activity

Extracts made from different cell preparations can vary in activity over a 5- to 10-fold range, with about two in three extracts exhibiting activity within 2-fold of the observed maximum. Extracts should be compared for their activity using a runoff assay from a standard polymerase II promoter, such as the major late promoter of adenovirus 2. With optimal DNA and extract concentrations, a good extract (20 μl) will yield 10^6 dpm, or 20 ng of a 2200 nucleotide runoff transcript from the Ad2 late promoter, in 1 hr ([α-^{32}P]UTP at 100 Ci/mmol). This represents the synthesis of one RNA molecule per 10 DNA template molecules present. However, the extract may actually be utilizing a smaller fraction of templates with multiple rounds of initiation per active template.

DNA and Extract Concentrations

Titrations both of DNA and of extract yield nonlinear responses. At a constant extract concentration, measuring runoff transcription as a function of DNA concentration yields (a) a threshold DNA concentration below which no transcription occurs; and (b) an inhibitory effect of high DNA concentration.[5] The requirement for a minimal DNA concentration is nonspecific; i.e., by using a concentration of a promoter specific DNA that is below the threshold, carrier DNA such as pBR322 or *Escherichia coli* DNA can be added to stimulate specific transcription. The duplex alternating copolymers poly[d(I-C)]:poly[d(I-C)] and poly[d(A-T)]:poly[d(A-T)] will also act as carrier DNA, thereby demonstrating a total lack of sequence specificity in the bulk DNA requirement.[17] A further advantage of these copolymers as carrier DNA is that the transcribed RNA products of the carrier poly[d(I-C)]:poly[d(I-C)] and poly[d(A-T)]:poly[d(A-T)] contain only two nucleotides. Thus, poly[d(I-C)]:poly[d(I-C)] carrier in a reaction containing [α-^{32}P]UTP yields no radioactive background. The key aspect of bulk DNA dependence is that at a fixed total DNA concentration, the molar yield of transcripts per promoter is constant and independent of the source of carrier. In general, specific competition between promoter-containing fragments is not observed.

A critical dependence of transcription upon extract concentration is also observed.[5] In fact, DNA concentration dependence and extract protein concentration dependence are not independent.[18] Specific transcription can be obtained in a range of 4–18 mg of extract protein per milliliter. At low extract concentration the DNA optima tend to be much lower (in the range of 10 μg/ml). There is still a bulk DNA dependence, but it is less steep and the threshold concentrations are lower. At a high extract concentration the DNA titration becomes sharper, and the threshold becomes higher. Under such conditions it is often necessary to use 60 μg/ml of DNA in order to see any transcription. Thus, for each new extract it is

[17] U. Hansen, D. J. Tenen, D. M. Livingston, and P. A. Sharp, *Cell* **27**, 603 (1981).
[18] A. Fire, C. C. Baker, J. L. Manley, E. B. Ziff, and P. A. Sharp, *J. Virol.* **40**, 703 (1981).

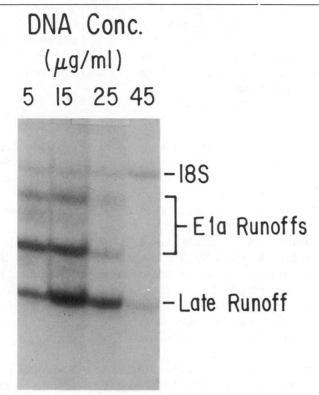

FIG. 2. Relative transcriptional activities of two adenovirus promoters as a function of bulk DNA concentration. Each reaction contained, per milliliter, 4 μg of a plasmid containing the adenovirus Ela promoter cleaved to give a 1220 n runoff and 1 μg of a plasmid containing the late promoter cleaved to give a 974 n runoff. Bulk DNA concentration was increased by addition of poly[d(I-C)]:poly[d(I-C)]. The transcription products were resolved on glyoxal gels as described above.

necessary to do careful DNA and extract titrations, to determine optimal conditions.

For a given promoter, very short runoff transcripts (<300 n) have a higher optimum DNA concentration than longer runoff transcripts.[18] This effect can be taken into account by measuring the synthesis of different length runoff products from the same promoter. No length dependence has been observed with runoff products between 400 and 4000 nucleotides.

To further complicate matters, the ratio of activity from two promoters can vary as much as 20-fold over a range of DNA and extract concentration.[17–20] An example of this is shown in Fig. 2 where the relative activ-

[19] D. Rio, A. Robbins, R. Myers, and R. Tjian, *Proc. Natl. Acad. Sci. U.S.A.* **77,** 5706 (1980).
[20] D. C. Lee and R. G. Roeder, *Mol. Cell. Biol.* **1,** 635 (1981).

ities of an early and a late Ad2 promoter are compared in an uninfected extract. The ratio of these activities varies 10-fold at different DNA concentrations. Comparison of promoter strengths in different extracts must thus be cautiously controlled and interpreted, a crucial point in assaying for regulatory phenomena.

Reaction Conditions

One unusual feature of the WCE is the temperature dependence of the reaction. Transcription is routinely done at 30°, where the *in vitro* synthesized RNA product is stable for 8 hr. Increasing the temperature to 37° greatly enhances the rate of RNA degradation; RNA made at 30° is degraded within 10 min after shifting to 37°. Transcription assayed at 23° yields the expected Arrhenius effect.

Specific transcription in the WCE is highly sensitive to ionic strength. Concentrations of KCl and NaCl above 60 mM significantly inhibit the reaction; concentrations in the 30–40 mM range are optimal. Reactions can also be performed in 15–30 mM $(NH_4)_2SO_4$. The divalent cations Ca^{2+}, Zn^{2+} and Mn^{2+} inhibit transcription and 0.5 mM EDTA is added to control their effect and the effect of other heavy metal contaminants. Reactions are done at pH 7.9, which is optimal for purified RNA polymerase II.[1]

Even after extensive dialysis, most WCEs seem to contain a free pool of 1 μM nucleotides.[18] The extract also contains creatine kinase (CPK) and other kinases and phosphatases so that the β and γ phosphates in nucleotides are labile.[21] For example, $[\gamma\text{-}^{32}P]ATP$ will rapidly exchange label with other triphosphates. Label at the α position of the nucleotide triphosphates does not exchange in the WCE, thus allowing RNA to be uniquely labeled with ^{32}P in the α position of each triphosphate. Addition of 5 mM creatine phosphate to the reaction mix ensures charging of the triphosphates and allows reduction in triphosphate concentrations, thus permitting the use of higher specific activities.[21] Extracts from some cell lines tested lack CPK activity and the enzyme must be added exogenously to maintain nucleotide concentrations.[22] One must also recall that $(NH_4)_2SO_4$ inhibits creatine kinase. In the presence of creatine phosphate, concentrations of UTP, CTP, and GTP as low as 5 μM saturate specific transcription; higher concentrations (up to 500 μM) do not inhibit specific transcription.[21] Because of endogenous pools, the transcription reaction is not fully dependent on addition of these three nucleotides.[18] A higher

---

[21] H. Handa, R. J. Kaufman, J. L. Manley, M. L. Gefter, and P. A. Sharp, *J. Biol. Chem.* **256**, 478 (1981).

[22] N. Crawford, unpublished results.

concentration of ATP is required for optimal activity (50 μM); ATP concentrations above 500 μM inhibit the reaction.[21]

The dialyzed extract also contains sufficient S-adenosylmethionine (SAM) to methylate the 5' ends of the *in vitro* transcripts.[5] Internal methylation has not been studied, however. Addition of exogenous SAM does not affect the reaction.[18,23]

Time Course

The rate of elongation in the WCE is approximately 300 nucleotides/min.[23] The rate *in vivo* is 10-fold higher, but one must recall the difference of 7° in temperature between the two. After DNA template, nucleotides, and extract are mixed, there is about a 5-min lag before specific transcription commences. The lag cannot be eliminated by preincubation of extract alone or with nucleotides, but is eliminated by preincubation of extract together with DNA (in the absence of nucleotides).[24] This suggests that the lag represents the time required for assembly on the DNA of factors required for initiation. The rate of accumulation of runoff transcripts is approximately linear for over an hour after the initial lag period.[5]

Some Other Properties of the WCE

The preparation procedure for the WCE was originally designed for solubilization of RNA polymerase II from mammalian cells. A standard 20 μl reaction mix typically contains 2–3 units of RNA polymerase II.[5] Under optimal conditions, at most one in ten polymerase II molecules gives rise to a specific transcript in a 1-hr reaction. Supplementation with excess purified polymerase has no significant effect on the WCE.[22] Endogenous RNA polymerase II in the WCE is inhibited by α-amanitin at 0.5 $\mu g/ml$; addition of a purified mutant enzyme resistant to α-amanitin reconstitutes specific transcription.[6,24,25]

RNA polymerase II preferentially initiates transcription at the termini of DNA fragments and at internal nicks.[26] The enzyme is also capable of end labeling DNA fragments with [α-^{32}P]NTPs to yield full-length labeled molecules, which are resistant to RNase digestion.[24] These reactions are each sensitive to α-amanitin (0.5 $\mu g/ml$), and are suppressed *in vitro* by the addition of a 110,000-dalton ADP-ribosyltransferase that may blockade nicks and ends.[27] This 110,000 dalton protein is present in large quantities in the WCE (up to 0.1% of total WCE protein).

[23] R. Jove, unpublished results.
[24] M. Samuels and A. Fire, unpublished results.
[25] C. J. Ingles, *Proc. Natl. Acad. Sci. U.S.A.* **75,** 405 (1978).
[26] M. K. Lewis and R. R. Burgess, *J. Biol. Chem.* **255,** 4928 (1980).
[27] E. Slattery, J. D. Dignam, and R. G. Roeder, unpublished results.

In Vitro TRANSCRIPTION STUDIES USING A WHOLE-CELL EXTRACT (WCE)

Template	Comments	References*
Adenovirus	Almost all *in vivo* promoters are recognized *in vitro*	2, 5, 7, 12, 18, 20
	Detailed 5' terminal analysis, dependence on upstream sequences	18, 20, *a–d*
	Inactivation of transcription in WCEs of poliovirus-infected cells	*e*
	Changes in transcriptional pattern in WCEs of adenovirus-infected cells	18
Globins	α-Globin and β-globin genes are recognized *in vitro*, dependence on upstream sequences	*f–i*
	Mutant α- and β-thalassemia globin genes are transcribed *in vitro*	*g, j–l*
SV40	Early and late promoters are recognized *in vitro*	19, 21
	Detailed 5'-terminal analysis of RNA from early promoter	17, *m, n*
	In vitro inhibition of transcription by T antigen	17, 19, *o*
	Cell-free translation of *in vitro* synthesized RNA	*p*
Conalbumin and ovalbumin	Promoters are recognized *in vitro*, dependence on upstream sequences, effects of altering TATA box	7, *a, q, r*
	Transcription in a homologous system	29
Type C retroviruses	Promoters in the long terminal repeat (LTR) of several RNA tumor viruses are recognized *in vitro*, dependence on upstream sequences	*s–u*
Fibroin	Promoter is recognized *in vitro* by WCEs of HeLa cells and of silk worm glands, dependence on upstream sequence in both extracts	*v, w*
Herpes simplex virus	Early promoters are recognized *in vitro* in uninfected cell WCEs	*x*

Template	Comments	References*
Histone H2A	Promoter is recognized *in vitro*, dependence on upstream sequences using linear or circular DNA template	y
Adeno-associated virus	Identification of a new promoter, detailed 5' terminal analysis	z

* *Key to references:* Numbers refer to text footnotes. Letters refer to the following.
  a. J. Corden, B. Wasylyk, A. Buchwalder, P. Sassone-Corsi, C. Kedinger, and P. Chambon, *Science* **209,** 1406 (1980).
  b. S.-L. Hu and J. L. Manley, *Proc. Natl. Acad. Sci. U.S.A.* **78,** 820 (1981).
  c. O. Hagenbuchle and U. Schibler, *Proc. Natl. Acad. Sci. U.S.A.* **78,** 2283 (1981).
  d. D. J. Mathis, R. Elkaim, C. Kedinger, P. Sassone-Corsi, and P. Chambon, *Proc. Natl. Acad. Sci. U.S.A.* **78,** 7383 (1981).
  e. N. Crawford, A. Fire, M. Samuels, P. A. Sharp, and D. Baltimore, *Cell* **27,** 555 (1981).
  f. D. S. Luse and R. G. Roeder, *Cell* **20,** 691 (1980).
  g. N. J. Proudfoot, M. H. M. Shander, J. L. Manley, M. L. Gefter, and T. Maniatis, *Science* **209,** 1329 (1980).
  h. C. A. Talkington, Y. Nishioka, and P. Leder, *Proc. Natl. Acad. Sci. U.S.A.* **77,** 7132 (1980).
  i. G. C. Grosveld, C. K. Shewmaker, P. Jat, and R. A. Flavell, *Cell* **25,** 215 (1981).
  j. R. A. Spritz, P. Jagadeeswaran, P. V. Choudary, P. A. Biro, J. T. Elder, J. K. DeRiel, J. L. Manley, M. L. Gefter, B. G. Forget, and S. M. Weissman, *Proc. Natl. Acad. Sci. U.S.A.* **78,** 2455 (1981).
  k. S. H. Orkin, S. C. Goff, and R. L. Hechtman, *Proc. Natl. Acad. Sci. U.S.A.* **78,** 5041 (1981).
  l. S. H. Orkin and S. C. Goff, *J. Biol. Chem.* **756,** 9782 (1981).
  m. D. J. Mathis and P. Chambon, *Nature (London)* **290,** 310 (1981).
  n. P. Lebowitz and P. K. Ghosh, *J. Virol.* **41,** 449 (1982).
  o. R. M. Myers, D. C. Rio, A. K. Robbins, and R. Tjian, *Cell* **25,** 373 (1981).
  p. C. L. Cepko, U. Hansen, H. Handa, and P. A. Sharp, *Mol. Cell. Biol.* **1,** 919 (1981).
  q. S. Tsai, M.-J. Tsai, and B. W. O'Malley, *Proc. Natl. Acad. Sci. U.S.A.* **78,** 879 (1981).
  r. B. Wasylyk and P. Chambon, *Nucleic Acids Res.* **9,** 1813 (1981).
  s. T. Yamamoto, B. deCrombrugghe, and I. Pastan, *Cell* **22,** 787 (1980).
  t. M. C. Ostrowski, D. Benard, and G. L. Hager, *Proc. Natl. Acad. Sci. U.S.A.* **78,** 4485 (1981).
  u. L. A. Fuhrman, C. Van Beveren, and I. M. Verma, *Proc. Natl. Acad. Sci. U.S.A.* **78,** 5411 (1981).
  v. Y. Tsujimoto, J. Hirose, M. Tsuda, and Y. Suzuki, *Proc. Natl. Acad. Sci. U.S.A.* **78,** 4838 (1981).
  w. M. Tsuda and Y. Suzuki, *Cell* **27,** 175 (1981).
  x. R. J. Frink, K. G. Draper, and E. K. Wagner, *Proc. Natl. Acad. Sci. U.S.A.* **78,** 6139 (1981).
  y. R. Grosschedl and M. L. Birnstiel, *Proc. Natl. Acad. Sci. U.S.A.* **79,** 297 (1982).
  z. M. R. Green and R. G. Roeder, *Cell* **22,** 231 (1980).

The WCE should contain most of the soluble proteins in the cell. Most of these are of no concern; however, some can interfere with interesting experiments. Most extracts have high levels of topoisomerase type I and II activities as well as DNA ligase. Thus DNA topologies can change rapidly in the reaction mix, preventing, for instance, studies of supercoiled DNA. The extract also contains RNA polymerases I and III.[13] Their contribution to the background pattern can be assessed with α-amanitin. Most template DNAs do not contain promoters for these enzymes, and their contribution to background incorporation is small. Some genomic clones contain dispersed repetitive elements, which often contain polymerase III genes.

Partial Fractionation of the WCE

A number of inhibitory activities can be removed by fractionation on phosphocellulose, yielding a more efficient transcription extract.[24,28] After dialysis to remove Mg^{2+} and dilution to 40 mM KCl, a WCE is chromatographed on phosphocellulose (Whatman P-11) yielding a breakthrough fraction, and two higher salt washes (0.35 and 1.0 M KCl). Reconstitution of the breakthrough and the dialyzed 1.0 M wash with purified RNA polymerase II in optimal ratios yields a mixture capable of specifically transcribing DNA at 10 times the efficiency of the original WCE. Tsai *et al.* have used a similar protocol to remove inhibitors from an extract of hen oviduct.[29]

Summary of Results

Since the first demonstration of transcription by RNA polymerase II in a soluble system and development of the WCE procedure a number of investigators have studied transcription of promoters using these systems. The preceding table represents a moderately comprehensive listing of such studies.

Acknowledgments

The work of R. Jove, H. Handa, U. Hansen, C. Cepko, N. Crawford, and A. Cano in helping to characterize the WCE and the advice and guidance of M. Gefter are gratefully acknowledged. This work was supported by Grant PCM78-23230 from the National Science Foundation, by Public Health Service Grants CA26717 (Program Project Grant) to P. A. S. and partially by CA14051 Center for Cancer Biology at MIT (Core). J. L. M. gratefully acknowledges support from Public Health Service Grant GM28983.

[28] T. Matsui, J. Segall, P. A. Weil, and R. G. Roeder, *J. Biol. Chem.* **255,** 11992 (1980).
[29] S. Y. Tsai, M.-J. Tsai, L. E. Kops, P. P. Minghetti, and B. W. O'Malley, *J. Biol. Chem.* **256,** 13055.

[37] Translation of Exogenous mRNAs in Reticulocyte Lysates

By WILLIAM C. MERRICK

In 1976, Pelham and Jackson first reported on the preparation of an active protein-synthesizing system for the translation of exogenous mRNAs.[1] By simply treating reticulocyte lysates with a Ca^{2+}-dependent nuclease and subsequently inactivating the nuclease with a chelator of Ca^{2+}, EGTA, they were able to convert the most efficient eukaryotic cell-free protein synthesis system[2] into one that would readily translate a variety of different mRNAs. In general this system is deemed superior to a Krebs II ascites or wheat germ cell-free system or frog oocytes because of the greater mRNA sensitivity (due to low backgrounds and high synthetic rates), low nuclease activity and/or low protease activity.[1] The availability of such an active translation system has been significant for molecular biologists. More than ever before it has become possible to assay for small amounts of mRNAs in order to monitor the purification of an mRNA, to quantitate levels of a specific mRNA during a developmental period, or to characterize recombinant DNA clones.

Solutions and Reagents

Phenylhydrazine, 2.5%: Phenylhydrazine (25 g) is dissolved in 1000 ml of H_2O that has been deoxygenated by bubbling with N_2. This solution is neutralized to pH 7 with 1 N NaOH. Single-use aliquots are frozen and protected from decomposition by storage in amber, airtight containers at $-20°$. Once thawed for use, residual amounts of the 2.5% phenylhydrazine solution are discarded, not refrozen for subsequent use.

Heparin solution, 500 units/ml

Physiological saline: 140 mM NaCl, 5 mM KCl, 7.5 mM magnesium acetate. This solution should be autoclaved to reduce the possibility of nuclease contamination.

Amino acid mixture: This solution contains all the nonradioactive amino acids at a concentration of 1 mM except for glutamate, glutamine, aspartate, asparagine, alanine, and glycine, which would be 2 mM.

Hemin hydrochloride solution: A 1 mM solution of hemin is prepared

[1] H. R. B. Pelham and R. J. Jackson, *Eur. J. Biochem.* **67**, 247 (1976).

[2] T. Hunt and R. J. Jackson, "Modern Trends in Human Leukemia" (R. Neth, R. C. Gallo, S. Spiegelman, and F. Stohlman, eds.), p. 300. Lehmanns Verlag, Munich, 1974.

as described by Ranu and London.[3] Initially 6.5 mg of hemin hydrochloride (Calbiochem-Behring) is dissolved in 0.25 ml of 1 M KOH. Subsequently, the solution is brought up to 10 ml final volume by the addition of 0.55 ml of distilled H_2O, 0.1 ml of 1 mM Tris-HCl (pH 7.9), 8.9 ml of ethylene glycol, and 0.2 ml of 1 M HC1. The resulting 1 mM solution of hemin can be stored at $-20°$ and is stable for more than a year.

Creatine phosphokinase solution: Ten milligrams of creatine phosphokinase (Sigma Chemical Co.) are dissolved in 1 ml of a buffer that is 20 mM HEPES–KOH (pH 7.5) and 50% glycerol. This enzyme solution is stable for more than 6 months when stored at $-20°$.

Master cocktail: To simplify the addition of small volumes of items that will normally be common to each assay, a number of stable reagents are mixed together to be subsequently added to the protein synthesis reaction mixture. The generation of a master cocktail not only saves time, but it also safeguards against the omission of a single component. The master cocktail is made up as follows:

100 μl	amino acid mixture
20 μl	100 mM GTP
50 μl	100 mM ATP
200 μl	2 M KCH$_3$CO$_2$, pH 7.5
10 μl	1 M Mg (CH$_3$CO$_2$)$_2$
75 μl	1 M HEPES–KOH, pH 7.5
X μl	radiolabeled amino acid (to be from 10 to 80 μM in the master cocktail, depending on expense)
(545-X) μl H$_2$O	
1000 μl	

It should be noted that depending on the amount of radiolabeled amino acid added to the master cocktail, it is possible that the rate and/or extent of protein synthesis will be limited by this component (especially if one uses high specific activity [^{35}S]methionine or [^{3}H]leucine: sp. act. 100 to 1500 Ci/mmol).

Micrococcal nuclease: Dissolve in H_2O at a concentration of 1 mg/ml and store frozen in aliquots.

Procedures

Preparation of Lysate

Immature New Zealand white rabbits (2.5–5 lb) are injected with 2.5% phenylhydrazine (0.10 ml/lb) for 5 consecutive days to cause reticu-

[3] R. S. Ranu and I. M. London, this series, Vol. 60, p. 459.

locytosis. During the last 1 or 2 days, the rabbits will become quite anemic and usually develop dark, watery stools. Subsequently the rabbits are allowed to recover for 2 days, then are bled on day 8. A few minutes prior to bleeding, the animals are anesthetized by an intraperitoneal injection of 30 mg of sodium pentobarbital. Care should be taken not to overdose the animals with the pentobarbital, as this may cause death of the animal prior to bleeding. Blood is taken by cardiac puncture with a No. 14 Huber-point needle (Becton-Dickinson) on a 60-ml plastic disposable syringe that contains 0.1 ml of heparin solution. Approximately 60–80 ml of blood can be obtained from each animal; however, some care in bleeding must be taken to avoid puncturing the heart too often, which leads to either loss of blood or death of the animal prior to removal of a satisfactory volume of blood.

After the blood has been withdrawn, it is transferred to chilled, plastic centrifuge tubes (250 ml) that contain 0.5 ml of the heparin solution. To avoid clotting and to increase cooling, the bottle is swirled gently with each fresh addition of blood and then kept on ice. After all the animals have been bled, the reticulocytes are collected by centrifugation in a Sorvall GSA rotor (8000 g, 10 min at 4°). The plasma is discarded and the reticulocytes are washed three times with chilled physiological saline. Washing of the cells involves (a) gentle but complete resuspension of the reticulocytes in 3–4 volumes of chilled physiological saline; (b) collection of the cells by centrifugation (8000 g, 10 min at 4°); (c) aspiration of the wash solution and the buffy coat (leukocytes), which sediment on the top of the reticulocyte pellet. Finally, the reticulocytes are osmotically lysed by the addition of chilled, distilled water (1.5–2.0 volumes of H_2O per volume of packed cells). Lysis is maximized by stirring the cells in the distilled water for 3–5 min on ice. Subsequently, the lysate is cleared of cell debris and mitochondria by centrifugation for 20 min at 16,000 g at 4°. The supernatant is carefully decanted off, taking care not to include any cell debris (the pellet formed is not very firm and does not remain attached to the bottom of the centrifuge tube). The supernatant is then immediately frozen in 0.5–2 ml aliquots and stored at the vapor temperature of liquid nitrogen.

Characterization of Reticulocyte Lysate

Prior to treatment with nuclease, it is worthwhile to check several parameters of the reticulocyte lysate to ensure that subsequent time and expense are not wasted on inactive lysate. Five relatively easily checked characteristics are as follows: (a) the A_{415} should be about 800–1000, which corresponds to a hemoglobin concentration of 200–250 mg/ml; (b) high speed centrifugation (100,000 g for 3 hr) should yield about 20–

25 A_{260} units of polysomes per milliliter of lysate; (c) using intact lysate, a 50-μl reaction mixture that is one-half lysate should direct the incorporation of about 300–400 pmol of radiolabeled leucine into hot trichloroacetic acid-precipitable product in 30 min; (d) in the absence of added hemin, protein synthesis should proceed at about 5–10% of the rate indicated above; (e) protein synthesis should be linear for the first 30 min and continue for 45–60 min. Although the list of characteristics may seem excessive, it should be noted that preparing active lysate is what is most important. The above characteristics provide a useful yardstick to judge the preparation. Another possibility is to compare the preparation with one of the commercially available "reticulocyte lysate translation kits" (see other considerations).

Assay for Hemoglobin Synthesis

The standard protein synthesis assay would contain the following components in a total of 100 μl: 20 μl of master cocktail, 50 μl of reticulocyte lysate, 1 μl of creatine phosphokinase solution, 5 μl of 100 mM creatine phosphate, 2 μl of 1 mM hemin hydrochloride solution, X μl of poly(A)$^+$ mRNA (0.3–5 μg), (22-X) μl of H$_2$O

For most assays the variable of interest will be the added mRNA. In such instances all the ingredients except the mRNA and lysate should be mixed together. Next, this mixture is then added to the unthawed lysate (ratio, 28 μl of mixture:50 μl lysate) and the lysate is thawed immediately. This thawing of the lysate in the presence of hemin minimizes any loss of activity due to apparent hemin deficiency in the lysate.[3] As should be apparent, in order to maximize the use of available lysate, aliquots should be made of appropriate size depending on need (100–1000 μl). The unused material should be discarded. Once the ingredients have been mixed on ice, they are transferred to a 30° water bath and incubated for the desired period of time. Reactions are terminated by dilution of a small aliquot, 2–10 μl, into 1 ml of H$_2$O and protein, then quantitated as (a) hot, TCA-precipitable radioactivity; or (b) alkali resistant, cold TCA-precipitable radioactivity (see below). For the initial untreated lysate, the investigator should optimize the assay conditions for added hemin (5–30 μM) and for added Mg(CH$_3$CO$_2$)$_2$ (0–4 mM). In most instances, those conditions optimal for globin mRNA translation will be similar to the conditions optimal for the translation of other mRNAs.

Hot TCA-Precipitable Radioactivity. To the 1 ml of distilled H$_2$O that contains 2–20 μl of the 100 μl reaction mixture is added in equal volume of 20% TCA. The solution is heated at 90° for 20 min to degrade aminoacyl-tRNA and then cooled on ice for 10 min. Precipitated radioactive protein is collected by vacuum filtration using glass fiber filters (Whatman GF/C) or nitrocellulose filters (type HA, Millipore). Glass fiber filters tend to have lower backgrounds when sulfur-containing or aromatic radiolabeled amino acids are used, but nitrocellulose filters give slightly better retention (background problems can be reduced by inclusion of 1 mM of the appropriate unlabeled amino acid in the TCA solution and rinsing of the filter prior to sample collection). The tube containing the 2-ml of sample is poured onto the filter and then rinsed three times with 5% TCA. The filters are then dried under an infrared lamp and transferred to scintillation vials; counting scintillant is added, and then radioactivity is determined using a liquid scintillation spectrometer.

Alkali-Resistant, TCA-Precipitable Radioactivity. This method of determination is primarily used for those few proteins that are not stable to hot TCA (i.e., collagen) or when ^{3}H-labeled amino acids are used (the precipitated hemoglobin will markedly quench ^{3}H but has little effect on ^{35}S or ^{14}C). In this instance, 0.5 ml of 1 M NaOH containing 0.5 M H_2O_2 (to decolorize) is added to the 1-ml sample (aliquot plus 1 ml of H_2O), and the mixture is incubated at 37° for 15 min. Then 1 ml of cold 25% TCA is added, and the sample is mixed and kept on ice for 10 min. Subsequently the radiolabeled, precipitated protein is collected by vacuum filtration, and radioactivity is quantitated as described above for hot TCA-precipitable radioactivity.

Treatment of Lysate with Micrococcal Nuclease

Having established that one has an active reticulocyte lysate, the next step is to remove the endogenous globin mRNA. This is done by treatment with a calcium-dependent RNase, micrococcal nuclease. Hemin (2–4 μl per 100 μl of lysate, as optimized in hemoglobin synthesis) is added to thawing lysate. This solution is then made 1 mM in $CaCl_2$ by the addition of 1 μl of 100 mM $CaCl_2$ per 100 μl of lysate and 10 μg/ml micrococcal nuclease by the addition of 1 μl of 1 μg/ml micrococcal nuclease per 100 μl of lysate. The lysate is then incubated for varying lengths of time and the action of the nuclease is stopped by the addition of the calcium chelator EGTA (1 μl of 100 mM EGTA per 100 μl of lysate). Subsequently the remaining ingredients may be added and protein synthesis tested. Usually, incubation of the lysate with nuclease for about 10 min gives optimal results (i.e., a low background and full activity). Underdigestion leaves residual globin mRNA, and consequently high backgrounds, in the absence of added mRNA. Overdigestion yields low backgrounds but the level of activity in the presence of added mRNA is reduced. Using the conditions indicated above, satisfactory lysate should be obtained with nuclease digestion times of 5–20 min.

Translation of Exogenous mRNA

The first mRNA to be tested should be one that is known to be active in directing protein synthesis. Having access to such an mRNA depends on whether or not one has a colleague from whom one might borrow such an mRNA. Lacking this, the preparation of globin mRNA using oligo(dT) cellulose is quite routine.[4,5] The advantage of using globin mRNA as an original test mRNA is that it should yield exactly the same incorporation as the untreated lysate, and the proteins (α and β globin) are readily characterized. Globin mRNA should saturate the assay system at about 20 μg/ml.[1] The major disadvantage in using globin mRNA is that one cannot determine how well larger mRNAs will be translated.

Having established an active, mRNA-dependent lysate, the mRNA is now to be analyzed. Listed below are a few considerations that should be reviewed prior to testing the mRNA.

1. It takes about 2 min to synthesize an intact globin chain at 30°. Using this ratio it will take about 10–15 min to synthesize a 75,000 dalton peptide, 20–30 min for a 150,000 dalton peptide, and 40–60 min for a 300,000 dalton peptide.

2. The reticulocyte tRNAs are ideally suited for the translation of globin mRNA, as the amino acid acceptance distribution very nearly matches the amino acid composition of the α and β globin chains.[6] Quite often the addition of a more general tRNA (i.e., liver tRNA) or tRNA from the same tissue will greatly enhance the translation of a specific mRNA. This is readily understandable for proteins of unique amino acid composition, such as collagen (high in glycine, serine, and proline), but some apparently normal proteins may also be dramatically responsive to added tRNA. A good example is the synthesis of fatty acid synthetase, where the presence of calf liver tRNA increased fatty acid synthetase production nine-fold while total protein synthesis was stimulated only slightly.[7] The preparation of any tRNA should include phenol extraction, ethanol precipitation, and Sephadex G-100 chromatography[8] to ensure that additional mRNA or mRNA fragments are not added.

3. The purity of the mRNA is not critical; however, poly(A)⁺ mRNA should be used. A total RNA preparation that is approximately 90–95% rRNA translates poorly, usually no better than 30–50% of maximum. A poly (A)⁺ mRNA preparation that is 20–60% rRNA translates well. The

[4] A. Krystosek, M. L. Cawthon, and D. Kabat, *J. Biol. Chem.* **250**, 6077 (1975).
[5] H. Aviv and P. Leder, *Proc. Natl. Acad. Sci. U.S.A.* **69**, 1408 (1972).
[6] D. W. E. Smith and A. L. McNamara, *Science* **171**, 577 (1971).
[7] A. G. Goodridge, S. M. Morris, Jr., and T. Foldflam, this series, Vol. 71, p. 139.
[8] W. C. Merrick, this series, Vol. 60, p. 108.

presence of some rRNA may help as an alternative substrate for nucleases and will aid in any precipitation steps as carrier RNA.

4. For most translations, one will use either [^{35}S]methionine (sp. act. 1000 Ci/mmol) or [^{35}S]cysteine (sp. act. 600 Ci/mmol) because of the very high specific activities and the higher energy of ^{35}S relative to ^{3}H (0.167 MeV vs 0.0186 MeV). However, although [3,4,5-^{3}H]leucine, [4,5,-^{3}H]isoleucine, or [4,5,-^{3}H]lysine are of lower specific activity (sp. act. about 120 Ci/mmol, 105 Ci/mmol, and 80 Ci/mmol, respectively), these amino acids constitute 6–10% of the amino acid composition of most proteins compared to the average 1% contribution of methionine or cysteine. The difference in specific activity between ^{35}S- and ^{3}H-labeled amino acids can thus be negated by the increased abundance of the three ^{3}H-labeled amino acids indicated above. This will not, however, compensate for the difference in the energy of decay between the two isotopes.

5. As indicated above, in general those conditions that are optimal for the translation of total poly(A)$^{+}$ mRNA (or even globin mRNA) should be useful in identifying the optimal conditions for translation of specific mRNAs. However, if mRNA is available, it would be worthwhile to titrate spermidine (0, 0.5, 1.0, and 1.5 mM), Mg(CH$_3$CO$_2$)$_2$ (1.0, 1.5, 2.0, 2.5, and 3.0) and KCH$_3$CO$_2$ (50, 100, 150, and 200 mM) for the specific enhancement of translation of the mRNA under study.

Analysis of Translation Products

A major part of using nuclease-treated lysates is the subsequent analysis of translation products, because only for a pure mRNA will TCA-precipitable radioactivity reflect the optimal translation conditions for a given mRNA There are two basic methodologies used to characterize translation products, and they are presented below.

1. *Polyacrylamide Gel Electrophoresis and Autoradiography.* Aliquots (2–10 μl) of sample reaction mixtures are mixed with 15 μl sample buffer [60 mM Tris-HCl, pH 6.8, 10% glycerol, 1% sodium dodecyl sulfate (SDS), 1% dithiothreitol, and 0.002% bromophenol blue] and then heated to 100° for 5 min to achieve uniform coating of the proteins with SDS. Subsequently, the 20–25-μl sample is subjected to electrophoresis in 7.5–12% polyacrylamide slab gels (100 × 140 × 0.75 mm; 20 wells per slab) as described by Laemmli.[9] The advantage of this gel system is that it is easy to prepare, readily reproducible and has high resolving power due to the discontinuous buffer system. The main variable is the percentage of acrylamide that yields best resolution of the molecular weight product expected.

[9] U. K. Laemmli, *Nature (London)* **227**, 680 (1970).

After electrophoresis, the proteins are visualized by staining with Coomassie Blue in 50% methanol, 7.5% acetic acid and destained with 5% methanol, 7.5% acetic acid.[10] Once the gel is destained, it is soaked in 10% acetic acid, 1% glycerol for 30–60 min and then dried onto filter paper (Whatman 3 MM or Bio-Rad filter paper backing). The dried gel is then exposed to X-ray film (Kodak X-Omat AR). Usually 1000 cpm of a [14]C- or ^{35}S-labeled protein can be visualized in 24 hr. However, for a mixture where it is not possible to anticipate the number of discrete polypeptides, it may take several days to visualize all the radiolabeled bands.

There are two common alternatives to the above procedure. The first is not to stain the gel, as staining is useful only in visualizing proteins used as molecular weight standards. In this case, after electrophoresis the gel is soaked in cold 20% TCA for 60 min to fix proteins, rinsed in the 10% acetic acid, 1% glycerol solution for 30–60 min, and then dried and exposed to X-ray film as above. The second alternative is the use of fluorography instead of autoradiography. In essence, this technique involves impregnating the gel with scintillation chemicals and then exposure to X-ray film.[11] In this process, the film is exposed by energy in the radioactive decay via the light emitted by the scintillation chemicals rather than by a photoconversion induced directly by the radioactive decay. This type of analysis is mandatory if ^{3}H-labeled amino acids are used, owing to the low energy in the ^{3}H decay, and gives about a 10-fold enhancement if [14]C- or ^{35}S-labeled amino acids are used. While greater sensitivity always sounds good, the use of the Bonner and Laskey procedure[11] or a commercial substitute (i.e., EN[3]HANCE, New England Nuclear) is expensive, a bit messy, and involves the use of hazardous solvents (i.e., dimethyl sulfoxide).

2. *Immunoprecipitation.* For many specific mRNAs to be studied, the protein for that mRNA and an antibody against the protein are available. It is possible to monitor the synthesis of a specific protein in an unfractionated mRNA population by immunoprecipitation. In this case, aliquots of the reaction mixture are diluted into an Eppendorf centrifuge tube that contains 1 ml of 150 mM NaCl, 10 mM EDTA, 20 mM Tris-HCl, pH 7.5, and 1% Nonidet P-40. To this solution is added 1 μl of a 200 mM solution of phenylmethylsulfonyl fluoride in absolute ethanol (phenylmethylsulfonyl fluoride is a general protease inhibitor that breaks down in H_2O and consequently must be added fresh; it is stable as a 200 mM solution in ethanol when stored at 4°). To quantitate total protein synthesis, a small aliquot can be taken, and hot TCA-precipitable radioactivity determined

[10] K. Weber and M. Osborne, *J. Biol. Chem.* **244,** 4406 (1969).
[11] W. M. Bonner and R. A. Laskey, *Eur. J. Biochem.* **46,** 83 (1974).

as above. To the remaining solution is added 10 μg of carrier protein and an appropriate amount of antibody to effect precipitation (this must be determined independently). The immunoprecipitate is allowed to form by incubating the tubes for 1 hr at room temperature and then overnight at 4°. The immunoprecipitate is collected by centrifugation for 5 min (Beckman or Brinkmann microfuge) and washed four times with 1 ml of 0.15 M NaCl, 1% Nonidet P-40 with vigorous mixing to resuspend the pellet uniformly. The pellet is then solubilized in SDS sample buffer as described above. At this point an aliquot is taken for determination of radioactivity. This radioactivity represents both completed chains as well as shorter peptides released by the EDTA present in the original buffer (see above). To quantitate the radioactivity in complete chains vs total radioactivity, it is necessary to analyze the immunoprecipitate by SDS polyacrylamide slab gel electrophoresis as described above. Such an analysis will indicate not only the percentage of radioactivity that represents complete chain synthesis, but also check the specificity of the immunoprecipitation process. As incomplete products will probably be randomly distributed throughout the low molecular weight region of the gel, it may be necessary to cut the dried gel into strips to quantitate the relative amount of radioactivity as full-length product.

As may be appreciated, the use of immunoprecipitation and SDS polyacrylamide gel electrophoresis should greatly increase the ability to identify and quantitate a specific radiolabeled peptide from the mixture. This is especially true for proteins that are minor representatives of the total due either to low mRNA abundance or relatively high molecular weight (all lysates are biased to yield a disproportionate amount of low molecular weight peptides because of low levels of nuclease or protease in the lysate and the greater length of time required to synthesize high molecular weight polypeptides). This procedure also allows resolution from more abundant proteins of similar molecular weight. However, as this process is more time consuming and requires additional materials, it is worthwhile to attempt identification just by SDS polyacrylamide gel electrophoresis[9] or as an alternative two-dimensional polyacrylamide gel electrophoresis.[12]

Other Considerations

Use of "The Kit." There are several commercial sources of nuclease-treated reticulocyte lysate that are guaranteed active and are usually accompanied by a proven test mRNA. If one is well-funded or plans only a limited series of translation experiments, the commercial materials (espe-

[12] P. H. O'Farrell, *J. Biol. Chem.* **250**, 4007 (1975).

cially from New England Nuclear) will readily prove cost-effective as the translation materials (lysate, mRNA, salts, etc.) are about the same price as the amount of radiolabeled amino acid required for the translations. In addition, most companies will give volume discounts (up to 40%) for an individual or group. Although optimal conditions for the test mRNA are provided, it is still necessary to ensure that these conditions are optimal for the mRNA being examined (most especially for high molecular weight polypeptides).

Messenger RNA Studies. The nuclease-treated lysate provides an excellent system for the examination of competition between different mRNAs for translation, especially as translation occurs at *in vivo* rates. Such studies can be extended by the use of inhibitors of initiation (i.e., ediene), elongation (i.e., cycloheximide), or initiation of capped mRNAs (i.e., m^7GDP).[13-15] The major asset of this system is the rapid rate of translation, which is usually 8- to 15-fold greater than most *in vitro* translation systems. The major limitation is that this is a system that would normally synthesize only α and β globin and as such is not prepared for many cellular responses (i.e., new mRNA synthesis, stimulation by hormones, cell division, viral infection, etc.).

Posttranslation Studies. Nuclease-treated lysates may be used to examine posttranslational modification. A number of studies, only a very few of which are referenced,[16-21] have examined *in vitro* the posttranslational modification of secreted proteins or proteins synthesized in the cytosol, but destined to be in mitochondria or chloroplasts. While it has not been demonstrated that the reticulocyte lysate system will be compatible with chloroplasts, it has been possible to demonstrate processing with microsomal membranes[16,17,19] or uptake into mitochondria of proteins synthesized *in vitro*.[20,21] At present a commercial kit for processing proteins with microsomal membranes is available (New England Nuclear), and by the time this chapter is published, one for the mitochondrial uptake of *in vitro* synthesized proteins will probably also be available.

[13] T. Brendler, T. Godefroy-Colburn, R. D. Carlill, and R. E. Thach, *J. Biol. Chem.* **256**, 11747 (1981).

[14] T. Godefroy-Colburn and R. E. Thach, *J. Biol. Chem.* **256**, 11762 (1981).

[15] D. Skup and S. Millward, *Proc. Natl. Acad. Sci. U.S.A.* **77**, 152 (1980).

[16] R. C. Jackson and G. Blobel, *Proc. Natl. Acad. Sci. U.S.A.* **74**, 5598 (1977).

[17] J. A. Majzoub, P. C. Dee, and J. F. Habener, *J. Biol. Chem.* **257**, 3581 (1982).

[18] A. R. Grossman, S. G. Bartlett, G. W. Schmidt, J. E. Mullet, and N. Chua, *J. Biol. Chem.* **257**, 1558 (1982).

[19] G. Rogers, J. Gruenebaum, and I. Boime, *J. Biol. Chem.* **257**, 4179 (1982).

[20] S. M. Gasser, A. Ohashi, G. Daum, P. Bohni, J. Gibson, G. A. Reid, T. Yonetani, and G. Schatz, *Proc. Natl. Acad. Sci. U.S.A.* **79**, 267 (1982).

[21] T. Morita, S. Miura, M. Mori, and M. Tatibana, *Eur. J. Biochem.* **122**, 501 (1982).

[38] Preparation of a Cell-Free Protein-Synthesizing System from Wheat Germ

By CARL W. ANDERSON, J. WILLIAM STRAUS, and BERNARD S. DUDOCK

Crude extracts of commercial wheat germ are known to be capable of translating a wide variety of messenge RNAs. The basic wheat germ cell-free system used in most laboratories was initially described by Roberts and Paterson.[1] However, several significant improvements have been introduced since that publication, such as the use of polyamines and the substitution of acetate for chloride ions in buffering systems.[2,3] The purpose of this communication is to describe in detail the preparation of a wheat germ cell-free system capable of synthesizing high molecular weight polypeptides and to enumerate some of the difficulties that may be encountered in preparing and using such a system. It should be noted that cell-free protein-synthesizing systems have been prepared also from wheat embryos.[4] Such systems are reported to have lower endogenous ribonuclease activity, but have not been as popular owing to the added procedure of isolating the embryos.

Preparation of Wheat Germ Extract

Materials. Wheat germ extract is commercially available from Bethesda Research Laboratories (BRL) (Gaithersburg, Maryland) and Miles Laboratories (Elkhart, Indiana), but can be easily and inexpensively prepared in any modern laboratory. Untoasted wheat germ can be obtained in small quantities (ca 1 pound) from General Mills (contact Director, Quality Control, P.O. Box 113, Minneapolis, Minnesota 05426; telephone 612-540-2354). General Mills has been setting aside parts of certain lots for research. We have not tested a sufficient number to comment on the relative activity of different lots. Alternatively, untoasted wheat germ can be purchased from local health food stores; however, only two or three of over a dozen lots tested gave satisfactory activity. Wheat germ can be stored for several years in a vacuum desiccator over silica gel at 4°. Exposure to moisture, heat, or freezing will result in a severe loss of activity.

[1] B. E. Roberts and B. M. Paterson, *Proc. Natl. Acad. Sci. U.S.A.* **70**, 2330 (1973).
[2] K. Marcu and B. Dudock, *Nucleic Acids Res.* **1**, 1385 (1974).
[3] J. F. Atkins, J. B. Lewis, C. W. Anderson, and R. F. Gesteland, *J. Biol. Chem.* **250**, 5688 (1975).
[4] A. Marcus, D. Efron, and D. P. Weeks, this series, Vol. 30, p. 749.

RECOMBINANT DNA
METHODOLOGY

Prior to extraction, assemble the following materials:

Acid-washed sand or powdered glass (oven sterilized)

Medium-sized mortar (ca. 10 cm in diameter) and pestle, rubber police-man, and spatula

U.S. Standard Sieve Series No. 25 screen (opening = 0.71 mm)

Corex centrifuge tubes, 30 ml (new, unscratched); refrigerated centri-fuge

Sephadex G-25 medium or fine column (2.5 × 30 cm) equilibrated with "column buffer," set up in a cold room at 4°; fraction collector and tubes

Spectrophotometer

Plastic microcentrifuge tubes, 500 μl (ca 150 per extract)

Liquid nitrogen in a small Dewar flask

Stock solutions:

    a. 1.0 *M* HEPES buffer, adjusted to pH 7.60 (at 20°) with KOH. Hepes is not stable to autoclaving and will support growth of mold and bacteria. Store as 5-ml aliquots at −20°.

    b. 1.0 *M* KOAc (acetate), adjusted to pH 7 with acetic acid. Auto-clave in 100-ml bottles.

    c. 0.1 *M* Mg(OAc)$_2$, adjusted to pH 7.0 with KOH if necessary. Au-toclave in 100-ml bottles.

    d. 0.1 *M* CaCl$_2$. Autoclave in 100-ml bottles.

    e. 1.0 *M* dithiothreitol (DTT). Store at −70° in 1.0 ml aliquots.

Extraction buffer (make fresh, keep ice cold):

	Volume	Final conc.
1.0 *M* HEPES, pH 7.6	1.0 ml	20 mM
1.0 *M* KOAc	5.0 ml	100 mM
0.1 *M* Mg(OAc)$_2$	0.5 ml	1 mM
0.1 *M* CaCl$_2$	1.0 ml	2 mM
1.0 *M* DTT	0.05 ml	1 mM
Distilled H$_2$O to final vol. 50.0 ml		—

Column buffer (make fresh, keep at 4°):

	Volume	Final conc.
1.0 HEPES, pH 7.6	10.0 ml	20 mM
1.0 *M* KOAc	60.0 ml	120 mM
0.1 *M* Mg(OAc)$_2$	25.0 ml	5 mM
1.0 *M* DTT	0.5 ml	1 mM
Distilled H$_2$O to final vol. 500.0 ml		—

The extraction should be performed as rapidly as possible and all ma-terials be kept at 0–4°. It is preferable to work in a cold room. All glass-ware should be scrupulously clean and oven-sterilized to minimize con-tamination from exogenous nucleases. Plastic gloves should be worn throughout. If nuclease contamination is a problem, then placental ribonuclease inhibitor[5] can be used to reduce degradation of the RNA.[6]

Extraction Procedure. Sieve about 13 g of wheat germ with moderate vigor for 3–5 min. The germ is retained on the screen while about 20% of the weight (mostly endosperm and chaff) passes through.

Vigorously grind 10 g of sieved wheat germ with an equal weight of acid-washed sand or powdered glass for 1 min in a prechilled mortar. Note that overgrinding will result in a loss of activity. Then mix in 20 ml of cold "extraction buffer" (see above) in 4–5 ml aliquots, with a rubber policeman or spatula, to obtain a thick paste. Transfer the paste with a spatula to a 30-ml Corex centrifuge tube and centrifuge at 30,000 g for 10 min at 2° (e.g., 16,000 rpm in a Sorvall SS-34 rotor). After centrifugation, remove the brownish supernatant ($\sim$12 ml) with a Pasteur pipette, taking care to avoid contamination from the yellow surface layer of fatty material and from the pellicle that covers the sand. A $\frac{1}{800}$ dilution of this supernatant into water should give approximate spectrophotometric readings at A_{260} of 0.5, and at A_{280} of 0.4.

Low molecular weight material is then removed from the extract by gel filtration. Load the extract on a Sephadex G-25 column (see above) and collect 4-ml fractions at a flow rate of 2–3 ml/min. The extract can be observed as a light brown band that elutes in the void volume. A yellow band is retained on the column and if desired, can be eluted in 1–2 column volumes. The void volume peak should be evaluated spectrophotometrically by diluting 10-μl aliquots of each fraction into 2 ml of water and determining the A_{260} values. The void volume should contain about one-third of the applied optical density at 260 nm. Pool appropriate portions of the void volume peak (i.e., discard the last third of the void volume peak) and centrifuge (15,000 g for 10 min at 2°) to remove any insoluble material. Note that the concentration of pooled extract can be changed by adding column buffer. We find it convenient to adjust the A_{260} to 100 for uniformity between different extracts.

The pooled extract is immediately aliquoted into 500-μl plastic tubes (e.g., 100–200 μl/tube) and dropped into a small Dewar flask of liquid nitrogen. The frozen tubes are then transferred to an appropriately labeled box and stored in liquid nitrogen or in a freezer at −70°. Extracts are stable at −70° for at least 6 months and often up to several years. From 10 g of wheat germ, we obtain about 25 ml of extract. The preparation can be scaled up or down as required.

Wheat Germ Protein Synthesis

Materials. Assemble reaction tubes, water bath (30°), wheat germ extract, mRNA, any test compounds, solutions, and assay materials (see

[5] P. Blackburn, G. Wilson, and S. Moore, *J. Biol. Chem.* **252,** 5904 (1977).
[6] G. Scheel and P. Blackburn, *Proc. Natl. Acad. Sci. U.S.A.* **76,** 4898 (1979).

below). The reactions can be conveniently performed in 1.5-ml plastic microcentrifuge tubes. The wheat germ extract, mRNA, and solutions should be kept ice cold prior to incubation.

The required stock solutions are listed below.

Hepes, 1.0 M, pH 7.60 (at 20°), adjusted with 5 M KOH, (do not autoclave)

$Mg(OAc)_2$, 0.25 M, adjusted to pH 7 with KOH, autoclaved

KOAc, 1.0 M, autoclaved (KOAc is very hydroscopic), adjusted to pH 7.0 with acetic acid

Stock solutions, 0.1 M, each of 20 amino acids (see Table I)

Working amino acid mixture (minus radioactive methionine, leucine, or other amino acids as required), 5 mM each amino acid ($\frac{1}{20}$ dilution of 0.1 M stock)

Creatine phosphokinase (CPK) (Sigma), 10 mg/ml in H_2O, 40–170 IU/mg

Creatine phosphate (CP) disodium salt (Sigma), 0.4 M, adjusted to pH 7

ATP, 0.1 M, adjusted to pH 7 with KOH

GTP + $Mg(OAc)_2$ (each 40 mM), adjusted to pH 7

DTT, 0.1 M

Spermine tetrahydrochloride, 1.5 mg/ml, adjusted to pH 7

TABLE I

AMINO ACIDS FOR STOCK SOLUTIONS

Acid	Molecular weight	0.1 M (mg/ml)
L-Alanine	89.09	8.9
L-Arginine-HCl	210.7	21.1
L-Asparagine-H_2O	150.10	15.0 suspension
L-Aspartic acid	133.10	13.3 suspension
L-Cysteine	121.16	12.1
L-Glutamic acid	147.13	14.7 suspension
L-Glutamine	146.16	14.6
Glycine	75.07	7.5
L-Histidine-HCl	209.6	21.0
L-Isoleucine	131.17	13.1
L-Leucine	131.17	13.1
L-Lysine-HCl (hydrate)	182.7	18.3
L-Methionine	149.12	14.9
L-Phenylalanine	165.19	16.5
L-Proline	115.13	11.5
L-Serine	105.09	10.5
L-Tryptophan	204.22	20.4 suspension
L-Threonine	119.12	11.9
L-Tyrosine	181.19	18.1 suspension
L-Valine	117.15	11.7

All solutions are stored at −20° in aliquots of 1−5 ml in glass vials or polypropylene plastic tubes. It may prove to be convenient to premix several components.

Procedure. Decide on the amount and type of radioactive label and lyophilize (if necessary). It is not usually necessary to lyophilize radioactive amino acids unless they are stored in acid (HCl) or unless large volumes are to be used. In general, [^{14}C]leucine is used for kinetic measurements and [^{35}S]methionine is used when products are to be analyzed by polyacrylamide gel electrophoresis. Typical assays contain: [^{14}C]leucine (ca 300 mCi/mmol, 100 μCi/ml) 1.5 μl/assay = 0.15 μCi/assay (0.5 nmol/assay = 20μM); [^{35}S]methionine (ca 350 Ci/mmol, 5 mCi/ml) 3 μl assay = 15 μCi/assay (42.9 pmol/assay = 1.7 μM).

Note that if substantially less than 15 μCi per assay of [^{35}S]methionine is used, cold methionine should be added to provide approximately 2 μM methionine. For maximum incorporation of isotope it is worthwhile to titrate the [^{35}S]methionine to see that it is not inhibitory at high concentrations.

Write up the protocol, thaw components (on ice), and set up reaction tubes in an ice water bath. Assemble the "master mix" (see Table II) in the tube containing the radioactive label, remembering to dilute the stock solution of 40 mM GTP + Mg^{2+} 1:20. Mix stock solutions before making additions, since some components (i.e., GTP, amino acids) are suspensions. Sufficient water may be added to the "master mix" to minimize pipetting of the reaction components into the individual reaction tubes.

Add to the individual reaction tubes appropriate volumes of water and master mix, any test solutions, and mRNA (approximately 0.1−1.0 μg) in order. Approximately 3 μg of rRNA or tRNA should be added to the background control tube to protect endogenous protein synthesis from endogenous nucleases so as to provide an adequate indication of the endogenous background (this RNA should be tested for inhibition of translation). Be careful to minimize bubble formations when pipetting and to mix tubes gently before incubation. Then incubate each tube for 90 min at 30°.

Assays for Protein Synthesis. If the entire sample is to be assayed for incorporated radioactivity, add 50 μl of bovine serum albumin (5 mg/ml) and 100 μl of 2 M KOH to each tube at the end of the incubation period (to provide carrier during precipitation and to hydrolyze aminoacylated tRNA). Incubate for 15 min at 30°, and then add 1.5 ml of cold 5% trichloroacetic acid (TCA) containing amino acids (20 mM each) or 2 g of "casamino acids" (Difco Laboratories) per liter and put on ice for 10 min. Filter the reaction mixtures through glass fiber filters (Whatman GF/C), and thoroughly rinse both the tubes and filters with additional TCA. The filters are then labeled (e.g., with a No. 1 pencil) and dried for 10−15 min under a heat lamp. Dried filters may then be placed in shell-vials and counted for radioactivity in about 3 ml of toluene-based scintillation fluid,

TABLE II

COMPONENTS FOR PROTEIN SYNTHESIS[c] BY WHEAT GERM EXTRACTS

Master mix[a]	Number of 25-μl reactions					Concentration		
	10 (μl)	20 (μl)	30 (μl)	40 (μl)	50 (μl)	Extract	Component	Total
1. 1.0 M HEPES, pH 7.6[b]	3.5	7	10.5	14	17.5	6	14	20 mM
2. 0.1 M ATP	2.5	5	7.5	10	12.5	—	1	1 mM
3. 0.4 M CP (creatine phosphate)	5	10	15	20	25	—	8	8 mM
4. 10 mg/ml CPK (creatine phosphokinase)	1	2	3	4	5	—	40	40 μg/ml
5. 1.5 mg/ml spermine, pH 7	5	10	15	20	25	—	30	30 μg/ml
6. 0.1 M DTT	4.25	8.5	12.75	17	21.25	0.3	1.7	2 mM
7. 2 mM GTP + Mg^{2+} (1/20 of stock)	2.5	5	7.5	10	12.5	—	20	20 μM
8. 5 mM each, 19 amino acid mix	1.25	2.5	3.75	5	6.25	—	25	25 μM
9. 0.25 M Mg(OAc)$_2$	1	2	3	4	5	1.5	1	2.5 mM
10. 1.0 M KOAc	21	42	63	84	105	36	84	120 mM
11. Wheat germ extract ($A_{260} \simeq 100$)	75	150	225	300	375	—	—	
Volume of "Master Mix"	122	244	366	488	610			
Total assay volume	250	500	750	1000	1250			
Remaining volume	128	256	384	512	640			

Additional components (for remaining volume)
12. Radioactive amino acid
13. Messenger RNA
14. Other test substances
15. Water

[a] Master mix optimized for translation of total cytoplasmic RNA from adenovirus-infected cells. Optimal translation of other RNAs may require slight adjustments of some concentrations.

[b] Tris-acetate, 1.0 M pH 7.2, will substitute for HEPES buffer.

[c] Incubate individual reactions at 30° for 90 min.

such as Econofluor (NEN). Discard TCA wash in radioactive waste.

If samples are to be analyzed by polyacrylamide gel electrophoresis,[7] remove 2–5-μl aliquots for analysis as above (scintillation counting). The remainder of the sample is incubated with 5 μl per assay of 200 μg/ml RNase A, 200 mM EDTA for 15 min at 30° prior to being precipitated with 0.5 ml of 90% aqueous acetone for about 30 min at 0–4°, and centrifuged (2000 rpm, 10 min at 4° in a clinical centrifuge). The supernatant, which is discarded, contains most of the unincorporated amino acid, and the precipitate is dissolved in 50 μl of standard SDS sample buffer. It is essential that the sample be well suspended in SDS sample buffer before it is heated. This is conveniently accomplished by placing the samples in an ultrasonic cleaning bath for 30 min prior to heating. Samples may be stored frozen at −20° until they are assayed. For best resolution do not apply more than 10 μl to a well of a standard 25-slot slab gel. It is desirable to stain gels or to soak gels in destaining solution to remove residual unincorporated radioactive amino acids prior to drying for autoradiography.

If the protein under study is soluble in acetone, alternative procedures are to dissolve the sample in an equal volume of sample buffer directly, or to precipitate the sample with 10% TCA and wash three times with ethanol–ether (1:3) before dissolving in sample buffer. If ethanol–ether washes are not thorough, residual TCA will be revealed by the sample turning yellow during heating. This is a pH change of bromophenol blue and is reversible.

When the reaction product is to be subjected to chemical or sequence analysis, it should be noted that the wheat germ extract can contribute significant quantities of endogenous amino acids, presumably from aminoacylated tRNA.[8] Furthermore, terminal amino acids in the polypeptide chains may be partially acetylated. Acetylation can be prevented by pretreating the reaction mixture with citrate synthetase and oxaloacetate as described by Palmiter.[9]

Troubleshooting. There are many ways in which a protein synthesis assay can fail, either totally or partially. Components should be reasonably fresh and handled and stored carefully. This is particularly true of the wheat germ extract and the mRNA. If it is not possible to obtain satisfactory results, then it may be necessary to test each individual component. It is preferable to test components one at a time in a system that is known to function, since it is otherwise difficult to identify the problem if two or more components are bad.

[7] U. K. Laemmli. *Nature (London)* **227,** 680 (1970).
[8] C. W. Anderson, *in* "Genetic Engineering: Principals and Methods," Vol. 4 (J. K. Setlow and A. Hollaender, eds.), p. 147–167. Plenum, New York, 1982.
[9] R. D. Palmiter, *J. Biol. Chem.* **252,** 8781 (1977).

Commercially available untoasted wheat germ may range in activity from very good to no activity at all. Extracts should be tested occasionally with an mRNA that is known to be efficiently translated. The 9 S RNA (globin mRNA) from rabbit reticulocytes (BRL) makes a useful standard with which to compare different extracts or different translation conditions. For maximal activity, the optimal amount of extract per 25-μl reaction should be determined. With extracts at $A_{260} = 100$, we have usually obtained satisfactory activity at 7.5 μl of extract per 25-μl reaction mixture. One cannot exceed 12 μl of extract per assay without exceeding the optimal Mg^{2+} concentration of 2.5 mM for most eukaryotic messages.

Most wheat germ extracts have very low endogenous activity. If necessary, endogenous activity may be further reduced by a preincubation step with creatine phosphokinase[1] or by treatment with micrococcal nuclease.[10] In practice we have generally found either treatment to be unnecessary (the overall activity is reduced as much or more than the endogenous activity). Even with extracts that have a relatively high endogenous activity as determined by the incorporation of radioactive amino acids, few if any endogenous protein "bands" are observed on SDS–polyacrylamide gels.

Pure mRNAs are extremely sensitive to degradation and should be stored frozen at $-20°$ in small aliquots or as an alcohol precipitate (do not use a frost-free refrigerator). Plastic gloves should be worn when handling RNA, and all solutions and glassware or plasticware must be nuclease free.

With regard to radioactive amino acids, [^{14}C]leucine is reasonably stable and can be stored frozen for several years without appreciable loss of activity. [^{35}S]Methionine is very sensitive to oxidation and should be stored frozen at $-70°$ in small aliquots in 25 mM mercaptoethanol. It should be used as soon as possible, certainly within 3 months of manufacture. The quality of [^{35}S]methionine can be readily checked in a few hours by thin-layer chromatography on cellulose sheets using n-butanol–pyridine–acetic acid–water (300:60:200:240) as a solvent. [^{35}S]Methionine should be mixed with a solution of cold methionine (ca 10 mM) to reduce oxidation during spotting and to provide a marker that can be detected by ninhydrin staining.

The assay system has a very sharp pH optimum. Furthermore, the pH of HEPES buffer changes significantly upon dilution. The value given above is for the 1.0 M stock solution, which is made from the acid by titration to pH 7.6 with 5 M KOH. The actual pH of the reaction mix is

---

[10] H. R. B. Pelham and R. J. Jackson, Eur. J. Biochem. **67,** 247 (1967).

considerably below pH 7.6 (i.e., pH 6.9–7.0). One should make up a series of HEPES solutions between pH 7.2 and pH 8 to determine the optimum for each new extract. HEPES should be stored frozen in small aliquots; it supports microbial growth, and it is not stable to autoclaving. One should also remember that the buffer supplies some potassium that has not been accounted for in the concentrations given above. Note that 1.0 M Tris-acetate buffer, pH 7.20 (at 20°) can be substituted for HEPES, pH 7.6, without apparent loss of activity.

The system has a very broad potassium optimum (using KOAc) with a maximum around 120 mM with Ad2 RNA. The potassium optimum reported by different groups (using different mRNAs) is often quite different. The Mg^{2+} [as $Mg(OAc)_2$] optimum is rather sharp but is close to 2.2 mM in the presence of 20 μg of spermine per milliliter for all mRNAs we have tested, including BMV, TMV, Ad2, KB cell, globin, and T7. The Mg^{2+} optimum is dependent on the concentration of polyamine used; spermidine or putrescine at the appropriate concentration will substitute for spermine. ATP, GTP, DTT, CP, and CPK are all somewhat unstable and are suspect when all else fails. Occasionally one may receive a bad lot of CPK.

[39] Direct Gene Transfer to Protoplasts of Dicotyledonous and Monocotyledonous Plants by a Number of Methods, Including Electroporation

By RAYMOND D. SHILLITO and INGO POTRYKUS

Introduction

Recent progress in cell and molecular biology has made available a great variety of new techniques for modifying the information content of plant cells. Moreover, in the last few years it has become possible to genetically engineer the plant cells by introduction of defined DNA sequences. This has been achieved mainly by exploiting the natural gene transfer system of the pathogen *Agrobacterium tumefaciens* (for a review, see Ref. 1). This is a convenient and efficient technique, but it can be used only with plants which fall inside its host range. This includes most herbaceous dicots, but only a limited number of monocots,[2] and, as yet, not the graminaceous monocots which make up the bulk of our crop species.

The introduction of genes by DNA-mediated transformation is a well-established procedure for bacterial, fungal, and animal systems and has proved to be a very powerful technique in the analysis of gene function. It has recently been shown that DNA can be introduced into plant protoplasts and integrated into the chromosomal DNA without intervention of a bacterial intermediary (direct gene transfer).[3-7] It is also possible, using this technique, to transform graminaceous cells.[8,9] A number of other

[1] G. Gheysen, P. Dhaese, M. Van Montagu, and J. Schell, *in* "Genetic Flux in Plants" (B. Hohn and E. S. Dennis, eds.), p. 12. Springer-Verlag, Vienna, Austria, 1985.

[2] M. De Cleene and J. De Ley, *Bot. Rev.* **42**, (1976).

[3] M. R. Davey, E. C. Cocking, J. Freeman, N. Pearce, and I. Tudor, *Plant. Sci. Lett.* **18**, 307 (1980).

[4] J. Draper, M. R. Davey, J. P. Freeman, E. C. Cocking, and B. G. Cox, *Plant Cell Physiol.* **23**, 451 (1982).

[5] F. A. Krens, L. Molendijk, G. J. Wullems, and R. A. Schilperoort, *Nature (London)* **296**, 72 (1982).

[6] J. Paszkowski, R. D. Shillito, M. Saul, V. Mandak, T. Hohn, B. Hohn, and I. Potrykus, *EMBO J.* **3**, 2717 (1985).

[7] R. Hain, P. Stabel, A. P. Czernilofsky, H.-H. Steinbiss, L. Herrera-Estrella, and J. Schell, *Mol. Gen. Genet.* **199**, 161 (1985).

[8] I. Potrykus, M. Saul, J. Petruska, J. Paszkowski, and R. D. Shillito, *Mol. Gen. Genet.* **199**, 183 (1985).

[9] H. Loerz, B. Baker, and J. Schell, *Mol. Gen. Genet.* **199**, 178 (1985).

techniques have also been developed recently, employing a viral vector,[10] liposomes,[11] bacterial spheroplasts,[12] and microinjection[13] to deliver the DNA. Thus a number of routes are available for introduction of defined pieces of DNA into plant cells, from which one can choose, on the basis of the culture system available, a suitable procedure for the particular species or cell type under study. Transformation of plant cells has the added advantage that in many cases the totipotency of the transformed cells allows regeneration of large numbers of whole plants and genetic and molecular analysis of progeny, which is not available in such an easily accessible manner with transformed animal cells.

Methods for the direct introduction of genes into plant protoplasts are presented, along with methods for the subsequent selection of transformed colonies, regeneration of the genetically altered fertile plants, and characterization of the introduced DNA by molecular, biological, and genetic techniques. The methods are drawn from two complementary fields: recombinant DNA and plant tissue culture. In the interests of continuity, methods from both these fields will be presented mixed with one another in the chronological order in which they are needed in a transformation experiment.

The frequency previously obtained by these methods was low (10^{-5}–10^{-4} of recoverable colonies).[5,6] Such low frequencies seriously hampered the application of the technique of direct gene transfer as a general method for introducing genes into plant cells. However, recent advances have increased frequencies into the percentage range.[14] Protocols for such efficient transformation, and improvements of the original protocols, are given in this publication.

Abbreviations

MES: 2(*N*-morpholino)ethanesulfonic acid
NAA: Naphthaleneacetic acid
2,4-D: 2,4-Dichlorophenoxyacetic acid
BAP: 6-Benzylaminopurine
SDS: Sodium dodecyl sulfate
TE: 10 mM Tris–HCl, 5 mM EDTA, pH 7.5
dsDNA: Double-stranded DNA

[10] N. Brisson, J. Paszkowski, J. R. Penswick, B. Gronenborn, I. Potrykus, and T. Hohn, *Nature (London)* **310,** 511 (1984).
[11] A. Deshayes, L. Herrera-Estrella, and M. Caboche, *EMBO. J.* **4,** 2731 (1985).
[12] R. Hain, H.-H. Steinbiss, and J. Schell, *Plant Cell Rep.* **3,** 60 (1984).
[13] T. Reich, V. N. Iyer, and B. Miki, *Proc. Congr. Plant Mol. Biol., 1st, 1985* p. 28 (Abstr.).
[14] R. D. Shillito, M. W. Saul, J. Paszkowski, M. Mueller, and I. Potrykus, *Bio/Technology* **3,** 1099 (1985).

Media

Bacterial media, as specified in Ref. 15
Plant culture media are shown in Table I

Materials

E. coli strain DH1[15]: Sources for the plant material are given with each
   individual protocol
Table-top centrifuge: Universal 2, (Hettich Zentrifugen, Tuttlingen,
   West Germany)
Osmometer: Roebling Micro-Osmometer (Infochroma AG, Zug, Swit-
   zerland)
Rocking table: Heidolph Reax 3 rocking table (Salvis AG, Reussbuehl,
   Switzerland)
Stainless steel sieves: Saulas and Company (Paris, France)
The 10-cm-diameter containers used for the "bead-type" culture are
   obtained from Semadani AG (Ostermundigen, Switzerland)
Petri dishes: These are obtained from a range of suppliers
Electroporator: "DIA-LOG" Elektroporator: DIA-LOG GmbH.,
   (Dusseldorf, West Germany)
Resistance meter: AVO B183 LCR meter (Thorn-EMI Ltd., Dover,
   England)
SeaPlaque agarose: Marine Colloids, FMC Corporation (Rockland,
   Maine)
Cleaned agar: This is prepared by washing with water, acetone, and
   ethanol in succession[16,17]
Tween 80: ICI (Runcorn, England) or Merck-Schuchardt (Munich,
   West Germany)
Greenzit[R]: Ciba-Geigy AG (Basel, Switzerland)
Restriction enzymes (ligase, etc.) can be obtained from a number of
   commercial sources
Cellulase "Onozuka" R10 and macerozyme R10: Yakult Pharmaceuti-
   cal Industries Company, Ltd. (Nishinomiya, Japan)
Driselase: Fluka AG (Chemische Fabrik, Buchs, Switzerland)
Hemicellulase: Sigma Chemical Company (St. Louis, MO)
Pectinol: Roehm GmbH (Chemische Fabrik, Darmstadt, FRG)
Antibiotics: Kanamycin sulfate: Serva, (Heidelberg, FRG); ampicillin:
   "Penbritin" Beecham SA (Bern, Switzerland); G418: Gibco
Polyethylene glycol (PEG): Merck PEG 6000 and PEG 4000

[15] T. Maniatis, E. F. Fritsch, and J. Sambrook, "Molecular Cloning: A Laboratory Manual." Cold Spring Harbor Lab., Cold Spring Harbor, New York, 1982.
[16] A. C. Braun and H. N. Wood, *Proc. Natl. Acad. Sci. U.S.A.* **48**, 1776 (1962).
[17] R. D. Shillito, J. Paszkowski, and I. Potrykus, *Plant Cell Rep.* **2**, 244 (1983).

TABLE I

COMPOSITION OF THE MEDIA USED[a]

A. Inorganic salts used in tissue culture media

Medium[b]:	T[1]	MS[2]LS[3]	K3[4]	H/J[5]/K0[6]	CC[7]	MDS[8]	RPZ[9]	NT[10]
Macroelements[c]: Final concentration (mg/ml)								
KCl								
KNO_3	950	1900	2500	1900	212	100	273	950
KH_2PO_4	68	170		170	136	85	271	680
$NaNO_3$								
NH_4NO_3	720	1650	250	600	640			825
$NaH_2PO_4 \cdot H_2O$			150					
$CaCl_2 \cdot 2H_2O$	220	440	900	600	588	220	57	438
$MgSO_4 \cdot 7H_2O$	185	370	250	300	247	185	233	123
$(NH_4)_2SO_4$			134					
$Ca(NO_3)_2 \cdot 4H_2O$							416	
$Mg(NO_3)_2 \cdot 6H_2O$							392	
Microelements[c]: Final concentration (mg/liter)								
Na_2EDTA	74.6	74.6	74.6	74.6	37.3	37.3	74.6	74.6
$FeSO_4 \cdot 7H_2O$	27.0	27.0	27.0	27.0	27.8	13.5	27.0	27.0
$FeCl_3 \cdot 6H_2O$								
H_3BO_3	10.0	6.2	3.0	3.0	3.1	0.06	0.12	6.2
KI		0.83	0.75	0.75	0.83	0.15	0.166	0.83
$MnCl_4 \cdot 4H_2O$	17.25							
$MnSO_4 \cdot H_2O$	10.0	16.9	10.0	10.0	11.15	2.0	3.38	16.9
$ZnSO_4 \cdot 7H_2O$		8.6	2.0	2.0	5.76	0.4	1.72	8.6
$CuSO_4 \cdot 5H_2O$	0.025	0.025	0.025	0.025	0.025	0.005	0.005	0.025
$Na_2MoO_4 \cdot 2H_2O$	0.25	0.25	0.25	0.25	0.24	0.05	0.05	0.25
H_2Mo_4								
$CoCl_2 \cdot 6H_2O$					0.028			
$CoSO_4 \cdot 7H_2O$	0.03	0.03	0.025	0.025		0.005	0.006	0.03

B. Vitamins and other organics used in culture media: Concentration (mg/liter)

Medium:	T	LS	K3	H/J^def	KO^de	CC	MDS^h	RPZ	NT
myo-Inositol	100	100	100	100	100	90	100	100	100
Vitamins^g									
Biotin	0.05				0.01			0.05	
Pyridoxine–HCl	0.50		0.10	1.00	1.00	1.0	0.5	0.5	
Thiamin–HCl	0.50	0.04	1.00	10.00	10.00	8.5	5.0	0.5	1
Nicotinamide				1.00	1.00				
Nicotinic acid	5.00		0.10	0.40	0.40	6.0	0.5	5.0	
Folic acid	0.50			1.00	1.00			0.5	
D-Calcium panto-thenate									
p-Aminobenzoic acid				0.02	0.02				
Choline chloride				1.00	1.00				
Riboflavin				0.20	0.20				
Ascorbic acid				2.00	2.00				
Vitamin A				0.01	0.01				
Vitamin D3				0.01	0.01				
Vitamin B12				0.02	0.02				
Glycine	2.0			0.10		2.0		2.0	
Coconut water (%v/v)						10.0			

C. Carbohydrates and phytohormones

Medium:	T	LS	K3A	K3C	K3E	H	J	KO	CC	MDS	RPZ	NT
Sugars and sugar alcohols (g/liter)												
Sucrose	10.0	30.0	102.96	102.96	36.0	0.25	0.25	68.4	20.0	30.0	0.5	10
Glucose						68.40	21.0	68.4	36.43			
Mannitol						0.25	0.25	0.25		50.0	18.0	
Sorbitol						0.25	0.25	0.25				

(continued)

TABLE I (continued)

Cellobiose						0.25	0.25	
Fructose						0.25	0.25	
Mannose						0.25	0.25	
Rhamnose						0.25	0.25	
Ribose						0.25	0.25	
Xylose				0.25	0.25	0.25	0.25	
Hormones (mg/liter)								
2,4-D	0.05	0.10	0.05	0.05	0.10			
p-CPA						2.0		2.0
NAA	2.00	1.00	2.00	2.00	1.00		0.05	
BAP	0.10	0.20	0.10	0.10	0.25	0.5	0.1	
Kinetin	0.10	0.20	0.10	0.10	1.0			
Zeatin							0.25	
Final pH	5.5	5.8	5.8	5.8	6.0	5.7	5.6	5.8

[a] Where the inorganic component is common, i.e., media H/J/K0, these have been given together.

[b] Key to the reference source: [1] J. P. Nitsch and C. Nitsch, Science 163, 85 (1969); [2] T. Murashige and F. Skoog, Physiol. Plant. 15, 473 (1962); [3] E. M. Linsmaier and F. Skoog, Physiol. Plant. 18, 100 (1965); [4] J. I. Nagy and P. Maliga, Z. Pflanzenphysiol. 78, 453 (1976); [5] K. N. Kao and M. R. Michayluk, Planta 126, 105 (1975); [6] H. Koblitz and D. Koblitz, Plant Cell Rep. 1, 147 (1982); [7] I. Potrykus, C. T. Harms, and H. Lörz, Theor. Appl. Genet. 54, 209 (1979); [8] I. Negrutiu, R. Dirks, and M. Jacobs, Theor. Appl. Genet. 66, 341 (1983); [9] P. Installé, I. Negrutiu, and M. Jacobs, J. Plant Physiol. 119, 443 (1985); [10] T. Nagata and I. Takebe, Planta 99, 12 (1971).

[c] Macroelements are usually made up as a 10× concentrated stock solution and microelements as a 1000× concentrated stock solution.

[d] Citric, fumaric, and malic acid (each 40 mg/liter final concentration) and sodium pyruvate (20 mg/liter) are prepared as a 100× concentrated stock solution, adjusted to pH 6.5 with NH4OH, and added to these media.

[e] Adenine (0.1 mg/liter), and guanine, thymine, uracil, hypoxanthine, and cytosine (0.03 mg/liter) are prepared as a 1000× concentrated stock solution, adjusted to pH 6.5 as above, and added to these media.

[f] The following amino acids are added to this medium using a 10× stock solution (pH 6.5 with NH4OH) to yield the given final concentrations: Glutamine (5.6 mg/liter), alanine, glutamic acid (0.6 mg/liter), cysteine (0.2 mg/liter), asparagine, aspartic acid, cystine, histidine, isoleucine, leucine, lysine, methionine, phenylalanine, proline, serine, threonine, tryptophan, tyrosine, and valine (0.1 mg/liter).

[g] Vitamin stock solution is normally prepared 100× concentrated.

[h] Ammonium succinate (770 mg/liter) is added to this medium as a source of reduced nitrogen.

All other organic and inorganic substances are of the highest purity available from the usual commercial sources.

Protocols for the Preparation, Transformation, and Culture of Protoplasts and Regeneration of Plants

We describe here protocols which are in everyday use in our laboratory for four plant species.

A number of factors are common to these protocols: (1) Centrifugations are carried out at 60 g except where otherwise stated. (2) Washing solutions (osmoticum) in all protocols are buffered with 0.5% (w/v) MES and adjusted to pH with KOH except where otherwise stated. (3) Counting of protoplasts is carried out by placing a drop of a 1 : 10 dilution of the suspension in the wash solution (where protoplasts will sediment) or in 0.17 M calcium chloride in a hemocytometer, counting, and estimating the density in the original suspension.

1. Source of the Hybrid-Selectable Gene

Several hybrid marker genes for use in plant cell transformation, using *A. tumefaciens*-mediated transformation, have been described in the last 2 years.[18–21] The elements necessary in such a construction can be summarized as follows:

1. Plant gene expression signals, i.e., promoter and terminator regions for an RNA, which are best derived from a constitutively and highly expressed plant or plant viral gene

2. A protein-coding region joined precisely to the above expression signals which when expressed will give an active product which allows easy selection at the plant cell level: e.g., detoxification of antibiotics lethal for plant cells

3. For DNA-mediated transformation, a region on the bacterial vector plasmid which allows recombination into the plant genome without disruption of the selectable gene

---

[18] M. W. Bevan, R. B. Flavell, and M. D. Chilton, *Nature (London)* **304,** 184 (1983).
[19] R. T. Fraley, S. G. Rogers, R. B. Horsch, P. R. Sanders, J. S. Flick, S. P. Adams, M. L. Bittner, L. A. Brand, C. L. Fink, J. S. Fry, G. R. Gallupi, S. B. Goldberg, N. L. Hoffman, and S. C. Woo, *Proc. Natl. Acad. Sci. U.S.A.* **80,** 4803 (1983).
[20] L. Herrera-Estrella, M. De Block, E. Messens, J.-P. Hernalstens, M. Van Montagu, and J. Schell, *EMBO J.* **2,** 987 (1983).
[21] L. Herrera-Estrella, A. Depicker, M. Van Montagu, and J. Schell, *Nature (London)* **303,** 209 (1983).

A plasmid fulfilling the above requirements has been constructed (pABD1). Details of the construction are given elsewhere.[6] The expression signals used are derived from gene VI of the plant dsDNA virus cauliflower mosaic virus (CaMV).[22] The selectable marker gene joined to these sequences is aminoglycoside phosphotransferase type II [APH(3')II][23] and the bacterial plasmid containing this construction is pUC8.[24] Before using the construction in direct DNA transformation experiments it was tested for biological activity and for the ability to be selected by introduction into *Nicotiana tabacum* cells via the *A. tumefaciens* method. Any suitable construction which satisfies the above criteria can be used for direct gene transfer.

Preparation of the DNA for Protoplast Transformation

Purification. The plasmid pABD1 is grown in *Escherichia coli* strain DH1 in the presence of 50 μg/ml ampicillin and isolated by a cleared lysate method.[25] After lysis in Triton X-100 containing lytic mix, supercoiled DNA is purified by a single round of CsCl/ethidium bromide gradient centrifugation. Ethidium bromide is removed by repeated extraction with CsCl-saturated 2-propanol solution. The DNA is then precipitated with ethanol (1 vol DNA solution plus 2 vol TE plus 6 vol 96% ethanol) at $-20°$ overnight. The precipitate is collected by centrifugation at 5000 g for 10 min, washed in 70% ethanol, dried briefly in an air stream, and redissolved in TE buffer. After spectrophotometric determination of the DNA concentration this is adjusted to 1 mg/ml. DNA for transformation is linearized by digestion with *Sma*I or *Bgl*II overnight and precipitated by addition of one-tenth volume of 3 M potassium acetate followed by 3 vol of ethanol. The precipitate is collected by centrifugation as above, washed in 70% ethanol and 100% (v/v) ethanol in succession, dried in an air stream, and redissolved in sterile double-distilled water at 0.4 mg/ml. All manipulations with the DNA after this sterilization step are carried out under aseptic conditions in a laminar flow cabinet.

Physical form of the transforming DNA: All of our early transformation experiments, which have already been well analyzed, were carried out with supercoiled plasmid DNA. We can conclude at this time that both linear and supercoiled molecules can be successfully taken up into plant protoplasts and integrated into the plant genome. However, linear molecules are clearly superior in the efficiency of transformation, amounting to a factor of 3–10 depending on the precise conditions used.

[22] H. Guilley, R. G. Dudley, G. Jonard, E. Balazs, and K. Richards, *Cell* **30**, 763 (1982).
[23] S. J. Rothstein and W. S. Reznikoff, *Cell* **23**, 191 (1981).
[24] J. Messing and J. Vieira, *Gene* **19**, 269 (1982).
[25] Y. M. Kuperstock and D. Helsinki, *Biochem. Biophys. Res. Commun.* **54**, 1451 (1973).

Carrier DNA: Experiments are generally carried out using high-molecular-weight calf thymus DNA (Sigma) as carrier, as described by Krens *et al.*[5] for experiments involving transformation of protoplasts with isolated Ti plasmid. Calf thymus DNA is dissolved in water, precipitated in 70% and washed with 70 and 100% (v/v) ethanol, dried, and redissolved at 2 mg/ml in sterile double-distilled water. The carrier DNA is mixed at a ratio of five times the amount of pABD1 DNA (equal volumes of the two DNA solutions as given). Trials with carrier DNA of other types have shown that salmon sperm DNA and *N. tabacum* DNA give comparable results but also that transformation is possible at reduced efficiency without any carrier DNA.

2. Preparation, Transformation, and Culture of Protoplasts from a Sterile Shoot Culture of N. tabacum, and Regeneration of Plants

The example given is for protoplasts from shoot cultures of the widely used genotype of *N. tabacum* cv. "Petit Havana," SR1.[26] This material is grown as sterile axenic shoot cultures.

The protocol for protoplast isolation is modified from that of Nagy and Maliga.[27] The culture method uses complex media based on that of Kao and Michayluk[28] (Table I) and the agarose "bead-type" system[17] to obtain high division frequencies and rates of conversion of protoplasts to colonies. Transformation is carried out using one of the four methods given.

Colonies are transferred to agar-solidified medium for one subculture and then placed on regeneration medium to promote the formation of shoots. Regenerated shoots are cultured on the original shoot culture medium.

Source of Material. Shoot cultures are established from seeds which are sterilized using mercury chloride (see protocol 2) or sodium hypochlorite [5 min, 1.4% (w/v) containing 0.05% (w/v) Tween 80]. The plants arising are serially subcultured every 6 weeks as cuttings on T medium[29] (Table I) solidified with 0.8% (w/v) Difco Bacto agar and grown at 26° with 16 hr of light (1000–2000 lx) per day in a growth chamber.

Preparation of Protoplasts. Just fully expanded leaves of 6-week-old shoot cultures are removed under sterile conditions and wetted thoroughly with enzyme solution. The leaves are then cut into 1- to 2-cm squares and floated on enzyme solution [1.2% (w/v) cellulase "Onozuka" R10, 0.4% (w/v) macerozyme R10 in K3A medium with 0.4 M sucrose] in

[26] P. Maliga, A. Breznovitz, and L. Marton, *Nature (London) New Biol.* **244,** 29 (1973).
[27] J. I. Nagy and P. Maliga, *Z. Pflanzenphysiol.* **78,** 453 (1976).
[28] K. N. Kao and M. R. Michayluk, *Planta* **126,** 105 (1975).
[29] J. P. Nitsch and C. Nitsch, *Science* **163,** 85 (1969).

Petri dishes (~1 g leaves in 12 ml enzyme solution in a 9-cm-diameter Petri dish). These are sealed and incubated overnight at 26° in the dark.

The digest is gently agitated and then left for a further half-hour to complete digestion. The solution is filtered through a 100-μm stainless steel mesh sieve and washed through with one-half volume of 0.6 M sucrose (MES, pH 5.6), distributed into capped centrifuge tubes and centrifuged for 10 min.

The protoplasts collect at the upper surface of the medium. The medium is then removed from under the protoplasts. A simple method of doing this uses a sterilized cannula (A. R. Howell, Ltd., London, England) attached to a 20-ml disposable plastic syringe. This must be done slowly so as to avoid disturbing the layer of protoplasts excessively. Alternatively, the protoplasts can be collected using a pipet (with a medium orifice).

The protoplasts are resuspended in K3A medium (Table I) containing 0.4 M sucrose. Washing of the protoplasts is carried out by repeated (3×) flotation and replacing of the medium in this way. A sample is taken for counting before the last centrifugation, and the protoplasts resuspended for the last time in the appropriate medium for the transformation protocol to be used.

Transformation

Method 1. "F medium" method: This method is a modification of the original method described by Paszkowski *et al.*[6,30] for transformation of protoplasts of *N. tabacum,* which was in turn a modification of the method of Krens *et al.*[5] We have added a heat shock step and changed the order of addition of DNA and PEG.

After counting, the protoplasts are adjusted to 2×10^6/ml in K3A medium and aliquots of 1 ml are distributed to 10 ml sterile plastic tubes. Heat shock is then administered by immersing the tubes for 5 min in a water bath at 45°, followed by cooling to room temperature on ice. Then DNA solution is added to the samples (10 μg of pABD1 plus 50 μg of calf thymus DNA in 50 μl sterile distilled water), followed by gentle mixing. Finally, 0.5 ml of PEG solution [40% (w/v) PEG 6000 in F medium (Table I)] is added with shaking.

The protoplasts are incubated with DNA and PEG for 30 min at room temperature with occasional gentle mixing. Then five aliquots of 2 ml of F medium are added at intervals of 5 min. We have noted that the pH of F medium drops to 4.3–4.6 after autoclaving. Since this is likely to be harm-

---

[30] J. Paszkowski and M. W. Saul, this series, Vol. 118, p. 668.

ful to many protoplast systems we recommend adjustment of the pH after autoclaving with KOH to 5.8.

Following transformation, the protoplasts are sedimented by centrifugation for 5 min, resuspended in 2 ml of K3A culture medium, and transferred in 1-ml aliquots to 9-cm Petri dishes. To each dish is added 10 ml of a 1 : 1 mixture of K3A and H media (Table I) containing 0.6% (w/v) Sea-Plaque agarose, and the protoplasts dispersed by gentle swirling. This protocol gives transformation efficiencies in the range of 10^{-4} to 10^{-3}.

Method 2. Electroporation of protoplasts[14]: Electroporation is a process in which cells or protoplasts are treated with high-voltage electric fields for short periods in order to induce the formation of pores across the membrane.[31,32] In this way it is possible to induce uptake of DNA into animal cells or plant protoplasts, leading to transient expression[33] or to stable transformation.[14,32]

Samples of protoplasts are pulsed with high-voltage pulses in the chamber of a "DIA-LOG" Elektroporator. This chamber is cylindrical in form with a distance of 1 cm between parallel steel electrodes and has a pulsed volume of 0.32 ml.[32] The pulse is applied by discharge of a capacitor across the cell. The decay constant of the pulse (time taken to decay to $1/e$ of the initial voltage) is in the order of 10 μsec using a capacitor of 10 nF and a chamber resistance of 1 kΩ. The resistance across the chamber is measured using an alternating current multimeter operating at 1 kHz.

The protocol given is for leaf mesophyll protoplasts of *N. tabacum*. These have an average diameter of 42 μm. The field strength required is inversely proportional to the diameter of the protoplasts being treated, and may vary a little from species to species in the field strength required for a given size of protoplast. In addition, protoplasts originating from suspension cultures generally require a slightly higher field strength than leaf mesophyll protoplasts.

Protoplasts are resuspended, following the last flotation step, in 0.4 M mannitol containing 1 g/liter MES (pH 5.6 with KOH), and containing 6 mM magnesium chloride to stabilize the protoplasts, at a population density of 1.6×10^6/ml. An aliquot of 0.37 ml is transferred to the chamber of the electroporator and the resistance measured.

In order to bring the resistance of the mannitol solution in which the protoplasts are suspended to the correct value, it is necessary to add ionic salts. Magnesium chloride is used to adjust the resistance to a value of 1–1.1 kΩ, by adding 1–3% (v/v) of a 0.3 M solution to the protoplast suspension.

[31] R. Benz, F. Beckers, and U. Zimmermann, *J. Membr. Biol.* **48**, 181 (1979).
[32] E. Neumann, M. Schaeffer-Ridder, Y. Wang, and P. H. Hofschneider, *EMBO J.* **1**, 841 (1982).
[33] M. Fromm, L. P. Taylor, and V. Walbot, *Proc. Natl. Acad. Sci. U.S.A.* **82**, 5824 (1985).

Heat shock is carried out before distributing the preparation into tubes, by treating the protoplasts for 5 min at 45°, followed by cooling to room temperature on ice.

Aliquots of 0.25 ml of protoplast suspension are placed in 5-ml-capacity polycarbonate tubes and DNA [4 μg pABDI linearized with SmaI and 20 μg of calf thymus DNA (Sigma) in 20 μl water/aliquot] and one-half volume of the PEG solution [24% (w/v) PEG 6000: Merck] added. This PEG is prepared in mannitol with sufficient magnesium chloride added (~30 mM) to bring the resistance when measured in the chamber of the electroporator into the region of 1.2 kΩ.

Ten minutes after addition of the DNA and PEG, samples are transferred to the chamber of the electroporator and pulsed three times at 10-sec intervals with pulses of an initial field strength of 1.4–1.5 kV/cm. They are then returned to a 6-cm-diameter Petri dish and held at 20° for 10 min before addition of 3 ml of a 1 : 1 mixture of K3A and H media (Table I) containing 0.6% (w/v) SeaPlaque agarose. This is the optimized protocol. It gives the highest frequencies we have as yet obtained—in the region of 1 to 3% for all colonies recoverable without selection. It is not necessary when using this method to cool the protoplasts on ice, due to the low heating effects when using such short pulses, in contrast to when using the method of Langridge et al.[34,35]

Method 3. Quick method for transformation without electroporation: Protoplasts are treated in an identical manner to that described for method 2 above. However, there is no need to adjust the resistance of the mannitol with magnesium sulfate, and 40% PEG is added to the protoplasts in place of the 24% (v/v) PEG. The protoplasts are transferred to the Petri dishes 10 min after addition of the PEG, and agarose added 10 min later. This protocol gives efficiencies in the range of 10^{-4} to 10^{-3}.

Method 4. Cotransformation: This is carried out using method 2 given above. The gene to be cotransformed into the protoplasts is added instead of the carrier DNA, i.e., at 50 μg/ml (20 μg/sample). This may be linearized before use, as this appears to give a higher rate of cotransformation. In experiments using a zein gene as the nonselected gene, 88% of the kanamycin-resistant colonies recovered contained sequences hybridizing to the zein gene sequences, with 27% containing a full copy of the zein gene clone used.[36]

Protoplast Culture, Selection of Transformed Lines, and Regeneration of Plants. Selection in the agarose bead-type culture system[17,37] has

[34] W. H. R. Langridge, B. J. Li, and A. A. Szalay, *Plant Cell Rep.* **4,** 355 (1985).
[35] W. H. R. Langridge, B. J. Li, and A. A. Szalay, this series, Vol. 153, p. 336.
[36] R. Scocher, R. D. Shillito, M. W. Saul, J. Paszkowski, and I. Potrykus, *Bio/Technology* **4,** 1093 (1985).
[37] I. Potrykus and R. D. Shillito, this series, Vol. 118, p. 549.

been found to be superior to selection in other culture systems tested. In this way a nearly constant selection pressure is maintained during the first 4 weeks of culture, thus suppressing any possibility of background colonies arising due to reduction in the selection pressure by the decay of the drug.

The dishes containing the protoplasts are sealed with Parafilm and incubated at 24° for 1 day in the dark followed by 6 days in continuous dim light (500 lx, cool fluorescent Sylvania "daylight" tubes). The agarose containing the protoplasts is then cut into quadrants and cultured in a 1 : 1 mixture of K3A and H media in the bead-type culture system[17] using one container with 30 ml medium for each 6-cm Petri dish from methods 2 and 3, and three containers with 30 ml of medium/10-cm dish from method 1. Kanamycin sulfate (50 μg/ml) is added to this and to all subsequent media for selection of transformants. Aliquots of the agarose containing the protoplasts can be cultured as a nonselected control in medium lacking kanamycin. After 1 week one-half of the medium is replaced with a 1 : 1 mixture of KC3 and J media (Table I), and thereafter one-half the medium is replaced weekly with a 1 : 1 mixture of K3E and J media (Table I).

Resistant clones are first seen 3–4 weeks after the start of the experiment, and after a total time of 5–6 weeks (when 2–3 mm in diameter) they are transferred to LS medium[38] (Table I) solidified with 0.8% (w/v) cleaned agar.

Regeneration of Plants. After 3–5 weeks, depending on the size of the original colony, these should reach 1 cm in diameter. Each colony is then split into four parts and two placed on fresh LS medium as above, and two on LS medium with 0.2 mg/liter BAP as the only phytohormone (regeneration medium). These latter dishes are incubated in the dark for 1 week and thereafter in the light (3000–5000 lx).

Shoots arising from the callus on regeneration medium are cut off when 1–2 cm long and placed on LS medium as above, but without hormones, where they produce shoots. When the shoots reach 3–5 cm in length they are transferred to T medium and treated as shoot cultures (see above). They can be transferred to soil once they have an established root system: the agar is gently washed away and the plantlets potted up. They require a humid atmosphere for the first week and can then be hardened off and grown under normal greenhouse conditions.

Genetic Analysis. When regeneration of fertile plants from tissue culture is possible then the introduced trait can be followed in its transmission to progeny. The earliest opportunity to observe this is by culture of

---

[38] E. M. Linsmaier and F. Skoog, *Physiol. Plant.* **18**, 100 (1965).

the male gametes via another culture.[39,40] Haploid plantlets developed from microspores can be tested under selective conditions by transfer to kanamycin-containing media (300 mg/liter kanamycin sulfate) when they are at the "seedling" stage. For a single dominant gene one expects approximately 50% of the plants to be resistant due to the segregation of the trait during meiosis.

For genetic analysis, plants of *N. tabacum* are grown in the greenhouse to flowering. They are then selfed and crossed in both directions with wild-type SR1 plants. Seeds are collected from individual capsules, and stored in a dry place at room temperature. Nonsterilized seed is sown on half-strength NN69 medium[29] (Table I) containing 300 μg/ml kanamycin sulfate and solidified with 0.8% cleaned agar.[41] Seeds germinate at a high frequency and, after 1 week, seedlings can be scored for resistance to kanamycin. Green seedlings are resistant, white sensitive. The counts are compared to the expected segregation ratios for one or more independent loci for the resistance character using the chi-square test.

A high proportion of the regenerated plants are fertile and pass the introduced genes to their progeny in a normal dominant Mendelian fashion, but a small proportion of the plants are disturbed in their fertility.

3. Preparation, Transformation, and Culture of Leaf Mesophyll Protoplasts from Greenhouse-Grown Plants (Petunia hybrida)

The method described for protoplast isolation is based on that of Durand *et al.*[42] and modified in our laboratory for use with the "cyanidin type,"[43] Mitchell,[44] and other petunias. A high division frequency and rate of conversion to colonies of protoplasts from a number of *Petunia* species is achieved by this method.

Source of Material. Plants of *P. hybrida* and other *Petunia* species are clonally propagated via cuttings, grown in clay pots in a controlled environment [12 hr light (5000 lx), 27/20° day/night, 50/70% relative humidity], and watered morning and evening with commercial fertilizer [0.01% (v/v) Greenzit[R]].

Preparation of Protoplasts. Young leaves at approximately two-thirds to three-quarters of their final size from 4- to 6-week-old plants are

[39] N. Sunderland and J. M. Dunwell, *in* "Plant Cell and Tissue Culture" (H. E. Street, ed.), p. 233. Univ. of California Press, Berkeley, 1977.

[40] E. Heberle-Bors, *Theor. Appl. Genet.* **71,** 361 (1986).

[41] I. Potrykus, J. Paszkowski, M. Saul, J. Petruska, and R. D. Shillito, *Mol. Gen. Genet.* **199,** 169 (1985).

[42] J. Durand, I. Potrykus, and G. Donn, *Z. Pflanzenphysiol.* **69,** 26 (1973).

[43] D. Hess, *Planta* **59,** 567 (1963).

[44] A. Z. Mitchell, B.A. thesis. Harvard Univ., Cambridge, Massachusetts.

washed with tap water and sterilized by immersion for 8 min in 0.01% (w/v) $HgCl_2$ solution containing 0.05% (w/v) Tween 80, and then washed carefully with five changes of sterile distilled water (each change 5 min).

Leaf halves without midribs are wet with osmoticum P1 [0.375 M mannitol, 0.05 M $CaCl_2$, 0.2% (w/v) MES, pH 5.7] and arranged in a stack of six on the lid of a 9-cm Petri dish ready for cutting. They are cut diagonally into clean sections 0.5 mm wide, transferred into a small screw-top flask containing 10 ml of enzyme solution [2% (w/v) cellulase "Onozuka" R10, 1% (w/v) hemicellulase, and 1% (w/v) pectinol in 0.3 M mannitol, 0.04 M $CaCl_2$, 0.2% (w/v) MES, pH 5.7], and vacuum infiltrated until the leaf tissue is translucent. The vacuum infiltration is carried out by placing the screw-top flask containing the leaf pieces and enzyme in a larger chamber (a small desiccator can be used), and applying a vacuum ($\sim$700 mm Hg) while gently shaking the material. The air present in the intracellular spaces in the leaf pieces will expand and bubble out. The pressure is then gently allowed back through a sterile filter. It may be necessary to repeat the infiltration one or two times to completely remove the air from the leaf spaces.

The leaf slices are placed in fresh enzyme solution in a Petri dish (0.5 g/10 ml in a 9-cm Petri dish), which is sealed with Parafilm and incubated at 28° for $\sim$3 hr. The incubation mixture is checked periodically under the inverted microscope for the release of protoplasts. The time required may vary, especially with greenhouse-grown material.

The digest is gently agitated, filtered through a 100-μm mesh stainless steel sieve, and transferred in 5-ml aliquots into 10- to 15-ml centrifuge tubes. Osmoticum P1 is added (5 ml) to each tube and, after gentle mixing, these are centrifuged for 5 min to sediment the protoplasts.

The supernatant is carefully pipetted off, and the sediment is gently shaken to free the protoplasts before resuspension in 10 ml of osmoticum P1. Washing by sedimentation is repeated two times. If necessary, the suspension is overlaid on 0.6 M sucrose to remove debris and the protoplasts collecting at the interface recovered and resuspended in osmoticum P1. A sample is taken and diluted in osmoticum for counting and the protoplasts sedimented once more and resuspended in medium (K0,[45] Table I) at 1×10^6/ml.

Transformation

Method 1. "F medium" method: Transformation is carried out as described for *N. tabacum* protoplasts above (protocol 2; method 1). After addition of the F medium, the protoplasts are resuspended in K0 culture

[45] H. Koblitz and D. Koblitz, *Plant Cell Rep.* **1,** 147 (1982).

medium at 2×10^5/ml. To 5 ml of this culture medium in a 9-cm-diameter Petri dish is added 5 ml of liquefied K0 medium containing 1.2% (w/v) SeaPlaque agarose. The protoplasts are dispersed by gentle swirling, and the agarose allowed to solidify. The dishes are sealed with Parafilm and cultured as described below.

Method 2. Quick method: The protoplasts are suspended following the last purification step in osmoticum P1 at 10^6/ml. Aliquots of 1 ml are placed in 10-ml polycarbonate centrifuge tubes, the protoplasts subjected to a heat shock as described above, and DNA (10 μg pABD1 plus 50 μg calf thymus DNA in 50 μl water) added. After 1 min, 0.5 ml of the PEG solution [40% (w/v) PEG 6000 in osmoticum P1) is added, followed 10 min later by 10 ml osmoticum P1. The protoplasts are collected by centrifugation, and plated in culture medium as described in method 1 above at 1×10^5/ml.

Method 3. Electroporation: The protoplasts are resuspended after purification at a density of 1.2×10^6/ml in 0.4 M mannitol containing 6 mM MgCl$_2$ as for electroporation of *N. tabacum,* and treated as described for *N. tabacum* above (protocol 2, method 2). A pulse voltage of 1.2 kV is delivered from a capacitor of 10 nF. They are then embedded, in the same way as used for *N. tabacum,* in K0 medium solidified with 0.6% (w/v) SeaPlaque agarose at 1×10^5/ml.

Culture and Selection of Transformants. Agarose-solidified cultures are incubated for 6 days in the dark at 26°. For plates containing 10 ml of protoplasts, half the protoplast-containing agarose gel is transferred to each of two 10-cm-diameter containers each containing 40 ml of liquid K0 medium containing 0.2 mg/liter, 2,4-D, 0.5 mg/liter BAP, and 2% (v/v) coconut milk. These are incubated at 26° on a gyratory shaker (60 rpm, 1.2 cm throw) in the light (500–1000 lx). The liquid medium is replaced weekly, each time reducing the glucose concentration in the original medium by one-quarter so as to reduce the osmotic pressure.

After 5–6 weeks the colonies are 1–2 mm in size and can be cultured further.

Regeneration of Plants. Colonies are transferred directly to the regeneration medium (NT,[46] Table I) containing 16 mg/liter zeatin and solidified with 1.0% (w/v) SeaPlaque agarose (16 colonies/6-cm-diameter Petri dish). These are cultured in the conditions used for the growth of plants.

The dark green calli which arise after 2–4 weeks are transferred to the same medium containing only 2 mg/liter zeatin. Those showing shooting morphology are placed on hormone-free medium to allow outgrowth of these and subcultured further as cuttings. In general, three such passages are necessary before shoots showing a "normal" morphology are ob-

[46] T. Nagata and I. Takebe, *Planta* **99,** 12 (1971).

tained and can develop a strong enough root system to be transferred to soil in the greenhouse.

4. Preparation, Transformation, and Culture of Protoplasts from a Nonmorphogenic Suspension Culture of the Graminaceous Monocot Species Lolium multiflorum

Protoplasts from leaf or other whole plant tissues of grasses do not in general divide in culture although there have been exceptions to this rule.[47] However, there are a number of suspension cultures of graminaceous species available, and these have been used to produce protoplasts which divide. There have been reports of division and colony formation of protoplasts from morphogenic suspension cultures of these species,[48] but these have not yet proved to be repeatable.

We describe a protocol developed in our laboratory for the isolation, transformation, and culture of protoplasts from *Lolium multiflorum* (Italian ryegrass) suspension culture cells.

Source of Material. The cell line was originally established by P. J. Dale, who has also used it for protoplast culture.[49]

Suspension cultures are maintained by weekly serial transfer (1:7 dilution) in CC medium[50] (Table I) without mannitol and with 2 mg/liter 2,4-D on a gyratory shaker (110 rpm, 2 cm throw) in low light levels (500 lx).

Preparation of Protoplasts. Cultures are used for protoplasting, 4, 5, or 6 days after subculture. Cells (10 ml) are sedimented by centrifugation (5 min) and resuspended in the same volume of enzyme solution [4% (w/v) Driselase in 0.38 M mannitol, 8 mM $CaCl_2$, MES, pH 5.6].

The solution is poured into a 9-cm Petri dish and this is sealed and placed on a rocking table for 1 hr at 20° before being incubated overnight (15 hr) without agitation at the same temperature. The preparation is then placed on the rocking table for an hour followed by another hour without agitation.

The protoplasts are filtered through a 100-μm mesh stainless steel sieve, an equal volume of 0.2 M $CaCl_2$ (MES, pH 5.8) added, and the suspension distributed into two centrifuge tubes. After centrifugation to sediment the protoplasts (10 min) they are taken up in 3 ml osmoticum L1 (0.25 M mannitol, 0.1 M $CaCl_2$, MES, pH 5.8) and overlayered on a 5-ml sucrose cushion (0.6 M sucrose, MES, pH 5.8).

[47] I. Potrykus, in "Advances in Protoplast Research" (L. Ferenczy and G. L. Farkas, eds.), p. 243. Hungarian Academy of Sciences, Budapest, 1980.

[48] V. Vasil and I. K. Vasil, *Theor. Appl. Genet.* **56,** 97 (1980).

[49] M. G. K. Jones and P. J. Dale, *Z. Pflanzenphysiol.* **105,** 267 (1982).

[50] I. Potrykus, C. T. Harms, and H. Loerz, *Theor. Appl. Genet.* **54,** 209 (1979).

Protoplasts collecting at the interface after centrifugation are carefully removed and washed twice with osmoticum L1, counted, and resuspended in CC medium (Table I) at a density of 2×10^6/ml.

Transformation

Method 1. "F medium" method: One-milliliter aliquots of the protoplasts, in CC medium, are heat shocked and treated with DNA as described for *N. tabacum* in protocol 2, method 1 above. Following the addition of the "F" medium, they are sedimented by centrifugation (5 min), and resuspended in 2 ml of CC culture medium.

Culture and Selection of Transformants. The protoplasts are cultured in 3.5-cm-diameter Petri dishes (2 ml/dish) at 26° in the dark. Fresh CC medium (1 ml) with 0.2 *M* mannitol is added after 7 days to dilute the culture and reduce the osmotic pressure. Two weeks after the DNA treatment, the cultures are sedimented and the cells taken up in the 3 ml of CC medium with 0.2 *M* mannitol containing 25 mg/liter G418. After a further 2 weeks these cultures are diluted 5-fold with CC medium without mannitol, and containing G418 as before. After a total of 6 weeks, calli arising from the cultures are transferred to CC medium as used for the suspension cultures solidified with 0.8% (w/v) cleaned agar and grown at 24° in the dark or in the light (2000 lx).

5. Preparation, Transformation, and Culture of Protoplasts from a Sterile Shoot Culture of N. plumbaginifolia, and Regeneration of Plants

Source of Material. This material is grown as sterile axenic shoot cultures. They are grown as described for *N. tabacum* above except that the medium used is Murashige and Skoog's medium[51] with 10 g/liter sucrose, as described by Negrutiu.[52]

Preparation of Protoplasts. For isolation of protoplasts, sterile leaves from shoot cultures are sliced carefully into 2-mm-wide strips using a sharp razor blade and these incubated overnight at 26° in enzyme solution [0.5% (w/v) Driselase in 0.5 *M* sucrose, 0.005 *M* CaCl$_2$, pH 5.5, with KOH]. The released protoplasts are filtered through 100- and 66-μm stainless steel sieves, and centrifuged for 5 min. The protoplasts at the surface are collected and washed two times with W5 salt solution (154 m*M* NaCl, 125 m*M* CaCl$_2$, 5 m*M* KCl, 5 m*M* glucose, pH 5.5, with KOH[53]). They are resuspended in W5 solution at a final density of $1-1.6 \times 10^6$ and used

[51] T. Murashige and F. Skoog, *Physiol. Plant.* **15,** 473 (1962).
[52] I. Negrutiu, *Z. Pflanzenphysiol.* **104,** 431 (1981).
[53] L. Menczel, G. Galiba, F. Nagy, and P. Maliga, *Genetics* **100,** 487 (1982).

immediately for transformation. Alternatively, they can be stored in the W5 solution for up to 6 hr at 6–8° before use.

Transformation

Method 1. PEG method: Aliquots (1 ml) of protoplasts are placed in centrifuge tubes, and a heat shock applied as described for *N. tabacum.* A mixture of linearized plasmid DNA (10 μg/ml) and carrier DNA (50 μg/ml calf thymus DNA) is added, followed by 1.0–1.5 ml PEG solution [45% (w/v) PEG 4000, 2% (w/v) $Ca(NO_3)_2 \cdot 4H_2O$, 0.4 M mannitol, pH adjusted to 9 with KOH repeatedly over a period of 4 hr and autoclaved; stored at −20°, final pH 7.5–8.5] with gentle shaking to give a final concentration of 22–27% PEG.

After incubation with the PEG for 20–30 min, W5 solution is added 3× in 1-ml aliquots at 2- to 5-min intervals, and the protoplasts centrifuged for 5 min. The pellet is resuspended in 10 ml culture medium (K3 medium with 0.4 M glucose[27]) and cultured in a 10-cm-diameter dish. Alternatively, the protoplasts can be embedded in the same culture medium solidified with 0.6% (w/v) SeaPlaque agarose as described above for *N. tabacum.*

Method 2. Electroporation: The protoplasts can be transformed by electroporation as described for *N. tabacum* above. Following the enzyme incubation, the released protoplasts are cleaned and transformed exactly as described for *N. tabacum* in protocol 2.

Culture and Selection of Transformants. At the 2- to 4-cell stage, the developing cultures are diluted 6- to 8 fold (to ~2000 surviving colonies/ ml) in low-hormone medium MDS[54] with 25 mg/liter kanamycin sulfate. In the case of agarose-embedded cells, these are suspended at a similar dilution factor as used for liquid medium in liquid MDS medium with (30 ml) or without (10 ml) shaking. The dishes are incubated at 1000–1500 lx at 26° for 3–5 weeks, when colonies should become visible.

Transformation frequencies of 10^{-4} to 10^{-3} should be obtained. It has not yet been possible to attain higher frequencies with this species despite many experiments to this end.

Calli are transferred to MDS medium solidified with 0.8% (w/v) cleaned agar containing 50 mg/liter kanamycin. After a further 2–4 weeks they can be transferred to RPZ medium[55] for regeneration, and shoots arising are cultured as shoot cultures as described for the source material, and transferred to the greenhouse.

[54] I. Negrutiu, R. Dirks, and M. Jacobs, *Theor. Appl. Genet.* **66**, 341 (1983).
[55] P. Installé, I. Negrutiu, and M. Jacobs, *J. Plant Physiol.* **119**, 443 (1985).

Evaluation of Results

Colonies which are selected as being resistant should be further analyzed for proof of transformation. Although with the system described we
have never observed resistant colonies from control cultures with *N.*
tabacum or *P. hybrida,* variations in the many factors present, particularly when using other protoplast systems, may lead to misinterpretation
of apparently resistant colonies as transformants.

We feel that the minimal criteria for confirmation of a transformation
event should be the following:

1. A phenotypic change to resistance, or growth under other selective
conditions, in a selection scheme which is proved to be "clean"
2. The presence for the transforming DNA in the selected lines in
form expected for transformed DNA (integrated in the genome or autonomously replicating)
3. Expression of the foreign DNA at the RNA/protein level

In addition, if the plant cell tissue culture system being used is capable
of regenerating plants, then genetic data should be obtained.

Phenotypic Change

The assumption of direct selection on kanamycin (or G418) is that only
transformed cell lines will be phenotypically resistant to the drug. Ideally
the level of selection should permit recovery of only transformed clones.
Therefore resistant clones should only appear after transformation with
the correct vector and not in any control treatments. Selection conditions
should be adjusted to produce this situation. The resistant phenotype
should be rechecked at later stages in culture by comparison with wild
type, for instance at the callus level.

In the case of *N. tabacum,* shoots can be regenerated from transformed callus under selective (100 mg/liter kanamycin sulfate) or nonselective conditions. These shoots can be rooted in medium containing 150
mg/liter kanamycin sulfate. Wild-type SR1 shoots regenerated in the absence of kanamycin never form roots, bleach, and die when cultured in
the presence of kanamycin. In order to confirm the resistant phenotype of
the cells of such regenerated transformed plants or to show the possible
loss of the introduced trait during plant development, mesophyll protoplasts from these plants can be isolated and their resistance checked by
culture in kanamycin-containing media.

Molecular Analysis of the DNA of Transformed Clones and Regenerated Plants

Analysis of DNA. DNA of transformed cell lines, regenerated plants, and their progeny should be analyzed using standard Southern blot techniques.[56] We will omit the exact procedures used since they are now standard in molecular biology and the choice of restriction enzymes, etc., will depend on the transforming DNA being analyzed. It should, however, be possible to prove unequivocally the presence of the transforming DNA in high-molecular-weight nuclear DNA of transformed lines in an integrated form in the absence of any hybridization to DNA from control untransformed lines. We shall concentrate here only on a method for the efficient isolation of DNA from small amounts of callus or plant tissue. The relatively slow growth rate of plant cell tissue cultures (~1 week doubling time for callus on solid media) means that the adaption of DNA extraction procedures to small amounts of tissue allows a significant shortening in the length of an experiment. We have adapted a method of Thanh Huynh (personal communication, Dept. of Biochemistry, Standford University), which allows the extraction of pure DNA from ~0.5 g of tissue:

Samples of 0.5 g of callus or leaf tissue are homogenized in a Dounce homogenizer in 3 ml of a buffer containing 15% sucrose, 50 mM EDTA, 0.25 M NaCl, 50 mM Tris–HCl, pH 8.0. Centrifugation of the homogenate for 5 min at 1000 g results in a crude nuclear pellet which is resuspended in 2 ml of a buffer containing 15% sucrose, 50 mM EDTA, 50 mM Tris–HCl, pH 8.0. SDS is added to a final concentration 0.2% (w/v). Samples are heated for 10 min at 70°. After cooling to room temperature, potassium acetate is added to a final concentration of 0.5 M. After incubation for 1 hr at 0° the precipitate formed is sedimented for 15 min in an Eppendorf centrifuge at 4°. The DNA in the supernatant is precipitated with 2.5 vol of ethanol at room temperature and redissolved in 10 mM Tris–HCl, pH 7.5, 5 mM EDTA. The DNA samples are then run in a cesium chloride/ethidium bromide gradient in the vertical rotor (Beckman VTi 65) for 17 hr at 48,000 rpm. The DNA is removed from the gradient with a widebore hypodermic syringe needle and the ethidium bromide extracted as for plasmid DNA (see above). The DNA obtained is of high molecular weight and susceptible to various restriction enzyme.

For Southern analysis 5–10 μg DNA is electrophoresed in a 1% agarose gel, transferred to a nitrocellulose membrane, and hybridized

[56] E. M. Southern, *J. Mol. Biol.* **98,** 503 (1975).

with nick-translated DNA[57] (5–10 × 10^8 cpm/μg). Filters are washed with 2× SSC at 65°, 3 × for 1 hr, and subsequently exposed to X-ray film with intensifying screens for 24–48 hr.

Activity Assay for the Product of the Transforming Gene

The assay for activity of the transformed gene will of course depend on the expected product. In the case described here, a method developed by Reiss et al.[58] allowed us to detect activity of the APH(3′)II gene product in transformed lines.

Callus or leaf pieces (100–200 mg) are crushed in an Eppendorf centrifuge tube with 20 μl extraction buffer [100% (v/v) glycerol, 0.1% (w/v) SDS, 5% mercaptoethanol, 0.005% (w/v) bromphenol blue, 0.06 M Tris, pH 6.8]. Extracts are centrifuged for 5 min at 12,000 g. Proteins in 35 μl of the supernatant are separated on a 10% nondenaturing polyacrylamide gel. The gel is incubated with kanamycin and [^{32}P]ATP and then blotted onto Whatman P81 phosphocellulose paper. The paper is washed five times with deionized water, wrapped in plastic film, and exposed to X-ray film with intensifying screens for 24–48 hr. Kanamycin binds to this paper but ATP does not, therefore radioactive bands on the paper reveal bands on the gel with an activity which transfers radiolabeled phosphate from the [^{32}P]ATP to the kanamycin, i.e., aminoglycoside phosphotransferases. Other transformant nonspecific bands are seen at the top of the gel. These are protein phosphorylases, and the bands can be removed from the P81 paper by treatment with protease if required.[59]

General Comments

Comparison with Other Gene Transfer Systems

A. tumefaciens-mediated gene transfer was, for a while, the only possible way to introduce foreign genes into plants. Although it is a well-established method, there has been little attempt until recently to study the fate of genes introduced in this way. There is also increasing evidence that the integration patterns found when using this method can be complex and show rearrangements of the original T-DNA. One of the great drawbacks of the method is the limited host range, which excludes, for instance, the graminaceous monocots. However, this will remain the method of choice for introduction of foreign DNA in most cases.

---

[57] W. J. Rigby, M. Dieckmann, C. Rhodes, and P. Berg, J. Mol. Biol. 113, 237 (1977).
[58] B. Reiss, R. Sprengel, M. Willi, and H. Schaller, Gene 30, 217 (1984).
[59] P. H. Schreier, E. A. Seftor, J. Schell, and H. J. Bonnert, EMBO J. 4, 25 (1985).

DNA viruses have been suggested as possible gene vectors. To date there is one example of this, in which cauliflower mosaic virus carrying a modified bacterial methotrexate-resistance gene was used to infect a plant. The foreign gene was thus transported into and systemically spread in the plant.[10,60] The advantages of this system are the ease of infection, systemic spread within the plant, and multiple copies of the gene being present per cell. The disadvantages are the narrow host range, lack of transfer to sexual offspring and the limited space available for passenger DNA.

Liposome fusion has been shown to be a method for transformation of plant cells.[11] As with direct gene transfer, this requires protoplasts. It will be of interest to see whether the pattern of integration of DNA by this method is different than that seen when using other methods.

Microinjection is an efficient means of transforming plant protoplasts.[13] However, the number of transformants that one can obtain is limited by the number which can be injected by the operator. The main use of this technique may be that it could be possible eventually to inject cells, thus circumventing the need for protoplasts which is inherent in other direct gene transfer techniques such as described above.

Conclusions

The procedures described above require a protoplast system which allows regeneration of callus from protoplasts and a suitable selectable marker gene. If these requirements are satisfied the methods for DNA delivery and selection of transformants are flexible. The protocols described here are certainly applicable to other systems and are reproducible in our hands.

Acknowledgments

The authors thank all the members of our laboratory, particularly M. W. Saul, J. Paszkowski, I. Negrutiu, and S. KrugerLebus for help in supplying information and protocols for this publication.

---

[60] N. Brisson and T. Hohn, this series, Vol. 118, p. 659.

[40] A Rapid Single-Stranded Cloning, Sequencing, Insertion, and Deletion Strategy

By RODERIC M. K. DALE and AMY ARROW

Introduction

A new procedure for producing a sequential series of overlapping clones for use in DNA sequencing uses single-stranded DNA and complementary DNA oligomers to form specific cleavage and ligation substrates.[1] This technique requires no restriction map information, is free from artifacts resulting from ligation of noncontiguous pieces of DNA found in double-stranded procedures, and is extremely easy and rapid (up to 2–4 kb of DNA can be subcloned and sequenced in one week). The same basic technology also allows one to do insertions and deletions.

The first step involves hybridization of an oligomer to single-stranded recombinant M13 or pBX DNA at the 3′ end of the cloned insert (see Fig. 1) creating a double-stranded region at either the EcoRI site of the odd-numbered M13 or pBX series or the HindIII site of the even-numbered series. Digestion with the appropriate restriction enzyme linearizes the circular molecule.

The second step employs the 3′- to 5′-specific exonuclease activity of T4 polymerase to digest the linearized DNA from the 3′ end. The polymerase will digest the DNA at a rate of approximately 40–60 bases/min. The enzyme rapidly heat inactivates and, if aliquots of the reaction products are removed at intervals, a series of overlapping deletions may be generated. The deletion products are tailed with either dAs for the even-numbered vector series or dGs for odd-numbered vectors using terminal deoxyribonucleotidyltransferase (TdT), thus producing a short homopolymer tail at the 3′ end. Fresh oligomer is annealed to the deletion products joining the two ends of the molecule. The remaining nick is sealed with T4 DNA ligase. The product is then used to transform competent Escherichia coli cells. Clones containing deleted DNA are identified and screened on agarose gels and templates are prepared for sequencing.

The procedure can easily be carried through to the point of plating transformed cells in about half a day. Clones can be picked and grown up on the second day and sizing gels run overnight. Templates can be prepared on the third day and sequencing begun on the fourth day. The rapid deletion subcloning procedure makes it easier to generate the necessary templates and to reduce the number of templates required to obtain a

---

[1] R. Dale, B. McClure, and J. Houchins, Plasmid 13, 31 (1985).

RECOMBINANT DNA
METHODOLOGY

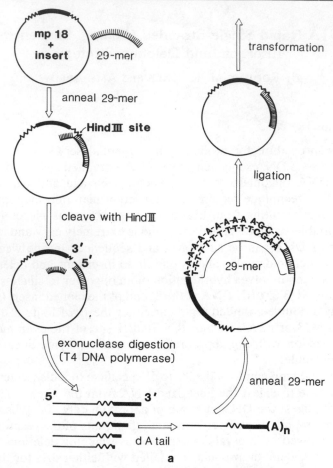

FIG. 1. Outline of the single-stranded cloning procedure for (a) mp8, mp10, or mp18, and (b) mp9, mp11, or mp 19. The solid bar represents the insert and the sawtooth line the polylinker region of M13.

complete sequence than do either shotgun procedures or other deletion protocols.

Methods

Template Preparation

It is imperative to have very clean single-stranded DNAs if the procedure is to work well. First, 10 μl of 2YT in a 125-ml flask was inoculated with 50 μl of an overnight culture of JM101, MV1190, or suitable alternate and grown for 1 hr at 37°, with shaking at 150 rpm. Then 50 μl of phage supernatant was added and grown for 6–12 hr. The 10 ml was centri-

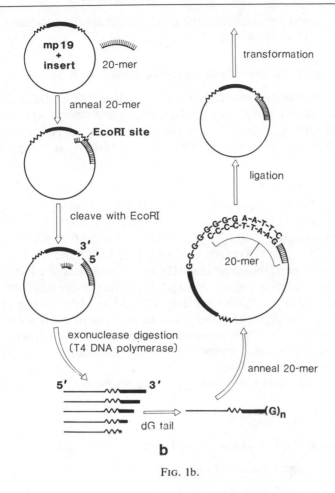

FIG. 1b.

fuged at 17,000 *g* for 15 min. The supernatant was collected and centrifuged again for 15 min at 17,000 *g*; 2.5 ml of a 20% PEG/3.5 *M* ammonium acetate solution was added to the cleared supernatant. The tubes were vortexed and placed in ice for 30 min and then centrifuged at 17,000 *g* for 15 min. The supernatant was then poured off. The tubes were allowed to drain and excess liquid was carefully removed with cotton swabs. The precipitates were resuspended in 600 μl of 20 m*M* Tris · HCl (pH 7.2), 1 m*M* EDTA. The samples were extracted three times with an equal volume of a 1:1 mixture of phenol and chloroform equilibrated with 1 *M* Tris · HCl (pH 8.0) and extracted once with an equal volume of chloroform. DNA was precipitated by the addition of 0.5 volume of 7.5 *M* ammonium acetate and 2 volumes of 95% ethanol for either 30 min at −70° or overnight at −20°. The DNA was then pelleted, washed once with 70% ethanol, dried, and resuspended in 60 μl of water or 10 m*M* Tris · HCl (pH

7.5). The A_{260} was measured and the DNA concentration was determined. The average yield was 40 μg.

Linearization of Circular Single-Stranded DNA

Most problems encountered in the procedure can be traced back to incomplete linearization at step I, which can almost always be traced back to problems with the DNA. Before proceeding, it is advisable to confirm that the DNA of interest can be digested with restriction enzymes.

Test Cut

In a reaction tube, combine 1 μg of DNA, 2.5 μl of 10× T4 DNA polymerase buffer [330 mM Tris–acetate (pH 7.8), 660 mM KOAc, and 100 mM magnesium acetate], 1 μl of RD-22 (5'-CGACGGCCAGTGAA-TTCCCCCC-3') at 0.5 A_{260}/ml, 1 μl of 0.1 M DTT, and 1 μl of *Eco*RI (for even-numbered DNAs) or *Hin*dIII (for the odd-numbered series) and water to a final reaction volume of 20 μl. Incubate the *Eco*RI reactions for 1 hr and the *Hin*dIII digests for 2 hr at 42–50°. Add a stop dye and electrophorese the digested sample and an undigested control on a 2.0% TBE agarose gel. The digestion should be at least 90% complete. If cutting is essentially complete, proceed with the protocol; if not, re-extract or reprepare the starting DNA.

If the test cut works well, a larger scale reaction can be set up.

Step I. Restriction Digestion

Cleavage at the *Eco*RI site of mp9, mp11, mp19, and um21:

1. In a reaction tube, combine
   4 μg DNA
   2.5 μl 10× reaction buffer
   4 μl RD-22-mer (20 μg/ml)
   1 μl DTT (100 mM)
   2 μl of *Eco*RI (10–20 units/μl)
   Add water to 20 μl final volume.
2. Incubate the reaction at 42–50° for 1 hr.
3. Inactivate the *Eco*RI by heating the reaction mixture at 65° for 10 min.
4. Allow the reaction to cool to room temperature.

Cleavage at the *Hin*dIII site of mp8, mp10, mp18, and um20:

1. In a reaction tube, combine
   4 μg DNA
   2.5 μl 10× reaction buffer
   8 μl RD-29-mer (20 μg/ml)

1 μl DTT (100 mM)
2 μl of HindIII (10–20 units/μl)
Add water to 20-μl final volume.
2. Incubate at 50° for 2 hr.
3. Inactivate the HindIII by heating the reaction mixture at 85° for 5 min.
4. Allow the reaction to cool to room temperature.

It is recommended that verification of complete digestion be performed by running a sample on a 1% agarose minigel at 100 V for 2 to 3 hr in TBE buffer. The digestion should be at least 90% complete. This DNA has been shown to give greater than 90% digestion following the above protocol; 250 ng is sufficient to see on a gel.

Step II. Exonuclease Digestion

The rate of digestion is proportional to the amount of T4 DNA polymerase added.

Units of T4 DNA polymerase per microgram of DNA	Approximate rate of digestion
1.5	45 bases/min
2	60 bases/min
2.5	75 bases/min
3	90 bases/min

1. To the above reaction mixture, add the following reagents:
2.5 μl DTT (200 mM)
1 μl BSA (10 mg/ml)
6–12 units of T4 DNA polymerase (1–2 μl)
2. Incubate at 37° for the appropriate period of time.
3. Remove aliquots at appropriate intervals. Digestion occurs at a rate of approximately 50 bp/min.
4. Terminate the reaction at each time point by heating the reaction mixture at 65° for 20 min. If multiple time points are collected, they may be stored on ice after heat denaturation. After the final time point, all the digests are combined before going to the next step.

Step III. Tailing Reaction

1. To the above reaction mixture, add the following reagents:
3.0 μl 50 μM dGTP for inserts in mp9, mp11, mp19, or um21
        OR
3.0 μl 50 μM dATP for inserts in mp8, mp10, mp18, or um20
1.0 μl of water

1.0 μl of TdT (at 15 units/μl)
2. Incubate at 37° for 10 min.
3. Heat inactivate for 10 min at 65°.
4. Remove 10 μl for use as an unligated control in the transformation step.

Step IV. Ligation

1. Add 1 μl of the appropriate oligomer (RD-22 for inserts in mp9, mp11, mp19, or um21 and RD-29 for inserts in mp8, mp10, mp18, or um20).

2. Anneal by placing the tube(s) in a heat block at 65° for 10 min, removing the heat block and allowing the reaction to cool to 40° (about 30 min).

3. Add the following reagents:
3.0 μl 10 mM ATP
4.0 μl of water
1.0 μl of T4 DNA ligase (1–3 units)

4. Incubate at room temperature for 1 hr or longer. Note: a control reaction without ligase should be prepared using the 10 μl saved earlier (Step III.4).

5. Transform competent *E. coli* JM101 or other cell line with 2–4 μl of the ligation mix. Let the balance of the ligation reaction sit at room temperature overnight.

Results: Transformation efficiency in the range of 10^4 to 10^5 transformants per microgram of DNA is routinely observed.

Step V. Sizing Subcloned Fragments

1. Pick isolated plaques with sterile toothpicks and grow in 2 ml of 2× YT media in 18 × 150-mm test tubes, shaking in a 37° incubator at a 45° angle at 150 rpm for 8–12 hr.

2. Transfer cultures to a 1.5-ml centrifuge tube.

3. Pellet cells in a microfuge at 12,000 g for 2 min.

4. Transfer supernatants to new tubes and store at 4° for phage preparations. The cell pellet may be saved if you wish to examine the double-stranded RF form of the phage DNA.

5. For phage, combine 25 μl of the supernatant with 5 μl of the SDS loading dye. Electrophoresis is carried out in 0.7% agarose gels run in 1× Tris–borate–EDTA gel buffer. Large 20-cm gels may be double combed and run for 12–14 hr at 100 V. The outside lanes on either side should contain markers consisting of the initial nondeleted single-stranded DNA as well as some single-stranded vector DNA without an insert.

6. Measure the migration distance of the two markers and plot distance migrated versus insert size on regular graph paper. Plotting the distance migrated of the DNAs containing various deletions will give a good estimate of the size of fragment remaining in each clone.

7. Select a group of clones showing an ordered decrease in size. Prepare the templates and sequence them.

Extension to Single-Stranded Plasmids

The same basic procedure can be applied to single-stranded DNA derived from plasmids such as the pEMBL plasmids. These plasmids contain an F1 origin of replication in addition to a ColE1 origin. When cells containing one of these plasmids are infected with a helper phage, single-stranded DNA is produced, packaged, and secreted into the media like M13 or F1 phage. Using the helper phage M13 K407 developed by Jeff Vieria (personal communication), a ratio of one helper phage to 50 copies of the single-stranded plasmid is found. After preparing the single-stranded plasmid DNA, the deletion subcloning procedure can be followed exactly as it was with M13. The only differences come in the plating/screening aspects of the procedure. After transformation, the cells should be spread on B-broth plates containing ampicillin. Cells infected with only helper phage will die; those cells transformed with the plasmid will be resistant to ampicillin. Colonies can be picked, grown up, superinfected with helper phage, and the single-stranded DNA sized on agarose gels as with M13.

Discussion and Conclusions

Oligomers have been used to serve two basic functions in the procedure: (1) to create a double-stranded region, thereby directing enzymes to a *specific* portion of the molecule, and (2) as a bandaid to direct the religation of *only* the desired fragments of the molecule. These principles can easily be applied to facilitate the construction of vectors, and for *in vitro* mutagenesis.

A modification of this procedure has been used by the authors to insert additional cloning sites into the multicloning region of M13. In addition, the origins of replication for T3 and T7 have been inserted to allow for RNA synthesis of cloned genes.[2] Synthetic oligomers were used to create a double-stranded region at the desired location for insertion, and the appropriate restriction enzymes were utilized to open the molecule at that

[2] J. Bailey, J. Klement, and W. McAllister, *Proc. Natl. Acad. Sci. U.S.A.* **80,** 2814 (1983).

region. Because the oligomers are specific and hybridize to only one site, any restriction enzyme can be used as long as the recognition site for the enzyme of choice exists in the double-stranded region that has been created. One need not worry about how many other sites there are in the vector, because only the double-stranded section (created by the oligomer) will be recognized by the enzyme. The oligomer fragments are removed by heat denaturation, cooling, and precipitation. Subsequent to the cleavage step, sequences complementary to the desired insert, in this case T3 or T7, are added with a second oligomer, which also serves to aid in the ligation reaction. This oligomer has a middle region complementary to whatever sequence is to be inserted, and ends complementary to the vector sequences which border the insert (Fig. 2a). Following addition of this oligomer, dNTPs, Klenow, and ligase are added. The resultant molecule is the original vector with the desired insert at the location of choice within the vector, in the desired reading frame.

A variation of this protocol can be utilized for large insert regions in order to decrease the cost inherent in the synthesis of the large oligomer that would be required for this procedure. In the modified procedure, after the directed cleavage step, the insert is added and a second set of oligomers, which serve to aid in the ligation reaction, is also added. These "bandaid" oligomers are designed such that only those molecules containing the desired insert will religate since the oligomers contain sequences complementary to both the insert and the vector (Fig. 2b). Re-

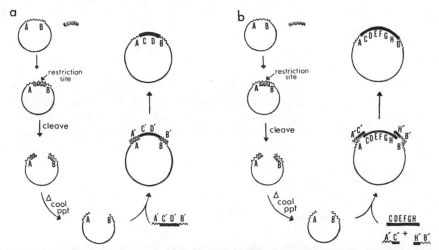

FIG. 2. (a) Use of oligomers to create a double-stranded region, directing enzymes to a specific portion of the molecule. (b) Use of oligomers as a bandaid to direct the religation of only the desired fragments of the molecule in the creation of a double-stranded region.

gardless of which insertion protocol is followed (a or b), this system has the same advantage as the subcloning procedure[2] in that all reactions can be performed in a single test tube. The percentage of positive clones found with either procedure has been 30–50%.

The basic subcloning procedure can also be modified for use in *in vitro* mutagenesis. A difficulty normally associated with deletion mutagenesis is the inability to define precisely the size of the deletion.[3] Most protocols are limited to generating deletions which are only capable of initiating at one end of the fragment, as well as resulting in deletions of varying extent and imprecise ends.[4] This procedure enables one to begin at any restriction site within, or outside, the gene and precisely define the end points of the deletion.

A synthetic oligomer, which hybridizes to a portion of the gene of interest at a restriction site, is used to create a double-stranded region within the molecule. This should present little difficulty since there are many restriction enzymes available, and the only possible site for cleavage will be in the gene fragment. The only double-stranded region is that created, and defined by, the oligomer; consequently, potential sites in the remaining single-stranded vector do not pose a problem. After digestion with the appropriate enzyme and removal of the oligomer, the result is a linear single-stranded molecule. A second set of oligomers, whose sequence is critical to the procedures, is then added. These oligomers will serve as borders for the deletion process. Therefore, the 3'-most base of the 5' oligomer, and the 5'-most base of the 3' oligomer (see Fig. 3) should correspond to those bases in the sequence which are just outside of the deleted region. One will have created a single-stranded deletion region which is bordered by oligomers. Hybridization of these oligomers to the molecule is followed by the addition of exonuclease VII. This enzyme will move along the single-stranded molecule in both a 3' and a 5' direction until it reaches the double-stranded border.[5] The extent of the deletion is then precisely defined by the positioning of the oligomers on either side of the single-stranded region. The border oligomers can be removed and a final oligomer added to act as a bandaid, facilitating the ligation reaction. The resultant molecule is the original vector with a precisely defined deletion in the gene of interest. Since the ends of the deletion are known, the bandaid oligomer is designed such that it will only allow for ligation of molecules with the appropriate ends. Another level of complexity can be generated by manipulation of the sequence of the bandaid oligomer. This

[3] D. Botstein and D. Shortlas, *Science* **229**, 1193 (1985).
[4] I. Nisbet and M. Beilharz, *Gene Anal. Tech.* **2**, 23 (1985).
[5] J. Chase and C. Richardson, *J. Biol. Chem.* **249**, 4545 and 4553 (1974).

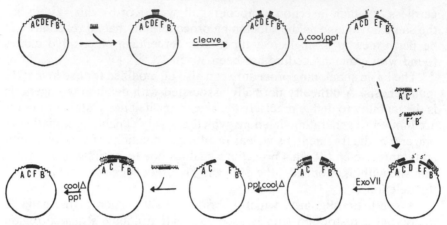

FIG. 3. A single-stranded deletion region bordered by oligomers. For detail, see text.

oligomer can be constructed to contain a unique restriction site. The resultant vectors can then be screened for those molecules which have successfully completed the deletion and religation process by virtue of a new restriction site.

With the appropriate vector, e.g., pBX10, 11 or pEMBLs, one which contains an F1 or M13 origin of replication and a T3 and/or T7 origin of replication, and using the basic protocols outlined, it should be possible to clone a gene, to sequence it using the deletion subcloning procedure, or to generate deletion or site-specific mutants and transcribe the respective RNAs for the wild-type and mutant constructs. The RNA can then be used for *in vitro* translation or injection into *Xenopus* oocytes,[6] so that one can readily study structure/function relationships of cloned and sequenced genes using this overall approach.

[6] T. Sakata, M. Tamaki, H. Teraoka, and K. Tanaka, *Gene Anal. Tech.* **2**, 57 (1985).

[41] Hydroxyl Radical Footprinting: A High-Resolution Method for Mapping Protein–DNA Contacts[1]

By Thomas D. Tullius, Beth A. Dombroski, Mair E. A. Churchill, and Laurance Kam

Proteins that bind to specific regions of DNA are leading players in the cast of molecules that obtains information from the genome. To begin to understand how a protein recognizes its binding site among the vast number of other sequences in a long DNA molecule, it is necessary first to find the DNA base sequence covered by the protein. Galas and Schmitz introduced an elegant method, called "footprinting," to detect contacts between DNA and protein.[2] A protein is allowed to bind to a radioactively labeled DNA molecule containing the sequence that the protein recognizes. The DNA–protein complex is digested by a nuclease, usually deoxyribonuclease I (DNase I).[3] The regions of the DNA molecule covered by the bound protein are protected from digestion while the rest of the DNA backbone is cut normally. If the products of this reaction are separated on an electrophoresis gel such as is used for DNA sequencing, the amount of digestion at each position in the sequence can be seen. A blank region on the autoradiograph of the gel, aptly called the footprint, is found at the sequence where the protein binds specifically to the DNA.

DNase I footprinting experiments have been invaluable in defining protein–DNA contacts.[2] But there are several disadvantages to the technique that other footprinting methods overcome only partially. Chief among these limitations is the preference of a footprinting reagent for cutting DNA only at particular bases or sequences of bases. Nucleases tend to cut at only a subset of the backbone positions along a DNA molecule.[4] Chemical footprinting reagents usually react with only one or two of the DNA bases, giving no information on protein contacts with the other bases. Dimethyl sulfate, for example, methylates at guanine and adenine,[5] so that contacts with cytosine and thymine cannot be observed.

[1] This research was supported by grants from the Searle Scholars Program of the Chicago Community Trust, the Research Corporation, the Biomedical Research Support Grant Program (S07 RR07041), and the National Cancer Institute (CA37444) of the National Institutes of Health.

[2] D. J. Galas and A. Schmitz, *Nucleic Acids Res.* **5,** 3157 (1978).
[3] Abbreviations: bp, base pairs; Cu(phen)$_2^+$, bis(1,10-phenanthroline)copper(I); DNase I, deoxyribonuclease I; MPE · Fe(II), methidiumpropyl-EDTA · iron(II).
[4] H. R. Drew and A. A. Travers, *Cell* **37,** 491 (1984).
[5] B. Singer, *Prog. Nucleic Acid Res. Mol. Biol.* **15,** 219 (1975).

DNase I has the further disadvantage that since it is roughly the size of the DNA-binding protein it is detecting, the DNA region covered by the protein often is overestimated by DNase I footprinting. Chemical reagents generally produce smaller footprints. For example, Sawadogo and Roeder found that the upstream stimulatory factor (USF), isolated from uninfected HeLa cells, protects 18–20 bases of the adenovirus major late promoter from DNase I digestion, but only 10 bases from cleavage by methidiumpropyl–EDTA · iron(II).[6]

We have developed a method, hydroxyl radical footprinting,[7] which overcomes these limitations. Hydroxyl radical cleaves DNA by abstracting a hydrogen atom from the deoxyribose sugars along the DNA backbone.[8] Since hydroxyl radical is exceedingly short lived and reactive, and attacks sites on the surface of the DNA molecule, there is almost no sequence dependence or base dependence in the cleavage reaction.[9] Every position along the backbone is cleaved nearly equally. We have used this chemistry before to determine the helical periodicity of a DNA restriction fragment bound to an inorganic crystal,[10] because hydroxyl radical could cut efficiently only the DNA backbone sugars that were directed away from the inorganic surface. We show here that the same principle can be applied to produce footprints of proteins bound to DNA,[7] since cutting of the DNA backbone by hydroxyl radical is blocked by bound protein.

To generate hydroxyl radical we make use of a venerable reaction in inorganic chemistry, the Fenton reaction,[11] in which iron(II) reduces hydrogen peroxide to give hydroxyl radical (I).

$$Fe(II) + H_2O_2 \rightarrow Fe(III) + OH\cdot + OH^-$$
$$(I)$$

We find that the Udenfriend system (II) is a convenient way to use this reaction to fragment DNA. Udenfriend et al. showed[12] that the EDTA complex of iron(II) [$Fe(EDTA)^{2-}$] would produce hydroxyl radical from hydrogen peroxide, and in turn the iron(III) product of the reaction could be reduced by ascorbate ion back to iron(II) to continue the process.

$$Fe(EDTA)^{2-} + H_2O_2 \rightarrow Fe(EDTA)^- + OH\cdot + OH^-$$

ascorbate
$$(II)$$

[6] M. Sawadogo and R. G. Roeder, Cell 43, 165 (1985).

[7] T. D. Tullius and B. A. Dombroski, Proc. Natl. Acad. Sci. U.S.A. 83, 5469 (1986).

[8] R. P. Hertzberg and P. B. Dervan, Biochemistry 23, 3934 (1984).

[9] W. D. Henner, S. M. Grunberg, and W. A. Haseltine, J. Biol. Chem. 257, 11750 (1982).

[10] T. D. Tullius and B. A. Dombroski, Science 230, 679 (1985).

[11] H. J. H. Fenton, J. Chem. Soc. 65, 899 (1894).

[12] S. Udenfriend, C. T. Clark, J. Axelrod, and B. B. Brodie, J. Biol. Chem. 208, 731 (1954).

A negatively charged complex of iron(II) is an advantage for footprinting since the inorganic reagent will not associate electrostatically with DNA. Only the hydroxyl radical interacts directly with the DNA, giving a very small (but powerfully reactive) probe of the DNA–protein complex.

In this article we present four aspects of the footprinting method. We first illustrate hydroxyl radical footprinting by its application to a well-studied DNA–protein complex, that of the bacteriophage λ repressor with the O_R1 operator DNA sequence. Next, we compare the footprints produced by hydroxyl radical with footprints made by several other reagents, to point out the different information available from the various methods. We then show how the reaction conditions for hydroxyl radical footprinting may be altered to accommodate other DNA–protein complexes. Finally, we apply the method to a more complex problem: the determination of the contacts made by the transcription factor TFIIIA with the internal control region of the 5 S ribosomal RNA gene of *Xenopus*.

Materials and Methods

DNA Preparation and Purification

Since hydroxyl radical is generated at some distance from the DNA molecule by negatively charged $Fe(EDTA)^{2-}$, the cutting reaction is rather inefficient. Although this has the advantage of making it unlikely that more than one break is introduced into a particular DNA molecule, it also demands that the DNA strands be intact before the reaction so that the footprinting "signal" is clear above the background. We have found that care must be taken in preparing a DNA fragment to be used for hydroxyl radical footprinting in order that the DNA molecule be as free as possible from nicks. Also to this end we use water purified through a Milli-Q system (Millipore) for all solutions.

We isolate crude plasmid DNA by a modification of a standard cleared-lysate procedure.[13] Final purification is achieved by chromatographing the plasmid over Sephacryl S-500 (1.6 × 40 cm; 1 ml/min) and RPC-5 (1 × 10 cm; Pharmacia HR-10/10 column; 1 ml/min) columns, using a FPLC system (Pharmacia). The starting buffer (buffer A) for the RPC-5 column is 10 mM Tris · Cl, 1 mM EDTA, and 0.4 M NaCl (pH 8.0). A complicated gradient from 0 to 100% buffer B is run, using 10 mM Tris · Cl, 1 mM EDTA, and 2 M NaCl (pH 8.0) as buffer B. Plasmid DNA elutes at around 15% buffer B.

S-500 chromatography separates plasmid DNA by size from residual chromosomal DNA and small RNA. This step also ensures that the RPC-5

[13] T. Maniatis, E. F. Fritsch, and J. Sambrook, "Molecular Cloning: A Laboratory Manual." Cold Spring Harbor Lab., Cold Spring Harbor, New York, 1982.

column is not overloaded with nucleic acid. RPC-5 chromatography[14] removes any remaining RNA and contaminating DNA, and often separates supercoiled from relaxed (nicked) plasmid DNA. By isolating mostly supercoiled plasmid we ensure that as few nicks as possible are present in DNA that is to be used for footprinting reactions. We find that DNA purified in this way is superior for hydroxyl radical footprinting to plasmid DNA prepared by standard methods (such as CsCl density gradient centrifugation). Plasmid DNA is stored at 4° in a buffer consisting of 10 mM Tris · Cl and 1 mM EDTA (pH 8.0) (TE buffer).

Labeling with Radioactive Phosphorus

DNA is labeled at the 3' end by standard methods,[13] using the Klenow fragment of DNA polymerase I and [α-^{32}P]deoxynucleoside triphosphates to fill in the nucleotides at a 5'-overhanging restriction endonuclease cut. When using the Klenow fragment to add several radioactive nucleotides, we find it advantageous to incorporate a dideoxynucleotide (either radioactive or nonradioactive) at the last position, since the 3'–5' exonuclease activity of the Klenow fragment cannot use this deoxyribose analog as a substrate. This procedure gives very high incorporation of radioactive label. DNA fragments are labeled at the 5' end by standard procedures[13] using T4 polynucleotide kinase and [α-^{32}P]ATP.

Hydroxyl Radical Footprinting

In this section, we present our protocol[7] for making footprints of λ repressor on the O_R1 DNA sequence. In later sections we discuss how changes in the reaction conditions affect the cutting reaction, so that the conditions can be adapted to accommodate other DNA–protein complexes.

Hydroxyl Radical Footprinting of the λ Repressor–O_R1 Complex

A typical sample contains end-labeled DNA (the 120-bp EcoRI–BglII restriction fragment of plasmid pOR1[15]) (30,000 to 40,000 cpm, 10 fmol) in 147 μl of repressor buffer, which consists of 10 mM bis-Tris · HCl (pH 7.0), 50 mM KCl, 1 mM CaCl$_2$, and 0.5 μg nonspecific (nucleosomal) DNA. A dilution of λ repressor (the generous gift of Michael Brenowitz and Gary Ackers) is made in this buffer, and 20 μl of the repressor solution (18 pmol of repressor monomer) is added to the DNA solution. The

[14] R. D. Wells, S. C. Hardies, G. T. Horn, B. Klein, J. E. Larson, S. K. Neuendorf, N. Panayotatos, R.K. Patient, and E. Selsing, this series, Vol. 65, p. 327.

[15] A. D. Johnson, Ph.D. dissertation. Harvard University, Cambridge, Massachusetts, 1980.

mixture is then incubated at room temperature to allow complexation of repressor with operator DNA.

A colorless aqueous solution of 0.4 mM iron(II) is prepared fresh before each experiment by dissolution of ferrous ammonium sulfate [$(NH_4)_2Fe(SO_4)_2 \cdot 6H_2O$] (Aldrich). Iron(II) is unstable in aerated solution,[16] oxidizing to iron(III). Iron solutions that turn orange and deposit a precipitate should be discarded. (Note: glassware containing iron solutions should ideally be used only for that purpose, and should be cleaned thoroughly with HCl before using for any other solution. Residual iron can cause nicking of DNA samples that come into contact with solutions prepared in such vessels.) The complex of iron(II) with EDTA[17] is prepared by mixing equal volumes of 0.4 mM iron(II) and 0.8 mM EDTA (Aldrich, Gold Label) solutions. Solutions of 20 mM sodium ascorbate (Sigma) and 0.6% H_2O_2 (diluted from a 30% solution, J.T. Baker) also are prepared for each experiment.

The iron(II) EDTA solution (10 μl), 0.6% H_2O_2 (10 μl), and 20 mM sodium ascorbate (10 μl) are placed on the inside wall of the 1.5-ml Eppendorf reaction tube above the solution of the repressor–operator complex. To initiate the cutting reaction, the reagents are mixed together on the side of the reaction tube, added to the repressor–DNA solution, and mixed well by tapping the tube. The final concentrations in the reaction mixture of iron(II), EDTA, H_2O_2, and ascorbate are 10 μM, 20 μM, 0.03%, and 1 mM, respectively. The reaction is allowed to proceed for 2 min at room temperature.

The reaction is quenched by adding 0.1 M thiourea (20 μl) (an efficient scavenger of hydroxyl radical) and 0.2 M EDTA (2 μl). To remove protein, the solution is extracted with 225 μl of a phenol:chloroform:isoamyl alcohol mixture (49:49:2). DNA is recovered from the aqueous phase by addition of 3 M sodium acetate (25 μl), tRNA (15 μg), and absolute ethanol (750 μl), and centrifugation in an Eppendorf microcentrifuge at 4° for 30 min. The DNA pellet is rinsed with 1 ml of cold 70% ethanol, dried in a Speed Vac concentrator (Savant Instruments), and dissolved in formamide–dye mixture.[18] (Note: if the DNA pellet is to be stored overnight before electrophoresis, it should be stored dry or as a suspension in ethanol. We have observed nicking of the DNA after storage in the formamide–dye mixture for times as short as overnight.)

[16] F. A. Cotton and G. Wilkinson, "Advanced Inorganic Chemistry," 4th Ed., p. 754. Wiley, New York, 1980.
[17] C. Bull, G. J. McClune, and J. A. Fee, *J. Am. Chem. Soc.* **105**, 5290 (1983).
[18] A. M. Maxam and W. Gilbert, this series, Vol. 65, p. 499.

Other Methods

DNase I Footprinting

Samples to be digested with DNase I[19] contain end-labeled DNA in repressor buffer in a total volume of 180 μl. Appropriate dilutions of λ repressor are added to the samples to bring the final volume to 200 μl. DNase I (10 ng) is added to the repressor–DNA mixture. Digestions are allowed to proceed for 2 min and are stopped as described above for the iron(II) EDTA-digested samples.

MPE · Fe(II) Footprinting[20]

Methidiumpropyl–EDTA[8] (MPE) was kindly provided by Peter Dervan. A solution of the λ repressor–DNA complex is prepared exactly as described above for hydroxyl radical footprinting. An aqueous solution containing 5 μM MPE and 4 μM iron(II) is made; 20 μl is added to the repressor–DNA mixture. This mixture is incubated at 37° for 30 min. The cutting reaction is initiated by addition of 10 μl of 80 mM dithiothreitol and is allowed to proceed for 15 min. The final concentrations of MPE and iron(II) in the reaction mixture are 500 and 400 nM, respectively. The reaction is stopped and worked up as described above for hydroxyl radical footprinting.

Cu(phen)₂⁺ Footprinting[21]

To a solution containing the λ repressor–DNA complex, prepared as described above, we add 20 μl of an aqueous solution consisting of 7.5 μM Cu(II) [prepared by dissolution of copper(II) sulfate (Aldrich, Gold Label)] and 50 μM 1,10-phenanthroline (Aldrich, Gold Label). To start the reaction, 0.1 M mercaptopropionic acid (Aldrich) (10 μl) is added. The final concentrations in the reaction mixture of copper(II), 1,10-phenanthroline, and mercaptopropionic acid are 750 nM, 5 μM, and 5 mM, respectively. The reaction is allowed to proceed for 2 min at room temperature, then is stopped as described above.

Gel Electrophoresis

DNA fragments are electrophoresed (after denaturation at 90° for 2 min) on a 6% polyacrylamide, 50% urea denaturing gel (0.35 mm thick)[22] at 55 W constant power. Gels are dried onto Whatman 3MM paper. Ko-

[19] A. Johnson, B. J. Meyer, and M. Ptashne, *Proc. Natl. Acad. Sci. U.S.A.* **76,** 5061 (1979).
[20] M. W. Van Dyke and P. B. Dervan, *Nucleic Acids Res.* **11,** 5555 (1983).
[21] A. Spassky and D. S. Sigman, *Biochemistry* **24,** 8050 (1985).
[22] F. Sanger and A. R. Coulson, *FEBS Lett.* **87,** 107 (1978).

dak XAR-5 film is preflashed[23] and the dried gel is autoradiographed either at $-70°$ with a Dupont Cronex Lightning Plus intensifying screen, or at room temperature without a screen. Autoradiographs are scanned with a Joyce-Loebl Chromoscan 3 densitometer using an aperture width of 0.05 cm.

Effects of Experimental Conditions on Hydroxyl Radical Cutting of DNA

DNA–protein complexes are made under a wide variety of conditions of temperature, buffer, pH, salt, and ionic strength. To assess the effects of the milieu on the hydroxyl radical footprinting method, we did a series of experiments to measure the degree of cutting of DNA by the iron reagent under conditions that might be used to stabilize a DNA–protein complex. We also studied the effects of changing the concentrations of the components of the iron cutting reagent, so that the reagent could be modified if one or more of its components were found to adversely affect a particular DNA–protein complex.

The standard cutting reagent for these experiments is 10 μM iron(II), 20 μM EDTA, 0.03% H_2O_2, and 1 mM sodium ascorbate. We incubate the reagent with end-labeled DNA under the conditions to be tested and separate the products of the cutting reaction by denaturing gel electrophoresis. We measure the amount of cutting by densitometry of a set of bands corresponding to cut and to uncut DNA, for each experimental condition examined. Absolute cutting frequencies of DNA for several buffers are presented in Table I. For the other conditions studied (Tables II–IV), the amount of cutting is normalized to the cutting observed with the standard cutting reagent in 10 mM Tris · Cl (pH 8.0), and the results are presented in schematic form.

Hydroxyl Radical Footprinting of TFIIIA

5 S DNA

Plasmid pXbs201,[24] which contains the *Xenopus* somatic 5 S ribosomal RNA gene, was kindly provided by Kent Vrana and Donald Brown. We used this plasmid to transform *Escherichia coli* strain HB101. The plasmid was isolated as described above, linearized by digestion with *Bam*HI, 5' or 3' end-labeled, and cut with *Hind*III to give the 249-bp *Bam*HI–*Hind*III fragment, which contains the 5 S gene and some surrounding sequences. The labeled fragment was purified by electrophoresis on a polyacrylamide gel.

[23] R. Laskey and D. A. Mills, *FEBS Lett.* **82,** 314 (1977).
[24] D. R. Smith, I. J. Jackson, and D. D. Brown, *Cell* **37,** 645 (1984).

TABLE I
EFFECT OF BUFFER AND pH ON HYDROXYL RADICAL
CUTTING OF DNA[a]

Buffer	pH	Uncut DNA[b] (%)	
		50 mM^c	10 mM^c
Tris · Cl⁻	8.5	98	77
Tris · Cl⁻	8.0	88.5	71
Na⁺-HEPES	7.5	91.5	78
Na⁺-HEPES	7.0	94	64
Sodium phosphate	7.0	18	33.5
Bis-Tris · Cl⁻	6.5	88	88
Sodium cacodylate	6.5	71	70.5
Sodium citrate	5.0	67	79.5
Sodium acetate	4.5	84.5	78.5

[a] DNA, dissolved in the indicated buffers, was cut at 22° for 2 min using 10 μM iron(II), 20 μM EDTA, 0.03% H_2O_2, and 1 mM sodium ascorbate.

[b] The percentage of uncut DNA for each sample was determining by densitometry of the gel band corresponding to uncut DNA, and comparison of the integral of this band to the integral of the corresponding band from untreated DNA.

[c] Buffer concentration.

TFIIIA

TFIIIA was isolated by Kent Vrana from the 7 S TFIIIA-ribosomal RNA particle by enzymatic digestion and BioRex 70 chromatography.[24] The protein was stored at −70° at a concentration of 1 mg/ml in a buffer consisting of 50 mM HEPES (pH 7.5), 5 mM $MgCl_2$, 10 M $ZnCl_2$, 1 mM dithiothreitol, 20% glycerol, and 1 M KCl. Further dilutions were made with TFIIIA binding buffer (see below), which contains no glycerol.

Footprinting Experiments

The 5 S DNA solution contains 35 fmol of end-labeled BamHI–HindIII fragment and 0.1 μg nonspecific (sonicated calf thymus) DNA, in TFIIIA binding buffer: 20 mM HEPES (pH 7.5), 70 mM NH₄Cl, 7 mM $MgCl_2$, 10 μM $ZnSO_4$, 0.02% NP-40, and 100 μg/ml bovine serum albumin (note: glycerol has been omitted from the protein binding buffer used by Smith et al.[24]). TFIIIA (0.7 pmol), in 20-fold molar excess over the labeled DNA fragment, is added to the DNA mixture. The protein–DNA mixture is incubated at room temperature for 15 min. The volume of the TFIIIA–DNA solution is 35 μl.

TABLE II

EFFECT OF REAGENT CONCENTRATIONS AND REACTION
CONDITIONS ON HYDROXYL RADICAL
CUTTING OF DNA[a]

Reaction condition	Cutting[b]
Temperature	
0°	−
22°	●
37°	●
65°	++
90°	++
Time	
1 min	●
2 min	●
5 min	●
Iron	
1 μM Fe^{2+}, 2 μM EDTA	−
100 μM Fe^{2+}, 200 μM EDTA	+
Ascorbate	
None	− −
0.5 mM	−
EDTA	
1 μM Fe^{2+}, 20 μM EDTA	−
10 μM Fe^{2+}, 200 μM EDTA	●
H_2O_2	
0.003%	●
0.3%	+
100 μM Fe^{2+}, 200 μM EDTA, no H_2O_2	− −
100 μM Fe^{2+}, 200 μM EDTA, 0.003% H_2O_2	●

[a] End-labeled DNA was dissolved in 10 mM Tris · Cl
(pH 8.0) and cut under the indicated conditions. The
DNA was cut at 22° for 2 min with 10 μM iron(II), 20
μM EDTA, 0.03% H_2O_2, and 1 mM sodium ascorbate, except in experiments where these conditions
were varied.

[b] We define DNA cutting by the following symbols: ●,
optimum cutting (20–40% of the DNA molecules are
cut); + or −, somewhat higher or lower than optimum cutting, but the degree of cutting is still acceptable; ++ or − −, unacceptably high or low degree of
cutting.

DNase I Footprinting.[24] Pancreatic DNase I (4 ng) is added and the
solution is mixed gently. The reaction mixture is incubated at room temperature for 1 min.

Hydroxyl Radical Footprinting. The cutting reagent consists of 5 μl
each of 1 mM iron(II)–2 mM EDTA, 0.03% H_2O_2, and 10 mM sodium
ascorbate. These three solutions are placed on the inner wall of the reac-

TABLE III

EFFECT OF SALTS ON HYDROXYL RADICAL
CUTTING OF DNA[a]

Salt/concentration	Cutting[b]
Sodium chloride	
50 mM	●
100 mM	●
500 mM	●
Ammonium chloride	
50 mM	●
200 mM	●
Potassium acetate	
50 mM	●
200 mM	●
Magnesium chloride	
1 mM	●
2 mM	●
5 mM	●
10 mM	●
20 mM	●
50 mM	−
100 mM	−
Calcium chloride	
1 mM	●
10 mM	●
Zinc chloride	
10 μM	●
100 μM	●

[a] Labeled DNA was dissolved in 10 mM Tris · Cl (pH 8.0) and the indicated concentration of salt and was cut at 22° for 2 min with 10 μM iron(II), 20 μM EDTA, 0.03% H_2O_2, and 1 mM sodium ascorbate.

[b] We define DNA cutting by the following symbols: ●, optimum cutting (20–40% of the DNA molecules are cut); −, somewhat lower than optimum cutting, but the degree of cutting is still acceptable.

tion vessel above the DNA–protein solution and then mixed gently with the TFIIIA–DNA solution. The final concentrations in the reaction mixture of iron(II), EDTA, H_2O_2, and ascorbate are 100 μM, 200 μM, 0.003%, and 1 mM, respectively. The reaction mixture is incubated at room temperature for 1 min.

Both footprinting reactions are stopped by addition of 21 μl of a solution containing 0.1 M EDTA, 0.05 M thiourea, and 250 μg/ml tRNA, and

TABLE IV
EFFECT OF ADDITIVES ON HYDROXYL RADICAL
CUTTING OF DNA[a]

Additive/concentration	Cutting[b]
Glycerol	
0.5%	−
1%	− −
2.5%	− −
5%	− −
10%	− −
BSA	
100 μg/ml	●
200 μg/ml	●
Nonspecific DNA	
2 μg/ml	●
20 μg/ml	●
Dithiothreitol	
1 mM	●
2-Mercaptoethanol	
5 mM	−

[a] Labeled DNA was dissolved in 10 mM Tris · Cl (pH 8.0) and the indicated concentration of additive and was cut at 22° for 2 min with 10 μM iron(II), 20 μM EDTA, 0.03% H_2O_2, and 1 mM sodium ascorbate.
[b] We define DNA cutting by the following symbols: ●, optimum cutting (20–40% of the DNA molecules are cut); −, somewhat lower than optimum cutting, but the degree of cutting is still acceptable; − −, unacceptably low degree of cutting.

extraction with phenol followed by ether. Addition of sodium acetate and ethanol and centrifugation in an Eppendorf microcentrifuge at 4° for 30 min precipitates the DNA. The DNA pellet is washed with 70% EtOH, dried, dissolved in formamide–dye mixture,[18] heated at 90°, loaded on a 7% DNA sequencing gel,[22] and electrophoresed for 3 hr at 55 W constant power.

DNase I Footprinting Assay for Compatibility of the Iron EDTA Cutting Reagent with a DNA–Protein Complex

We make use of DNase I footprinting to test whether a particular component of the iron cutting reagent affects the ability of the protein to bind to DNA. To perform the assay, we incubate the DNA with the protein for 15 min. We next add the component to be tested and incubate for an additional 5 min. DNase I digestion then shows whether the protein remains bound to the DNA in the presence of the added component. We used this assay to test the effect on TFIIIA binding to 5 S DNA of 0.03% H_2O_2, 0.003% H_2O_2, and 1 mM ascorbate plus 100 μM iron(II) EDTA. (Note: if all three components of the iron(II) EDTA reagent are tested simultaneously, the DNA will be cleaved.)

Comments on the Method

Structural Information from Hydroxyl Radical Footprints

The hydroxyl radical footprint of λ repressor bound to the O_R1 operator DNA sequence is shown in Fig. 1. What can we learn about the structure of the repressor–DNA complex from this footprint? From computer graphics model-building studies,[25] as well as from the crystal structure of a related protein–DNA complex,[26] it is well established that the repressor binds as a dimer in a symmetrical fashion to only one side of the DNA molecule. The DNA sequence to which repressor binds has approximate inversion symmetry around the ninth base pair of the 17-bp operator consensus sequence. Each monomer of the repressor dimer makes sequence-specific contacts with DNA base pairs in the major groove.[25]

The symmetry of the repressor–DNA complex clearly is present in the hydroxyl radical footprint. Near the dyad of the operator sequence there

---

[25] C. O. Pabo and M. Lewis, *Nature (London)* **298,** 443 (1982).
[26] J. E. Anderson, M. Ptashne, and S. C. Harrison, *Nature (London)* **316,** 596 (1985).

FIG. 1. Hydroxyl radical footprinting of the λ repressor–O_R1 DNA complex. (A) Autoradiograph of an electrophoresis gel on which was run the products of the footprinting reaction. Lanes 1–4, DNA labeled on the 5' end of the noncoding strand. Lanes 5–8, DNA labeled on the 3' end of the coding strand. The DNA strands were labeled in this way (on the same end of the molecule) so that bands corresponding to cutting at each base pair would run at the same position on the gel. The symmetry of the repressor–DNA complex is thus apparent from the pattern of hydroxyl radical footprints on the two strands. Lanes 1 and 8, untreated DNA. Lanes 2 and 7, products of hydroxyl radical cutting of DNA in the absence of repressor. Lanes 3 and 6, products of the Maxam–Gilbert G-specific sequencing reaction. Lanes 4 and 5, products of hydroxyl radical cutting of DNA in the presence of 90 nM λ repressor. (B) Densitometer scans of lanes 4 and 5 of the autoradiograph shown in part A. The broken vertical line passes through bp 9 in the O_R1 operator, the center of symmetry of the sequence. The 3-base offset of the central footprints on each strand can be seen around the dyad axis.

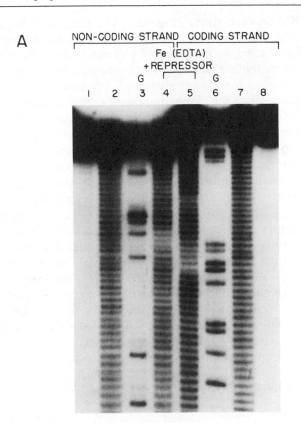

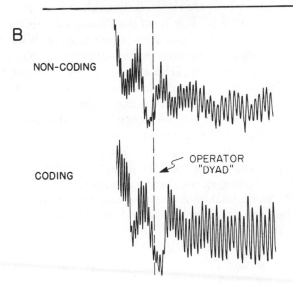

is a footprint on each strand (Fig. 1B). The minima of these two central footprints are offset from each other by 3 bp. The geometry of B-DNA tells us that these two protein–DNA contacts must be across a minor groove from each other, since the backbone positions which are closest to each other across a minor groove are 3 bp apart in the sequence. The corresponding offset for closest approach across a major groove is 7 bp. Each of the protein–DNA contacts on one strand revealed by hydroxyl radical are related by inversion symmetry, about the dyad of the operator sequence, to a corresponding contact on the other strand.

Besides reflecting the symmetry of the repressor–DNA complex, the hydroxyl radical footprint also shows directly that the protein binds to only one side of the DNA molecule.[25] This conclusion follows from the presence of backbone positions, between the central footprint and the two outer footprints on each strand, that are cut by hydroxyl radical. The outer footprints are 10 bases from the central footprint on the same strand, one turn of the helix away (see Fig. 1B). Between these protected backbone positions the DNA strand is on the "backside" of the repressor–DNA complex and is not covered by protein. Hydroxyl radical is small enough to be able to cut these exposed positions, thus showing that the repressor binds to only one face of the helix. Hydroxyl radical footprinting gives a clear picture of this structure, amounting to a "contact print" of the repressor protein bound to the helix.

Comparison with DNase I

There are fundamental differences between DNase I and hydroxyl radical footprints, which are particularly evident in densitometer scans of footprints. A scan of a gel lane from the DNase I footprint of λ repressor (such as lane 3 in Fig. 2) would show a perfectly flat "valley" (the footprint) between bands corresponding to normally cut DNA sites outside the protein binding site. The DNase I footprint of repressor is like a step function. Either a position is protected by bound protein from DNase I digestion, or it is not, with no intermediate degrees of cutting evident. It is impossible to tell from the DNase I footprint that the repressor is bound to only one side of the DNA molecule.

The hydroxyl radical footprint is more like a sine wave (Fig. 1B). Between minima (maximally protected backbone deoxyriboses) and maxima (unprotected deoxyriboses) are positions which are cut to an intermediate extent. Some backbone positions are thus "shadowed" by bound repressor, but not completely shielded. Highly ordered cocrystals of λ repressor with DNA recently have been obtained.[27] Detailed comparison

[27] S. R. Jordan, T. V. Whitcombe, J. M. Berg, and C. O. Pabo, *Science* **230,** 1383 (1985).

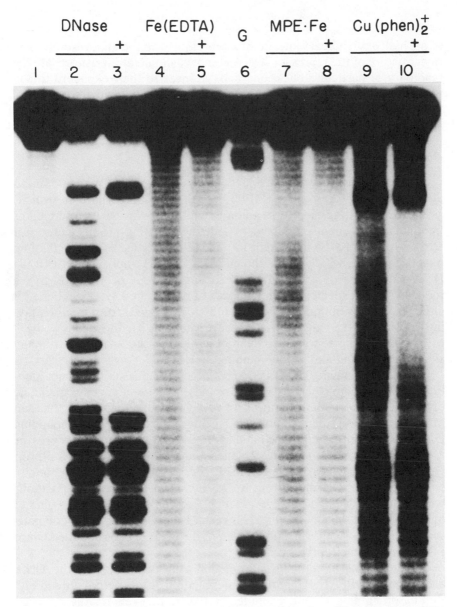

FIG. 2. Comparison of footprints of λ repressor made by four different reagents. DNA containing the O_R1 operator sequence, labeled on the 3′ end of the coding strand, was allowed to react with each of the four DNA-cutting reagents in the presence (lanes marked with +) or absence of λ repressor (90 nM). Lane 1, untreated DNA. Lanes 2 and 3, products of DNA cut by DNase I. Lanes 4 and 5, products of DNA cut by hydroxyl radical produced by the iron(II) EDTA reagent. Lane 6, products of the Maxam–Gilbert G-specific sequencing reaction. Lanes 7 and 8, products of DNA cut by MPE · Fe(II). Lanes 9 and 10, products by DNA cut by Cu(phen)$_2^+$.

of the hydroxyl radical footprint of λ repressor with the high-resolution X-ray structure of the repressor–DNA complex should allow us to relate the reactivity of particular backbone deoxyriboses to their exposure to solvent. This information will be valuable in interpreting the hydroxyl radical footprints of protein–DNA complexes for which little high-resolution structural data are available.

Comparison with MPE · Fe(II)

Hertzberg and Dervan[28] have developed a footprinting reagent, methidiumpropyl–EDTA · iron(II) [MPE · Fe(II)], which incorporates iron(II) as the DNA-cutting moiety in a compound consisting of an intercalator (methidium) covalently coupled to an iron chelating group. Cutting of DNA by MPE · Fe(II) is thought to depend on binding of the methidium moiety to DNA, which brings the chelated iron(II) in proximity with the DNA backbone.[28] The iron(II) then reacts with O_2 or H_2O_2 to give a reduced oxygen species which attacks and breaks the DNA backbone.[8,28] Other molecules bound to DNA exclude MPE · Fe(II) from binding, giving a blank in the cutting pattern corresponding to the binding site.[29] MPE · Fe(II) has been used to define the binding sites on DNA of many small intercalating and groove-binding molecules.[29]

Since the chemistry of MPE · Fe(II) footprinting[8] is similar (if not identical) to the Fenton chemistry we employ, the two reagents might be expected to give similar footprints for proteins bound to DNA. In fact, an MPE · Fe(II) footprint resembles much more closely the corresponding DNase I footprint. The DNase I and MPE · Fe(II) footprints of λ repressor are similar in appearance (Fig. 2, lanes 3 and 8). The DNase I footprint encompasses 26 bases, while the MPE · Fe(II) footprint covers about 20 bases. In particular, no cutting by MPE · Fe(II) of backbone positions within the footprint is seen, even though λ repressor is thought to bind to only one side of the DNA helix. This result is very similar to the comparison reported by Van Dyke and Dervan of the MPE · Fe(II) and the DNase I footprints of *lac* repressor.[20]

These results fit the suggestion[28] that efficient cutting of the DNA backbone by MPE · Fe(II) requires intercalation of the methidium moiety. Repressor binding should exclude intercalation, even on the exposed backside of the DNA–protein complex, so that no cutting would be seen. Iron(II) EDTA, in contrast, does not bind to the DNA but produces hydroxyl radical at some distance from the helix. This "spray" of radical

[28] R. P. Hertzberg and P. B. Dervan, *J. Am. Chem. Soc.* **104,** 313 (1982).

[29] M. W. Van Dyke, R. P. Hertzberg, and P. B. Dervan, *Proc. Natl. Acad. Sci. U.S.A.* **79,** 5470 (1982).

can reach the exposed DNA backbone on the backside of the repressor–DNA complex.

Comparison with Cu(phen)$_2^+$

Spassky and Sigman[21] have used the simple inorganic complex bis(1,10-phenanthroline)copper(I) [Cu(phen)$_2^+$] to make footprints of proteins bound to DNA. The chemistry used by this reagent to cleave DNA also is related to Fenton chemistry, but it appears that the copper complex binds to DNA and positions the copper near to the deoxyribose hydrogen that is abstracted. It is thought that Cu(phen)$_2^+$ binds along the minor groove, not by intercalation. Cutting of the DNA backbone by the copper complex is much less random than by either MPE · Fe(II) or iron(II) EDTA. Several positions of high cutting frequency can be seen (Fig. 2, lane 9) for Cu(phen)$_2^+$ reaction with free DNA.

The footprint of λ repressor made by Cu(phen)$_2^+$ (Fig. 2, lane 10) also resembles the DNase I footprint more than the hydroxyl radical footprint. Bound repressor blocks copper-induced cutting over 21 bases, and there is no evidence for backside accessibility of the DNA strands to the copper reagent. Cu(phen)$_2^+$ is thus another footprinting reagent [like DNase I and MPE · Fe(II)] that needs to bind to DNA before it can cleave the backbone. A bound protein can exclude binding of such a cutting reagent even on the side of the helix not contacted by the protein.

Compatibility of Other Conditions with Hydroxyl Radical Footprinting

The buffer system we used for footprinting λ repressor is one in which the repressor was known to bind to the operator DNA sequence. What about footprinting other protein–DNA complexes? There are two types of compatibility to consider. One is the compatibility of the protein–DNA complex with the conditions for generating hydroxyl radical. The other is the compatibility of Fenton chemistry with the conditions necessary for stabilizing a particular DNA–protein complex.

We have addressed these questions by performing the cutting reaction, in the absence of protein, under various conditions and measuring DNA cutting via densitometry. Our aim is to nick no DNA molecule more than once. The Poisson distribution shows that this condition is satisfied if less than around 30% of the DNA molecules is cut.

We have found that the iron(II) EDTA/H_2O_2/ascorbate reagent described above is relatively insensitive to buffer components, pH, temperature, cation concentration, sulfhydryl compounds, and concentration of nonspecific DNA. The buffers and pH conditions we have examined are displayed in Table I. Of the buffers we have tried, only sodium phosphate

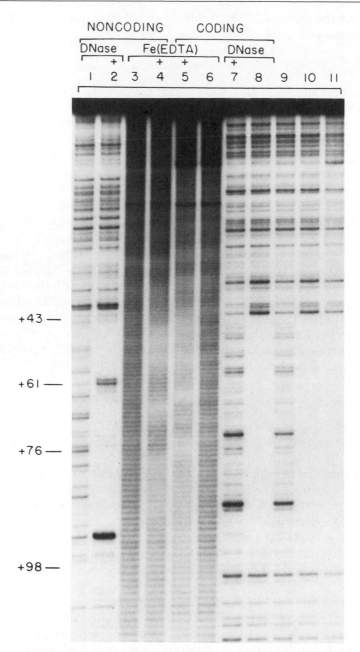

Fig. 3. Hydroxyl radical footprinting of TFIIIA bound to the *Xenopus* 5 S RNA gene. The 249-bp *Bam*HI–*Hin*dIII restriction fragment of 5 S DNA, labeled at the *Bam*HI end, was used. The numbers to the left of the autoradiograph indicate base positions relative to

buffer (pH 7.0) drastically affects the reaction, by causing a large increase in cutting of the DNA. We suspect that this is the result of a specific complex, formed between iron(II) and phosphate, which is particularly efficient in the production of hydroxyl radical. For the other buffers, the cutting reaction proceeds to the same extent regardless of pH. With a 2-min reaction time the amount of cutting is within the range required for ensuring that only single nicks are produced. Lower concentrations of Tris and HEPES buffers cause a small increase in cutting, while the concentrations of the other buffers do not affect the extent of cutting.

We have found that the concentrations of iron(II), EDTA, H_2O_2, and sodium ascorbate may be changed if a particular DNA–protein complex is sensitive to the standard reagent concentrations (see Table II). For example, in our initial experiments TFIIIA gave no footprint. We found that this protein is especially sensitive to hydrogen peroxide, and we were then able to modify the concentrations of the reactants so that the protein–DNA complex would be stable during the cutting reaction. We present the details of the TFIIIA footprinting experiment below as an example of how to adjust the reaction conditions to accommodate a different DNA–protein complex.

Monovalent cations have no effect on the cutting reaction (see Table III). Divalent cations will, at sufficiently high concentration, displace iron(II) from its complex with EDTA. Free iron(II) is not as stable as EDTA-chelated iron(II), so divalent cations can thus inhibit the cutting reaction.

The major adverse effect we have seen comes from glycerol (see Table IV), which is sometimes added to buffers as a preservative for proteins. Glycerol is a scavenger of hydroxyl radical. It greatly decreases DNA cutting by hydroxyl radical at concentrations as low as 0.5% in the footprinting reaction mixture. Other hydroxylated compounds (such as alcohols and sugars) similarly should be avoided, since they often are hydroxyl radical scavengers. Dithiothreitol and 2-mercaptoethanol inhibit cutting only slightly and are compatible with the footprinting chemistry.

---

the start of the 5 S gene. Lanes 1–4, DNA labeled at the 3' end of the noncoding strand. Lanes 5–11, DNA labeled at the 5' end of the coding strand. Lanes 1, 2, 7, and 8, products of DNase I digestion of DNA in the presence (+) or absence of TFIIIA. Lanes 3–6, products of hydroxyl radical cutting of DNA in the presence (+) or absence of TFIIIA. Lanes 9–11, assay for determining the effect on TFIIIA binding to DNA of components added to the DNA–protein mixture. TFIIIA was allowed to bind to DNA, and then 0.03% H_2O_2 (lane 9), 0.003% H_2O_2 (lane 10), or 100 μM iron(II) EDTA plus 1 mM ascorbate (lane 11) was added. DNase I then was used to digest the DNA. Lane 9 shows that TFIIIA does not bind to 5 S DNA in the presence of 0.03% H_2O_2, while lanes 10 and 11 show that the TFIIIA footprint is stable in the presence of either 0.003% H_2O_2 or iron(II) EDTA plus ascorbate.

Hydroxyl Radical Footprinting of the TFIIIA–DNA Complex

We have been successful in using hydroxyl radical to make a footprint of TFIIIA on the 5 S ribosomal RNA gene of *Xenopus* by introducing a few modifications to the method developed for footprinting λ repressor. These modifications were necessary for two reasons. First, cutting by the iron(II) EDTA reagent is severely inhibited by glycerol, which is found in buffers used to store TFIIIA. Second, the concentration of H_2O_2 normally used in the iron reagent is high enough to inactivate the protein, causing it to lose its ability to bind to the DNA.

The first problem was overcome by omitting glycerol from the protein dilution and reaction buffers. TFIIIA binding to 5 S DNA was not affected by this adjustment, as assayed by DNase I footprinting under these buffer conditions. The second problem was discovered using an assay illustrated in Fig. 3, lanes 9–11. A DNA–protein complex displays its sensitivity to a component added to the mixture either by maintaining the bound state, or by dissociation of the protein from DNA. Binding of protein to DNA is assayed by performing a DNase I footprinting reaction in the presence of the component to be tested. The concentration of H_2O_2 normally used in the cutting reagent, 0.03%, caused loss of protein binding (Fig. 3, lane 9), while 0.003% H_2O_2 did not (lane 10). To compensate for the lower peroxide concentration the concentration of iron(II) was increased to 100 μM. These modified conditions resulted in the DNA being cut to an acceptable level in 1 min, the same length of time as the DNase I digest. When subjected to the protein binding assay, iron(II) EDTA and sodium ascorbate were compatible with TFIIIA binding to DNA (lane 11).

The hydroxyl radical footprint of TFIIIA (Fig. 3, lanes 4 and 5) has features that are similar to those of the λ repressor footprint and other features that are distinctly different. Near the center of the region protected from DNase I digestion (roughly positions +42 to +92 in the 5 S gene),[24] we find a footprint on each strand, offset from each other by 3–4 bp. This is the hallmark of a protein bound to one side of the DNA helix across the minor groove, similar to what we observe for λ repressor. Two more footprints are found on each strand, one on each side of the two central footprints, with initial minima around 12 bp from the central footprint on the strand. These four outer footprints are different from the central footprints, in that they have broad minima around 10–12 bases wide. This kind of footprint also is unlike any of the λ repressor footprints, even those at the edges of the operator sequence.

We interpret the hydroxyl radical footprint as showing that TFIIIA binds across the minor groove near position +65 on one side of the DNA helix, extends across the major groove to each side of this central binding site, and then covers a whole turn of the DNA helix on each end of the control region. The center of the internal control region (roughly positions 56–77) is covered on only one side of the DNA helix by TFIIIA, while at the ends of the control region (roughly positions 44–55 and 78–88) TFIIIA surrounds the helix.

Index

C

F

F medium transformation method
Lolium multiflorum suspension culture
protoplasts, 704
N. tabacum protoplasts, 696–698
Petunia hybrida protoplasts, 701–702
Fragmentation, high-molecular-weight
DNA, 153–154
Freezing, competent *E. coli* cells, 261–
262
Fusion, *see* Gene fusion

G

β-Galactosidase
assays for regulated yeast genes, 349
gene fusions for gene expression in *E.
coli* and yeast, 413–415
alternative insertion fragments, 420–
422
choice of fragment, 422–424
materials, 415–418
pMC1403 vectors, 418–420
selection of fusion clones, 424–426
Gapped duplex DNA, oligonucleotide-
directed mutagenesis
construction of gdDNA, 577
DNA polymerase/DNA ligase reaction,
577–578
features, 571
flexibility, 585
hybridization of mutagenic primer to
gdDNA, 577–578
maintenance of bacterial strains, 580–
581
materials, 572–573
M13mp9rev RF-DNA, 575–576
M13 phage stocks, 579–580
schematic, 569–572, 570
side reactions, 584–585
structure of synthetic oligonucleotide,
582–584
structure of target region, 582
template DNA for nucleotide sequence
analysis, 579–580
transfection and segregation, 578–579
virion DNA from recombinant M13mp9,
573

Gel electrophoresis
alkali-extracted plasmid DNA, characterization, 181–187
DNA cleaved at 8-, 9-, and 10-bp sequences by *Dpn*I, 68–69
isolation of DNA fragments, 87
labeled *Xenopus* RNA, 495–496
λ repressor–O_R1 complex, 726–727
orthogonal-field-alternation, 163–166
apparatus construction, 166–168
apparatus safety, 168
microbead-protocol, 162–163
sample preparation, 159–160
solid-plug protocol, 160–162
troubleshooting
apparatus, 168–171
electrophoresis conditions, 168–
171
pulsed-field, *see* Pulsed-field gel electrophoresis
restriction fragments, 113–114
detection of DNA fragments, 115
gel dimensions, 114–115
hybridization, 120–122
mobility measurements, 124–125
photography, 115–116
preparative gel apparatus, 128–133
analysis of fractions, 135–137
gel casting, 133
recovery and purification of fragments, 137
set up, 133–135
radioautography of ^{3}P and ^{125}I, 117–
118
recovery of DNA from gel slices, 126–
127
size
calculation from mobility, 122–
123
estimation from mobility, 122
standards, 123–124
specific sequences, 118–119
transfer of DNA from agarose gels,
119–120
vertical apparatus, 114
unlabeled *Xenopus* RNA, 496
Gels, double-decker, preparation and
applications, 69–71
Gel slices, recovery of DNA from, 126–
127